Lecture Notes in Civil Engineering

Volume 352

Lecture Notes in Civil Engineering (LNCE) publishes the latest developments in Civil Engineering—quickly, informally and in top quality. Though original research reported in proceedings and post-proceedings represents the core of LNCE, edited volumes of exceptionally high quality and interest may also be considered for publication. Volumes published in LNCE embrace all aspects and subfields of, as well as new challenges in, Civil Engineering. Topics in the series include:

- Construction and Structural Mechanics
- Building Materials
- Concrete, Steel and Timber Structures
- Geotechnical Engineering
- Earthquake Engineering
- Coastal Engineering
- Ocean and Offshore Engineering; Ships and Floating Structures
- Hydraulics, Hydrology and Water Resources Engineering
- Environmental Engineering and Sustainability
- Structural Health and Monitoring
- Surveying and Geographical Information Systems
- Indoor Environments
- Transportation and Traffic
- Risk Analysis
- Safety and Security

To submit a proposal or request further information, please contact the appropriate Springer Editor:

- Pierpaolo Riva at pierpaolo.riva@springer.com (Europe and Americas);
- Swati Meherishi at swati.meherishi@springer.com (Asia—except China, Australia, and New Zealand);
- Wayne Hu at wayne.hu@springer.com (China).

All books in the series now indexed by Scopus and EI Compendex database!

Krishna Kant Pathak · J. M. S. J. Bandara ·
Ramakant Agrawal
Editors

Latest Developments in Civil Engineering

Proceedings from the International Conference on Recent Advances in Civil Engineering, RACE 2022

Springer

Editors
Krishna Kant Pathak
Department of Civil Engineering
Indian Institute of Technology BHU
Varanasi, Uttar Pradesh, India

J. M. S. J. Bandara
Department of Civil Engineering
University of Moratuwa
Moratuwa, Sri Lanka

Ramakant Agrawal
Faculty of Engineering
VNS Group of Institutions (AICTE
approved Integrated Campus)
Bhopal, Madhya Pradesh, India

ISSN 2366-2557 ISSN 2366-2565 (electronic)
Lecture Notes in Civil Engineering
ISBN 978-981-99-2675-6 ISBN 978-981-99-2676-3 (eBook)
https://doi.org/10.1007/978-981-99-2676-3

This Springer imprint is published by the registered company Springer Nature Singapore Pte Ltd.
The registered company address is: 152 Beach Road, #21-01/04 Gateway East, Singapore 189721, Singapore

Conference Committee

Chief Patron

Shri Praveen Thakral, Chairman, Oriental Group of Institutions, Bhopal and Chancellor, Oriental University, Indore.

Patron

Shri S K Sahni, CEO, Oriental Group of Institutions, Bhopal.

General Chair

Prof. (Dr.) Rajesh K Shukla, Director, OIST, Bhopal, (MP), India.

Organizing Committee Chair

Dr. Ramakant Agrawal, Head, Civil Engineering, OIST, Bhopal.

Program Chair/Volume Editor

Dr. K. K. Pathak, IIT (BHU), Varanasi, India
Dr. J. M. S. J. Bandara, University of Moratuwa, Colombo, Sri Lanka
Dr. Ramakant Agrawal, Oriental Institute of Science and Technology, Bhopal, India.

Program Committee

Dr. Sandeep Chaudhary	Professor, IIT Indore
Dr. S. S. Bhadauria	Professor and Director, University Institute of Technology, RGPV Bhopal
Dr. J. S. Chouhan	Professor, SATI Vidisha
Dr. M.S. Hora	Professor, MANIT, Bhopal
Dr. H. K. Mahiyar	Professor, SGSITS Indore
Dr. Vivek B.	Professor, BITS Pilani, Dubai campus
Dr. P.K. Jain	Professor, MANIT, Bhopal
Dr. Manish Mudgal	Principal Scientist, CSIR-AMPRI, Bhopal
Dr. Sanjeev Saxena	Principal Scientist, CSIR-AMPRI, Bhopal
Dr. Prachand Man Pradhan	Associate Professor, Kathmandu University, Nepal Dean of School of Engineering at Manmohan Technical University, Morang, Nepal
Dr. B. B. Das	Associate Professor, NITK, Surathkal

Publicity Chair

Ms. Akansha Rupal Nath, Assistant Professor, Civil Engineering, OIST Bhopal
Mr. Abhishek Vishwakarma, Assistant Professor, Civil Engineering, OIST Bhopal.

Publication Chair

Dr. Ramakant Agrawal, Head, Civil Engineering, OIST, Bhopal
Mr. Parshottam Sarathe, Assistant Professor, Civil Engineering, OIST Bhopal
Ms. Padma Pandey, Assistant Professor, Civil Engineering, OIST Bhopal.

Registration Chair

Mr. Deepak Garg, Assistant Professor, Civil Engineering, OIST Bhopal
Ms. Mumtaz Begum, Assistant Professor, Civil Engineering, OIST Bhopal.

Session Management Chair

Mr. Pramod Gour, Assistant Professor, Civil Engineering, OIST Bhopal

Mr. Vivek Rangnekar, Assistant Professor, Civil Engineering, OIST Bhopal.

Local Arrangement Chair

Mr. Abhishek Dubey, Assistant Professor, Civil Engineering, OIST Bhopal
Ms. Shivani Singh Baghel, Assistant Professor, Civil Engineering, OIST Bhopal.

Members

Mr. Pankaj Kumar Dwivedi
Mr. Deepak Kumar Uikey
Mr. Abhishek Rajpoot.

Preface

This Lecture Notes in Civil Engineering volume contains documented versions of the papers accepted at the International Conference on Recent Advancements in Civil Engineering, 2022 (RACE-2022). The conference was held during October 14–15, 2022 at the Oriental Institute of Science and Technology, Bhopal (Madhya Pradesh), India.

This conference was a platform for academicians, researchers, and industry delegates to present their research and contributions. The conference highlighted emerging research on different disciplines of Civil Engineering. The objective of this International Conference was to provide opportunities for the participants to interact and exchange ideas, experiences, and expertise in the recent technological trends. Along with sharing, an array of lectures from eminent personalities in this field was delivered to bring value to the conference.

The inauguration was held in the presence of Dr. Neeraj Saxena (AICTE New Delhi) and Dr. C. S. Verma, (AICTE Bhopal) on October 14 with their enlightening talks. The keynote talk was delivered by Dr. Sandeep Chaudhary (IIT Indore). The conference had been a good opportunity for participants from across the country. The sessions were a perfect learning place with speakers from diverse expertise.

The sessions were mentored by academic leaders from NITs, CSIR-Advanced Materials Process Research Institute and other Institutes like Dr. J. P. Shukla, Dr. Sanjeev Saxena, Dr. P. K. Jain, Dr. P. K. Agrawal, Dr. Rakesh Khare, Dr. Rana Pathak, Dr. Amit Vishwakarma, Dr. Aruna Rawat and Er. Shiv Singh Patel. The areas covered in the sessions included Structural Engineering, Transportation Engineering, Geotechnical engineering, Concrete Technology. Water Resources Engineering, Environmental Engineering, Construction Technology and Management and recent technical topics that align with the theme of the conference. There were 84 papers in 09 sessions that filled the gaps in the recent research and suggested new measures and tools for improvising the existing state of research and applications of the new techniques and innovations.

A committee of external and internal reviewers was formed for a rigorous peer review of submitted papers which were 122 in number. For maintaining the quality of the conference, the committee took full efforts and helped to shortlist 84 papers for the presentation. We are thankful to all the reviewers. Our acknowledgements are due also to Ms. Priya Vyas and Mr. Fermine Shaly, who were a constant support for communications with the Springer publications.

Finally, we take the privilege to thank all sponsors, committee members, volunteers, participants, press, print and electronic media for the success of the conference.

Varanasi, India Krishna Kant Pathak
Colombo, Sri Lanka J. M. S. J. Bandara
Bhopal, India Ramakant Agrawal

Contents

About the Editors

Dr. Krishna Kant Pathak is Professor in the Department of Civil Engineering IIT (BHU) Varanasi since April 2016. Before that he was a professor in the Department of Civil and Environmental Engineering, NITTTR Bhopal. He has also served as Scientist in CSIR from 1996 to 2011. Dr. Pathak received his B.Tech. and M.Tech. from KNIT Sultanpur and MNNIT Allahabad, in 1991 and 1993, respectively. He obtained his Ph.D. from IIT Delhi in the area of computational solid mechanics in 2001. He has guided 25 Ph.D. thesis. He has published more than 300 papers in journals and conference proceedings. He is a Fellow of Institution of Engineers (India). He is a recipient of the prestigious George Oomen Memorial Prize, The Metallurgical and Materials Engineering Division Prize and E. P. Nicoles Prize by the Institution of Engineers (India) in 2004, 2009 and 2011, respectively. He was the editor of Journal of Modelling and Simulation in Design and Manufacturing and Journal of Engineering. Science and Management Education. Dr. Pathak has delivered many invited talks in Conferences and STTP programs. His research interests include structural analysis and design, structural shape optimization, computer simulation of manufacturing processes, material characterization using miniature testing methods, finite element analysis, artificial intelligence and software development.

Dr. J. M. S. J. Bandara is a Senior Professor in Civil Engineering, University of Moratuwa. He graduated from the University of Moratuwa as a Civil Engineer and obtained his Ph.D. in Transportation Engineering with specialization in Airport Planning from The University of Calgary, Canada. He has nearly 40 years of experience in teaching and research. Currently, he is serving as the Director of Intelligent Transport Systems Research Center. Dr. Bandara is a Chartered Engineer and a Fellow of Chartered Institute of Logistic and Transport, Sri Lanka. He is also the Immediate Past President of Sri Lanka Evaluation Association, Immediate Past President of Highway Engineering Society, Sri Lanka and Past Chairman of Chartered Institute of Logistics and Transport, Sri Lanka. Dr. Bandara has served in a number of national level committees in the areas of public transport, traffic planning, airport development, road safety, highway engineering, monitoring & evaluation, and environmental assessment and has served as the Chairman of the Road Development Authority Sri

Lanka. There are over 200 national and international research publications in the above areas and joint patent for UniRoad Traffic Signal System to his credit.

Dr. Ramakant Agrawal is Principal/Associate Director, Faculty of Engineering, VNS Group of Institutions (an AICTE Approved Integrated Campus) Bhopal. He received his B.E. and M. Tech. from Maulana Azad College of Technology (REC), Bhopal, and obtained his Ph.D. in structural engineering from Maulana Azad National Institute of Technology, Bhopal. He has more than 24 years of teaching experience. His research interests include soil-structure interaction, high-performance concrete, and structural health monitoring. He has published 25 research papers in reputed journals. He is Life Member of Indian Society of Technical Education.

CTM

To Develop Geopolymer Concrete by Using Pozzolanic Industrial Waste and to Study Their Durability and Mechanical Property

Siddharth Jain, Sanjeev Singh, Pranshu Tiwari, Vijayesh Singh Chauhan, Nikhil Rai, Honey Kishore, and Utkarsh Mishra

Abstract In this study, the experimental development of the geopolymer concrete mix design process is done using the hit and trial method. By changing the proportion of different ingredients of geopolymer concrete that was established by the previous researchers (Vijaya Rangan B in Indian J, 2008), we found out their compressive strength and modelled the interrelation between them. From the various factors, we study the relationship between the fly ash content and alkaline activator content to fly ash content ratio with compressive strength of concrete after curing the specimen of 15 cm * 15 cm * 15 cm at 80 °C for 48 h, thereby making a model curve to depict the relation. Further by using the same procedure and the IS10262:2019, we make an M30 concrete and study its mechanical properties and durability to justify and validate the process of mix design that was modelled. It was analyzed from the result that the Indian standard code 10262:2019 and model curves themselves can be used for the development of geopolymer concrete with some design modification and quality regulations.

Keywords Geopolymer concrete · Green concrete · Sustainable concrete · Fly ash · Mix design

1 Introduction

The incorporation of geopolymer concrete is not new in civil engineering. Geopolymer Concrete is extensively used by Engineers since the Roman period but its importance and requirement become more critical in recent times since recent times we are facing more critical and challenging environmental impact, and one of the main constituents of environmental degradation are air pollutants and greenhouse gases like SOx, NOx, and COx. As we know that the manufacturing of conventional

S. Jain (✉) · S. Singh · P. Tiwari · V. S. Chauhan · N. Rai · H. Kishore · U. Mishra
Department of Civil Engineering, KIET Group of Institutions, Delhi-NCR, Ghaziabad, India
e-mail: siddharth.jain@kiet.edu

K. K. Pathak et al. (eds.), *Latest Developments in Civil Engineering*, Lecture Notes in Civil Engineering 352, https://doi.org/10.1007/978-981-99-2676-3_1

concrete requires the burning of clinker using coal and natural gases, and we are also aware of the fact that burning of these fossil fuels releases a high amount of air pollutants and greenhouse gases. Cement is the source of about 8% of the world's carbon dioxide (CO_2) emissions, according to the think tank Chatham House, if the cement industry were a country, it would be the third-largest emitter in the world—behind China and the US [1].

The development of geopolymer in a more scientific way started in the late twentieth century. The chemistry and terminology of inorganic polymers were introduced in detail by Davidovits. The term Geopolymer was first mentioned by Dr. Davidovits [2]. Geopolymers form three-dimensional disordered frameworks of the tecto-alumino-silicate type with the general empirical formula Mn[-(SiO_2)z-AlO_2]n.wH_2O, in which n is the degree of poly-condensation, and M is predominantly a monovalent cation (K^+, Na^+), although Ca^{2+} may replace two monovalent cations in the structure [3].

In this study, the experimental development of the geopolymer concrete mix design process is done using the hit and trial process by altering the proportions of different ingredients of geopolymer concrete that was established by previous researchers [6], we find their compressive strength and modelled the interrelation between them. From the various factors, we study the relationship between the fly ash content and alkaline activator content to fly ash content ratio with compressive strength of concrete after curing the specimen of 15 cm * 15 cm * 15 cm at 80 °C for 48 h, thereby making a model curve to depict the relation. Further as per codal provisions of IS10262:2019, we make an M30 concrete and study their mechanical properties and durability to justify and validate the process of mix design that was modelled by us.

2 Material Used in Mixed Design

The geopolymer concrete mix is broadly divided into three groups:
Cementitious material, Alkaline activator, and Aggregates.

2.1 Cementitious Material

The cementitious material used in this study was ASTM Type F fly ash (ASTM C 618) of low calcium content (less than 5%) which mostly consists of Alumina and Silica dioxide. These kinds of fly ashes are mostly obtained from anthracite and bituminous coal. For this study, fly ash is obtained from NTPC Dadri thermal power plant. The specific gravity of fly ash used is 2.4, Grey in colour, and available in powdered consistency which means the particle shape was spherical.

Table 1 Physical properties of sodium hydroxide

Property	Description/Value
Appearance	White deliquescent pellets
Solubility	10% solution in water is clear
Specific gravity of solution	1.54

Table 2 Chemical properties of sodium hydroxide

Material	Content
Molecular weight	40 g/mol
Purity	97%
Carbonate	2.0%
Chloride	0.01%
Phosphate	0.001%
Silicate	0.02%
Sulphate	0.01%
Arsenic (As)	0.0001%
Iron (Fe)	0.005%
Zinc	0.02%

2.2 *Alkaline Activator*

The alkaline activator is a mixture of sodium silicate solution and sodium hydroxide solution, which is mixed together at a particular proportion at 25 °C. When it mixes with fly ash, it starts a polymerization reaction of silicon and oxygen which makes a long chain of silicon and oxygen bonded with a covalent bond, which further imparts cohesion and adhesion properties in the mixture.

2.2.1 Sodium Hydroxide

Sodium hydroxide is available in the form of pellets and also in powdered form. This is further mixed with water to make a solution of particular molarity according to the requirement. The cost of sodium hydroxide mainly depends on its purity, it is available at 95% to 99% pure form. So it is recommended to use less pure sodium hydroxide pellets to save money. Physical properties of sodium hydroxide used are tabulated in Table 1 and chemical properties are given in Table 2.

2.2.2 Sodium Silicate

Sodium silicate solution is also known as water glass. It is a transparent solution made by dissolving sodium silicate crystals in water. Sodium silicates are stable in neutral

Table 3 Properties of sodium silicate solution

Property/Material	Description/Value
Purity	99.9%
Molar mass	122 g/mol
Density	2.4 g/cm^3
Chemical formula	Na_2SiO_3
Appearance	Colour Less

and alkaline solutions, and in acidic solutions, the silicate ions react with hydrogen ions to form silicic, which tend to decompose into hydrated silicon dioxide gel [4]. The properties of the sodium silicate solution used in this study are tabulated in Table 3.

2.3 *Aggregate*

The coarse aggregate and fine aggregate are used to fill the vacant space in the matrix of geopolymer concrete, which also imparts considerable strength and workability. The coarse aggregates used in this study were locally available crushed stoned aggregates of specific gravity 2.69, percentage water absorption 0.5% (%w/w), and their fineness modulus was 7.6. It was prepared by passing it through a 20 mm sieve and retained at a 4.75 mm sieve. The fine aggregate used in this study was crushed stone which is locally available aggregates of specific gravity 2.55, percentage water absorption 5% (%w/w), and their fineness modulus was 3.7. All the aggregates were used and tested in saturated surface dry conditions.

2.4 *Water*

The water usage in geopolymer concrete is different than in conventional OPC concrete, here in geopolymer concrete instead of hydration of cement. Water is mostly used to dissolve the sodium hydroxide and sodium silicate solution to make it more workable and homogeneous. The other purpose of water is to provide lubrication between coarse and fine aggregates and produce workable and economical concrete.

3 Research Methodology and Development of Mixed Design Process

3.1 Design Specifications and Other Data Required for Mixed Design

The process of mixed design is an iterative process in which the strength requirement needed from the geopolymer concrete and properties of ingredients involved in the manufacturing of geopolymer concrete act as an input while using different variables which decide the mechanical properties and serviceability of geopolymer concrete. We can depict the content of different ingredients involved in the manufacturing of concrete which is nothing but the output. Following is the basic list of data required to make a mixed design:

1. Characteristic compressive strength of geopolymer concrete requires curing at 80 °C for 48 h, followed by 24 h cooling at room temperature; say Fck.
2. Maximum size of the coarse aggregate and zone of the fine aggregate (sand) used.
3. Specific gravity of coarse aggregate, fine aggregate, sodium hydroxide solution, sodium silicate solution, alkaline activator solution, fly ash, and water.
4. Percentage Water absorption of fine aggregate and coarse aggregate.
5. Required workability and slump value to find consistency.
6. Fineness modulus of aggregates to make a well-graded mix.
7. Molarity of sodium hydroxide solution.
8. Suitable content of fly ash in kg/m^3 will be depicted using the content of fly ash to strength of concrete curve, in which curve will be found out experimentally.
9. Suitable alkaline activator to fly ash content ratio will be depicted using (alkaline activator to fly ash content ratio) versus (strength of concrete) curve, in which curve will be found out experimentally.
10. The suitable ratio in which sodium silicate solution to sodium hydroxide solution is added to make an alkaline activator solution will be depicted using the ratio in which sodium silicate solution to sodium hydroxide solution added versus strength of concrete curve, which will be found out experimentally.

3.2 Estimation of Entrapped Air Using IS10262:2019

The approximate value of entrapped air to be present in non-air-entrained concrete is obtained using Table 3 (clause 5.2) given in IS10626:2019 [5]. The same value as given in IS10262:2019 is tabulated in Table 4.

Table 4 Approximate air content

Nominal max size of aggregate (mm)	Entrapped air, as percentage of vol of concrete
10	1.5
20	1.0
30	0.8

3.3 *Process of Development of Content of Fly Ash to Strength Versus Strength of Concrete Curve*

The process of development of Content of Fly Ash Versus Strength of Concrete Curve is based upon the hit and trial method, in which, by varying the content of fly ash in concrete, we find out the compressive strength of concrete while considering all the other factors stated in 3.1, which was taken constant in all the mixed designs. The content and value of all the ingredients and requirements other than fly ash content are tabulated in Table 5. The varying value of fly ash content and resultant value of compressive strength of concrete is tabulated in Table 6. The resultant curve between fly ash content versus strength of concrete is shown in Fig. 1. Further manufacturing of concrete will be done by using the same curve, and results will be discussed in the subsequent chapter.

Table 5 Data for mixed design

Material/Description	Content/Value
Characteristic strength	Fck
Target strength	Fck + 1.65 * 5 (calculated using IS456:2000)
Max size of aggregate	20 mm
Zone of sand	Zone 3
Na_2SiO_4 to NAOH ratio	2.0
Alkaline activator to fly ash content ratio	0.62
Molarity of sodium hydroxide solution	12
Specific gravity of coarse aggregate	2.69
Specific gravity of fine aggregate	2.55
Specific gravity of alkaline activator solution	
Percentage water absorption of coarse and fine aggregate	0.5 and 5%
Coarse aggregate to total aggregate ratio	0.552 (taken from IS10262:2019)
Dimension of specimen	15 cm * 15 cm * 15 cm
No. of specimen tested per mixed design and volume of concrete accordingly	3 and 0.010125 m^3

Table 6 Compressive strength of concrete obtained at particular fly ash content

Content of fly ash (kg/m^3)	Compressive strength (MPA)
450	24.30
475	25.89
500	26.93
525	27.77
550	28.79
575	30.38
600	32.90

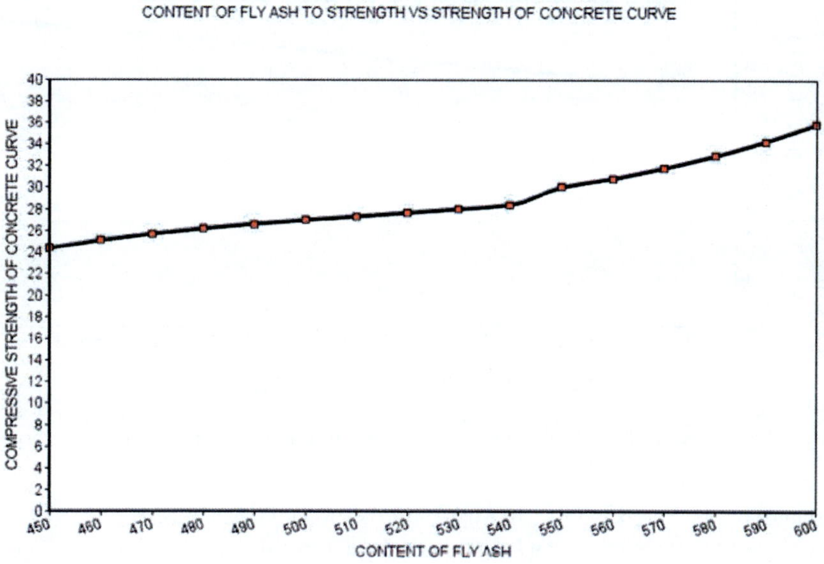

Fig. 1 Curve Between fly ash content versus compressive strength of concrete

3.4 Process of Development of (Alkaline Activator to Fly Ash Ratio) Versus Strength of Concrete Curve

The process of development of this curve is based upon the hit and trial method, in which, by varying the content of alkaline activator to fly ash ratio in concrete, we find out the compressive strength of concrete while considering all the other factors stated in 3.1 as constant for all the mixed design under this category. The content and value of all the ingredients and requirements other than the content of alkaline activator to fly ash ratio is tabulated in Table 5. The varying value of fly ash content and resultant value of compressive strength of concrete is tabulated in Table 7. The resultant curve between the content of alkaline activator to fly ash content versus

Table 7 Compressive strength of concrete obtained at particular alkaline activator to fly ash content ratio

Alkaline activator to fly ash content ratio	Compressive strength of concrete (MPA)
0.50	16.21
0.55	17.71
0.60	26.58
0.65	41.43

Note Mean of 3 cubes of size 15 cm * 15 cm * 15 cm were taken to calculate it

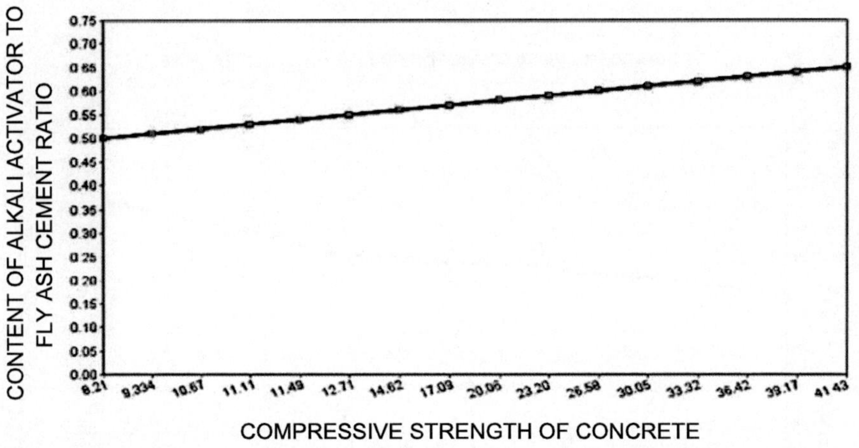

Fig. 2 Curve between the content of alkaline activator to fly ash content versus strength of concrete

strength of concrete is shown in Fig. 2. Further manufacturing of concrete will be done using the same curve, and results will be discussed in the subsequent chapter.

3.5 *Estimation of Coarse and Fine Aggregate Content*

1. The amount of total volume of aggregates in the concrete will be calculated by using mathematical operation, in which, the volume of all the ingredients other than aggregates will be subtracted from the total volume of manufactured concrete. The following formula will be used to calculate the total volume of aggregate:

   From

$$V = S^{Na_2SiO_4}/D^{Na_2SO_4} + S^{NAOH}/D^{NAOH} + S^{Fly\ ash}/D^{Fly\ ash} + V^{TA} + V^{a}$$

We get

$$\boxed{V^{TA} = V - (S^{Na_2SiO_4}/D^{Na_2SO_4} + S^{NAOH}/D^{NAOH} + S^{Fly\,ash}/D^{Fly\,ash} - V^{A})} \quad (1)$$

where

$$V = \text{Complete volume of green concrete}$$

$$S^{Na_2SiO_4} = \text{Amount sodium silicate in kg}$$

$$D^{Na_2SO_4} = \text{Density of sodium silicate in kg/m}^3$$

$$S^{NAOH} = \text{Amount sodium hydroxide in kg}$$

$$D^{NAOH} = \text{Density of sodium silicate in kg/m}^3$$

$$S^{Fly\,ash} = \text{Amount of fly ash in kg}$$

$$D^{Fly\,ash} = \text{Density of fly ash in kg/m}^3$$

$$V^{TA} = \text{Volume of total aggregate in m}^3$$

$$V^{a} = \text{Volume of entrapped air}$$

2. After the claculation of the total volume of aggregate using Eq. 1, calculation of the amount of coarse aggregate and fine aggregate will be done by using IS10262:2019, In which by using the nominal max size of coarse aggregate and zone of sand we can find the ratio between the volume of coarse aggregate and volume of total aggregate using Table 5 (clause 5.5) given in IS10262:2000 [5]. The following formula will be used to find it:

   From

$$P = V^{CA}/V^{TA}$$

$$V^{CA} = P * V^{TA}$$

   We get

$$S^{CA} = V^{CA} * D^{CA} \quad (2)$$

And

$$V^{FA} = V^{TA} - V^{CA}$$

We get

$$\boxed{S^{FA} = V^{FA} * D^{FA}} \quad (3)$$

where

P = ratio between the volume of coarse aggregate and the volume of total aggregate

$$V^{TA} = \text{Volume of total aggregate in m}^3$$

$$V^{CA} = \text{Volume of coarse aggregate in m}^3$$

$$V^{FA} = \text{Volume of fine aggregate in m}^3$$

$$D^{CA} = \text{Density of coarse aggregate in kg/m}^3$$

$$D^{FA} = \text{Density of fine aggregate in kg/m}^3$$

$$S^{CA} = \text{Amount coarse aggregate in kg}$$

$$S^{FA} = \text{Amount fine aggregate in kg}$$

4 Illustrative Development of M30 Concrete and Its Mechanical Property Analysis

1. Following are the step involved in the development of mix design:

Step 1: Calculation of target strength by selecting Fck (Characteristic strength) and standard deviation.

Step 2: Find the volume of required green concrete by analyzing the requirement.

Step 3: Find the content of fly ash using a curve between fly ash content versus strength of concrete given in Fig. 1. A further amount of fly ash will be calculated by multiplying the volume of green concrete and the content of fly ash.

Step 4: Mixing of NaOH solution and Na_2SiO_4 solution to make an alkaline activator solution. In this study, Na_2SiO_3 to NAOH ratio is taken as 2.

Step 5: For the calculation of alkaline activator content, the amount of alkaline activator solution is calculated by using the ratio between alkaline activator content and fly ash content i.e., by multiplying the amount of fly ash calculated in step 3 with the alkaline activator to fly ash ratio. Thus, we get the total amount of alkaline activator solution in the green concrete. The curve for the calculation of the ratio between alkaline activator content and fly ash content is given in Fig. 2.

Step 6: Calculation of the amount of coarse aggregate and fine aggregate.

2. Calculation involved in the development of M30 mix designing.

Following are the design data for the required concrete.

Material/Description	Content/Value
Characteristic strength (Fck)	30 MPA
Standard deviation	5
Target strength	Fck + 1.65 * 5 (calculated using IS10262:2000)
Max size of aggregate	20 mm
Zone of sand	Zone 3
Na_2SiO_4 to NAOH ratio	2.0
Molarity of sodium hydroxide solution	12
Specific gravity of coarse aggregate	2.69
Specific gravity of fine aggregate	2.55
Specific gravity of NaOH	1.52
Specific gravity of Na_2Sio_3 (water glass solution)	2.4
Volume of entrapped air as PERIS10262:2000	1% of volume of concrete (for 20 mm aggregate)
Percentage water absorption of coarse and fine aggregate	0.5 and 5%
Dimension of specimen	15 cm * 15 cm * 15 cm
No. of specimens tested per mixed design and volume of concrete accordingly	3 and 0.010125 m^3

(a) Calculation of target strength: $F_t = 30 + 1.65*5 = 38.25$ MPA
(b) Volume of concrete $= 3*0.15*0.15*0.15 = 0.010125$ m^3
(c) Content of fly ash for 38.25 MPa strength: 640 kg/m^3
(d) Amount of fly ash = 6.48 kg
(e) Ratio of alkaline activator to fly ash for 38.25 MPa strength = 0.64
(f) Amount of alkaline activator = 4.15 kg
(g) Amount of NAOH solution = 1.0368 kg
(h) Amount of NA2SIO3 solution: 3.110 kg

(i) Coarse aggregate to the total aggregate ratio: 0.572

Note 1. For every decrease in water/cement ratio by 0.05 from 0.5, the ratio of vol of coarse to total aggregate is increased by 0.01 (As given in IS10262:2019).

Note 2: Because alkaline liquid to fly ash is taken as 0.64, so ratio of volume of coarse aggregate to total aggregate is decreased by 0.028 from the value given in IS456:2000.

(j) Volume of entrapped air = 1% of volume of concrete = 0.00010125 m^3
(k) Calculation of amount total aggregate:

From,

$$V^{TA} = V - (S^{Na_2SiO_4}/D^{Na_2SO_4} + S^{NAOH}/D^{NAOH} + S^{Fly\ ash}/D^{Fly\ ash}) - V^A$$

$$V^{TA} = 0.010125 - (3.110/2400 + 1.0368/1520 + 6.48/2500) = 0.0054538\ m^3$$

(l) Volume of Coarse aggregate: 0.572*0.0054538 = 0.003119
(m) Amount of coarse aggregate in kg = 8.38 kg
(n) Volume of fine aggregate = 0.0023348
(o) Amount of fine aggregate: 5.95 kg

Result of Mixed Design

Following final design data was used for manufacturing of M30 concrete.

Material	Amount (kg)
Fly ash	6.48
Alkaline activator solution	4.15
Coarse aggregate	8.38
Fine aggregate	5.95
Final ratio (fly ash: fine aggregate: coarse aggregate	1:0.92:1.29
Alkaline solution to fly ash ratio	0.64

Following is the mechanical properties obtained after 48 h of curing at 80 °C.

Property	Cube 1	Cube 2	Cube 3	Average	Comment
Compressive strength (MPA)	37.6	36.5	34.4	36.16	Pass
Flexure tensile strength (MPA)	4.29	4.21	4.10	4.20	Pass
Workability (SLUMP)	67 mm				Pass

5 Conclusion and Discussion

1. Need for Geopolymer concrete starts with concerning environmental issues and it was seen that it can contribute to lower down the negative impact of conventional concrete on the environment.
2. In this study, with the help of previously available information provided by various researchers and IS10262:2019, we developed the steps to make a geopolymer concrete and also make an M30 concrete and test its mechanical properties and durability to justify the mixed design process. From the result, it was analyzed that the Indian standard code 10262:2019 and the model curves themselves can be used for the development of geopolymer concrete with some design modification and quality regulations.
3. It was seen that Type F fly ash (ASTM C 618) of low calcium content (less than 5%) provides very good compressive strength in concrete. The curve formed between the content of fly ash versus compressive strength concrete after curing the specimens at 80 °C for 48 h seems to be concave upward, In which at low fly ash content we find that strength of concrete looks less sensitive but at a higher level it looks more sensitive, but after crossing fly ash content 650 the increase in strength stagnate with an increase in fly ash content, so here we conclude that to increase the strength further we have to use silica fumes or high silica fly ash as suggested in many pieces of research [7]. The actual relationship is concluded in the curve given in Fig. 1.
4. The alkaline solution to fly ash content ratio is another variable that is tested in this study. We found that the alkaline solution to fly ash ratio is directly proportional to the compressive strength of concrete. The actual relationship is concluded in the curve given in Fig. 2.
5. The M30 concrete formed using the same mix design process that was established in the same study was seen to be passed in all aspects in which it was tested. Which also provides sufficient data and proof to validate the study.
6. So, geopolymer has some excellent benefits which are suitable for structural applications. And seeing the recent scenario and taking into consideration the environmental impacts of conventional concrete, it becomes more important to introduce such innovative material in the construction industry.

References

1. Making concrete change: innovation in low-carbon cement and concrete, Chatham House Report, 13 June 2018. ISBN: 978 1 78413 272 9. https://www.chathamhouse.org/2018/06/making-concrete-change-innovation-low-carbon-cement-and-concrete
2. Davidovits J (1988) Structural characterization of geopolymeric materials with X ray diffractometry and MAS NMR spectroscopy. In: Geopolymer '8: first European conference on soft mineralogy, Compiegne, France, vol 2, pp 149–166

3. Hardjito D, Rangan BV (2005) Development and properties of low-calcium fly ash-based geopolymer concrete, Research Report GC, Faculty of Engineering, Curtin University of Technology, Perth, Australia, pp 1–130
4. Gelpi C, Feaver P, Reifler J (2005). Replication data for: success matters: casualty sensitivity and the war in Iraq. OCLC 795918959
5. IS10262:2019, Recommended guidelines for concrete mix design
6. Vijaya Rangan B (2008) Mix design and production of fly ash based geopolymer concrete. Indian J
7. Fernandez J, Palomo A (2004) Activation of fly ashes: a general view, fly ash, silica fume, slag, and natural pozzolans in concrete. In: Proceedings eighth international conference, V.M. Malhotra editors, Las Vegas, USA, pp 351–366

Effect of Untreated Recycled Aggregate on Properties of GPC

Tanuja Gupta and Meesala Chakradhara Rao

Abstract This research article provides a comparison of various properties of geopolymer concrete on the complete substitution of natural aggregate with untreated recycled aggregate. In total, 4 design mixes were prepared, out of which 2 design mixes were of traditional concrete, while the other 2 were of geopolymer concrete (GPC). In GPC, curing temperatures of 60 and 90 °C and three curing periods (i.e., 24, 48, and 72 h) were adopted. The mix composition of aggregates was decided using the particle packing method. The specimen testing was done after 7 and 28 days of casting. The GPC density is about 3% less than that of OPC. As the curing temperature rose for both natural and recycled aggregate concrete, an improvement in mechanical properties was observed. The study further compared the UPV value of OPC and GPC.

Keywords Geopolymer concrete · Recycled aggregate · Compressive strength · Ultrasonic pulse velocity (UPV)

1 Introduction

Concrete is the most durable material. It is being used second only to water. The binder material used in traditional concrete adversely affects the environment as CO_2 emission takes place during the manufacturing of cement. About 900 kg of CO_2 is emitted for the production of 1000 kg of cement [1]. Researchers have suggested various alternatives to overcome this advantage. Some of the alternatives are the use of blended cement [2], reducing the cement clinker by the addition of additives [3], use of sustainable cement [4], and alkali-activated binders [5, 6]. "Geopolymer binders" is the commercial name for alkali-activated binders, which was coined by Prof. Davidovits in the late 1970s. The polycondensation reaction between silica and aluminum oxides (which are present in abundance in industrial wastes such as fly ash and ground granulated blast furnace slag) in the presence of alkaline medium such

T. Gupta (✉) · M. C. Rao
Guru Ghasidas University (A Central University), Bilaspur 495009, Chhattisgarh, India
e-mail: tanujagupta1994@gmail.com

K. K. Pathak et al. (eds.), *Latest Developments in Civil Engineering*, Lecture Notes in Civil Engineering 352, https://doi.org/10.1007/978-981-99-2676-3_2

as Sodium hydroxide (NaOH), Potassium Hydroxide (KOH), and Sodium Silicate (Na_2SiO_3) results in the formation of geopolymeric compound. Various studies have reported a reduction in the emission of CO_2 in the case of Geopolymeric binders. According to one of these studies, GPC concrete emits 9% less CO_2 than concrete which is 100% OPC [7]. In addition to being used for local projects like retaining walls, water tanks, and boat ramps, GPC has also been used for large-scale production. One such instance is the construction of pavements at Brisbane Airport in Australia, which comprised an apron and taxiways.

In the past decade, researchers have attempted to develop geopolymer binders, so that the use of traditional cement could be reduced. In addition to this, due to the high demand for aggregates, there is a decrease in the availability of aggregates, also there is an increase in construction and demolition waste (C and D). A study suggested that there is a 40% replacement of recycled aggregate by natural aggregate in GPC [8, 9]. Accumulating evidence from multiple researches demonstrates that the decrease in strength of GPC is not so significant and it is within the limits when natural aggregates were replaced by recycled aggregate [10]. The research recommended using recycled aggregate made from lightweight blocks to make geopolymer concrete, which can be utilized to build partition walls because its density is about 860–1400 kg/m^3 [11].

Recycled aggregates have found usage not just in buildings but also in pavements, where they have taken the place of natural aggregates. In comparison to roller-compacted concrete with 12% OPC, a study suggests that geopolymer concrete with 100% recycled aggregate should be used [12]. The impact of recycled aggregate in GPC has been studied by replacing natural coarse aggregates with recycled aggregates. Additionally, the impact of variables including the curing temperature, duration, and amount of recycled aggregate on the workability of fresh concrete as well as the density, compressive strength (7 and 28 days), and ultrasonic pulse velocity of hardened concrete has been addressed.

2 Experimental Program

This experimental study is intended to examine how 100% untreated recycled coarse aggregate affects M25 grade GPC. Three different curing times (24, 48, and 72 h) in the oven—and two different curing temperatures (60 and 90 °C) were used. Density and Ultrasonic Pulse Velocity (UPV) testing were conducted at 28 days, whereas compression strength tests were performed on the 7th and 28th day.

2.1 Materials

In this analysis, the following source materials were used:

a. **Fly Ash (Class-F)** collected from the National Thermal Power Plant, Sipat, with a specific gravity of 2.2
b. Aggregate:
   i. **Fine Aggregates (FA)**: Zone II Conforming fine aggregate as per IS 2386 (Part 1) with a specific gravity of 2.648 and fineness modulus of 2.92
   ii. **Natural Coarse Aggregates (NCA)**: Locally available Crushed aggregate as per IS 383:2016 with a specific gravity of 2.683 (blended aggregate)
   iii. **Untreated Recycled Concrete Aggregate (UTRA)**: Laboratory produced and tested OPC concrete coarse aggregate which are crushed with the help of a jaw crusher whose Specific gravity is 2.458

   The particle packing method recommended by Thunuguntla et al. [13] was used in this work; Table 1 shows the quantity of aggregates and their sizes.

c. **Alkaline Solution**
   i. Sodium Hydroxide (NaOH)—12 molarity (12 M) solution,
   ii. Sodium Silicate (Na_2SiO_3) (commercial grade) of specific gravity 1.41, with a molar ratio (SiO_2: Na_2O) as 3.22, Water % = 61.4%, $Na_2O + SiO_2 =$ 38.6%, procured from a local supplier.
   iii. Ratio of Na_2SiO_3 to NaOH = 2.0.

   To prepare the NaOH solution of 12 molarity (12 M), 480 gm of NaOH was first dissolved in 250 ml of distilled water, then the solution was made to 1000 ml by adding distilled water. Since the mixture of NaOH and Na_2SiO_3 releases a lot of heat during the geo-polymerization process, it was combined 24 hours before actual casting [14, 15].
d. **Superplasticizer**—2% by weight of fly ash

Table 1 % Optimum aggregate proportion for monosized untreated recycled concrete aggregates (100% replacement of natural aggregate) for mix design

Size of aggregate	100% NCA (%)	100% UTRA (%)
20–16 mm	11.55	11
16–12.5 mm	11.55	15
12.5–10 mm	18.9	14
10–4.75 mm	18	20
4.75–2.36 mm	9.75	9.75
2.36–1.18 mm	9.75	9.75
1.18–0.6 mm	10.5	10.5
0.6–0.15 mm	10	10

2.2 Mix Design

The conventional concrete was prepared with OPC-53. The water–cement ratio used was 0.45. This concrete was used for comparison purposes. Table 2 lists the mix design for OPC concrete. Fly ash was used to bind four different mix designs, and 100% natural coarse aggregates were replaced with 100% untreated recycled aggregates. The alkaline activator to binder ratio, superplasticizer, and binder content were all kept constant at 340 kg/m^3, 0.45, and 2%, respectively. The mix composition of conventional concrete was designed as per the guidelines of IS 10262: 2019 and IS 456: 2000. The composition of coarse and fine aggregate was determined using the particle packing method. A standard mix design is not available for GPC, so the same guidelines as conventional concrete were adopted. Additionally, the design procedure suggested by Anuradha et al. [16] and Patankar et al. [17] was referred to. Mix ID number has been assigned to the various mix proportions, the details of which are presented in Table 3.

Table 2 Mix proportion (in kg/m^3) of OPC and GPC with natural and recycled aggregate concrete

Mix ID	Cement	Fly ash	FA	NCA	RA	AA/B	S
100N-OPC	340	–	776.2	1179.33	–	0.45	6.8
100N-GPC	–	340	776.2	1179.33	–	0.45	6.8
100R-OPC	340	–	776.2	–	1080	0.45	6.8
100R-GPC	–	340	776.2	–	1080	0.45	6.8

Note 1. 100N-OPC: OPC concrete with 100% NCA, 2. 100N-GPC: GP concrete with 100% NCA, 3. 100R-OPC: OPC Concrete with 100% RCA, 4. 100R-GPC: GP Concrete with 100% RCA

Table 3 Mix no. for OPC and GPC

Mix no.	Mix ID	Aggregate	Curing temperature (°C)	Curing Period (hr)
M1	100N-OPC	NA	–	–
M2	100N-GPC	NA	60 °C	24, 48, 72
M3	100N-GPC	NA	90 °C	24, 48, 72
M4	100R-OPC	RA	–	–
M5	100R-GPC	RA	60 °C	24, 48, 72
M6	100R-GPC	RA	90 °C	24, 48, 72

Note NA—Natural Aggregate, RA—Recycled Coarse Aggregate (untreated)

2.3 *Preparation of GPC Specimens*

Surface-saturated dry condition aggregates were used. 24 h before the actual mixing of GPC, the alkaline solution was mixed. The two-stage mixing process [18] recommended by Tam et al. was used in this research. For about a minute, the dry aggregates (coarse and fine) were mixed. After that, the superplasticizer and half of the water (or alkaline solution) were mixed, and the mixing was continued for an additional minute. After adding the cementitious material (fly ash for GPC concrete and cement for OPC), the mixture was then mixed for an additional 30 s. After adding the leftover liquid, the mixture was well blended to create a uniform composition. The mixture was then poured into the molds and compacted for two to three minutes on a vibrating table. The compressive strength test was performed on 100 mm cube specimens.

After casting, the specimens were cured in an oven at 60 and 90 °C for 24, 48, and 72 h. The specimens were demoulded after oven curing and were kept at room temperature until the day of testing.

3 Results and Discussions

3.1 *Workability*

As the workability of GPC was low and no slump was observed, the compaction factor test was performed. The workability of concrete with natural aggregate is more than that of recycled aggregate concrete. Concrete with natural aggregate is more workable than concrete with recycled aggregate. Additionally, the workability of GPC is lower than that of OPC (Fig. 1).

3.2 *Density*

The concrete's density was tested on the 28th day. The density of GPC with natural aggregates varied within the range of 2290–2390 kg/m^3. The density of GPC with 100% natural aggregate (or 0% RCA) is about 5 and 7% lower than the OPC respectively for 60 and 90 °C cured concrete. However, GPC with 100% UT RA demonstrated a reduction of 1% (cured @ 60 °C) and 4% (cured @ 90 °C) in density against that of OPC. The density reduction of 90 °C cured GPC is more than 60 °C cured specimens. The variation of density is represented in Figs. 2 and 3.

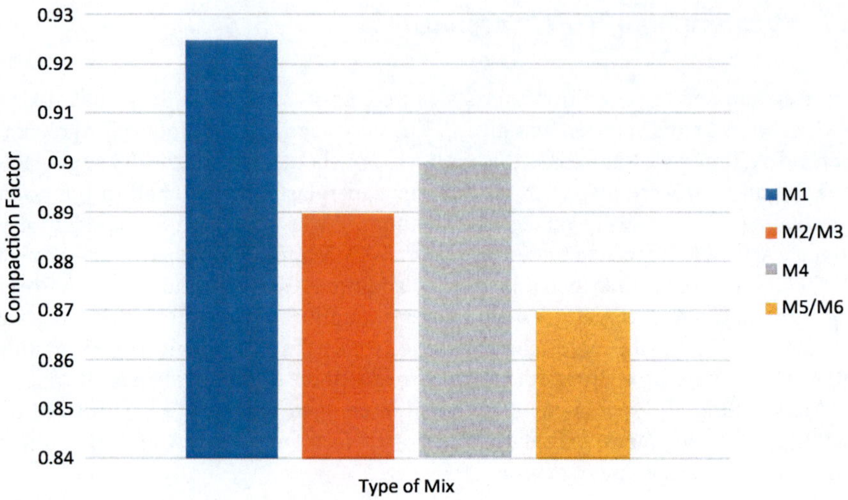

Fig. 1 Compaction factor of various mixes of concrete

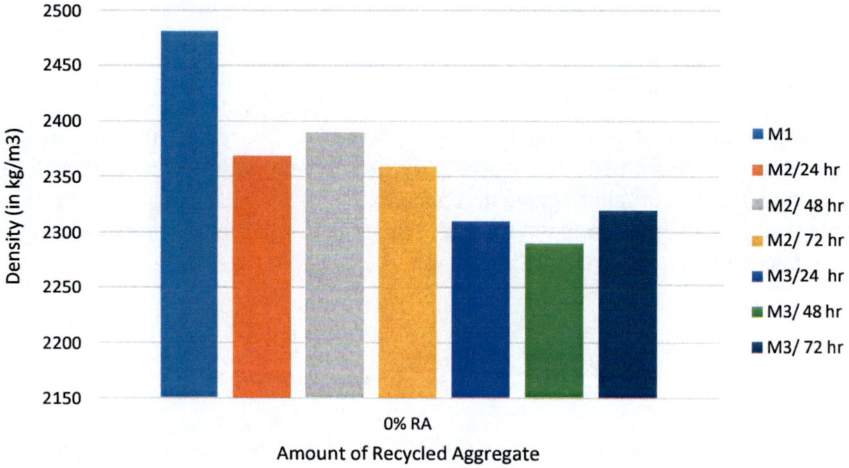

Fig. 2 Density variation of OPC and GPC with 100% NA

3.3 Compressive Strength

The compressive strength variation of OPC and GPC are graphically represented in Figs. 4 and 5. Figure 4 displays that in the case of GPC with natural aggregate; an increase in strength takes place with an increase in curing temperature and curing period strength, but the amount of increment of strength after 7 days at higher temperature (i.e., 90 °C) is less as compared to 60 °C cured specimens. Additionally, the

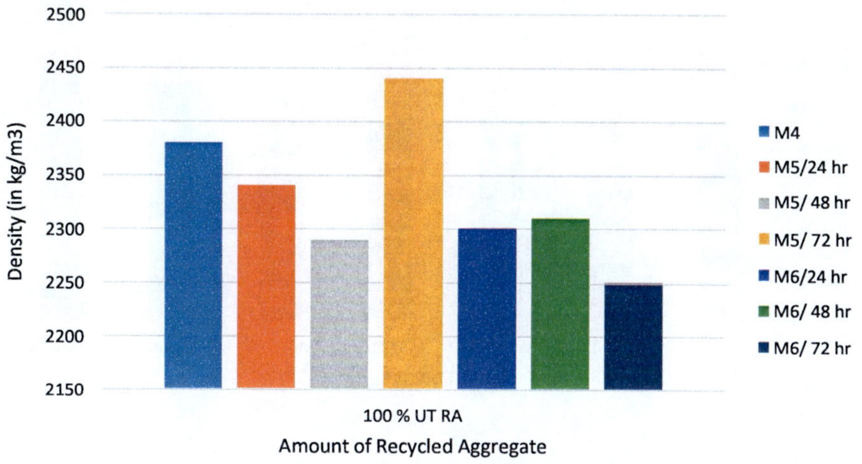

Fig. 3 Density variation plot of OPC and GPC with 100% RCA

GPC cured at 90 °C for 72 h exhibited higher strength than OPC. A similar variation of strength was observed in GPC with 100% untreated recycled concrete aggregate (refer to Fig. 5). Figure 6 displays the variation of the curing regime on GPC. The best curing regime of 90 °C for 72 h is selected. The variation of compressive strength (optimum parameters) of GPC and OPC is plotted in Fig. 7 for both concrete with natural and recycled coarse aggregate.

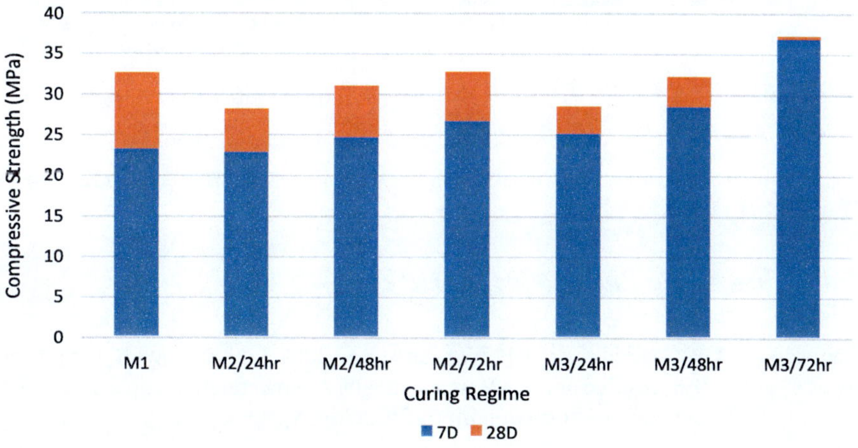

Fig. 4 Compressive strength—OPC and GPC with 100% NA

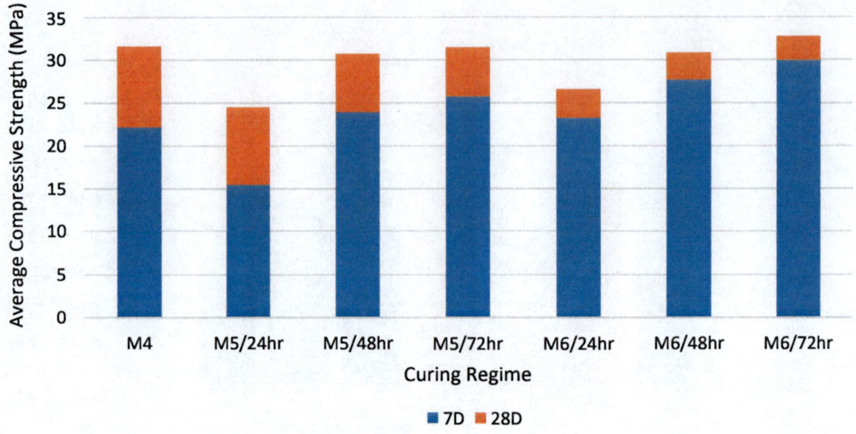

Fig. 5 Compressive strength—OPC and GPC with 100% UTRA

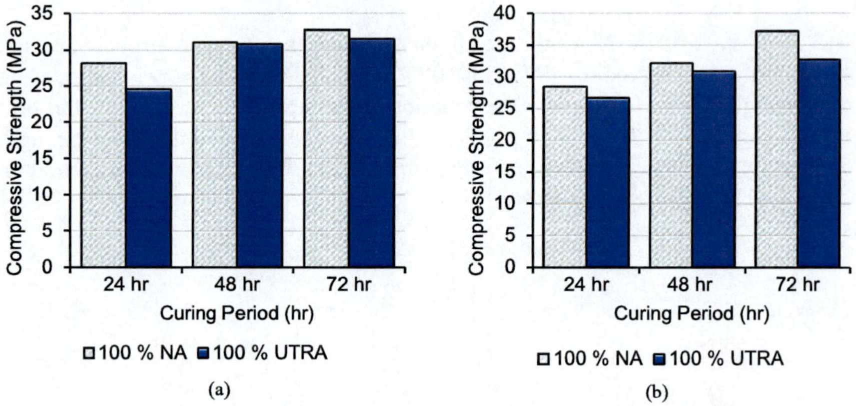

Fig. 6 28 day compressive strength of GPC with 100% NA and 100% UTRA (**a**) 60 °C and (**b**) 90 °C

3.4 Ultra Sonic Pulse Velocity (UPV)

The value of UPV varied from 3.1–3.75 km/s, and 2.88–3.43 km/s with natural (or 0% RA) and 100% recycled aggregate, respectively. A direct relationship was noticed between UPV and compressive strength (refer to Fig. 8).

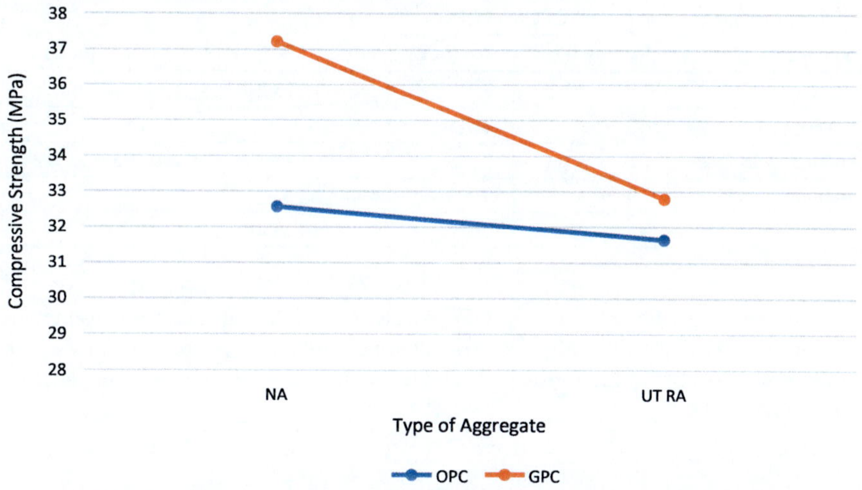

Fig. 7 Optimum compressive strength in both OPC and GPC with and without RA

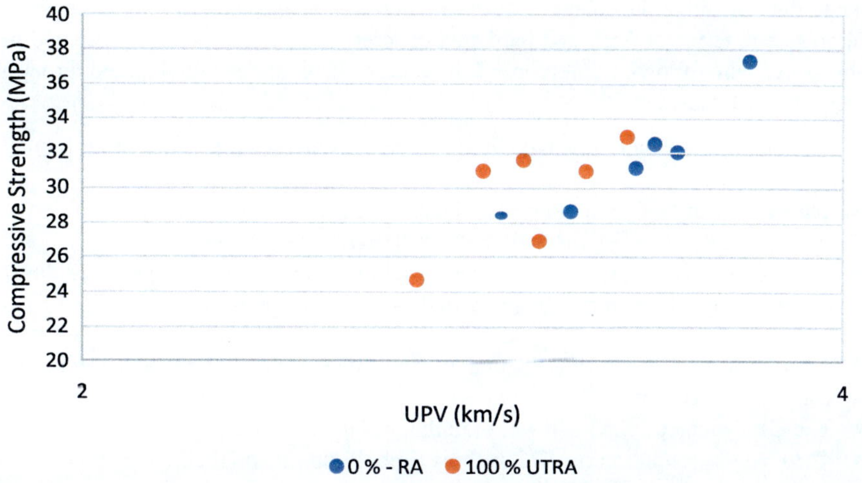

Fig. 8 The association between UPV and compressive strength for both natural and recycled concrete aggregate GPC

4 Cost Analysis

To assess the manufacturing cost of fly-ash geopolymer concrete, a preliminary analysis based on market prices was performed. The cost evaluation of 1 m^3 of M25 OPC and GPC is shown in Table 4.

Table 4 Cost estimation of 1 m^3 GPC and OPC

Ingredients	Quantity (kg)	Cost (in Rs.) per kg	OPC (in Rs.)	GPC (in Rs.)
Cement	340	7	2380	–
Fly ash	340	2	–	680
Sodium hydroxide	21	50	–	1050
Sodium silicate	110	50	–	5500
		Total	2380	7230

A similar production cost of 1 m^3 GPC was reported [19, 20]. The high production cost of GPC is due to the high cost of alkaline activators. A study estimated an approximate cost of production as 1.7% (lower grade) and 11% (higher grade) against that of OPC [21].

5 Conclusions

The effect of the replacement of 100% natural aggregates with untreated recycled aggregate in terms of fresh and hardened concrete properties has been presented in this paper. The following inferences can be drawn from the result obtained through experimental analysis:

- The incorporation of recycled aggregate decreases the workability of both OPC and GPC.
- The density of GPC with natural and untreated recycled aggregate lies within the range of 2290–2390 kg/m^3 and 2250–2440 kg/m^3, respectively.
- The comparative analysis of GPC specimens cured at 60 °C (7 days → 28 days) exhibited an enhanced rate of strength development against that of 90 °C (7 days → 28 days).
- An increase in the curing regime demonstrated a significant increase in the compressive strength of GPC.
- The UPV value of GPC is lower than that of OPC.
- The cost of production of GPC is three times higher than OPC.

References

1. Mahasenan N, Smith S, Humphreys K (2003) The cement industry and global climate change current and potential future cement industry CO_2 emissions. Greenh Gas Control Technol—6th Int Conf 995–1000
2. Gilani V (2013) The cBalance Blog. In: cBalance Blog. http://cbalance.in/2013/12/carbon-emissions-in-the-cement-sector-in-india/#.XnSrhhBR1PY

3. Ali MB, Saidur R, Hossain MS (2011) A review on emission analysis in cement industries. Renew Sustain Energy Rev 15:2252–2261. https://doi.org/10.1016/j.rser.2011.02.014
4. van Deventer JSJ, Nicolas RS, Ismail I et al (2014) Microstructure and durability of alkali-activated materials as key parameters for standardization. J Sustain Cem Mater 4:116–128. https://doi.org/10.1080/21650373.2014.979265
5. Shi C, Jiménez AF, Palomo A (2011) New cements for the 21st century: the pursuit of an alternative to Portland cement. Cem Concr Res 41:750–763. https://doi.org/10.1016/j.cemconres.2011.03.016
6. Juenger MCG, Winnefeld F, Provis JL, Ideker JH (2011) Advances in alternative cementitious binders. Cem Concr Res 41:1232–1243. https://doi.org/10.1016/j.cemconres.2010.11.012
7. Turner LK, Collins FG (2013) Carbon dioxide equivalent (CO2-e) emissions: a comparison between geopolymer and OPC cement concrete. Constr Build Mater 43:125–130. https://doi.org/10.1016/j.conbuildmat.2013.01.023
8. Srinivas T, Abhignya G (2021) Behaviour of structural elements made of geopolymer concrete with recycled aggregates. IOP Conf Ser Mater Sci Eng (ICIRMCT 2021) 1091:012030. https://doi.org/10.1088/1757-899x/1091/1/012030
9. Raj SD, Bhoopesh J (2017) Strength and behaviour of recycled aggregate geopolymer concrete beams. Adv Concr Constr 5:145–154. https://doi.org/10.12989/acc.2017.5.2.145
10. Ojha A, Gupta L (2020) Comparative study on mechanical properties of conventional and geopolymer concrete with recycled coarse aggregate. Mater Today Proc 28:1403–1406. https://doi.org/10.1016/j.matpr.2020.04.811
11. Posi P, Teerachanwit C, Tanutong C et al (2013) Lightweight geopolymer concrete containing aggregate from recycle lightweight block. Mater Des 52:580–586. https://doi.org/10.1016/j.matdes.2013.06.001
12. Rahman SS, Khattak MJ (2021) Roller compacted geopolymer concrete using recycled concrete aggregate. Constr Build Mater 283:122624. https://doi.org/10.1016/j.conbuildmat.2021.122624
13. Thunuguntla CS, Gunneswara Rao TD (2018) Mix design procedure for alkali-activated slag concrete using particle packing theory. J Mater Civ Eng 30:04018113. https://doi.org/10.1061/(asce)mt.1943-5533.0002296
14. Poojidha M, Kumar KN (2017) Study on strength of geopolymer concrete in ambient curing. Int Res J Eng Technol 3:318–321
15. Duxson P, Fernández-Jiménez A, Provis JL et al (2007) Geopolymer technology: the current state of the art. J Mater Sci 42:2917–2933. https://doi.org/10.1007/s10853-006-0637-z
16. Anuradha R, Sreevidyaa V, Venkatasubramania R, Rangan BV (2012) Modified guidelines for geopolymer concrete mix design using Indian standard. Asian J Civ Eng (Building Housing) 13:353–364. https://doi.org/10.1007/s10853-006-0523-8
17. Patankar SV, Ghugal YM, Jamkar SS (2015) Mix design of fly ash based geopolymer concrete. Adv Struct Eng 3:1619–1634. https://doi.org/10.1007/978-81-322-2187-6
18. Tam VWY, Gao XF, Tam CM (2005) Microstructural analysis of recycled aggregate concrete produced from two-stage mixing approach. Cem Concr Res 35:1195–1203. https://doi.org/10.1016/j.cemconres.2004.10.025
19. Abbas R, Khereby MA, Ghorab HY, Elkhoshkhany N (2020) Preparation of geopolymer concrete using Egyptian kaolin clay and the study of its environmental effects and economic cost. Clean Technol Environ Policy 22:669–687. https://doi.org/10.1007/s10098-020-01811-4
20. Tempest B, Snell C, Gentry T et al (2015) Manufacture of full-scale geopolymer cement concrete components: a case study to highlight opportunities and challenges. PCI J 60:39–50. https://doi.org/10.15554/pcij.11012015.39.50
21. Thaarrini J, Dhivya S (2016) Comparative study on the production cost of geopolymer and conventional concretes. Int J Civ Eng Res 7:117–124

A Critical Appraisal on Green Building Design by Utilizing New Materials and Techniques

Nitu, Rajesh Kumar, Vanita Aggarwal, and Surinder M. Gupta

Abstract Green construction advancement is perhaps the most moving subject all over the planet, advancing to decrease the significant impact of the improvement business on the environment, society, and economy. The world urgently needs down-to-earth and sharp improvement as pollution and a risky barometrical deviation are hastily increasing. Developing countries such as India, Sri Lanka, Pakistan, and others are far behind in terms of creating effective new development that is not harmful to biological system advancement. Several scholars and specialists have conducted extensive research on green structure over the years. This paper means to basically evaluate the several types of research related to green structures materials and advancements. A basic examination of writing has been conveyed to investigate Green Building Construction (GBC) execution in India, which comprises pertinent manageability works in development materials, building rating apparatuses, and comparative studies to upgrade plan efforts. As energy assumes an imperative part in green structures, one of the most important arrangements is to achieve sustainable natural engineering. Sustainable engineering attempts to manage nature to beat it and use renewable energy rather than fossil fuels. Thus, the usage of intelligent materials diminishes energy utilization in green structures with a sustainable improvement approach. This work can help scientists and students better appreciate the green structure drifts and discern constraints in green structure development and clear the way for future research. This paper addresses G-IoT innovations that form a green and smart world, reducing contamination and diminishing energy consumption.

Keywords Green structure · Sustainable · Environment · Smart materials · Sustainability · Green technology

Nitu (✉) · R. Kumar
M.M. Engineering College, M.M (Deemed to Be University), Mullana, Ambala 133207, Haryana, India
e-mail: nbalhara884@gmail.com

V. Aggarwal
Civil Department, M.M. Engineering College, M.M (Deemed to Be University), Mullana, Ambala 133207, Haryana, India

S. M. Gupta
Department of Civil Engineering, NIT Kurukshetra, Kurukshetra 136119, Haryana, India

K. K. Pathak et al. (eds.), *Latest Developments in Civil Engineering*, Lecture Notes in Civil Engineering 352, https://doi.org/10.1007/978-981-99-2676-3_3

1 Introduction

A Green Building, otherwise called a Sustainable Building, is a design that is intended to be built, changed or reused economically and asset scrupulously. They achieve specific goals such as occupant well-being security, water, energy, and other asset protection, and overall environmental impact reduction.

Green development techniques advance sound developments that benefit the two tenants and the environment. These procedures, regardless of whether they are sustainable, can possibly radically diminish the pessimistic effect of building plan and development on the two individuals and the environment. The utilization of energy and materials in development can significantly affect worldwide ecological change.

A major piece of the green structure idea is improving individuals' lives and addressing the requirements of the future. Natural insurance, general well-being, and social advantages are focused on. Private energy utilization is the most elevated, everything being equal, representing 61% of complete power utilization; this proportion is supposed to ascend by 5.5% throughout the following five years. From that point onward, it's indispensable to zero in on home energy productivity by establishing agreeable conditions that limit the interest in mechanical cooling and heating. The government has expected different rules as an underlying move towards guaranteeing compliance with the environment-cognizant plan.

Nonetheless, the weight on the construction industry for working at a quicker pace and with modest materials frequently prompts forfeiting adherence to rules. The cost of energy has amplified due to the decrease in fossil fuel supply all over the planet. Accordingly, nations all over the planet have started sustainable techniques through the formation of strategy instruments.

With regard to the planning stage, the draftsman plans the structure through advanced apparatuses that anticipate, compute, and gauge ecological execution attributes [1]. The environmental assessment instruments for structures have been created to assess indoor natural quality, asset usage, biological loadings, and so on [2].

These devices present different techniques to characterize the rules of green structures. Those issues tended to by the apparatuses may impact ecological policies, designs, and building rehearses. The evaluation approaches play a few roles; they work with figuring out the impact of structures on normal frameworks, promoting green buildings, and tending to manageability [2].

Likewise, the development area turns into the potential contributor to manageable improvement at a huge level. The essential diagram behind green structure embraces effective decrease in consumption of water, energy, and different assets, reusing waste, and bracing tenants' well-being. The green structure utilizes elective energy sources like sun-oriented energy to restore electrical energy during the day time.

The water utilizes an example that likewise varies in green working as water gathering frameworks are utilized, or the water supply system is planned so that water is productively reused before definite removal. A Green building is an energy-proficient structure that enhances water use, to some degree depends on natural wellsprings of energy, produces less waste, and gives superior indoor air quality to its inhabitants. It by and large seems like a traditional structure.

Decreased use of concrete in the construction, advancing sustainable power, façade, and envelope are some of the prerequisites of green structures. It brings about improved solace level and productivity of its inhabitants. The green structure idea has become more important in the business field to get more effective representatives' working.

Through different specialists, the elements of the green structure can be summed up as:

1. Resourceful utilization of water, energy, and different assets.
2. Use of sustainable power, like sun-based energy and wind energy.
3. Sufficient indoor natural air quality.
4. Use of non-poisonous, moral, and feasible materials.
5. Taking into account the natural impacts during the planning, development, and activity stages.
6. Considering the distinction of life of inhabitants during the planning, development, and operation stages.
7. It should be conceptualized as a plan that permits variation in an evolving environment.

Going to different lengths to ration natural assets, diminish the carbon impression, and advance productive energy utilization methods is one reason for changing to Green IoT. Green IoT centres around diminishing energy. Green IoT is characterized as "the review and practice of planning, utilizing, producing, and discarding servers, PCs, and related subsystems, for example, screens, capacity gadgets, printers, and correspondence network frameworks productively and really with insignificant or no effect on Environment." Due to the advancement of green IoT innovations, green IoT is turning out to be more proficient through risky emanations, diminished asset utilization, and decreased contamination. In this manner, green IoT centres around green creation, green use, green plan, and green removal.

2 Scenario of Green Building in India

Like other non-industrial nations, India is going through an uprising in the development area, which results from a developing populace and urbanization. Subsequently, there should be an expansion in mindfulness about saving the environment. India has a huge part of metropolitan areas with populations of more than 1,000,000. The most populated metropolitan cities, i.e., Mumbai, Delhi, and Kolkata, have an extended populace of in excess of 18 million, 16 million, and 14 million, separately (Registrar General and Census Commissioner, India, 2011).

Moreover, it has been exhibited that the materials used before times, which were promptly accessible privately, were ecologically positive. Various new materials and building innovations, then again, present significant natural dangers and impacts because of their inadequate arrangement without designing judgement.

In India, administrative associations have even commanded that structures accomplish a base degree of green rating and offered motivators, for example, upgraded "Floor Area Ratio" (FAR). The census lodging figures, then again, portray the present status of the lodging stock. From a logical, mechanical, and cultural point of view, the designed, semi-designed, and non-designed structures decay with age and natural factors and may not endure dangers like a quake, as clear from the preliminary performance contextual investigations.

Some of them are totally reliant upon human mistakes; other peculiarities are a characteristic rot of the material yet can be made worse by man's carelessness. These conclusions validate the request and proper engagement of the AEC business to develop green and evaluate the ongoing structure stock both for underlying condition/well-being and greenness.

Indeed, even the programmes add to the improvement of administrations in certain urban communities. Under these missions, city improvement and redevelopment and greenfield enhancement are of prime concentration. The public authority has proposed new, effective waste administration practices for metropolitan areas that could promote a healthy environment.

Efficient power energy hall undertaking will keep up with energy prerequisites, and in this manner, help in accomplishing a sustainable environment. Vehicle Emission force declines with the execution of Bharat Standard VI vehicles, which could likewise assist with accomplishing supportability [3].

3 Alternative Materials and Technologies

Buildings use a significant amount of nonrenewable energy and impressively add to GHG outflows. Due to the intricacies of the issues, practical development is seen as a possible future arrangement. Since it represents 80–90% of energy utilization all through a construction's life cycle, the functional stage is more significant.

Wang et al. [4] explained that the expansion and distribution of green construction materials are critical in the field of green building due to sustainable resources and energy. As a fundamental component of a structure, wall materials can create expansive views in a greenway. Their analysis indicated that nations' green building growth strategies were organized around four pillars: establishing the policy framework, implementing primary education, expanding partnerships, and developing economic incentives to support the aims and implementation approach. Additionally, they identified a few materials that may be substituted to increase structural sustainability. For instance, recyclable building materials can be used to manage construction waste in an environmentally responsible manner. Natural insulating material is an excellent option for agricultural waste because it is environmentally friendly and performs incredibly well. Photochromic glass may be utilized to conserve energy in cities and provides a vitreous solution for improving indoor air quality and creating a healthy environment. The development and implementation of green construction materials continue to face several obstacles, but they will be more widely used in the near future.

Siksnelyte-Butkiene et al. [5] discussed the European Commission's newly implemented Renovation Wave Strategy, which aims to improve buildings' energy performance. This could be achieved by appropriately choosing the correct thermal insulation materials. Also, it becomes the utmost widespread strategy that efficiently decreases the energy demand of buildings [6]. The researchers found that choosing a suitable material is one of the most significant challenges and complex stages of a building project. The researchers tried to bring about an exhaustive assessment of prevailing applications for choosing insulation materials. The existing MCDM provides significant insights to abridge the approaches' procedure and choose the criteria for future research. The MCDM was earlier also studied by Streimikiene et al. in the year 2020. The study provided the chief advantages and disadvantages of various techniques determining which MCDM technique is the most suitable for diverse questions.

Sustainable materials are used in the construction of green buildings as energy harvesting concrete, martian concrete, bacterial concrete, light emitting cement, wood foam, and thermally insulating gypsum composites incorporating aerosol and metal roofing. It possessed good compressive and flexural strength.

Green IoT Technology interfaces everybody in the smart world, so the utilization of energy by IoT innovation is a difficult and appealing exploration region. The G-IoT is seen as the IoT of the future that is not harmful to the ecology. Greening ICT innovation plays an important role in G-IoT and provides numerous benefits to society, such as proficient creation and reducing the energy used to develop and distribute ICT gadgets and supplies. Green advances are becoming essential tools for practical development because they can be used to ensure that people have access to clean water and perfect and reasonable energy, live in less hazardous conditions, manage natural assets more productively and honestly, and have proficient systems. Recently, there has been an increase in the use of eco-friendly objects, green development, green rooftops, the use of sustainable energy sources, energy reserve funds at home, awareness of the need of reusing, and so on. Consequently, natural innovations empower the acknowledgement of the advantages presented by new advancements, while limiting the effect on climate and energy utilization. After various conversations by certain creators on green ICT and green IoT, as a rule, depend on green correspondence innovations and green shrewd frameworks, and the end is that the innovations that make up green ICT are: Green RFID, Green remote sensor organization, Green distributed computing, Green M2M, and Green Data Center. Data and correspondence advances are imbued in our regular routines that supporting a general public without them would be troublesome. Numerous things need an extraordinary concentration in the space of normalization, security, and administration for the smooth working of the Internet of Things that can help society in general. These advancements mean lower energy consumption through more productive utilization of available green foundations. Following several discussions by various creators about what green IoT comprises in regard to green communication improvements and green intelligent grids, it is assumed that the developments that build green ICT are recorded as beneath in Fig. 1 (Table 1).

4 Green Building Concept and Design

Akadiri et al. [7] established a theoretical foundation for integrating sustainability ideas into architectural design. Following a review of the literature, each topic pertaining to the techniques and procedures to be used throughout the life cycle of construction projects is presented, along with a few case studies to illustrate the approaches. The framework aided design teams in striking a balance between economic, social, and environmental concerns, influencing how construction practitioners view the statistics they use to evaluate construction projects and promoting the building industry's sustainability.

Ahmed et al. [8] conducted a critical analysis of the green building concept's implementation in Ghana. The research established the exercise's then-current state in Ghana and the drivers responsible for it. The recommendations of Ghana's construction industry's drivers were thoroughly explored. Additionally, major initiatives to guarantee rapid uptake for improved perception agreement in Ghana and Sub-Saharan Africa were advocated (SSA). Among the methods are a national education, awareness, and publicity campaign, the enactment of required government rules and policies, the provision of financial and market-based incentives, and a variety of additional measures.

Varma et al. [9] examined green design as a dynamic transformation of contemporary architecture in developing countries. The researchers did research and presented their findings on the fundamentals and aspirations of green architecture, as well as the ideas and opinions stated in the field, and on successful examples of environmentally friendly buildings in India.

Nazer and Rodrigues [10] evaluated the thermal capacity of a typical residential unit. They offered interferences that would reduce the company's dependency on motorized space conditioning during cooling seasons. Through the use of dynamic simulation modelling, a parametric analysis was constructed, which involved multiple iterations analysing alternative fenestration designs and wall thermal transmittance values. The simulation assumptions were validated by enhancing the model's accuracy using the findings of a longitudinal survey of tenants in 145 identical flats. The survey collected data on residents' thermal perception and behaviour, socioeconomic status, and the physical qualities and function of the building. The researchers concluded that small design tweaks, such as adding natural ventilation during certain seasons, could enhance thermal performance by up to 45%. In Jordan, the optimal glass-to-wall ratio for more energy-efficient residential buildings was set at 15–20% in all orientations. According to the researchers, this could be a better

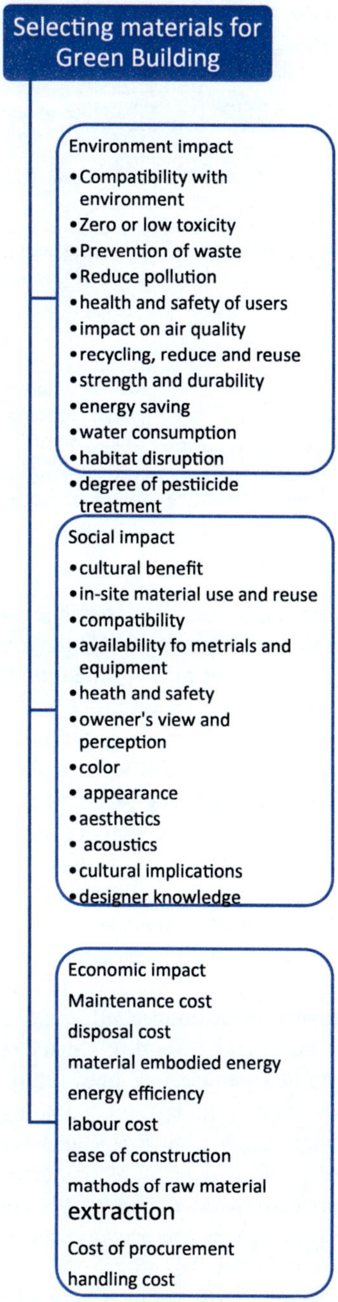

Fig. 1 Factors for selecting suitable and sustainable materials in construction

Table 1 Green mechanism and strategies for G-IoT

Green technologies	Energy saving mechanisms	Energy saving strategies
Green RFID	Existence of active tags and passive sensors	Energy productive Calculations and conventions, decreasing the size of RFID labels because of reusing Creation of marks on reused paper substrate
Green WSN	Reducing communication between nodes-sleep mode	Smart methods of activity through powerful energy the board procedures New energy productive directing calculations
Green cloud computing	Making equipment arrangements pointed towards delivering gadgets that consume less energy	Utilization of assets that are eco-accommodating and keep up with registering execution
Green machine to machine (M2M)	Normal energy-saving instruments, use of productive correspondence conventions	Group-based strategies low-mobility-based optimizations

approach to assisting designers in selecting the ideal building component qualities and orientation for each function at the early design stages.

Hwang et al. [11] identified flaws in Taiwan's present building envelope and energy conservation index practises for adjusting to a changing environment. They proposed a new metric for thermal transfer efficiency known as the total thermal transfer value (OTTV). Based on the simulation results, the new OTTV updates the prior baseline criterion. This research also proposed a way for updating the OTTV index in response to environmental change. They discovered that as the temperature rises, the coefficient in the OTTV equation must be changed over time to adapt to the changing environment. The method described in this article shows how an environment-responsive index could be established in the future to achieve a more energy-efficient construction.

Wakefield-Rann et al. [12] conducted an in-depth investigation on an exemplary profile of positioned information that will serve as the foundation for future, integrative, and cross-disciplinary research into indoor ecologies. Among the subjects studied were design, architecture, social and human environmental science, environmental psychology, biotechnology, spatial sciences, sociology, mycology, statistics, engineering, philosophy, and geography. The research found a number of difficulties and provided useful recommendations for overlapping epistemic kinds and areas of concern indoors. The study's objective was to identify and emphasize critical sites of divergence and convergence in order to develop a new transdisciplinary technique for indoor exploration.

Among the subjects studied were design, architecture, social and human environmental science, environmental psychology, biotechnology, spatial sciences, sociology, mycology, statistics, engineering, philosophy, and geography. The research found a number of difficulties and provided useful recommendations for overlapping epistemic kinds and areas of concern indoors. The study's objective was to identify and emphasize critical sites of divergence and convergence in order to develop a new transdisciplinary technique for indoor exploration. They discovered that players in the building construction business are hesitant to deploy green construction approaches on a discrete basis. Specifically, using game theory, the administrative trials of various players in the building construction industry predict self-centred choices. Individual preferences were incorporated into the group choice, resulting in a preference for non-adoption over adoption among stakeholders in the building construction sector.

The end result was a case study of a lakeside resort built employing green building and design with nature principles, boosting the lake's potential for aesthetic delight and environmental sustainability.

Baheta et al. [13] reviewed new advancements in thermoelectric air conditioning systems for building applications, such as thermoelectric radiant panel ceilings, thermoelectric air duct systems, and thermoelectric cooling facades.

Manso et al. [14] reviewed the literature to identify the benefits and costs of green roof and wall systems. They demonstrated a different method of carrying out the same operation. Green roofs, also known as eco roofs, living roofs, or vegetated/vegetative roofs, are any method of greening roofs that permits a variety of vegetation species to grow on the tops of structures [15–17]. Their research anticipated an examination of the present benefits and costs associated with various types of green roofs and walls. The investigation uncovered scant data on negligible benefits, such as improved quality of life and well-being. Additionally, there are few researches that quantify the advantages and costs associated with green walls. Diverse system characteristics, the surrounding environment, local weather conditions, and the building envelope all contribute to a significant degree of statistical irregularity [18].

Battista et al. [19] evaluated the impact of cool pavements and green roofs on building energy performance. The urban heat island effect has been shown to enhance the effect of global warming, with practical implications for building energy usage. The heating effect increases the need for air conditioners, whereas the cooling effect aids in chilling homes and expels hot air from outside, decreasing the cooling effect of night time air. The researchers wanted to know how urban heat island mitigation measures like cool pavements and green roofs affected building energy performance. Temperature data were used to calibrate an ENVI-met model of a densely populated Rome neighbourhood. Then, in order to define the boundary conditions, an energy analysis was performed using TRNSYS to investigate the energy behaviour of a reference building in its urban setting.

Shree et al. [20] conducted a study to determine the carbon footprint of an academic building in India. They defined the entire process of carbon footprinting more skillfully in their paper, which assesses the consequences of carbon emission assessment in a specific process from cradle to grave. As cities expand at a breakneck pace, vast volumes of pollutants are created across all industries, most notably the building industry. The researchers attempted to quantify the emissions linked with an academic building in this work by utilizing the life cycle assessment technique included in GaBi education software. Following an examination of the process's carbon footprint, various corrective steps were proposed to help lower the process's carbon footprint and make it more sustainable. Additionally, the study aided in determining the effect of various architectural factors on emission levels and made recommendations for modifying the structure to generate fewer pollutants.

Das et al. [21] investigated fibre-reinforced polymers that were discovered to be constructed of a thermosetting petrochemical resin or matrix and synthetic fibres such as glass or paper. Additionally, the study discovered a developing demand for high-performance natural fibres, notably bast fibres (flax, hemp, and jute). As a result, the sector incorporates a number of noteworthy novelties at each stage of the composite manufacturing process, from fibres and their precursors or preforms through manufacturing techniques and allied services. Finally, the study summarized current trends in the production and characterization of green polymer-based composite materials.

Through the creation of a new wall and roof panel system, Khan et al. [22] examined the energy efficiency of a building envelope. Ferro Cellular lightweight-concrete Insulated Panel Assembly was the researchers' name for a structure (FCIPA). The study discovered that improving a building's thermal comfort subtly reduces its energy consumption throughout the year. Numerous aspects, such as the building's orientation and the kind of window glass, are inextricably linked to reducing the building's cooling and heating loads. The most energy-efficient building, in theory, is north–south oriented, has a longer east–west axis, and uses grey glass for window/door openings.

5 Conclusion and Future Recommendation

To a greater or lesser extent, sustainability requirements are integrally related. The challenge for designers is to blend these various manageability criteria creatively. The expansion of green buildings also brings a plethora of obstacles. While green design appears to be more environmentally benign than conventional construction, the prices are much higher. Similarly, a few experts have explored the astonishing prevalence of green architecture in cosy retreats. According to Paul and Taylor [23], after the detailed literature survey on green building materials and techniques, it is seen that even after much research has been carried out, there are some limitations. There are minimal suggestions for alternate construction materials as per the literature studied. Limited or no study was available on indoor environmental quality and productivity. Though there are a few comparative studies for various green building tools, no concrete outcome has been presented. Also, there is a lack of knowledge and methodologies for assessing the cost analysis of converting old buildings to new green buildings. Thus, using more efficient renewable sources of energy for the construction/maintenance of buildings is required. Thus, the present work suggests distinctive green building design practices that have a consolidative tactic, that is, progressing in design while employing green building systems and techniques as future scope of work.

Significant perspectives are connected with sustainability where the quick improvement of IoT innovations should be painstakingly referenced according to the perspective of assets and climate to guarantee adjusted and economical advancement of IoT products. The advantages of this innovation are most prominent in a reduction in energy consumption as well as a reduction in carbon dioxide emissions, which directly effects climate change and reduces global contamination while also reducing costs. It promotes economic development.

References

1. Morledge R, Jackson F (2001) Reducing environmental pollution caused by construction plant. Environ Manag Health 12(2):191–206
2. Cole RJ (2000) Editorial: cost and value in building green. Build Res Inf 28(5–6):304–309
3. Patel MS, Rangrej S (2021) Experiences of green infrastructure in urban planning. Int J Res Eng Sci 9(1):11–13
4. Wang H, Chiang P-C, Li YCC, Wang X, Chen T-L, Wei S, Huang Q (2018) Application of wall and insulation materials on green building: a review. Sustainability 10:3331. https://doi.org/10.3390/su10093331
5. Siksnelyte-Butkiene I, Streimikiene D, Balezentis T, Skulskis V (2021) A systematic literature review of multi-criteria decision-making methods for sustainable selection of insulation materials in buildings. Sustainability 13:737. https://doi.org/10.3390/su13020737
6. Amani N, Kiaee E (2020) Developing a two-criteria framework to rank thermal insulation materials in nearly zero energy buildings using multi-objective optimization approach. J Clean Prod 276:122592
7. Akadiri PO, Chinyio EA, Olomolaiye PO (2012) Design of a sustainable building: a conceptual framework for implementing sustainability in the building sector. Buildings 2:126–152. https://doi.org/10.3390/buildings2020126
8. Ahmed K, Hatira L, Valva P (2014) How can the construction industry in Ghana become sustainable?
9. Varma K, Chaurasia M, Shukla P, Ahmed T (2014) Green building architecture: a literature review on designing techniques. Int J Sci Res Publ 4(2). ISSN 2250-3153
10. Nazer H, Rodrigues L (2015) Optimizing housing design to improve energy efficiency in Jordan. In: 14th International conference on sustainable energy technologies SET 2015—25th to 27th August 2015, Nottingham UK
11. Hwang RL, Shih WM, Lin TP, Huang KT (2018) Simplification and adjustment of the energy consumption indices of office building envelopes in response to environment change. Appl Energy 230:460–470
12. Wakefield-Rann R, Fam D (2018) Initiating a transdisciplinary conversation to improve indoor ecologies. Human Ecol Rev 24(2)
13. Baheta AT, Looi KK, Oumer AN, Habib K (2019) Thermoelectric air-conditioning system: building applications and enhancement techniques. Int J Air-Cond Refrig 27(02):1930002
14. Manso M, Teotonio I, Silva CM, Cruz CO (2021) Green roof and green wall benefits and costs: a review of the quantitative evidence. Renew Sustain Energy Rev 135:110111. https://doi.org/10.1016/j.rser.2020.110111
15. Shafique M, Kim R, Rafiq M (2018) Green roof benefits, opportunities and challenges—a review. Renew Sustain Energy Rev 90:757–773
16. Vijayaraghavan K (2016) Green roofs: a critical review on the role of components, benefits, limitations and trends. Renew Sustain Energy Rev 57:740–752
17. Besir AB, Cuce E (2018) Green roofs and facades: a comprehensive review. Renew Sustain Energy Rev 82:915–939

18. Seiwert A, Robler S (2020) Understanding the term green infrastructure: origins, rationales, semantic content and purposes as well as its relevance for application in spatial planning. Land Use Policy 97(June):104785. https://doi.org/10.1016/j.landusepol.2020.104785
19. Battista G, de Lieto Vollaro E, de Lieto Vollaro R (2021) How cool pavements and green roof affect building energy performances. Heat Transf Eng. https://doi.org/10.1080/01457632.2021.1874667
20. Shree V, Nautiyal H, Goel V (2021) Carbon footprint estimation for academic building in India. In: Muthu SS (eds) LCA based carbon footprint assessment. Environmental footprints and eco-design of products and processes. Springer, Singapore. https://doi.org/10.1007/978-981-33-4373-3_3
21. Das PP, Chaudhary V, Mishra S (2021) Emerging trends in green polymer based composite materials: properties, fabrication and applications. In: Sharma B, Jain P (eds) Graphene based biopolymer nanocomposites. Composites science and technology. Springer, Singapore. https://doi.org/10.1007/978-981-15-9180-8_1
22. Khan MY, Baqi A, Talib A (2021) Energy efficiency analysis of a building envelope. In: Bose M, Modi A (eds) Proceedings of the 7th international conference on advances in energy research. Springer proceedings in energy. Springer, Singapore. https://doi.org/10.1007/978-981-15-5955-6_160
23. Paul WL, Taylor PA (2008) A comparison of occupant comfort and satisfaction between a green building and a conventional building. Build Environ 43:1858–1870

Design of Pre-engineered Building (PEB) Structural Frame

Kiran Hubli and R. J. Fernandes

Abstract In recent years, priority has been given to structures that are cost-effective, energy efficient, quality efficient, and has a short construction period. The current technology allows the structural engineer to design the section and profile as per the strength requirement. Many researchers have designed Pre-Engineered Buildings (PEB) using a direct approach to structural software, where the cross-section of PEB components was randomly assumed without any validations. In such a case, the drawback is that the time consumption to arrive at an effective section is high. This paper focuses on drawbacks and a method has been evolved to design the components of PEB which is a prior requirement to be considered for analysis in any structural software. The design analysis of the PEB frame is been carried out through plastic analysis and the validation is been done using Indian Standard codes. The obtained design section was analyzed in STAAD Pro software with the optimization of the sectional profile which reduces the steel quantity.

Keywords Pre-engineered buildings · Design analysis · Plastic analysis method · STAAD Pro V8i · Indian standard codes

1 Introduction

Pre-Engineered Buildings (PEB) are steel structures whose primary components are pre-fabricated in the factory and then shipped to the site; assembled using bolting and fastenings. In general, PEB has tapered rigid columns and rafters which are built-up welded sections and are joined or linked through connection plates using

K. Hubli (✉)
Civil Engineering Department, KLE Technological University, Dr. M. S. Sheshgiri College of Engineering and Technology, Belagavi 590008, Karnataka, India
e-mail: kiranhubli07@klescet.ac.in

R. J. Fernandes
Civil Engineering Department, SDM College of Engineering and Technology, Dharwad 580002, Karnataka, India

K. K. Pathak et al. (eds.), *Latest Developments in Civil Engineering*, Lecture Notes in Civil Engineering 352, https://doi.org/10.1007/978-981-99-2676-3_4

high-strength bolts. These structures are generally adopted for large-scale infrastructures like industrial buildings, metro stations, railway workshops, warehouses, supermarkets, airports, factory buildings, and others. This structure proves to be economical and easy in fabrication and construction as opposed to traditional steel buildings when higher span widths are adopted.

This paper elaborates on the manual design of PEB simple framed structure subjected to all loads under consideration. The design includes each component such as the design of purlin, sag rod design, and built-up frame design. Purlin's design was met based on the load delivered by the roof covering sheet and the wind force acting on it. The selection of the purlin cross-section was as per IS811:1987. Further, the section was checked for various aspects as per BS5950-5:1998 and IS 801:1975. The sag rod was designed using IS 800:2007 in consideration of the axial load expected in it. The combination of columns and rafters forms a frame for PEB whose cross-section was assumed initially for self-weight, live load, and wind load with their combinations traced as per IS 800:2007. Further, the plastic analysis was followed for different load combinations among DL, LL, and WL. The analysis of the frame was accomplished through 6 possibilities which are Beam Mechanism, Sway Mechanism, Gable Mechanism, Combined Beam and Gable Mechanism, Combined Sway and Gable Mechanism, and Combined Beam and Sway Mechanism to extract the design moments. The highest design moment was considered for the design of the built-up section using the limit state method approach. The obtained and assumed sections were compared and validated through various checks like check for shear, check for bending moment, and combined axial force as per the guidelines of IS 800:2007.

The manual design of components was further validated through STAAD Pro software by optimizing the frame, i.e., through tapering the longitudinal section. On tapering the PEB frame section, a considerable reduction in the weight of the structure was found.

2 Literature Review

Patil et al. [1] have compared the PEB with traditional steel buildings with structure weight, analysis, design, and cost. In comparison to CSB, they disclosed that the PEB is lighter, rapid, and efficient in drafting, analyzing, and optimizing; easy and rapid procedure in software; lower cost as approach leads to appreciable savings while designing, fabricating, and erecting. Patil and Deshpande [2] conceived and designed the PEB structure having different geometries and parameters and presented the results based on self-weight, cost, and time of construction of different models. They concluded that in comparison with a traditional steel building, PEB is lighter in structural weight and economical for large span structures, also on average, its construction period is up to 30% lesser than a traditional steel structure. Chavanke and Tolani [3] have carried out an investigation and designed the PEB and traditional steel building for spans widths like 15, 20, 25, and 36 m using structural software.

They reported that PEB is economical for larger spans preferably more than 25 m at a 10–20% rate, whereas uneconomical for smaller spans (<25 m) at a 30–50% rate. Jayaraman et al. [4] carried out a comparison study between limit state and working stress methods for the design and economics of purlins and roof trusses. They stated that the limit state method is less efficient than the working stress method in load-carrying capacity, stability, structure weight, and cost. Nakum et al. [5] compared analyzing the CSB and PEB structural material specifications using STAAD-pro and designed the same using IS: 800:2007. They concluded that with a rise in structure buildup area, the cost is effectively less against the small area, they also mentioned that the use of fy345 material leads to effectiveness in cost. Goswami and Shende [6] have examined and drafted a typical frame for an Industrial shed using the software STAAD. Pro and compared CSB with PEB.

They concluded that traditional steel building shows less displacement against PEB due to the low structural weight of PEB. The support reaction is more for CSB (Portal frame) against PEB, and in the case of CSB (Truss frame), the loading is nodal, and hence the maximum load is taken care of by the member itself leading to less reaction at support for CSB (Truss frame). PEB building is 45% lighter than that of CSB (Portal Frame) and 27% lighter than the CSB (Truss Frame) structures.. Other studies have described that PEB is lighter, economical, and easy in adaptability and construction than CSB. They also found that the theoretical results of bending moment and load-carrying capacity are much higher in the limit state method of design against the working stress method of design. The authors also showcased that the yield strength in steel 345 MPa is cost-effective when compared to Fy250. On overall glancing of the literature study, PEB is cost-effective, durable, design flexible, adaptable, recyclable, and imparts good strength. In consideration of these literature studies, the problem is defined and work is brought off accordingly.

3 Design Problem Trails

Design an open terrain, general class simple PEB structure having a design life of 50 years for the span of 35 m (c/c) **[Trail 1]**, 25 m (c/c) **[Trail 2]**, and 20 m (c/c) **[Trail 3]**, length of 60 m (c/c), and eave height of 8 m. Adopt roof slope 1:10 (5.71°), Yield Strength of Steel (fy) 345 N/mm^2, Roofing sheet shall be 0.5 mm thick at 50 N/m^2 weight. The percentage of openings shall be less than 20. Assume purlin spacing is 1.5 m c/c, main column spacing 6.0 m, and purlin material's Young's modulus E = 203,000 N/mm^2 [7].

3.1 Design of Purlin

Load Calculations

Dead Load (DL) = 52.5 N/m^2 [8];
Live Load (LL) = 746.28 N/m^2 [9] and
Wind Pressure = 476.33 N/m^2 [10].
For Internal Pressure co-efficient (Cpi) and Percentage of opening ≤ 20%
Internal pressure co-efficient is Cpi = ±0.5.
For External Pressure co-efficient (Cpe).
Height to Width Ratio = 0.23 ≤ 0.5 and Length to Width Ratio = 1.7 > 1.5.

On preparing the wind force chart (F) = [Cp. Pd. A] (kN) in reference with Table 1, considering the highest Wind Load i.e. in X direction, W, wx = 3.43 kN and in Y direction W, wy = 2.86 kN.

Maximum UDL among Load Combinations of 1.5(DL + LL), 1.5(DL + WL), and 1.2(DL + LL + WL), along X axis, Wx = 5222.71 N/m and along Y axis, Wy = 4294.79 N/m.

Bending Moment (M) along x-axis = 19.33 kNm and along y-axis = 5.88 kNm.
Shear Force (V) along x-axis = 15.67 kN and along y-axis = 6.44 kN.
Allowable Tensile or Compressive Stress [11] = fb = 0.6 fy = 207 N/mm^2.
Allowable Shear Stress = fv = 0.4 fy = 0.4 X 345 = 138 N/mm^2.
Section Modulus Required along X and Y axis: Zxx = 93.37 cm^3 and Zyy = 42.58 cm^3.
Considering **Z 260 x 75 x 20 x 3.15 @ 9.37 kg/m** can be adopted from IS 811:1987 [12].

Check for Z Purlin Section [13]

The following were the checks performed as per the recommendation of BS5950-5:1998 [14] and IS 801:1975 [11] which fetched satisfactory results.

Table 1 Wind load co-efficient for wall and roof [10]

Wind angle (θ)	Cpe for wall as per IS 875:2015 (Part 3), Table 5			
	Left (A)	Right (B)	Rear (C)	Front (D)
0°	+0.7	−0.25	−0.6	−0.6
90°	−0.5	−0.5	+0.7	−0.1
Wind angle (θ)	Cpe for roof[a] as per IS 875:2015 (Part 3), Table 6			
	Left (EF)		Right (GH)	
0°	−0.94		−0.4	
90°	−0.8		−0.4	

[a] Interpolation for 5.71° roof slope is taken

Overall Depth, Overall width of compression flange, Width of the lip, Total width over both flanges, Zxx of purlin, Minimum depth of lip, check for stresses on purlin like, check for allowable compressive stress v/s lateral buckling of the flange, check for allowable shear stress and bending stress in web, combined bending and shear stress in web, and deflection check.

3.2 *Design of Sag Rod*

Design Calculation

Axial Force in Sag Rods (P) = 0.15Vx = 2350.5 N.
Sag Rod Length (KL) and Rod Diameter = 1.5 m and 20 mm.

Check on Sag Rod Section [15]

The stress and slenderness ratio check was performed as per IS 800:2007 which was found satisfactory.

3.3 *Design of Built-Up Frame*

Load Calculations

The dead load is considered with the assumption of roof sheet load, purlin weight, and cross-section of column and rafter Web Plate: 1300 mm X 10 mm and Flange Plate: 280 mm × 12 mm is been assumed initially.

Total Dead Load, DL = 8.08 kN/m.
Total Live Load, LL = 2.99 kN/m.
Total horizontal force at eave height. (WL/Eave Height), WL = 14.29 kN.

Plastic Analysis 6 possibilities [16] were carried out for 3 Load Combinations (LC): **LC1** = 1.5DL + 1.5LL, **LC2 =** 1.5DL + 1.5WL, and **LC3 =** 0.9DL + 1.2LL + 1.2WL to arrive design moments as summarized in Table 2.

The highest design moment (Mp) for limit state design is **1148.43 kNm.**

Limit State Design for Built-Up Sections [15]

We have,

$$Z_p = \frac{\gamma_{mo} M_p}{\beta_b f_y} = 4160978.26\,\text{mm}^3$$

where

$\beta_b = 1.0$ for plastic and compact section,

Table 2 Plastic analysis possibilities result

LC1 moment (Mp)	1148.43 kNm	
Moment value (Mp) for	LC2 (kNm)	LC3 (kNm)
Beam mechanism	247.31	221.60
Sway mechanism	85.72	68.6
Gable mechanism	824.64	738.91
Combined beam and gable mechanism	534.29	478.74
Combined sway and gable mechanism	862.74	769.40
Combined beam and sway mechanism	602.87	533.62

Z_p = Plastic section moduli of cross-section,
f_y = yield stress of the material, and
γ_{mo} = partial safety factor [Shop Fabrication—welds].
Assumed Web plate dimension was 1300 × 10 mm.
Therefore, the area required for a flange plate is 3200.75 mm^2.
Flange plate dimensions as 280 × 12 mm can be used.
Hence, Section Moduli provided,
$Z_{p,pro} = 8{,}633{,}320\ mm^3 > Z_p$ OK Safe.
Safe factor percentage $= 100 - \left[\frac{4160978.26}{8633320} \times 100\right] = 51.80\%$.

This value should be greater than 50% from the structural safety point of view. Therefore, provide a **web plate of 1300 X 10 mm and flange plates of 280 x 12 mm** as **trail 1** result.

Check for the Frame Section

The check for shear, Nominal plastic shear resistance under pure shear, check for combined axial force and bending moment, and other geometric shape checks were performed and were found satisfactory.

Trails 2 and 3 Results

Similar manual analysis was performed for **trail 2** (width span of frame 25 m) and **trail 3** (width span of structure 20 m), the arrived sections are **web plates of 850 x 10 mm and flange plates of 190 x 12 mm,** and **web plate of 675 x 10 mm and flange plates of 150 x 12 mm,** respectively.

4 STAAD Pro Software Validation

It is the basic Civil Engineering software to conceive and design the structures like buildings, towers, bridges, and infrastructures. Most engineers use this software for analysis and design of its user-friendly modeling and analysis. Four major steps are performed in this platform, they are modeling, analysis, check and verification of results, and forwarding the analysis result to steel design for designing purposes. The current study is based on steel industrial structural analysis and design, which is manually per-formed for a single 2D portal frame and the same frame is analyzed, validated, and optimized further in 3D form.

The 3D model of the PEB structure was modeled, examined, and designed in consideration with DL, LL, and WL arrived in manual design works, and also the cross-section and longitudinal sections of PEB mainframes were optimized based on the trial-and-error approach and practical concept of PEB structure in discussion with PEB manufactures for all 3 trails as shown in Fig. 1. As a result, Table 3 shows the optimized safe section of the PEB frame of different trails. The steel take-off was reduced at a considerable rate in comparison with manual design.

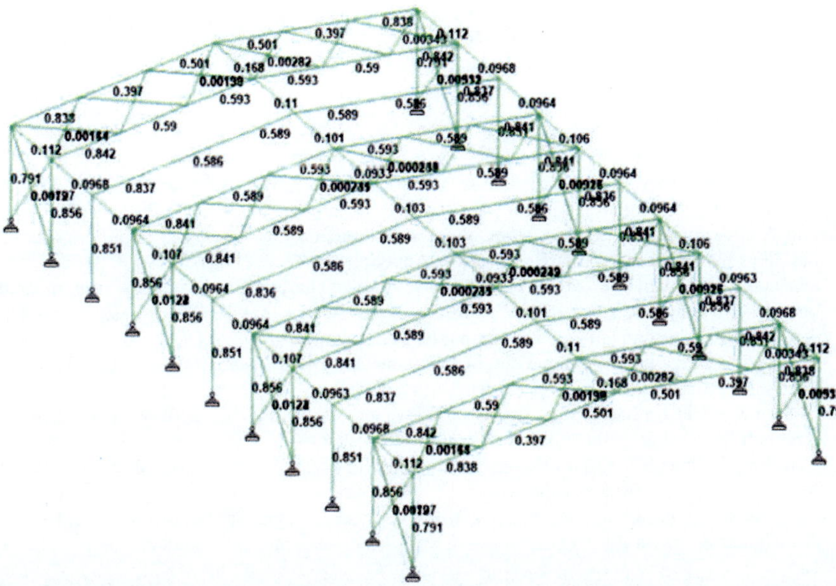

Fig. 1 STAAD pro section passed view

Table 3 Final section list validated through software

Sl. No.	Component	Trail 1	Trail 2	Trail 3
1	Main column	**Web**: 1300–700 × 10 **Flange**: 280 × 12	**Web**: 825–595 × 10 **Flange**: 180 × 12	**Web**: 651–525 × 10 **Flange**: 150 × 12
2	Main/ridge rafter	**Web**: 1300–700 × 10 **Flange**: 280 × 12	**Web**: 825–595 × 10 **Flange**: 180 × 12	**Web**: 651–525 × 10 **Flange**: 150 × 12
3	Main mid rafter	**Web**: 700 × 10 **Flange**: 280 × 12	**Web**: 595 × 10 **Flange**: 180 × 12	**Web**: 525 × 10 **Flange**: 150 × 12

5 Conclusion

This study describes a conformable procedure for the design of a PEB frame that is analyzed in steel structural software for validation and optimization. This method of approach (manual design and software validation) to design a PEB building was found to be less time-consuming.

The total weight of the structure in trail 1 (span width 35 m) is 87218.543 kgs through manual design, whereas, by software analysis, the steel take-off was reduced by 23.11%. Similarly, the total weight of the structure in trail 2 (span width 25 m) is 46462.517 kgs with a reduction of 17.79% and the total weight of the structure in trail 3 (Span width 20 m) is 32156.136 kgs with a 14.20% reduction through software analysis. There is a decrease in steel tonnage with the decrease in span width which is obvious. Further, in terms of per unit length, trail 1 frame has a 15.21% of reduction, trail 2 has a 10.72% reduction, and trail 3 has an 8.41% reduction in weight between straight sections and tapered sections.

References

1. Patil SS, Jadhav RV, Mali PA, Bhanuse MM, Katti MR (2017) Analysis & design of pre-engineered building of an industrial warehouse. Int J Curr Eng Sci Res (IJCESR) 4(12):39–43
2. Patil SS, Deshpande S (2018) A study on the structural analysis and design of pre-engineered buildings for different geometries. Int J Res Appl Sci Eng Technol 6(6):184–190
3. Chavanke SB, Tolani KK (2017) A comparative study of analysis and design of pre-engineered building and conventional steel building for a polymer factory. Int J Sci Res Dev 5(05):883–886
4. Jayaraman A, Geethamani R, Sathyakumar N, Shenbagam NK (2014) Design and economical of roof trusses & purlins (comparison of limit state and working stress method. Int J Res Eng Technol 3(10):199–207
5. Nakum M, Zala J, Shah D (2017) Cost effectiveness of multi storey CSB & PEB industrial steel building. J Emerg Technol Innov Res 4(4):264–268
6. Goswami A, Shende T (2018) Pre-engineered building design of an industrial warehouse. Int Res J Eng Technol 5(6):1484–1488
7. Kirby Technical Handbook (2016) Kirby Building System, Alghanlm Industries, Safat, Kuwait
8. Arunkumar S, Purlin Z (2011) Design—Excel Sheet, Madurai ES Consultancy Services Pvt. Ltd.
9. IS 875: Part 2: 1987: code of practice for design loads (other than earthquake) for buildings and structures, imposed loads

10. IS 875: Part 3: 2015: design loads (other than earthquake) for buildings and structures—code of practice, wind loads
11. IS 801: 1975: code of practice for use of cold-formed light gauge steel structural members in general building construction
12. IS 811: 1987: cold formed light gauge structural steel sections
13. Sankeshware NM, Kulkarni RB (2014) Parametric study and comparison of Indian Standard Code with British Standard Code for the design of light gauge cold formed flexural members for inverted hat section. Int Res J Eng Technol 7(6):3689–3695
14. BS 5950-5: 1998: structural use of steelwork in building, code of practice for the design of cold-formed thin gauge sections
15. IS 800: 2007: General construction in steel—code of practice
16. Shiyekar MR (2010) Limit state design in structural steel. PHI Learning Pvt. Ltd., New Delhi

Experimental Study on Low-Fines Self-compacting Concrete Using Fly Ash Replacement

M. Shinde Jaydeep and K. Ravi

Abstract Smart Dynamic Concrete has a low cementitious material and can have self-compacting properties. Low-Fines Self-Compacting Concrete (LFSCC) was developed in this experimental investigation using cementitious content less than 380 kg/m^3. To achieve economy and to create ecological concrete, by replacement of fly ash with cement which ranges from 10 to 50%. The fresh flow (flowable) properties of LFSCC were investigated as per EFNRAC specifications. Compressive strength, strength for flexural, and strength for splitting tensile are evaluated as per Indian standards specifications. From experimental results, all the mix combinations satisfied ENFRAC specifications in the fresh state of LFSCC. The 32.12 MPa highest compressive strength was observed when fly ash was replaced by 30%. The flexural strength and the splitting tensile strength of the LFSCC decreased as the fly ash content increased. Overall findings suggest that LFSCC can be developed for everyday concrete applications by incorporating fly ash.

Keywords Low-fines self-compacting concrete · Fly ash · Workability · Compressive strength

1 Introduction

Self-Compacting Concrete (SCC) has been distinguished by its low yield stress, moderate viscosity, and high deformability without external compaction. These properties are essential for ensuring the suspension of uniform solid particles throughout the transportation and placement processes without external compaction. This form of concrete enables the casting of substantially reinforced sections in regions where compacting vibrators are inaccessible and in elaborate formwork designs that would

M. S. Jaydeep (✉)
VTU, Belagavi, Karnataka, India
e-mail: jaydeep3002@gmail.com

K. Ravi
AGMR College of Engineering, Varur, Karnataka, India

K. K. Pathak et al. (eds.), *Latest Developments in Civil Engineering*, Lecture Notes in Civil Engineering 352, https://doi.org/10.1007/978-981-99-2676-3_5

be incredible to cast other type of concrete. It produces a surface that is much superior to standard concrete [1].

Since the introduction of SSC in the 1980s in Japan, traditional vibrate concrete should have been replaced to a greater extent by this technology's significant benefits. However, traditional SCC has drawbacks in the form of a high amount of fines and an imbalance between constancy and fluidity due to the mix's sensitivity to changes in concrete ingredients. In addition, it observes that the unit cost of SCC was 30–50% more than that of standard concrete material [2]. The extra cement content and fines required for the SCC, in addition to logistical costs (additional silos, additional mixing period, severer quality control, etc.), increase manufacturing costs. Therefore, the cost per volume of SCC becomes a constraint. Because of this, SCC will never be able to replace traditional concrete [3].

Also, using SSC like everyday concrete isn't cost-effective, especially in Ready Mix Concrete (RMC) industry. This is due to the rise in the cost of binder content as a more refined material used to achieve desirable self-compacting properties. With the incorporation of a modern Viscosity-Modifying Agent (VMA) into a Polycarboxylate Ether (PCE)-based hyper plasticizer, it is possible to lower the overall fines content, as a result of which the unit cost of Low Fines Self Compacting Concrete (LFSCC) is significantly reduced (compared to conventional SCC). Additionally, essential SCC properties have been retained. Thus, it is conceivable to realize the dream of compatible concrete in the range of 25–40 MPa for ordinary concrete. In this direction, a new type of SCC has been introduced to the building construction industry in recent years. A lower quantity of cementitious substantial has been used to obtain self-compacting properties in this Smart Dynamic Concrete [4]. In this LFSCC, the unit cost has been reduced as the total fines content is reduced and also maintaining the self-compacting concrete [1].

Some researchers have developed SCC by substituting cement content with blast furnace slag, powder of limestone, fly ash, powder of lime content, silica fume, etc. Sustainable improvements in fresh as well as hardened characteristics are anticipated when cement is substituted by fly ash [5–7]. Lower grades of SCC can be produced by a low percentage of cementitious content in the range of 340–380 kg/m^3 [8]. Low-fines self-consolidating concrete employs a new synthetic Viscosity-Modifying Agent (VMA) in a hyper plasticizer-based on Polycarboxylate Ether (PCE).

1.1 Importance of Low-Fines Self-compacting Concrete

The self-compacting with low fines will assist engineers, owners, and concrete producers achieve their objectives by increasing production and efficiency. Costly fines SCC learns many good things, like how to save money on fines, increased durability, reduced impact on the enviroment, and reduced user fatigue due to vibrations.

2 Review of Literature

Khatib et al. [9] researched the investigation of incorporating fly ash (FA) on the characteristics of SCC, and he discovered that a large content of FA could be employed in SCC. He concluded that substituting 40% of the cement with FA will increase the strength up to 65 N/mm^2. Naik et al. [10] studied the improvement of strength, cost-effective SCC with 35% to 55% fly ash replacement. They have observed that VMA and PCE are reduced as compared to the usual dosage. They concluded that 62 MPa inexpensive self-consolidating concrete might be constructed with 35% high-volume fly ash. Blends of Metakaolin (MK) and Fly Ash (FA) have been tried as substitutes for cement by Murthy et al. [11]. They created a simple excel-based model for designing SCC mixes with 29% coarse aggregate, cement substitution with FA and MK, mixtures of the two, and regulated SCC mixes with a 0.36 water/cementitious ratio. Dinkar et al. [12] investigated SCC made with cement and varying amounts of FA. They have replaced FA in between 10% to 70% of cement. They noticed that a thirty percent replacement of FA resulted in 100 MPa strength at 56 days.

According to the literature review, there has been a lot of research done on making SSC. Many of the SCC mixes being made now have higher amounts of fines, which has problems and is also not economical [13]. In this direction, there is a necessity for developing low-fines self-compacting concrete for day-to-day applications in the construction industry. The result of low-fines self-compacting concrete is possible at present by using PCE. With the innovation of low-fines SSC, it is possible to develop concretes in grades of 20–40 MPa. Since enough literature is not available at present on low-fines self-compacting concrete, hence further investigation is required.

3 Materials Used

The experiment used cement of grade 43, having a specific gravity of 3.15 (IS: 8112-2013). Artificial sand with a specific gravity of 2.60 was utilized (IS: 383-2002). Crushed stone aggregate has been collected from a nearby quarry. A 12.5 mm and smaller coarse aggregates were used with a specific gravity of 2.74 (IS: 383-1970). Pozzocreat 60, with 2.2 specific gravity was used. The superplasticizer MasterGlenium Sky 8654 PCE was used.

4 Mix Proportions

Table 1 shows the mixed content of proportion for LFSCC mixes with 0, 10, 20, 30, 40, and 50% fly ash as a replacement to cement as a cementitious mate rial. The characteristic strength considered is M20. The mix proportions inwards as per the design of the mix refers to IS:10262-2019 [14]. The total powder content of the six

Table 1 Mix proportions of LFSCC mixes

Mix id	Cement (kg/m^3)	Fly ash (kg/m^3)	Fly ash (%)	Fine aggregate (kg/m^3)	Coarse aggregate (kg/m^3)	Water (kg/m^3)	SP (kg/m^3)	W/P
S20F00	382.87	0	0	1150.26	647.78	180	3.82	1.03
S20F10	344.68	38.29	10	1150.26	635.62	180	3.82	1.1
S20F20	306.38	76.59	20	1150.26	621.18	180	3.82	1.07
S20F30	268.04	114.89	30	1150.26	606.75	180	3.82	1.03
S20F40	229.78	153.19	40	1150.26	592.31	180	3.82	1
S20F50	191.48	191.48	50	1150.26	577.87	180	3.82	0.97

concrete mixtures (cement + fly ash) was limited to 375 kg/m^3. The coarse aggregate content varying from 577.87 to 647.78 kg/m^3 had been maintained by volume. While fine aggregate content of 1150.26 kg/m^3 was maintained by volume. The Water to Powder content (W/P) ratio varies from 0.97 to 1.03 has been adopted. The initial mixture consisted of water content of 180 kg/m^3.

5 Test Specimens Preparation and Casting

In order to thoroughly combine all of the LFSCC's constituent parts, a tilting mixer was put to use. The needed quantities of aggregates, the content of cement, and fly ash were first completely mixed before proceeding to the next step. In order to achieve a good mixing, water and the additive were added. The testing of fresh flow properties of the LFSCC was carried out first, followed by the casting of test specimens for the hardened properties. After leaving all of the cast specimens in their moulds for 24 h at regular temperature, the specimens were removed from the mould and placed for curing until they reached the appropriate age for testing.

Compressive strength tests were carried out with a 150 mm × 150 mm × 150 mm cube size. Beams of 100 mm × 100 mm × 500 mm sizes were used for flexural strength [15]. For split tensile strength, 150 mm diameter and 300 mm length of cylinder specimens were used [16]. Strength for compression, splitting tensile strength, and strength for flexural have been studied as per IS specifications.

6 Testing of Specimens

With reference to EFNARC specification [17], the Slump test (flow test), J-ring flow test, L-box test, V-funnel flow, and U-box test were performed. A compressive strength test was performed at 7 days, 28 days, and 90 days. Flexural and tensile strength tests were performed at 28 days and 90 days of ageing; according to the IS Specification.

7 Results and Discussion

7.1 Properties of Fresh LFSCC

Table 2 summarizes the results of several flow properties for various mix compositions. Figures 1, 2, 3 and 4 demonstrate the diversity in test findings. Fresh flow properties observations indicate that all mixtures with different cement substitutions by fly ash (0–50%) satisfy ENFRAC specifications for LFSCC. Also, it has been noticed that as the proportion of percentage of replacement increases, fresh flow qualities improve in all circumstances.

Table 2 Properties of fresh flow of LFSCC mix

Mix id	Slump flow	V-funnel	J-ring	L-box
	Dia. (mm)	Flow (sec)	Dia (mm)	(H2/H1)
S20F0	625	16	564	0.83
S20F10	638	15	586	0.84
S20F20	660	13.5	601	0.86
S20F30	672	13	605	0.88
S20F40	680	12	618	0.91
S20F50	690	11.5	623	0.93

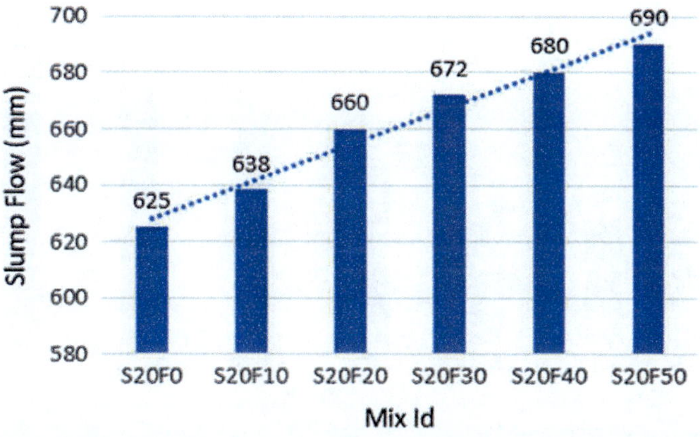

Fig. 1 Slump values of LFSCC mix

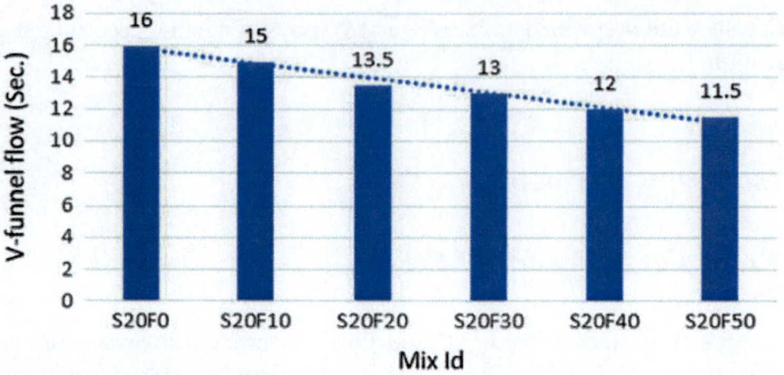

Fig. 2 V-funnel test of LFSCC mix

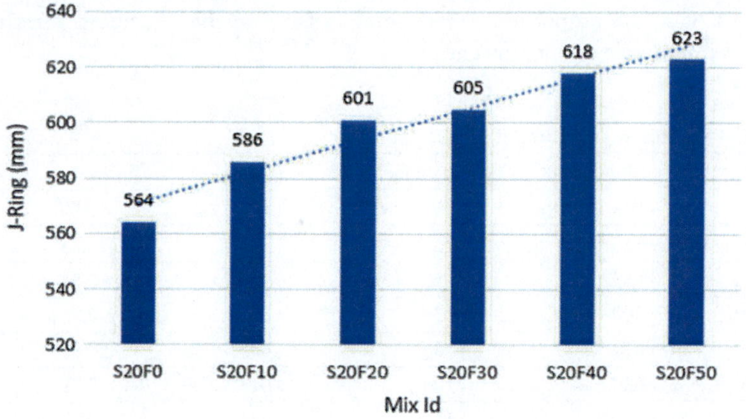

Fig. 3 J-ring test of LFSCC mix

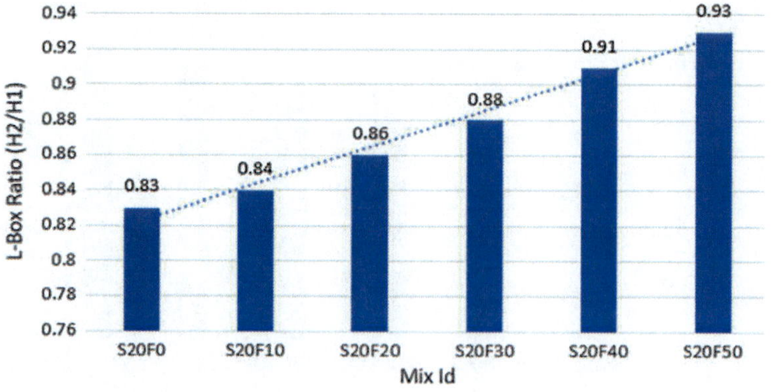

Fig. 4 L-box test of LFSCC mix

All mixes with varying percentages of cement replacements by fly ash are found to produce fresh flow charactristics and satisfy EFNARC standards for LFSCC. Additionally, it has been found that fresh flow qualities always improve as the percentage of replacement rises.

7.2 *Properties of Hardened LFSCC*

7.2.1 Compressive Strength

Table 3 summarizes the results of various strength parameters for different mix compositions. Figures 4, 5 and 6 show the variation in test results. According to the compressive strength test results (Fig. 4), all compressive strength values rise with age. A similar finding was made by Khayat [18] for tests conducted up to 90 days after curing.

Table 3 Strength properties of LFSCC mix

Mix id	Compressive strength (MPa)			Tensile strength (MPa)		Flexural strength (MPa)	
	7 days	28 days	90 days	28 days	90 days	28 days	90 days
S20F0	21.15	26.12	43.35	2.9	3.19	5.32	6.22
S20F10	22.53	27.15	43.44	2.75	3.27	4.72	5.5
S20F20	24.80	29.18	41.14	2.4	3.08	4.17	5.02
S20F30	25.05	32.12	44.32	2.12	2.8	3.87	4.75
S20F40	21.35	28.1	38.22	1.92	2.41	3.66	4.52
S20F50	20.53	25.67	35.93	1.78	2.08	3.37	4.12

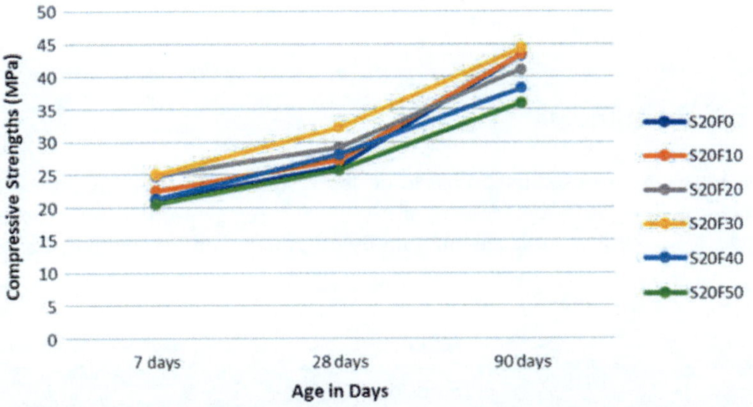

Fig. 5 Compressive strengths of LFSCC mix

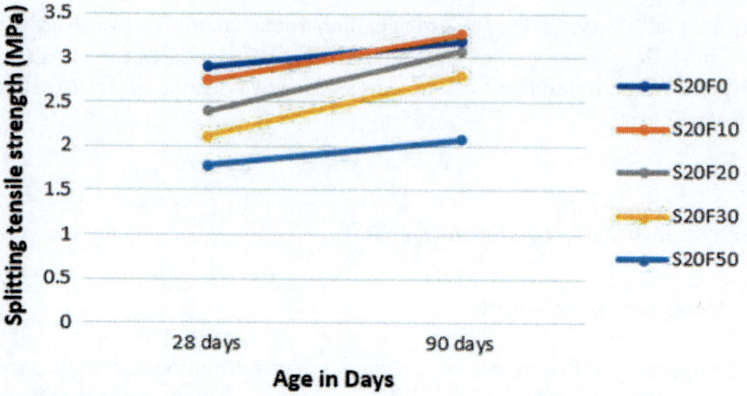

Fig. 6 Splitting tensile strengths of LFSCC mix

Additionally, it has been found that the strength values rise up to a 30% replacement of fly ash as the percentage of fly ash replacement increases. With more fly ash, strength decreases. Up to 30% fly ash replacement, it may have participated in pozzolanic activity; above that, it may have operated as filler. Lachmi et al. [19] found maximum strength at 40% replacement.

7.2.2 Splitting Tensile Strength

For LFSCC, the splitting tensile strength test was conducted on 28 days to 90 days cured specimens (Fig. 6). From test results, it is observed that tensile strength increases when the days of the curing period increase. However, as the fraction of fly ash replacement increases, the split tensile strength is observed to decrease. Comparing 30% replacement with fly ash to 0% replacement, the decrease in split tensile strength is just about 12%. In the case of 50% replacement by FA, the observed reduction in strength was around 39%.

7.2.3 Flexural Strength

The observation was established based on the results of flexural strength tests, as shown in Fig. 7. At 30% replacement of FA, the percentage drop in flexural strength is around 27%. At 50% replacement with fly ash, the flexural strength decreases by 36%.

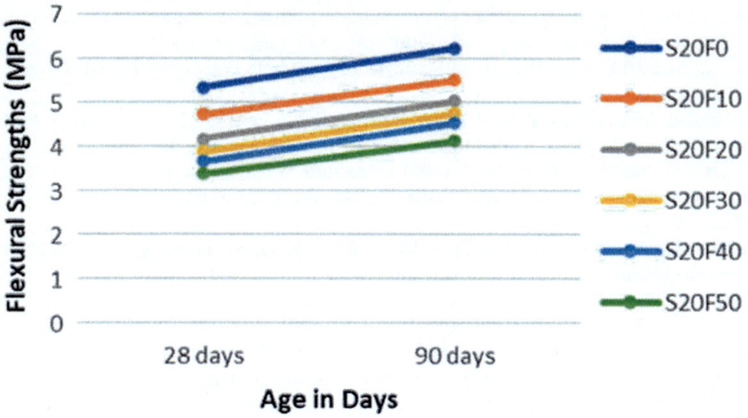

Fig. 7 Flexural strengths of LFSCC mix

8 Conclusion

It is essential to design self-compacting concrete with a range of 20–40 MPa at this time to meet the requirements of concrete applications. This can be accomplished by utilizing a category of concrete known as low-fines self-compacting concrete, which has a low cementitious content (LFSCC). The use of low-fines SCC results in various benefits, including economical, long-lasting, ecological, and ergonomic. To attain the needed SCC properties in the manufacturing of LFSCC, class F fly ash can be utilized in place of cement. In their fresh state, all test results are satisfied as per requirements of the EFNARC specifications, regardless of the percentage of cement replaced by fly ash. The portion of cement replaced by FA results in a decline in both the split tensile strength and the flexural strength of the concrete. The highest compressive strength was measured when fly ash replaced cement at 30%.

References

1. Okamura H, Ouchi M (2003) Self-compacting concrete. J Adv Concr Technol 1(1):5–15
2. Huat SK, Kar N, Qiuling F (2011) The Asian experience in low fines self consolidating concrete (SCC) in everyday applications. In: 36th conference on our world in concrete & structures, Singapore, 14–16 August
3. Kar N, Firth D (2012) Smart dynamic concrete—economic, ecological and ergonomic, In: Proceeding of conference on IQA-CC construction materials industry conference, CMIC12, Melbourne, Victoria, 19–22 September
4. D'Souza B, Yamamiya H (2013) Applications of smart dynamic concrete. In: Proceeding of 3rd international conference on sustainable construction materials and technologies, 18–21 August 2013, Kyoto Research Park, Kyoto, Japan, pp 1–7
5. Jain A, Gupta R, Chaudhary S (2020) Sustainable development of self-compacting concrete by using granite waste and fly ash. Construct Build Mater 262:120516

6. Xie Y, Liu B, Yin J, Zhou S (2002) Optimum mix parameters of high–strength selfcompacting concrete with ultrapulverised fly ash. Cem Concr Res 32(3):477–480
7. Siddique R (2011) Properties of self-compacting concrete containing class F fly ash. Mater Design 32:1501–1507
8. Ashish DK, Verma SK (2019) An overview on mixture design of self-compacting concrete. Struct Concrete 20(1):371–395. https://doi.org/10.1002/suco.201700279
9. Khayat KH, Paultre P, Tremblay S (2001) Structural performance and in-place properties of self-consolidating concrete used for casting highly reinforced columns. ACI Mater J 98(5):371–378
10. Naik TR, Kumar R, Ramme BW, Canpolat F (2012) Development of high-strength economical self-consolidating concrete. Construct Build Mater 30:463–469
11. Murthy NK, Rao N, Reddy IR, Reddy MVS (2012) Mix design procedure for self compacting concrete. IOSR J Eng (IOSRJEN) 2(9):33–41
12. Dinakar P, Reddy MK, Sharma M (2013) Behaviour of self compacting concrete using Portland pozzolana cement with different levels of fly ash. Mater Design 43:161–169
13. Krieg W (2003) Self-compacting concrete: definition, development, and applications, A technical paper presented in the meeting of the ACI, Saudi Arabia Chapter, Eastern Province, October
14. IS 10262-1982 (1989) Indian Standard recommended guidelines for concrete mix design. Bureau of Indian Standards, New Delhi
15. IS: 516-2002, Flexural strength of concrete method of test, Bureau of Indian Standards, New Delhi, India.
16. IS: 5816-1999, Splitting tensile strength of concrete method of test, Bureau of Indian Standards, New Delhi, India
17. EFNARC-Specification and guidelines for self compacting concrete surrey GU 9 7EN, UK February 2002
18. Khayat KH, Bickley J, Lessard M (2000) Performance of self-consolidating concrete for casting basement and foundation walls. ACI Mater J 97(5):374–380
19. Lachemi M, Hossain KMA, Lambros V, Bouzoubaa N (2003) Development of cost effective self-consolidating concrete incorporating fly ash, slag cement, or viscosity-modifying admixtures. ACI Mater J 100(5):419–425
20. Nuruddin MF, Chang KY, Azmee NM (2014) Workability and compressive strength of ductile self compacting concrete (DSCC) with various cement replacement materials. Construct Build Mater 55, pp 153–157
21. Corradi M, Khurana R, Magarotto R (2007) Low fines content self compacting concrete. In: 5th international RILEM symposium on self compacting concrete, 3–5 September, Ghent, Belgium
22. Roncero J, Corradi M, Khurana RS (2007) New admixture system for low fines self compacting concrete. In: 5th international RILEM symposium on self compacting concrete, 3–5 September, Ghent, Belgium
23. Khayat KH, Manai K (1996) Les betons autonivlants: proprietes, caracterisation et applications, colloque sur les betons autonivlants. Universite de Sherbrooke. November 28, Canada
24. IS: 516-1959. Indian Standard for methods of tests for strength of concrete. Bureau of Indian Standards, New Delhi (India)
25. Nagataki S, Fujiwara H (1995) Self-compacting property of highly-flowable concrete. In: Malhotra VM (ed) Am Concr Inst SP 154:301–314
26. Khayat KH (2000) Optimization and performance of air-entrained, self-consolidating concrete. ACI Mater J 97(5):526–535

Utilization of Coal Dust Waste in Bricks an Experimental Approach

Ramakant Agrawal, Deepak Garg, and Pramod Gour

Abstract For developing nations like India, the major part of electricity is produced from thermal power stations where coal is used for burning. In such a process, a large amount of waste is generated which can be classified into three parts, namely fly ash, bottom ash and coal dust waste; out of these wastes, there are no specific efforts to utilize the coal dust waste. Disposal of coal dust waste is not easy as it can harm the living hood of the surrounding area. Coal Dust waste can be a good option for preparing bricks or blocks for low-cost construction. An attempt to use coal dust waste in the preparation of bricks has been made in this study. Coal dust waste is collected from the Korba area of Chhattisgarh and brick samples are prepared with different configurations. Field and lab tests are performed on different samples as per Indian standards. After summarizing the test results, it can be concluded that CDW bricks are cheap and suitable alternatives to conventional clay bricks.

Keywords Bottom ash · Clay bricks · Coal dust waste · Field test · Fly ash

1 Introduction

The population is increasing at a rapid rate in our country; now India is next to China in terms of the population, therefore, there is a huge demand for basic facilities like food, shelter and employment. Providing shelter involves the construction of new houses, apartments and offices. It is very well known that a building requires various partition walls or masonry units, and if we look around us, the most common construction material used for such work is brick or block. Conventional bricks are made of clay or soil that is having a high amount of silica, many of us have visited a brick production plant which requires a lot of freshly excavated soil or clay.

To construct numerous houses, a huge quantity of bricks is needed, especially for a developing country like India. Most of the construction is needed in rural areas, and precast ad volumetric construction technology is not so popular in rural areas.

R. Agrawal (✉) · D. Garg · P. Gour
Department of Civil Engineering, OIST Bhopal, Bhopal, India
e-mail: rkagrawal0017@gmail.com

K. K. Pathak et al. (eds.), *Latest Developments in Civil Engineering*, Lecture Notes in Civil Engineering 352, https://doi.org/10.1007/978-981-99-2676-3_6

Government schemes like PM awas yojna, Indira awas yojna, etc., are running for providing low-cost houses to the poor, which encourages civil engineers to utilize some locally available waste materials in construction and develop an approach for a low-cost sustainable construction practice.

Another problem with conventional bricks is the raw ingredient itself which is fresh soil or clay. There is a huge problem with the availability of suitable clay that is fit for making bricks. Excavation of fresh soil from fields is not a good practice as there is a severe shortage of good soil available for agricultural activities.

One more disadvantage of using clay brick is that for good strength, bricks should be burnt at a very high temperature above a thousand degree centigrade. Burning requires fuel and the emission of large amounts of greenhouse gases like carbon dioxide also takes place. The complete production process of brick takes 4–5 months and a lot of manpower too. On the basis of the above points, it can be easily concluded that for a good sustainable practice, clay brick should be replaced by some other waste material-based block or brick which is locally available at low cost.

As we know in developing countries like India, the major part of electricity is produced from thermal power technology, where coal is burnt and then power is produced. Coal is a mineral obtained from mines. During this entire process, some amount of coal is wasted in the form of powder which is called coal dust waste. Coal dust waste is formed during the following operations performed on coal [1]:

- Coal cutting
- Drilling
- Blasting
- Loading of coal
- Transporting of coal (including conveyor loading and transfer points, tub loading points tippler, bunkers)
- Movement of mobile machinery
- Coal handling/Screening

Coal dust waste, which is a waste material produced during coal mining and transportation activities, has a high potential to be used as the main ingredient for making bricks [2].

The following points justify the above statement:

- It will reduce the demand for clay and hence indirectly it can help in increasing the agricultural yield.
- It has several issues with disposal on land; makes the land barren and reduces its productivity.
- It increases the carbon content in the air and renders the air polluted.
- Manufacturing bricks from CDW is a comparatively easy and fast process.
- Manufacturing of CDW bricks needs very less burning at comparatively low temperature for more strength of bricks. Hence, it is eco-friendly.

In this work, coal dust waste is collected from a dump yard in the Korba area of Chhattisgarh (Fig. 1).

Fig. 1 CDW bunkers near coal mine

2 Literature Review

- Dubale et al. [3] intended to use the waste of ceramic tiles after demolition in the production of bricks. The demolished waste is mixed with laterite soil in different proportions (from 5 to 45%) and the bricks were burnt at different temperatures. The authors concluded that waste can be utilized up to 40% by weight into fired brick.
- Vasić et al. [4] the idea of this study was to use the coal waste in hollow block manufacturing of size 250 × 190 × 190 mm^3 the blocks were burnt at 180 °C temperature ad tested as per European standards.it was concluded that 3% waste is suitable for making hollow blocks.
- Mohammed et al. [5] performed a review study of the characteristics of coal ash to be used as a construction material. They studied the physical properties, such as specific gravity, water absorption, moisture content, fineness modulus etc., and chemical composition. Authors concluded that coal ash is a class F pozollona containing more than 70% silica in its composition.
- Singh and Singh [6] studied the properties of coal dust waste in the Kobra area of Chhattisgarh. They manufactured bricks from this coal dust waste. The manufactured brick has good strength and completely satisfies the requirement of the Bureau of Indian Standard code for common burnt clay bricks for construction purposes.

Table 1 Properties of soil taken

S. No.	Properties	Values
1	Type of soil	Alfisols (alluvial)
2	Colour	Yellowish brown
3	Specific gravity	2.70
4	Bulk density	1.1 g/cc
5	Grain size	0.002 mm

Table 2 Properties of sand taken

S. No.	Properties	Values
1	Specific gravity	2.63
2	Bulk density	1750 kg/m^3
3	Natural moisture content	0.95%
4	Appearance	Rounded (microscope)
5	Sand type	River sand from River, zone-II

Table 3 Characteristics of CDW used

S. No.	Properties	Value
1	Size	1–100 microns
2	Colour	Brownish-yellow
3	Specific gravity	1.32
4	Moisture content	75%
5	Carbon content	40–60%

Source SECL

3 Material Used

The following materials are used during the research work:

- Soil—Collected from river bed; mostly alluvial soil
- Fine aggregates (Sand)—Collected from a local supplier
- Coal dust waste—Collected from dump yard near Korba district
- Water (Tables 1, 2 and 3)

4 Methodology

For the experimental study, the different materials are mixed together. To provide uniformity, the condition of the different samples is kept constant. Five different samples, each containing 20 bricks, were prepared, and the size of the specimen was kept at 190 mm × 105 mm × 75 mm; all bricks were prepared using mould by hand moulding only (Table 4).

The following procedure has been adopted for the manufacturing of CDW bricks:

1. Procurement of materials
2. Preparation of mix sample
3. Moulding
4. Drying
5. Burning

They were further tested for quality and then dispatched to the sites (Figs. 2, 3 and 4).

The following tests are conducted on different samples to assess the quality of the sample prepared [7–10]:

- Compressive strength test
- Water absorption test
- Efflorescence test

Table 4 Proposed mix proportioning

Sample ID	Clay (%)	Fine aggregate (%)	Coal dust waste (%)
S 1	40	10	50
S 2	30	10	60
S 3	20	10	70
S 4	15	10	75
S 5	10	10	80

Fig. 2 Preparation of mix sample

Fig. 3 Preparation of mix sample

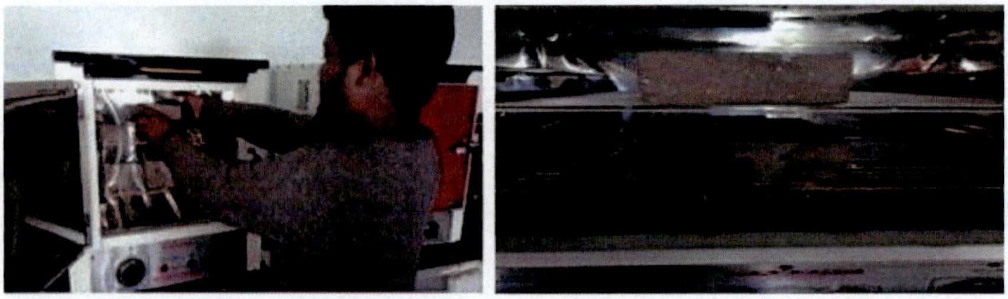

Fig. 4 Placing brick in the oven for drying

- Shape and size test
- Colour Test
- Sound Test
- Hardness test
- Warping Test

5 Results

The above graph shows that compressive strength increases gradually with the increase in the percentage of coal dust waste. This trend is recorded up to sample S 3, then it starts decreasing on further increase of coal dust waste percentage (Table 5 and Figs. 5, 6).

Table 5 Results of compressive strength test

Sample id	Baked bricks	Baked bricks	Burnt bricks	Burnt bricks
	15 days MPa	28 days MPa	15 days MPa	28 days MPa
S 1	2.50	3.40	4.31	5.90
S 2	2.63	3.86	4.54	6.36
S 3	2.86	4.09	4.77	6.59
S 4	2.72	3.72	4.09	6.13
S 5	2.36	3.27	3.86	5.68

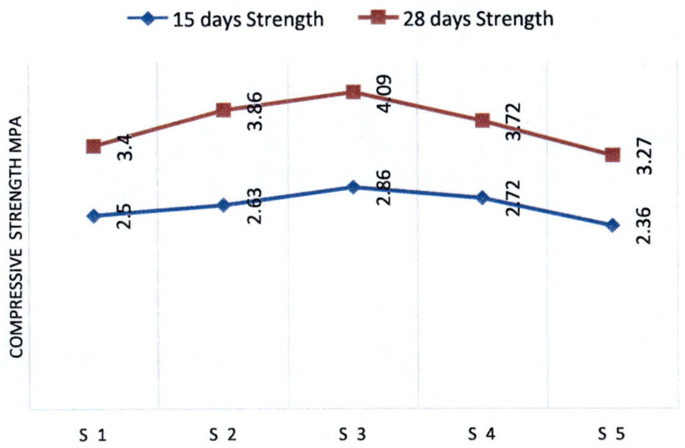

Fig. 5 Compressive strength of baked bricks at 15 and 28 days

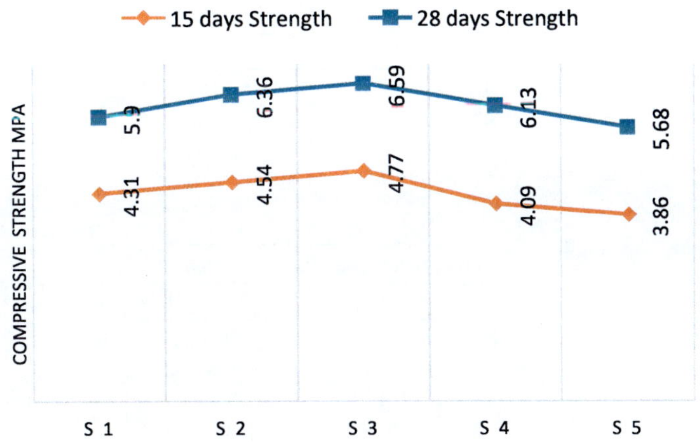

Fig. 6 Compressive strength of burnt bricks at 15 and 28 days

From Fig. 7, it is observed that the increase in the percentage of coal dust waste reduces the water absorption capacity of the brick sample. The continuous drop in water absorption is recorded up to S 5 Sample.

For all the samples casted, there is less efflorescence observed. It can be easily seen in Table 6 that the degree of efflorescence is a sight for all samples: S1, S2, S3, S4 and S5 (Table 7).

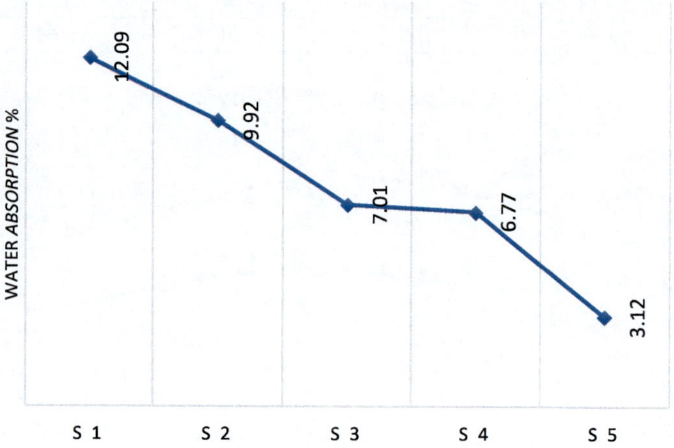

Fig. 7 Water absorption of burnt bricks at 28 Days

Table 6 Efflorescence test results

Burnt brick sample id	Degree of efflorescence
S1	Slight
S2	Slight
S3	Slight
S4	Slight
S5	Slight

Table 7 Sound test results

Burnt brick sample ID	Sound
S1	Unsatisfactory sound and breaking
S2	Slightly metallic sound without breaking
S3	Good metallic sound without breaking
S4	Slightly ringing sound without breaking
S5	Unsatisfactory sound and breaking

Table 8 Results of colour test

Sample id	Colour of (unburnt bricks)	Colour of (burnt bricks)	Remarks
S 1	Light grey	Dark greyish	Colour changed
S 2	Light grey	Dark greyish	Colour changed
S 3	Greyish	Dark greyish	Colour changed
S 4	Greyish	Greyish	Unchanged
S 5	Greyish	Greyish	Unchanged

Table 9 Warpage test results

Sample id	Concave (mm)	Convex (mm)
S 1	3.5	4.80
S 2	4.75	4.95
S 3	4.95	5.10
S 4	5.25	5.55
S 5	5.50	6

Table 10 Calculated cost of materials for 1 brick

Sample id	CDW (Rs)	Sand (Rs)	Clay (Rs)	Water (Rs)	Total cost of material/brick (Rs)
S 1	0.30	0.03	0.72	0.1	1.15
S 2	0.36	0.03	0.54	0.1	1.03
S 3	0.42	0.03	0.36	0.1	0.91
S 4	0.45	0.03	0.27	0.1	0.85
S 5	0.48	0.03	0.18	0.1	0.79

For samples S1 andS5, a nonmetallic sound was observed after striking two bricks. For samples S2 and S4, a little bit of metallic sound was observed; but for sample S3, pure metallic sound was recorded when two bricks were stroked together (Table 8).

Initially, the colour of all specimens was light grey or greyish, but after burning, it turns into dark greyish for samples S1, S2 and S3 (Tables 9 and 10).

The cost of material required for making 1 brick is minimum for sample S5. This is because of the utilization of more amount of coal dust waste in it as compared to the other sample (Table 11).

Table 11 Cost analysis per brick

Sample id	Total cost of material/brick	Total cost of material/brick including damage @ 5%	Cost including man-power, fuel and marketing Rs.
S 1	1.15	1.2075	2.3575
S 2	1.03	1.0815	2.2315
S 3	0.91	0.9555	2.1055
S 4	0.85	0.8925	2.0425
S 5	0.79	0.8295	1.9795

6 Conclusion

From the results, we can conclude that:

- Initially, with the increase in the percentage of CDW, the compressive strength of brick increases up to 70% of CDW addition in it, but beyond that, i.e., 75% and 80% addition of CDW, this value goes on decreasing.
- Suitable compressive strength has been recorded for the S3 sample (70% CDW waste, 20% clay and 10% sand).
- Water absorption capacity of brick sample has been deceased and is reduced by 72% as compare to the maximum water absorption specified for third class brick in IS code as coal dust waste content is increased.
- CDW-based bricks are 18% lighter than conventional clay bricks for the same dimensions.
- CDW-based brick offers the required hardness to be used in construction. For the same strength bricks, CDW bricks are cheaper and more economical than conventional clay bricks by 35% and fly ash bricks by 15%.
- CDW brick consumes very less fuel for baking. It does not require high temperature burning of 1000–1200 °C like conventional clay bricks. They give satisfactory results after baking even at 180 °C with the normal laboratory oven. Thus, they are eco-friendly bricks.
- CDW brick baked at 180 °C comes under 'third class' brick as per the strength classification of bricks specified by the IS code.

References

1. Yao ZT, Ji XS, Sarker PK, Tang JH, Ge LQ, Xia MS, Xi YQ (2015) A comprehensive review on the applications of coal fly ash. Earth Sci Rev 141:105–121
2. Senapati MR (2011) Fly ash from thermal power plants waste management and overview 100(12):1791–1794
3. Dubale et al (2022) An investigation of demolished floor and wall ceramic tile waste utilization in fired brick production. Environ Technol Innov 25:10228

4. Vasić et al (2021) Recycling of waste coal dust for the energy-efficient fabrication of bricks: a laboratory to industrial-scale study. Environ Technol Innov 21:101350
5. Mohammed et al (2021) A review of the utilization of coal bottom ash (CBA) in the construction industry. Sustainability 13:8031
6. Singh SK, Singh RK (2020) Manufacturing of bricks from coal mine overburden dump of Korba. Sustainable Environment Research, Chhattisgarh, India
7. IS: 3495:1992 (Part 1), "Methods of test of burnt clay building bricks. Determination of compressive strength, New Delhi
8. IS: 3495:1992 (Part 2), Methods of test of burnt clay building bricks. Determination of water absorption, New Delhi
9. IS: 3495:1992 (Part 3), Methods of test of burnt clay building bricks. Determination of efflorescence, New Delhi
10. IS: 3495:1992, (Part 4), Methods of test of burnt clay building bricks. Determination of warpage in bricks, New Delhi

Lightweight Concrete by Using Waste Materials

Ayush Mittal, Akhilesh Singh, Aman Kumar Chaudhary, and Avinash Kumar

Abstract With the growing needs of the population, the need for plastic and bottles of glass is increasing tremendously. After their use, they are not good for the environment as they decompose very slowly. Therefore, to solve these types of problems, many researchers are looking for effective ways of their disposal. One such way is to use these materials in the field of construction where crushed glasses are used in place of sand and melted plastic bottles are used in place of coarse aggregate. In today's scenario, lightweight concrete is formed by replacing the ingredients in various proportions in M20 grade. Cube compressive strength tests are conducted after a curing period of 7, 14, and 28 days. On the basis of weight analysis, cost comparison, and non-destructive tests like ultrasonic pulse results, the most optimum mix is found to be 50% sand replacement and coarse aggregate with crushed glass and melted plastic, respectively.

Keywords Crushed glass · Melted waste plastic · Lightweight concrete

1 Introduction

Currently, around 3.5 million tonnes of plastic waste and 3 million tonnes of glass waste are generated in India annually [1]. The effective removal of such a huge quantity is a challenging task for environmental engineers. These wastes are great threats to water and soil if remain untreated. Currently, these wastes are burned to produce hazardous fumes, and they are also buried in the ground or in landfills, rendering the soil barren. To overcome such situations, these wastes can be effectively used in the field of construction in place of sand and coarse aggregates.

In this study, the weight of the concrete building is primarily reduced, and the strength of the concrete structure is increased by using waste glass powder and plastic aggregate, as well as decreasing waste pollution issues and removing the demand for natural aggregate from the concrete.

A. Mittal · A. Singh (✉) · A. K. Chaudhary · A. Kumar
REC Ambedkar Nagar, Akbarpur, U.P. 224122, India
e-mail: amittal@recabn.ac.in

K. K. Pathak et al. (eds.), *Latest Developments in Civil Engineering*, Lecture Notes in Civil Engineering 352, https://doi.org/10.1007/978-981-99-2676-3_7

2 Literature Review

Various studies have been conducted using plastic waste in place of aggregate and crushed glass in place of sand. Plastic aggregate is used in various proportions in making concrete blocks as well as crushed glass is used in place of sand.

Modern waste management has led to the development of lightweight construction materials, such as plastic aggregate generated from the moulding of plastic objects into aggregate and glass sand produced from the crushing of glass waste products. Glass powder is replaced with cement. The main purpose of this work is to improve the green area as well as the concept of building with no harmful effect on the environment. The purpose of this study is to assess the feasibility of using waste glass powder. If you use less amount of glass powder in cement, concrete congestion and carbon-dioxide emissions can be reduced. Lightweight concrete is upgraded with plastic aggregate and foam agent, and lightweight aggregate recently upgraded waste such as plastic objects [2].

Premalathaet al. (2019) reported that a slight modification of the glass powder is approx. river sand. The main objective of this work is to develop compressive strength and flexibility. The strength and absorption of water in the same grade of concrete are different for different percentage of glass powder used. Addition of garbage powder causes a progressive increase in pressure force, strong dividing force, and flexible force of up to 30%, as well as a conversion of equivalent energy values of 40% and 50% for regulating samples. In concrete, cory can be reduced with glass powder. Chloride entry testing is performed by RCPT [3].

Khajuria et al. (2019) reported that flexural strength of lightweight concrete gets enhanced by the gradual replacement of waste plastic aggregates. Plastic aggregates in various percentages (i.e. 1%, 2.5%, 5%, and 10%) are used as replacement for conventional aggregates. 2.5% was found to the most optimum in terms of strength, workability. Recycling plastic aggregate helps us cut down on pollution [4].

According to Brito et al. (2012), the 2012 study on plastic in cement and concrete mud was applied. They found that the usage of angular plastic aggregates decreases the performance, while the use of smooth aggregates boosts performance. No matter what kind of plastic was used, there was a reduction in pressure, but the reduction in flexibility and durability was minimum [5].

Plastic trash was studied by Amaluet al. (2016) as a potential aggregate for concrete. Instead of paying fines at rates of 10%, 15%, 2%, and 25%, they utilised plastic. Although they saw a drop in concrete's strength, they nevertheless encourage its usage in non-structural concrete because it offers great performance and a reduction in natural waste [6].

The effect of sand with waste plastic aggregates on compressive strength values of concrete was studied by Peter et al. (2016) [7].

In sizes ranging from 0 to 5%, they use plastic bottles and bags to turn fine particles into concrete.

Plastic was used along with concrete in various percentages (i.e. 5, 10 and 20%) to dispose of plastic in Hossain et al. (2016)'s work. They found modified concrete to light in weight, but compared to regular concrete, the compression force was slightly lower. Additionally, they found that concrete containing 10% plastic particles demonstrated nearly identical strength to regular concrete [8].

3 Literature Gap

Based on comprehensive literature review, following gaps are identified.

(a) Some researcher has used 50 percent or less plastic waste in place of aggregate in concrete.
(b) Some researcher has used crushed glass in place of sand in concrete block. They used very fine glass powder in place of sand. But very fine glass gives very low strength and makes it uneconomical because the making of glass powder is very costly.
(c) Less studies have been seen on the use of both materials (crushed glass in place of sand and plastic waste in place of aggregate) for making the concrete mix.

4 Materials

4.1 Cement

It sets and solidifies when it dries, reacts with carbon-dioxide in the air through dependency, and can bind to other substances. Cement is a substance that is utilised as a binding medium around the world. Ordinary Portland Cement of 43 is the type of cement utilised in the investigation and is commonly employed in the construction industry. It is also readily available in local markets. With a specific gravity of 3.0, the cement is examined for a number of qualities in accordance with Indian Standard Code: IS 8112-2013. Tables 1 and 2 list the physical and chemical characteristics of cement, respectively.

Table 1 Physical properties of OPC

Physical properties of OPC cement	Values
Initial setting time (minutes)	30
Final setting time (minutes)	600
Standard consistency	31.0
Specific gravity	3.15
The fineness of cement (m^2/Kg)	Not less than 225

Table 2 Chemical properties of OPC

Chemical properties of cement	Values (%)
Lime CaO	62–65
Silica SiO_2	17–25
Alumina Al_2O_3	3–8
Iron oxide Fe_2O_3	1–5
Magnesia MgO	1–3
Sulphur trioxide	1–2

4.2 Glass Powder (Fine Aggregate)

Utilising waste glass in the concrete building sector provides advantages because it lowers the cost of making concrete. Because glass items are used so frequently, waste glass progressively grows over time. Topcuet et al. (2004) found that effective use of crushed glass in concrete products decreases their cost significantly. If sized and processed properly, crushed glass or cullet can yield properties like sand [9].

Glass bottles that had been abandoned were employed in our investigation; these bottles were gathered from various sites and crushed in the laboratory by impact test. After breaking down every bottle, we found fine particles in the form of glass powder that had been filtered through a 2.36 mm sieve to produce such fine particles as fine mixing.

Concrete made with glass powder can have features ranging from a colourful terrazzo floor to a granite- or marble-finished surface to concrete that reflects light like a mirror. In fact, glossy concrete can be created using glass as fine aggregate [10].

The physical and chemical properties of glass are shown in Fig. 1, Tables 3, and 4.

4.3 Plastic Aggregate

With the fast growth of the population, the usage of plastics is additionally growing dramatically. With time, researchers and environmentalists are looking for a strategy to control plastic waste disposal. Many plastic recycling strategies had been used; however, the ones strategies had been now no longer enough to control the quantity of generated plastic waste. So, putting plastic waste in concrete may be another manner to lessen the waste plastic trouble. It can't be most effective but it helps to lessen the trouble of waste.

The presence of a strong aggregate reduces drying and other size changes that occur due to moisture movement. In our investigation, we used an aggregate over 20 mm IS sieve and it is retained in a12.5 mm sieve. Instead, we melted the plastic bottles into a solid mixture (20 mm). While plastic is a robust compound, it is able

Fig. 1 Crushed glass powder

Table 3 Physical properties of glass

Physical properties	Values
Fineness modulus	2.99
Specific gravity	2.50
Water absorption	0.40
Colour	Light grey
Density (kg/m^3)	1680

Source Olofinnade et al. (2018)

Table 4 Chemical properties of glass

Composition	Value
SiO_2	64.32%
Al_2O_3	2.90%
Fe_2O_3	6.25%
CaO	18.18%
MgO	0.63%
SO_3	0.25%
K_2O	1.56%
Na_2O	13.03%
TiO_2	0.61%
Cr_2O_3	0.02%
P_2O_5	0.01%
Density	2555 m^3

Source Climent et al. (2018)

Fig. 2 Plastic aggregate

Table 5 Physical and chemical property of plastic aggregate

Physical and chemical property of plastic aggregate	Values
Fineness modulus	4.45
Specific gravity (g/m^3)	2.18
Aggregate maximum size (mm)	20
Bulk density (kg/m^3)	1510
Crushing value	0.83
Water absorption	0.05%
Apparent	1.13
Saturated surface dry basis	0.87

Source [11]

to lessen the load of concrete slabs. So, in our project, we use plastic aggregate. The physical and chemical properties of plastic aggregate are proven in Fig. 2 and Table 5.

5 Experimental Procedure

The mixing proportion is set for a grade of M20 concrete for conventional concrete and the size of the cube is (150 × 150 × 150) mm. The cubes are designed for testing compressive power which is tested for 7 days, 14 days, and 28 days for each piece of coloured glass powder, and a combination of the disposable plastic aggregate mixture. Filled with 18 cubes are unmolded after 24 h, and curing is done in water for 7 days, 14 days, and 28 days.

6 Preparing Lightweight Concrete

6.1 *Preparation of Glass Fine Aggregate*

Bottles of garbage glass are collected at various locations. Then, firstly, they were crushed with an impact machine and filtered with a 2.36 IS sieve shown in Fig. 3.

6.2 *Making of Plastic Aggregates*

Bottles of plastic waste are collected at a dumping location and chemical treatment is done to clean the bottle. For melting it, we used a microwave oven and a gas stove to create an unusual shape (20 mm). Plastic bottles melt at approx. 600 °C and will be strong, rough, and uneven in shape as shown in Fig. 4 [10].

7 Result and Discussion

7.1 *Concrete Compressive Strength*

This test of 7 days, 14 days, and 28 days of concrete cubes is performed in accordance with IS 516-1959 by using a 40-tonne compressive test machine (CTM). The results are shown in Fig. 5, 6, and Table 6.

The pressing capacity of 7 days, 14 days, and 28 days of disposable concrete waste (glass powder + plastic aggregate) is reduced in comparison to conventional concrete.

7.2 *Weight Comparison*

The results of weight reduction of concrete due to use of glass powder and plastic aggregates are shown in Table 7.

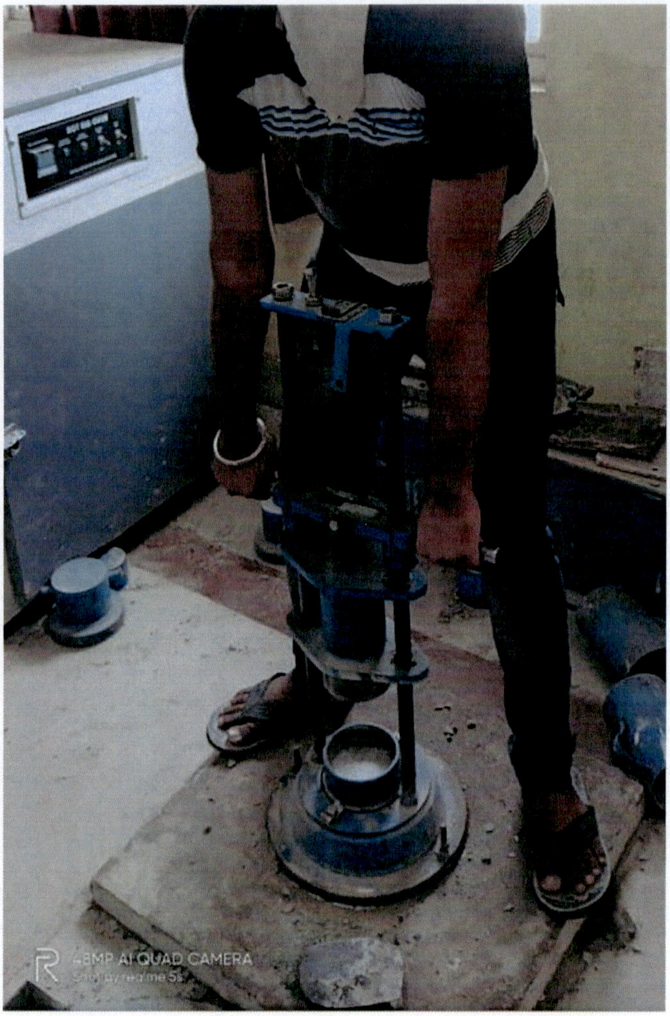

Fig. 3 Preparation of glass fine aggregate

7.3 Cost Analysis of Concrete

The cost analysis of concrete cube with and without replacement with glass powder and plastic aggregates are conducted. These consequences have been acquired via way of means of studying the fee of glass scrap and waste plastic which determined that the rate of glass scrap became four Rs/kg and a bad plastic bottle became thirteen Rs/kg at the market [12]. Results are proven in Table 8.

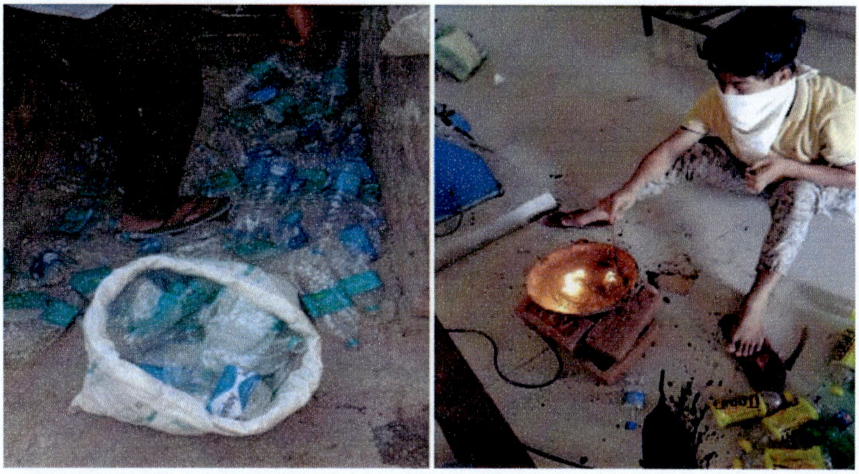

Fig. 4 Making of plastic aggregate

Fig. 5 Concrete compressive strength

7.4 Ultrasonic Pulse Velocity Test (NDT)

Ultrasonic pulse velocity checking out is a non-invasive take-a-look-at approach and it is used to test the inner shape of an object. With a heartbeat, it could be ensured that the inner shape of the cube is easy and well-integrated. If there are cement-based compounds, it is helpful to check their similarity. It is also known that life-saving techniques are applied to concrete or mud. There are various levels available for this

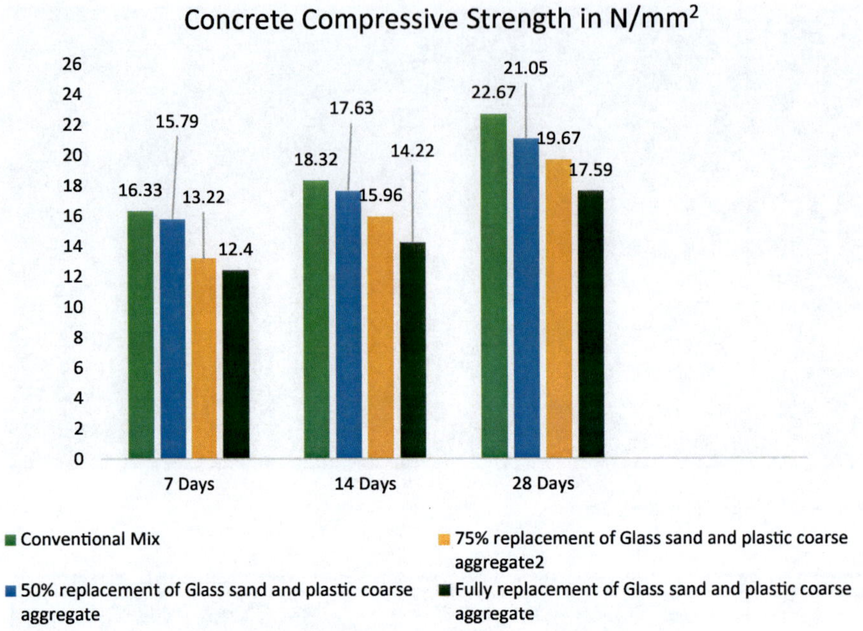

Fig. 6 Concrete compressive strength in N/mm^2

Table 6 Concrete compressive strength

	Conventional mix	50% of sand of glass sand and coarse aggregate with plastic replaced	75% of sand of glass and coarse aggregate with plastic replaced	Fully sand of glass sand and coarse aggregate with plastic replaced
7 days	16.33	15.79	13.22	12.40
14 days	18.32	17.63	15.96	14.22
28 days	22.67	21.05	19.67	17.59

Table 7 Weight comparison

	Conventional mix	50% of sand of glass sand coarse aggregate with plastic replaced	75% of sand of glass sand and coarse aggregate with plastic replaced	Fully sand of glass sand and coarse aggregate with plastic replaced
Weight (Kg)	8.98	8.033	7.23	6.02

Table 8 Cost analysis of concrete

	Conventional mix	50% of sand of glass sand and coarse aggregate with plastic replaced	75% of sand of glass sand and coarse aggregate with plastic replaced	Fully sand of glass sand and coarse aggregate with plastic replaced
Cost (in rupees)	78.299	63.062	55.578	47.825

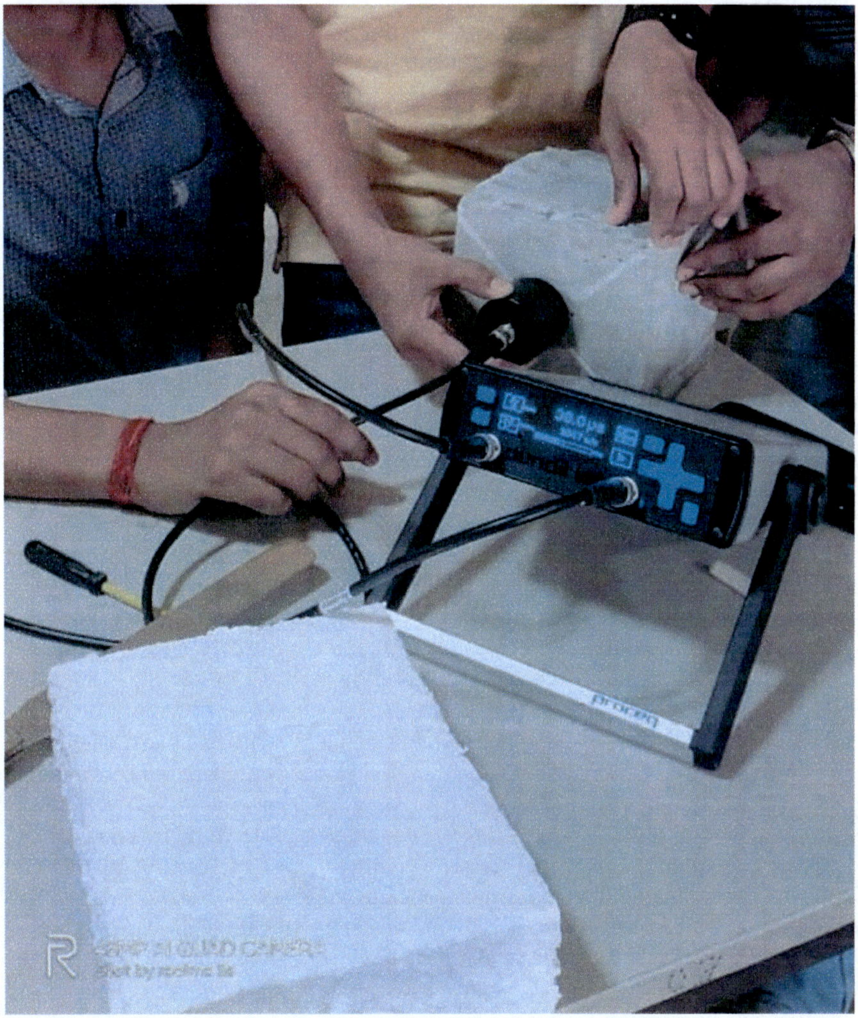

Fig. 7 Ultra pluse velocity test on M20 cube

Table 9 Velocity of ultrasonic pulse

	Conventional mix	50% of sand of glass sand coarse aggregate with plastic replaced	75% of sand of glass sand and coarse aggregate with plastic replaced	Fully sand of Glass sand and coarse aggregate with plastic replaced
Velocity of ultrasonic pulse (m/s)	4257	4106	3897	3811

test to test the behaviour of a concrete sample [13]. The result of these tests is shown in Fig. 7 and Table 9.

8 Conclusion

- Investigating the outcomes of dirty plastic and glass waste on concrete, 3 test mixes had been designed. Each sample was examined with a compressive test, weight evaluation testing, cost analysis, and ultrasonic pulse velocity test. The following conclusions are drawn from the existing work:
- The prepared concrete cubical blocks were tested for compressive strength, weight comparison, and cost analysis, and the results were obtained by comparing the data obtained from a standard cube with a cube prepared using glass powder (sand) and plastic aggregate.
- Compressive test data show that the compressive strength of 50% of sand of glass sand and coarse aggregate with plastic replaced cube, 75% of sand of glass sand and coarse aggregate with plastic replaced cube, and full replacement of Glass and plastic coarse aggregate cube decreased by 11.27%, 16.27%, 21.46% in 7 days, 7.43%, 12.30%, 19.34% in 14 days, and 6.29%, 8.16%, 18.69% in 28 days, respectively.
- These results show that 50% replacement of glass sand and plastic aggregate cubes have given better strength; they are lightweight compared to conventional concrete.
- Replacement of 50%, 75% and 100% sand and aggregate with glass powder and plastic aggregates respectively causes weight reduction of 10.54%, 19.48% and 32.96% as compared to conventional concrete cube.
- Therefore, 100% sand of glass sand and coarse aggregate with plastic replaced blocks are relatively lightweight and have low overall strength.
- In the cost analysis, we found that the cost of replacing 50% of sand of glass sand and coarse aggregate with plastic replaced cube, 75% of sand of glass sand and coarse aggregate with plastic replaced cube, and complete replacement of Glass sand and the coarse plastic aggregate cube was reduced by 19.46%. 29.01%, and 38.92% than the cost of a conventional mix cube, respectively.
- In ultrasonic pulse velocity tests, no significant differences were observed in the results.

- In the future, more explicit research can be done to replace 50% glass sand and plastic aggregate blocks by increasing the test period, as well as block exposure to natural conditions.

References

1. Nandy B, Sharma G, Garg S, Kumari S, George T, Sunanda Y, Sinha B (2015) Recovery of consumer waste in India—a mass flow analysis for paper, plastic and glass and the contribution of households and the informal sector. Resour Conserv Recycl 101:167–181
2. Sudharsan N, Saravanaganesh S (2019) Feasibility studies on waste glass powder. Int J Innov Technol Explor Eng 8:1644–1647
3. Khajuria A, Sharma P (2019) Use of plastic aggregates in concrete. Int J Innov Technol Explor Eng 9:4406–4412
4. Premalatha J, Srinivasan R (2019) Properties of concrete with waste glass powder (Gp) as fine aggregate replacement. Int J Recent Technol Eng 8:2308–2314
5. Saikia N, Brito JD, Use of plastic waste as aggregate in cement mortar and concrete preparation 34:385–401
6. Amula RG, Ashraf A, Hussain M, Rejith KU, Vijitha V (2016) Use of waste plastic as aggregates in Concrete. Int J Innov Technol Explor Eng 9:4406–4412
7. Jibrael MA, Peter F (2016) Strength and behavior of concrete contains waste plastic. J Ecosyst Ecography 6:1–4
8. Hossian MB, Bhowmik P, Shaad KM (2016) Use of waste plastic aggregation in concrete as a constituent material. Progress Agric 3:383–391
9. Topcu IB, Canbaz M (2004) Properties of concrete containing waste glass. Cem Concr Res 34:267–274
10. Tamang LWT, Wangmo T, Darjay KT, Phuntsho KS, Namgyal P, Wangchuk U (2017) Use of plastics in concrete as coarse aggregate. Int J Educ Appl Res 7:9–13
11. Jayaraman A, Gowtham SM, Praveen N, Hariharan V, Afrith MR (2020) Light transmitting concrete using eco-friendly materials (waste materials). Int J Chemtech Res 13:240–250
12. Lasiyal N, Pawar GS, Dixit M (2016) Effect of plastic waste as partial replacement of fine aggregate in concrete and cost analysis. Int J Eng Res Technol 4:1–4
13. Lorenzi A, Tisbierek F, Filho LCPDS, Ultrasonic pulse velocity analysis in concrete specimens. https://www.researchgate.net/publication/237322793

Improving Pedestrian Movements in Congested Urban Areas

C. S. Punchihewa and J. M. S. J. Bandara

Abstract Pedestrians are one of the main components of the urban traffic environment. Improved corridors for vehicle movements, but the lack of pedestrian facilities are indifferent identifications of a conventional urban setting. Unplanned and uncontrolled pedestrian movements result in delays and safety risks at town centers. Often, there are lots of public requests for a proper network of pedestrian pathways including amenity development. However, it is very difficult to plan and design an efficient pedestrian network without understanding pedestrian movement behavior in such a vicinity. This study is focused on developing a methodology to identify pedestrian movement behavior in critical areas and make necessary adoptions to develop such facilities to encourage a walkable city environment. Pedestrian movement has a high degree of freedom in selecting Origin–Destination pair than any mode of transportation. Household or occupational purpose utility-related trips are commonly identified in such urban environment and it directly relates to the land use pattern of a town area. This study identifies specific land uses that serve as trip generators or attractors, generated pedestrian trips, and potential pedestrian paths within an urban territory. Collecting vehicle speed data using Google Maps to identify the heavy use crosswalks, data verification using field surveys, and developing a GIS-based land use model with pedestrian paths are also under the framework of the study. Shortest path Origin–Destination matrix development for pedestrian networks is one objective of this study. According to the OD matrix, the frequency of sidewalk or crosswalk usage in each OD pair is counted and ranked. A prioritized list is prepared according to the rank and level of interacting traffic.

Keywords Pedestrian trips · Connectivity · Land use · Prioritization of pedestrian amenities

C. S. Punchihewa (✉)
Road Development Authority, Battaramulla, Sri Lanka
e-mail: chamithaph@yahoo.com

J. M. S. J. Bandara
Department of Civil Engineering, University of Moratuwa, Moratuwa, Sri Lanka
e-mail: bandara@uom.lk

K. K. Pathak et al. (eds.), *Latest Developments in Civil Engineering*, Lecture Notes in Civil Engineering 352, https://doi.org/10.1007/978-981-99-2676-3_8

1 Introduction

1.1 General

The town center is the core area of a city associated with various commercial and retail activities. Urban development in a city is interpreting number of human factors and physical factors. "Economic growth" is a simple explanation for these human and physical factors for urbanization [1].

Distribution of land use is determined by human activities such as education, employment, leisure, shopping, etc. Those human activities create banks, schools, shopping areas, and government institutes in the spatial space in an urban area. These activities require spatial interaction or trips. The distribution of transportation systems feeds that demand and that opportunity can be measured as accessibility [2]. The result would be the complex transport network and diversified land use area over urban space.

Pedestrians are a main component of different kinds of land use activities in urban spaces. So, improving pedestrian movement is a necessary action to optimize traffic movement in a city environment.

Individual facility improvement without analyzing the entire study area may impact another part of the city and it could create delays and safety risks. Often, there are lots of public requests for a proper network of pedestrian pathways including amenity development. However, it is very difficult to plan and design an efficient pedestrian network without understanding pedestrian movement behavior in such a vicinity [3].

1.2 Research Problem

The analysis of the pedestrian network, the identification of optimum walking paths, and the correct positioning of pedestrian crossings are primary methods for improving traffic flow in the urban environment. In many parts of the world, main cities have inadequate facilities for pedestrians. Urban traffic optimization requires a methodology to identify network conflicts and improvements for the pedestrian environment. The study is focusing on this problem, and it proposes a method to optimize critical pedestrian amenities for the reduction of traffic conflicts and related congestion in the town center.

1.3 *Research Objective*

The main goal of this study is to develop a methodology to identify the most used pedestrian amenity at a specific urban center. The specific objectives are:

- Identify the Pedestrian Behavior and Walking Pattern in a Town Center.
- Develop a methodology to identify the locations of traffic conflicts and congestion due to pedestrian movement.
- Propose a methodology to prioritize existing sidewalk links for improvements, and identify the best locations for crossings to optimize pedestrian movement and traffic flow.

2 Literature Review

2.1 *Walkability*

Walking is a basic requirement of mobility, and it promotes a healthy lifestyle and social interactions. Overall, walking is sustainable, environmentally friendly transport mode, and increases the quality of life in society. Generally, 95% of trips below 100 m are completed by foot [4].

Walkability defines in many ways at the urban environment. Basically, walkability forces on pedestrian comfortability for walking at urban environment. Walkability describes the quality of pedestrian facilities, roadway conditions, land use patterns, community support, security, and comfort of walking in an area.

The walkability index project explains four types of basic scenes: the safety, security, economy, and convenience of traveling by foot. Healthier worksite initiative described walkability as a measurement of the transportation and recreation opportunities for pedestrians while considering safety, convenience, and route aesthetic for pedestrians [5].

2.2 *Acceptable Walking Distance*

Critical walking distance is identified as the distance from any kind of transportation facility (Railway station, Bus terminal, Parking lot) at a commercial capital or a town center to main trip generators [6]. In the South Asian context, office premises, schools, tuition classes, banks and other government institutions, and commercial areas such as markets and shops are the main trip generators.

According to Seneviratne [6], these kinds of trips in a town center would depend on some external factors such as the type of trip, the purpose of trip, and the type of day. It may directly influence walking distance. Utility-related walking trips may indirectly influence some factors such as congestion, parking fee, one-way roads, parking availability, etc. Even if there is no parking fee or subjectively low cost in the town center, people including workers are willing to walk up to 610 m. The threshold limit of walking distance in a town center is identified as 610 m in this study [6].

Walking distance depends mainly on the trip's purpose. However, a threshold value for average walking distance is required to identify the study area. According to another study, the mean value of walking distance is 1120 m and the median value is 804 m. Their purpose varies with the income category. Relatively lower income population walks longer distances to work and shorter for recreation. But higher household income population walks longer distances for recreation and shorter to work [7].

Roddin's work regarding determining the benefits of separating pedestrians and vehicles stated the general statement about pedestrian walking length. Most of the time, pedestrian route length is less than 3,000 ft (915 m) in length [8]. Highway capacity manual Chap. 13 quotes Rodin's method [8] to calculate total travel time in pedestrian networks.

Total travel time = Number of pedestrian $\times$ ((Route length/Walking speed) + Signal Delay).

In here, Rodin introduced a maximum walking route length is less than 915 m. Furthermore, an ideal pedestrian route length is defined as less than 1.2 times the straight-line distance between the trip origin and destination. This assumption is important to avoid long-distance alternative routes which reduced the complexity of the OD matrix [8].

2010 and 2016 editions of the Highway Capacity Manual mention the threshold length of walking as 3000 ft (912 m) in Chap. 23 in 2010 and Chap. 24 in 2016 versions. A recent medical study [9] reveals that walking outdoors at a usual pace was an average speed of 1.31 m/s, a cadence of 116.65 steps/min [9]. This value is important to use as an assumption for walking speed on the above equation proposed by Rodin.

There are several studies regarding the threshold value for walking in an urban center. Rongrong Yang [7] introduced the threshold value for pedestrians at transit stations in China as 400 m to 800 m according to Krishi's logit Price Sensitivity Model [7]. Another study was conducted in Spain regarding the threshold distance to walk from home to school by Carlos Rodríguez-López [10] and it was 875 m for young people's community.

The average acceptable walking distance for the study of a typical city in Sri Lanka is selected as 750 m by considering the literature.

2.3 *Pedestrian Route Identification*

Proper data collection methods should be identified to find possible pedestrian routes. A broad range of methods for data collection were found in the literature. Raford et al. [17] proposed a method where respondents can draw their route on paper, but this approach would be difficult in an urban context. Snizek et al. [18] developed a criterion regarding web-based questionnaires. This method is quite advanced; connected to web API, Google Maps, and Open StreetMaps. Application of web-based methodology would be difficult in the Sri Lankan city because most of the pedestrian categories have low levels of computer literacy. It would result to omit the majority of pedestrian categories from the survey.

A Hybrid data collection including web-based methods and personal interviews have covered all categories of pedestrians.

2.4 Pedestrian Route Choice

There are many studies about pedestrian route choice. Pedestrians are enjoying a high degree of freedom of movement in a congested city environment in comparison to other travel modes.

Hill [11] found important factors regarding pedestrian route choice according to four different attributes: age, gender, trip purpose, and environment. Across all factors, responders considered minimization of distance in the first place while selecting a route. Only 1 out of 211 observations deviated from the shortest distance [11].

In another study conducted in a town center in Sweden, private vehicle users who parked their vehicles in the central parking area were observed to typically visit the most distant location first and then gradually walk back to their vehicle over the course of their visit. Nearly 70% of shoppers attempt to minimize the distance while on their trip [12]. In another research conducted in a UK urban context, out of the 820 sampled pedestrians, 616 (75% of the total) chose the shortest route. The remaining 25% of trips were slightly longer than the shortest route [13].

Zielstra's study [14], conducted in more pedestrian-friendly environment in the USA and Germany, examined the usage of smart devices among pedestrians. The study found that the popularity of pedestrian-related routing applications has increased due to the widespread use of smart devices such as tablets and smartphones in such environments. Nonrecreational and multimodal trip pedestrians have used that kind of application because pedestrians typically aim to reduce their walking distance. Pedestrian network shortest path algorithm program is really helping to find the shortest way and thus to avoid unnecessary detours [14].

2.5 Shortest Path

The shortest path generally saves time, cost, and energy. Pedestrians are choosing the shortest path to avoid unnecessary detours on utility-based trips most of the time [11, 13]. There are some other considerations for pedestrian route choice. There would be a low walk score in the shortest path, due to the poor condition of walkability measures, the probability of selecting the next available connectivity path would be high for many categories of pedestrians [5, 21].

2.6 Application of Dijkstra's Algorithm

Dijkstra's Algorithm is the fundamental shortest path algorithm that gives the optimum route on the network by counting method. It computes the shortest distance from one node to every other node. Hence, it is used in many researches and is very efficient in shortest-distance problems. Dijkstra algorithm is used to find the list of shortest routes in the OD pair list. At the end of the iteration process, each node consists of (x, y) values. y value at the destination node (x, y), represents the shortest distance from the origin node.

If the y value > the threshold value of walking distance (750 m), the probability of occurrence of the trip using pedestrian mode in particular OD pair is very less.

2.7 Long Walking Trips

Straight line distance between the destination and city center will correspond to the Rayleigh distribution [23]. Rayleigh distribution functions as follows:

$$F(x) = 1 - e^{-\frac{x^2}{2\sigma^2}} \tag{1}$$

where

x—Straight line distance between destination and city center.

σ—Rayleigh distribution parameter concerning the variable x [24]

This distribution is implied to find the probability of trips for long-distance OD paths.

2.8 Pedestrian Demand in Medium Size City

Unlike vehicle traffic demand, pedestrian traffic demand is very difficult to estimate. In this methodology, the first step is to understand pedestrian trip origin nodes.

Moulden's study in 1997 describes apartments and grocery stores creating high number of pedestrian trips in the urban area. It means that the apartments and residential units generate pedestrian trips in general [15].

Based on the literature, the following key points highlight the assessment of pedestrian demand in urban town centers:

- Bus stations, bus stops, and train stations are the main nodes of pedestrian trip generators.
- Vehicle parking area serve as the secondary nodes for pedestrian trip generation.
- Residential areas such as flats and apartments in the urban vicinity are secondary trip generation nodes.
- Trip generation nodes are acting as trip attractors on the return trip.

3 Methodology

3.1 Framework Development

- Step 1

  The first step in understanding pedestrian activities in a medium-sized city's urban center is to identify trip origin nodes including the threshold limit of average walking distance and setting boundaries using a buffer area around trip origins. Threshold walking distance is used to determine the buffer radius. In addition to that, geographical barriers and built environment barriers are to be considered to set the spatial limits for pedestrian movement.
- Step 2

  GIS-based methodology is developed based on pedestrian movement network. The best connectivity path between a trip origin node and the destination node is considered. The best connectivity path is the minimum cost path that consumes less energy to reach the destination. It would be the shortest distance path with a minimum number of road crossings in an urban network. The minimum cost path is made from a series of sidewalk segments and crosswalk segments. OD path matrix is developed by manipulating the sequence of each origin and destination connectivity path.
- Step 3

  Elements of OD path matrix consist of a set of sidewalk and crosswalk segments. Each segment's frequency at the urban network OD matrix is counted and ranked according to descending order. Higher rank means that a particular crosswalk or sidewalk belongs to many numbers of pedestrian OD paths. So, the priority list for pedestrian amenities is developed according to that ranked list.
- Step 4

  Road segment-wise Google traffic speed data is extracted using Google script. Continuous series of low-speed segments represent the vehicle queue at junctions or at pedestrian crossing locations. Data collection is performed using a Google sheets script, enabling the automatic extraction of speed data for each segment at 10-minute intervals. After collecting data, spatio-temporal graph, which represents the average speed data matrix, is developed according to location vs time. This data analysis was conducted to identify congested locations and normal traffic behavior in the city environment.

- Step 5

  Pedestrian counts are conducted at selected links to compare the OD matrix results. The selected path should include a mix of higher- and lower-ranked sidewalk and crossing locations to verify the pattern of pedestrian variation with the model result vs real-world scenarios. Conducting manual pedestrian spot counts for 15 minutes during a selected peak time is a simple and straightforward survey method. Pedestrian count using bird eye view video from a drone camera is highly effective in this kind of verification.

OD matrix results and pedestrian count results are plotted on the XY graph according to the rank and compared with the R^2 value in linear regression to validate the model results. Prioritize Road segments can be short-listed as per budget constrain and proceed with improvement projects for better utilization of public funds.

Study area identification and methodology implementation were done as a case study to identify the prioritized pedestrian amenities in Rathnapura town. The municipal council of Rathnapura, Sri Lanka, was allocated funds for the development of pedestrian facilities at the town center to improve pedestrian movement in the town area.

3.2 Study Area Identification

Walkable distance is the main influencing factor in pedestrian mode choice. Based on the literature, a 750 m is selected as the convenient distance for walking. Walkable distance is the main factor to be considered in the selection of Origin–Destination locations. Pedestrians will not choose walking as their mode for longer distances, especially in urban centers where commuter buses, three-wheelers, and taxis are available. Hence, the study area should not extend beyond the city center because pedestrian trip does not occur between those O-Ds.

Geographical barriers are another influencing factor for pedestrian trips between OD pairs. This has plus and minus points. Natural barriers like streams, marshy land, etc., limit the walking trips between a OD pair. In hilly environments with steps may encourage walking as a mode of transportation over vehicles. This is because steps with high gradients can shorten the distance compared to vehicular routes, even for motorcycles. This is another important factor to be considered in selecting the study area.

The economic status of the population category also influences pedestrian trips to a town center. Low-income population, school children use walking mode without using three-wheelers or taxis for short trips.

3.3 Study Area Identification Using ArcGIS

The following nodes are considered trip generation nodes:

- Main bus stand
- Train station
- Bus stops
- Parking outlets
- Apartment/Flats exit gates

Acceptable walking distance from the trip origin point is considered 750 m at the town center. It means the probability of a trip length of more than 750 m is very rare for pedestrians in the urban environment.

According to Rodin [8], 625 m (750 m/1.2) will be considered as the maximum radius in the pedestrian trip zone for a generating node. After identifying all trip generating nodes, a collection of 500 m radius buffer defines the most probable pedestrian zone for each origin node. This pedestrian movement zone is restricted from geographical barriers such as rivers, water streams without a bridge or any other cross path, physical barriers, environmental barriers, marshy land, etc. Those geographical features are extracted as a geographical boundary layer to fine-tune the pedestrian movement zone. A close polygon is created by connecting boundaries of barriers and it is named a barrier polygon layer. After identifying environmental barriers, a polygon is created as a barrier polygon.

The study area is extracted after merging those geographical barrier layers and the node pedestrian movement buffer zone layer.

3.4 Pedestrian Route Choice

A route is required to connect an Origin and Destination pair of the network.

Sidewalks and crosswalks are two main physical components in the typical urban pedestrian network. The next step is the identification of possible pedestrian routes which connects all origins and destinations in the town center.

As discussed in the literature review, the hybrid data collection method was carried out to identify the possible pedestrian routes. Open street maps and Google satellite maps are used to identify possible pedestrian paths. Pedestrian paths on both sides of roads, crosswalks, steps, and footpaths were identified as an initial step. Pedestrian interviews were conducted at trip origin locations, and site visits to the study area were undertaken to verify pedestrian routes and identify preferred alternative routes within the urban network.

3.5 Origin–Destination Matrix Development

The shortest distance results for all of OD pairs in the network are extracted from the Arc GIS OD matrix solver. Each OD pair has a distance value and it consists of a series of sidewalk lengths and crosswalk lengths.

Arc GIS solver works in two ways. Reduce distance and time of the journey. In the pedestrian network, the time factor affects only major road crossings. There would be a waiting time at main road crosswalks. So, the additional length modification applies at such crosswalks as follows.

The average walking speed for pedestrians is 1.3 m/s [9]. Correction should be added to the graded crossing locations as mentioned in the literature [22].

If the average waiting time at a crosswalk location is t.

Then, the length correction at the crosswalk location = 1.3 t in meters (Up to 5% grade section).

Representation of the sample network illustrated in Fig. 1 and results of the shortest distance in OD matrix are tabulated in Table 1.

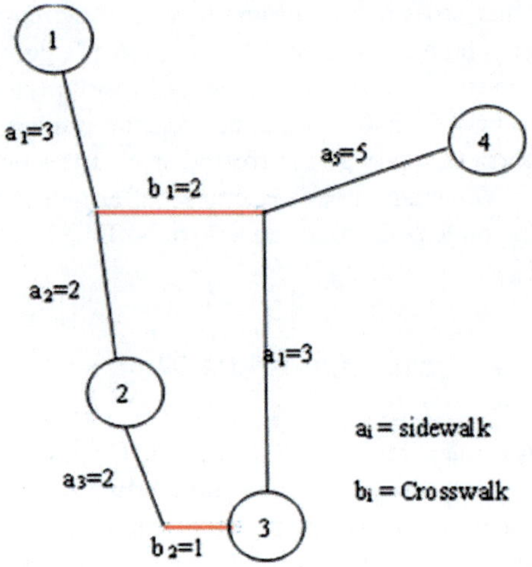

Fig. 1 Sample pedestrian network

Table 1 Shortest path and shortest distance for each OD pair

Origin	Destination	Shortest path	Distance
1	2	a1-a2	5
1	3	a1-a2-a3-b2	8
1	4	a1-b1-a5	10
2	3	a3-b2	3
2	4	a2-b1-a5	9
3	4	a4-a5	10

After finding the shortest paths between all pairs of nodes, add each Origin–Destination pair to the OD matrix. Origin–Destination pair of shortest paths consists of a sidewalk section and crosswalk section series. OD pair i-j, shortest path *l*, sidewalk section a, crosswalk section c;

$$l(i,j)=\{a_1,a_2...a_n,c_1,c_2,..c_n\} \text{ where } n \in \mathbb{N} \tag{2}$$

Origin–Destination matrix (in tens of components of shortest path);

$$[l(I,j)] = \begin{bmatrix} l(1,1) & l(1,2) & \dots & l(1,j) \\ l(2,1) & l(2,2) & \dots & l(2,j) \\ \vdots & \vdots & \vdots & \vdots \\ l(i,1) & l(i,2) & \dots & l(i,j) \end{bmatrix} \tag{3}$$

OD path matrix can be obtained from the above data for the shortest distance between each OD pair. If the shortest distance for an OD pair is greater than the threshold value of convenient walking distance, pedestrian trip generation probability would be very small, pedestrians would move into another mode such as a taxi or commuter bus on those occasions unless there is a special reason to walk between that OD pair. According to the literature, the probability of longer walking trip occurrence can be obtained from the Rayleigh distribution.

Table 2 shows the example of the cost OD matrix for the sample network in Fig. 1. All 4 nodes on the network act as trip origins and trip destinations in the network. Destination nodes are represented in columns and origin nodes are represented in rows.

Road element notations such as sidewalk segments and crosswalk segments are denoted as matrix elements. Hence, there are 4 nodes, each node has 3 destination routes, e.g., for origin node 1, routes are 1–2, 1–3, and 1–4.

Table 2 OD matrix for the shortest distance

O \ D	1	2	3	4
1	-	a1 a2	a1 a2 a3 b2	a1 b1 a5
2	a1 a2	-	a3 b2	a2 b1 a5
3	a1 a2 a3 b2	a3 b2	-	a4 a5
4	a1 b1 a5	a2 b1 a5	a4 a5	-

Table 3 Frequencies of pedestrian amenities usage on the network

Segment	Frequency
a1	6
a2	6
a3	4
a4	2
a5	6
b1	4
b2	4

3.6 Analysis of the Model and Results

Two sets of amenities can be extracted from the OD matrix. The sidewalk segment set and crosswalk segment set are separate sets of elements on the network. Sidewalk set includes the frequent occurrence of

a_n; where n ∈ ℕ.

max $|a_n|$ gives the most repetitive element in the network. The most repetitive crosswalk section was also identified by the same counting method.

b_n; where n ∈ ℕ, max $|b_n|$ gives the most repetitive crosswalk segment.

Table 3 illustrates the frequency of usage of both sidewalks and crosswalks in the network. This kind of frequency table helps to identify the priority elements in a network, and it reinforced the optimum utilization of public funds in development projects. According to this example, a1, a2, and a5 segments are frequently used in the network. Developing the walkability of those elements would improve the overall efficiency of the network.

4 Application of Google Speed Data Analysis

Google application gathers crowdsource data, such as location from smart devices, and processes those location displacements with time as traffic data on roads. This algorithm is denoted as the distance matrix API, which provides travel distance and travel time from Origin to Destination on Google Maps.

4.1 Google Map Travel Time and Distance Data Extraction to Google Sheet

Google directional finder class on the Google app script allows finding the direction from the Origin location to the Destination. On the direction data, this class returns travel time between OD, and distance between OD according to real-time traffic

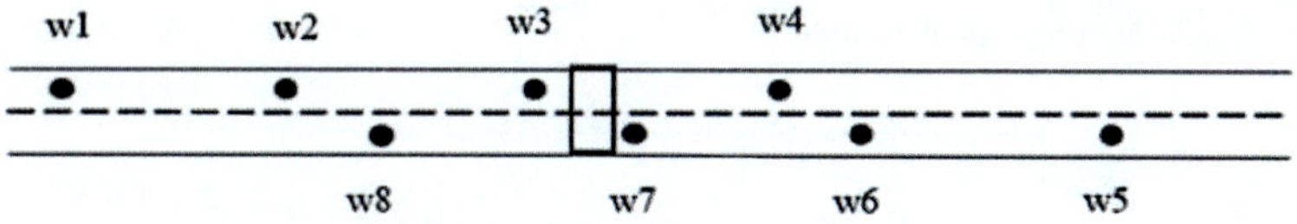

Fig. 2 Way point arrangement near the crosswalk

conditions. Google sheet script editor allows us to automate this basic process in specific time intervals.

Traffic behavior study near the crosswalk requires some modification on this basic process. The latitude and longitude of each waypoint were extracted from Google Maps to check the speed variation as illustrated in Fig. 2.

Google script is set up to trigger every 10 min and it collects the live duration to pass the segment. The duration output file contains the estimated duration, trip length, date, and time. This raw output file is modified as separate attributes of the date, time, speed (length/time), and section ID. From that four attributes, a simple pivot table is generated with y-axis as time, and x-axis as section IDs. Speed data variation can be identified from the graph as denoted in Table 4.

This pivot table is spatio-temporal graph and speed variation illustrates the congestion at any crosswalk.

The logic of speed data analysis at a crosswalk is as follows:

- If the road section before the crosswalk displays a red, and the section after the crosswalk displays a green, it indicates a congested crosswalk.
- If sections before and after show red means heavy congestion on the network despite of crosswalk location
- Green on each side means no congestion on the network.

4.2 Results Validation

Site visits and spot pedestrian counts are required to validate the OD matrix results. The pedestrian movement behavior at the congested crosswalk location is also monitored to identify possible countermeasures for avoiding conflicts between vehicular traffic and pedestrian flow.

Pedestrian route, including sidewalks and crosswalk segments, is selected to test the validity of OD matrix results. The sample route section would contain a mix of higher-rank and lower-rank pedestrian amenities.

Simultaneous 15 min pedestrian counts at all pedestrian amenities on selected routes and comparison with OD matrix results is the basic method of data validation. Bird's eye view video from a drone camera is a very cheap and practical application to capture simultaneous pedestrian counts on specific routes.

OD matrix results and pedestrian count results are plotted on XY graph according to the rank of the segment and compared R2 value in linear regression to validate OD

Table 4 Spacio-temporal graph sample

Time	Section ID					
	w1-w2	w2-w3	w3-w4	w5-w6	w6-w7	w7-w8
6:10	38.00	20.00	36.00	32.00	25.71	50.00
6:20	38.00	15.78	21.20	20.00	18.00	50.00
6:30	38.00	15.78	21.20	18.60	14.00	50.00
6:40	38.00	13.68	15.90	15.94	18.00	50.00
6:50	38.00	13.68	15.90	15.94	18.00	50.00
7:00	38.00	13.68	15.90	15.94	18.00	50.00
7:10	32.00	12.00	21.20	15.94	12.00	50.00
7:20	32.00	12.00	21.20	15.94	12.00	30.00
7:30	32.00	12.00	21.20	15.94	12.00	30.00
7:40	32.00	12.00	23.85	15.94	12.00	30.00
7:50	32.00	15.78	23.85	15.94	12.00	30.00
8:00	38.00	15.78	23.85	15.94	12.00	30.00
8:10	38.00	17.10	23.85	22.32	12.00	30.00
8:20	38.00	17.10	23.85	22.32	12.00	27.00
8:30	38.00	17.10	23.85	22.32	12.00	27.00
8:40	38.00	15.78	23.85	22.32	12.00	27.00
8:50	38.00	15.78	23.85	22.32	20.00	50.00
9:00	38.00	15.78	23.85	22.32	20.00	50.00

matrix results and actual values. Figure 3 represent the flow chart of the methodology which is described in the above section.

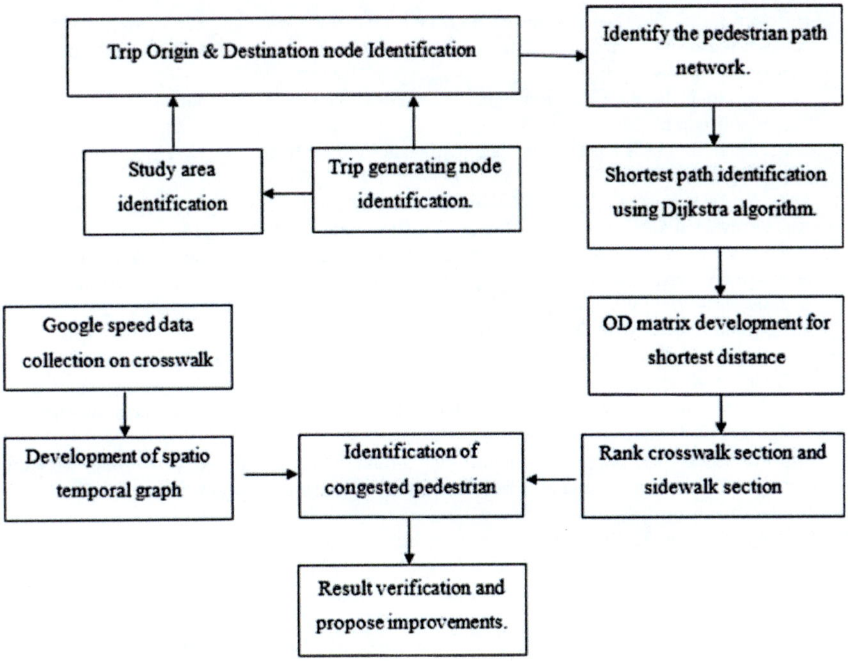

Fig. 3 Flow chart of methodology

5 Conclusion

Pedestrian amenity development project is required to identify and prioritize the order of segments at urban center. Identification of pedestrian paths and Origin–Destination connectivity is important in this process. OD matrix, considering the shortest path, is a powerful tool for identifying frequently used homogeneous segments in the urban pedestrian network. Improving walkability in those segments is a good observation when utilizing public funds for the development of the city environment in an efficient manner.

References

1. Black D, Henderson V (1999) A theory of urban growth. J Polit Econ 107:252–284. https://doi.org/10.1086/250060
2. Wegener M (2004) Overview of land use transport models. J Discrete Algorithms—JDA 5
3. Zheng Y, Elefteriadou L, Chase T, Schroeder B, Sisiopiku V (2016) Pedestrian traffic operations in urban networks. In: International symposium on enhancing highway performance ISEHP June 14–16 2016 Berlin, vol 15, pp 137–149. https://doi.org/10.1016/j.trpro.2016.06.012

4. Dörrzapf L, Kovács-Győri A, Resch B, Zeile P (2019) Defining and assessing walkability: a concept for an integrated approach using surveys, biosensors and geospatial analysis. Urban Dev Issues 62:5–15. https://doi.org/10.2478/udi-2019-0008
5. Hewawasam C, Bandara S, Wirasinghe S (2013) Analysis of factors affecting pedestrian route choice. J Chem Inf Model 53:1689–1699
6. Seneviratne P (1985) Acceptable walking distances in central areas. J Transp Eng-Asce 111. https://doi.org/10.1061/(ASCE)0733-947X(1985)111:4(365)
7. Yang Y, Diez-Roux AV (2012) Walking distance by trip purpose and population subgroups. Am J Prev Med 43:11–19. https://doi.org/10.1016/j.amepre.2012.03.015
8. Roddin MF (1981) A manual to determine benefits of separating pedestrians and vehicles
9. Murtagh EM, Mair JL, Aguiar E, Tudor-Locke C, Murphy MH (2021) Outdoor walking speeds of apparently healthy adults: a systematic review and meta-analysis. Sports Med 51:125–141. https://doi.org/10.1007/s40279-020-01351-3
10. Rodríguez-López C, Salas-Fariña ZM, Villa-González E, Borges-Cosic M, Herrador-Colmenero M, Medina-Casaubón J, Ortega FB, Chillón P (2017) The threshold distance associated with walking from home to school. Health Educ Behav 44:857–866. https://doi.org/10.1177/1090198116688429
11. Hill MR (1982) Spatial structure and decision-making aspects of pedestrian route selection through an urban environment. PhD dissertation, Department of Geography, University of Nebraska-Lincoln
12. Gärling T, Gärling E (1988) Distance minimization in downtown pedestrian shopping. Environ Plan Econ Space 20:547–554. https://doi.org/10.1068/a200547
13. Verlander N (1997) Pedestrian route choice: an empirical study. PTRC Education and Research Services, London, England, pp 39–50
14. Zielstra D, Hochmair HH (2012) Using free and proprietary data to compare shortest-path lengths for effective pedestrian routing in street networks. Transp Res Rec 2299:41–47. https://doi.org/10.3141/2299-05
15. Moudon A, Hess P, Snyder M, Stanilov K (1997) Effects of site design on pedestrian travel in mixed-use medium-density environments. Transp Res Rec 1578:48–55. https://doi.org/10.3141/1578-07
16. Ariffin RNR, Zahari RK (2013) Perceptions of the urban walking environments. AicE-Bs 2013 Lond Asia Pac Int Conf Environ-Behav Stud 4–6(105) Sept 2013 105:589–597. https://doi.org/10.1016/j.sbspro.2013.11.062
17. Raford N, Chiaradia A, Gil J (2005) Critical mass: emergent cyclist route choice in central London
18. Snizek B, Sick Nielsen TA, Skov-Petersen H (2013) Mapping bicyclists' experiences in Copenhagen. J Transp Geogr 30:227–233. https://doi.org/10.1016/j.jtrangeo.2013.02.001
19. Baillie BH (2008) Shared space: reconciling people, places and traffic. Built Environ 34:161–181
20. Kaparias I, Wang R (2020) Vehicle and pedestrian level of service in street designs with elements of shared space. Transp Res Rec 2674:1084–1096. https://doi.org/10.1177/0361198120933627
21. D'Orso G, Migliore M (2020) A GIS-based method for evaluating the walkability of a pedestrian environment and prioritised investments. J Transp Geogr 82:102555. https://doi.org/10.1016/j.jtrangeo.2019.102555
22. Board TR (2016) Highway capacity manual 6th edition: a guide for multimodal mobility analysis. The National Academies Press, Washington, DC
23. Horbachov P, Svichynskyi S (2018) Theoretical substantiation of trip length distribution for home-based work trips in urban transit systems. J Transp Land Use 11. https://doi.org/10.5198/jtlu.2018.916
24. Bunke O, Feller F (1969) An introduction to probability theory and its applications, vol II, John Wiley & Sons, Inc., New York-London-Sydney, 1966. XVIII + 626 S., 3 Abb., 2 Tab., Preis $ 12,00. Biom Z 11:428–428. https://doi.org/10.1002/bimj.19690110615

Study on the Strength Behavior of Clay Blended with Nonwoven Surgical Masks Fabrics

Pallavi Badry

Abstract Amid the Covid-19 pandemic, it is declared by WHO to wear masks in public places. As per the guideline recieved from WHO in 2021, the people were instructed to wear 2 masks which surgical mask recommended to be more effective considering the situation in the second wave. Considering this scenario, for the next 8 to 12 months, we need to wear surgical masks compulsorily and need to dispose of them every alternate day. Considering the huge population in India, if this habit and the manner of wearing masks continue till the near future, the waste created with these masks will be voluminous. Thus, the need to recycle becomes obvious and need to address effectively. There are many techniques to recycle biomedical wastes. But if constructive recycling can be achieved it will help effectively in economic project execution in the civil construction industry. Thus, in this study, an attempt is made to use such nonwoven fabrics in soil improvement techniques. Thus, in this study, an attempt is made to use such nonwoven fabrics in soil strength improvement techniques. Before using the mask, it's well sanitized and then mixed with the clay soil. The different aspect ratios are selected for fabrics and several tests like Atterberg's limit (LL,PL), permeability, Triaxial test (Undrained), consolidation test, and compaction test are carried out. The result so obtained with the clay with mask fabrics is compared with the test result of virgin clay. The study found that the accountable strength is achieved in the clay soil with some limitations.

Keywords Covid-19 · Clay improvement · Mask · Non-woven fabric · Triaxial test · Atterberg's limit

P. Badry (✉)
Department of Civil Engineering, Vidya Jyothi Institute of Technology, Aziznagar Gate, Chilkur Road, Hyderabad 500075, Telangana, India
e-mail: pallavibadrim@vjit.ac.in; pallavi.badry@gmail.com

K. K. Pathak et al. (eds.), *Latest Developments in Civil Engineering*, Lecture Notes in Civil Engineering 352, https://doi.org/10.1007/978-981-99-2676-3_9

1 Introduction and Literature Review

Human health is severely affected by the recent pandemic caused due to COVID-19 disease which disturbed human life at every pace. The virus spreading can be arrested by using the face mask extensively and primary personal protective equipment (PPE) as per the severity of the places. Thus, the manufacturing and utilization of face masks to protect at a personal level notably increase as the COVID-19 pandemic is not concluded completely and it is still shooting up with time. It is a matter of fact that every surgical mask is manufactured with plastics or plastic derivatives. This substantial utilization of masks produces millions of tons of plastic waste on Earth with a considerable short time span [2]. This present pandemic situation compelled everyone to wear protective masks to avoid the virus infection. There are several varieties of masks such as surgical, N95, and commercial fabric/cloth masks used to control the present pandemic situation [1].

The statistics on surgical mask utilization and habits of disposing of it were reviewed in detailed through the available literature studies. To meet the demand of the mask it is realized by several countries like USA, China, Russia etc. that there is need of enhancement of production of the masks. As per the study attempted by Xiang et al., in 2020, it is stated that World Health Organization (WHO) mentioned a statistics that, COVID-19 will probably will be extended for some time span and to manage this situation about 89 million medical masks are foreseen to be manufactured [10]. Further, it is observed by the plastic innovation hub that the household insistence for the mask in the UK is around 24.37 billion per year [7]. As of February 2020, China has raised its daily production of medical masks to 14.8 million. It is also recorded by the Japanese Ministry of finance, trade, and industry that 600 million plus face masks were dispensed in the month of April 2020 [2, 7]. The rise in mask utilization due to mandatory practice to wear it in public for self protection from virus, it is obvious that the production of mask releases huge order. A research attempted by Klemeš et al., 2020a,b reveals that a mask manufacturing process emancipates about 59 g of carbon dioxide gas per unit of production. Further, ever-increasing uses of face mask induces waste management burdens and thus requires to increase the landfill area to manage such medical waste [7].

During this apex situation of the first wave of the COVID-19 crisis, several counties imposed lockdown. Due to the imposed lockdowns, the pool and public transportation, individual vehicles, and walking were much reduced with the remarkable count, which impacted the rate of mask usage considering the weekly rate. Extending these statistics to the global scale with the whole population when the lockdown is removed partially and fully in some countries, it is anticipated to attain the highest level of mask wastage in a short span of time, which will, in due time, generate a noteworthy quantity of plastic waste to the mother nature (Fig. 1).

The surgical mask in priority consists of plastic forms like polypropylene and/or polyethylene, polyurethane, polystyrene, polycarbonate, and polyacrylonitrile. The compound adds plastic or microplastic pollution to the environment [1] (Fig. 2).

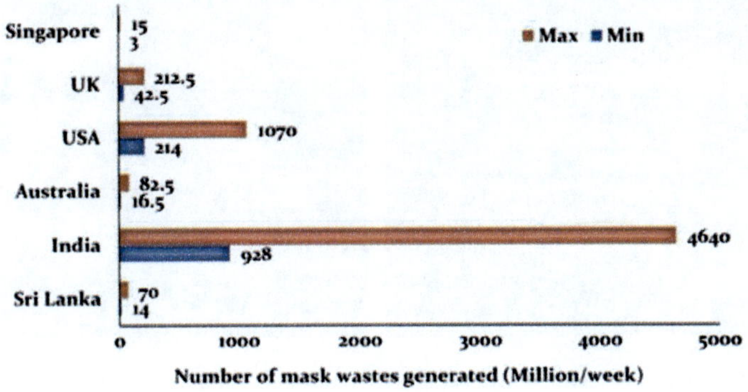

Fig. 1 Countrywise statistics on mask waste per week [2]

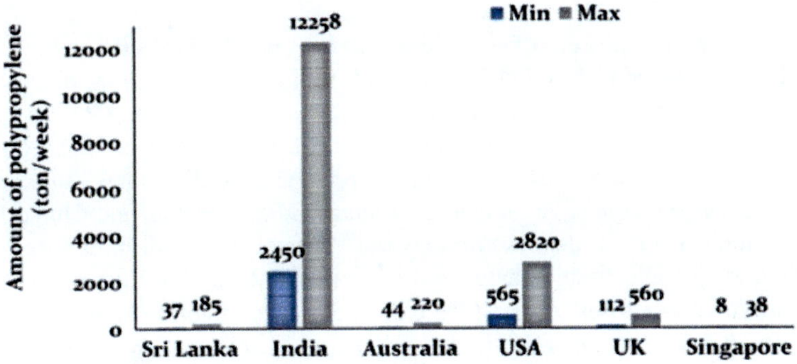

Fig. 2 Waste generation at a global scale due to the use of masks [7]

2 Material Used in the Production of the Masks

It is a matter of fact that literature findings inferred that in dense local areas, people are habituated to use surgical masks to protect themselves from the COVID-19 virus. The very vivid reason to use nonwoven fabrics in surgical face masks is to filter the bacteria and maintain air permeability (Fig. 3). Also, these nonwoven fabrics help in retaining their position on the skin by resisting slippage. Considering these blockade properties and expiratory resistance, the ASTM F2100 (ASTM, 2007) and EN ISO 15223–1 (BS EN ISO 15223–1, 2016) segregate this mask into several categories such as low barrier, moderate barrier, and high barrier [7].

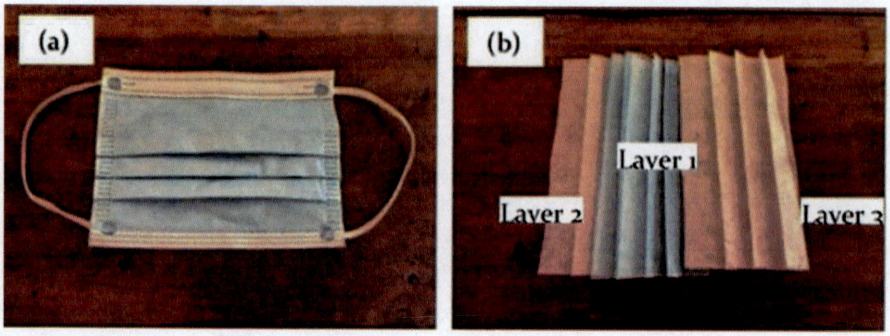

Fig. 3 Surgical mask structure: **a** surgical mask **b** three layers in the surgical mask

2.1 Impact of Mask Waste on the Environment

Due to the lack of awareness of suitable methods of disposal for the used mask, the wastes have increased tremendously. This leads to the challenges in the disposal and protection of the environment in this COVID crisis condition. In addition to this, there is no suitable mask or nondegradable waste accumulation method specified in any region of the country. This situation is found in almost all countries in the world and a few are mentioned as reference (Sri Lanka, India, Pakistan, and China) [2, 7]. This practice leads to an increase in the great volumetric growth in nondegradable plastic and its particle waste in nature, which causes unhealthy visuals on the streets and increases the stresses on landfills. Considering its effect on water bodies and waterways, it is observed that the mask waste disposed of in it reaches the fresh water and seawater. This increases the existence of plastics in the water. The health and environmental results of plastic and its substances due to the inappropriate disposal of facemasks were also presented in detail by a number of literature studies [3, 7].

It is also noted that the carbon dioxide emission will be increased after each step of mask manufacturing, which will certainly contribute to global warming [10].

It is observed that the masks are mostly catapulted without proper scientific disposal on the paths and or accumulated as mixed waste.

In waste disposal management techniques for urban solid waste and hazardous medical waste, the pandemic has led to great challenges. The face masks collected from hospitals and other clinical mixed waste are sent to the incineration and landfill. However, due to the existence of the plastics in the mask, such methodologies often have the potential to cause ill effects on the environment. Usually, plastics are chemically stable, resistant to Corrosion, and difficult to degrade by microorganisms [10]. Yet they prefer to endure in the soil and pose environmental threats.

This indicates that the current ongoing pandemic increases environmental pollution and negative impact on human and animal health. Also, it is very vivid that the population and awareness of the pandemic are the functions of the quantity of waste generation, thus India needs to be provided with a proper disposal mechanism for such waste.

It is quite obvious that it is impossible to cease the usage of masks considering public safety but need to identify several techniques to dispose of them in a meaningful manner. This study focused on the utilization of mask waste to treat the clayey soil in foundation constructions.

3 Methodology

It is understood from the literature that it is the need of the hour to dispose of the waste generated due to surgical masks in a proper mechanism. Also, the civil construction industry is one of the sizable industries in the global scenarios and thus these masks waste can be diverted toward it.

It is noted that the cost of the foundation construction is about 15 to 20% of the total cost of construction for any project and this cost is more for the foundations where the bearing resistance of the soil is very less. The clayey soils always offered the differential settlement and thus required to be treated the soil either with chemicals or provide a raft foundation which will be recommended as per the standard foundation design aspects. Treating or improving the soil strata for better bearing capacity is one of the recommended practices to avoid the failures of the foundations for several reasons. The use of fiber reinforcement with site soils has been recognized as a recommended technique for ground improvement in numerous geotechnical engineering applications where shear strength needs to be improved for soils. The short staple fiber or sometimes continuous filament yarn is used to fabricate the nonwoven geosynthetics [9]. The manufacturing process of nonwoven fabrics consist of thermal, chemical or mechanical techniques or a combination of techniques [8]. Generally, the thermal bonding technique is used in order to adopt a wide range of opening sizes and a typical thickness of about 0.5–1 mm of nonwoven fabrics. To have a proper chemical bond between soil particles and fibre the greater thickness fibers ranging from 2 to 3 mm is preferred. On the contrary, mechanically fastened nonwoven geosynthetics have a typical thickness in the range of 2–5 mm [4, 6]. It is recommended in practice not to entrap stones or moisture in the nonwoven geotextile during blending into the soils, especially in clayey types of soils.

In this study, the clay soil is assorted with a nonwoven fabric prepared from the masks. The filaments of masks in the form 10, 15, 20, and 30 mm are prepared mechanically with a thickness of 0.5–1 mm. The fabrics are added into the soil with the percentages of 0.5%, 2%, and 5% with the volume of dry clay [5, 11] (Fig. 4).

For all percentages added in the clay, several tests including Atterberg's limit (LL, PL, and SL), permeability, Triaxial test (Undrained), and compaction test are carried out. And the results are tested against virgin clay (Fig. 5).

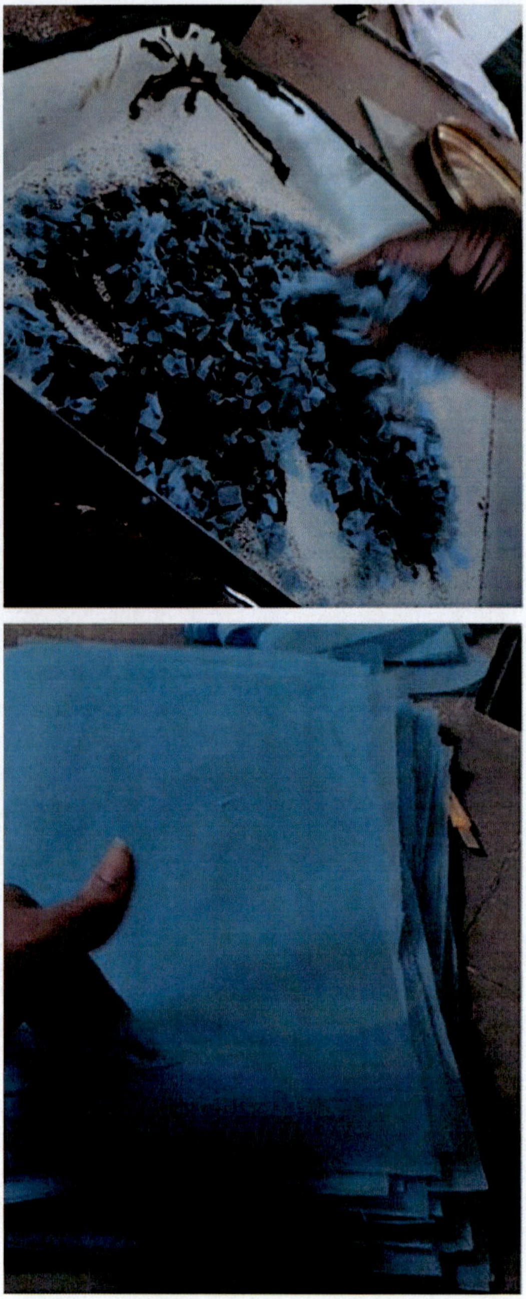

Fig. 4 Mixing of nonwoven mask fabrics in a clay

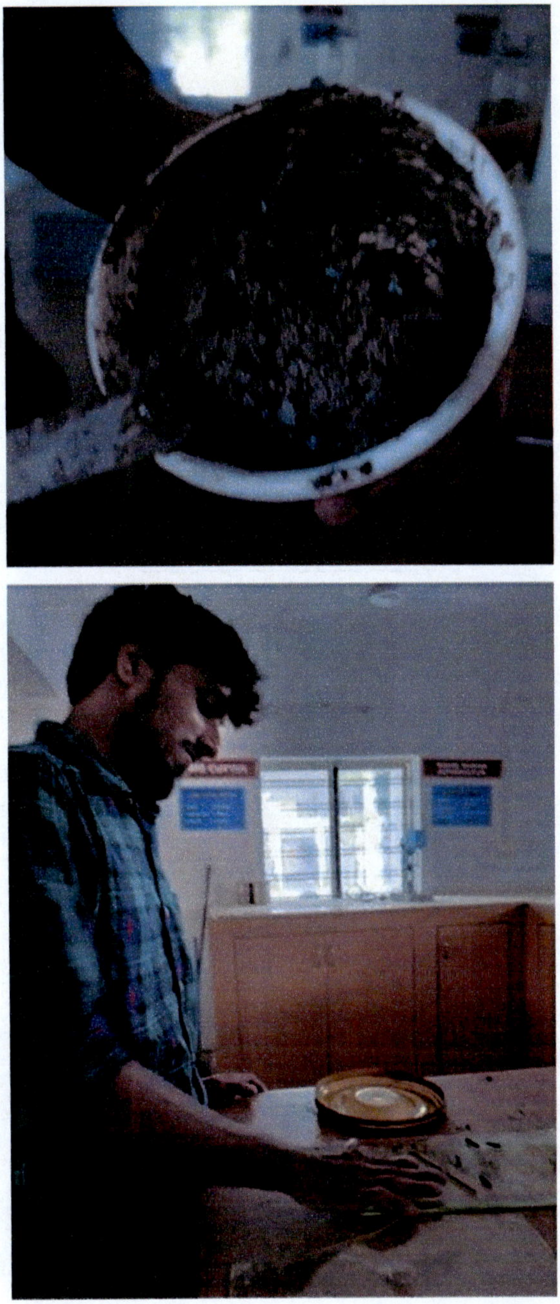

Fig. 5 Test on clay and assorted clay in the laboratory

4 Result and Discussion

The virgin clay is first dried and clumps are removed by ramming it gently, the sample is prepared as per the IS guidelines for each test.

The several tests include grain size analysis, Atterberg's limit (PL, LL), Modified Compaction test, variable head permeability test, and undrained triaxial test.

The results are obtained for different asortation percentages of mask segment of thickness 1 mm (avg.) and aspect ratio 1:2 with the average length of each piece of fibre were kept 20 mm. The assortation percentages with dry clay varied from 0.5%, 2%, and 5% by volume to understand the impact of the material on the engineering properties of the clay. The above percentages are decided from the ground improvement techniques guidelines recommended in the past literature studies [5, 8, and 9]. The test results are compared with virgin clay results. The following section shows a detailed discussion on the test results obtained in the present study (Fig. 6).

The grain size distribution is carried out on virgin clay samples and the different mask fiber assortation percentages. It is observed that the retention is 0% for 4.75 and 2.36 mm aperture sieve sizes; hence the results are presented from the 1.5 mm sieve size in order to capture the detailed profile of the curve. It is observed that more than 0% of particle size falls below 0.3 mm for virgin clay. But when the clay is assorted with fiber, the average grain size increased to some extent as % finer started reducing and this change of weel graded behaviour is more prominent when the clay has 2% mixed fiber a compared to other percentage mixes (Fig. 7).

The plastic limit and liquid limit tests have been carried out on the clay sample. The test procedure is followed strictly to prepare soil samples as per the guidelines given in an IS code. The liquid limit of the virgin clay is found to be 41% which is a common range of clay soils. After adding the fibers, an improvement in LL is

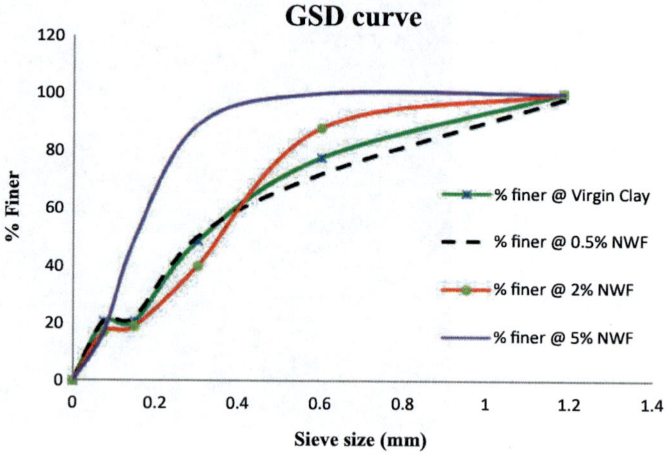

Fig. 6 GSD for nonwoven mask fabrics in clay and virgin clay

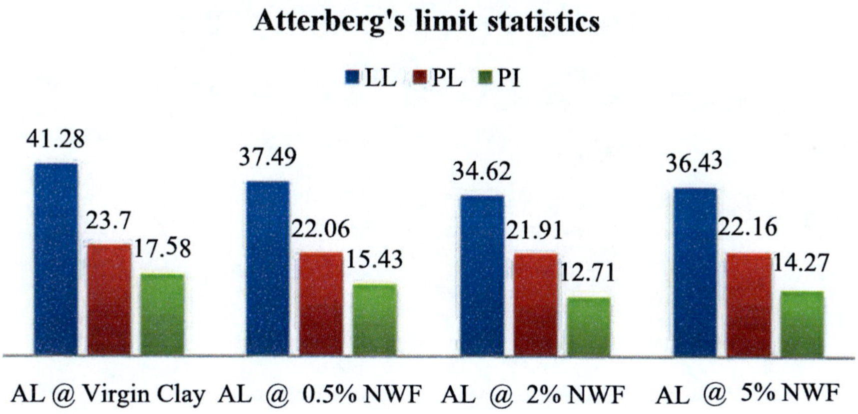

Fig. 7 Atterberg's limit for non-woven mask fabrics in clay and virgin clay

observed. This shows the LL is decreased by 12% on average. A similar observation is held for plastic limit. Among all the mixes, the 2% mix shows a decrease in both the limits, i.e., around 15% (Fig. 8).

The modified compaction test has been carried out on the virgin clay as well as different assortation percentages with dry clay varied from 0.5%, 2%, and 5% by volume. It is observed that the OMC and DD profile of 5% fiber mixed clay is in synchronization with virgin clay results with the same value of OMC, i.e., 21% but a slight drop of dry density magnitude. Also, it is noted that the mixing of fibers in clay shifts the OMC and DD profile showing the slow compactions and attaining the maximum dry density with some delay. The results reveals that the 2% mix shows

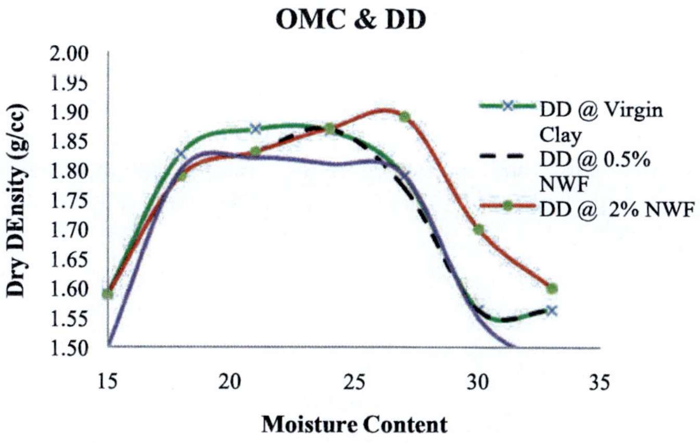

Fig. 8 Compaction test results for non-woven mask fabrics in clay and virgin clay

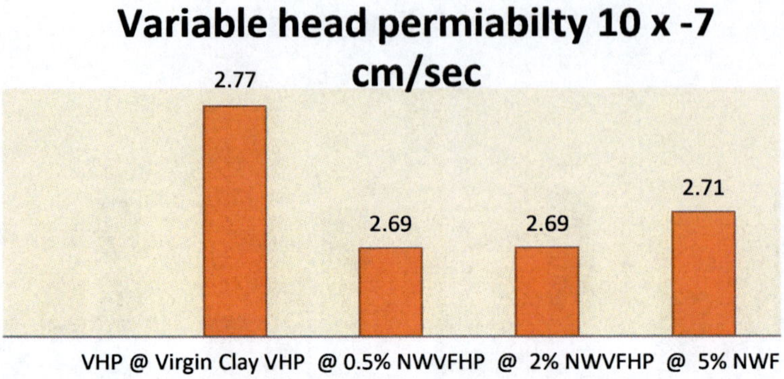

Fig. 9 Permeability test results for non-woven mask fabrics in clay and virgin clay

a countable change in dry density value in comparison with the virgin clay. This improvement is nearly 5% less than the virgin clay (Fig. 9).

Also, the hydraulic behavior of clay is studied by performing the variable head permeability test which is the recommended test for the clayey soils. The permeability characteristic of clay reduces with the addition of fibers in the soil mass. This is due to the fiber exhibiting the water holding capacity and opposing the free flow of the water through a soil mass. It is a fact that clayey soils have always poor permeability and thus can be treated separately to enhance this characteristic. But it is vivid from the result that the addition of fibers in clay does not affect the hydraulic characteristic of the clay.

Typical results of the UU test on saturated cohesive clay are performed for normally consolidated clay. It was observed that the increase in cell pressure only results in an equal increase in pore water pressure, since no drainage is allowed, and the subsequent increase in the total major principal stress also results in the same change in pore water pressure (Fig. 10).

The result shows that there is a countable increase in cohesion and friction resistance values for all fiber mixes including 0.5%, 2%, and 5%. This enhancement in shear strength is more in the 2% mix (Avg 15%) than the rest other mixes. This result shows that there is an improvement in the shear strength of the clay as additional cohesion will be offered by the fiber pieces, thus improving the overall shear resistance of the clay.

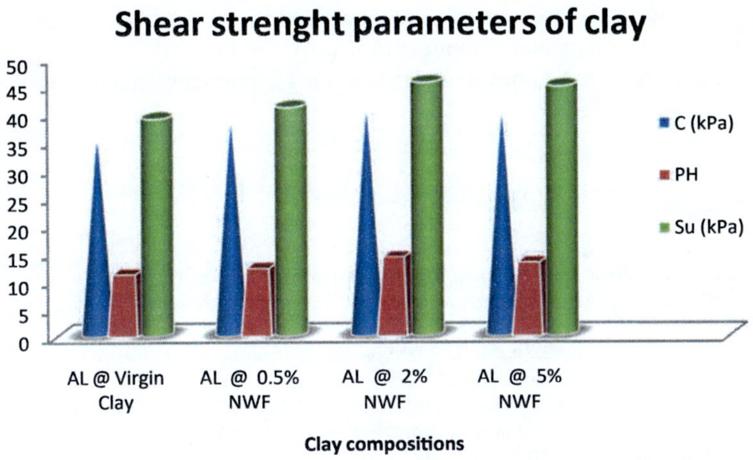

Fig. 10 Unconsolidated-Undrained test results for nonwoven mask fabrics in a clay and virgin clay

5 Conclusions

From several test results including grain size analysis, Atterberg's limit (PL, LL), Modified Compaction test, variable head permeability test, and undrained triaxial test, the following salient conclusions have been drawn:

1. It is observed that certainly mixing the mask fibers improves the engineering characteristics of the clay. In some cases, the deviation is remarkable, and in some tests, it is negligible.
2. It is inferred from the study that 2% mixed fiber shows a uniform gradation curve within the limits of the grain sizes of clay soils and can be recommended to use it for effective means of ground improvement techniques.
3. Adding the fibers increases the average grain size of the clay and thus provides an additional area for grain to grain contact and shows a remarkable improvement in shear strength of the soil.
4. The clay shows a decrease in LL and PL when fibers are added to it and this change is more prominent for the 2% mix
5. The OMC and DD profile shows slow compactions when mask fabrics are assorted with clay than virgin clay. Also, it is found that the former case attains the maximum dry density with some delay than virgin clay condition. The results show that the 2% mix exhibits a countable change in dry density value in comparison with the virgin clay. This improvement is nearly 5% to the virgin clay
6. The hydraulic characteristics are found to be unaffected much with a countable range for fiber mixed clay. Thus, some other mechanism needs to work out to improve this characteristic of clay.

7. The present study offers an overall conclusion that the 2% mixed fibers prove to be a more recommended solution to improve the clay strength although a few more trials need to be done to understand the improvement in permeability.

References

1. Barasheed O, Alfelali M, Mushta S, Bokhary H (2016) Uptake and effectiveness of facemask against respiratory infections at mass gatherings: a systematic review. Int J Infect Dis 47:105–111
2. Fadare OO, Wan B, Guo L, Zhao L (2020) Microplastics from consumer plastic food containers: are we consuming it ? Chemosphere 253
3. Fakhrabadi A, Ghadakpour M, Choobbasti AJ, Kutanaei SS (2021) Influence of the non-woven geotextile (NWG) on the engineering properties of clayey-sand treated with copper slag-based geopolymer. Int J Constr Build Mater 23
4. Mitra Ashis (2015) Application of geotextiles in coastal protection and coastal engineering works: an overview. Int Res J Environ Sci 4(4):96–103
5. Mohammadi M, Habibagahi G, Hataf N (2021) A bioinspired technique for improving the interaction between cohesive soil and geotextile reinforcements. Int J Geosynth Ground Eng 22
6. Nguyen M-D, Yang K-H, Yalew WM (2020) Compaction behavior of nonwoven geotextile-reinforced clay. Geosynth Int 2
7. Sangkham S (2020) Face mask and medical waste disposal during the novel COVID-19 pandemic in Asia. Case Stud Chem Environ Eng 2
8. Trauner L, Dolinar B, Mišič M (2005) Relationship between the undrained shear strength, water content, and mineralogical properties of fine-grained soils. Int J Geomech 5(4):350–355
9. Umesh S (2014) Application of geotextiles in pavement drainage systems. Int J Civ Eng Res 5(4):385–390
10. Xiang Y, Song Q, Gu W (2020) Decontamination of surgical face masks and N95 respirators by dry heat pasteurization for one hour at 70°C. Am J Infect Control 48:880–882
11. Zuzulova A, Hodakova D, Capayova S, Schlosser T (2019) Design of pavement structures in tunnels. In: Proceedings of the 19th international multidisciplinary scientific geoconference SGEM, vol 30. pp 1–13

Experimental Study on the Use of Pond Ash and Walnut Shell in Concrete

Parshottam Sarathe, Ramakant Agrawal, Pramod Gour, and Vinay Kumar

Abstract The main objective of the present study is to find out a suitable, effective and alternative material for partial replacement of cement and coarse aggregate, to find out the possible utilization of waste materials in the construction industry that in turn considerably minimizes the usage of cement and coarse aggregate and ultimately reduce construction cost, to explore possibilities of improving mechanical properties of concrete using pond ash instead of cement and Walnut Shell instead of coarse aggregate partially, to evaluate the effect of using pond ash and Walnut Shell in concrete and to investigate the strength of replaced concrete with that of conventional concrete. This project is mainly undertaken to study the behaviour and performance of concrete using waste materials such as pond ash and Walnut shell. This type of use of waste material can solve problems of lack of aggregate in various construction sites and reduce environmental problems related to aggregate mining and waste disposal. The use of waste aggregates can also reduce the cost of concrete production and increase the workability.

Keywords Concrete mix · Pond ash · Strength parameter · Sustainable construction · Walnut shell · Workability

1 Introduction

In today's world, the importance of concrete should not be underestimated. The concrete structures are everywhere like buildings, roads, bridges and dams. Concrete is a composite of binder and filler. Typical concrete is a mixture of materials of fine and coarse aggregates and water. Cement is commonly used as binders and sand fillers are mixed as fine aggregate and crushed stone, gravel and crushed stone bricks. Clinker is used as coarse aggregate. Cement concrete is concrete that contains, in addition to water, a certain proportion of binder (cement), fine aggregate (sand) and coarse aggregate (crushed stone or gravel).

P. Sarathe (✉) · R. Agrawal · P. Gour · V. Kumar
Department of Civil Engineering, OIST, Bhopal, India
e-mail: parshottam.sarathe@gmail.com

K. K. Pathak et al. (eds.), *Latest Developments in Civil Engineering*, Lecture Notes in Civil Engineering 352, https://doi.org/10.1007/978-981-99-2676-3_10

In the 1824, the Portland cement was introduced by British stone-mason John Aspdin, who heated the mixture of clay and ground limestone on stove.

The name Portland came from the Isle of Portland which is in Britain, the limestone was quarried from this place.

Over the past years, global warming and environmental degradation have become major problems. Greenhouse gas emissions from the cement industry contribute to climate change. Preventing the depletion of natural resources and improving the use of waste are challenges for scientists and engineers. A lot of research has been done on the conservation of natural resources, the prevention of pollution and the contribution of this waste to economic efficiency.

The first factory of the cement was built in 1904 in Chennai in India, the name of the cement factory was South India Industrial Limited which was established in Madras, after this, India went into the Cement Era in 1914 and started the cement company named India Cement Company Limited at Porbandar, Gujarat; fter that, many more companies were established in India. The cement production of India was 267,000 tonnes in 1924. Nowadays, India is the second largest cement manufacturing country in the world after China. In 2020, the cement production of India was 329 Million Metric Tonnes.

Pond ash is a by-product of the work in power plants. It has several applications in civil engineering; mainly in the construction sector. In this blog, we will go through the basic details, properties, applications and limitations.

About 20% of the total ash produced is discharged to the embankment as bottom ash and about 80% of the total ash is collected in dry form in the ESP/silo system and shipped to cement manufacturers and various fly ash users/traders. will be not transported dry ash is mixed with water and dumped into the ash dike. The ash accumulated on the ash embankment is called pond ash. Ash ponds, also known as coal ash ponds or surface dams, are man-made structures used in coal-fired power plants to dispose of two types of coal combustion products: bottom ash and fly ash. Concrete will be worth nearly $7.5 billion by 2050, and m3 (about 18 billion tonnes) per year will continue to grow. At least three quarters of the total volume of concrete consists of coarse and fine aggregates. Natural resources such as river sand are increasingly scarce. The country's electricity demand is growing rapidly in India, coal-fired power plants supply about 65% of the country's electricity demand. The coal-fired power plant will burn approximately 407 million tonnes of coal annually and produce approximately 131 million tonnes of coal ash. Forms fine particles less than 4.75 mm. Accumulation of pond ash (PA) around thermal power plants poses an environmental threat, making its proper management increasingly important.

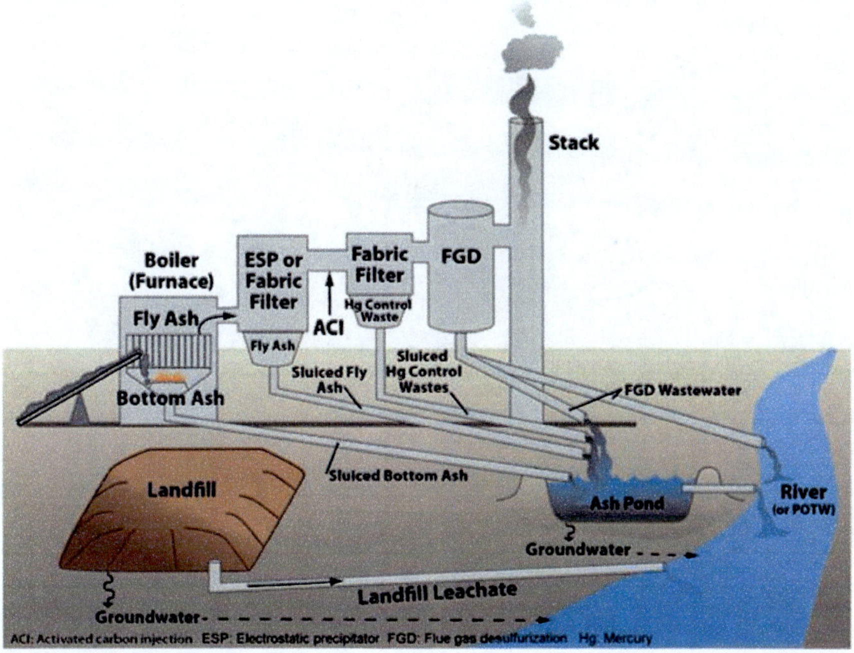

Fig. 1 Process of generation of Pond Ash

The following points justify the problems caused due to pond ash:

- Cause severe respiratory problems
- Visual and aesthetic problems in almost all the major industrial cities in India
- Degradation in the environment
- Depletion of the natural resources
- Ash pond areas lose all vegetation and the area never can be reclaimed for irrigation
- The ash bunds collapse and contaminates freshwater resources

In this work, pond ash waste is collected from a dump yard from NTPC Bhusawal (Figs. 1, 2).

.

2 Literature Review

- Nusrat et al. [1] focuses on the use of two agricultural wastes, egg shells and walnut shells to produce unfired clay blocks. Tests were conducted on 3 sample series, first incorporating eggshell (10–50%) and walnut shell (5–20%) separately,

Fig. 2 Pond Ash (NTPC Bhusawal)

then by combining the mixture (5% walnut, 10–30% eggshell). Determination of their effects to evaluate the physical and mechanical properties of unfired clay blocks. The results showed that when the ingredients were used individually, the eggshell increased strength compared to the control, while the walnut shell decreased strength. In addition, in a mixer, he combined both the ingredients, which further reduced the strength of the sample. Nevertheless, waste entrapment reduced sample density, capillary water absorption coefficient and linear shrinkage. This result indicates that eggshell has great potential for the production of unfired clay blocks. However, the integration of walnut shells requires further research.

- Sreelakshmi and Reshmi (2018) dealt with the strength and durability performance of concrete in which pond ash is replaced as fine aggregate and also details the effect of triethanolamine, an integrated inhibitor. The weight fractions of the added pond ash fine aggregate were 0.10, 20, 30, 40 and 50, and the fractions of triethanolamine were 1%, 2% and 3% by weight of the cement: 4 wt% and 5 wt%. Test results show that replacing sand with pond ash increases the strength of concrete, while the addition of integrated inhibitors reduces permeability and increases density, thereby improving corrosion resistance and durability in adverse environments.

- Kamal et al. (2017) studied the use of walnut shells, the agro waste, as partial replacement of fine aggregate in concrete for the purpose of reducing natural resource exploitation and associated costs, minimizing waste landfill and modifying the concrete properties. The walnut-concrete samples seemed to have lower density, and lower initial and total surface water absorption compared to control samples. Similar findings were obtained for compressive strength. However, it was confirmed that walnut shell can be used as light replacement material instead of fine aggregate up to 30% at 0.38 water/cement ratio without adversely affecting the acceptable compressive strength for structural Portland cement concrete.
- Cheng et al. (2017) studied the use of agricultural waste as a partial replacement for fine aggregate in concrete with the aim of reducing natural resource depletion and associated costs, minimizing waste landfills and altering concrete properties. We are using walnut shells from The walnut-concrete samples appeared to be less dense and had lower initial and total surface water absorption compared to the control samples. Similar results were obtained for compressive strength. However, it has been determined that walnut shells can be used as a lightweight alternative for up to 30% fine aggregate at a water/cement ratio of 0.38 without adversely affecting the allowable compressive strength of structural Portland cement concrete.
- Bagwan and Kulkarni (2015) present an experimental investigation conducted to investigate the use of pond ash in concrete. It is now important to think about the efficient use of pond ash in order to conserve natural resources and achieve sustainable development. Concrete was made with different percentages of pond ash (15, 25, 35, 45 and 55%) and tested at different ages (3, 7, 28, 56, 90 and 180 days). The results of pond ash concrete were compared with control concrete. Maintained a slump in the range of 100–120 mm in all parts.

3 Material Used

The following materials are used during the research work (Tables 1, 2, 3):

- Cement
- Fine aggregates (Sand)

Table 1 Physical properties of cement (OPC 53)

S. no	Properties	Values
1	Grade of cement	OPC 53
2	Normal consistency	33%
3	Initial setting time	35 min
4	Final setting time	385 min
5	Fineness of cement	8%

Table 2 Properties of sand taken

S. No	Properties	Values
1	Specific gravity	2.63
2	Bulk density	1750 kg/m^3
3	Fineness modulus	2.80
4	Silt content	2.93
5	Sand type	River sand from River, Zone-II

Table 3 Characteristics of pond ash used

S. No	Properties	Value
1	Size	1 to 100 microns
2	Colour	Grey
3	Specific gravity	2.38
4	Moisture content	75%
5	Silica content	40–60%

- Coarse Aggregates(12–20 mm)
- Walnut chips
- Pond Ash (dried)
- Super Plasticizer
- Water

4 Methodology

For the experimental study, the different materials are mixed together. To provide suitable strength and uniformity amount of sand is kept constant. The slump cone test was used for the determination of the workability of the concrete. For determining compressive strength, different samples, each containing 3 cubes, were prepared. The size of the specimen was kept at 150 mm × 150 mm × 150 mm; all cubes were prepared using a mould through hand moulding only. Flexural strength was determined using one point loading test through UTM.

The following procedure has been adopted for the work (Figs. 3, 4 and Table 4):

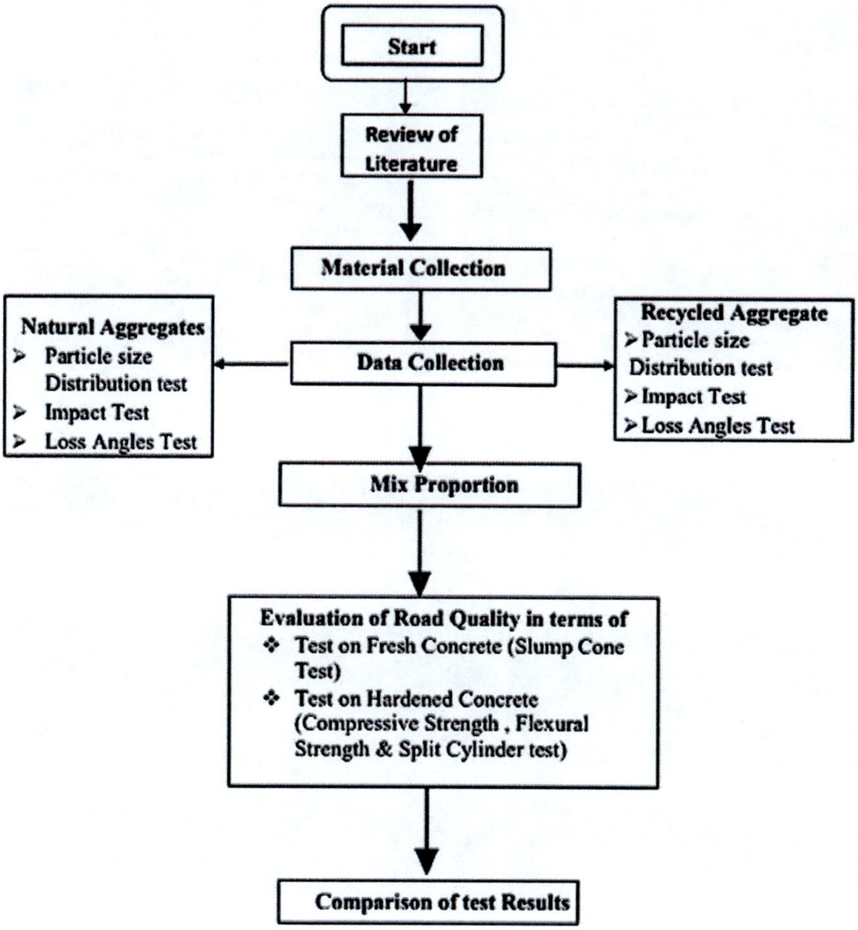

5 Results

5.1 Slump Cone Test Results

See Tables 5, 6 and Figs. 5, 6.

The above graph shows that the slump value decreases gradually with the increase in the percentage of Pond ash and Walnut Shell; this trend was recorded for all samples.

Fig. 3 Preparation of mix sample

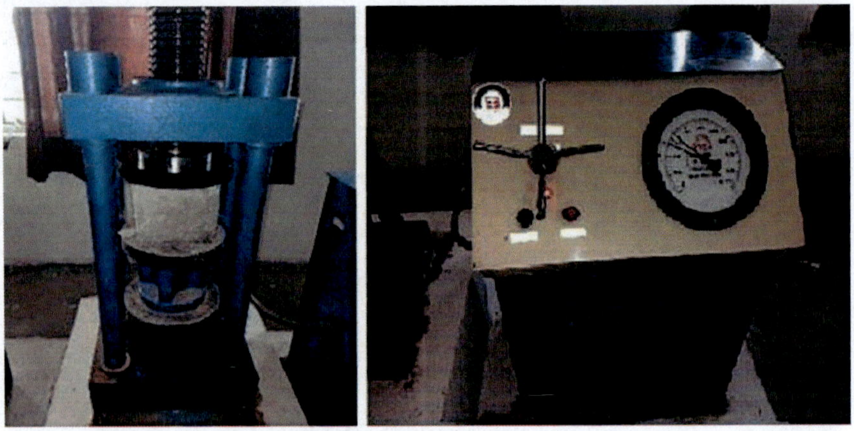

Fig. 4 Placing on compressive testing machine

5.2 Compressive Strength Test Results

The above graph shows that Compressive Strength decreases gradually with the increase in the percentage of Pond ash; this trend was recorded for all samples (Figs. 7, 8).

5.3 Flexural Strength Test Results

See Fig. 9.

Table 4 Cases considered for the study

Phase 1 Using pond ash as cement replacement		
Sample ID	Cement %	Pond ash %
CONC 0	100	0
CR 5	95	5
CR 10	90	10
CR 15	85	15
CR 20	80	20
Phase 2 Using Walnut shell as coarse aggregate replacement		
Sample ID	Coarse aggregate %	Walnut shell %
CONC 0	100	0
AR 5	95	5
AR 10	90	10
AR 15	85	15
AR 20	80	20
AR 25	75	25
Phase 3 Using pond ash and walnut shell		

Table 5 Slump in mm of Concrete on cement replacement

Pond ash used	0%	5%	10%	15%	20%
Slump in mm	76	77	70	65	60

Table 6 Slump in mm of Concrete on coarse aggregate replacement

Walnut shell used	0%	5%	10%	15%	20%	25%
Slump in mm	76	68	68	65	60	58

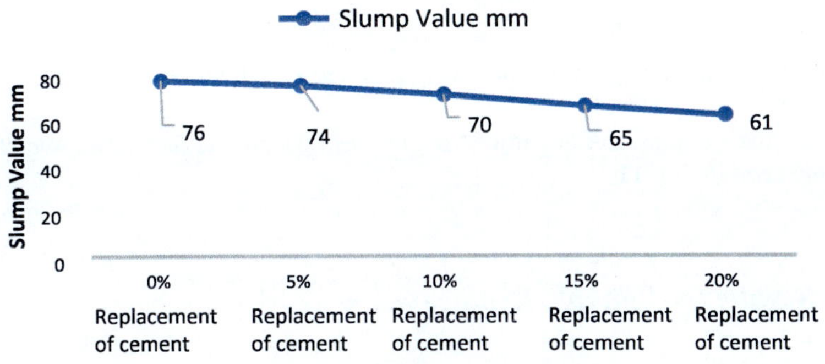

Fig. 5 Slump Values observed on cement replacement

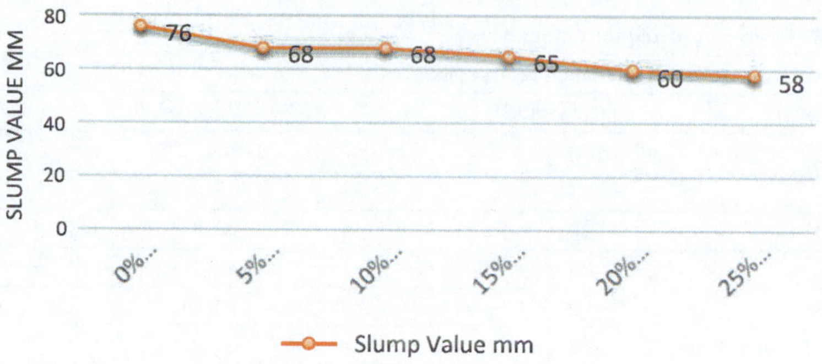

Fig. 6 Slump Values observed on coarse aggregate replacement

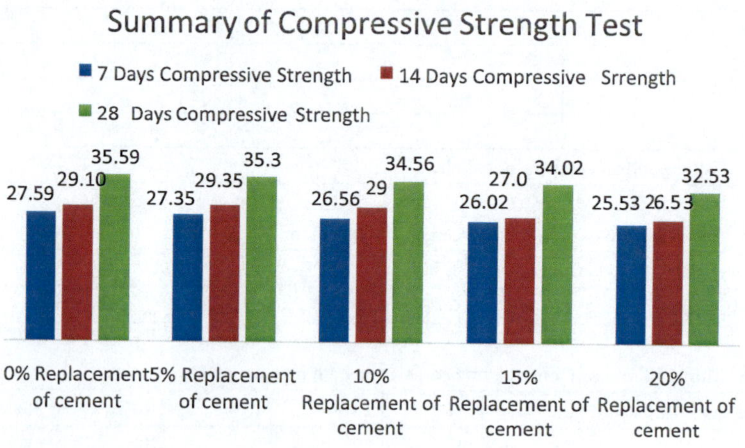

Fig. 7 Compressive Strength of M30 Grade Contain of Pond Ash

For all the samples, the flexural Strength is decreased, which can be easily seen in Fig. 10.

For all the samples, the flexural Strength was observed is decreased, which can be easily seen in Fig. 11

5.4 *Results for Suitable Replacement*

See Table 7 and Fig. 11.

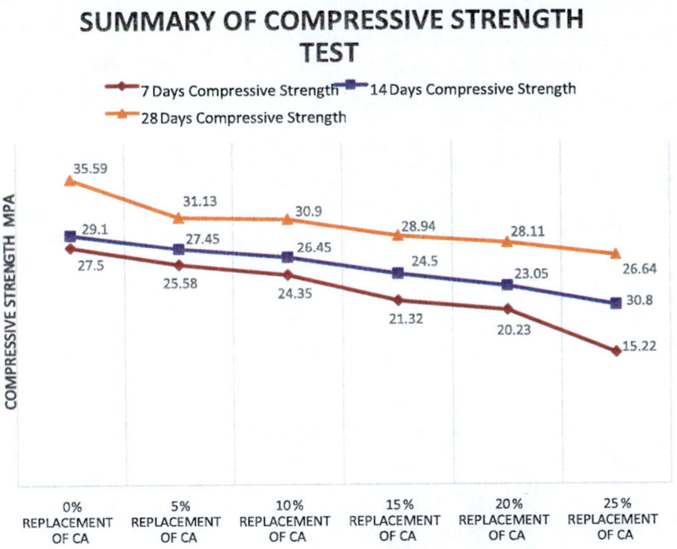

Fig. 8 Compressive Strength of M30 Grade Contain of walnut shell

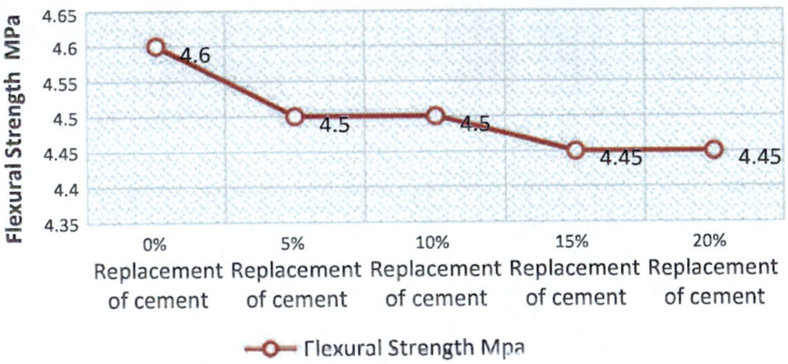

Fig. 9 Flexural Strength of M30 having pond ash

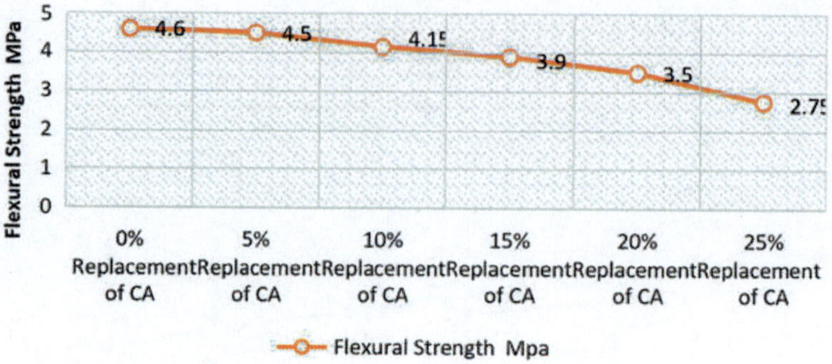

Fig. 10 Flexural strength of M30 having walnut shell

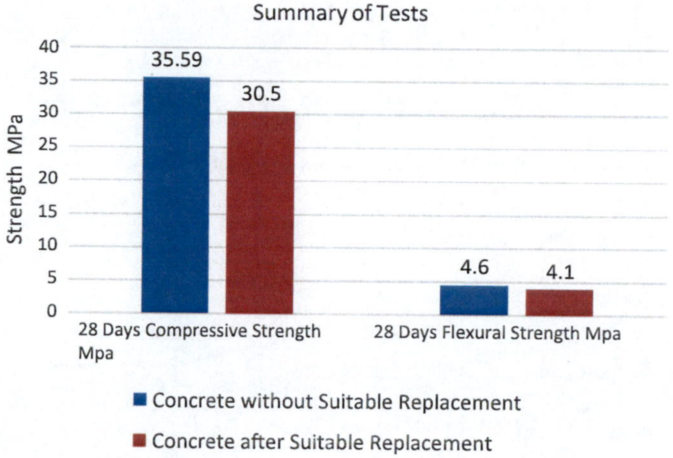

Fig. 11 Results for suitable replacement

Table 7 Results for suitable replacement

suitable %, coarse aggregate with walnut shell at 28 days	suitable %, cement with pond ash at 28 days	compressive strength at 28 days (N/mm^2)	flexural strength at 28 days (N/mm^2)
10%	10%	30.50	4.10

6 Conclusion

From the results, we can conclude that:

- Slump test values show that the Workability of concrete is decreased as the percentage replacement of coarse aggregate with the walnut shell is increased. However, the workability of all concrete mixes up to 25% coarse aggregate replacement was suitable for structural uses.
- When the replacement of cement is increased up to 20% by Pond ash in the concrete mix, there is a decrease in slump value which shows the workability of the concrete is decreased.
- Both the compressive strength and flexural strength of concrete decrease with the increase in pond ash.
- Compressive and flexural strength of concrete was decreased with an increase in walnut shells as a partial replacement of coarse aggregate.
- The maximum level for the replacement (by weight) of cement for severe exposure condition, i.e. M30 grade, is found to be 5–10%, and for the walnut shell as partial replacement of coarse aggregate, it is found to be 10%.

Reference

1. Nusrat et al (2022) Eggshell and walnut shell in unburnt clay blocks. Civil Eng 3:263–276

Process Management of Project Activities in a High-Rise Building Using Primavera

Md. Imteyaz and Ravindra Gautam

Abstract High-rise building architecture has seen significant evolution in the last several decades. Buildings are increasingly being modelled using three-dimensional finite element software. The usual construction activities in a "high-rise structure" have been reviewed in this paper. The conventional ways of scheduling the construction activities have been shifted over to a software called Primavera which provides better and more accurate time management of the activities. The timeline of activities has been detailed using a "Gantt chart" and a total time duration of 398 working days has been obtained through the process. Moreover, the stress analysis of "high-rise structure" has been also attempted in this paper using the Staadpro software. A software model of the G + 3 structure has been prepared "and then subjected to various loads as per Indian standards. It has been observed that the maximum load is bored by the ground floor, i.e. 0.018 N/mm^2, which goes on decreasing gradually towards the upward floors and is minimum for the topmost floor, i.e. 0.123 N/mm^2. While the beams show "maximum bending moment" in the z-direction.

Keywords High-rise buildings · Gantt chart · Primavera · StaadPro · Stress analysis

1 Introduction

The use of cast iron as a construction material, which was lighter but better than masonry at the time of its development in the 1880s, marked the beginning of the twentieth century's advancements in the "high-rise building structure" design. Elevators helped pave the way for skyscrapers to increase in popularity over time. "High-rise structures" have seen a resurgence in popularity in urban areas with limited land supply in recent decades due to their economics, environmental friendliness, and other advantages. Everywhere we look, skyscrapers are being erected to increasingly greater heights. Disasters of many dangerous types have become more common,

Md. Imteyaz (✉) · R. Gautam
Department of Civil Engineering, Technocrats Institute of Technology Excellence, Bhopal, India
e-mail: mdimteyaz98@gmail.com

K. K. Pathak et al. (eds.), *Latest Developments in Civil Engineering*, Lecture Notes in Civil Engineering 352, https://doi.org/10.1007/978-981-99-2676-3_11

making the task of designing robust structures more difficult. Because of this, it is critical that the behaviour of these buildings under several hazards be understood in order to utilise this information in the design process [1].

In current high-rise structures, there are numerous examples where enormous spans are necessary to obtain spacial freedom on normal levels and broad atria to provide continuity with the exterior areas on the lower floors. In order to accomplish these spaces, it is required to offer great strength in the structural parts that compose the building structure, notably the columns. It is feasible to prevent overly large volume columns by employing proper mixes of high-strength materials. (1) New design of strong lines and rectangles, where hefty constructions made out of steel, indicate power and grandeur. (2) The merger of inner rooms and outer façade as one entity. (3) Use of concrete and steel as structural members [2].

The "natural frequencies" are the frequencies at which a structure starts to resonate, and they are found in all constructions. The "basic frequency" is the lowest possible natural frequency. When doing a dynamic analysis of a structure, it is critical to consider the "natural frequencies" [3, 4]. In the event of an earthquake or a windstorm, for instance, buildings might wobble dramatically and even collapse. Due to wind-induced oscillation [5] in 1940, the "Tacoma Narrows Bridge" collapsed in Washington state. Modes are used to characterise the natural frequencies; each mode has a "natural frequency" and a form [3]. The x and y-axis sway, and the torsional sway, are usually the first 3 natural frequencies for a structure (around the z-axis). No matter where the structure is situated, wind loads must be taken into account. Pressure variations in the atmosphere move air, resulting in wind. High above the earth's surface, airflow is mostly unaffected; but, near the ground, the earth's surface and man-made buildings generate turbulence and eddies. An eddie is a vortex that forms when the wind flows past an obstruction, such as a structure or rugged terrain. The serviceability limit state is one of the most critical considerations in the construction of "high-rise structures". The horizontal deflection, as well as building motion, are the two most important considerations in the comfort design. Maximum deflection and storey drift are examples of the building's lateral motions [6]. Non-structural materials, like cladding, may be damaged if there is a significant variation in deflection between floors of a building owing to "story drift." [7]. Generally, a limit of H/450H/500 is used to calculate horizontal deflection using equivalent static loads. The acceleration is taken into account when analysing the building's motions. Because of its complexity, the topic of comfort in motion will be further explored.

One definition of "progressive collapse" is the failure of one load-bearing element to cause the collapse of an entire section of a building [8, 9, 10]. For instance, a tiny gas explosion on the 18th floor ripped away a concrete panel that supported the structure, leading to its gradual collapse. This resulted in the whole side of the building falling due to the floors above. There are fewer walls on the lower floors of high-rise structures, allowing for more open space, and as a result, a reduced degree of stiffness. It's possible that lateral loads can cause the first story of a building to collapse, while the rest of it stands firm [11, 12, 13].

2 Literature Review

Babu et al. [14] Attractive architecture and skyscrapers are in demand, and everyone on the planet desires a structure that will elevate his or her social position. As a consequence, the construction sector is seeing an increase in investment every day, and this trend does not seem to be slowing down any time soon. The construction business is always evolving, allowing new and sophisticated technologies to be used. There is a global desire for new technologies, tools, and machinery. In addition, we relied on well-known programmes such as STAAD.Pro, AutoCAD, and Bentley's RCDC to complete our investigation. STAAD has been utilised to design a building, that has a unique form, while AutoCAD was used to write out its layout plan. Bentley's RCDC was used for pro and cost estimates. According to the findings, the software has an important impact on the structural design, as well as the study's primary goal was to support an uncommon idea of designing an unevenly shaped but visually arresting building, as well as to put that idea into ractice using multiple civil engineering software programmes, in particular, in the structural design.

Elhegazy et al. [12] The structural designer's number one aim is to find the best structural system at the lowest possible cost. Even though previous study has focused extensively on this topic, it has not yet been fully addressed. A "multi-story reinforced concrete" (RC) structure is the subject of this study, which attempts to provide suggestions for the most cost-effective gravity and lateral systems. Accordingly, 72 RC structures with many storeys ranging from 5 to 50 floors and grid spacing of 6.0 to 12.0 m were used in the parametric analysis. Solid, two-way ribbed, waffle, and flat slabs were all evaluated. In addition, shear walls, intermediate moment resisting the frames, and a dual system were all investigated for resisting lateral loads. There was research on low, medium, and high-rise structures. A combination of solid slabs for short spans and ribbed slabs for the medium and long spans was shown to be the most appropriate dual system for the "medium and high-rise structures".

Zhu et al. [15] There is a possibility of considerable across-wind produced vibrations in 2 buildings due to the aerodynamic interference between their closely spaced links. TMDI (tuned mass-damper-inerter) is used to reduce the effects of wind on the connected structures. TMDI-connected buildings make use of the considerable relative motion between the 2 terminals of inerter to decrease the wind-induced reactions of the individual structure. TMDI motion equation for the connected structures under "wind loads" is first provided. "Wind tunnel experiments" for two connected structures with heights of 268 m (the building-1), as well as 210.2 m (the building-2), were carried out in order to determine their relative wind loads. For this purpose, we evaluated and compared the reactions of the connected buildings with TMDI, along with those without any control device, to highlight the benefits of TMDI over TMDs.

Aradhanarao et al. [16] Cost estimating, negotiated price, hard bid, conventional, construction management, management contracting, at risk, design, and design-build-bridging are some of the precise estimate approaches used in apartment building construction (APARTMENT BUILDING). A commercial building with many floors is referred to as a multi-story building. In our project, they estimate and price out a multi-story apartment complex that meets all of the client's needs while staying within their budget. According to considerations such as the customer's budget, the site's circumstances, and the availability of materials, an estimator may be able to alter the building design's parameters. Quantity surveying, as well as volumetric analysis, are both used as part of our project to provide an estimate and cost for the (G + 5) apartment complex. The "tandard schedule rate sheet" (SSR) 2013 was used in our project for the volumetric analysis as well as quantity surveying.

Mubarak et al. [17] Because structural dimensions fluctuate, seismic loads must be considered while estimating costs. When it comes to constructing structures, the seismic stresses they are subjected to might vary greatly from one place to another. "Reinforced concrete" (RC) beam, as well as column costs, may vary depending on the kind of seismic load that a structure receives, and that is the subject of this research. A two-story building prototype was used in this investigation. An earthquake-based dynamic study of the structure was used, with different seismic design categories applied to each of the eight seismic zones that were identified. Three indices of seismic important factors were used in a building prototype to indicate the group of occupants. There is a 0.68 percent, 1.70 percent, and 1.54 percent increase in the overall cost of two RCC components when the seismic significance factor indices are 1.00, 1.25, and 1.54.

Vorotyntseva et al. [18] Construction price standards are used to determine the cost of capital construction items for pre-design workings out as well as initial maximum initial price determinations on bids (CPNs). High-rise property values have not been taken into account while developing modern CPNs. Modern CPNs need to be adapted so that high-rise development costs may be taken into consideration with additional considerations. The most common methods are: applying further corrective correction factors or selecting new representative objects.

Chrysanidis et al. [19] Currently, a "high-rise building" design idea is being studied. This "braced shear wall core system" is proposed as the structural structure for the high-rise skyscraper. To begin, a rough sketch of the design concept is created. Using the member-sections produced from the final detailed design of the structure, a "parametrical cost analysis" of the structural system is performed. As a final point, various conclusions can be drawn from this preliminary design and the parametrical cost analysis based on its usefulness and the numerous aspects that impact the calculation of construction costs.

Palmisano [20] Researchers hope to demonstrate that simple measures like activating the "elastoplastic catenary behaviour" of slab reinforcement, might be very effective to increase the building's robustness without significantly increasing the cost of a structural system if the primary goal of the design is to protect the human lives in such extreme conditions.

3 Methodology

3.1 Process Management

Process management includes the management of time duration and the calculation of an estimated completion period. It includes "resource management" and all the activities that will happen during the whole project. It is managed using "Oracle Primavera p6 software", which provides the management of activities and resources. This has been done in accordance with the guideline of IS 1200-1 (1992): Methods of measurement of building and civil engineering works [21].

Primavera software manages a project in 5 main parameters which are as follows:

- "Work breakdown structure (WBS)
- Activities
- Calendar
- Estimated Duration of activities
- Gantt chart"

For scheduling this project, a 6-day working calendar is selected with Sunday as a nonworking day. The project start date is 3rd of June 2022 and the end date is analysed using Primavera p6. Working hours for each day are set as 8 h from 8 am to 4 pm. The calendar which is used in Primavera is described below.

Figure 1 shows the calendar which is used in Primavera for time tracking and the tracking of working hours which can be seen in the white area in work hours. The blue colour shows the "nonworking days" and the nonworking hours within the day.

3.2 Stress Analysis

The StaadPro. software has been used for carrying out stress analysis of the floor plans and the elevation. Figure 2a, b show the dimensions taken for stress analysis in Staadpro. Figure 2a shows the base plan, whereas Fig. 2b shows the elevation plan. An elevation height of 3 m is taken for each storey.

Figure 3a shows the connection between each column with implying beams at each storey. The base of this building is fixed for analysing the effect of load on building. Figure 3b shows the placement of the plate for each storey. Plates are defined by violet-colour in the above figure. The value of loads is taken as standard values for G + 3 storey buildings according to Indian standards [23].

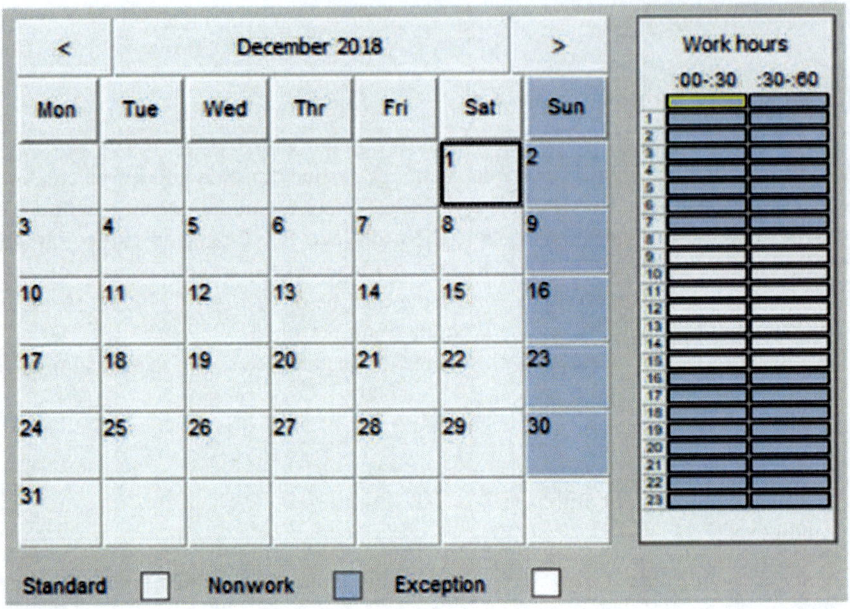

Fig. 1 Calendar

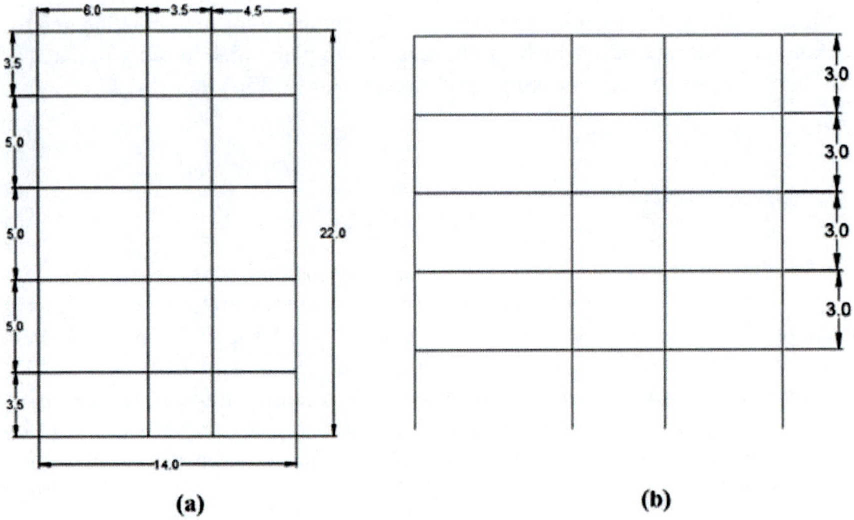

Fig. 2 **a** Floor plan for stress analysis **b** Elevation height for each floor Column having a size of 300 * 400 mm is used with concrete as column material. Here the 150 mm thickness plate is used across all storeys with concrete as plate material [22]

Fig. 3 **a** Connection between each column **b** Plate placement

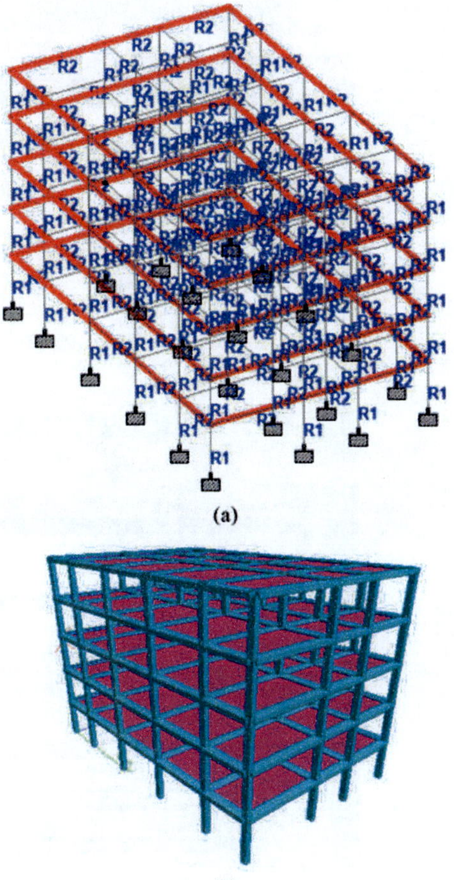

(a)

(b)

4 Results and Discussions

4.1 Duration of Activities

The required duration for each activity in days is provided in primavera according to which it concludes the final date of completion for every activity and the final completion of the project. The activity timings have been detailed in primavera as shown in Fig. 4.

Layout: Classic Schedule Layout — Filter All: Not Started

Activity ID	Activity Name	Original Duration	Start	Finish
I imtiyaaz		341d	03-Jun-22 08:00 AM	06-Jul-23 08:00 AM
I.1 start milestone		0d	03-Jun-22 08:00 AM	03-Jun-22 08:00 AM
A1000	project started	0d	03-Jun-22 08:00 AM	03-Jun-22 08:00 AM
I.2 engineering		16d	03-Jun-22 08:00 AM	21-Jun-22 04:00 PM
A1010	finishing works drawing	8d	03-Jun-22 08:00 AM	11-Jun-22 04:00 PM
A1020	MEP works drawing	7d	03-Jun-22 08:00 AM	10-Jun-22 04:00 PM
A1030	structural drawing	7d	03-Jun-22 08:00 AM	10-Jun-22 04:00 PM
A1040	structural analysis using STAADPRO	9d	11-Jun-22 08:00 AM	21-Jun-22 04:00 PM
I.3 procurement		16d	22-Jun-22 08:00 AM	09-Jul-22 04:00 PM
I.3.1 Material submission & approval		8d	22-Jun-22 08:00 AM	30-Jun-22 04:00 PM
A1050	Finishing materials	6d	22-Jun-22 08:00 AM	28-Jun-22 04:00 PM
A1060	MEP materials	6d	22-Jun-22 08:00 AM	28-Jun-22 04:00 PM
A1070	structural materials	8d	22-Jun-22 08:00 AM	30-Jun-22 04:00 PM
I.3.2 Material orders & arrivals		8d	01-Jul-22 08:00 AM	09-Jul-22 04:00 PM
A1080	Finishing materials	8d	01-Jul-22 08:00 AM	09-Jul-22 04:00 PM
A1090	MEP materials	8d	01-Jul-22 08:00 AM	09-Jul-22 04:00 PM
A1100	structural materials	8d	01-Jul-22 08:00 AM	09-Jul-22 04:00 PM
I.4 construction		254d	11-Jul-22 08:00 AM	02-May-23 04:00 PM
I.4.1 earth work		14d	11-Jul-22 08:00 AM	26-Jul-22 04:00 PM
A1110	excavation of all soil	11d	11-Jul-22 08:00 AM	22-Jul-22 04:00 PM
A1120	PCC	3d	23-Jul-22 08:00 AM	26-Jul-22 04:00 PM
I.4.2 substructure		10d	27-Jul-22 08:00 AM	06-Aug-22 04:00 PM
I.4.2.1 footing		6d	27-Jul-22 08:00 AM	02-Aug-22 04:00 PM
A1130	concreting of footing	3d	29-Jul-22 08:00 AM	01-Aug-22 04:00 PM
A1140	reinforcement of footing	2d	27-Jul-22 08:00 AM	28-Jul-22 04:00 PM
A1150	shuttering of footing	6d	27-Jul-22 08:00 AM	02-Aug-22 04:00 PM
I.4.2.2 plinth beam		5d	02-Aug-22 08:00 AM	06-Aug-22 04:00 PM
A1160	concreting of plinth beam	3d	04-Aug-22 08:00 AM	06-Aug-22 04:00 PM
A1170	reinforcement of plinth beam	2d	02-Aug-22 08:00 AM	03-Aug-22 04:00 PM
A1180	shuttering of plinth beam	2d	02-Aug-22 08:00 AM	03-Aug-22 04:00 PM

Layout: Classic Schedule Layout — Filter All: Not Started

Activity ID	Activity Name	Original Duration	Start	Finish
I.4.2.2 plinth beam		5d	02-Aug-22 08:00 AM	06-Aug-22 04:00 PM
A1160	concreting of plinth beam	3d	04-Aug-22 08:00 AM	06-Aug-22 04:00 PM
A1170	reinforcement of plinth beam	2d	02-Aug-22 08:00 AM	03-Aug-22 04:00 PM
A1180	shuttering of plinth beam	2d	02-Aug-22 08:00 AM	03-Aug-22 04:00 PM
I.4.3 super structure		230d	08-Aug-22 08:00 AM	02-May-23 04:00 PM
I.4.3.1 columns		80d	08-Aug-22 08:00 AM	08-Nov-22 04:00 PM
A1190	concreting of column	20d	17-Oct-22 08:00 AM	08-Nov-22 04:00 PM
A1200	reinforcement of column	60d	08-Aug-22 08:00 AM	15-Oct-22 04:00 PM
A1210	shuttering of column	15d	08-Aug-22 08:00 AM	24-Aug-22 04:00 PM
I.4.3.2 slab/beams		70d	09-Nov-22 08:00 AM	28-Jan-23 04:00 PM
A1220	concreting of slab/beams	20d	06-Jan-23 08:00 AM	28-Jan-23 04:00 PM
A1230	reinforcement of slab/beams	50d	09-Nov-22 08:00 AM	05-Jan-23 04:00 PM
A1240	shuttering of slab/beams	15d	09-Nov-22 08:00 AM	25-Nov-22 04:00 PM
I.4.3.3 brickwork		50d	30-Jan-23 08:00 AM	28-Mar-23 04:00 PM
A1250	main walls	25d	30-Jan-23 08:00 AM	27-Feb-23 04:00 PM
A1260	partition walls	25d	28-Feb-23 08:00 AM	28-Mar-23 04:00 PM
I.4.3.4 terrace floor		30d	29-Mar-23 08:00 AM	02-May-23 04:00 PM
A1270	exterior walls/parapet walls	30d	29-Mar-23 08:00 AM	02-May-23 04:00 PM
I.5 finishing work		55d	03-May-23 08:00 AM	05-Jul-23 04:00 PM
A1280	electrical work	12d	03-May-23 08:00 AM	16-May-23 04:00 PM
A1290	plumnbing work	15d	03-May-23 08:00 AM	19-May-23 04:00 PM
A1300	plastering work	21d	20-May-23 08:00 AM	13-Jun-23 04:00 PM
A1310	flooring work	16d	20-May-23 08:00 AM	07-Jun-23 04:00 PM
A1320	doors/windows work	13d	08-Jun-23 08:00 AM	22-Jun-23 04:00 PM
A1330	painting work	14d	08-Jun-23 08:00 AM	23-Jun-23 04:00 PM
A1340	fixture work	10d	24-Jun-23 08:00 AM	05-Jul-23 04:00 PM
I.6 finish milestone		0d	06-Jul-23 08:00 AM	06-Jul-23 08:00 AM
A1350	finish work	0d	06-Jul-23 08:00 AM	06-Jul-23 08:00 AM

Fig. 4 Activities

4.2 Gantt Chart

Gantt chart shows the timeline on which the project will be performed. It shows the relationship between 2 activities of the project. It is decided that the next activity should start after the completion of the previous activities or if they can work simultaneously, they both can be started as start to start function which will reduce the overall time taken by the project (Fig. 4).

4.3 Timeline According to WBS Section

The start as well as end date of each section are described in Table 1 for the baseline structure of scheduling. And the duration of each WBS is also mentioned in Table 2.

Table 2 shows the overall time which project will take with nonworking days and without nonworking days. A total time of 398 days will be used in this project to complete fully. Further process of the project can be started according to the timeline decided with the Gannt chart. With the help of the Gannt chart, the project can be tracked throughout the project processing.

4.4 Stress Analysis Results

- Beam bending direction

  Figure 6 below shows the "bending moment diagram" of every beam in the structure. This has been evaluated as per the Indian Standards [23].
- Plate load

  As can be seen in Fig. 7, the maximum load is bored by the ground floor, i.e. 0.018 N/mm^2, which decreases gradually towards the upward floors and is minimum for the topmost floor, i.e. 0.123 N/mm^2.

Figure 8 shows the "beam diagrams" obtained in the analysis for the forces applied in different directions. In the z-direction, the beam shows a polynomial load displacement, in the y-direction, it's linear, while in the x-direction, the beam shows a uniform displacement.

5 Conclusion

In this paper, the process management of construction activity of a "high-rise building" has been attempted in detail. Conventional management activities have been performed using software called Primavera which further simplifies the process and

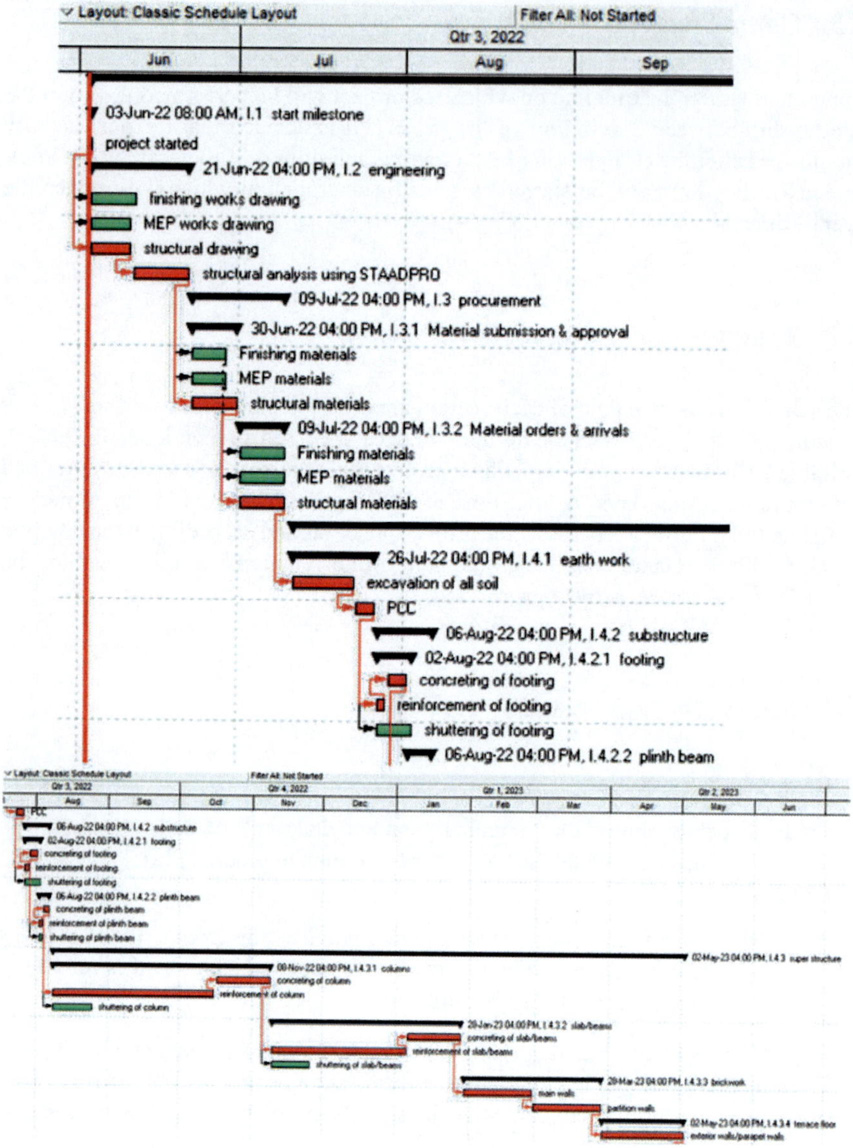

Fig. 5 Gantt chart

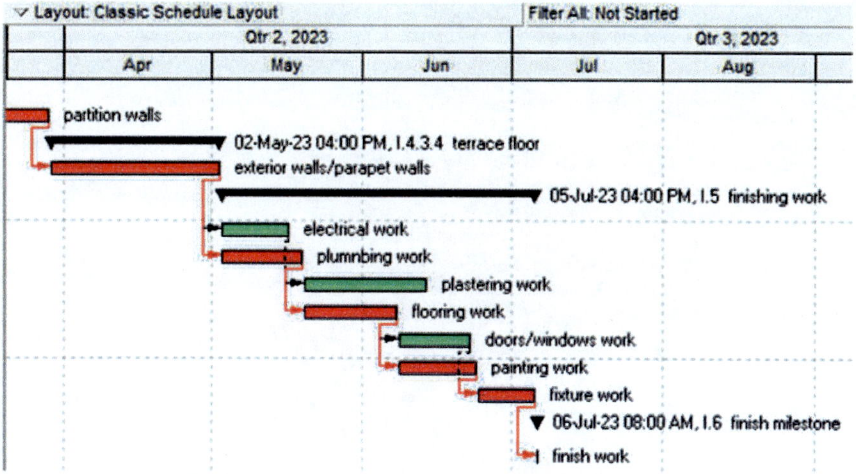

Fig. 5 (continued)

Table 1 Duration for each WBS

WBS	Duration (days)	Start date	End date
Starting milestone	0	03–06–2022	03–06–2022
Engineering	16	03–06–2022	21–06–2022
Procurement	16	22–06–2022	09–07–2022
Construction	254	11–07–2022	02–05–2023
Finish work	55	03–05–2023	05–07–2023
Finish milestone	0	06–07–2023	

Table 2 Total number of days

Excluding nonworking days	Total working days
341 days	398 days

saves time with the least error. The timeline of activities has been detailed using a Gantt Chart in addition to their duration periods. A total of 398 working days have been calculated by the Primavera in the set activities timeline. Furthermore, the stress analysis of the "high-rise structure" has also been carried out. It has been observed that the maximum load is borne by the ground floor, i.e. 0.018 N/mm^2, which decreases gradually towards the upward floors and is minimum for the topmost floor, i.e. 0.123 N/mm^2.

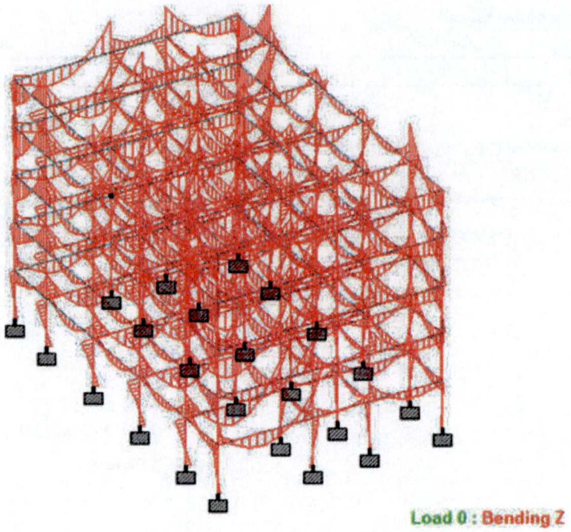

Fig. 6 Beam bending

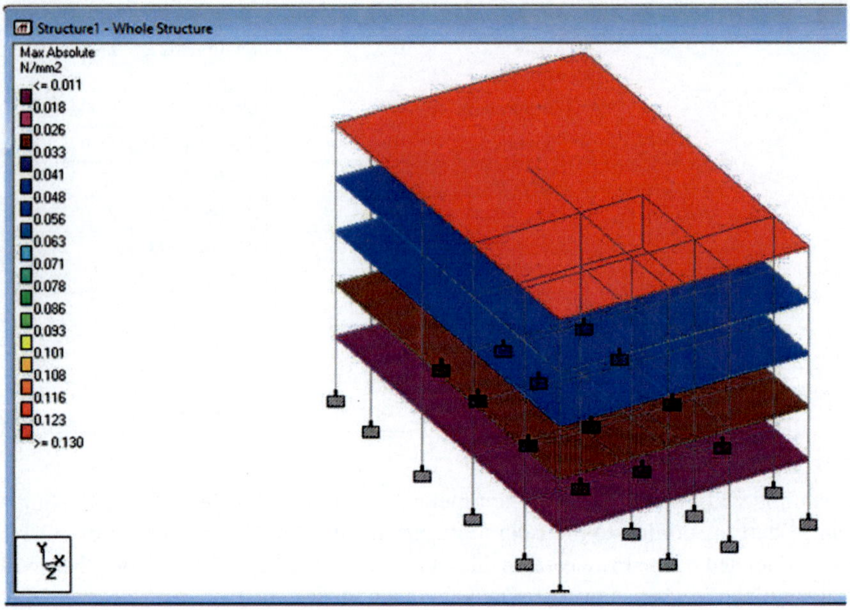

Fig. 7 Structural load across whole storeys

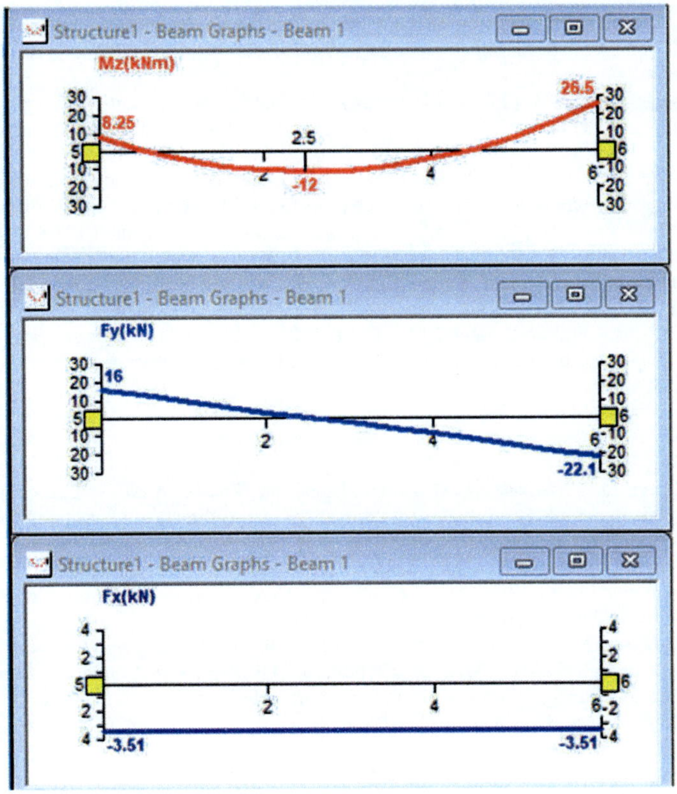

Fig. 8 Beam graphs

References

1. Ovsiannikov A, Vorotyntseva A, Bolgov V (2018) Features of calculations of the limit cost of high-rise constructions for housing and civil purposes with the use of consolidated standards. MATEC Web Conf 193. https://doi.org/10.1051/matecconf/201819305037
2. Pu B, Zhou X, Liu Y, Liu B, Jiang L (2021) Mechanical behavior of concrete-filled rectangular steel tubular composite truss bridge in the negative moment region. J Traffic Transp Eng (English Ed.) 8(5):795–814. https://doi.org/10.1016/j.jtte.2021.09.002
3. Nazarimofrad E, Zahrai SM (2016) Seismic control of irregular multistory buildings using active tendons considering soil–structure interaction effect. Soil Dyn Earthq Eng 89:100–115. https://doi.org/10.1016/j.soildyn.2016.07.005
4. Konar T, Ghosh A (2021) Development of a novel tuned liquid damper with floating base for converting deep tanks into effective vibration control devices. Adv Struct Eng 24(2):401–407. https://doi.org/10.1177/1369433220953539
5. Chen B, Yang D, Zheng Y, Feng K, Ouyang Y (2018) Response control of a high-rise television tower under seismic excitations by friction dampers. Int J Struct Stab Dyn 18(11):1–25. https://doi.org/10.1142/S0219455418501407

6. Anjali B, Raji M (2015) Seismic analysis and soil structure interaction of multistoried building with different types of footing. Int J Eng Res V4(09): 820–824. https://doi.org/10.17577/ijertv4is090690
7. Foraboschi P, Vanin A (2013) Non-linear static analysis of masonry buildings based on a strut-and-tie modeling. Soil Dyn Earthq Eng 55:44–58. https://doi.org/10.1016/j.soildyn.2013.08.005
8. Prasad AS, Ramakrishna S, Madhavi M (2018) Experimental investigations on static and dynamic parameters of steel and composite propeller shafts with an integrated metallic joints. Mater Today Proc 5(13):26925–26933. https://doi.org/10.1016/j.matpr.2018.08.180
9. Jaiswal S, Chauhan VB (2021) Ultimate bearing capacity of strip footing resting on rock mass using adaptive finite element method. J King Saud Univ Eng Sci. https://doi.org/10.1016/j.jksues.2021.09.004
10. Cheng Z, Zhang Q, Bao Y, Deng P, Wei C, Li M (2021) Flexural behavior of corrugated steel-UHPC composite bridge decks. Eng Struct 246:113066. https://doi.org/10.1016/j.engstruct.2021.113066
11. Peng J, Hou C, Shen L (2021) Numerical analysis of corner-supported composite modular buildings under wind actions. J Constr Steel Res 187:106942. https://doi.org/10.1016/j.jcsr.2021.106942
12. Elhegazy H, Ebid AM, Mahdi IM, Aboul Haggag SY, Rashid IA (2019) Selecting optimum structural system for R.C. multi-story buildings considering direct cost. Structures 24:296–303. https://doi.org/10.1016/j.istruc.2020.01.039
13. Carpinteri A, Lacidogna G, Cammarano S (2013) Structural analysis of high-rise buildings under horizontal loads: a study on the Intesa Sanpaolo Tower in Turin. Eng Struct 56:1362–1371. https://doi.org/10.1016/j.engstruct.2013.07.009
14. Babu G, Bawa S, Almheri S, Mahajan A (2021) Detailed planning, designing and cost-estimation of high rise (G + 3) residential building. J Emerg Technol Innov Res 8(5):951–959
15. Zhu Z, Lei W, Wang Q, Tiwari N, Hazra B (2020) Study on wind-induced vibration control of linked high-rise buildings by using TMDI. J Wind Eng Ind Aerodyn 205:104306. https://doi.org/10.1016/j.jweia.2020.104306
16. Aradhanarao M, Mohanrao P, Sombabu Y (2019) A case study on estimation and costing of multi-storey building. Int J Adv Res Sci Eng Technol 6:455–462
17. Mubarak M, Abdullah A, Azmeri A, Hayati Y (2019) Cost estimation of structural components of a building by considering the seismic load on different regions. Adv Civ Eng 2019. https://doi.org/10.1155/2019/7357913
18. Vorotyntseva A, Ovsiannikov A, Bolgov V (2018) Formation of costs of high-rise objects of housing and civil purpose based on enlarged norms. E3S Web Conf 33. https://doi.org/10.1051/e3sconf/20183303036
19. Chrysanidis T, Panoskaltsis V, Tegos I (2016) Preliminary design and analysis of cost parameters of a high-rise building: braced shear wall core system. Int J Civ Eng Technol 7(5):137–152
20. Palmisano F (2014) Mitigation of progressive collapse by the activation of the elasto-plastic catenary behaviour of R.C. slab structures. Open Constr Build Technol J 8(1):122–131. https://doi.org/10.2174/1874836801408010122
21. Kisan M, Sangathan S, Nehru J, Pitroda SG (1992) IS-1200. Bur Indian Stand Dehli
22. Bureau of Indian Standards (2017) IS 16700:2017 criteria for structural safety of tall concrete buildings. Bur Indian Stand Dehli 6
23. IS 456 (2000) Plain Concrete and Reinforced. Bur Indian Stand Dehli 1–114

An Experimental Study on Partial Replacement of Cement in Concrete with Sugarcane Bagasse Ash Using Magnetized Water

Sontam Lakshminarayana, Andhi Shilpika, Nakkala Kumar, Paramayala Kranthireddy, Kemsaram Deepak, Sutari Usha, and Miryala Vijayakumar

Abstract In this investigation, the "SUGARCANE BAGASSE ASH (SBA)" is used as the replacement of cement partially in concrete with proportions of 0, 5, 10 and15% using magnetized water. The silica present in the SBA gives strength for concrete and reduces the concrete's permeability. And magnetized water will give the additional strength and also helps to decrease the weight of the concrete. The SBA contains silica which interacts with cement components in the hydration process and additional properties are also imparted such as chloride resistance, corrosion resistance etc. Tests were conducted on the fresh concrete are slump cone test and compaction factor test (Cf). The compressive strength test (Cs) and tensile strength test were conducted on concrete that is in hard state for cubes that were casted and cured in normal water for ages of 7, 14, 21 and 28 days. In this investigation we were using M25 grade concrete.

Keywords Sugarcane bagasse ash (SBA) · Magnetized water · M25 grade concrete · Compressive strength · And Split tensile test

1 Introduction

The Research shows that for every ton of cement manufacturing will expose half a ton of CO_2 and also SBA waste disposal is also becoming a big problem which causes environmental damage. So there is an immediate need to control both the problems. However, the SCBA is high in silica which helps in strengthening the concrete. In our investigation, we are using magnetized water for the concrete cube preparation which further increases the strength. By using the SCBA as a replacement partially

S. Lakshminarayana · A. Shilpika · N. Kumar · P. Kranthireddy · K. Deepak · S. Usha · M. Vijayakumar (✉)
KG Reddy College of Engineering and Technology, Hyderabad, India
e-mail: miryalavijaykumar757@gmail.com

K. K. Pathak et al. (eds.), *Latest Developments in Civil Engineering*, Lecture Notes in Civil Engineering 352, https://doi.org/10.1007/978-981-99-2676-3_12

in cement for concrete will help us to control wastage and to protect the environment and also it becomes very economical when compared to pure cement concrete.

This investigation is adopted to reduce the carbon dioxide content in the environment which results in global warming and also depletion of ozone layer. Because the manufacturing every ton of cement exposes half a ton of co_2 and recycling of SCBA is also a big task. So, in order to reduce the pollution in the environment, which shows the very serious impact on the earth we are conducting the experiments on the concrete which is prepared by replacing partially SCBA in cement.

2 Literature Review

Abdel-Magid et al. [1], used concrete with magnetized water, which increases in concrete workability up to 400% and also it reduces the concrete cubes' weight by 3% and Cs is increased by 10%.

Shruthi et al. [2], In their research investigation, they concluded that up to 5 and 10% of increased the concrete strength by SCBA and further it showed strength is decreased

Singh et al. [3], This study finalized that the fly ash uses up to 20%, compressive strength increased and beyond that sudden decrease in the workability and concrete strength.

Madhusudan Reddy et al. [4], This study explores that HCl effect on fly ash and silica fume blended cement and their concretes. They concluded that the compressive strengths reduction of fly ash blended concrete and silica fume blended concrete was 2 to 19% for 28 days and 90 days.

Al-Tamimi and Sonebi [5], studied and discussed in this paper Self-Compacting Concrete properties when it is submerged in acidic solutions. They conclude that self-compacting concrete was performing better than control concrete when revealed to 1% H2SO4 and hydrochloric acid.

3 Motivation

As the cement manufacturing increases the carbon dioxide content which results in an increase in global warming and in the other hand the recycling of SBA becoming hazardous to environment. So based on the literature reviews, we came to know that SBA increase the strength in cement and magnetized water also reduces the weight of the cement, the strength of the concrete is also increases. So, this is motivated us to do an investigation.

4 Problem Domain

As we know that SBA increases the strength in concrete so, we partially replaced the SBA by 5, 10 and 15% along with the Magnetized water of water-cement ratio of 0.5% to the concrete.

5 Problem Definition

To reduce the environmental hazards caused by the cement manufacturing and SBA and to convert SBA into useful manner in concrete.

6 Statement

Replacing of SCBA partially to the cement increases the strength and we can use as alternative for the cement up to some percentages.

7 Materials

7.1 Cement

In this investigation, we have used the Portland pozzolana cement of 53 grade [11]. We have adopted this cement because there is no need of any special requirements for our investigation and it is widely available.

7.2 Fine Aggregates

In this investigation, we have used sand which is locally available and specific gravity of fine aggregate was 2.60. The sand is available of river sand conforming to Grading Zone I is: 383-1970 [12]. It is dried well and cleaned thoroughly before use.

7.3 Coarse Aggregates

Generally, the coarse aggregate occupies 70 to 80% of the volume of the concrete. So, it is mandatory to have keen knowledge on the properties of the aggregates [12]

Fig. 1 Change in surface tension (viscosity) of the water

that we are using. We use aggregates with different sizes in order to avoid improper filling of the voids in the concrete. Basically, the aggregates which are above 4.75 mm in size are termed as coarse aggregates and we have used up to 20 mm in size of aggregates.

7.4 *Magnetized Water*

Magnetized water will act as a lubricant in the sand and coarse-grained aggregate. Sand reacts with cement chemically to form the binding material for the aggregate and reinforcement. It increases the strength by 10 to 22% compared to normal water (Fig. 1).

7.5 *Sugarcane Bagasse*

Bagasse is a product that is obtained after burning the raw material after drying. The waste sugarcane obtained after use is taken and dried until its losses the moisture content. It is rich in pozzolanic properties that would potentially be used as a cement replacement material (Figs. 2, 3 and Tables 1, 2).

Fig. 2 Sugarcane waste

Fig. 3 SBA

Table 1 SBA properties

S. no	Component	%
1	Cellulose	50
2	Hemi cellulose	25
3	Sugar	10

Table 2 Chemical properties of SBA

S. no	Components	%
1	Silica	63.00
2	Alumina	31.50
3	Ferric oxide	1.79
4	Manganese oxide	0.004
5	Calcium oxide	0.48
6	Magnesium oxide	0.39

Table 3 M25 grade MIX DESIGN [1:1:2]

Cement	1900 gm
Coarse aggregate	1930 gm
Fine aggregate	4106 gm
Magnetized water	950 gm

8 Methodology

In this investigation, we have adopted a grade concrete mix of M25 (1:1:2) [IS:10262-2009]. Tests on the cement were done and all the requirements for the making of cubes were satisfied by the OPC cement. Proportions were adopted accurately for the desired quantities while making the cubes (Table 3).

9 Results

The tests were conducted on the different proportions of 0%, 5%, 10%, 15%, at different ages of 7 days, 14 days and 28 days the cubes (Table 4).

Graphical Identification:

See Table 5.

Table 4 Concrete compressive strength (N/mm^2)

% SBA	7 days	14 days	28 days
0	20.44	23.11	30.22
5	21.77	26.22	32.88
10	24.44	27.55	36.44
15	23.55	26.6	34.66

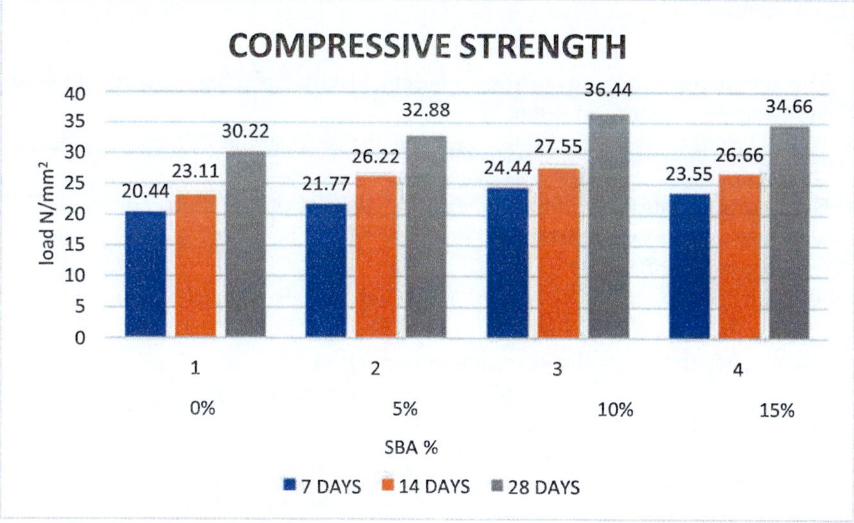

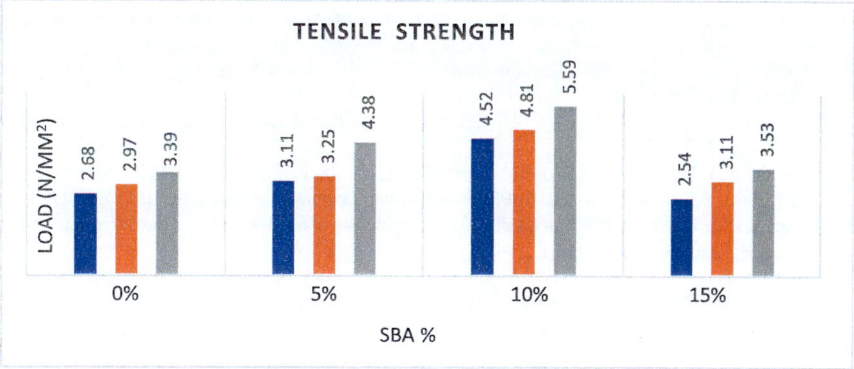

Table 5 Spilt tensile strength of the concrete (N/mm²)

SBA %	7 days	14 days	28 days
0	2.68	2.97	3.39
5	3.11	3.25	4.38
10	4.52	4.81	5.59
15	2.54	3.11	3.53

10 Conclusion

- The maximum compressive strength is achieved by replacing cement with waste SCBA using magnetized water.
- 10% of SCBA blended mix has attained the M25 grade CC 28 days desired strength.
- By the investigation, we finalized that the 10% of partial replacement of SBA with magnetized water also increases the workability, compressive strength and tensile strength of concrete.
- And further increase of SBA with magnetized water with 15% will reduce the strength and also workability.
- The magnetized water also increased its strength by 6%.
- Finally, we conclude that Cs of concrete will increase by using the SCBA of optimum of 10% with magnetized water.

References

1. Abdel-Magid TIM (2017) Effect of magnetized water on workability and compressive strength of concrete. Sudan University of Science and Technology, Almarata 61 street
2. Shruthi HR, Yashwant MK, Keerthi Gowda BS (2019) A study on bagasse ash replaced in concrete. IJATES 02(08)
3. Singh VK, Srivastava V, Agarwal VC (2018) Effect of fly ash as partial replacement of cement in concrete. IJIRSET 4(7)
4. Madhusudan Reddy B, et al (2019) Effect of HCL on blended cement. IJIRSET 6(4)
5. AL-Tamimi AK, Sonebi M (2016) Properties of self-compacting concrete. IJIRSET 6(2)
6. Modania PO, Vyawaharebnology MR (2013) Utilization of bagasse ash as a partial replacement of fine aggregate in concrete. 51:25–29
7. SivaKumar M, Mahindra N (2013) Experimental studies of strength and cost analysis of concrete using bagasse ash. Int J Eng Res Technol 2(4):1–8
8. Hailu B, Dinku A, Application of sugarcane bagasse ash as a partial cement replacement material. 1–21
9. Abbasi A, Zargar A (2013) Using bagasse ash in concrete as properties of sugarcane bagasse ash and its potential as cement-pozzolana binder. Ain Shams University
10. Noor Ul Amin (2010) Chemical activation of bagasse ash in cementitious system and its impact on strength development. Abdul Wali Khan University, Pakistan
11. IS 8112 (1989) Indian standard, 53 grade ordinary Portland cement—specification||. Bureau of Indian Standards, New Delhi
12. IS 383 (1970) Indian standard, specification for coarse and fine aggregates from natural sources for concrete||. Bureau of Indian Standards, New Delhi
13. IS 2386 (1963) Indian standard, Methods of test for aggregates for concrete. Bureau of Indian Standards, New Delhi
14. IS 10262 (2009) Indian standard, recommended guidelines for concrete mix design. Bureau of Indian Standards, New Delhi

Cost Benefit Analysis of Adoption of Renewable Energy (Solar) in a Residential Building—A Case Study on Residential Apartment in Pune

Arjita Biswas and J. S. Sudarsan

Abstract The climate catastrophe is primarily caused by greenhouse gas (GHG) emissions from buildings. Going beyond a building's functioning energy use and associated GHG emissions, one must consider a building's entire life cycle to address climate change mitigation concerns. But these can only be fruitful when the existing buildings are targeted, as the percentage of conventional building are much more than existing green building. These existing buildings consume huge resources and thus the challenge is to convert the existing residential buildings into reasonably energy-efficient green buildings. But whether it is viable and practicable. This study is an attempt for the same. Solar roofing system has increased significantly around the world as a result of a significant drop in equipment costs for photovoltaic (PV) systems, an increase in electricity costs, and mounting environmental pressure to use green energy sources rather than fossil fuels, which pollute the environment more severely. This study investigated the economic validity of using a PV system at the existing residential buildings. The study has quantified that the installation of PV solar panels will be beneficial and the payback period is quite less as compared to the long-term benefits. The study will motivate the residents to switch to the sustainable method.

Keywords Green building · Carbon credits · Solar panels

1 Introduction

In the recent past green buildings have been in discussion and both real estate developers and public are taking attention in the technology. If people start thinking in non-conventional ways and start harnessing green energy sources, they could actually be benefited in the long run as their expenditure on operation and maintenance

A. Biswas (✉) · J. S. Sudarsan
National Institute of Construction Management and Research (NICMAR) University, Pune, India
e-mail: abiswas@nicmar.ac.in

K. K. Pathak et al. (eds.), *Latest Developments in Civil Engineering*, Lecture Notes in Civil Engineering 352, https://doi.org/10.1007/978-981-99-2676-3_13

of their house as well as entire building would get reduced and they would move towards a sustainable lifestyle with minimal problems. Christopher.

Scheuer [1] presented the methods for implementation of suburban green building practices & understanding the role of awareness in the University of Michigan in 2007. This could be done by conversion of existing buildings into green buildings or constructing new buildings based on green technology. Buildings use a lot of energy over the course of their existence. The key source of environmental contamination is the production of energy, which mostly relies on traditional sources.

All of the factors that affect a building., building's energy efficiency must be considered in order to optimize its environmental performance [2, 3]. Carbon footprints are a worry for the entire world and they are assessed in a variety of different ways [4]. Although the methods for calculating carbon footprints are still being developed, they are becoming a crucial tool for managing greenhouse gas emissions [5]. The carbon footprint primarily affects individuals, products, organizations, cities, and countries [4, 6].

India's housing sector is one of the highest contributors to country's carbon dioxide emissions, contributing about 22%. According to Indian Green Building Council (IGBC), by the year 2017, only 2200 (i.e., 2%) buildings have been certified as green buildings. Parikh et al. [7] examines the carbon dioxide (CO_2) emissions of various industries. Impact on Indian economy as a result of final consumer spending in the producing sectors and households are identified by the authors. Apart from these, the buildings continue to report consumption of maximum amount of energy and thus the challenge is to convert the existing residential buildings into reasonably energy-efficient green buildings. As of present, the amount of such projects being undertaken are piece meal. People find it difficult to analyze the paybacks of green building over its cost.

All phases of a building's life cycle result in carbon emissions, which are often divided into operating and embodied carbon. Based on the properties of the building, each type of carbon makes a different contribution to the life cycle of carbon [8]. As per Dakwale et al. [2], all of the factors that affect a building's energy efficiency must be considered to optimize its environmental performance. Ren et al. [9] emphasized on saving energy consumption to meet the standards for national energy and environmental criteria of any nation. There are many ways by which conventional structure can be converted to energy-efficient building. One very common approach suitable for Indian conditions is residential photovoltaic (PV) systems, which absorb and convert sunlight into electricity using solar PV panels. Housing PV systems allow residents to generate emissions-free electricity, removing the dependency on external agency to supply electricity [10]. Kapure and Jain [11] documented the parameters of Upgrading Existing Buildings into Green Buildings.

They said that from a microeconomic perspective, studies have shown that energy-efficient and certified green buildings merit higher market values, greater rents, and higher occupancies as well as from a corporate sustainability viewpoint, greening existing buildings is a direct way to reduce a company's carbon footprint. Kavani and Pathak [12], discussed how a Present building can be converted into a green building. This research paper is related to the conversion of existing structure into a green building. Kataria and Salunkhe [13], observed the retrofitting of existing building as per IGBC existing building norms. According to them, retrofitting or renovation of an existing facility is most of the time more cost-effective and economical than investing in a new facility.

In 2020, the US installed 19.2 GW of solar PV capacity, an increase of 43% from 2019. In 2020, there was 97.3 GW of installed PV capacity, which was sufficient to power 17.7 million American homes [14]. For Asia to achieve net zero energy/ carbon for buildings, there are still numerous obstacles to overcome. Blindly replicating building and construction methods from Europe or North America increases consumption more than traditional design (Trevor and Ng 2013). There are three main rating systems in India for Green building, although they all only consider very large or tiny commercial structures.

Lakshmi [15] rated an existing residential building using the SVAGRIHA (Simple Versatile Affordable Green Rating for Integrated Habitat Assessment) green building rating system. The study shows that it is feasible and practical to convert small-scale existing residential buildings to green buildings. However, there isn't a reliable rating system that can handle this conversion.

If a conventional building in India can be partially transformed into a green building by adding or changing a few things, this will not only reduce the structure's operating costs but also advance the country's sustainability goals. The goal of the current study is to comprehend the practical challenges that inhabitants of typical buildings face and explore solutions. In the study, installation of solar panels is the major change suggested and cost benefit analysis are also presented in detail in addition to other smaller recommendations.

The objectives of the study are:

- To compare the existing power demand of the entire building with the potential supply which could be delivered after installation of solar panels for power generation and to estimate the cost benefit analysis.
- To reduce the effect of external heat by providing vertical gardens on the outer wall surfaces wherever possible.

Use of soft tools such as BIM using REVIT, Form. It is to prepare and analyze the 3D model and establish required reports using Green Building Studio.

2 Methodology

The following methodology was adopted to derive the outcome for the research study using the following methodology as depicted in Fig. 1. The study started with introducing the topic and reviewing the literature. The study area selected is Kothrud which is an upmarket suburban neighborhood in Pune city of Maharashtra, India. Kothrud has a latitude of 18.5 North and Longitude of 73.82 East and an altitude of 611 m above MSL. The area receives approximately 12 h of sunlight on sunny days. Then a Questionnaire survey form was floated to understand the

Fig. 1 Methodology

people's perspective on green building and their problems on a daily basis related to conventional building. After understanding the difficulties faced by Middle income people with conventional building, a building was selected and a complete study was done on that particular building. Market research was conducted to find out different types of solar panels available in the market, their specifications suitability, cost etc. A building was selected from the study area and detailed study was conducted. In the last part of the study, detailed calculations and analysis are discussed.

3 Data Collection

To start with a detailed questionnaire in the form of a Google form was prepared to collect data from people regarding the parameters of their existing building and their understanding about Green building. Respondents were all from same locality specifically from the study area i.e., Kothrud, Pune and respondents belong to middle income group.

This form was shared through social media like WhatsApp, Facebook, e-mails etc. and 43 residents/families were interviewed personally. From the study area, one building was selected as the prominent location. All the apartments were 2BHK approximately of the same area, so it was assumed the residents are from similar income groups. The Chairman of the society was known to the Researcher and it would have been possible to collect all information required from the building and respondents were enthusiastic to share all the details.

After analyzing the responses received, the following observations were made: Approximately 60% people are found to live in residential buildings and societies. About 20% people are facing water and electricity problems and about 55–60% are facing trouble due to increasing heat. Moreover, 40% people face the problems of waste management. 40–45% people have reported problems as far as parking facilities and play areas are concerned. 76% of the houses have to spend more than Rs.500 on electricity. 81% population is facing the problem of managing the storm water in monsoon. Around 70–85% buildings do not treat gray water or sewage. 82–90% buildings do not have green roofs. 40–45% people are still disposing their household waste directly without separation. 55–60% people are facing parking problems either with 2-wheeler/4 wheeler or both. About 70% of the people have no or very less idea about govt. policies and subsidies. 72–75% population is aware of green building technology. 91–97% people believe that sustainable approach towards living will help them have a better life style. Only 35–40% of the people are not ready to invest for use of green technology. About 70–80% people find difficulty in implementing green building technology because of its high initial cost and its complexity.

After having done all the data analysis and interpretation, it was observed that the problem of increasing heat, power cuts in summer and any other season, and high electricity bill amounts were of major concern to most of the people.

Two problems were observed—the people were reluctant to use air conditioners as it would cost them too much at the end of the month and the other problem was that their efficiency was noticeably reduced due to the outside temperature. Solution to this could be the maximum use of natural cooling of the buildings and the use of increasing solar intensity for harnessing energy as shown in Fig. 2.

4 Case Study

For the case study, a building in Pune city, Maharashtra, India was selected. Sneh Paradise society, near MIT Engineering College, Rambaug Colony, Paud Road, Kothrud was chosen as the area for study and building 'M' of the same society was selected for carrying out the further analysis of the residential building.

Measurements were taken by conducting manual surveys in the society and a building plan was prepared using AutoCAD.

5 Market Research

The information obtained from solar panel manufacturers in the local market, installers and solar energy consultants are as follows.

One small rectangular panel of size 1 m * 2 m can produce around 320–350 W. Hence, around 3 such panels constitute 1 KW of solar power generation. 1 KW solar

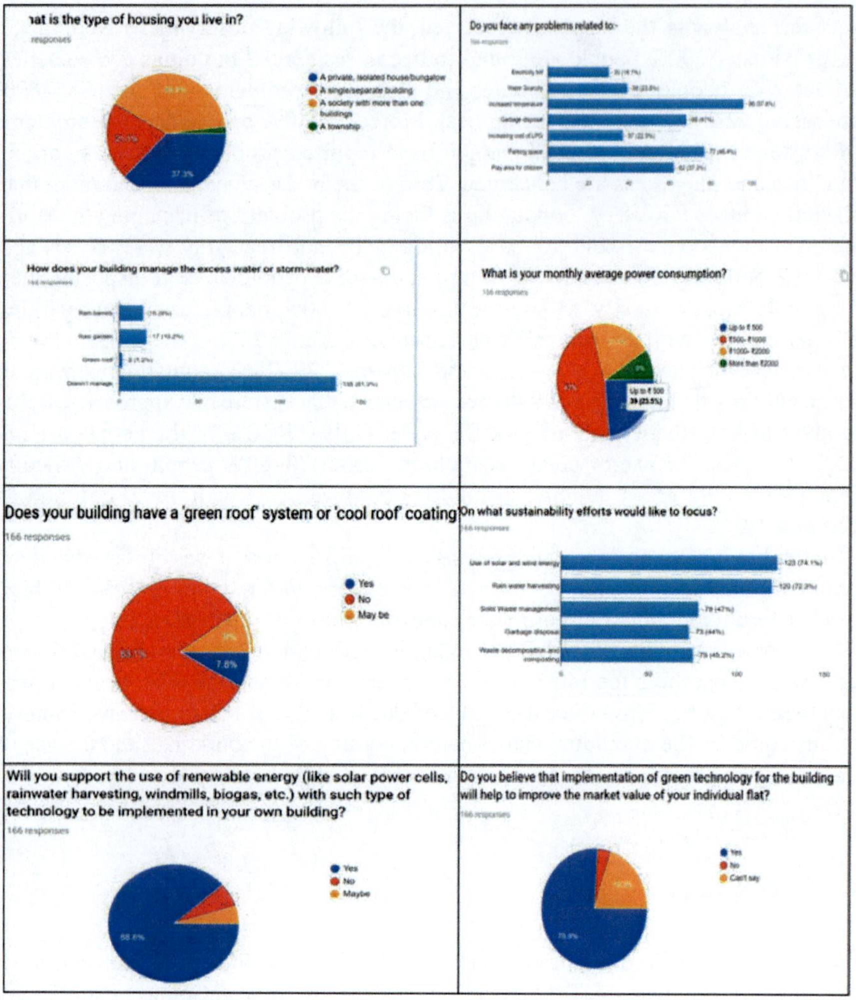

Fig. 2 Pie chart showing different responses

panel costs around Rs. 1,50,000 to Rs. 1,80,000. 2–2.5 KW panel may cost around 2.5–3 lakhs rupees. Further 5 KW panel will cost around 3.75–4.25 lakhs rupees.

Total cost of solar system includes (panel cost + structure cost + cable cost + inverter + labor charges + net meter cost + AMC). Inverter is to be installed for converting DC into AC and one inverter may be sufficient for power generation of 1–5 KW. Number of inverters required is equal to the number of meters. Suppose the inverter required for 1KW solar panel costs 80,000 to 90,000, inverter for 4–5 KW panel will cost 120–150 thousand at maximum.

Maintenance of solar system includes periodic checking/inspection of solar panels, cables and structure, periodic cleaning of panels, checking the condition of inverter, meter and all electrical connections. Small window/opening is provided in big solar panels to reduce the effect of wind pressure on the panels. Also, cleaning of panels is its secondary benefit.

Household where monthly bill for electricity is around Rs. 400–600 would require 1 KW solar panel. Solar panel works on negative temperature coefficient i.e., when the panels are less heated, cooled, they work more efficiently. Hence in summer, though the sunrays/ radiations are continuous and intense, the heating of these panels reduces its efficiency up to 70–80%.

But in monsoon, though the sunrays may not be available for the entire day, whenever they are present, the cool panels have increased efficiency to generate power. Hence, the overall efficiency of solar panels, even in monsoon is 40–60%.

The Warranty for the complete solar system is around 18–20 years, it increases up to 25 years for solar panels only, whereas the inverter has only about 7 years warranty. From the market research it was explored that there are different solar types available in India with varied advantages and disadvantages. The most effective one is Concentrated PV Cell (CVP) but solar tracker and cooling system are needed to reach high efficiency rate. Monocrystalline solar panels (Mono-SI) have a good efficiency rate, high life-time value but are expensive.

On the other hand, Polycrystalline solar panels (p-Si) are lower in price but are sensitive to high temperatures with lower life span. Many times people prefer Amorphous Silicon Solar Panels-(A-SI) for its relatively lower costs but they actually have shorter warranties and lifespan as represented in Table 1.

The most typical size used for residential installations is 65 in. by 39 in., while the common size for commercial applications is 77 in. by 39 in.

Numerous researchers have studied the properties of climate changes on the efficiency of PV systems under different levels of solar irradiation and air temperature [16–18].

Many solar panel manufacturing companies in India provide optimum tilt of solar panels by month for particular areas based on the directions of sun movement. In general, as a thumb rule, optimum angle of 48° in winter, 72° angle for spring/autumn and 96° for summer are preferably followed.

Table 1 General specifications of solar panels

System size (kW)	Approx. roof space (m^2)	Typical annual output (kWh)
1	8	850
2	14	1,700
3	21	2,550
4	28	3,400

6 Results and Discussion

For understanding the practical implications of solar panels, two different types of solar panels were studied and compared through their techno-commercial specifications. The details of these panels are given below:

1. Monocrystalline Panels
2. Polycrystalline Panels

The general, technical and commercial details were obtained from the manufacturers and were studied thoroughly. After comparing all the features and technical details, it was decided that the use of polycrystalline solar panels proved to be economical with greater technical benefits for the residential buildings.

After acquiring all these techno-commercial data and specifications, it was further required to get the present power consumption data of all the households as it would give the annual or monthly or daily demand of any particular flat/apartment and moreover the entire building.

The detailed calculations were carried out to calculate the segregation in annual bills and monthly bills as shown in below Figs. 3 and 4. Figure 3 gives the average monthly bill of all households in the building. Figure 4 shows the annual total bills.

The total annual power consumption of the entire building was calculated as 81,480 kW/hr. The output of the selected solar panel (Canadian Solar CS6X-325P 325 W) is 325 Watts per hour. Assuming some loss, it's predicted that the actual output would be somewhere around 300 Watts per hour. The sunrays are assumed

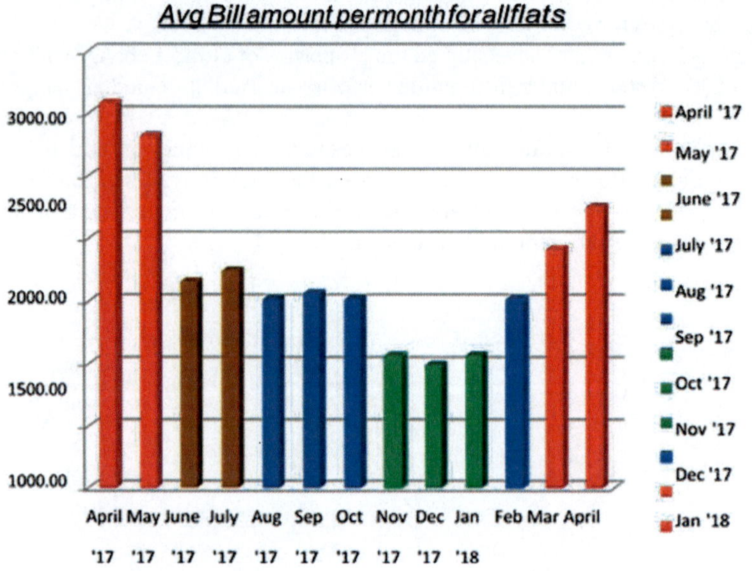

Fig. 3 Monthly electricity bill amount

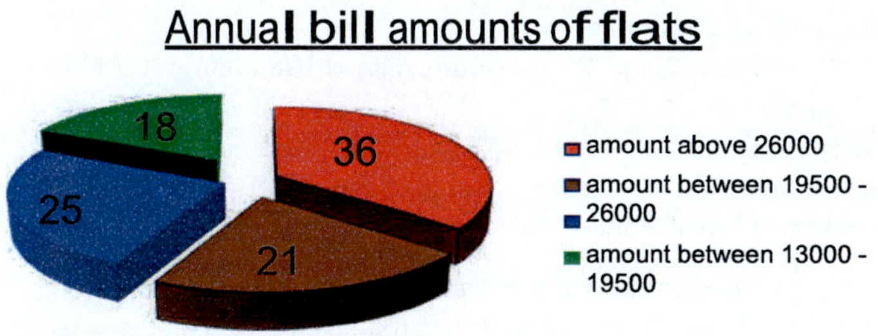

Fig. 4 A pie chart representing annual bill amounts of flats

to be intense and capable of generating electricity only for 6 h. a day. Therefore, the annual output of the above-mentioned solar panel would be = 300 * 6 * 365 = 6,57,000 Watts i.e., 657 kW.

Therefore, the required number of solar panels for the entire building is as follows:

(1) Only for common areas: 9482/657 = 14.43 (approx. 15) panels.
(2) Only for all households (28 flats): 72,000/657 = 109.5 (approx. 110) panels.
(3) Both for common areas along with all flats: 81,500/657 = 124.04 (approx. 125) panels.

The area required for the required number of solar panels to be installed is

(1) Only for common areas: 2 * 15 = 30 m^2
(2) Only for all households (28 flats): 2 * 110 = 220 m^2
(3) Both for common areas along with all flats: 2 * 125 = 250 m^2
(4) the Cost of selected solar panel (Canadian Solar CS6X-325P 325 W) is Rs. 10,739 for a set of 2 panel.

Therefore, the cost of required number of solar panels for the entire building is as follows:

1. Only for common areas: (10,739 * 15)/2 = Rs. 80,550.
2. Only for all households (28 flats): (10,739 * 110)/2 = Rs. 5,90,645.

Both for common areas along with all flats: (10,739*125)/2 = Rs 6,71,190 including Installation charges.

- **Government subsidy**:
  The Central Government pays 40% of the benchmarked installation cost as a subsidy for installing rooftop solar at home in states falling under the general category, according to the Ministry of New and Renewable Energy.

  Therefore, the cost of solar panel for the entire building after deducting the government subsidy is as follow

  1. Only for common areas: (80,550 * 0.7) = Rs. 56,400
  2. Only for all households (28 flats): (5,90,645 * 0.7) = Rs.4,13,450
  3. Both for common areas along with all flats: (6,71,190 * 0.7) = Rs.4,69,830

  :

- **Payback period (annually)**

  Overall payback period for common areas = 108,805/56400 = 1.92 years i.e., (1 yr 11 months).

  For all households (28 flats) = 642,351.10/413450 = 1.5 yrs i.e., (1 yr 6 month).

  Both for common areas along with all flats = 751,156.72/469830 = 1.59 yrs i.e.,19 months (1 year 7 months).

The above calculation outcome represented in Table 2 highlights the total area required, total no of panels and approximate total cost based on the subsidy provided by the government and it was assumed in this case as 40% and the payback period was calculated accordingly.

It is observed that power consumption varies as per season and as per the temperature. April and May being the hottest and the requirement of AC and November and December being the coldest and the requirement of heaters. In both scenarios, solar panels can be a savior as the sunlight received is maximum in these 4 months and can be used effectively.

Figure 5 shows the bifurcation of power consumption in the common areas which reveals lifts take the major contribution. From the pie chart it is also known that the maximum consumption is for lifts followed by pumps. If solar energy can suffix these areas, it will be a great relief in power consumption and subsequently it will help in reducing the society's electric bill and in turn it will be saving for the residents.

Table 2 Solar panel and cost estimation of the selected building

S. no	Location	Total area (m^2)	Total cost	No of solar panels	Payback period
1	Common areas	30	Rs. 56,400	15	(1 year 11 months)
2	Households	220	Rs. 4,13,450	101	(1 year 6 monthss)

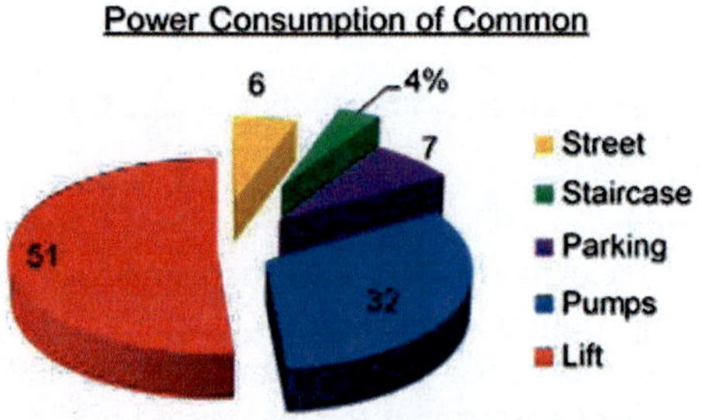

Fig. 5 Pie chart representing the % of power consumption by common units

7 Future Scope of Work

(1) Doing detailed study of green roof system or green wall system or vertical gardening system.
(2) Checking whether solar water heaters could be installed to suffice the needs of all the households in the entire building.
(3) Modifying the layout of the society and efficiently planning the open and free spaces.
(4) Maximum utilization of the terrace spaces and roof tops to be done for provision of green cover.
(5) Rain Water Harvesting System to be implemented for one building and then for the entire society.

8 Conclusion

It was evident from the research study carried out using the different sources of data it was observed that the people who were living in the building suffered from the problem of high electricity bills and maintenance of their households due to ever increasing costs of renewable energy sources and thus there it is a need of harnessing renewable energy resources such as solar energy. It was also observed that the payback period for installation of solar panel was found to be 2 years approximately and it depends on the topography, electricity tariff and other factors. To enhance the implementation of renewable energy in the built environment the Government encourages more subsidies to boost the use of renewable energy sources and 40% subsidy needs to be provided to solar energy users. The efficiency and availability of solar cell and ancillary components need to be improved and the government at all levels must give rebate in taxes to promote the adoption of renewable energy sources. Also, calculating the heating and cooling loads, and defining the orientation of the house and the direction of the wind could be helpful in making certain changes in the existing building which would facilitate to convert the existing structure into an efficient structure.

References

1. Scheuer CW (2007) Adoption of residential green building practices. Natural Resources and Environment in the University of Michigan
2. Dakwale V, Ralegaonkar RV, Mandavgane SA (2011) Improving environmental performance of building through increased energy efficiency: a review. City Soc 1
3. Yim S, Ng S, Hossain Md, Wong J (2018) Comprehensive evaluation of carbon emissions for the development of high-rise residential building, buildings, 8:147
4. Gao T, Liu Q, Wang J (2013) A comparative study of carbon footprint and assessment standards. Int J Low-Carbon Technol 9:237–243

5. Pandey D, Agrawal M, Pandey JS, Carbon Footprint (2011) Current methods of estimation, environmental monitoring and assessment, 178:135–160
6. Finkbeiner M (2009) Carbon footprinting—opportunities and threats. Int J Life Cycle Assess 14:91–94. https://doi.org/10.1007/s11367-009-0064-x
7. Parikh J, Panda M, Ganesh Kumar A, Singh V (2009) CO2 emission structure of Indian economy, energy, 34:1024–1031
8. Akbarnezhad A, Xiao J (2017) Estimation and minimization embodied carbon of building. Academic Editor: Manuel Duarte Pinheiro
9. Ren H, Gao W, Ruan Y (2009) Economic optimization and sensitivity analysis of photovoltaic system in residential buildings. Renew Energy 34(3):883–889. https://doi.org/10.1016/j.renene.2008.06.011
10. Burns JE, Kang J-S (2012) Comparative economic analysis of supporting policies for residential solar PV in the United States: solar renewable energy credit (SREC) potential. Energy Policy 44(May):217–225. https://doi.org/10.1016/j.enpol.2012.01.045
11. Kapur R, Jain RK (2014) Parameters of upgrading existing building into a green building. Int J Eng Res Appl 4(2). ISSN: 2248-9622 (Version 1)
12. Kavani N, Pathak F (2015) An existing building into a green building. IJRET: Int J Res Eng Technol 3(6)
13. Kataria V, Salunkhe H (2016) Retrofitting of existing building as per IGBC existing building norms. Int J Civil Struct Eng Res 4(1):250–253
14. Feldman D, Ramasamy V, Margolis R (2021) US Solar photovoltaic BESS system cost Benchmark Q1 2020 report. Rep.No. DE-AC36–08GO28308. National Renewable Energy Laboratory-Data, Golden, CO
15. Lakshmi R (2021) Conversion of existing conventional building to green building using simple versatile affordable green rating for integrated habitat assessment and cost analysis. EPRA Int J Res Dev (IJRD) 6(12):December 2021
16. Feron S, Cordero RR, Damiani A, Jackson RB (2021) Climate change extremes and photovoltaic power output. Nat Sustain 4(3):270–276. https://doi.org/10.1038/s41893-020-00643-w
17. Jerez S, Tobin I, Vautard R, Montávez JP, López-Romero JM, Thais F, Bartok B, Christensen OB, Colette A, Déqué M (2015) The impact of climate change on photovoltaic power generation in Europe. Nat Commun 6(1):1–8. https://doi.org/10.1038/ncomms10014
18. Zhao X, Huang G, Lu C, Zhou X, Li Y (2020) Impacts of climate change on photovoltaic energy potential: a case study of China. Appl Energy 280(December):115888. https://doi.org/10.1016/j.apenergy.2020

"Study on Binary and Ternary Blended Concrete"

Krishna Thakur, Manendra Pratap Verma, and Vinay Kumar Singh Chandrakar

Abstract The scope of the current investigation is to determine whether it is feasible to use fly ash and rice bran ash as admixtures in material and to look into how high temperatures affect the characteristics of the material. The amounts of water, cement, fine and coarse aggregate that we can make control mix binary mix (77% cement + 23% fly ash) and (80% cement + 20% rice husk ash) and ternary material (71% cement + 20% rice husk ash + 9% silica fume) and status of material is M20, M30, and M40. We undergo testing at room temperature in this study. Following casting, all Samples were moist cured for 28 days. Compressive strength tests were performed on the samples. Utilizing fly ash and rice husk ash in material not only lowers costs but also increases durability and tolerance to high temperatures. The use of Rice bran Ash (RHA) in a material is environmentally benign and lowers carbon dioxide emissions after exposure to greater temperatures was raised by more fly ash and rice bran ash.

Keywords Ternary material · Compressive strength · Binary blended material · RHA · Control material · Temperature

1 Introduction

1.1 General

Mixtures of three extraordinary cementations components makes up ternary material. This document discusses those mixtures that include a third textile, slag cement, and Portland cement. Fly ash is often the third element, however, silica fume might also be present.

K. Thakur (✉)
School of Engineering, Madhyanchal Professional University, Bhopal, M.P, India
e-mail: krishnathakur15034@gmail.com

M. P. Verma · V. K. S. Chandrakar
Department of Civil Engineering, School of Engineering and Technology, Bhopal, India

K. K. Pathak et al. (eds.), *Latest Developments in Civil Engineering*, Lecture Notes in Civil Engineering 352, https://doi.org/10.1007/978-981-99-2676-3_14

A ternary mixture also includes binary (ASTM C595) mixed hydraulic cement used in conjunction with a third cementation material. Other components in conjunction with Portland and slag cement, such as met kaolin or rice bran ash, are not now in a general application and will not be mentioned here.

The binary (ASTM C595) ternary mixed cement can also be used to create ternary mixtures. Slag cement has long been used in ternary mixtures. For instance, the Empire State Building was reconstructed using a mixtures of slag cement.

Benefits

- Good strength
- More corrosion resistance
- Average sulfur resistance
- Prevention of thermal cracking
- Minimum permeability
- Maximum Sulfate resistance
- Diminution of carbon footprint

Ternary mixed have been utilized in material showing to freezing, thawing, de-icing agents and still are today. Any material's capacity to survive freezing, thawing, and de-icing chemicals will be maximized with optimum air entrainment, adequate curing, and precise material finishing techniques. With great success, numerous unique mixture proportions have been tried. Paving projects utilizing ternary mixtures have worked for decades under demanding conditions without clearly losing durability.

Ternary Mixtures Can Be Used

After warming, material's mechanical qualities, such as strength, elastic modulus, and volume twisting, dramatically deteriorate, which results in a change in the fundamental makeup of cement. One of the principal real weakening cycles that affect the stability of considerable constructions and may result in unintended initial disappointments is high temperature. In light of this, precautions like selecting the proper materials should be taken to limit the dangerous effects of high temperatures on the material (Bahar 2008).

Objectives of the Study

- To expand binary mixed Grade M20, grade M30 and grade M40 ordinary material via way of means of the use of most fulfilling alternative stage of FA and RHA
- To expand ternary mixed Grade M20, grade M30 and grade M40 material via way of means of carrying out investigations on workability, placing time and along with compressive strength, splitting tensile strength, via way of means of different percent of SF and to examine the overall performance of ternary mixed material with manipulate and FA and RHA primarily based binary mixed material.
- To compare the compressive strength of control mix, binary material and ternary material at different elevated temperatures.

- To compare the price of fly ash based and rice bran based binary and ternary material with plain cement material for M20, M30 and M40.

2 Literature Review

Alonso et al. (2002) have announced that potential qualities are subject to the conductivity of the framed consumption items and on the period of cement. Stray flows and age of intersection possibilities can likewise affect the expected readings. Different variables affecting potential estimations incorporate reference terminal position, material sort and the presence of breaks. In any case, the strategy is generally utilized and functions as the first way to deal with consumption recognition. Notwithstanding, quantification and solid expectation of the erosion rate needs the utilization of other electrochemical methods.

Anwar (2002) introduced the outcomes from research facility studies on properties of substantial that contain ternary mixes of Portland material, Micro Silica and Fly Ash. Chosen four substantial blends were ready with water to cementitious material proportion of 0.4.

Austin et al. (2001) have revealed that the dazzled current procedure is affirmed to be a powerful and speedy technique for speeding up chloride-actuated consumption. Be that as it may, the electrochemistry behind the component contrasts from normally instigated erosion. The essential electrochemical contrast of the dazzled current procedure to normally eroding frameworks is the raising of the possibility to a worth more noteworthy than the transpassive potential, where the consumption rate doesn't compare with a harmonious blended potential or a potential realistic under common assistance conditions without applying an outer wellspring of energy. Henceforth, forcing such a potential will result in the rebar being in a misleadingly enraptured state.

Auyeung et al. (2010) have observed that the hypothetical and genuine consumption mass misfortune are not equivalent due to the different variables, for example, need for electrical energy to start the erosion, resistivity of cement, creation of the bar, electrical properties of minerals in material.

Basha and Muntohar (2010) (Electronic Journal of Geotechnical Engineering, 2003) RHA can be added to soil to support compatibility, the versatility of soil is decreased when rice bran debris as well as material is added, just like the greatest dry thickness, and the ideal

3 Materials and Methods

The detailed experimental programmer design of in this chapter.

3.1 Materials

The material used for study are cement, Sand, Aggregate, Rice bran ash, Silica Fume, Fly ash and Water (Table 1).

C.A: Water = 1: 1.66: 2.27: 0.45

Mix Proportions for Binary Blended Material

The binary combined material Samples the usage of low calcium FA and RHA had been organized and the partial substitute of cement by FA and RHA was taken into consideration as 5, 10, 15, 20, 21, 22, 23, 24 and 25% by way of the mass of cement. The different mix mixtures of binary mixed material shown in Table 3.19 and 3.20.

Mix Proportions for Ternary Blended Material

For preparing ternary blended material, the replacement of cement by SF was considered as 3, 6, 9 and 12% of total powder content by weight. The different mix mixtures of ternary blended material are shown in the Tables 2 and 3.

Table 1 Mix proportions for control material

Material grade	Materials required 1 m^3 of material			
	Cement in kg	Sand in kg	Aggregate in kg	Water in liter
M 20	422	680	1137	190
M 30	440	736	1032	197
M 40	445	742	1011	200

Table 2 Cementations material mixtures for FA and SF based ternary blended material

Mix	Cement content (kg/m^3)				Mineral admixture content (kg/m^3) Silica fume			
	%	M20	M30	M40	%	M20	M30	M40
PCC	100	422	440	445				
BFA23	77	324.94	338.8	342.65	0	–	–	–
SF3	74	315.19	328.64	332.37	3	9.75	10.16	10.28
SF6	71	305.44	318.47	322.09	6	19.50	20.33	20.56
SF9	68	295.7	308.31	311.82	9	29.24	30.49	30.83
SF12	65	285.95	298.14	301.53	12	38.99	40.66	41.12

Table 3 Cementations material mixtures for RH and SF based ternary blended material

Mix	Cement content (kg/m^3)				Admixture content (kg/m^3) Silica fume			
	%	M20	M30	M40	%	M20	M30	M40
PCC	100	422	440	445				
BRH20	80	337.6	352	356	0	–	–	–
SF3	77	327.47	341.44	345.32	3	10.13	10.56	10.68
SF6	74	317.34	330.88	334.64	6	20.26	21.12	21.36
SF9	71	307.22	320.32	323.96	9	30.38	31.68	32.04
SF12	68	297.09	309.76	313.28	12	40.51	42.24	42.72

4 Results and Discussion

Binary blended material was prepared by adding mineral admixtures to partially replace cement, along with other ingredients. FA-based and RHA based binary blended material Samples were prepared for grade M20, grade M30 and grade M40 of material.

Compressive Strength

Compressive energy is the capability to hold masses of material and shape on its floor with no cracking or deformation. An object below compression will reduce in length and, underneath anxiety, the size will maintain to extend (Table 4).

$$\text{Compressive energy} = \text{Load/Cross - sectional region}$$

Table 4 Compressive strength for binary blended material (fly ash + cement)

Mix	Compressive strength (N/mm^2)					
	M20		M30		M40	
	7 days	28 days	7 days	28 days	7 days	28 days
PCC	20.08	29.91	28.43	39.82	38.16	54.23
BFA-5	19.86	29.29	28.06	41.72	38.66	53.14
BFA-10	19.53	28.87	27.82	41.92	38.17	52.17
BFA-15	18.79	28.43	27.43	43.25	37.92	52.79
BFA-20	16.98	28.12	26.17	44.27	37.42	51.63
BFA-21	15.45	27.92	25.84	42.13	36.52	50.92
BFA-22	14.42	27.12	24.16	41.46	35.12	50.43
BFA-23	13.23	27.32	22.30	38.23	32.63	49.62
BFA-24	12.91	24.32	21.18	37.13	31.25	47.65
BFA-25	11.83	22.35	21.10	36.47	30.32	45.42

- We checked the compressive strength of all grades of material at 7 days and 28 days and we found that the strength of M20 grade of control mix material after 28 days was 29.91 Mpa and after 7 days 20.08 Mpa (Table 5).
- The consequences of 7 and 28 days cured M20 grade material and the usage of numerous cement replacement stages by using RHA are shown in diagram 4.4 while cement become changed by RHA from 5 to 25% for developing the RHA based binary mixed material.
- The compressive electricity decreased from 18.13 MPa at 5% substitute to 11.43 MPa at 25% replacement after 7 days of curing because of the slower pozzolanic response of RHA particle.

Workability of Material

Performance of material is the assets of newly mixed material which determines the convenience and homogeneity with which it could be mixed as defined with the aid of ACI trendy 116R-90 (ACI 1990b) (Tables 6, 7, 8).

Table 5 Compressive strength for binary blended material (rice bran ash + cement)

Mix	Compressive strength (N/mm^2)					
	M20		M30		M40	
	7 days	28 days	7 days	28 days	7 days	28 days
PCC	20.08	29.91	28.43	39.82	38.16	54.23
BRH-5	18.13	30.12	27.29	41.30	37.91	54.70
BRH-10	17.73	28.32	26.81	42.70	37.13	54.93
BRH-15	16.95	27.78	26.13	41.30	36.83	52.13
BRH-20	12.23	26.12	21.43	37.82	31.48	48.80
BRH-21	12.13	24.73	20.30	35.82	30.70	47.80
BRH-22	12.43	24.12	20.16	34.92	29.81	45.19
BRH-23	11.91	24.23	20.91	36.48	28.23	45.34
BRH-24	11.80	24.15	20.54	35.12	27.69	44.79
BRH-25	11.45	24.10	20.10	35.16	26.12	43.29

Table 6 Workability for Binary and Ternary blended material (cement + fly ash + silica fume)

Mix	Workability (mm)		
	M20	M30	M40
PCC	34	30	27
BFA-23	37	34	29
SF3	32	27	22
SF6	25	22	16
SF9	23	20	15
SF12	16	12	8

Table 7 Workability for binary and ternary blended material (rice bran ash + cement + silica fume)

Mix	Workability (mm)		
	M20	M30	M40
PCC	34	30	27
BRH-20	35	30	37
SF3	30	25	23
SF6	26	22	21
SF9	20	17	14
SF12	16	13	6

Table 8 Initial setting time for Binary and Ternary blended material (fly ash + cement + silica fume)

Mix	Time (minutes)		
	M20	M30	M40
PCC	315	285	270
BFA-23	435	340	308
SF3	321	298	264
SF6	298	281	245
SF9	285	265	234
SF12	345	335	270

- FA-based M23 grade binary mixed material took 5 h 15 min to set initially and 7 h 30 min to set completely.
- When 23% Fly ash is utilized as a partial replacement for cement in material, the initial setting time is increased by about 27.58% compared to control material's preliminary setting time and the finishing setting time of binary blended material is also increased compared to control material's preliminary setting time and the finishing.
- As a result of the findings, it is concluded that replacing cement with FA prolongs the setting qualities of material. The results show that replacing cement with FA significantly extends the setting qualities of material, which is mostly owing to the FA particle's slower pozzolanic action (Mirza et al. 2002). Material Samples of the M30 and M40 grades yielded similar results (Figs. 1, 2).

When 20% RHA is added to M20 grade binary mixed material, the initial and final setting times increase by about 20–25% when compared to the control material's initial and final setting times.

When mineral admixtures such as FA/RHA are used to partially replace cement, the relative cement content in the material is lowered, which slows down the hydration process. Additionally, FA and RHA do not react in the first few hours, increasing the setting time of blended material.

Figure 3 show that adding 3 and 9% SF as a second mineral admixture reduces both the preliminary and finishing setting time of grade M20, grade M30, and grade M40 grade Fly ash/rice bran ternary blended material when compared to the initial

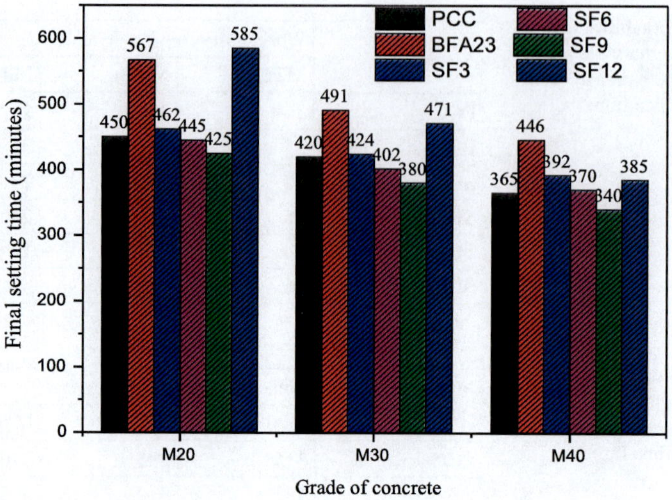

Fig. 1 Show the results of the beginning and final setting times of grade M20, grade M30, and grade M40 control, binary, and ternary mixed material samples

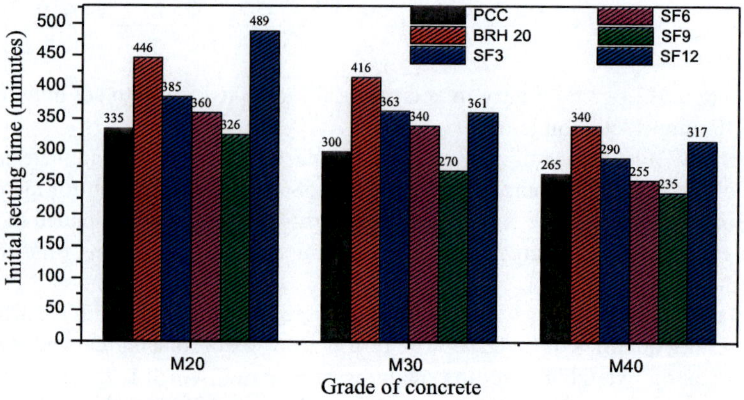

Fig. 2 Initial setting time for binary and ternary blended material (rice bran ash + cement + silica fume)

and final setting time of binary blended material, whereas adding 12% SF increases both the initial and final setting time of FA/RHA ternary blended material.

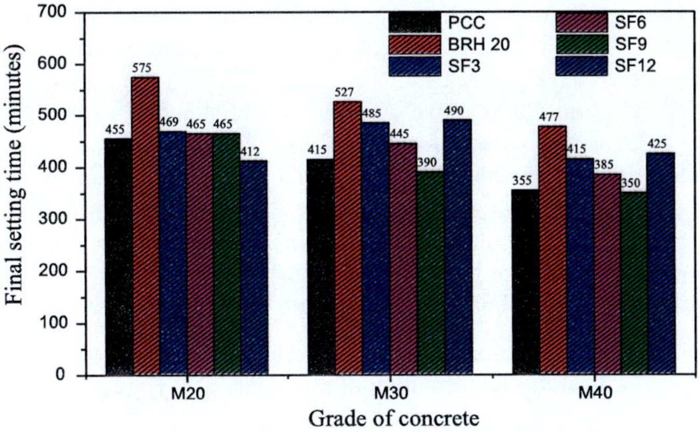

Fig. 3 Show that adding 3 and 9% SF as a second mineral admixture reduces both the initial and final setting time of M20, M30, and M40 grade FA/RHA ternary blended concrete

5 Hardened Properties of Ternary Blended Material

There are many characteristics of hardened material, including compressive strength, splitting tensile strength, bond strength and elastic modulus. These characteristics show relationships between compressive strength and splitting tensile strength of binary and ternary blended material (Table 9).

- The addition of 9% SF to FA-based blended material resulted in a compressive strength of 20.21 MPa after 7 days of curing, which is the approx same as the compressive strength of control material at 7 days,
- The addition of 12% SF to FA-based ternary blended material resulted in a compressive strength of 17.23 MPa, which is less than the compressive strength of control material (20.08 MPa).

Table 9 Compressive strength for ternary blended material (fly ash + cement + silica fume)

Mix	Compressive strength (N/mm^2)					
	M20		M30		M40	
	7 days	28 days	7 days	28 days	7 days	28 days
PCC	20.08	29.91	28.43	39.82	38.16	54.23
BFA-23	13.23	27.32	22.30	38.23	32.63	49.62
SF3	14.26	26.56	23.69	38.06	33.41	48.95
SF6	16.23	27.46	25.21	38.96	36.25	49.13
SF9	20.21	29.62	29.23	40.12	39.62	53.79
SF12	17.23	26.12	27.06	37.82	37.86	48.12

Figure 4 show that the 9% SF based M20, M30 and M40 grade ternary blended material had higher compressive strength than the other mix mixtures in all curing periods of 7 and 28 days (Table 10).

Figure 5 show the compressive strength development of control, RHAbased binary, and RHA and SF based ternary blended material Samples.

Figure also depicts the strength of 7-day-old control, binary, and ternary blended material.

The strength of M20 grade control material was found to be 20.08 MPa after 7 days, whereas RHA based binary blended material was found to be 12.23 MPa, which is 40% less than the compressive strength of the control material.

- It demonstrates that the binary blended material reached its maximum strength at later ages as a result of the formation of secondary C-S-H gel.

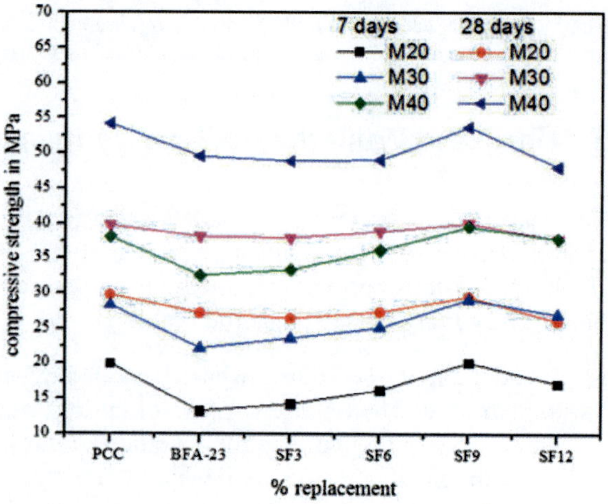

Fig. 4 Compressive strength for ternary blended material (fly ash + cement + silica fume)

Table 10 Compressive strength for ternary blended material (rice bran ash + cement + silica fume)

Mix	Compressive strength (N/mm^2)					
	M 20		M 30		M40	
	7 days	28 days	7 days	28 days	7 days	28 days
PCC	20.08	29.91	28.43	39.82	38.16	54.23
BRH–20	12.23	26.12	21.43	37.82	31.48	48.80
SF3	14.23	26.84	22.13	36.82	31.97	48.71
SF6	16.23	27.21	25.31	37.83	34.80	50.12
SF9	19.98	29.80	29.11	40.08	37.82	53.17
SF12	15.23	26.12	26.25	37.18	35.71	47.28

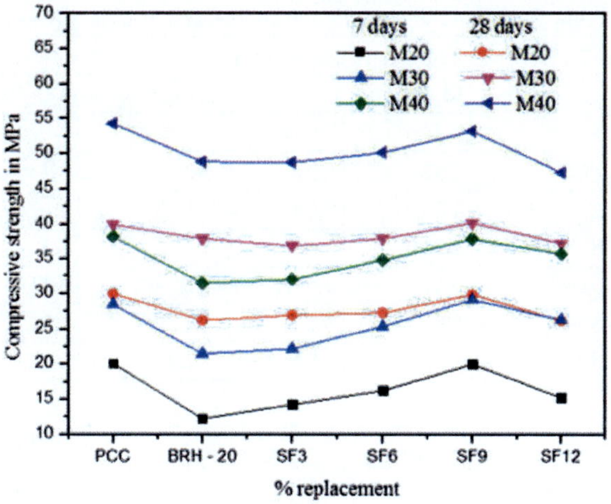

Fig. 5 Compressive strength for ternary blended material (rice bran ash + cement + silica fume)

- The increased compressive strength of ternary blended material in both early and late ages is due to an overall improvement in material homogeneity (Table 11, Fig. 6).
- The parting rigidity of control material, FA-based binary mixed material (BFC 23) and FA and SF based ternary mixed material are displayed in Figs. 4.13.
- The parting elasticity at the period of 7 days control, parallel and ternary mixed material are likewise displayed in Fig. 4.13.
- From Fig. 4.13, the parting rigidity of M 20 grade control material at 7 years old days was viewed as 2.52 MPa while FA-based binary mixed material was viewed as 1.53 MPa which is around 40% lesser than the parting elasticity of control material.
- The expansion of 9% SF in FA-based ternary mixed material showed the parting rigidity of 2.87 MPa in 7 days restoring which is somewhat higher than the parting elasticity of control material.
- The expansion of 12% Silica fume in fly ash based ternary mixed material showed the compressive strength of 1.67 MPa which is lesser than the parting rigidity of control material (2.52 MPa) at 7 years old days (Table 12, Fig. 7).

Cast Analysis

The cost is the main important factor in any research. The invented material should be cost effective so that it can be easily adopted in modern construction industry. For construction of any project 60% of the cost is incurred in the material and in which most costly among all three of construction material is cement. Our material in which we have used Fly ash, Rice bran ash and silica fume as a fractional replacement of cement by 5–25%, 5–20%. In this way less cement will be used for every m^3 of material and we can save the direct material cost. The details of the cost economics of the material is described briefly in following table 14.

Table 11 Tensile strength for ternary blended material (fly ash, cement and silica fume)

Mix	Tensile strength (N/mm^2)					
	M 20		M 30		M40	
	7 days	28 days	7 days	28 days	7 days	28 days
PCC	2.52	2.91	2.71	3.11	3.09	3.65
BFA-23	1.53	2.06	1.73	2.6	2.11	2.93
SF3	2.31	2.68	2.43	3.17	2.86	3.6
SF6	2.57	2.91	2.89	3.31	3.08	3.9
SF9	2.87	3.21	3.40	4.01	3.91	4.61
SF12	1.67	1.98	1.91	2.23	2.30	2.64

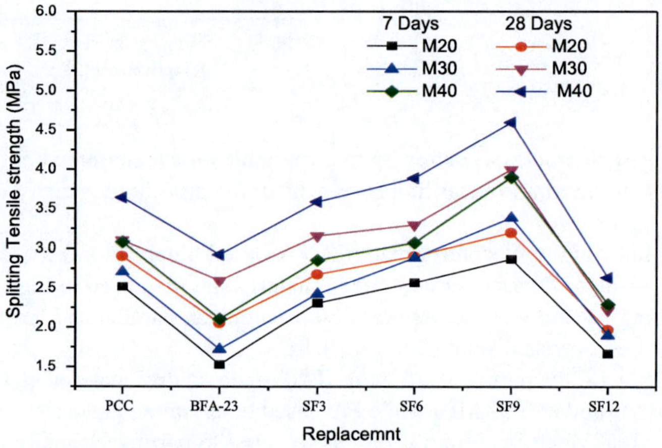

Fig. 6 Tensile strength for ternary blended material (fly ash + cement + silica fume)

Table 12 Tensile strength for Ternary blended material (Rice Bran Ash + cement + silica fume)

Mix	Tensile strength (N/mm^2)					
	M 20		M 30		M40	
	7 days	28 days	7 days	28 days	7 days	28 days
PCC	2.52	2.91	2.71	3.11	3.09	3.65
BRH–20	1.26	1.89	1.69	2.17	1.96	2.46
SF3	1.81	2.6	2.31	2.85	2.41	2.93
SF6	2.14	2.91	2.72	3.10	3.21	4.05
SF9	2.89	3.02	3.06	3.56	3.67	4.25
SF12	1.18	1.63	1.72	2.38	2.18	2.33

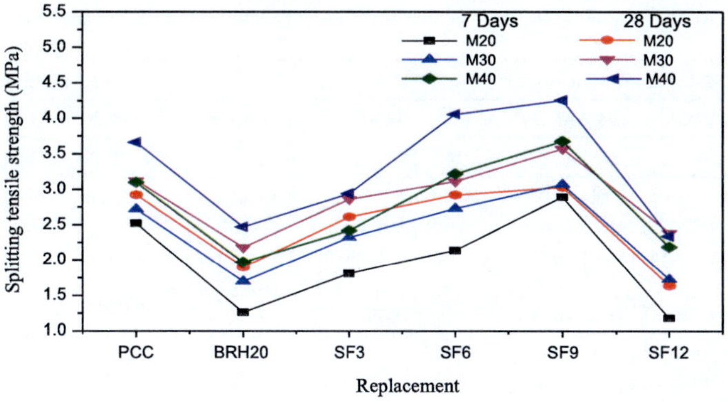

Fig. 7 Tensile strength for ternary blended concrete (rice bran ash + cement + silica fume)

Material Rate in Bhopal 8-03-2022
Cement—355 Per Bag
1 bag = 50 kg
So rate of 1 kg cement = 355/50 = 7.1 Rs
Sand—43 cu ft
1 cu ft = 28.32 kg
So rate of 1 kg sand = 43/28.32 = 1.52 Rs
Aggregate—25 cu ft
So rate of 1 kg aggregate = 25/28.32 = 0.88 Rs
Table 13 shown the individual cast of materials.

Table 14 show the fresh material, fly ash binary material and fly ash ternary material.

- If we can use binary blended material to save the cast 14% in per cubic meter and use ternary material to save the cast of 11% of the control mix material.
- Ternary material rate 3% higher than the binary material (Table 15).
- Table 4.21 show the cast of rice bran based binary and ternary material/cubic meter for grade M20, grade M30 and grade M40 grade of material
- Plain cement material cast for all grade of material is Rs 5030 for M20 Rs 5150 for M30 and Rs. 5178 for M40.

Table 13 Market rate of all material

S. No	Material	Rate (Rs./Kg)
1	Cement	7.1
2	Fine Aggregate	1.52
3	Coarse Aggregate	0.88
4	Fly ash	0.2
5	Rice bran ash	1
6	Silica Fume	12

Table 14 Cast of Cementations material mixtures for fly ash and silica fume based ternary blended material

Mix	Material content (kg/m^3)				Rate of Material in Rs		
	% of C	M 20	M30	M 40	M 20	M30	M 40
PCC	100	422	440	445	2996	3124	3160
BFA23	77 + 23	324.94 + 97.06	338.8 + 101.2	342.65 + 102.35	2326	2426	2437
SF9	68 + 23 + 9	295.7 + 97.06 + 29 24	308.31 + 101.2 + 3 0.49	311.82 + 102.35 + 30 83	2469	2575	2604

Table 15 Cast of Cementations material mixtures for RH and SF based ternary blended material

Mix	Cement content (kg/m^3)				Rate of Material in Rs./m^3		
	%	M20	M30	M40	M20	M30	M40
PCC	100	422	440	445	2996	3124	3160
BRH20	80 + 20	337.6 + 84.4	352 + 88	356 + 89	2481	2587	2617
SF9	71 + 20 + 9	307.22 + 84.4 + 30.38	320.32 + 88 + 31.68	323.15 + 89 + 32.85	2629	2742	2777

- When cement is placed with rice bran ash, it forms a binary material whose cast is for different grades of material is Rs 4515, Rs 4613, Rs 4635 for M20, M30 and M40.
- After binary material, we added silica fume in the same way and made ternary material whose rate is for all three grades Rs 4663, Rs 4768, Rs 4795 in per meter cube.
- If we can use binary blended material to save the cast 10% in per cubic meter and use ternary material to save the cast of 8% of the control mix material.
- Ternary material rate 3% higher than the binary material.

6 Conclusions

From the extensive experimental investigations, the following mi poratntare arrived:

The optimum replacement of cement for developing the grade M20, grade M30 and grade M40 grade binary blended material by fly ash and rice bran was found to be 23% and 20% respectively based on the 28 days required mean strength.

- The slump value of 23% fly ash grade M20, grade M30 and grade M40 grade binary blended material was found to be slightly higher whereas 20% rice bran grade M20, grade M30 and grade M40 binary blended material slightly compared to control material.

- The addition of silica fume for developing grade M20, grade M30 and grade M40 grade ternary blended material reduced slump value compared to control and binary blended material. Addition of 12% SF drastically reduced the hunch value due to the more water demand of silica fume particles.
- The addition of 23% fly ash increased the initial setting time approximately 30% and also increased the final setting time approximately 20% compared to control material. The addition of 20% rice bran increased the initial setting time of binary blended material approximately 35% and also increased the final setting time approximately 30% compared to control fresh material.
- The addition of 9% SF along with 23% FA/20% RHA enhanced the setting time characteristics of blended material.
- The compressive strength of 7 days cured 23% FA and 20% RHA based grade M20, grade M30 and grade M40 grade binary blended material was found to be less compared to control material. The compressive strength of 9% SF mixed FA/RHA based ternary blended material was found to be more compared to the control, binary blended material and 3% and 12% SF based ternary blended material.

  – The splitting tensile strength of FA/RHA based grade M20, grade M30 and grade M40 grade binary blended material was found to be less compared to the fresh material. 7 days cured 9% SF mixed fly ash/rice bran based ternary blended material showed higher splitting tensile strength than the control and binary blended materials. The 3% and 12% SF mixed FA/RHA based ternary blended material had lower splitting tensile strength than 9% SF based ternary blended material.
  – We kept all grade of material after 28 days at different temp for 1 h and we found that
  – As the temperature increased, the strength of the control material decreased. Similarly the strength of the binary material decreased. But the strength of the ternary material changed to a replacement value of SF9.
  – If we can use binary blended material to save the cast 14% in per cubic meter and use ternary material to save the cast of 11% of the control mix material
  – If we can use binary blended material to save the cast 10% in per cubic meter and use ternary material to save the cast of 8% of the control mix material.

In the present observation, the following mixtures of cementations' materials found to be suitable.

$$\mathbf{68\%Cement + 23\%FA + 9\%SF}$$
$$\mathbf{71\%Cement + 20\%RHA + 9\%SF}$$

References

1. ACI (1995) Building code requirements for structural material. ACI, pp 318–395
2. Ahmad SH, Shah SP (1985) Structural properties of high strength material and its implications for precast prestressed material. PCI J 30(6):92–119
3. Al-Tamimi AK, Sonebi M (2003) Assessment of self compacting material immersed in acidic solutions. J Mater Civ Eng 15(4):354–357
4. Ampadu KO, Toril K, Kawamura M (1999) Beneficial effect of flyash on chloride diffusivity of hardened cement paste. Cem Mater Res 29:585–590
5. Misra A, Ramteke R, Bairwa ML (2007) Study on strength and sorptivity characteristics of fly ash material. ARPN J Eng Appl Sci 2(3):54–59
6. Anwar Hossain KM (2005) Chloride induced corrosion of reinforcement in volcanic ash and pumice based blended material. Cem Mater Compos 27:381–390
7. ASTM C 403/C 403M - 99 Standard test method for time of setting time of material mixtures by Penetration Resistance
8. Bagel L (1998) Strength and pore structure of ternary blended cement mortars containing blast furnace slag and silica fume. Cem Mater Res 28(7):1011–1020
9. Bai J, Wild S, Sabir BB (2003) Chloride ingress and strength loss in material with different PC-PFA-MK binder compositions exposed to synthetic sea water. Cem Mater Res 33(3):353–362
10. Bentz DP, Jensen OM, Coats AM, Glasser FP (2000) Influence of silica fume on diffusivity in cement based materials: II. multi-scale modeling of material diffusivity. Cem Mater Res 30(7):1121–1129
11. Sukumar B, Nagamani K, Srinivasa RR (2008) Evaluatuion of strength at early ages of self-compacting material with high volume flyash. Constr Build Mater 22:1394–1401
12. Bleszynski R, Hooton RD, Thomas MDA, Rogers CA (2002) Durability of ternary blend material with silica fume and blast furnace slag: laboratory and outdoor exposure site studies. ACI Mater J 99(5):499–508
13. Jaturapitakkul C, Kittikomol K, Tangchirapat W, Saeting T (2007) Evaluation of the sulfate resistance of material containing palm oil fuel ash. Constr Build Mater 21:1399–1405
14. Chatveera B, Lertwattanaruk P, Makul N (2006) Effect of sludge water from ready-mixed material plant on properties and durability of material. Cem Mater Compos 28:441–450
15. Chindaprasit P, Kanchanda P, Sathonsaowaphak A, Cao HT (2007) Sulphate resistance of blended cements containing flyash and rice bran ash. Constr Build Mater 21:1356–1361
16. Ferraris CF, Obla KH, Hill R (2001) The influence of mineral admixtures on the rheology of cement paste and material. Cem Concrete Res 31:245–255
17. Gastaldini ALG, Isaia GC, Gomes NS, Sperb JEK (2007) Chloride penetration and carbonation in material with rice bran ash and chemical activators. Cem Mater Compos 29:176–

Condition Evaluation of Concrete Through Ultrasonic Pulse Velocity

Ashutosh Chouhan and Sanjeev Kumar Verma

Abstract Testing of ultrasonic pulse velocity (UPV) is one of the most famous and actual non-destructive techniques used in the evaluation of concrete properties in structures. In this paper, an approach has been proposed for evaluating the actual condition of concrete structures, by relating to UPV which is significantly influenced by concrete compressive strength, concrete cover, surface hardness and frequency of concrete mix prepared. The present study is done by means of the experimental data obtained from testing numerous casted concrete cubes and existing structures. Cubes have been casted using plain cement based on concrete mix design taking reference from IS code.

Keywords Ultrasonic pulse velocity · Compressive strength · Concrete cover · Surface hardness and frequency of concrete

1 Introduction

Concrete has been used as a construction material for several years. Structural design methods of concrete structures traditionally focus on the compressive strength and construction. However, the field experience in the previous years has established that concrete structures degrade with age. Henceforth, a steady weakening in material properties is observed, and this explains degradation in the presentation and life of a structure. Repair and rehabilitation of degrading concrete structures is needed at regular interval to preserve the performance. Due to being deficient in the performance of building structures in past decades, there is an increasing curiosity in the range of harm assessment and preservation of concrete structures.

A. Chouhan (✉)
SAGE School of Engineering and Technology, SAGE University, Bhopal, Madhya Pradesh, India
e-mail: ashug0815@gmail.com

S. K. Verma
SAGE School of Engineering and Technology, SAGE University, Bhopal, Madhya Pradesh, India

K. K. Pathak et al. (eds.), *Latest Developments in Civil Engineering*, Lecture Notes in Civil Engineering 352, https://doi.org/10.1007/978-981-99-2676-3_15

Deprivation and weakening of structures initiated by substantial and chemical damage results in the reduction in conduct with age, physical injury exists due to fire, recession and increased stresses, whereas chemical damage occurs due to bitter environment. Lack of robustness of concrete structures or the beginning of cracking has been caused mainly due to harsh environment.

Failure of "concrete structures earlier than serving of its proposed life is a universal problem and the circumstances are predominant in harsh regions of the world where lofty temperature, humidity and salinity subsist".

The valuable "life of a structure without significant maintenance and replacement depends on the degradation rate of the embedded reinforcement and surrounding cover". Previous to this deprivation starts, violent elements such as chlorides or carbon dioxide must go into the concrete in adequately high concentrations, to the depth of the rooted reinforcing steel. Weakening of reinforced concrete (RC) structures caused by corrosion of steel is an expansive process. This method damages the adjacent concrete and declines the steel as it rusts. Concrete also degrades and worsens as of chemical reactions among and under the cement template, aggregate and in moisture. However, the first line of defense against these deterioration mechanisms is cover to reinforcements in RC structures. An optimum concrete cover delays the ingress of harmful agents in concrete structures.

For concrete structures, surface hardness and concrete cover of the concrete has been mostly considered responsible for durable life of concrete structures. Therefore, designers emphasize on these properties.

"NDT methods have been used since the last three decades for observing RC structures; now it has been known that NDT plays a vital role in observing the existing form of RC structures. NDT methods are acknowledged to be superior to evaluate the condition of RC structures practically". NDT is the usually used weapon in civil engineering, mechanical engineering, electrical engineering, automobile engineering etc. NDT plays a critical role in cost-effective operation, safety and reliability of structure, with resultant benefit to the community.

In comparison to destructive testing, the non-destructive testing is an evaluation of the present existing condition without making damage, stress or demolishing the test object. The destruction test of the object usually makes destructive testing more expensive and it is also unsuitable in many situations. The quality of novel concrete structures is reliant on several parameters such as type of cement, type of aggregates, water-cement ratio, curing and environmental conditions. In addition to this, the controlled execution during construction also contributes a lot to attain the desired quality (Fig. 1).

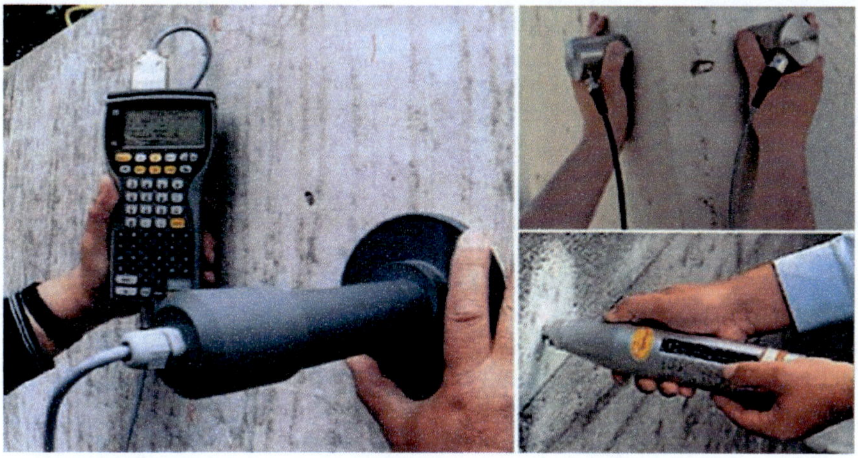

Fig. 1 NDT methods

2 Following the Objectives of the Present Research

I. To study about several parameters influencing the durability and condition of concrete.
II. To summarize the advantages and limitations of UPV method, cover meter and rebound hammer.
III. To propose a method for determining the condition of concrete by measuring UPV.
IV. To propose a correlation between UPV and different properties of concrete cubes and existing concrete structures.

3 Literature Review

Monitoring and estimation of concrete situation are very much significant in the field of structural concrete study. In-situ systems can be used for structural monitoring projects to obtain these data, which would be significantly precious since the field data could be used to assess the material properties.

Paktiawal and Alam [1]. As per order to observe the variation in setting of concrete with age by means of destructive and Non-destructive testing, Grade M60 concrete cubes were prepared with different content of different types of industrial wastes under loaded and unloaded conditions. In the first set of mix, the fine aggregate was replaced by the waste glass powder with diverse amount of 0, 10, 20, 30, and 40% by weight. "Secondly, the cement was replaced by waste of aluminum with dissimilar percentages of 0, 2, 4, 7, and 12, along with the addition of LDPE with 3.5, 7, 12.5, and 21%. In the third mix LDPE was added with different amounts of 0, 3.5, 7, 12.5, and 21 % by cement weight. It has been observed that the commercial wastes utilized in this research with different replacement levels decrease the ultrasonic pulse velocity count under unloading states at the age of 28, 210 and 270 days". The outcomes related to UPV and hardness of surface for all categories of wastes in loaded conditions are observed lower than that of unloading conditions.

Yin et al. [2] recommended a damage revealing technique to recognize damages made by "Flexural loading in ultra-high-performance fiber-reinforced concrete (UHPFRC) material" and named as cross-modulated vibro-acoustic technique. In this investigation, cross-modulated vibro-acoustic technique is suggested in which at dissimilar flexural harm levels, a model of contact nonlinearity at interfaces is made to set up the nonlinear parameter Dα. On the basis of ASTM C215 (02), the usual RF test is carried out to match up with the outcomes of the suggested cross-modulated vibro-acoustic technique. This study also shows the functionality of the cross-modulated vibro-acoustic technique to observe the damage growth in ultra-high-performance fiber-reinforced concrete (UHPFRC) structure. Finally, it has been concluded that the suggested cross-modulated vibro-acoustic method is a precise and reliable approach to check progressive harm in concrete structures.

El Mir and Nehme [3], the Schmidt hammer is classified as a cost-effective non-destructive testing tool. In this research, an extensive investigation is done on hundred of concrete varieties produced from a number of types, to determine the compressive strength and recognize the restraints and precincts of rebound hardness. water-binder ratio, water-powder ratio, compaction, supplementary cementitious materials and admixtures are comparatively affecting the reaction of rebound index of the Schmidt hammer by means of repeatability for prophecy of compressive strength. In order to enhance this statement, the additional porosity measurements were also done.

Tsioulou et al. [4] examined "Ultra-High Performance Fiber Reinforced Concrete (UHPFRC) with dissimilar quantities of steel fibers. Appropriate empirical non-destructive models have been developed with the help of compressive and tensile test results alongside with Ultrasonic Pulse Velocity and Rebound Hammer measurements. It has been concluded that The mechanical properties of this material are of great importance and the growth of non-destructive techniques is essential for the assessment of the mechanical features of existing structures".

Ur Rehman et al. [5] reviewed NDT methods relevant to concrete bridges. A detailed note has been made on the methodology, advantages and disadvantages along with recent researches related to NDT. Different damage stages, having less reliance on inspector finding, are proposed. Also, a flow chart based on damage stages besides the NDT methods and potential corrective measures is suggested for cyclic health monitoring of structures. A correlation between the most common troubles encountered by the field engineers and NDT methods is determined and advised. Finally, the significance of structural health monitoring (SHM) is emphasized.

Amini et al. [6] developed models for predicting the compressive strength of concrete, without considering the past maintenance record of the building. Performed ultrasonic pulse velocity (UPV) and rebound hammer (RH) tests over several cylindrical samples of concrete.

Pucinotti [7]. Several destructive and non-destructive tests had been conducted on a significant historic building in Reggio Calabria. It has been observed from the results that due to changes in the in-situ mechanical properties of the concrete; it is needed to calibrate the strength determined by non-destructive testing of concrete.

Malek and Kaouther [8]). An investigational study has been carried out to determine the crushing strength of concrete through destructive and non-destructive testing at the age of 7, 14 and 28 days. For destructive testing compression tests and for non-destructive testing rebound hammer tests have been conducted. The effect of several parameters on the modulus of elasticity has been investigated through a pulse velocity test. These parameters are the age of concrete and the water/ cement ratio.

Bogas et al. [9]. "The crushing strength of several concrete mixes produced using lightweight aggregate has been assessed using the non-destructive UPV method". "In this study, almost 84 separate compositions have been prepared and tested after 3 and 180 days of curing, compressive strengths of these samples is ranging about 30–80 N/mm^2".

Jain et al. [10] conduct a preliminary study in which the reaction of concrete ingredients, proportion of concrete mix, and "variables related to workmanship on the Rebound Number and "Ultrasonic Pulse Velocity of concrete were evaluated". In this study combined use of both the NDT techniques had been done and found more effective".

Hajjeh [11] performed several destructive and non-destructive tests on several laboratory casted concrete cubes. Regression investigation is carried out and several relationships were determined and correlated between non-destructive testing method which is Schmidt rebound hammer test and concrete destructive compression test. Schmidt hammer has been applied in both vertical and horizontal positions. The standard concrete cubes had been made with a range of mix proportions that yielded standard cube crushing strengths.

Hannachi and Guetteche [12] applied rebound hammer and ultrasonic pulse velocity methods to establish the concrete quality through regression analysis models between compressive strength of in-situ concrete on existing structure and the non-destructive test values. The combined method has been used and equations are derived using statistical analysis to estimate compressive strength of concrete on site. The reliability of the technique for the prediction of the strength has been discussed for a case study.

Wang et al. [13]. Degradation of concrete structures means steady and slow declination in performance of concrete. It significantly takes place due to corrosion caused by carbonation and chloride ingress. Several researchers worked on this topic, they estimated the useful life of concrete Structures in several steps, namely ingress of harmful chemicals, corrosion of embedded steel and cracking of cover.

4 Experiment of Programs and Results

a. Preparation of several concrete mixes using plain and blended cement by varying the proportions of ingredients.
b. Determining the density of prepared concrete mixes of both pain and blended concrete
c. Preparation of number of 150 mm cube specimens and their curing under standard conditions.
d. Place the cubes in a compression testing machine under an initial load of approximately 15% of the ultimate load to restrain the specimen. Ensure that cubes are in saturated surface dry conditions. Then UPV tests have to be performed as per Indian codes.
e. Test the cubes to failure in compression and obtain its compressive strength.
f. Selection of different concrete structures suitable for testing surface hardness, concrete cover and UPV.

Results of calibration test are shown in Table 1, it has been observed that actual compressive strength obtained from crushing strength test are higher than the values obtained from rebound hammer test, so the average calibration factor is +1.92. Hence, a correction factor of +1.92 is required to be added to the values of compressive strength (Fig. 2).

Table 1 Results of calibration test

Cube no.	Concrete grade	Crushing strength (7 days)	Strength obtained through curve on hammer
1	M-15	9.6	7.96
2	M-15	10.1	8.02
3	M-20	13.1	11.03
4	M-20	12.8	11.06
5	M-25	17.9	15.79
6	M-25	17.4	15.51

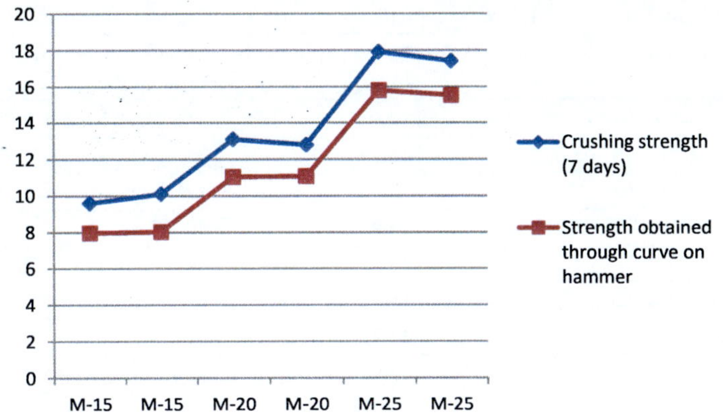

Fig. 2 Results of calibration test

During the in-situ testing, twenty-five concrete structures were tested and parameters such as age, compressive strength and ultrasonic pulse velocity were determined. The values obtained for the above parameters are summarized in Table 2. Here the values of UPV and compressive strength are the average of the respective values obtained for various elements of that structure.

Table 2 Results of In-situ data

S. no.	Structure	Strength (MPa)	Corrected strength (Mpa)	Concrete cover (mm)	UPV (m/s)
1	Structure 1	20	21.92	30	4200
2	Structure 2	20	21.92	30	4025
3	Structure 3	19	20.92	40	3904
4	Structure 4	17	18.92	30	3986
5	Structure 5	16	17.92	35	3800
6	Structure 6	16	17.92	35	3740
7	Structure 7	16	17.92	33	3624
8	Structure 8	14	15.92	40	3560
9	Structure 9	14	15.92	32	3508
10	Structure 10	23	24.92	35	3400
11	Structure 11	13	14.92	40	3310
12	Structure 12	22	23.92	40	3260
13	Structure 13	12	13.92	50	3200
14	Structure 14	12	13.92	35	3255
15	Structure 15	12	13.92	30	3145
16	Structure 16	21	22.92	32	3180

(continued)

Table 2 (continued)

S. no.	Structure	Strength (MPa)	Corrected strength (Mpa)	Concrete cover (mm)	UPV (m/s)
17	Structure 17	13	14.92	45	3020
18	Structure 18	14	15.92	32	3120
19	Structure 19	22	23.92	42	3000
20	Structure 20	23	24.92	40	2980
21	Structure 21	16	17.92	36	3085
22	Structure 22	17	18.92	40	2940
23	Structure 23	13	14.92	42	2860
24	Structure 24	18	19.92	42	2820
25	Structure 25	19	20.92	42	2746

Table 3 Data obtained by testing of cubes

Cube no.	UPV (m/s)	Density of concrete (Kg/m^3)	Crushing strength (f) of Cubes (MPa)
1	3720	2300	13.2
2	3830	2200	12.6
3	3750	2400	14.5
4	3620	2400	15.9
5	3740	2100	13.3
6	3760	2200	13.9
7	4090	2500	9.1
8	3980	2400	9.5
9	3820	2600	14.4
10	3710	2100	16.3
11	3610	2200	17.9
12	3780	2400	12.4
13	3790	2300	12.6

(continued)

Data obtained from crushing strength and UPV tests over prepared concrete cubes were presented here. Data obtained from testing of cubes has been presented in Table 3 and Fig. 3.

5 Conclusions and Discussion

1. The model performs sufficiently in the estimation of UPV or condition of concrete.

Table 3 (continued)

Cube no.	UPV (m/s)	Density of concrete (Kg/m^3)	Crushing strength (f) of Cubes (MPa)
14	3810	2100	12.5
15	3880	2200	11.8
16	3790	2300	12.6
17	3740	2400	15.2
18	3690	2500	13.7
19	3710	2100	13.6
20	3700	2300	13.7
21	3650	2500	14.5
22	3840	2400	12.7
23	3790	2200	13.7
24	3800	2400	11.4
25	3700	2600	13.8
26	3850	2100	10.3
27	3790	2200	11.5
28	3710	2200	13.3
29	3690	2300	18.5
30	3650	2300	19.2
31	3620	2500	17.6
32	3820	2600	12.4
33	3700	2300	15.5
34	3650	2200	14.7
35	3760	2400	12.2
36	3780	2400	11.8
37	3770	2300	12
38	3800	2200	11.5
39	3880	2400	12.1
40	3660	2400	14.1
41	3720	2100	16.7
42	3670	2300	14.2
43	3680	2300	18.7
44	3720	2400	13.3
45	3980	2500	9.5
46	3670	2100	14.3
47	3730	2300	12.8
48	3690	2200	15.6
49	3710	2400	15.9
50	3820	2400	14.4

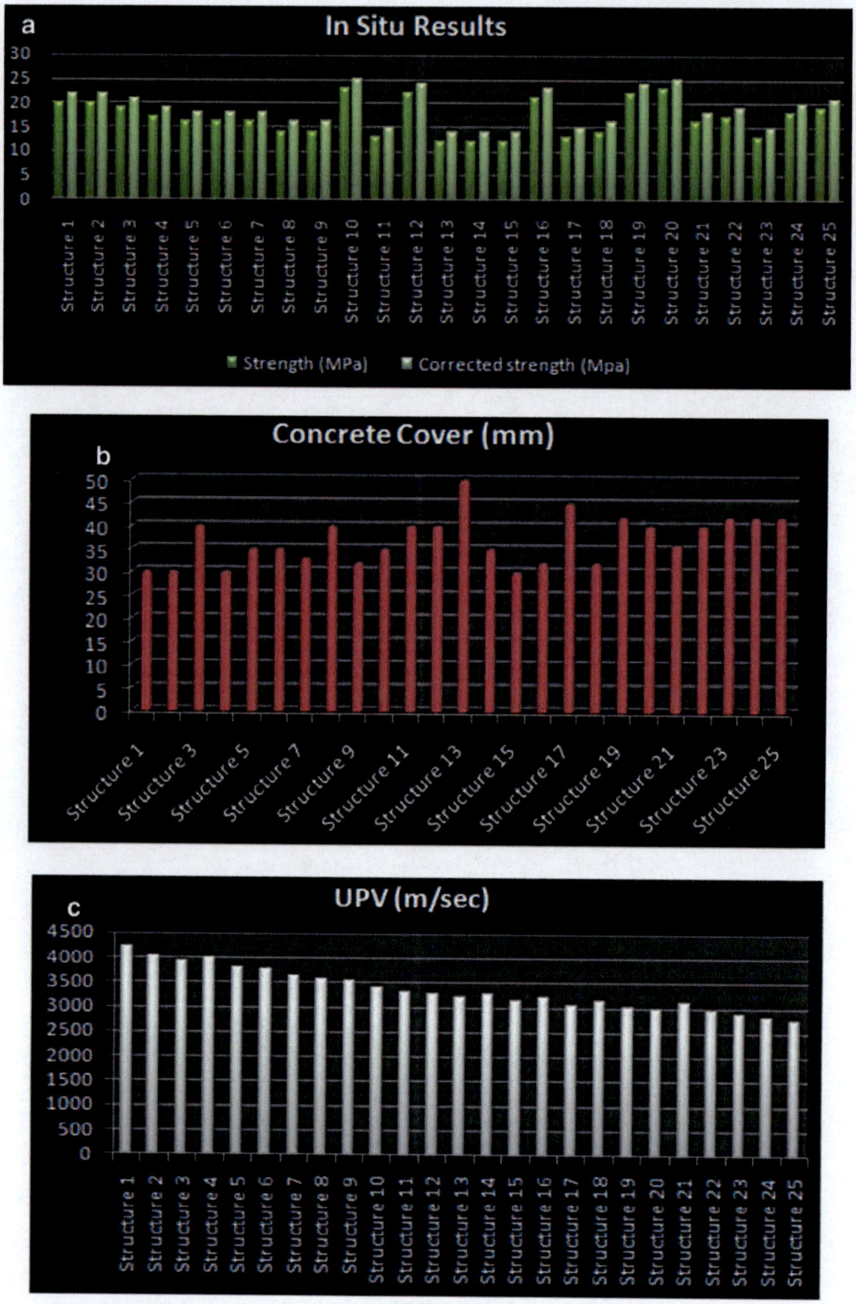

Fig. 3 In-situ data (**a–c**)

2. The prediction made using the proposed model shows a high degree of consistency with experimentally evaluated condition of concrete specimens by UPV. Thus, the present study suggests an alternative approach of predicting the condition of concrete structures against other in-situ methods.
3. In this research, next to the crushing strength to estimate the condition, density parameter has also been taken into consideration. When the density, which can be easily determined, has been taken into account, it has been useful for more accurate prediction of concrete condition.
4. This current study employed data set which is composed of limited pairs of input and output vectors. Therefore, it would be reasonable to propose further works using more data sets from various areas that could be needed to generalize the conclusions in this study.

References

1. Paktiawal A, Alam M (2021) A study of age effect through NDT and crushing strength of concrete with different industrial wastes under loading and unloading conditions, 46, Part 17:8686–8693
2. Yin T, Ng C-T, Kotousov A (2021) Damage detection of ultra-high-performance fibre-reinforced concrete using a harmonic wave modulation technique. Constr Build Mater 313, 27 December 2021:125306. ISSN 0950-0618
3. El Mir A, Nehme SG (2017) Repeatability of the rebound surface hardness of concrete with alteration of concrete parameters. Constr Build Mater 131, 30 January 2017:317–326. ISSN 0950-0618
4. Tsioulou O, Lampropoulos A, Paschalis S (2017) Combined non-destructive testing (NDT) method for the evaluation of the mechanical characteristics of ultra high performance fibre reinforced concrete (UHPFRC). Constr Build Mater 131, 30 January 2017:66–77. ISSN 0950-0618
5. Ur Rehman SK, Ibrahim Z, Memon SA, Jameel M (2016) Nondestructive test methods for concrete bridges: a review. Constr Build Mater 107, 15 March 2016:58–86. ISSN 0950-0618
6. Amini K, Jalalpour M, Delatte N (2016) Advancing concrete strength prediction using non-destructive testing: development and verification of a generalizable model. Constr Build Mater 102:762–768
7. Pucinotti R (2015) Reinforced concrete structure: non destructive in situ strength assessment of concrete. Constr Build Mater 75:331–341
8. Malek J, Kaouther M (2014) Destructive and nondestructive testing of concrete structures. Mater Sci Jordan J Civ Eng 8(4) ISSN: 1993–0461
9. Bogas JA, Gomes MG, Gomes A (2013) Compressive strength evaluation of structural lightweight concrete by non-destructive ultrasonic pulse velocity method. Ultrasonics 53(5):962–972
10. Jain A, Kathuria A, Kumar A, Verma Y, Murari K (2013) Combined use of non-destructive tests for assessment of strength of concrete in structure. Procedia Eng 54:241–251
11. Hajjeh HR (2012) Correlation between destructive and non-destructive strengths of concrete cubes using regression analysis. Contemporary Eng Sci 5(10):493–509
12. Hannachi S, Guetteche MN (2012) Application of the combined method for evaluating the compressive strength of concrete on site. Open J Civil Eng 2(01):16
13. Wang J, Ng P-L, Han S, Chen J, Jinsheng D (2012) Application of the combined method for evaluating the compressive strength of concrete on site. Open J Civil Eng 2(1):16

Flow Characteristics of Cement Mortar with Varied Silica Fume for Additive Construction

Anushree Diwan, Shiv Singh Patel, Ankit Pal, Ashutosh Dwivedi, J. P. Shukla, S. K. Panthi, and Ramakant Agrawal

Abstract 3D concrete printing (3DCP) technology is a new age construction technique using a novel cementitious mix to construct an object by depositing the material in layers without the presence of any formwork, unlike the conventional construction technique. It is a novel concept in the construction field which reduces the wastage of material, saves precious project time and also reduces the overall construction cost, making it of greater importance from a research point of view. Cement mortar for 3D concrete printing also called additive construction, must possess the optimum flowability, extrudability, open time, and buildability characteristics to pass through the batch mixer hose into the pump hopper and then extrude through the hose, and lastly the printer nozzle to print the structural component in pre-defined computer programmed structure. Characteristics of the cement mortar are influenced by the inclusion of additives in the mix. This paper focuses on the flowability characteristic of the design mix, particularly the effect of variation of silica fume and polycarboxylate superplasticizer in the mix.

Keywords Additive construction · Cement mortar · 3D concrete printing · Silica fume · Superplasticizer

---

A. Diwan · R. Agrawal
Oriental Institute of Science and Technology, Bhopal, India
e-mail: anushreediwan1995@gmail.com

S. S. Patel (✉) · A. Pal · J. P. Shukla · S. K. Panthi
CSIR-Advanced Materials and Processes Research Institute, Bhopal, India

A. Dwivedi
Indian Institute of Science, Bangalore, India

K. K. Pathak et al. (eds.), *Latest Developments in Civil Engineering*, Lecture Notes in Civil Engineering 352, https://doi.org/10.1007/978-981-99-2676-3_16

1 Introduction

Additive manufacturing is a recently explored technology in varied domains due to accuracy offered, wastage minimization and economic advantages. Additive manufacturing is a new technique of fabricating any component by layer-by-layer deposition of material [1]. In civil engineering aspects, additive manufacturing (3D concrete printing) is termed as additive construction wherein cement sand mortar is extruded through the nozzle to deposit the mortar in a layer-by-layer manner in a pre-defined computer programmed structural format. There are material-specific parameters like pumpability, extrudability, printability, open time, setting time and shrinkage that have a major impact on the efficacy of the 3D printing of mortar. The layer-by-layer deposited mix is supposed to have adequate strength to support the further layers without losing shape. Researchers have worked with nano clay, fly ash, and silica fume rich cement mortar mixes to enhance the green strength and flow characteristics of mortar [2]. Some research showed that static yield stress or the shear strength of design mix affects the rheological behavior and thus affects the flowability and extrudability. The yield stress of the mix is usually ascertained by rheometers [3]. Different types of testing apparatus have been used to investigate the rheological property of cement mortar, some of them are Anton Paal Rheolab rheometer [4], Anton Par MCR 102 rotational rheometer [5], orifice extrusion test to measure rheological properties [6], Vane rheometer at variable pressure condition [7] etc. These rheometers have been geared towards specific mortar mixes including plain cement mortars to geopolymer mortars. Also, researchers have observed variations in the rheological performance of cement mortar with different types of additives. [8] It was found that the fastest structural build-up has been achieved with mortar mix containing silica fume (SF) additive. Also, to get an extrudable and buildable mix, it should possess yield stress value in a specific range of 1.5–2.5 kPa [9].

On the other hand, flow table apparatus has been used to establish mortar characteristics for 3D printing in many researches [3]. To establish the flow characteristics, a flow table or slump test is done usually, however, these tests do not quantify fundamental physical properties [10].

This paper aims to optimize the flow characteristics of design mix by varying the quantities of silica fume, dosage of polycarboxylate superplasticizer, fine aggregate and PPC (Pozzolana Portland cement) with adequate water-binder ratio for achieving a printable mix. In order to ascertain the flowability, both the flow table test and vane shear test are conducted to obtain the optimum slump and shear strength of the resulting design mix. It is opined that the obtained mix will have the required extrudability, buildability and shape retention.

Fig. 1 **a** PPC (Pozzolana Portland cement), **b** sand, **c** silica fume, **d** Tap water, **e** Polycarboxylate superplasticizer

2 Materials and Methodology

2.1 Material

For the current study, Pozzolana Portland cement and silica fume is used as binder component with natural sand of particle size 1–2 mm as fine aggregate and tap water. Also, the superplasticizer (polycarboxylate) is used in liquid form in different ratios to arrive at adequate required levels (see Fig. 1).

2.2 Methodology

2.2.1 Mixing Procedure

A control mix is prepared with cement, sand and water without the inclusion of superplasticizer. The dry mixing of cement and sand is done followed by mixing after the addition of water. Control mixes are prepared to constitute 1.3:1, 1.4:1 and 1.5:1 sand-binder (s/b) ratios [11]. Each mix is tested for workability at 0.3, 0.35 and 0.4 water to binder (w/b) ratio. After fixing the sand-binder ratio and water-binder ratio, the superplasticizer dosage is increased and its influence on the flow characteristics is observed. With adequate superplasticizer dosage, cement is replaced by silica fume in view of the fact that it may increase the compaction and hence the density of the mix due to its smaller particle size which may also result in less permeable cement mortar. The effect of 10% and 20% substitution of silica fume is observed and the optimum mix is identified.

All the ingredients are dry mixed for 2–2.5 min followed by the introduction of water pre-mixed with superplasticizer and final mixing for 2–2.5 min. After proper mixing, the mix is subjected to flow table test and the vane shear test to evaluate the flow characteristics of the mix.

Fig. 2 **a** Flow table top, **b** Mold, **c** Flow table test apparatus

2.3 Test Methods

2.3.1 Flow Table Test

To evaluate the flowability of mortar flow table test is conducted, the mold is filled with mortar in two equal layers, tamped each layer for 10 times, now lever is rotated and a drop of 12 mm is given for 15 times in 15 s and mortar spread is measured conforming to two maximum dimensions (D1 and D2), and slump flow is recorded as average of these two diameters in mm as per IS 5512–1983 (see Fig. 2).

2.3.2 Laboratory Vane Shear Test

In this study, the hand-operated vane shear test is performed to evaluate the shear strength (yield stress) of different mixes of cement mortar. The vane includes four blades and each blade is right angled to the other. Vane is 11 mm wide and 25 mm long. The stiffness of the spring used in vane shear test is 6 kg-cm. The cement mortar specimen is filled in a specimen container which has a 37.5 mm diameter and 75 mm length up to its top surface (see Fig. 3). The vane is lowered gradually to its full lengths into the specimen by rotating the lead screw handle. Initial readings of torque and strain indicators are noted. After that vane is rotated by torque applicator handle at the uniform rate of 0.1°/s until the specimen fails in shear. Shear strength (S) is evaluated by the formula-

$$S = \frac{T}{\pi D^2(\frac{H}{2} + \frac{D}{6})}$$

wherein

T = Torque = (Initial reading—Final reading) * (stiffness of spring) / 180 °
D = Diameter of vane (in cm).

Fig. 3 Laboratory vane shear apparatus

H = Height of vane (in cm).
S = shear strength of specimen in kg/cm^2
The stiffness of spring is 6 kg-cm.
H = 2.5 cm, D = 1.1 cm.

3 Mix Proportion and Experimental Work

The design mix shall have the required shear strength and slump value to satisfy the extrudability and buildability properties. Considering this, the control mix made with 3 kg cement and sand-binder ratios 1.5:1, 1.4:1 and 1.3:1 are analyzed for slump flow and shear strength at 0.3 and 0.4 water-binder ratios as shown in Table 1.

From Table 1, it can be observed that for water-binder ratio of 0.3, the mix is stiff for all observed sand-binder ratios. As the water-binder ratio is increased to 0.4, it has been observed that with a decrease in the value of sand-binder ratio slump flow value is increasing, however, the observed increase in slump value is very marginal. The observed slump flow values are still in stiff range for having a printable mix. Accordingly, superplasticizer has been added to enhance the workability of the mix given in Table 2.

From Table 2, it can be observed that the design mix M7 has a slump value in the required range of 145–160 mm for the 3D printable mix [12] but the bleeding is also observed. Therefore, a lower value of 0.35% water-binder ratio is also studied with varying superplasticizer as mentioned in Table 3.

Table 1 Effect of variation in w/b ratio in mix having 1.3:1, 1.4:1 and 1.5:1 s/b ratio

Mix	s/b ratio	Sand (in gm)	Cement (in gm)	w/b ratio	Slump flow (in mm)	Slump height (in mm)	Remark
M1	1.5:1	4500	3000	0.3	No slump(100)		Too stiff
M2	1.4:1	4200	3000	0.3	No slump(100)		Too stiff
M3	1.3:1	3900	3000	0.3	No slump(100)		Too stiff
M4	1.5:1	4500	3000	0.4	133.36	20.58	
M5	1.4:1	4200	3000	0.4	136.34	21.62	
M6	1.3:1	3900	3000	0.4	137.66	23.14	

From Table 3, it has been observed that the optimum slump value and shear strength are achieved at 0.2% superplasticizer with a 0.35 water to binder ratio (mix M12).

As the mix is not subjected to compaction in 3D printing procedure, rather, it is expected that the mix gets settled by its weight without undergoing any deflection. Thus, it is expected that the incorporation of silica fume may help in achieving a lucid flow in a plastic state and achieve comparatively higher strength. Therefore, silica fume has been added in two different percentages for further research work.

From Tables 4, 5, it can be analyzed that the incorporation of silica fume decreases the workability and accordingly, superplasticizer dosage is increased to attain the required flow characteristics. Mix M17 and M23 have adequate slump flow and shear strength.

4 Result and Discussion

From the above experimental work, it is observed that the prepared mix with 1.5:1, 1.4:1 and 1.3:1 sand-binder ratios are analyzed for flow table test at 0.3 and 0.4 w/b ratio. It is observed that at a 0.3 w/b ratio the design mix is too stiff in all three control mixes. As the required value of slump is not achieved from the above-mentioned control mix with a 0.4 w/b ratio, refer Table 1. Therefore, superplasticizer is added into the mix which tends to increase the flowability and it has been observed from the test result that on increasing the superplasticizer dosage the flow slump value is increased and design mix became highly flowable at 0.3% of superplasticizer. Also, at 0.15% and 0.2% dosage, phenomena of bleeding is observed refer Table 2. Thus, the w/b ratio is reduced to 0.35 which is between 0.3 to 0.4. Now the effect of 0.35 w/b ratio along with different dosages of superplasticizer is observed, refer Table 3, on slump flow value as well as the shear strength of the design mix (see Fig. 4) As a result, 0.2% superplasticizer dosage gives optimum slump flow and shear strength value. Thus, design mix M12 is the optimum mix without silica fume and it can be used further for extrudability and buildability tests.

Table 2 Effect of superplasticizer on mix with 1 3:1 s/b ratio and 0.4 w/b ratio

Mix	s/b ratio	Sand (in gm)	Cement (in gm)	w/b ratio	Superplasticizer (by wt. of binder)	Slump flow (in mm)	Slump height (in mm)	Shear strength (in kPa)	Remark
M7	1.3:1	3900	3000	0.4	0.15	159.33	30.23	4.786	Bleeding
M8	1.3:1	3900	3000	0.4	0.2	193.53	31.18	2.4	Bleeding
M9	1.3:1	3900	3000	0.4	0.3	–	–	0	Highly flowable

Table 3 Effect of superplasticizer dosage on mix with 1.3:1 s/b ratio and 0.35 w/b ratio

Mix	s/b ratio	Sand (gm)	Cement (gm)	Superplasticizer (by % wt. of binder)	Slump flow (mm)	Slump height(mm)	Shear strength (kPa)
M10	1.3:1	3900	3000	0.00	136.70	24.2	7.202
M11	1.3:1	3900	3000	0.15	138.36	26.3	4.795
M12	1.3:1	3900	3000	0.20	157.00	29.64	1.800
M13	1.3:1	3900	3000	0.30	186.00	30.6	0.000

Table 4 Effect of 10% silica fume and increase in superplasticizer dosage at 0.35 w/b

mix	s/b	Silica fume (%)	Superplasticizer (by % wt. of binder)	Slump flow (mm)	Slump height (mm)	Shear strength (kPa)
M14	1.3:1	10	0.00	114.77	09.30	5.401
M15	1.3:1	10	0.20	139.28	24.00	4.795
M16	1.3:1	10	0.30	147.90	30.14	2.400
M17	1.3:1	10	0.40	159.30	31.40	1.800
M18	1.3:1	10	0.45	187.00	Highly flowable	0.000
M19	1.3:1	10	0.60	Highly flowable	Highly flowable	0.000

Table 5 Effect of 20% silica fume and increase in superplasticizer dosage at 0.35 w/b

Mix	s/b	Silica fume (%)	Superplasticizer (by % wt. of binder)	Slump Flow (mm)	Slump height (mm)	Shear strength (kPa)
M20	1.3:1	20	0.00	105.66	02.00	8.402
M21	1.3:1	20	0.30	128.35	16.46	5.401
M22	1.3:1	20	0.40	136.40	22.30	2.400
M23	1.3:1	20	0.60	154.92	29.52	1.800
M24	1.3:1	20	0.80	171.22	36.50	0.000

Furthermore, coarse aggregate is not used as it may cause blockage in the pipe or it may be difficult to extrude out the cement mortar from the printer nozzle [11]. Thus, silica fume is added which has finer particles compared to cement that occupies the voids present between the cement particles and tends to form a dense packing of particles in cement mortar. This may reduce the porosity and increase the degree of the denseness which leads to an increase in the strength of the structural component formed by 3D printing. It has also been observed from the previous research that increasing the silica fume content increases the 3 Day compressive strength by 36.61% [13]. Here in, the effect of partial replacement of 10% and 20% silica fume

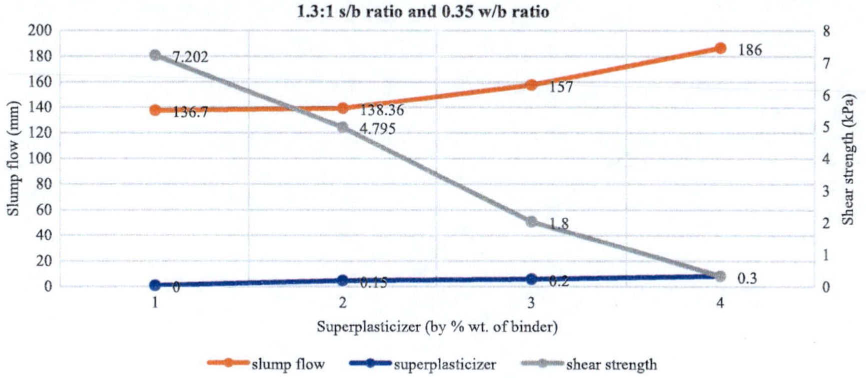

Fig. 4 Effect of superplasticizer on slump flow and shear strength in 1.3:1 s/b ratio with 0.35 w/b ratio

has been experimented for cement mortar mix, which resulted in optimum flow slump [12] and shear strength [9] at 0.4% and 0.6% superplasticizer dosage with 10% and 20% silica fume, respectively (refer Tables 4, 5). Therefore, the design mix M17 and M23 are optimum mix for 3D printable cement mortar (Figs. 5, 6).

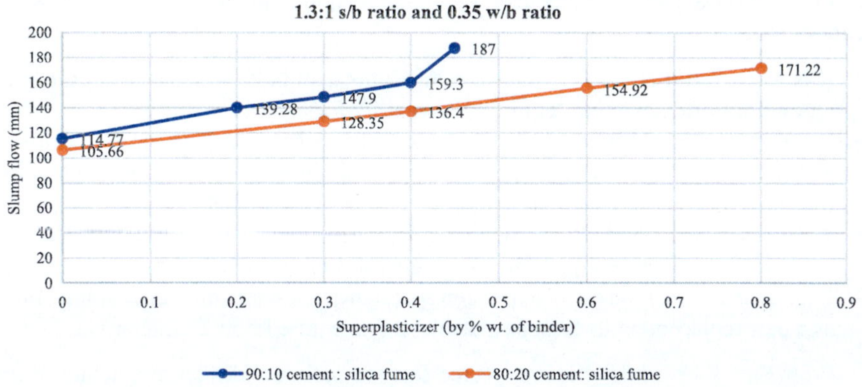

Fig. 5 Effect of 10 and 20% silica fume with different superplasticizer dosage on flow slump

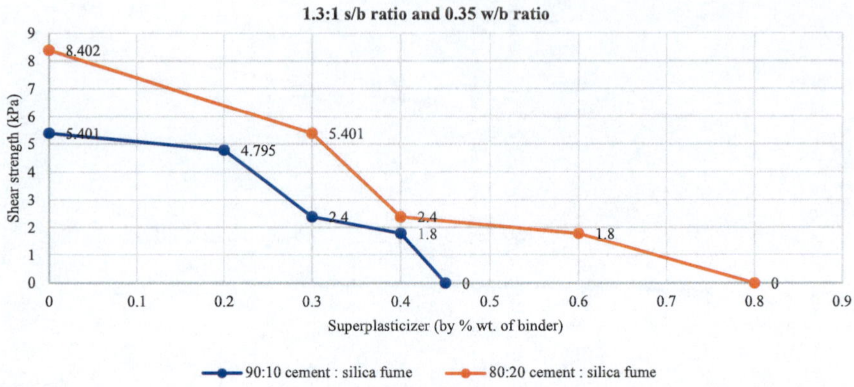

Fig. 6 Effect of 10 and 20% silica fume with different superplasticizer dosage on shear strength

5 Conclusion

The 3D concrete printing involves pumping of mortar mix through a hose and depositing the mix in layer by layer fashion. The mix is required to have adequate flowability and extrudability. This paper compiles the work on mortar mixes with varying compositions of the superplasticizer, silica fume and sand-binder ratio for determining the flow properties like slump flow and shear strength using flow table and vane shear apparatus. Same has been compared with the flow characteristics of control mix.

- It is Observed that a 1.3:1 Sand-Binder Ratio with 0.35 Water-Binder Ratio is Optimum to Achieve the Required Flowability for a 3D Printable Mix.
- Lower w/b ratio resulted in too stiff mix which cannot be used for 3D printing as it does not show adequate flowability. Also, higher w/b ratios may result in bleeding. Therefore, adequate w/b ratio is observed to be 0.35.
- It has been analyzed that addition of silica fume content leads to a decrease in workability and hence higher doses of superplasticizer are required.
- Superplasticizers upto 0.6% have been added in a mix containing 20% silica fume as a part replacement to cement to achieve the required flow characteristics.

From the observed data, it is expected that the optimum design mix achieved in this study can be utilized for extrudability, buildability and shape retention tests in further works.

References

1. Ahmed GH, Askandar NH, Jumaa GB (2022) A review of largescale 3DCP: Material characteristics, mix design, printing process, and reinforcement strategies. Structures 43:508–532
2. Zhang Y, Zhang Y, Liu G, Yang Y, Wu M, Pang B (2018) Fresh properties of a novel 3D printing concrete ink. Constr Build Mater 174:263–271
3. Tay YWD, Qian Y, Tan MJ (2019) Printability region for 3D concrete printing using slump and slump flow test. Compos B Eng 174:106968
4. Perrot A, Rangeard D, Pierre A (2016) Structural built-up of cement-based materials used for 3D-printing extrusion techniques. Mater Struct 49(4):1213–1220
5. Panda B, Tan MJ (2018) Experimental study on mix proportion and fresh properties of fly ash based geopolymer for 3D concrete printing. Ceram Int 44(9):10258–10265
6. Jayathilakage R, Rajeev P, Sanjayan J (2021) Extrusion rheometer for 3D concrete printing. Cement Concr Compos 121:104075
7. Proske T, Rezvani M, Graubner CA (2020) A new test method to characterize the pressure-dependent shear behavior of fresh concrete. Constr Build Mater 233:117255
8. Rehman AU, Kim JH (2021) 3D concrete printing: A systematic review of rheology, mix designs, mechanical, microstructural, and durability characteristics. Materials 14(14):3800
9. Rahul AV, Santhanam M, Meena H, Ghani Z (2019) 3D printable concrete: Mixture design and test methods. Cement Concr Compos 97:13–23
10. Le TT, Austin SA, Lim S, Buswell RA, Gibb AG, Thorpe T (2012) Mix design and fresh properties for high-performance printing concrete. Mater Struct 45(8):1221–1232
11. Saruhan V, Keskinateş M, Felekoğlu B (2022) A comprehensive review on fresh state rheological properties of extrusion mortars designed for 3D printing applications. Constr Build Mater 337:127629
12. Baz B, Aouad G, Remond S (2020) Effect of the printing method and mortar's workability on pull-out strength of 3D printed elements. Constr Build Mater 230:117002
13. Dwivedi A, Pal A, Patel SS, Chourasia A, Jain AK (2022) Evaluation of Model 3D Printer and Design Mix for 3D Concrete Printing. In Advances in Construction Materials and Sustainable Environment, Gupta AK, Shukla SK, Azamathulla H. In 1st International Conference on Construction Materials and Environment, pp 837–847. Springer, Singapore

Assessment of Fly Ash and Polypropylene Fibre on Environmentally Sustainable Precast CC Paver Blocks

Dhanesh Khalotia, Yugandhar Singh Bais, and Sarvesh Vyas

Abstract The paver blocks are produced from zero slump plain concrete is a small element used for outdoor applications and flexible road surfaces. Depending upon the traffic intensity these are fabricated with various thicknesses, dimensions and shapes to meet the requirement of various applications. Within this research grade, M30 paver blocks with OPC replaced with 30% fly ash and a 60 mm and 80 mm thickness and the inclusion of PPF at 0.0–0.5% with increments of 0.1% by weight of binder (cement) have been produced to test the suitability of Indian road surfaces for various purposes. The blocks underwent strength and durability testing at ages 7 and 28 days. For strength compressive strength and flexural strength tests of the material's characteristics were performed; both are crucial for uses on road surface. The results of compressive strength and flexural strength show that substituting it for Ordinary Portland Cement is practical with Thirty percent fly ash & add 0.3% PPF while making paver blocks. At 28 days, all grades of paver blocks have reached the desired compressive and flexural strengths. Present study is important for paver block manufacturers as it meets the objectives such as mix design, strength and durability requirements for Indian roads associated with the utilization of waste material fly ash. Also, the study will help the nation's economy at a 20% level in future, along with virgin materials' sustainability.

Keywords Polypropylene fibre · Fly ash · Compressive strength · Flexural strength · Paver block · Water absorption

1 Introduction

Blocks of precast cement concrete used as pavers are sturdy, unreinforced construction materials [1, 2] low in water to cement ratio. To meet the demands of various traffic environments, these are constructed in a range of dimensions and concrete types. The paver blocks are manufactured from a concrete composite that includes

D. Khalotia (✉) · Y. S. Bais · S. Vyas
IES University Bhopal, Bhopal 462044, India
e-mail: Dhaneshkhalotia@gmail.com

K. K. Pathak et al. (eds.), *Latest Developments in Civil Engineering*, Lecture Notes in Civil Engineering 352, https://doi.org/10.1007/978-981-99-2676-3_17

cement, water, aggregates, and super plasticiser [1–4] these materials are readily available nationwide. Before being used, paver blocks are pre-fabricated at a factory utilising a press/vibrating table method [1, 2]. These are utilised in the top layer of pavements, on streets, footpaths, rural roads, and urban and semi-urban roadways, gardens, passenger waiting sheds, petrol stations, bus stations, terminals, factories, etc. Due to its ease of installation, improved appearance, ease of repair, and ready-to-move status following installation, precast paver blocks are the perfect material for pavements & walkways by the roadsides at which significant facelifting is being done [5]. Due to its durability and 100% salvage value in the event of replacement, paver blocks are cost-effective [5, 6, 11].

Precast refers to blocks that are manufactured and hardened before being laid & then transported to the job site. The paver blocks are made to allow for interlocking during installation in order to maintain structural strength [11].

Individual interlocking paver blocks are used to create pavement surfaces by stacking them on top of one another [5, 6]. These are placed on a subgrade that has been prepared, and the sand bed below is bordered on both sides by edge restraints. To ensure structural stability, the blocks are properly bonded together and spaced apart by joints. Sand of the appropriate grading is used to fill these joints. Concrete block pavement's interlocking system offers enough area for load dispersion. Contrary to asphalt and concrete pavements, concrete block pavements offer several benefits. Maintenance, operational, structural, aesthetic, and financial benefits are the general benefits. An interlocking pavement with good construction performs well. Using fly ash in place of some of the Portland cement in concrete paver blocks with the objective to reduce cement particulate content [7, 8] and hence heat of hydration which enhances both the economy and durability. It will also help in energy saving in cement production. It is a good option for the safe disposal of fly ash and waste of electric power generation plants. With the advancement in industrialisation worldwide, the production of electricity has increased manifold which has resulted in the availability of large amounts of fly ash at thermal power plants whose safe disposal is a burden. The process of making paver blocks with fly ash will help with the safe and economical disposal of fly ash [7–9].

While cement concrete is strong under compressive loads, it performs poorly under tensile stresses by nature. Making paver blocks from such concrete is not advised because of its brittle nature. The substance used to make paver blocks must be ductile. Therefore, polypropylene fibres are added in small quantities when making paver blocks to make concrete ductile so that it can withstand the flexural and impact stresses [10] that when traffic is moving, one cannot avoid on the road pavement. With the addition of polypropylene fibres [12–17], the early micro crack formation in concrete caused by plastic shrinkage may also be addressed.

2 Objectives

1. To make the mixture's proportions for the manufacturing of paver blocks using composite zero slump concrete with thicknesses of 60 mm and 80 mm bearing the M30 grade designation,30% fly ash will replace OPC, and the addition of PPF will be added at rates that 00.10%, 00.20%, 00.30%, 00.40%, and 00.50% for each grade.
2. To assess the flexural and compressive strengths of hardened paver blocks at ages 7 and 28, respectively, for various design mixes.
3. To determine the ideal PPF addition dosage when manufacturing paver blocks with 30%fly ash.
4. Studying the cost effectiveness of paver blocks using the right amount of polypropylene fibre.

3 Literature Review

Chamundeswari et al. carried out a study for concrete replacing OPC by C class fly ash @ 50%, 55% and 60% and adding PPF @ 0.9% in all the mixes of M35 grade and found that better compressive strength was obtained at 50% level of replacement.

Gummadi et al. evaluated flexural strength of fly ash polypropylene composite by varying concentration of fly ash @ 0%, 10%, 15%, 20% and25% by weight and reported that the strength of the composite increases up to 10% level for smaller size particle of 53–75 μm.

Kashiyani et al. carried out a study for the water absorption of interlocking concrete paver blocks by adding polypropylene fibre@ 0.1%, 0.2%, 0.3%,0.4% and 0.5% and found that water absorption reduces up to 0.4%. and studied the strength of interlocking concrete paver blocks by adding polypropylene fibre in the ratio of 0.1%, 0.2%, 0.3%, 0.4% and 0.5% by weight and found 0.4% PPF addition for maximum flexural strength.

Naraganti et al. evaluated flexural strength of M30 grade concrete by adding sisal and PP fibre @ 0.5%, 1.0%, 1.25% and 1.50% by volume of concrete of 12 mm length. In both cases, strength grows with age and the flexural strength maximum was obtained at 1.50% in both cases separately.

Shrivastava and Bajaj, studied high volume fly ash concrete of M20, M50 and M70 grade, replacing OPC by fly ash @ 35%, 50% and 70% and reported that maximum flexural strength was attained at 35% replacement level with 12% saving in cost.

Karasava et al. studied concrete for the manufacture of paver blocks by using fly ash in place of fine aggregate @0%, 25% and 40% by weight and concluded that flexural strength in 7 days is 6 MPa satisfied the target value.

Table 1 OPC 43 grade physical properties

Physical property	Results obtained by observation
Normal consistency in %	30
(IST) Initial setting time	94 min
(FST)Final setting time	245 min
Fineness in % (using 90 μm IS sieve)	6%
Soundness	2 mm
S.G (Specific gravity)	3.15
Compressive strength in N/mm^2 03 days 07 days 28 days	25 35.5 44.5

4 Methodology for the Present Study

First, we understand about our study, then we go for literature review, then we collect material for testing after collecting material, conduct preliminary testing of materials and then prepare the mix design of concrete after this casting of paver block started. After casting of paver blocks curing of specimen is necessary after this process, conduct testing of prepared paver block, then we go for the result and get a conclusion about our study.

5 Experimental Programme

5.1 Used Materials

5.1.1 OPC (Ordinary Portland Cement)

Cement is a finely ground material which possess adhesive and cohesive properties. It is obtained by burning a mixture of argillaceous and calcareous materials at a high temperature of about 1450 °C as per Neville et.al. There are three main grades of Portland cement: OPC33, OPC43, and OPC53. According to Aggarwal et al., the classification of cement is determined by the cement's strength after 28 days. In the manufacturing of paver blocks, cement serves as a binder. For this research, 43 grade OPC that complies with IS: 8112-1989 and was purchased locally in Patna was used. Table 1 provides the results for the Physical-properties [4].

Procedure to obtain physical property of OPC

1. Normal consistency—we use Vicat plunger apparatus having 10 mm diameter and 50 mm length which can penetrate up to 33 to 55 mm from the top of the mould, then we get 30% normal consistency.

Table 2 The physical characteristics of coarse aggregate

Physical properties	Results obtained
Bulk density	1440 Kilogramme per m^3 as per IS: 2386(Part -III)-1963
S.G. (Specific gravity)	02.63
Water absorption	0.48%
Impact value	14%
Abrasion value	19%

2. Fineness test—we use Sieve method for this test in which we have to take 100gm of cement sample, the sample is placed on a 90 μ sieve and continuously sieved for 15 min after this we get 6% of consistency.
3. Soundness Test—we get soundness for OPC through the autoclave apparatus.
4. Compressive strength Test—Three cubes are tested for compressive strength at 3 days, 7 days and 28 days.

5.2 *Coarse Aggregates (CA)*

The aggregates most of which are retained on a 4.75 mm IS sieve are known as coarse aggregates. These can be crushed or uncrushed gravel. Stone cursed aggregates with up to 10 mm nominal size have been employed in the present research and was purchased from a nearby market in Maihar, Madhya Pradesh. The coarse aggregates underwent IS: 2386 testing. Results of a sieve analysis test and the physical characteristics of coarse aggregates (Table 2).

- We obtain Bulk Density of 1440 kilogramme per metre cube as per IS 2386(Part-III)- 1963.
- Impact value is obtained by the ratio of the weight of the fraction passing through sieve by the total weight.
- We get 19% of Abrasion value with the help of Los Angeles Test.

5.3 *FA (Fine Aggregate)*

Natural sand or Artificial sand (stone crushed sand) with a fraction passing 4.75 millimetre IS sieve that are called fine aggregates. For this research, river sand that complied with IS: 383 was purchased from the Satna Son River in M.P. The sand has undergone IS:2386 testing. The results of the sieve analysis test & the physical characteristics of the fine aggregates (Table 3).

Fly ash is used in various sectors. One such sector is concrete manufacturing. The fly ash concrete has good long-term strength but poor early strength. The

Table 3 Physical characteristics for fine aggregates

Properties	Results obtained by observation
Bulk density (loose)	1567 kilogramme/m^3 as per IS: 2386(Part-III)—1963
S.G	2.57
Water absorption	00.60%

Table 4 Physical characteristics of fly ash

Sr. no	Properties	Values obtained by observation
01	S.G	02.08 as per IS 3812–1981
02	Class	F-type

heat of hydration is low. It increases the life of the product by filling the gaps in concrete, producing more durable concrete products. Fly ash for the current study was purchased from Vijay tiles in Khramseda, Satna, M.P (Table 4).

5.4 Chemical Admixture (Super Plasticiser)

Semi-dry concrete, which is used to make paving blocks, has poor flow properties when subjected to vibration. According to the Concrete Institute in Midrand, the use of chemical admixtures enhances workability. For the creation of cement concrete inter locking paver blocks, there has been the use of the PCE (Poly-carboxylic Ether) chemical admixture-based superplasticiser BASF Master Glenium SKY 8233.

5.5 Water

The paver blocks were cast and dried using potable tap water. The water complies with IS: 456 specifications

5.6 PP (Polypropylene)

PPF is a form of PP that has been used. Recron 3s is the PPF product's brand name. The manufacturer advises using 125 gm per 50 kg bag of cement as the standard dosage. To achieve excellent dispersion, before being added to the concrete batch and mixed, the required amount of PPF is first briefly submerged in water. Recron

Table 5 Recron 3 s specifications

Properties	Value
Cut length (mm)	12
Fibre's shape	Triangular
S.G	00.91
Effective diameter	25 to 40 μm
Tensile strength	4×10^3 to 6000 kilogram per cm^2
Melting point	165 °C
Dosage rate	125 gm per 50 kg cement

Table 6 M30 grade concrete design mix with 30% fly ash and various PPF percentages

Mix design	Cement- based materials		W	F.A	C.A	SP	PPF
	C	Fly ash					
kg/m^3							
M30F30 P0.0%	269	116	152	953	879	2.08	0
M30F30 P0.1%	269	116	152	953	879	2.08	0.38
M30F30 P0.2%	269	116	152	953	879	2.08	0.77
M30F30 P0.3%	269	116	152	953	879	2.08	1.15
M30F30 P0.4%	269	116	152	953	879	2.08	1.54
M30F30 P0.5%	269	116	152	953	879	2.08	1.95

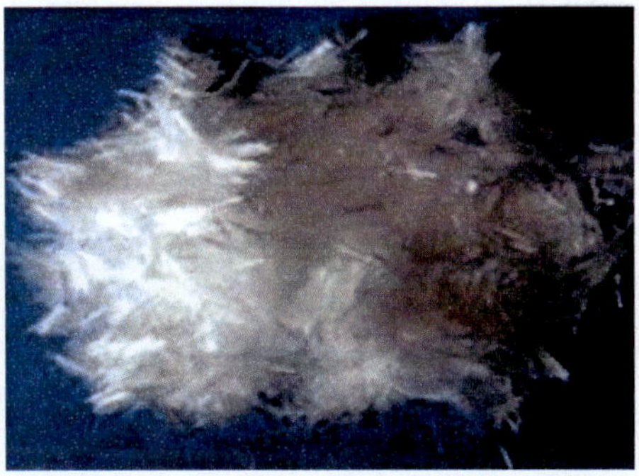

Fig. 1 PPF

3s' specifications as provided by the suppliers are given in the Tables 5, 6 [13–15] (Fig. 1).

5.7 Compressive Strength Test

The paver block specimens 4 in number shall be selected randomly and physically checked for observation of dimensions, aspect ratio and plan area before testing as per code. A 200-tonne capacity compressive testing machine was utilised during this test. The sample needs to be covered in 4 mm thick plywood sheets that larger the specimen in size, which must be placed in the middle between the compressive testing machine's bearing plates and manually tightened. Until the specimen fails, the load must be applied steadily at a rate of 153 N/mm2 per minute without any jerks. The number N indicates the failure of load. F (Comp. strength) = failure load/plan area in Newton/mm2 is the formula used to determine the clearly evident compressive strength of the paver block for each specimen. Size of the specimen for this test we take 200 × 100 × 60 mm and 200 × 100 × 80 mm [12].

5.8 Flexural Strength Test

The paver blocks manufactured will be used for road surfacing. When used on roads with moving traffic, the paver block's flexural property must be carefully considered. The test specimen's dimensions (length, width, thickness, and aspect ratio) must be examined. the testing equipment must comply with IS: 15658 & IS: 516. Supporting rollers for the machine need to have a diameter of 25 to 40 mm. To fix the specimen at 50mm, the rollers' centre-to-centre distance must be changed. Four paver blocks were chosen at random for testing and kept with capping material in accordance with IS: 15658. Without applying any shock, the load must be gradually increased at a rate of 6 kN per minute until the specimen fails [15].

6 Results and Discussion

6.1 Compressive Strength

Corrected Compressive Strength for 60 millimetre thick Paver Blocks with 30% Fly Ash in place of Ordinary Portland Cement (OPC) and Various PPF Proportions for M30 Grade Paver Blocks.

Table 7 lists the results of the corrected compressive strength tests of 60 millimetre thick paver blocks with Ordinary Portland Cement in place of 30% Fine Aggregate & addition of 0–0.5% PPF of various ages. The names of the paver blocks are based on their grade, Polypropylene fibre (PPF) addition and FA replacement proportion. For paver blocks in the M30 grade, the corrected compressive strength variation with age has been shown graphically (Fig. 2).

Table 7 Corrected compressive strength study results for M30 grade paver blocks, 60 mm thick, with various polypropylene fibre (PPF) content ratios

Grade	Water-cement Ratio (WCR)	SP	Polypropylene fibre (PPF)	Thickness	Corrected compressive strength (Newton per mm^2)	
					07 days	28 days
M30F30 P0.0	0.43	2.08	0.00	60	23.63	35.80
M30F30 P0.1	0.43	2.08	0.38	60	23.80	37.50
M30F30 P0.2	0.43	2.08	0.77	60	24.96	38.40
M30F30 P0.3	0.43	2.08	1.15	60	25.53	39.10
M30F30 P0.4	0.43	2.08	1.54	60	24.83	38.80
M30F30 P0.5	0.43	2.08	1.92	60	24.60	38.40

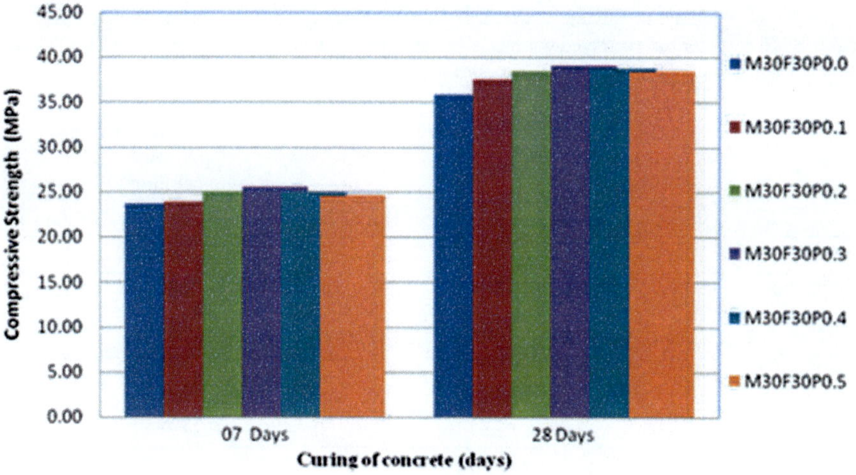

Fig. 2 Variation in corrected compressive strength for M30 grades with different Polypropylene fibre (PPF) proportions for 60 mm thick paver blocks

Paver Blocks' Corrected Compressive Strength with an 80 mm Thickness with 30% Fly Ash Replaced by OPC and Various PPF Proportions for M30 Grade Paver Blocks.

Table 4.9 summarises the corrected compressive strength study results for 80 millimetre thick paver blocks with Ordinary Portland Cement in place of 30% fine aggregates & addition of 0–0.5% polypropylene fibre (PPF) at various ages. The names of the paver blocks are based on their grade designation, PPF addition and FA replacement proportion. Fig. 3 depicts visually how corrected compressive strength for M30 classes of paver blocks varies with age (Table 8).

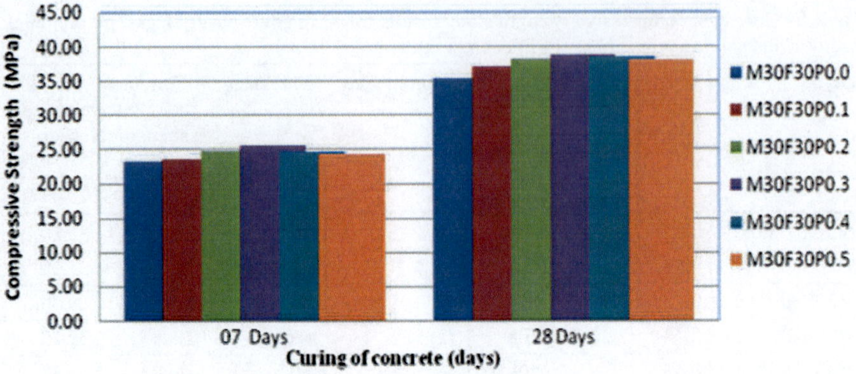

Fig. 3 Variation in corrected compressive strength for M30 grades with different polypropylene fibre (PPF) proportions for 80 mm thick paver blocks

Table 8 Results for the corrected compressive strength of paver blocks of the M30 grade that are 80 mm thick and contain various PPF proportions

Grade	Water-cement Ratio (WCR)	SP	Polypropylene fibre(PPF)	Thickness	Corrected compressive strength (Newton per mm^2)	
					07 day	28 days
M30F30P0.0	0.43	2.08	0.00	80	23.24	35.30
M30F30P0.1	0.43	2.08	0.38	80	23.54	37.10
M30F30P0.2	0.43	2.08	0.77	80	24.70	38.10
M30F30P0.3	0.43	2.08	1.15	80	25.50	38.70
M30F30P0.4	0.43	2.08	1.54	80	24.60	38.40
M30F30P0.5	0.43	2.08	1.92	80	24.36	38.06

6.2 Flexural Strength

Flexural strength of 60millimetre thick M30 grade paver blocks which have 30% fly ash in place of the OPC and 0.0–0.5% of PPF added over the course of 7 & 28 days was measured and noted in Table 9, which is represented graphically.

Flexural Strength of 60-millimetre-Thick Paver Blocks with 30% Fly Ash in Place of OPC And Varying PPF Contents for M30 Grade Paver Blocks (Fig. 4).

Flexural Strength of Paver Blocks 80 millimetre Thickness with 30% Fly Ash in place of Ordinary Portland Cement and different contents of PPF for M30 grade of paver blocks.

Flexural strength for M30 grade paver blocks which is 80 millimetre in thickness with 30% fly ash in place of OPC and 0.0–0.5% of PPF added for 7 and 28 days

Table 9 Results for paver blocks which are 60 millimetre thick and M30 in grade with different ratios of Polypropylene fibres in terms of flexural strength

Grade	Water-Cement Ratio	SP	Polypropylene fibre (PPF)	Thickness	Flexural Strength (Newton per mm^2)	
					07 days	28 days
M30F30P0.0	00.43	02.08	00.00	60	3.65	4.56
M30F30P0.1	00.43	02.08	00.38	60	3.80	4.76
M30F30P0.2	0.43	2.08	0.77	60	3.97	5.30
M30F30P0.3	0.43	2.08	1.15	60	4.24	5.94
M30F30P0.4	0.43	2.08	1.54	60	4.28	5.72
M30F30P0.5	0.43	2.08	1.92	60	4.08	5.48

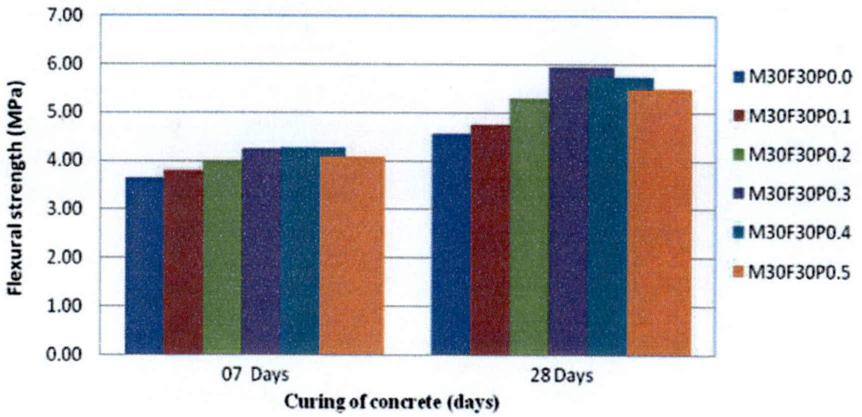

Fig. 4 Flexural strength variation with age for M30 grades with different ratios of polypropylene fibres for 60 mm thick paver block

was observed and tabulated in Table 10, which is depicted graphically in Fig 5. The names of the paver blocks correspond to their mix-in proportions.

6.3 *Cost Effectiveness*

The cost of manufacture of paver block with optimum addition of 0.3% PPF and 30% replacing OPC by fly ash has been calculated in Table 11.

Table 10 Results for 80 millimetre thick M30 grade paver blocks with different contents of PPF in terms of flexural strength

Grade	Water-cement ratio	SP	Polypropylene fibre (PPF)	Thickness	Flexural strength (Newton per mm^2)	
					07 days	28 days
M30F30 P0.0	0.43	2.08	0.00	80.00	3.58	4.48
M30F30 P0.1	0.43	2.08	0.38	80.00	3.73	4.66
M30F30 P0.2	0.43	2.08	0.77	80.00	3.88	5.18
M30F30 P0.3	0.43	2.08	1.15	80.00	4.35	5.80
M30F30 P0.4	0.43	2.08	1.54	80.00	4.14	5.56
M30F30 P0.5	0.43	2.08	1.92	80.00	3.98	5.34

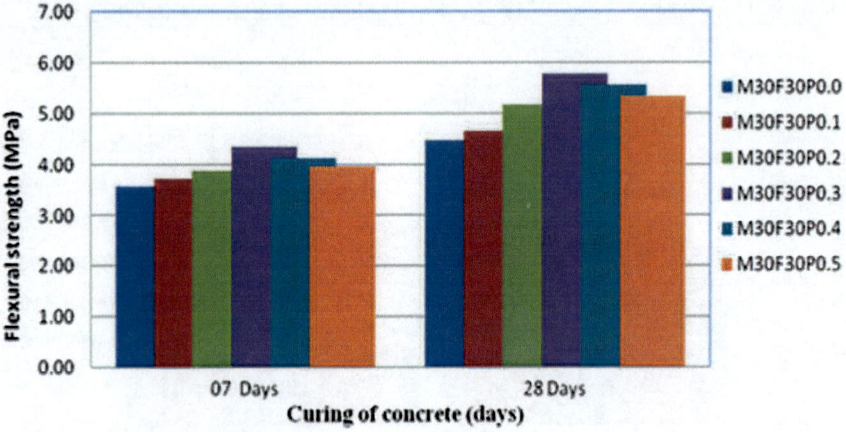

Fig. 5 Variation of Flexural strength with age for the grade of M30 with different Polypropylene Fibre proportions of 80 mm thick paver blocks

Table 11 Cost effectiveness

Paver blocks with 100% OPC	Paver blocks with OPC, fly ash, PPF composite
Manufacture of 100 pieces of paver blocks by using cement 50 kg (1 bag) Cost of 50 kg (1 bag) of OPC = Rs 300/-	Manufacture of 100 pieces of paver blocks by using (cement 35 kg + 15 kg fly ash + 0.125 kg PPF) Cost of 35 kg cement @ 300/50 kg = Rs 210/- Cost of fly ash (waste material) PPF0.125 kg Total cost
Saving in cost = 300–240 = Rs 60/-Percentage Saving = 20%	

7 Conclusions

- For reference mixes, corrected compressive strength in 60 mm and 80 mm thick paver blocks increases when age increased in M30 grade. After curing for 28 days, the strength has just barely surpassed the target strength.
- Maximum strength gains for all grades at 28 days were seen with the addition of 0.3% Polypropylene Fibre (PPF), which may be considered the ideal dose.
- Every mix that included PPF in varying amounts across all grades attained the desired strength after 28 days.
- Fly ash tends to reduce the strength of hardened paver blocks, & the strength was only slightly increased by the addition of Polypropylene Fibre (PPF).
- Flexural strength of the reference mixes in M30 grade 60 mm and 80 mm thick paver blocks increase with age. Reference mix reached the target strength after 28 days.
- The addition of 0.3% PPF at 28 days led to the greatest increase in flexural strength for all grades, which may be considered the optimum dose.
- Cost of paver blocks will increase in small amount due to the addition of PPF.
- Fly ash is a waste material to be used for the manufacture of paver blocks of different grades and thicknesses the resulting product will be economical, energy saving and eco-friendly.
- OPC replaced by fly ash will earn more economy resulting in an overall reduction in cost by 20% at an optimum level of 0.3% PPF in all the grades and thicknesses.

References

1. Mohod MV (2015) Performance of polypropylene fibre reinforced concrete. J Mech Civ Engg 12(1):28–36
2. Kolli R (2013) Strength properties of polypropylene fiber reinforced concrete. Inter J Inno Res Sci Engg Tech 2(8):3409–3412
3. Siddique R (2004) Performance characterstics of high-volume class F fly ash concrete. Cem Conc Res 34:487–493
4. PCA (2005) An engineer's guide to: building green with concrete. Portland Cement Association, Skokie, IL
5. Sachdeva SN, Aggarwal V, Gupta SM (2014) High volume fly ash concrete for paver blocks. World Acad Sci, Engg Tech, Inter J Civ, Arch, Struct Cons Engg 8(3):238–244
6. Singh G, Goel S (2016) Performance evaluation of pet-polypropylene hybrid fiber reinforced concrete in terms of workability, strength and cost effectiveness. Inter J Civil Structure Engg Res 3(2):85–94
7. Ondova M, Stevulova N, Sicakova A (2012) Assessment of selected indicators of portland cement containing fly ash in road concrete. Ad Alta: J Interdiscip Res 2:114–116
8. Karahan O, Atis CD (2011) The durability properties of polypropylene fiber reinforced fly ash concrete. Mater Des 32(2):1044–1049
9. Murahari K, Rao R (2013) Effects of polypropylene fibres on the strength properties of fly ash based concrete. Int J Eng Sci Inven 2(5):13–19
10. Zhang P, Li Q (2013) Effect of polypropylene fiber on durability of concrete composite containing fly ash. Composites: Part B 45:1587–1594

11. Gencel O, Ozel C, Koksa F, Erdogmus E. Barrera GM, Brostow W (2012) Properties of concrete paving blocks made with waste marble. J Clean Prod 21:62–70
12. Thirumurugan S, Sivakumar A (2013) Compressive strength index of crimped polypropylene fibers in high strength cementitious matrix. World AppSci J 24(6):698–702
13. Rao MVK, Murthy NRD, Kumar VS (2011) Behavior of polypropylene fiber reinforced fly ash concrete deep beams in flexure and shear. Asian J Civ Engg (Build Hous) 12(2):143–154
14. Mohod MV (2015) Performance of polypropylene fibre reinforced concrete. IOSR J Mech Civ Eng (IOSR-JMCE) 12(1):28–36
15. Nehvi R, Kumar P, Nahvi UZ (2016) Effect of different percentages of polypropylene fibre (recron 3s) on the compressive, tensile and flexural strength of concrete. Int J Eng Res Technol (IJERT) 5(11):124–130
16. Ahmed TW, Mohammed Ali AA, Zidan RS (2020) Properties of high strength polypropylene fiber concrete containing recycled aggregate. Constr Build Mater 241:118010
17. Muhammed R, Varkey D (2016) Experimental investigation on properties of geo-polymer concrete paver block with the inclusion of polypropylene fibers. Inter J Sci Engg Res 4(10):10–15
18. Haue MN, Langan B, Ward M (1988) High fly ash concretes. ACI Mater 8(1):54–60
19. Uygunglo T, Topcu IB, Gencel O, Brostow W (2012) The effect of fly ash content and types of aggregates on the properties of pre-fabricated concrete interlocking (pcibs). Const Build Mater 30:180–187
20. Patel I, Modhera CD (2011) Study effect of polyester fibers on engineering properties of high volume fly ash concrete. J Engg Res Stud 2(1):159–166
21. Raju JJ, John J (2014) Strength study of high volume fly ash concrete with fibers. Inter J Adv Struct Geotech Engg 3(1):60–64

Embodied Energy Analysis of Different Types of Wall Insulating Materials Used in Buildings

Rahul Patel, Suresh Singh Kushwah, and Aruna Rawat

Abstract Several approaches have emerged in recent years for Building energy analyzes from construction to demolition. These methodologies are often categorized into three stages: embodied energy, operational energy and demolition energy. During the life of a building, operational energy accounts for 40–60% of overall life cycle energy. However, this operational energy can be reduced by using advanced energy saving materials, but this increases the building's embodied energy. Accordingly based on various past research the increased embodied energy is minimized by conserving energy during the working phase of the building. The structure's envelope should be created in accordance with the environmental conditions of that site since the environment and topography of that area have a significant impact on how much energy the building consumes. In the present work, A residential house's (G + 1) outside wall embodied energy is around 60–68% of the overall structure's embodied energy. The embodied energy of six distinct types of wall insulating materials is calculated and compared with non-insulated (NI) walls. The insulating wall materials of varying thicknesses are considered. The increased embodied energy and percentage of embodied energy savings of several types of wall insulating material are also compared. It is observed that the outside wall's embodied energy increases as its insulation increases. 125 mm thick cellulose insulation (CI) has the highest embodied energy saving of 13.55% and the lowest increase in embodied energy of 104.13 MJ/m^2. 250 mm thick Polyurethane foam spray (PUI) has the lowest embodied energy saving of—16.75% and the highest increase in embodied energy of 1413.15 MJ/m^2.

Keywords Embodied energy · Insulating materials · Outside wall · Buildings

R. Patel (✉) · S. S. Kushwah · A. Rawat
Department of Civil Engineering, University Institute of Technology, RGPV, Bhopal, Madhya Pradesh 462 033, India
e-mail: rahulpatelcivilengg@gmail.com

K. K. Pathak et al. (eds.), *Latest Developments in Civil Engineering*, Lecture Notes in Civil Engineering 352, https://doi.org/10.1007/978-981-99-2676-3_18

1 Introduction

The utilization of expanded polystyrene, mineral wool, polyurethane foam, cellulose fibre, cork board insulation, glass wool/fiberglass batt, and other exterior cladding materials creates an external wall insulation system that is thermally insulated, protective, and appealing. The thickness of thermal insulation is determined by the type required to provide a partition with a $U = 0.25 - 0.3$ W/m^{2}k heat transmission factor. Through its ability to stop wind-driven rain, exterior wall insulation can assist in resolving problems with rain seeping through solid walls. By using insulated solid walls can significantly minimize the heating expenses and makes the livelihood more comfortable. Embodied energy (EE), which is the total amount of non-renewable energy used in the creation of a material, is crucial when choosing any kind of building material. This energy is calculated at every step in the production of a product or commodity, such as mining, manufacturing, and transportation. The levels of embodied energy in the various materials and construction methods will be typical. It is an important topic to consider when considering a building's life cycle because it is directly tied to the sustainability of the built environment. While choosing materials and construction techniques, to build a really low-energy home, embodied energy must be taken into consideration. Buildings currently contribute 39% of the world's greenhouse gas (GHG) emissions from energy use, and by 2050, the built environment is predicted to treble, largely due to the expansion of emerging economies.

In Central India, where many different types of meteorological conditions, such as summer, rainy, and winter weather prevail, temperatures range from 5 to 45 °C. In such situation, buildings' outer walls play a critical part in ensuring that the interior temperature stays constant independent of the outside temperature. A residential building's outside walls contribute somewhere between 60 and 68% of the building's overall embodied energy (G + 1). Buildings with minimal energy consumption, the reduction of 60–68% embodied energy outer walls to make them more environmentally friendly and resource efficient. Numerous studies have been carried out for this goal, but they are focused on a particular location or area since the environmental circumstances of that location have a considerable impact on the construction of a wall for that location. As a result, prior to building a wall, it is necessary to research the regional weather patterns.

The embodied energy of a residential building's (G + 1) outside walls made of six different types of wall insulation materials is calculated in the current work using Microsoft Edge software, with the thickness of each wall being adjusted separately. Several wall insulating material's embodied energy comparison is also carried out.

2 Related Previous Researches

The various researches were carried out in the past are discussed herewith: Sadeghifam et al. [1] showed building information modeling (BIM) implementation and assessment for a library building. The Ecotect program's aim of 8% electrical energy savings was achieved with the modification in the type of wall materials used. If the amount of energy utilized decreases, a year's worth of savings may be realized. A green campus with less environmental damage will result from the environmental benefits of utilizing less energy. The results of this study suggest that when choosing an appropriate material for buildings with bioclimatic design and optimal energy utilisation, designers should pay more attention to crucial aspects including U-value, admittance, solar absorption, and the thickness of wall materials.

Mishra et al. [2] calculated the optimal insulation thickness for the various wall materials, such as brick, concrete, and stone, utilized in building construction in India. Extruded polystyrene (XPS) and expanded polystyrene served as the study's insulating materials (XPS). In this investigation, the heat load was calculated using the De- gree-Day method (loss of heat). Because of this, depending on fuel prices, the appropriate insulation thickness ranges from 5.2 to 7.4 cm, and energy savings range from 2560 to 5510 Rs/m^2.

Carlier et al. [3] examined the possible energy savings by adding switchable insulation systems (SIS) to the walls of homes in Belgium and other parts of Europe. The study examines two low-energy prototype dwellings (an apartment and a detached house) that were typical of building and remodeling in Belgium around 2010. The effectiveness of the static and dynamic wall insulation systems in both homes was assessed using a 3R2C-based analytical tool. The switchable insulating system and the associated straightforward two-step rule-based control method were initially introduced. It was discovered that SIS-integrated walls in Belgium could reduce space heating and cooling costs by up to 98 percent and 3.7 percent, respectively.

Abey and Anand [4] evaluated the amount of energy that was used to build a house utilizing conventional in-place techniques and prefabricated materials. A sizable number of energy-demanding materials utilized in both types of construction account for 90% of the EE. Transport energy is the second-largest consumer since prefab producers, in contrast to ready-mixed concrete plants, were likely to be spread out over vast distances. Prefabricated manufacturers are likely to be located far away from ready-mixed concrete facilities, making transportation the second-largest energy consumer. Due to the use of energy-efficient building materials and ideal construction times, prefabricated buildings were found to have moderately (5.7 percent) higher energy consumption than conventional construction. Prefabricated residential structures made up a substantially larger portion of the infill partition wall materials than did commercial buildings.

Reilly et al. [5] compared in a moderate maritime climate, typical wall construction methods and heating systems. The transient energy ratio approach makes it possible to assess the effect of walls alone by extracting the heat flow through the walls (operational energy for heating). Investigations were conducted on the three retrofit possibilities for solid walls. The total amount of energy used can rise as the embedded energy surpasses the operating energy at very low U-values; given the short building lifespan and current best practises walls, this point might be reached soon. Given its relative importance to total energy consumption, the embedded energy measurements that were currently in use had a large amount of uncertainty.

Dissanayake et al. [6] Expanded polystyrene (EPS) that has been recycled 50% is used to manufacture lightweight foam concrete panels, which is a novel walling method. To determine the embodied energy of solar panels, a thorough investigation was done. A single-story house comparison research with other building materials showed that foam concrete precast panels might be an excellent option for walling.

Devi and Palaniappan [7] carried out a case study of the life cycle energy analysis of a residential complex in Southern India that contains 96 apartment types that are identical to one another. Construction site equipment use and the energyutilized to transport materials were also monitored. The sensitivity analysis was performed to examine the relationship between the relative importance of construction energy and operational energy as a function of building service life and monthly power usage per resident. It was discovered that the life cycle energy of residential buildings with little or no air conditioning was largely comprised of the energy used during construction. Construction energy now plays a similar role to operational energy in terms of the life cycle of a building due to increasing operational energy efficiency and a shorter building service life.

Dixit et al. [8] carried out a literature analysis of a group of factors that fluctuate and cause volatility and inconsistency in embodied energy estimates was also found and presented. They also addressed the existing state of conflicting interpretations of embodied energy and provided some insight into the underlying diversity. The feed-stock energy element was found to be the least discussed, despite the fact that the geographic location is generally highlighted in studies. An accurate, comparable database of the embodied energy of building materials can be created if these parameters are taken into account.

Grazieschi et al. [9] Buildings' operating energy consumption can be reduced and their environmental profile improved with the use of insulation materials. In comparison to natural and innovative insulation materials, conventional insulating materials like stone wool and glass wool have proven to be still quite competitive. They per- form well in terms of fire resistance, mould prevention, good water proofing, less constrained durability conditions, and low embodied energy values. The environmental costs of new, high-performance insulating materials, however, are still very costly. Ramesh et al. [10] Building embodied energy is estimated using building material embodied energy coefficients that are appropriate in an Indian environment. e-Quest energy simulation software is used to estimate the building's operating energy. According to the findings, embodied energy (11%), followed by operating energy (89%) of the building, is the highest contributor to the life cycle energy of the building. The overall life cycle energy demand of the structure is reduced by 9.7% when aerated concrete blocks are used to form walls and cover the roof, according to research on this energy saving approach. Additionally, it has been determined that integrated photovoltaic (PV) panels in buildings hold the most promise for a reduction (37%) in the building's life cycle energy use.

Ramesh et al. [11] Even if using passive and active technology results in a slight increase in embodied energy, the life cycle energy demand of a building can be greatly decreased. However, using passive and active characteristics of a structure excessively could have the opposite effect. In terms of life cycle performance, low-energy buildings outperform self-sufficient (zero operational energy) structures.

3 Present Methodology

A residential House (G + 1) located in Bhopal city in Central India is considered for present study. The built-up area is 1550 square-feet. The outside wall is made from normal clay brick wall along with six types of wall insulating materials as given in Table 1. The main factor in ensuring thermal comfort of outside wall in a building is done by using insulation materials. They must be wisely chosen based on our requirements and desire to minimize environmental effects. They come in a variety of forms, including mineral, synthetic, animal, and vegetable-based types. Microsoft Edge programme is used to compute the embodied energy for walls that are not insulated (NI) and walls that are insulated with various insulating materials. We take into account insulated walls of various thicknesses. It also looked into how different types of wall insulating material compare in terms of increased embodied energy and percentage embodied energy savings.

(i) Cellulose Insulation (CI)

Recycled paper is used to make cellulose insulation. Comparatively speaking, cellulose insulation has a far higher percentage of recycled materials than other insulations, with recycled newspaper making up about 85% of its composition. To restrict airflow, the tiny pieces of paper are fiberized and compactly packed. Following collecting, the material is treated with safe chemicals like the mineral borate, which serves as a flame retardant and deters insects and mold.

(ii) Cork Board Insulation (CBI)

Excellent insulating and damp-proofing qualities can be found in cork, an eco-friendly insulation material. There are many different insulating uses for this material. The two forms of cork are boards and granules. Cork boards are excellent for insulating floors, walls, and roofs.

(iii) Glass Wool/Fiberglass Board (GFI)

A roll of fiberglass insulation is a thin, compressible insulating material comprised of long, fine inorganic fibers that are joined by a high-temperature adhesive. Because there are numerous tiny air spaces in between the glass fibers, it provides excellent

Table 1 Insulating wall materials

Insulating wall materials	Picture	Types	Made from	Thickness
Cellulose insulation (CI)		Natural	Plant fiber	
Cork board insulation (CBI)		Natural	Cork	
Glass wool/ fiberglass board (GFI)		Natural inorganic	Silica and glass	
Mineral wool insulating board (MWI)		Natural inorganic	Volcanic rock	
Polystyrene foam spray (PSI) Polyurethane foam spray (PUI)	 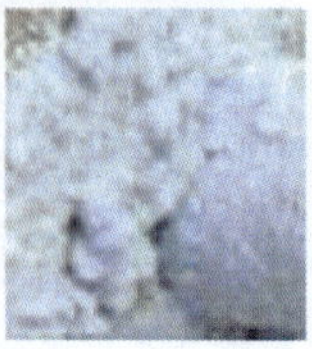	Synthetic Synthetic	Polymer made from molecules composed of carbon and hydrogen complex polymer made from molecules composed of nitrogen and oxygen as well as carbon and hydrogen	125 mm, 150 mm, 200 mm and 250 mm

thermal and acoustic insulation qualities. Rolls of fiber glass wool are produced, each with a unique set of mechanical and thermal characteristics.

(iv) Mineral wool Insulating Board (MWI)

High-temperature insulation called mineral wool board is created using inorganic fibers taken from volcanic rock. Due to its hydrophobic properties, mineral wool board helps prevent rot, rust, and mildew. Basement walls, commercial roofs, and other areas of the building envelope can all benefit from mineral wool board insulation.

(v) Polystyrene Foam Spray (PSI)

Any variety of expanded polystyrene foam is referred to as EPS foam. Insulation made of closed-cell, lightweight expanded polystyrene is available. It can be utilized as an insulated panel system for building facades, walls, roofs, and floors since it is strong and long-lasting.

(vi) Polyurethane Foam Spray (PUI)

One of the most popular materials for home furnishings, including furniture, bedding, and carpet underlay, is polyurethane, primarily in the form of flexible foam. An innovative and highly effective form of thermal insulation for buildings is polyurethane foam insulation. To make upholstered furniture more robust, comfortable, and supportive, flexible polyurethane foam is used as a cushioning material.

4 Results and Discussions

The embodied energy-non-insulated (NI) wall and wall insulated with different insulating materials are calculated with different thicknesses. The total embodied energy-G + 1 building is obtained as 3763.06 MJ/m^2. Comparative analysis of increased embodied energy- different types of wall insulating material is also obtained as described in Table 2.

Figure 1 demonstrates the variation in increased embodied energy for various insulating materials for various thicknesses under consideration. It can be observed that as the thickness of wall insulation increases embodied energy of the wall also increases. The increased embodied energy is lowest in the case of CI and highest in PUI.

Table 3 gives the percentage saving in the embodied energy for different insulated walls. It can be seen that cellulose insulation (CI) of thickness 125 mm has the highest embodied energy saving of 13.55% and the wall composed of polyurethane foam spray (PUI) of thickness 250 mm has the lowest embodied energy saving of -16.75% as compared to the non-insulated (NI) wall. Figure 2 shows the percentage of embodied energy saving for different insulating materials along the thicknesses. Furthermore, by comparing the embodied energy saving for wall insulation with varying thicknesses of 125 mm, 150 mm, 200 mm and 250 mm with non-insulated

Table 2 Increased embodied energy for different insulating materials

Insulating material	Increase embodied energy (MJ/m^2) for different thickness of wall			
	125 mm	150 mm	200 mm	250 mm
Cellulose insulation (CI)	104.13	106.82	112.19	117.57
Cork board insulation (CBI)	167.97	183.42	214.33	245.24
Glass wool/Fiber glass board (GFI)	190.16	210.05	249.84	289.63
Mineral wool insulating board (MWI)	190.2	210.1	249.84	289.63
Polystyrene foam spray (PSI)	368.22	423.72	534.73	645.75
Polyurethane foam spray (PUI)	751.92	884.17	1148.66	1413.15

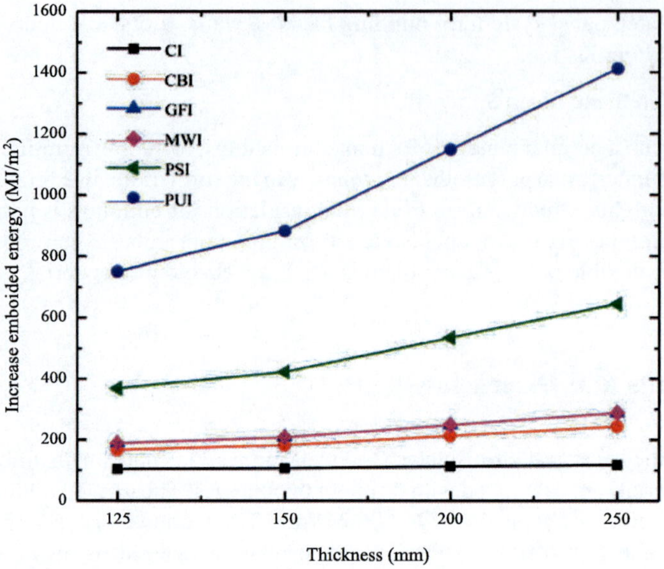

Fig. 1 Increased embodied energy

(NI) wall, it is observed that the percentage saving in embodied energy decreases with the increase in the thickness of the wall insulating materials.

Embodied Energy is the total non-renewable energy which is consumed during the manufacturing of these insulating wall materials from cradle to grave. Hence it is observed that as we increase the thickness of wall insulating material embodied energy is also increased. Thus the selection of which type of wall insulating material to use and also its thickness is the important parameter for creating a Low-Energy Building. These wall insulating material help in maintaining comfort condition inside the building, this will reduce the operational energy of building and that compensates the increase embodied energy during the life cycle energy of the building.

Table 3 Embodied energy saving for different insulating materials

Insulating material	Embodied energy saving (%) for different thickness of wall			
	125 mm	150 mm	200 mm	250 mm
Cellulose insulation (CI)	13.55	13.48	13.36	13.24
Cork board insulation (CBI)	12.07	11.71	11.00	10.28
Glass wool/Fiber glass board (GFI)	11.56	11.10	10.17	9.25
Mineral wool insulating board (MWI)	11.56	11.10	10.17	9.25
Polystyrene foam spray (PSI)	7.43	6.15	3.58	1.01
Polyurethane foam spray (PUI)	− 1.45	− 4.51	− 10.63	− 16.75

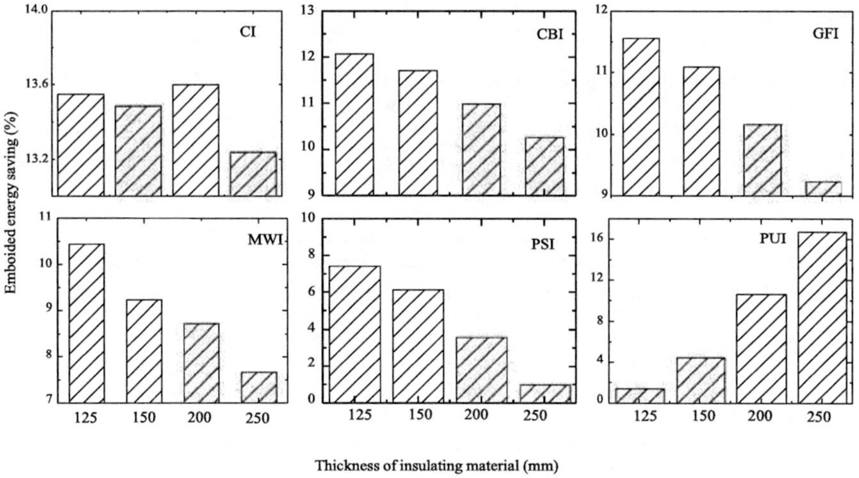

Fig. 2 Percentage embodied energy saving for different insulating materials

5 Conclusions

A residential House (G + 1) located in Bhopal city in Central India is taken into account for the current study. The embodied energy-outside wall is calculated from normal clay brick wall along with six different types of wall insulating materials is presented herein, the following conclusions are drawn:

1. The total embodied energy G + 1 House is obtained as 3763.06 MJ/m^2. It is observed that as the thickness of wall insulation increases embodied energy of wall also increases. The increased embodied energy is lowest in the case of CI and highest in PUI.
2. It can be seen that cellulose insulation (CI) of thickness 125 mm has the highest embodied energy saving of 13.55% and the wall composed of polyurethane foam spray (PUI) of thickness 250 mm has the lowest embodied energy saving of −16.75% as compared to the non-insulated (NI) wall
3. Cellulose insulation (CI) of thickness of 125 mm has the lowest increase in embodied energy by 104.13 MJ/m^2 and Polyurethane foam spray (PUI) of thickness 250 mm has the highest increase in embodied energy by 1413.15 MJ/m^2.
4. Furthermore, by comparing embodied energy saving for wall insulation with varying thicknesses of 125 mm, 150 mm, 200 mm and 250 mm with non-insulated (NI) wall, it is observed that the percentage saving in embodied energy decreases and embodied energy increases with the increasing thickness of wall insulating materials. Thus, the insulation must be wisely chosen based on project requirements and the desire to minimize environmental effects.

References

1. Nobahar SA, Kadir MA, Iman K, Umit I, Asghar BA, Sanaz T (2016) Energy analysis of wall materials using building information modeling (BIM) of public buildings in the tropical climate countries. Jurnal Teknologi 78(10):35–41
2. Subhash M, Usmani JA, Sanjeev V (2012) Energy saving analysis in building walls through thermal insulation system. Int J Eng Res Appl (IJERA) 2(5):128–135
3. Remy C, Mohammad D, Moncef K (2022) Energy performance of integrated wall and window switchable insulated systems for residential buildings. Energies 15(1056):1–22
4. Abey Sharon T, Anand KB (2019) Embodied energy comparison of prefabricated and conventional building construction. J Inst Eng (India) Ser A 100(4):777–790
5. Reilly A, Oliver K, Richard O (2020) Energy embodied in, and transmitted through, walls of different types when accounting for the dynamic effects of thermal mass. J Green Build 15(4)
6. .Dissanayakea DMKW, Jayasingheb C, Jayasinghec MTR (2017) A comparative embodied energy analysis of a house with recycled expanded polystyrene (EPS) based foam concrete wall panels, energy and buildings. Energy Build 135:85–94
7. Devi LP, Palaniappan S (2014) A case study on life cycle energy use of residential building in Southern India. Energy Build 80:247–259
8. Kumar DM, Fernández-Solís JL, Sarel L, Culp CH (2010) Identification of parameters for embodied energy measurement: a literature review. Energy Build 42(8):1238–1247
9. Gianluca G, Francesco A, Guilhem T (2021) Embodied energy and carbon of building insulating materials: a critical review. Clean Environ Syst. https://doi.org/10.1016/j.cesys.2021.100032
10. Ramesh T, Ravi P, Karunesh SK (2013) Life cycle energy analysis of a multifamily residential house: a case study in Indian context. Open J Energy Effi 2:34–41
11. Ramesh T, Ravi P, Shukla KK (2010) Life cycle energy analysis of buildings: an overview. Energy Build 42:1592–1600

Relationship Between Properties of Plastic Composite Construction Material

Devansh Jain, Sudhir Singh Bhadauria, and Suresh Singh Kushwah

Abstract Plastic waste generation is increasing rapidly, and using plastic waste to make construction materials is a good way to recycle the plastic waste. In this paper, a correlation has been established between the characteristics of the plastic composite construction material and the technique used to develop empirical relations between the properties are linear regression, where compressive strength is used as an explanatory variable and other properties are used as dependent variables. The result indicates that compressive strength is proportional to flexural and split tensile strengths.

Keywords Linear regression · Plastic composite construction material · Plastic waste · Recycle

1 Introduction

India has seen a significant increase in the consumption of plastics as well as the manufacture of plastics. The management of the trash produced by discarded old plastic goods, especially those used for packaging purposes, has become a hard problem in the absence of an appropriate waste collection and sorting mechanism. The management of solid waste has become a serious issue in the majority of nations, including India. The issue of solid waste, particularly plastic garbage, is continually becoming worse in India because of the frightening quantity of plastic waste that is produced every day and pollutes the environment. The general public in India has not recognised the threat that plastic trash poses to the planet, wildlife, and people. They are unaware that plastic takes a very long time to breakdown and is not biodegradable. Therefore, plastic garbage pollutes our ecosystem. In India, the typical method of handling solid waste which includes plastics is landfilling, which also has some long-term issues [1]. Recycling plastic waste into construction material can solve the

D. Jain (✉) · S. S. Bhadauria · S. S. Kushwah
Department of Civil Engineering, University Institute of Technology—Rajiv Gandhi Proudyogiki Vishwavidyalaya, Bhopal, MP, India
e-mail: devansh.jain@live.com

K. K. Pathak et al. (eds.), *Latest Developments in Civil Engineering*, Lecture Notes in Civil Engineering 352, https://doi.org/10.1007/978-981-99-2676-3_19

problem upto some extend. Ge et al. [2] examined the effects of aggregate gradation with different sand and PET ratio along with curing conditions on recycled PET mortar. Choi et al. [3] evaluate the effect of granulated blast furnace slag on the surface microstructure of waste PET bottle lightweight aggregate (WPLA). Wang et al. [4] developed concrete containing lightweight aggregate, light weight aggregate manufactured using PET bottles. María et al. [5] Using a glycolysis process, used PET bottles were utilised to produce an unsaturated polyester resin (UPER). UPER was utilised to manufacture polymer mortars. First, the ratio of sand and UPER was varied, then the ratio of sand, UPER and PET. Reis et al. [6] developed Non-biodegradable PET plastic aggregates from beverage containers that are used in mortar. Mahdi et al. [7] investigated the use of recycled PET plastic trash as an alternative binder to conventional Portland cement. In addition, PET plastic waste was glycolyzed to create UPER (unsaturated polyester resin). UPER was utilised to produce polymer concrete and mortar [8]. Mitchell and colleagues [9] investigated Most disposable cups are produced from (PPL) paper plastic laminates, which are comprised of cellulose with a thin polyethylene covering. Disposable cups were shredded to generate PPL flakes, which were used to strengthen polypropylene to make innovative paper plastic composites (PPCs). Yesilata et al. [10] used PET bottles and tyre parts, which may be found in the environment for free, shredded and mixed with concrete to test heat insulation. Ferreira et al. [11] Assess the effect on mechanical properties by different curing conditions of concrete containing waste plastic aggregate. Atis et al. [12] The utilisation of bottles containing PET fragments as lightweight aggregate in mortar was investigated. Two mortar samples were evaluated, one with PET aggregates alone and the other with PET aggregates and sand. In order to reduce cement usage and save money, 50% of blast furnace slag was substituted for cement. Jassim [13] made use of HDPE waste, this waste is mixed with Portland cement to produce cement containing plastic or plastic cement, and the effect of replacing sand with varied percentages of fine polyethylene waste is evaluated. Sellakutty et al. [14] used bags made from HDPE and polyethylene PE. These bags are cleaned and combined with aggregates (gravel and sand) in varying quantities to produce bricks with thermal and acoustic insulation properties [15].

Waste water sachets made of LDPE are melted and then mixed with sand to produce manufactured sand bricks bonded by LDPE. Sand particle size along with the ratio of ingredients affects the properties of bricks like density, compressive strength, water absorption, and flexural strength. Shanmugavalli et al. [16] examined the plastic waste reutilisation in paver blocks. The study shows plastic waste is substituted for cement in paver blocks to reduce the cost of paver blocks relative to conventional concrete paver blocks.

In this paper empirical relationship is developed between the properties of plastic composite construction material using liner regression technique. Compressive strength has been considered as explanatory variable and remaining properties like split tensile strength, density, flexural strength and water absorption have been the dependent variable.

2 Method and Data Collection

Waste plastic was collected and HDPE (high-density polythene), LDPE (low-density polythene) and PETE (polyethylene terephthalate) were sorted manually, then shredding of plastic was done using a machine. To manufacture plastic composite construction material waste plastics were softened in the pan. For softening of plastic used engine oil was heated upto 100 °C then waste plastic where added to it. Once reached to softening state then aggregates were added in it. Then specimen was casted and tested for various properties i.e. CS (compressive strength), FS (flexural strength), TS (split tensile strength), D (density) and WA (water absorption) [17].

3 Relationship Between Properties of Plastic Composite Construction Material

Relationship between properties of plastic composite construction material were obtained using linear regression. The obtained empirical relationships are given below. Figure plotted between dependent variable and independent variable, where x-axis is independent and y-axis is the dependent variable.

3.1 Compressive Strength Versus Flexural Strength

Figure 1 shows the relationship between (CS) and (FS) obtained from this study. The relevant empirical expression obtained from this study is given in Eq. (1).

$$\mathrm{fs} = 0.218\mathrm{cs} + 1.396. \tag{1}$$

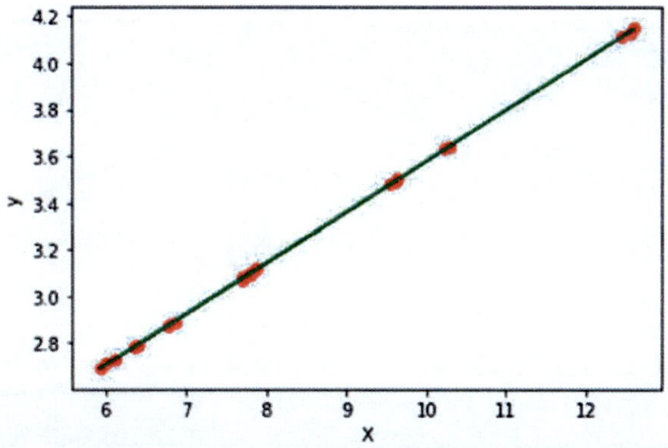

Fig. 1 Relationship between CS and FS

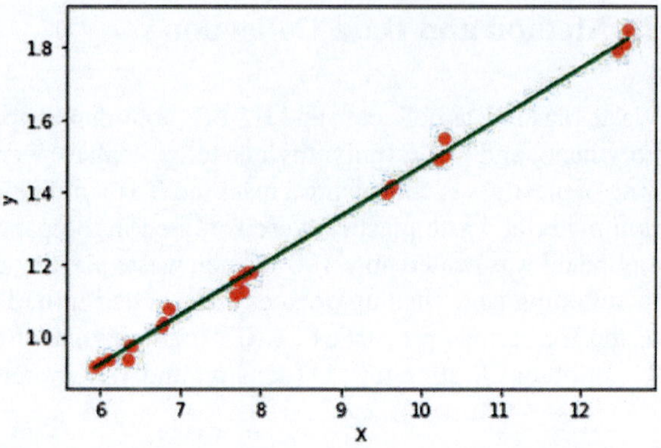

Fig. 2 Relationship between CS and TS

3.2 Compressive Strength Versus Split Tensile Strength

Figure 2 illustrates the relationship between CS and TS obtained from this study. The relevant empirical expression obtained from this study is given in Eq. (2).

$$ts = 0.137cs + 0.102. \tag{2}$$

3.3 Compressive Strength Versus Density

Figure 3 shows the relationship between CS and D obtained from this study. The relevant empirical expression obtained from this study is given in Eq. (3).

$$D = -40.638cs + 1598.88. \tag{3}$$

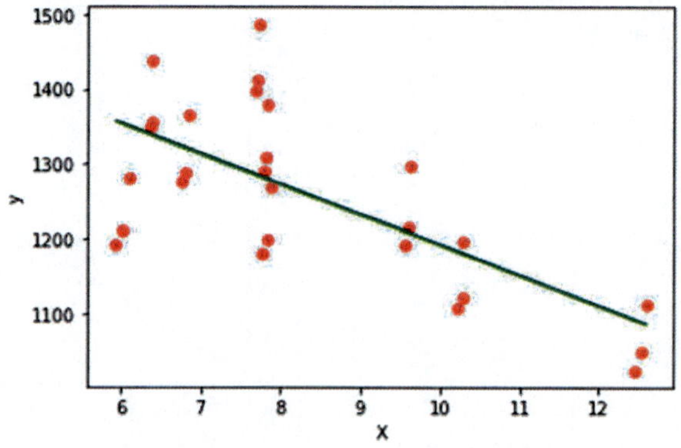

Fig. 3 Relationship between CS and D

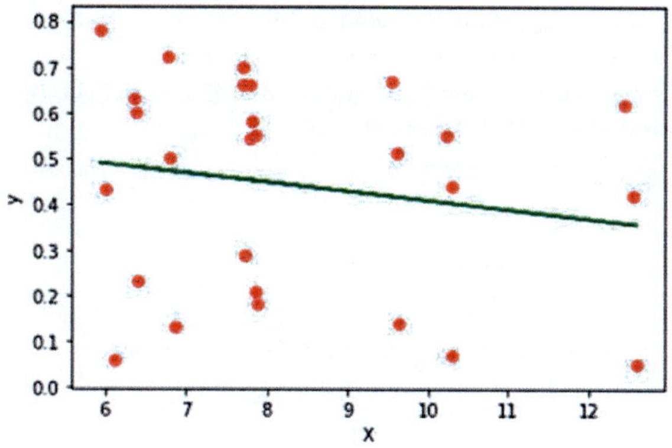

Fig. 4 Relationship between CS and WA

3.4 Compressive Strength Versus Water Absorption

Figure 4 shows the relationship between CS and WA obtained from this study. The relevant empirical expression obtained from this study is given in Eq. (4).

$$wa = -0.020cs + 0.608 \tag{4}$$

4 Evaluation Metrics

Three approaches, namely MSE, R^2, and RMSE, are used to evaluate the developed equation that demonstrates the relationship between the properties of Plastic composite construction material. The average of the squared difference between the actual and predicted values in a data set is known as the mean squared error (MSE). It measures the variation between residues. Root Mean Squared Error (RMSE) is equal to Mean Squared Error squared (MSE). It calculates the distance between the mean and the residuals. R-squared (R^2) shows how much of the variation in the dependent variable can be accounted for by the linear regression model. This indicates that regardless of how large or little the values are, the value of R square will always be less than one [18]. Values of the validation model are given in Table 1.

Table 1 Accuracy of the model

	MSE	RMSE	R^2
Compressive strength versus Flexural strength	7.162	0.00267	0.999
Compressive strength versus Split tensile strength	0.00030	0.0175	0.995
Compressive strength versus Density	7022.19	83.798	0.480
Compressive strength versus Water absorption	0.0310	0.222	0.049

The evaluation metrics used in this paper to find the precision and performance of Linear Regression model are Mean Squared Error (MSE), Root Mean Squared Error (RMSE), and R-Squared (R^2). The concise table for various properties and its comparative MSE, RMSE and R^2 is shown in Table 1.

5 Conclusion

An empirical relationship between the properties of plastic composite construction materials has been developed using the linear regression technique. The relationship between CS and FS has an accuracy of 99.9% and The relationship between CS and TS has an accuracy of 99.5%, while the error between the predicted and actual value is 0.0175. Relationship between CS and D, along with CS and WA, are less accurate and have a high error rate.

From the study, it has been concluded that FS and TS are proportional to CS and predict the values accurately, whereas water absorption and density are not considerably proportional to compressive strength and are unable to make an empirical relationship.

References

1. Banerjee T (2014) Plastics waste management in India: an integrated solid waste management, July 2014
2. Ge Z, Sun R, Zhang K, Gao Z, Li P (2013) Physical and mechanical properties of mortar using waste Polyethylne Terephthalate bottles. Constr Build Mater 44:81–86. https://doi.org/10.1016/j.conbuildmat.2013.02.073
3. Choi Y, Moon D, Chung J, Cho S (2005) Effects of waste PET bottles aggregate on the properties of concrete, 35:776–781. https://doi.org/10.1016/j.cemconres.2004.05.014
4. Wang Y, Joong D, Jic Y, Lachemi M (2009) Characteristics of mortar and concrete containing fine aggregate manufactured from recycled waste polyethylene terephthalate bottles. Constr Build Mater 23(8):2829–2835. https://doi.org/10.1016/j.conbuildmat.2009.02.036
5. María J et al (2014) Polymer mortars prepared using a polymeric resin and particles obtained from waste pet bottle, 65:376–383. https://doi.org/10.1016/j.conbuildmat.2014.04.114
6. Reis JML, Chianelli-junior R, Cardoso JL, Marinho FJV (2011) Effect of recycled PET in the fracture mechanics of polymer mortar. Constr Build Mater 25(6):2799–2804. https://doi.org/10.1016/j.conbuildmat.2010.12.056
7. Mahdi F, Abbas H, Ali A (2013) Flexural, shear and bond strength of polymer concrete utilizing recycled resin obtained from post consumer PET bottles. Constr Build Mater 44:798–811. https://doi.org/10.1016/j.conbuildmat.2013.03.081
8. Mahdi F, Abbas H, Khan AA (2010) Strength characteristics of polymer mortar and concrete using different compositions of resins derived from post-consumer PET bottles. Constr Build Mater 24(1):25–36. https://doi.org/10.1016/j.conbuildmat.2009.08.006
9. Mitchell J, Vandeperre L, Dvorak R, Kosior E, Tarverdi K, Cheeseman C (2014) Recycling disposable cups into paper plastic composites. Waste Manag 34(11):2113–2119. https://doi.org/10.1016/j.wasman.2014.05.020

10. Yesilata B, Isıker Y, Turgut P (2009) Thermal insulation enhancement in concretes by adding waste PET and rubber pieces. Constr Build Mater 23(5):1878–1882. https://doi.org/10.1016/j.conbuildmat.2008.09.014
11. Ferreira L, De Brito J, Saikia N (2012) Influence of curing conditions on the mechanical performance of concrete containing recycled plastic aggregate, 36:196–204. https://doi.org/10.1016/j.conbuildmat.2012.02.098
12. Atis CD (2010) An investigation on the use of shredded waste PET bottles as aggregate in lightweight concrete, 30:285–290. https://doi.org/10.1016/j.wasman.2009.09.033
13. Jassim AK (2017) Recycling of polyethylene waste to produce plastic cement. Procedia Manuf 8(October 2016):635–642. https://doi.org/10.1016/j.promfg.2017.02.081
14. Sellakutty D (2016) Utilisation of waste plastic in manufacturing of bricks and paver, January 2016
15. Jnr AK, Yunana D, Kamsouloum P, Webster M, Wilson DC, Cheeseman C (2018) Recycling waste plastics in developing countries: use of low-density polyethylene water sachets to form plastic bonded sand blocks. Waste Manag 80:112–118. https://doi.org/10.1016/j.wasman.2018.09.003
16. Shanmugavalli B (2017) Reuse of plastic waste in paver blocks, 6(2):313–315
17. Jain D, Bhadauria SS, Kushwah SS (2022) An experimental study of utilization of plastic waste for manufacturing of composite construction material. Int J Environ Sci Technol (0123456789). https://doi.org/10.1007/s13762-022-04447-7
18. Maiti and Bidinger (1981) Performance metrics in machine learning. J Chem Inf Model 53(9):1689–1699

Prediction of Compressive Strength of Plastic Composite Construction Material Using Artificial Neural Network

Devansh Jain, Sudhir Singh Bhadauria, and Suresh Singh Kushwah

Abstract Due to the huge amounts of plastic trash being produced, which seriously impact the ecosystem and its people, this threat to the environment is increasing. In order to find an effective way to manage these wastes and improve the sustainability of our environment and recycling waste plastic is a solution to this problem. Machine learning and deep learning have captured most of the prediction and classification problems in recent times. In this paper, deep learning Artificial Neural Network technique is used to predict the compressive strength of plastic composite construction material. Developed deep learning Artificial Neural Network model shows 88.59% accuracy.

Keywords Composite construction material · Artificial neural network · Deep learning · Plastic waste · Recycling · Waste management

1 Introduction

Based on the functioning of the brain of human beings that consists of hundreds of billions of interconnected neurons which process information. Artificial neural network mimics the human brain. Artificial neural network commonly known as ANN technique consists of three layers of neurons i.e. input layer, hidden layer and output layer. Standard architecture of ANN comprises of lines connecting neurons. [1]. A study by Moretti et al. studied the forecast of elastic modulus and compressive strength (CS) of various laboratory-prepared and tested concrete mixtures, using a feed-forward ANN technique along with back-propagation technique. It demonstrates how neural networks may be used to quantify the elasticity and strength of concrete samples made with various mix ratios [2]. Abdulla et al. observed the compressive stress and strain atthe highest stress of concrete were predicted using the artificial neural network (ANN) technique. Using a variety of statistical indices, the

D. Jain (✉) · S. S. Bhadauria · S. S. Kushwah
Department of Civil Engineering, University Institute of Technology – Rajiv Gandhi Proudyogiki Vishwavidyalaya, Bhopal, M.P., India
e-mail: devansh.jain@live.com

K. K. Pathak et al. (eds.), *Latest Developments in Civil Engineering*, Lecture Notes in Civil Engineering 352, https://doi.org/10.1007/978-981-99-2676-3_20

proposed equation's capability and performance are compared with actual experimental findings and forecasts from 53 empirical equations, including multiple design codes and different strain models for regular and high-strength concrete. The results demonstrated that the ANN technique has a high potential for predicting the concrete compressive strength and strain at maximum stress, with accurate predictions and significant agreement with the actual data [3]. Dantas et al. developed ANN models which predict concrete compressive strength. In this study waste from construction demolition is used as aggregate. Data from previous studies have been used to build the models. 1178 data points in total, 77.36% of which were used for training and 22.24% for testing, were used for modeling. The results from training and also from testing phases clearly demonstrate forecasting ability of ANN [4]. Using data from earlier experimental studies, Rajeshwari et al. constructed an ANN model for compressive strength (CS) of concrete which contains high-volume fly ash (HVFA). Experimental data sets and data from previous studies are collected to predict the CS of HVFA concrete with eight input parameters, having hidden layer nodes, weights, and biases which are then adjusted through trial and error to produce the best model, study shows that ANN can be used to forecast the strengths of HVFA concrete [5]. For the nonlinear mapping between the 28d compressive strength of seawater sea sand concrete and the concrete water-cement ratio, cement content, and sand ratio, Zheng et al. developed an ANN model for compressive strength of seawater sea sand concrete. Python was used to establish it. The outcomes demonstrated that the model training fitting was good with suitable network settings, and the prediction outcomes were satisfactory. The result shows that it is feasible to create ANN model for compressive strength prediction of seawater sea sand concrete [6]. Sakshi et al. demonstrates the use of ANN to create a model for forecasting the concrete compressive strength containing nano-silica as a partial replacement of cement. The correlation coefficient can be used to assess the model's performance and comparison metrics with the previous study data. The mean absolute error and root mean square error have been adopted, and the result shows that ANN can predict concrete compressive strength containing nano-silica [7]. The compressive strength (CS) of plastic composite construction material is predicted in this paper using a deep learning artificial neural network technique.

2 Experimental Dataset and Methods

The threat to the ecosystem is developing as a result of the vast volumes of plastic garbage being created, which harms both the environment and its inhabitants, and the rapid increase in plastic waste. Recycled plastic waste can be utilized to produce construction materials. To manufacture material containing waste plastic or plastic waste composite construction material (PCCM), three types of plastic waste i.e. HDPE, LDPE and PETE also known as high-density polyethylene, low density polyethylene, and polyethylene terephthalate respectively is melted or softened to semiliquid condition using UEO (used engine oil) and mixed with CA (coarse aggregate) and (FA) fine aggregate in different compositions. The compressive strength (CS) of this material has been measured [8].

A machine learning model capable of predicting CS has been developed with Six predictors, namely HDPE, LDPE, PETE, UEO, FA, CA and target variable CS. For training and testing, random data sets are used, and the data is also standardized. Standardizing the data accelerates model training and facilitates the identification of global minima. Seventy percent of the data in this study is used for training and thirty percent for testing. To create a deep learning artificial neural network, the tensor flow and keras libraries were utilized.

Figure 1 depicts the deep learning ANN architecture used in this paper. Each of the two hidden layers contains five neurons and there is only one neuron in the output layer. Changes may be made to the number of hidden layers and the number of neurons in each layer for optimal accuracy.

While implementing ANN architecture, specific parameters are used including the Sequential and Dense module of Keras, which is a library of Python language. The sequential module helps in creating a sequential stack of various layers of ANN whereas the Dense module defines the number and initial weight of neurons in each layer and the activation function. In this ANN model, five units of neurons have been used which will receive the input values or variables. The dimension of input is six This indicates that the input data contains six predictors, as predicted by the first layer. Because the sequential approach communicates this information to subsequent levels, this value is not defined for the second dense layer. The initializer for the kernel is set to "normal." When neurons begin their computation, an algorithm determines the weight value for each neuron. This states that you can select several values, such as 'normal' and 'glorot uniform'. The activation function "Relu" has been used. The output for Relu that has been given in Eq. 1 is that when the input value is negative, the output becomes zero, and when the input value is positive, the output becomes the input value.

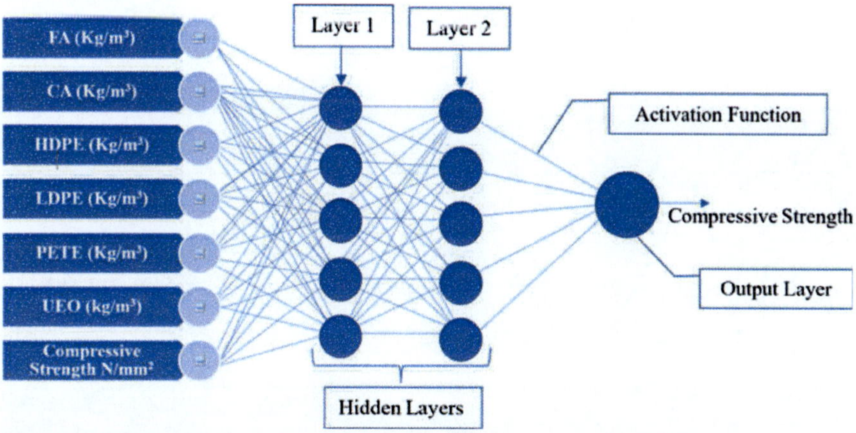

Fig. 1 ANN architecture

$$f(x) = \begin{cases} 0, if\, x\, is\, negative \\ x, if\, x\, is\, positive \end{cases} \quad (1)$$

where x = input value

The total rows that have to be passed in a single pass after that pass the computation of neural network starts where it calculates the sum of squares of error (SSE) and adjusts the weights of neurons based on errors. A single epoch completes when all rows are delivered in batches of defined rows which is called as mini-batch gradient descent. In order to predict the correct value of batch size and epochs, a hyperparameter modification has been used to establish the correct value. ANN hyperparameters are optimized by calculating the ideal batch size and epoch values. The number of epochs and batch size has been identified to be used as fifty and two from Fig. 2. which means ANN examines the whole training dataset fifty times and modifies its weights. Hyperparameter tunning has a direct impact on the performance of the model. Improper selection of values of hyperparameter causes overfitting or underfitting. In this paper, two parameters tunning namely epoch and batch size are presented.

Figure 2 depicts a scatterplot of the trial parameters for ANNs. The ideal batch size and epochs parameters are 2 and 50, respectively. Next, it is necessary to train the model using these parameters. The model is retrained using the optimal set of parameters to predict the compressive strength of the training data. Figure 3 depicts the anticipated values obtained from the ANN model plotted along with the actual data.

Using the final trained model and calculating the Absolute Percentage Error for each testing data row, the prediction error for each row has been computed. Mean Absolute Percentage Error is calculated by averaging all the rows' absolute percentage errors (MAPE), MAPE for this paper has been calculated as MAPE = 100 * (abs (Testing Data ['Compressive strength'] – Testing Data ['Predicted Compressive strength'])/Testing Data ['Compressive strength']) where abs is absolute. The accuracy is calculated as 100 – np.mean MAPE, where np is Non-Deterministic Polynomial-time and the calculated accuracy of deep learning ANN model is 88.59%.

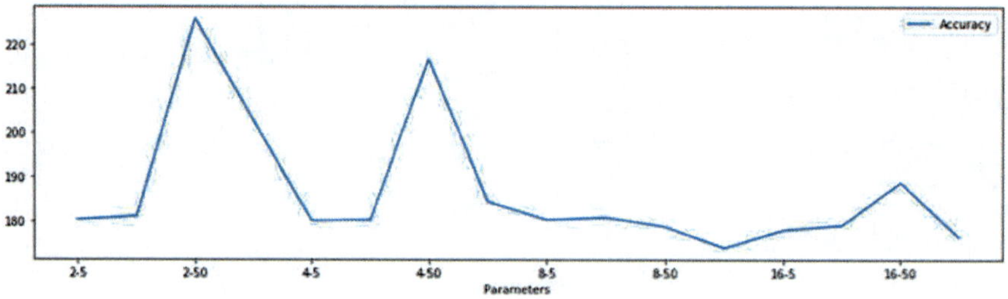

Fig. 2 Outcomes of parameter trials for ANN

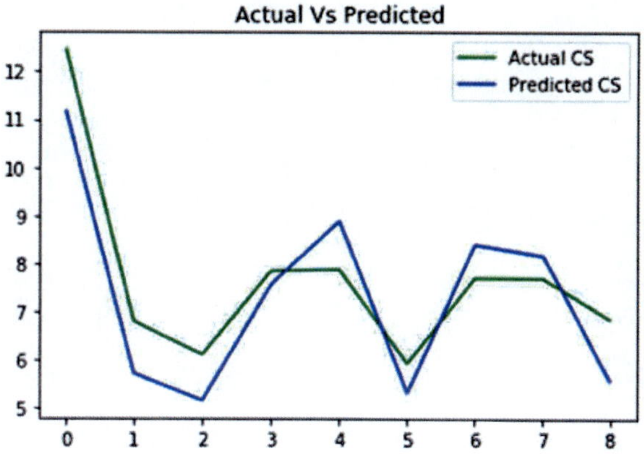

Fig. 3 Relationship between predicted and experimental CS

3 Conclusion

The study demonstrates that the compressive strength of plastic waste composite construction materials can be predicted with 88.59% accuracy using a Deep Learning ANN model. Current research demonstrates that developing a model using ANN shows a substantial correlation between input parameters i.e. FA, CA, HDPE, LDPE, PETE, UEO and the response variable i.e. compressive Strength of PCCM.

References

1. Bell J (2014) Chapter 5—artificial neural networks. Mach Learn Hands-On Dev Tech Prof 91–116
2. Moretti JF, Minussi CR, Akasaki JL, Fioriti CF, Melges JLP, Tashima MM (2016) Previsão do módulo de elasticidade e da resistência à compressão de corpos de prova de concreto por meio de redes neurais artificiais. Acta Sci Technol 38(1):65–70. https://doi.org/10.4025/actascitechnol.v38i1.27194
3. Abdulla NA (2022) Application of artificial neural networks for prediction of concrete properties. Mag Civ Eng 110(2). https://doi.org/10.34910/MCE.110.7
4. Dantas ATA, Batista Leite M, De Jesus Nagahama K (2013) Prediction of compressive strength of concrete containing construction and demolition waste using artificial neural networks. Constr Build Mater 38:717–722. https://doi.org/10.1016/j.conbuildmat.2012.09.026
5. Rajeshwari R, Mandal S (2019) Prediction of compressive strength of high-volume fly ash concrete using artificial neural network. Lect Notes Civ Eng 25:471–483. https://doi.org/10.1007/978-981-13-3317-0_42
6. Yu H, Zheng J, Lin Q (2022) Strength prediction of seawater sea sand concrete based on artificial neural network in python. Mater Res Express 9(3). https://doi.org/10.1088/2053-1591/ac5957
7. Gupta S (2013) Using artificial neural network to predict the compressive strength of concrete containing nano-silica. Civ Eng Archit 1(3):96–102. https://doi.org/10.13189/cea.2013.010306
8. Jain D, Bhadauria SS, Kushwah SS (2022) An experimental study of utilization of plastic waste for manufacturing of composite construction material. Int J Environ Sci Technol. https://doi.org/10.1007/s13762-022-04447-7

Estimating the Life Cycle Cost of a Structures–Case Study

Jaydeep Singh Sanodiya and Harsh Rathore

Abstract For some random structure, the "life cycle cost" is "the current worth of all assessed uses influencing development in addition to all costs connected to support and organization over the existence of the design. "A product's "life cycle cost" is the sum of all its repair and replacement expenses, as well as the costs associated with the risk of damage from environmental factors like corrosion and earthquakes. That's why it's so important to keep up with repairs on concrete buildings as they age to ensure they continue to work as intended. Existing concrete buildings in Bhopal, India were the focus of a case study reported here.

Keywords Concrete · Structures · Service life · Cost · Durability

1 Introduction

"If the monitoring system offers enough notice that remedial action can be performed before the damage grows to a failure level, only then can the life-safety and economic advantages associated with such a concept be achieved. A more complex monitoring hardware deployment on the system and a more complex data processing technique to query the collected data are necessary costs of adopting such a mindset. The "life cycle cost" of a structure is "the current worth of all normal consumptions connected with the development in addition to all expenses related to upkeep and organization over the existence of the design."

"The term "lifecycle cost" is often used to describe the overall price tag associated with a product's inevitable degeneration owing to wear and tear processes like corrosion and the danger of damage from natural disasters like wind and earthquake. "Total cost = Pc + Pi + Pm + Pr + other expenses" Construction cost (Pc), inspection cost (Pi), maintenance cost (Pm), and renewing/replacing cost (Pr) may all contribute to an estimate of the total life cycle cost (Pr). When comparing two alternatives for a project, life cycle costing is the strategy for decision assuming the two arrangements

J. S. Sanodiya (✉) · H. Rathore
SAGE School of Engineering and Technology, SAGE University, Bhopal, Madhya Pradesh, India
e-mail: jaydeepsanodiya@gmail.com

K. K. Pathak et al. (eds.), *Latest Developments in Civil Engineering*, Lecture Notes in Civil Engineering 352, https://doi.org/10.1007/978-981-99-2676-3_21

give basically similar degrees of administration and advantages to the undertaking's end client [1]. After deciding to improve or implement a new project, businesses often turn to LCCA to determine the most cost-effective means of doing so. Kong and Frangopol [2] are only two of many scholars all around the globe focusing on this topic. The proposed approach may be used for either brand-new or already-built civil infrastructures, and it is compatible with a wide range of upkeep plans. For their life cycle cost prediction model, Li and Guo [3] used 42 years' worth of data on repairs and upkeep. When it comes to predicting how well a building will perform, Kim and Frangopol [4] established a method that uses structural health monitoring to make such predictions."

2 Life Cycle Cost of a Concrete Structures

Case study of a concrete structure situated in the City of Bhopal has been conducted and presented here. Information of structural plan, compressive strength, concrete cover and reinforcement design is collected and the approximate cost of construction is calculated based on the above data. Compressive strength is found to be 20 MPa. Concrete cover is taken as 30 mm. The building has a 50-year expected lifespan. Corrosion initiation time is determined as 18.37 from different established relations such as $ti = (Cc/K)^2$ where Cc is concrete cover and K = coefficient of carbonation. Below, we examine these expenses in light of the period from corrosion onset, the original cost of construction, and how much of the part has to be replaced after its useful life is over.

Pr = replacement cost;

m = how often a part needs to be changed as a percentage = 50%= 0.5, mostly because it's just necessary to repair the concrete that's around the rebar's

I= Yearly rate of inflation i.e., 4% = 0.04

N= Expected Service life = 18.37 years = 18 years

Pc = Cost of building = 353000/-

The following relation may be used to estimate the cost of renewal or replacement:

$$Pr = Pc \text{ x } m\,(1+i)^N \tag{1}$$

So, $Pr = 353000 \text{ x } 0.5\,(1.04)^{18} = 357557$

It is possible to calculate the present value of this expenditure by utilizing,

Present Cost = To be modified costs/$(1+r)^t$

The net discount rate of money, r, is substituted for t, which is the number of years after the initial investment was made. Considering r = 0.05

Therefore, Pr = 148572 at the current time.

The total cost is—if maintenance is not performed when corrosion first appears and the beam is repaired or replaced when it is 40 years old and fails.

To delay the inevitable collapse of a building, cracked concrete should be replaced at regular intervals. Current findings indicate that the overall cost of replacement may be reduced by replacing 50%, 60%, and 75% of corroded concrete every 18 years (at ages 18, 36, and 54).

Pr = Pr (18) + Pr (36) + Pr (54);

Pr (18) has been calculated to be 148572

It is possible to calculate the initial building cost for Pr (36), given Pr (18) and the value of 353,000 at Pr (18) years of age.

Pr (36) = (353,000 * $(1.05)^{18}$ + 148,572) * 0.6 $(1.04)^{18}$) = 1,213,190.

Hence, current value of Pr (36) = 209,467/-

Pr (54) = (353,000*$(1.05)^{36}$ + 148,572*$(1.05)^{18}$ + 209,467) *0.75$(1.04)^{18}$) = 3,967,867/-

Consequently, current value of Pr (45) = 441,610/-

Therefore, Pr = 148,572 + 209,467 + 441,610 = 799,649.

When no damage is found during the first year after construction, a conservative estimate puts the cost of inspection at 0.7 percent of the overall price of the building. (Frangopol et al., 2), a fraction of the total cost if modern non-destructive techniques (NDT) are used to detect corrosion of rebars and the formation of cracks.

Therefore, the initial cost of inspections is.

0.07* 353,000 = 24,710

The price of future inspections may be estimated using

$$\mathrm{Pi} = \mathrm{Pin}\,(1 + \mathrm{r})^{\mathrm{D}} \tag{2}$$

Pi = the total amount spent on inspections throughout the structure's whole expected service life.

Pin = Annual inspection fees during the first year of a building's existence = 24,710.

D = structure's expected lifespan = 50 years.

S0, Pi = 283,360.

Protection of repair and upkeep expenses Pm may be gauged by how often checks are performed. Frangopol, et al. (2011) [4] state that

$$\mathrm{Cmain,\,t} = \mathrm{Cmain\ x\ t}; \tag{3}$$

If we divide the total cost of defense maintenance in year one by the concrete's age in years, we get the annual cost of defense maintenance, Cmain, for that year. For the sake of this analysis, the total cost of maintenance is calculated assuming that efences are maintained at 15, 30, and 45 years of age, respectively.

Pm = Cmain,15/ $(1+r)^{15}$ + Cmain,30/ $(1+r)^{30}$ + Cmain,45/ $(1+r)^{45}$ Cmain is estimated to account for 0.05% of total building costs = 17650

Pm = 127,349 + 122,514 + 88,397 = 338,260.

Therefore, the total cost of ownership for the beam is calculated to be.

Total cost = 353000 + 799649 + 283360+ 338260 = 17, 74,269.

3 Conclusions and Discussion

This study is a case study of determining the total cost of ownership for a concrete building. This technique optimizes the monetary costs while still taking into account the structural service life.

The beginning of corrosion due to carbonation is the basis for the limit state. Determining the service life of a building is important since it determines the total cost of ownership and the frequency of costs across the structure's design lifetime. To guarantee a constant comparison of expenses throughout time, the total cost during its useful life is determined by adding together the initial build price and the estimated repair costs in the future and discounting them to present day values.

References

1. (USDTFHAO) U.S. Department of transportation federal highway administration office of asset management. August 2002: Life-Cycle Cost Analysis Primer
2. Kong JS, Frangopol DM (2003) Evaluation Of Expected Life-Cycle maintenance cost of deteriorating structures. J Struc Eng 129(5):682–691
3. Li C, Guo S (2012) Life cycle cost analysis of maintenance costs and budgets for university buildings in Taiwan. J Asian Arch And Bild Eng 11(1):87–94
4. Kim S, Frangopol DM, Zhu B (2011) Probabilistic optimum inspection/repair planning to extend lifetime of deteriorating structures. J Perf Constr Fac 25(6):534–544

Environmental Engineering

Design of Grey Water Treatment Using Community Soak Pit Method

Sayed Shabaj M. Sagar and Sanjeev Sangami

Abstract The wastewater produced by homes is separated into two types grey water and black water, with grey water making up the majority of the flow. Grey water from residential apartments makes up 50–70%. Grey water's composition varies widely depending on where it comes from (e.g., bathroom, laundry, or kitchen grey water), and it is affected by the local water quality. Preliminary data has been collected and surveying has been done for these selected village Honnakalasapura village with a population of 509. After surveying these villages, community soak pits are provided at the end of drainage points prior to mixing with black water. If any houses are not connected with drainage, then the individual soak pits are given for a particular house. Villages with Grey water pipelines that are directly connected with sanitary lines are the villages not considered for treatment. Reusing and Recycling of Grey water can be achieved by providing a community soak pit as well as individual soak pit. The design of the community soak pit is according to the population of village. Based on this 24.48 KLD greywater is produced per day. Based on this the diameter of the soak pit is given as 1.80 m, depth = 4.3 m and 7 numbers of rings are provided, the main aspect of this method is cost effective and easy to use.

Keywords Grey water · Methods · Reduce · Reuse · Recycle and Recharge

1 Introduction

In order to improve public health in rural regions, grey water management is crucial. Grey water that is safely handled can help reduce exposure to water-borne and water-washed illnesses, as well as vector-borne diseases. Grey water management can significantly lessen the difficulties associated with freshwater consumption [1].

S. S. M. Sagar (✉) · S. Sangami
Department of Civil Engineering, Jain College of Engineering, Belagavi 590014, Karnataka, India
e-mail: 1997shabaz.s@gmail.com

S. Sangami
e-mail: sanjeevenv08@gmail.com

K. K. Pathak et al. (eds.), *Latest Developments in Civil Engineering*, Lecture Notes in Civil Engineering 352, https://doi.org/10.1007/978-981-99-2676-3_22

Grey water is frequently collected from residential neighborhoods, business districts, and institutional settings. The water is described as being grey due to its appearance in the foggy environment and the fact that it is neither fresh nor very contaminated (black water). To be precise, grey water is any water emitted from residential structures that are not toilet waste. Grey water may still be suitable for reuse despite the presence of grease, food particles, and other contaminants [2].

Specifically, grey water is wash water, water used for baths, dishes, and laundry that is free of garbage-grinder residue and excludes toilet waste. When handled appropriately, greywater might be a resource for both agricultural and horticultural crops as well as home gardeners [3].

Grey water and black water are the two main categories of wastewater. Grey water is a type of residential wastewater that is created during activities including bathing, laundry, clothes washing, and dishwashing. Between 55 and 75% of the home water discharges were wastewater. Different from black water is grey water. Grey water varies from black water in terms of both the kind and quantity of its biological and chemical pollutants (feces water to noxious chemicals) [4].

In addition to landscape architects, builders, developers, and contractors, on-site greywater treatment and management for design and landscaping has several benefits. Greywater is a source of phosphorus, potassium and potassium nitrogen pollutants for lakes, rivers, and groundwater. Also, make this specific kind of wastewater is a great supply of nutrients for plants when it is made accessible for irrigation [5]. Wastewater is any water that has been polluted. Use and human activity. Wastewater is used to clean up any mix of home, commercial, industrial, or activities including agriculture and any sewage inflow infiltration. Consequently, wastewater is a byproduct of residential, commercial, industrial, or agricultural pursuits [6].

Grey water reuse serves two purposes: First and foremost, it uses less freshwater than is required for domestic tasks and produces less wastewater that is collected in sewage tanks or septic tank systems. Grey water naturally includes nutrients that can be useful when utilized as fertilizer. When it is treated like sewage and discharged into ponds, lakes, and rivers, it becomes a pollutant [7].

As a resource, water is becoming increasingly scarce. By 2025, one third of Indians would be experiencing water scarcity, according to the worldwide water management institute. The disparity between water availability and demand is growing as the population grows. By reusing the cleaned water, the free technologies would undoubtedly save a significant amount of water [8].

1.1 Technology Options for Grey Water Treatment

(1) **On site (Decentralized)**

   (a) Soak pit
   (b) Leach pit
   (c) Kitchen garden

(d) Community Soak pit
(e) Community Leach pit
(f) Community Kitchen gardens

(2) **Off Site (Centralized)**

(a) Small Bore System
(b) Waste Stabilization Ponds
(c) Constructed Wetland
(d) Soil Bio-Technology
(e) Phyto rid Technology
(f) Anaerobic Baffled Reactor
(g) Moving Bed Bio-Film Reactor

When comparing the above methods and lots of studies, the best method suited for rural areas are Community soak pit method and also economical method when compared to other methods. Under the *Swaccha Bharath mission Grameen phase II Ministry of Jal Shakthi Government of Karnataka has raised the fund for this method due to being Economical and Cost Effective.*

1.2 Comparison Between Decentralized and Centralized

See Table 1.

2 Methodology

See Fig. 1.

Table 1 Comparison of both technologies

Decentralized	Centralized
O&M costs are low and borne by the house owner	O&M costs are more when compared to other methods
Requires less space	Requires more space
Unskilled manpower is required	Skilled manpower is required
Less capital cost	High capital cost depending on the price of land
Required less maintenance	Required periodic maintenance
No drains required	Required drains to convey the grey water from village to treatment unit

Fig. 1 Flow chart of Methodology

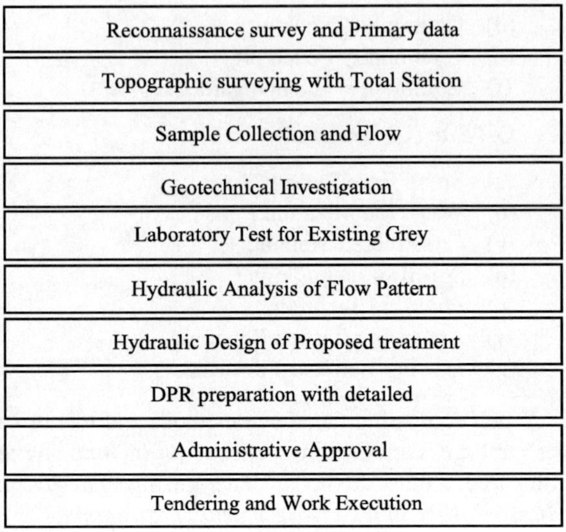

3 Result and Discussion

See Table 2.

Design methods are taken from the Grey water resource management Book Ministry of Jal Shakthi Government of India.

3.1 Population Method

Population = 509
Production Rate = 30%
Water Required = 55 LPCD
Grey water = 70 %

(Note: If we provide 100% of water in that 65–70% of the total water supply in rural areas is converted to grey water after use.)

$$= (509) + (509 \times 30\%) \times 55 \times 70\%$$

Table 2 Data of Honnakalsapura

Gram Panchayat	Surgajakkanahalli
Name of village	Honnakalasapura
Village population (present)	509
Total number of houses	80

$$= 25475.45\ \text{L}$$
$$= 25.48\ \text{KLD}$$

3.2 Design of Community Soak Pit

Number of community soak pit proposed = 5
Quantity of grey water for each community soak pit = 5.10 KLD
Percolation rate = 1 min (IS 2470-1985 part II)

(Note: Based on the test, the soil is found as clay soil, for this soil the percolation rate is taken as 1)

$$\text{Percolation area} = \frac{5.10 \times 1000}{204} = 25.32\ \text{m}^2$$

Diametere of community soak pit = 1.80 m

$$\text{Depth of community soak pit} = \textbf{Percolation area} - \pi/4 \times \mathbf{D}^2/(\mathbf{D}^2)$$
$$\text{D} = (24.98 - 2.55)/(5.66)$$
$$\text{D} = 3.97\ \text{m} \quad \text{D} = 4\ \text{m}$$
$$\text{Free Board} = 0.3\ \text{m} \quad \text{Total Depth} = 4.3\ \textbf{m} \tag{3.2.1}$$

3.3 Soak Pit Ring Design

See Fig. 2.

$$\text{Quantity of grey water} = 5.10\ \text{KLD}$$
$$\text{V} = (\pi\ d^2/4) \times h$$
$$\text{h} = \text{V}/(\pi\ d^2/4)$$
$$\text{h} = \frac{5.10}{3.142 \times 1.8 \times 1.8/4}$$
$$\text{h} = 2.00$$
$$\text{Depth of each ring} = 0.3$$
$$\text{Number of rings to be provided} = 2.00/0.3$$
$$= 7\ \textbf{numbers} \tag{3.3.1}$$

Fig. 2 Village map with road connections

As per the Census 2011 information the Village code of Honnakalasapura is 613310. Honnakalasapura village is situated in Anekal taluka Bangalore urban district in Karnataka India. It is located 35 km from the district headquarters in Bangalore. According to 2009 data Suragajakkanahalli is gram panchayat of honnakalasapura. The overall geographical area of the village is 143.08 ha. There are about 105 households in Honnakalasapura village. Pincode of the village locality is 56210 (Figs. 3, 4, 5, 6 and 7).

4 Conclusions

The benefits of grey water recycling by using community soak pit and Individual soak pit technologies will reduce Water/Environmental pollution and mainly it is a benefit to the ground water recharge. Most of the other technologies are there such as waste stabilization ponds, Phyto rid Method and DEWATS technologies but these technologies are required centralized land, required more area and also High-cost technologies. Hence, to conclude soak pit method is suitable in rural areas and semi-rural areas. Because it required less capital cost, maintenance free, easy to construct, comparatively it required very less area to construct and it will directly improve the ground water table.

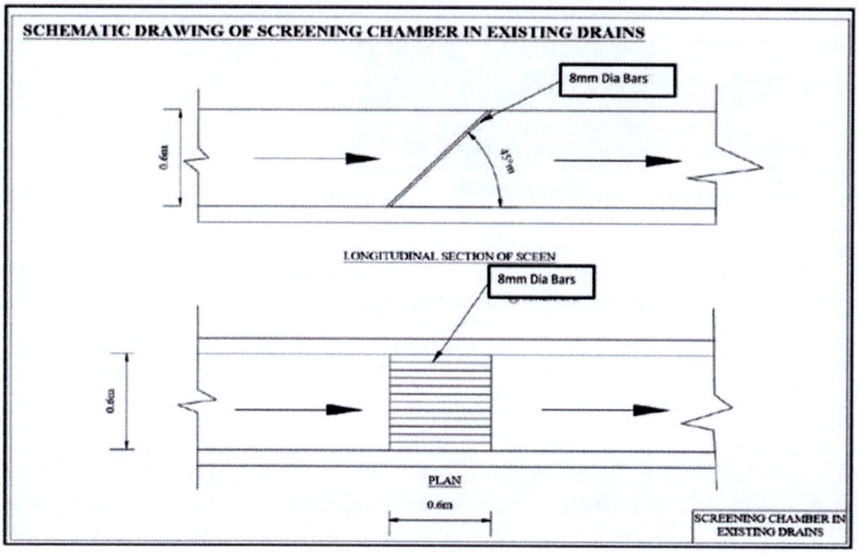

Fig. 3 Screen Chamber is provided at the end of drainage it is a type of mesh which eliminates the unwanted objects at the beginning of the treatment process

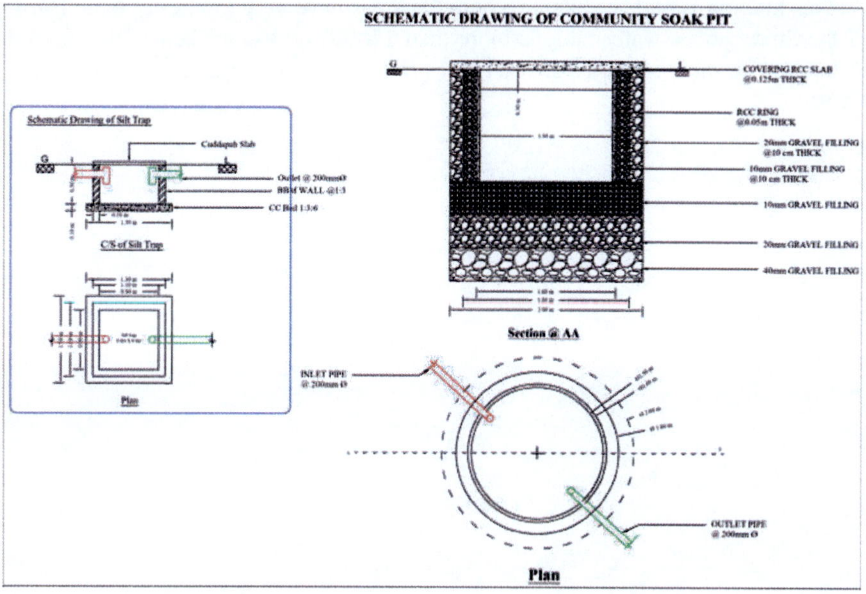

Fig. 4 Community soak pit, is given after the screening this is a pit where the grey water percolates in this pit then the water can be used as groundwater recharge or else it can be used for gardening purpose

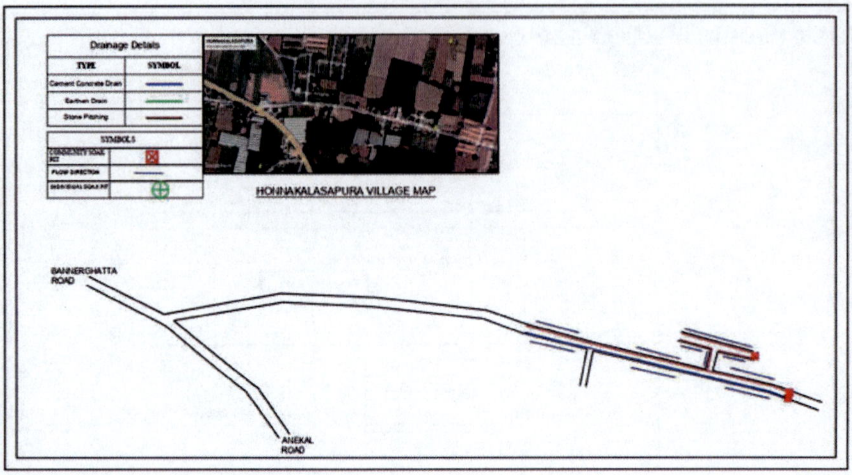

Fig. 5 Layout of Honnakalasapura Village Which comes under the Surgajakkanahalli GP after surveying and data collected from this village, three community soak pits are required according to the population of this village the red symbol shows the community soak pit at the end of drainage and blue arrow mark shows the flow directions

This method is highly suitable where the areas have very low water table (Most of the Study area's water table is more than 1200 ft or 400 m) hence the soak pit technology is most suitable for rural areas in the community level and house hold level.

Fig. 6 Outfall point 1 at the end of drainage. Where the community soak pit is to be propose

Fig. 7 Outfall point 2 at the end of drainage

References

1. Swachh Bharath mission (Grameen Phase II) guidelines for grey water management
2. Rakesh et al (2019) A reviews on Characteristics and treatment of grey water. IJCS 8(1):34–40
3. Sheikh et al (2018) A review on grey water: journal of contemporary water research & education article about the use of reclaimed water for agriculture in California (165):28–41, December 2018
4. Lambe et al (2013) Greywater—treatment and reuse by using house hold technology. IOSR-JMCE 20–26. ISSN: 2278-1684
5. Singh et al (2018) A review on feasibility study of treatment technologies for greywater to enhance water security, 13(6):4042–4048. ISSN 0973-4562
6. Shelar et al (2019) Review on treatment of grey water using cost effective technologies for Kushvartha Kund Water. IRJET 6(5), May 2019
7. Tandekar et al (2021) Research paper on grey water treatment by using coconut shell. IRJET 8:2 Feb 2021. e-ISSN: 2395-0056
8. Singh V et al (2018) A review on the possibility study of greywater treatment technologies to improve water security, 13(6):4042–4048

Experimental Analysis of Impact of Solar Chimney on Ventilation Rate in a Building

Jitendra Raghuwanshi and Vishvendra Nath Bartaria

Abstract The rapidly developing socio-economic conditions since last few decades are tremendously escalating the energy demand in all spheres of life. Accordingly, the energy needs in buildings have also increased significantly. The energy requirements are majorly fulfilled by the burning of fossil fuels. Huge energy consumption and burning of fuel are impacting environmental sustainability through global warming, climate change and changing weather conditions. Building energy use may be reduced by utilization of renewable energy sources such as solar energy. This paper presents the strategy for reducing building energy use through the application of solar energy in solar chimney to produce ventilation. The performance parameters and the concept of solar chimney have been discussed along with the estimation of flow. Analytical calculations using mathematical models are made. Experimental analysis on ventilation produced through solar chimney is also performed. The research facility at GWPC, Bhopal for carrying out this research work has been used with suitable modifications in instrumentation and flow arrangement for the determination of ventilation rate during winter sunny days. The experimental results are compared to those calculated from simple mathematical model. The ventilation rate as obtained from experiment varies between 0.2 to 0.4 m^3/s. Mathematical calculations on the rate of ventilation show little variations of up to 10% value arising due to the assumptions made in the model formulation. The results are indicative of application of solar chimney in the establishment of ventilation in buildings. It also shows that this passive ventilation technique may reduce the use of conventional energy in buildings significantly and without affecting the indoor thermal environment.

Keywords Passive ventilation · Indoor environment · Solar chimney

J. Raghuwanshi (✉) · V. N. Bartaria
Mechanical Engineering, LNCTU, Bhopal, M.P, India
e-mail: jitendrar@lnct.ac.in

V. N. Bartaria
e-mail: vnbartaria@lnct.ac.in

K. K. Pathak et al. (eds.), *Latest Developments in Civil Engineering*, Lecture Notes in Civil Engineering 352, https://doi.org/10.1007/978-981-99-2676-3_23

1 Introduction

1.1 Energy Demand in Buildings

The rapidly developing socio-economic conditions since last few decades are tremendously escalating the energy demand in all spheres of life. With the overall increase of energy needs, the energy requirements in a building is also increased as humans are spending most of their time indoors for most of the activities such as office work, academic affairs, indoor games, sleeping, cooking, eating and other related tasks. Due to long time human occupancy and increased indoor activities, the indoor environment is normally much more polluted than the outside atmosphere [1]. According to the available literature on the indoor environment, the need of improved indoor environmental conditions has continuously been increasing since last 150 years [2]. The poor or non-ventilated building spaces are often found filled with dust, smoke and odor. In improving indoor environment, mechanical means of cooling, heating, and ventilation are commonly being employed in buildings. Their operation relies on the availability of electricity [3]. Various building energy use are shown in Fig. 1.

Environmental Impact. The huge energy consumption in urban areas is responsible for creating environmental issues such as global warming, climate and weather change conditions [4]. The need of more and more energy in carrying out indoor activities in modern domestic, commercial and industrial spaces is increasing continuously. The increasing need of energy can be reduced up to a certain level by improving the efficiency of the systems and equipments being used in the building spaces. But, the modern practices of creating artificial indoor environment will still require huge energy until their use is reduced or replaced by other innovative systems and strategies which require no or low conventional energy [6]. Incorporation of sustainable strategies and the use of renewable sources of energy such as solar energy in buildings has a prospective solution to reduce energy usage and further degradation of environmental conditions [7].

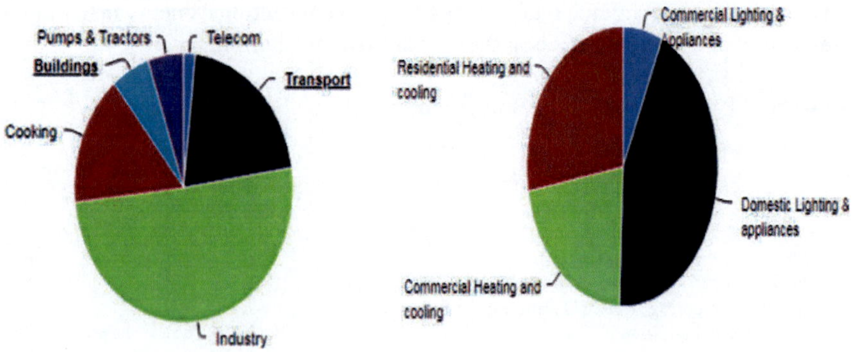

Fig. 1 Energy use in buildings (*Source* India energy security scenarios)

2 Future Energy Needs

With the economic growth, population and rapid urbanization as discussed, the energy needs in buildings have risen. According to an International Agency report (International Energy Outlook, 2017), the energy use patterns for the rural and urban populations are different and rising. With the rapid global development causing more industrial, commercial and economic activities concentrated near or within the urban areas, it is expected that the rate of commercial energy growth will be higher than the residential, but, the building energy consumption will be greater [8].

As per the statistics of the Government of India (GoI), in India, the energy consumption in residential and commercial buildings is about one-third of the total electricity consumption. It is estimated that the electricity consumption in commercial and residential buildings for heating, cooling, lighting and air conditioning will rise by 860% by the year 2047 [9].

Fulfilling energy requirements and effective utilization of energy in buildings can be ensured by its effective management and use of alternative energy sources. Effective utilization of building energy is also important in mitigating the climate change and preventing environmental pollution. This will help in reducing the energy needs in buildings and achieving the HVAC requirements. This is also required for a sustainable society and usage of non-renewable resources [11]. The conventional practices can be reduced with the construction of new buildings with energy efficient building designs. Use of green technologies and passive features in building designs and construction can reduce the energy application in HVAC [12]. Improvement in the efficiency of energy use will result in the reduction of both energy consumption and emissions of CO_2 in the atmosphere [13].

3 Parameters Affecting Energy Performance of Buildings

Energy consumption of buildings is related with the parameters of energy performance of the building. Building orientation, solar daylight parameters, construction materials used in building, and design parameters of buildings are some of the important parameters which affect the energy needs in buildings and indoor environments. These parameters are required to be carefully considered at the initial design stage of the building design and construction [14]. For any later modifications in buildings for efficiency improvement, only limited changes can be made at a higher cost. Modern tools and software will also help in building design where simulation of building energy consumption provides solutions for efficient energy use in buildings. EnergyPlus software is commonly being used for such purposes. Optimization techniques including Genetic Algorithms etc. are also useful and are being adopted in the optimal design of energy efficient buildings [15, 16].

4 Passive Ventilation in Buildings

The ventilation in buildings is a process of replacing the indoor air by the outside fresh air. In natural ventilation, this process takes place naturally by thermal buoyancy effect. It is an attractive technique involving the natural flow of air from indoor spaces. Mechanical ventilation as the most commonly used strategy needs consumption of large amount of conventional energy or electricity while natural ventilation use passive methods. Natural ventilation needs suitable arrangements in buildings which drive the air to flow outside. Natural ventilation with the help of solar energy is a centuries old technique famously used in Italy and known as "Scirocco rooms" [7]. Solar energy is a clean renewable energy source, abundant on the earth surface and its use is inevitable for us [17]. It must be explored in the natural ventilation.

5 Solar Chimney

Solar chimney is a possible means of driving air through available solar energy. It consists of an absorber plate with a transparent cover over it. When the solar radiations pass through the transparent cover and fall on the black painted absorber plate, the air in contact with the absorber plate heats up. Due to this heating effect, the air flows in the upward direction due to the buoyancy effect. The properly designed solar chimney may be used for heating or cooling and ventilation of building spaces [18, 19].

Solar chimney provides natural ventilation and the flow of air takes place by the use of solar energy. The principle of working of solar chimney is based on greenhouse effect and buoyancy forces are created which causes the flow of air as shown in Fig. 2. Solar chimney systems can be effectively used in creating natural ventilation by driving the indoor air to replace it by the fresh outside air.

6 Experimental and Analytical Investigations

6.1 Analytical Analysis

In a solar chimney solar energy drives the air. Heat transfer is involved in the operation of solar chimney in heating air above or below the absorber plate. The mathematical model used in the analysis of flow through the chimney is on the heat transfer equations. Considering the assumptions of steady state one-dimensional heat transfer and constant temperatures at various points of absorber plate and neglecting the heat losses due to conduction, convection and radiations. The equation for the flow rate of air Q in solar chimney is given by the relation [20, 21].

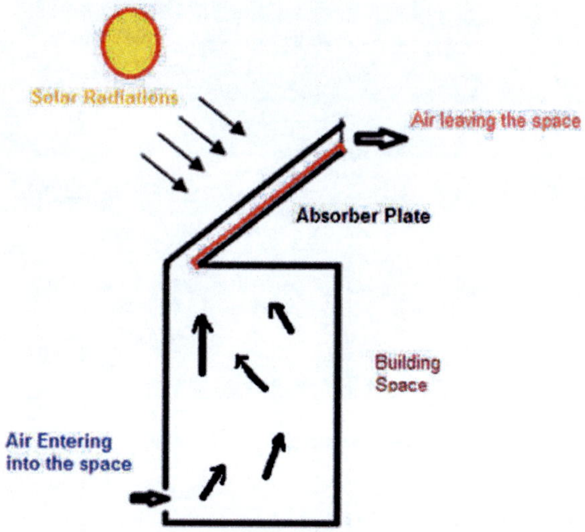

Fig. 2 Use of solar chimney in ventilation

$$Q = C_d A_o \sqrt{\frac{2g\Delta TL}{T_{a,i}\left\{1 + (A_o/A_i)^2\right\}}}$$

In this relation, Cd is the discharge coefficient., and the cross-sectional areas of chimney at inlet and outlet sections are Ai and Ao respectively. ΔT is the difference in temperature of air at inlet and outlet and Ta, i is the temperature of air at inlet. L is taken as the stake height of the chimney. The discharge coefficients are taken between 0.65 to 0.7 [22, 23] in the equation.

6.2 *Experimental Set-Up*

The established research facility at GWPC, Bhopal as shown in Fig. 3 is used in this experimental study with suitable modifications to capture temperature at flow rate at various places of test space. The solar chimney is placed on the roof of a building and the test space is connected with the solar chimney through PVC pies of 200 mm diameter. Measurements are made for the determination of air flow rate on a sunny day of Dec, 2021 in the ventilated test space. Hot wired anemometer records the air velocity and RTD measures the temperatures at the inlet and outlet of the pipe connected between the test space and the solar chimney.

Fig.3 Solar chimney at test facility

7 Results and Discussions

The analytical solutions are obtained for the determination of the rate of air flow by using the mathematical relations [20, 21]. Experiments are also performed on the research set-up. Flow of air leaving the solar chimney and the solar radiations falling on the chimney are recorded at an interval of 1 h. The temperatures at inlet and exit sections of the chimney are recorded and the theoretical calculations of air flow rate as shown in Table 1 are made.

Table 1 Air flow calculations from mathematical model

Time (hrs.)	To (°C)	Ta,I (°C)	L (m)	Q (m3/s)
9	20	18	4.82	0.3
10	22	18	4.82	0.4
11	25	21	4.82	0.4
12	26	24	4.82	0.2
13	30	25	4.82	0.4
14	32	26	4.82	0.4
15	32	26	4.82	0.4
16	28	27	4.82	0.2

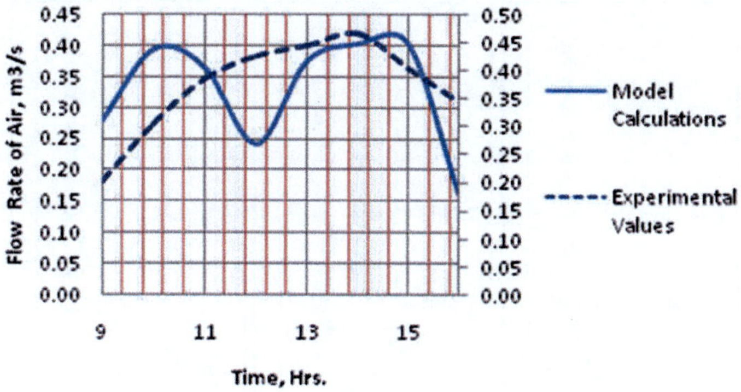

Fig.4 Comparative analysis of experimental and model calculation for air flow rate

Results on Ventilation Rate. The test results were observed on a sunny day of December, 2021from 09.00 h. to 16.00 h. Air flow rate of the solar radiations falling on the absorber plate of the chimney are recorded at an interval of one hr. The temperature of air at the inlet and exit section is recorded for the theoretical calculation of air flow rate. The experimental values show that the minimum air flow rate is 0.20 m3/s and maximum is 0.47 m3/s. The minimum value of air flow rate from analytical calculations is 0.16 m3/s and the maximum is 0.40 m3/s. These show a similar pattern of the air flow from analytical calculations and the experimental values, Fig. 4.

The experimental value of air flow rate shows good agreement with the analytically obtained values. The room air temperatures are also recorded to show the vertical temperature distributions at the set test locations. The test locations at the center and two opposite sides of the room are considered for the simulation of vertical temperature distributions as depicted in fog. The vertical temperature plates are as shown in Fig. 5a, b and c.

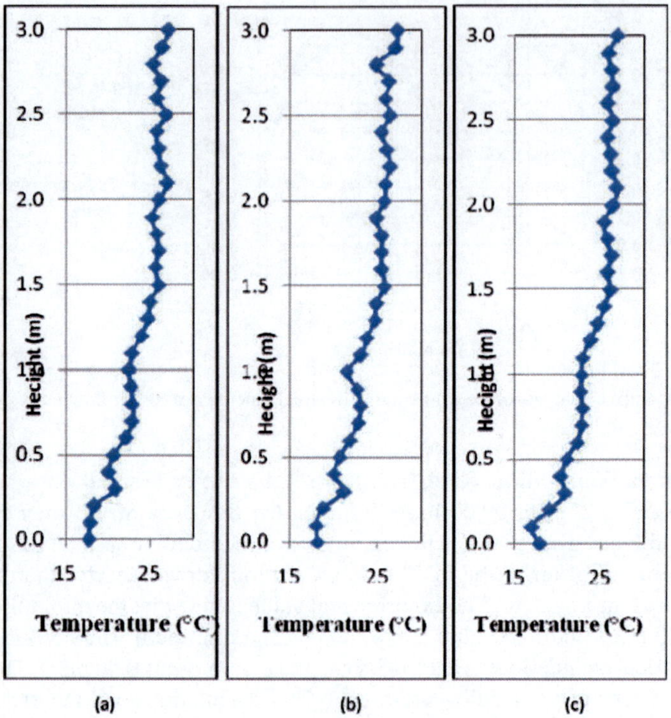

Fig. 5 The vertical temperature distributions at test locations **a** Near left wall location, **b** Central location and **c** right wall location of the room to show indoor thermal environment

8 Conclusion

Experimental and analytical study is carried out for the evaluation of natural ventilation through the solar chimney at the research facility with solar chimney. The test set-up used includes a solar chimney fitted test room and equipped with instruments needed to record flow and temperature parameters. The study shows the potential application of solar chimney strategy in producing natural ventilation in buildings. The impact of passive ventilation strategy on indoor environment is presented along with air flow rate through the solar chimney. The results are compared with the model calculation. The room thermal environment is simulated through vertical temperature plots at the three test locations.

The tests were performed on a bright sunny day of December, 2021from 09.00 h. to 16.00 h. Flow of air leaving the solar chimney and the solar radiations falling on the chimney are recorded at an interval of 1 h. The temperature of air at inlet and exit section is recorded for the theoretical calculation of air flow rate. The experimental values show that the minimum air flow rate is 0.20 m3/s and maximum is 0.47 m3/s. The minimum value of air flow rate from analytical calculations is 0.16 m3/s and the maximum is 0.40 m3/s. These show a similar pattern of the air flow from analytical calculations and the experimental values. The vertical temperature distributions show the indoor thermal environment temperature stratification.

References

1. Rajput KS, Bartaria VN (2022) Experimental study of vertical air temperature distribution and ventilation with energy saving using solar chimney. Materials Today. Proceedings 63:692–698. https://doi.org/10.1016/j.matpr.2022.04.914
2. Wei JS, Zhao JN, Chen QY (2010) Energy performance of a dual airflow window under different climates. Energy and Buildings. 42:111–122
3. World Health Organization (1983) Indoor air pollutants exposure and health effects. Copenhagen EURO Reports and studies. 78
4. Al Khidir T, Zailani S (2009) Going green in supply chain towards environmental sustainability. Glob J Environ Res, 3(3), pp 246–251
5. A.I.J Japan Architectural Institution (2005) Architecture for a sustainable future, IBEC
6. Puchkal V, Jurmanov B (2013) Stochastic model of the thermal regime and heat consumption of residential buildings for heating. World Appl Sci J 23(812):191–196
7. Ghalamchi M, Kasaeian A, Ghalamchi M (2015) Experimental study of geometrical and climate effects on the performance of a small solar chimney. Renew Sustain Energy Rev 43:425–443
8. International Energy Outlook (2017)
9. Building an energy efficient India, Urban Futures, July 26, 2018, https://www.orfonline.org/programme/energy.
10. Perez-Lombard L, Ortiz J, Pout C (2008) A review ´ on buildings energy consumption information. Energy and buildings 40(3):394–398
11. Yu Z, Haghighat F, Fung BC, Yoshino H (2010) A decision tree method for building energy demand modeling. Energy and Buildings 42(10):1637–1646
12. Santamouris M, Mihalakakou G, Argiriou A, Asimakopoulos DN (1995) On the performance of buildings coupled with earth to air heat exchangers. Sol Energy 54(6):375–380
13. Zheng X, Yu Y, Wang , Deng H (2014) Identifying the determinants and spatial nexus of provincial carbon intensity in China: A dynamic spatial panel approach. Regional environmental change, 14, pp 1651–1661
14. Magnier L, Haghighat F (2010) Multi objective optimization of building design using trnsys simulations, genetic algorithm, and artificial neural network. Build Environ 45(3):739–746
15. Lin S-HE, Gerber DJ (2014) Designing-in performance: A framework for evolutionary energy performance feedback in early stage design. Autom Constr 38:59–73
16. Cristofalo S, Orioli S, Silvestrini G (1989) Thermal behavior of "Scirocco rooms" in ancient Sicilian villas. TunnIng Undergr Space Technol 4(4):471–473
17. Bartaria VN et al (2018) CFD Analysis of room air ventilation through solar chimney. IJSR 7(8):16–18
18. Lal S, Kaushik SCR (2016) Hans Experimental investigation and CFD simulation studies of a laboratory scale solar chimney for power generation, Sustain Energy Technol & Assess 13 13–22
19. Vishvendra Nath Bartaria et.al. (2019) Experimental study on ventilation through solar chimney, Int J Civ Eng Technol (IJCIET), 12, 10(12) 545–549
20. P B Kommineni Venkateshwara Rao ,Vishvendra Nath Bartaria, Passive Ventilation and Solar Chimney: A Review, J Emerg Technol Innov Res 6(5), 2904–2908
21. Kommineni Venkateshwara Rao, Vishvendra Nath Bartaria (2021) Effective passive ventilation technique with solar chimney: A Review, PB, Indian Social Science Congress XLIV 2021, Vidisha Madhya Pradesh. March 15–19, pp 211

22. Kommineni Venkateshwara Rao (2019) Vishvendra Nath Bartaria, Prashant Bhave. passive ventilation and solar chimney: a review, JETIR 6(5):2904–2908
23. Kommineni Venkateshwara Rao (2019) Vishvendra Nath Bartaria, Prashant Bhave. Experimental study on ventilation through solar chimney, IJCIET 10(12):545–549

Microplastics in the Environment: Its Sources, Occurrence, Impact on Human Health and Environment

Dinesh Kumar Gupta, Amit Vishwakarma, and Archana Singh

Abstract Microplastics is gaining global attention because of recent encounter in aquatic and terrestrial environments and related ecological impacts. Microplastics have been encountered in drinking water, air, freshwater environment, biota, marine environment and table salt, presenting a very serious human exposure risk. However, detailed analytical and practical methods for quantification and characterization of microplastics still remain very limited. Studies about the very serious harmful effects on human health via dietary and respiratory exposures are also very limited. To address these issues, we have reviewed various studies related to its quantification, occurrence, microplastic encounter in ecosystem, and qualification methods of microplastics in the environment. Homo sapiens ingests microplastics through various mediums such as drinking water, inhalation and sea food. Total ingested particles ingested by homo-sapiens are estimated more than 100 particles/person annually. Floating Microplastics are very harmful for the digestive and respiratory systems through ingestion and respiratory. Considering the quantum of particles ingested by any organism in the environment, it is the need of the hour to understand the expected hazardous effects of microplastics related to human health.

Keywords Microplastics · Occurrence · Human health · Sources · Pollution

1 Introduction

The word "plastic" is derived from the Latin "plasticus" and the Ancient Greek plastikos, which means "capable of being molded". In modern age, Thompson and coworkers [1] stated that plastic may be defined according to its physical properties, which are lightweight, resistant, and durable while offering efficient thermal

D. K. Gupta (✉) · A. Vishwakarma
Department of Civil Engineering, University Institute of Technology RGPV, Bhopal 462033, India
e-mail: gupta.dinesh96@gmail.com

D. K. Gupta · A. Singh
Advanced Materials and Processes Research Institute, Hoshangabad Road, Bhopal 462026, India

K. K. Pathak et al. (eds.), *Latest Developments in Civil Engineering*, Lecture Notes in Civil Engineering 352, https://doi.org/10.1007/978-981-99-2676-3_24

and electrical insulation. Alternatively, plastic may also be defined according to its chemical nature and structure. Carpenter and coworkers [2]defined plastic as a synthetic organic polymer which is derived from the polymerization of monomers which are extracted from oil or gas. Plastic can also defined be on the basis of its polymeric structural characteristics and these characteristics provide different physical properties. Plastic Europe [3] defines plastics by employing resin identification code.

We are living in the plastic age, due to our continuous negligence towards waste management and the hazardous impacts of plastic pollution on our ecosystem. In the present world, plastics are present nearly in all the products used in regular lifestyles such as toothbrushes, clothes, cosmetic products, vehicles, computers, phones, shoes, furniture, home appliances, single use plastics and the list goes on. The life span of plastic is considered to be a closed cycle after recycling or a linear path from production to user and to the landfill or incineration based on the waste management system. In spite of these measures, plastic waste can be observed all around the globe which proves that there is a leak in the closed cycle and all these products directly or indirectly go into our ecosystem once discarded by us. It is well known fact that it takes hundreds of years for the degradation of plastics. Once this plastic reaches our ecosystem it stays there and gets degraded over time into fragments and gets mixed into various mediums such as water, air, earth etc.; these fragments are known as microplastics which are less than 5 mm. Thompson and coworkers [4, 5] defined the upper limit of microplastic to be particles less than or equal to 5 mm.

2 Sources of Microplastics

Microplastics are produced for various purposes which are cosmetics, resin pellets, microbeads, and scrubbers used in air blasting technology; these are termed as primary microplastics. The particles which get fragmented over time due to weathering are termed as secondary microplastics [6]. Microplastic is being ingested by all the organisms in the aquatic food chain from plankton to mammals since plastics waste was being dumped in marine environment since the beginning of plastic use and the thread of its bioaccumulation increases as the size of microplastic decreases [7].

In general definition two categories of plastic families exists, first one is thermoplastics which can be melted when heated and hardened when cooled, in other words, it can be frozen, reshaped and reheated simultaneously and the other category of plastic family is Thermosets which changes chemically if reheated, which creates a three dimensional network. Once these are heated and molded they cannot be reformed and reshaped.

As per plasticEurope [8] following are the different plastic products (which becomes the source of microplastic contamination) along with the percentage annual demand in the year 2018.

Table 1 Microplastic possible Sources based on demand [8]

Polymer type	Polymer name	Products	% Plastic demand
Thermoplastics	Polypropylene	Pipes, food packaging, automotive parts, packing wrappers (for sweets, tobacco and snacks), hinged caps, microwave, bank notes, containers, etc.	19.3
	Polyethylene (low density)	Food packaging film, Reusable bags, containers, trays agricultural film, etc.	17.5
	Polyethylene (high density)	House-ware, shampoo, milk bottles, bottles, pipes, toys, etc.	12.2
	Polyvinyl Chloride	Garden hoses, cable insulation, furniture, flooring sheets, inflatable pools, etc.	10
	Polyethylene Terephthalate	Disposable bottles, juice packets, packaging, soft drinks bottles etc	7.7
	Polystyrene	Fishery and dairy packaging, thermal insulation in buildings, fridges liners, electrical &electronic components, optical industry, etc.	6.4
	Others	ABS sheets, roofing sheets, cable coating in telecommunications, touch screens, medical industry, etc.	19
Thermosets	Polyurethane	Pillows and mattresses, insulation works in different areas, etc.	7.9

Microplastics can be described and categorized into various terms after its first review [5, 9, 10]. They categorized microplastics (MPs) into two categories "primary" and "secondary" by differentiating it on the basis of its sources which is a terminology from atmospheric sciences. They stated that, in primary sourcesMPs are produced intentionally by grinding, as a raw material for other products like pellets & for direct use like abrasives used in different cleaning products, and in secondary sources microplastics are produced in the ecosystem from discarded plastic products into small pieces of the size of micro and nanometer. Andrady and coworkers [11]categorized microplastic sources into direct and indirect sources [12]. They further stated that runoff water is a direct source as it becomes part of sewage or storm water and indirect sources are fragmentation of existing plastic waste in the ecosystem [13].

Browne and coworkers [12]reviewed that microplastic sources may be categorized into four different categories (i) Personal care products and cleaning products (ii) Larger Plastic waste (iii) Medicines and (iv) Textiles. Van and coworkers [14] stated that marine activities such as fishing, waterways transportation, oil exploration activities etc. generates large plastic debris. These discarded particles disintegrates during the course of time by weathering action into micro-size particles. The process of defragmentation of these large plastic fragments into microparticles is called fragmentation and this fragmentation becomes the pathway for microplastic contamination.

Lechner and coworkers [15] studied spatial correlation between microplastic particles discovered at any particular site and surrounding human activities. They further stated that the origin of microplastic at any site can be investigated by studying the concentration and polymer type found at that site. Eriksen and coworkers [16] observed that microplastics encountered in the North American Great Lakes are strongly similar in size, shape, color, and elemental composition. Carr and coworkers [17] studied the microplastics in the sewage treatment plant effluent and observed that it is analogous in color, size and shape with the categories of microplastics which are used in toothpaste manufacturing, concluding that the microbeads used in daily use products like toothpaste, facewash, scrub etc. are one of the leading sources of microplastic pollution in freshwater environments. Free and coworkers [18] observed great amounts of secondary microplastics near the shoreline of populated mountain lakes. Lechner and coworkers [15] observed traces of Industrial activities and the sources of microplastics in the Danube River. It is observed that isolated waterbodies like lakes, ponds and estuaries may accumulate more microplastics as compared to lakes [18, 19]. Industrial resin particles and microspheres were in abundance near the industrial zone of Huron Lake [16, 20]. Tourist activities also cause a huge amount of microplastic pollution, which may conclude from various studies conducted near tourist sites; first example is the study by [21] in China's Qinghai Lake where the concentration of microplastics was 5000–757,500 units km^{-2}.

Another method of tracking the history and origin of microplastic is by studying the amount of weathering on the surface of microplastics. Zbyszewski and coworkers [22] stated that surface features help in identifying whether the plastic waste went through some mechanical degradation, by the course of waves, photo-degradation from the exposure to ultraviolet radiation or biodegradation from the process of weather by hydrocarbon-degrading microorganisms [23]. So the identification of the reason for degradation of plastics in different environments becomes important as it may help us in finding the particle interaction with the environment and the impact of various factors on their life, transportation and the extent of effects to marine-organisms [24]. Various methods of degradation have been shown in Fig. 1.

In the past decades, plastic degradation was investigated for different outdoor applications under different conditions to increase its service life or to upgrade the disposal capacities of different categories of plastic debris generated [25]. It was studied for various outdoor conditions such as Photo-degradation, in this process

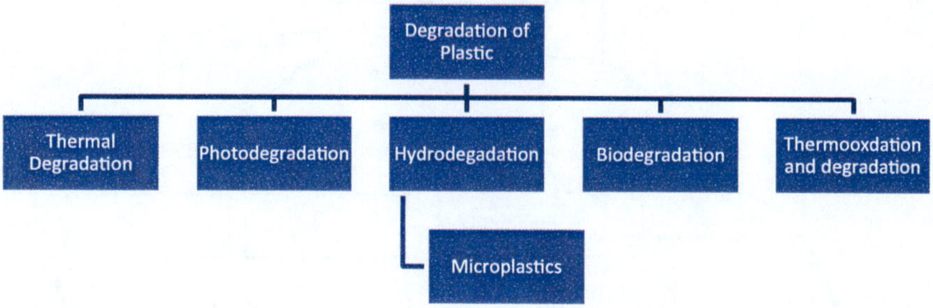

Fig. 1 A Schematic view of different degradation processes of plastic

polymers firstly absorb UV light for the formation of a polymer radical, and then with the reaction of oxygen and hydrogen abstraction happens, which concludes in the cleavage of polymer chains [26, 27]. Thermal degradation is another process of plastic weathering which is very similar to photo-degradation process, but the main difference in this process is that it requires heat to kick start the reactions [28] which is very rare. Use of Biodegradation is very limited for degradation of plastic [29, 30]. Initially, Abiotic decomposition of the plastic waste is required for subsequent biological degradation as only the abiotic degradation can break the polymer chain-link. Human activities have been the root cause of microplastic contamination and transportation in the environment (Fig. 2).

3 Occurrence

With the increase in the daily demand of plastic, microplastic is being observed in the environment all around the world; be it freshwater, marine water, sediments, soil, air, biota, marine organisms, humans, etc. Firstly Thompson and coworkers [4] reported the occurrence of microplastic in the shore line sediments, since then many researchers conducted studies on the presence of microplastic all around the globe in different mediums like water, soil, air, organism and most recently in the placenta of humans [31].

3.1 Microplastic in Aquatic Environment

In the past few decades, various global studies have been conducted to figure out the extent of the microplastics pollution [18, 32, 33] in various freshwater ecosystems around the globe. Li and coworkers [34] studied that microbeads and microplastics are one of the major pollutants in the North Atlantic region and Northern Hemisphere subtropical whorls. In case of microplastics special accumulation zones have been

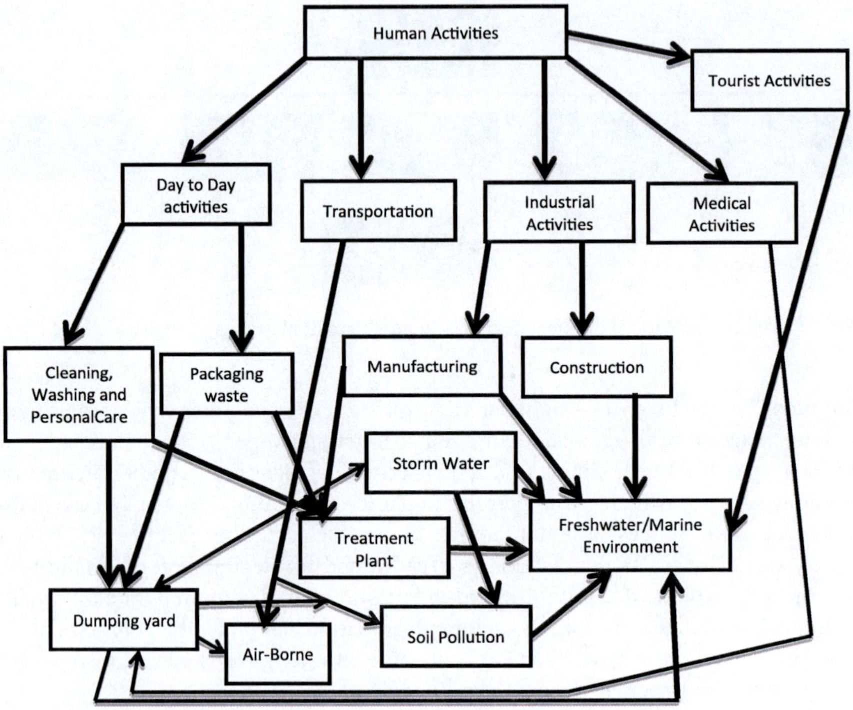

Fig. 2 A Schematic view of different plastic sources and its pathway for the microplastic transportation in the environment

formed due to the movement of plastic waste from different terrestrial and marine sources to the subtropical whorls [35]. Very high clusters of microplastics have been observed in the North Pacific central whorls by Moore and coworkers [36] for the first time. The term "ocean garbage patches" with reference to microplastic was firstly used by Kaiser [37]. Nearly 26,898 microplastic particles km^{-2} with a size range of 0.355–4.750 mm have been observed by Eriksen and coworkers [16] during their study for the quantification of microplastics in the South Pacific subtropical whorl. In a similar study by Law and coworkers [38] classification of microplastics was reported with the presence of Thin films, polypropylene and Polyethylene with an abundance of 33,000 microplastic particles km^{-2}. In a study by Free and coworkers [18] in the Mediterranean Sea and coastal Australia around 270,000 tons of microplastics have been observed in the past 5 years. Eriksen and coworkers [16] observed that different natural processes in the marine environment of Ecuador and Indonesia are one of the leading causes for the high concentration of microplastic in the South Pacific subtropical whorl. An additional garbage patch has been observed by Sebille and coworkers [39] in different zones of Barents Sea. Microplastic particles ranging from 38 to 234 particles per m^3 have been observed in the Arctic Ocean [40]. The presence of microplastics has been witnessed by entire freshwater ecosystem as well as Polar Regions and deep sea. It has been observed that polar areas are the new garbage patches with an abundance of plastic debris in the Arctic and Antarctic regions [41, 42] which may pose a serious threat to the entire ecosystem. Microplastic particles in the Northeast Atlantic Ocean were firstly discovered by Lusher and coworkers [42] with an abundance of 2.46 particles per m^{-3}. An abundance of 4,600 particles per cubic meter was observed in the subsurface seawaters of the northeastern Pacific Ocean in a study by Desforges and coworkers [43]. Various studies have indicated that the abundance of microplastic particles increases with increase in depth of sea and is highest in deep sea sediments.

The freshwater ecosystem has witnessed the presence of microplastic particles in different parts of the world such as North America (Five Lakes, Lake Ontario, St. Lawrence), Europe (Danube, Rhine, Seine), different parts of China like Tai Lake, Pearl River, Yangtze River, Wuhan River, in India, Brazil, South Africa, Mongolia [44]. Microplastics have been observed in the isolated lakes of Tibet plateau [45]. Su and coworkers [46] in their recent study reported the presence of microplastics in the Taihu Lake with an abundance ranging from 3.4 to 25.8 particles per litre. Ballent and coworkers [24] in their study stated that the most polluted area of Lake Ontario is Humber Bay shoreline as they studies the presence of microplastics at different locations of Lake Ontario. Zhang and coworkers [47] in their study have shown that the plentitude of microplastics in the Swiss river is on a much lower side as compared to the Three Gorges Reservoir area, due to the prevailing human activities nearby Three gorges reservoir area. Free and coworkers [18] reported the abundance of microplastics in a large isolated mountain lake Hovsgol present in the country of Mongolia with an abundance of 20,264 particles per square kilometer. They further stated that Lake Hovsgol is the most polluted lake as compared to the other more developed lakes which is due to its location. This lake is situated in a very remote location geographically, and the areal population is exiguous, it may be hypothesized that the environmental factors such as monsoon water, runoff water, and other atmospheric conditions, etc. may also have an influence on the microplastic pollution. The extent of transportation (generally by oceanic waves) of microplastic from aquatic to marine environment and further to arctic and Antarctic region depends on its properties such as shape, size and density [10, 24] as the presence of microplastics have been witnessed along the coastal area of various continents [12, 48], the Arctic region [40], sub-Antarctic waters [18], deep-sea sediments [49] and central Atlantic Islands [48]. Faure and coworkers [50] reported that the most polluted lake in Europe is Lake Geneva, Switzerland with the presence of highest microplastic particles amounting to nearly 48,146 units per square kilometer. Cole and coworkers [51] stated that some quantity of microplastics may get washed off from the land areas to the rivers with the surface runoff and further it may reach to marine environment travelling via rivers. There are a number of ways for microplastic to travel in between different ecosystems and this movement is not in the linear direction from territorial environment to freshwater environment to marine environment in one way direction. Bergmann and coworkers [52] in their study indicated that the particles in the freshwater environment might wash back to the territorial environment due to floods and storms. It is evident that the microplastic particles with a higher density than the marine water may heap up in marine sediments as they tend to sink over time due to the presence of gravity.

3.2 Microplastic in Sediments

The floating microplastic particles may get deposited with the shoreline sediments as they have a lower density than marine water and a high quantum of this plastic waste may stay at the coastline of the marine environment due to the forces of waves. Not until recent times studies on microplastic pollution were only conducted on the Lakes, Rivers, estuaries, islands and beaches, but recently various studies have been conducted on the remote lake sediments and deep sea sediments reflecting the spatial distribution of microplastic pollution in the ecosystem [53]. A study at eight different coastal beaches in Guangdong Province, China, was conducted to examine the particle size distribution of microplastics and very high quantum of fragmentation was observed [54]. Mathalon and coworkers [55] stated that Sediments in multiple environments are used for the verification of the severity of pollution. A study by Retama and coworkers [56] examined the causes for microplastic pollution in the tourist beaches of Huatulco Bay and it was observed that most of the particles were from tourist activities, However, after studying the quality, quantity and size of microplastic particles at various beaches of Slovenia Laglbauer and coworkers [57] concluded that tourism does not impact the microplastic quantity at coastal area. Van and coworkers [45] detected the presence of microplastic particles in deep sea sediments while collecting samples at different locations with different depths and confirmed the extensive presence of microplastic particles even at very remote depth in the sea. Fok and coworkers [58] studied the presence of microplastic in the marine sediments from 25 beaches form the coastal area of Hong Kong and observed that more than 90 percent of microplastic particles consisted of fragments, pellets and polystyrene and the mean abundance of microplastic particles was 5,595 particles per square meter. This data is very alarming for Hongkong beaches when compared to the average particle abundance from around the world. It also shows that the Hongkong is becoming the hotbed for marine microplastic pollution.

In a study at Belgian coast by Claessens and coworkers [59] for microplastic contamination in beach sediments the average abundance was 92.8 units per kilogram of dry sediments. They observed that most of the particles are fiber which might have originated from carpets, fishing nets and ropes and these are made up of polypropylene, nylon and polyvinyl alcohol. It has been observed on the beaches that the microplastic abundance increases rapidly in rainy season due to climatic conditions like wind currents, heavy rains and hurricanes which leads to the washing of plastic debris on the beaches [60]. Ivar and coworkers [61] investigated the presence of microplastics on the beaches of Fernando de Noronha, (Equatorial Western Atlantic). These beaches are generally divided into two types which are windward and leeward beaches. They observed a higher concentration of plastic pollution on windward beaches as compared to the leeward which may be due to the surface currents.

Various studies [62–64] have investigated and reported the extent of microplastic pollution in various non-industrial remote locations such as the Pitcairn Islands, Tonga, the Hawaiian and Chile Islands, Rarotonga and Fiji. Jayasiri and coworkers [65] investigated the presence of microplastic on the sandy beaches of Mumbai and an abundance of 7.49 g for dry sediments and 68.83 particles per square meter was reported. Their particle size distribution indicated the dominance of the size range 1–5 mm which poses a serious risk of ingestion by sea organisms resulting in the microplastic transfer to food web. Neto and Carvalho [66] studied the beaches of the Brazilian coastline of Guanabara Bay for the quantification of microplastic and observed a very high concentration of microplastics dominated by tiny fragments, Styrofoam, pellets and fibers with particle abundance ranging from 12 to 1300 units per square meter. These results led them to a conclusion that these beaches are the most polluted beaches in the Brazilian coastline. Turner and Holmes [67] reported a high concentration of plastic pellets (primary microplastic) on the beaches in the island of Malta. Microplastic debris having dominated contents of polyethylene, Polystyrene and polypropylene fragments have been observed in the sub-surface and surface layers of Singapore Coastline. The microplastic pollution in the Singapore coastal area may be due to the presence of various industries, shipping ports and recreational activities [68]. Browne and coworkers [69] studied the causes for the abundance and spatial pattern of microplastic particle distribution on the shoreline of English Channel and observed that the major abundance of microplastic particle fragments in the sediment of shoreline was due to the wind pressure and water waves. Microplastic particles and fragments have been observed in the various deep sea sediments from different locations [45, 70, 71] such as Indian Ocean, Mediterranean Sea and Atlantic Ocean with an abundance in the range from 100 to 10,000 particles.

3.3 Microplastic in Soil

Microplastic is one of the major emerging pollutant observed in soil. In some parts of the world such as Europe, North America etc. It has been observed that microplastic abundance in agricultural soil is dominant as compared to the freshwater environment which might enter the environment through various routes such as sewage sludge, runoff water, urban waste dumping yards, etc. [72]. As the soil structure has diversity and complexity, the microplastic sampling, separation and extraction processes become very hectic and cumbersome. There are very limited studies available related to microplastic detection and distribution characterization in the terrestrial environment. Distribution of microplastic contamination in the terrestrial environment is very different regionally for example in a study conducted by Peng and coworkers [73] in the various economically developed places like Shanghai microplastic particle distribution can be attributed to the dominant exploitation of personal care products. Zhang and Liu [74] attributed the main cause of microplastic pollution in the terrestrial environment to the continuous utilization of plastic films and long term use of land for storing sewage sludge during their study of Dianchi Basin area in China.

Another source of microplastic pollution in the terrestrial environment may be the continuous use of pesticides, industrial compost and chemical fertilizers in agriculture. Liu and o-worker [75] observed the microplastic particles on surface soil in the industrial zones of Sydney in the range of 6.7%.

Being an indispensable part of the terrestrial environment Soil bears very massive pollution pressure which may be due to direct or indirect influence of humans. Degradation of the plastic debris into smaller microplastic particles in aqua environment occurs mainly due to the direct influence of UV-radiation from sunlight and other climatic processes such as abrasion but same is not the case with terrestrial environment which may lead to a very slow fragmentation process in the soil. Some studies have been conducted on the soil reporting very low degradation in the terrestrial environment. In a study conducted by Albertsson and coworkers [76] weight loss reported ranged from 0.1% to 0.4% by weight of sample of polyethylene in soil after 800 days. After putting polypropylene in soil for one year only 0.4% of weight loss was observed and fragmentation was not observed even after 35 years in the polyvinyl chloride [77]. Rate of plastic fragmentation is greatly affected by the soil texture. In a study conducted by Cesar and coworkers [78] it was observed that fragmentation in clay soil was higher as compared to sandy soils. Microplastic particles have a very high impact on water holding capacity of soil, soil structure, microbial activities within soil texture and its bulk density which end up in disturbing the relationship between the soil and water [79].

3.4 Microplastics in Air

There is a scarcity of studies conducted on the detection, mapping and distribution of microplastics in the atmosphere. It is very difficult to detect, track and map microplastic distribution in the atmosphere because of various factors such as characteristics of microplastic particles and its behavior in air, different types of microplastic sources in the atmosphere, various environmental conditions affecting the movement of microplastic particles in the air such as atmospheric moisture, temperature, and pressure. Presence and transportation of microplastic particles in the atmosphere cannot be ignored just because of knowledge scarcity in this area. Cai and coworkers [80] studied the presence of microplastic particles in the air of Dongguan city in China and observed that the primary microplastics are dominant in the urban atmosphere of the city. Transportation of microplastic particles in the atmosphere may be influenced by wind pressure and airflow as it has very low weight and the transportation may lead to the movement of particles from its source [81]. Since this transportation cannot be restricted by environmental boundaries, the result might be very hazardous with the highest degree of air contamination. Presently, there are no reports indicating the impact of microplastic atmospheric contamination on water and terrestrial microplastic pollution, further studies are required in this area.

Presence of microplastic particles in the atmosphere started getting attention only after 2015. Dris and coworkers [82] detected the presence of microplastic particles in the atmosphere for the first time and evaluated the TAF (total atmospheric fallout) in the urban and suburban areas of Paris city of France. They used rain sampling method for the collection of atmospheric samples further the microplastic particles were filtered, quantified and qualified under a stereomicroscope.

Various researchers have used different sampling methods which can mostly be categorized into three categories which are; dray and wet deposition [80, 83, 84], dust collection [85, 86] and atmospheric sampling, [82, 87]. Since there are different sampling methods for collection of microplastic particles from air, it is impractical to compare various studies having different sampling techniques. Air samplers can be used to observe microplastic particles upto very small size range [88, 89] but it has been observed that the particle size distribution generally varies from 100 μm to 5 mm in the atmosphere [80, 82, 90]. The mean width size of microplastic fibers may vary from 1 μm to 10 μm [91, 92]. In a study conducted by Dris and coworkers [82] in the urban and sub-urban area of Paris city microplastic abundance was observed to be greater in former case as compared to the latter. Microplastic particles scattering and distribution quantum in the atmosphere are highly influenced by meteorological factors and climatic conditions. For example, the highest abundance was observed in rainy season and the lowest abundance of microplastic particles was witnessed during rainy season [84]. Dry and wet deposition method was used for the sampling of microplastic fibers during rainy season [82]. Further due to very weak air ventilation in urban area because of high population leads to a strong abundance of microplastic particles as compared to suburban and rural area where air ventilation is sufficient for continuous air movement [82]. In a study by Dris and coworkers [82] it has been observed that indoor atmosphere hosts more microplastic particles as compared to outdoor atmosphere which may be due to the factors like airflow and indoor partitions [93]. Abbasi and coworkers [88] identified road dust as one of the emerging source hosting the microplastic particles in the urban atmosphere and this may lead to the direct ingestion of microplastic by children playing in the contaminated area from the dust. They also calculated the annual microplastic particle ingestion by any child via dust to be around 900 units making it to be 200 mg per day in a normal scenario. Studies on microplastic contamination in the air have been mainly reported in two continents which are Europe and Asia. Browne and coworkers [94] stated that microplastic particles are easily noticeable in the downwind side in a high abundance since these particles are of small size having very low density and easily transferable by the air.

A study was conducted by Cai and coworkers [80] in Dongguan city of China for the calculation of microplastic abundance in fiber and non-fiber form in the total atmospheric fallout. They reported the microplastic abundance in the range from 175 to 313 units per square meter per day. They observed that most of the particles are fibers and other particles like fragments, film and foams were witnessed occasionally. Similar results were reported in another study by Zhou and coworkers [95] in Yantai (A coastal city) situated in Shandong Province. A study in Edinburg situated in United Kingdom was conducted by Catarino and coworkers [96] to access the household fiber fallout during food cooking and eating. They observed that a large quantum of microplastic particles are ingested via food in the evening during cooking time.

The particle deposition rate was ranging from 1666.8 to1671.84 particles per day per square meter. Most of the particles collected were less than 500 μm and consisted of Polyurethane and Polyethylene Terephthalate. Another study was conducted by Kaya and coworkers [97] in the Sakarya province situated in Turkey for the presence of microplastics in the atmosphere of an intercity terminal of a university campus. The sampling was done with the help of vacuum pump at the volumetric collection rate of 0.3 cubic meters per minute and the minimum duration for sample collection was kept 30 min. The samples were observed using micro FTIR followed by visual sorting using a microscope. Microplastic particle abundance was observed in the range from 259 to 12,895 units per liter with mainly two categories of microplastics which are fibers and fragments. Another study was conducted in Asaluyeh situated in Iran for the evaluation of microplastic particles in the suspended dust which was collected for an eight day period being collected every day [88] and it was observed that the quantum of microplastic particles in the atmosphere is highly influenced by anthropogenic activities. In a similar study conducted by Klein and coworker [98], identical deductions were done but they also observed Teflon in the atmospheric fallout which may be due to some nearby regional sources. Liu and coworkers [89] conducted an investigation in Shanghai city situated in China for the identification of microplastic particles from textile industry with the help of principal component analysis. They concluded that the dominant category of microplastic is fibers followed by fragments which clearly indicates the source of microplastic pollution in the area is the textile industry.

4 Impacts on Ecosystem

Some research have been conducted for understanding the extent of toxicity of microplastics after its ingestion by freshwater as well as marine organisms but no efforts have been made for characterization of its monotonous pathways and their toxicological effects on the aquatic organisms [43, 99]. There are three possible pathways via which microplastic particles may pose a threat and become toxic for aquatic organisms: (i) plastic additive leakages like that of a plasticizer, (ii) direct ingestion of microplastic leading to blockage inside the body of aquatic organisms and related expenditure of energy for its removal, (iii) manifestation and direct contact with the emerging contaminants (like persistent organic pollutants) hosted by plastic on its surface [10, 11]. There can be very serious repercussion of microplastic subjection to aquatic organisms which may be highly influenced by its translocation and aggregation inside the tissues of organisms and its potential to effuse the accumulated particles before very severe exposure [7].

Derraik and coworkers [100] stated the toxicological effects of plastic litter into various sea creatures like whales, turtles, polar bears and seals. In a study conducted by Fossi and coworkers [101] stated that the microplastic particles may have been ingested by Baleen whales which may happen via microplastic ingestion by sea water filtering organisms which enters their system continuously. Barvo and coworkers

[102] investigated the microplastic exposure and its ingestion in harbor seals and they observed the presence of microplastic particles in their intestines and stomach. Nerland and coworkers [103] investigated the exposure of microplastic particles into the sea turtles of Brazil and observed microplastic particles in the digestive tracts of more than 60% sea turtles. It can be easily interpreted that the large size organisms and animals like whales, turtles, seals etc. are more susceptible to microplastic ingestion as compared to small size organisms. A lot of whales have died and washed up ashore due to the ingestion of plastic debris in their body. Stimmelmayr and coworkers [104] examined polar bears from Alaska and observed that 13 out of 51 subjects had ingested plastic debris in the form of garbage bags, local plastic bags and some other particles of unknown origin.

Wright and coworkers [7] concluded that microplastic leakage may also cause chemical contamination in the environment apart from its physical nuisance. Various studies have already implicated and shown the toxicological effects of microplastic pollution on the ecosystem. The chemical contamination of the ecosystem may be a result from the various additive and plasticizers (very toxic in nature) used for the enhancement of the physical properties (such as corrosion resistance, heat resistance, durability improvement, surface softening etc.) of plastic during its production [105]. For example, Polybrominated diphenyl ethers are added to increase fire resistivity and for corrosion resistance, nonylphenol is added. Since these additives and plasticizers are applied as a coating on the plastic surface and are not the structural part of the polymers, they gets transported very comfortably into the ecosystem from the microplastic surfaces with the passage of time [72]. Exposure to these chemicals and additives poses a serious menace for the ecosystem and its organisms. Eriksen and coworkers [16] stated that during the production of plastic addition of the additive and plasticizers depends on its application. Since the different types of microplastics particles have been known to be dissipating all the variety of additives in the ecosystem have a very different impact on the biota [1]. Chances of these additives being released into the ecosystem are higher at the places with high UV exposure and having accumulated microplastic in very high abundance. Wang and coworkers [106] have observed that in places where plastic films are being used have higher concentrations of plasticizers in the environment. These observations clearly indicate that the microplastic particles may also dissipate the additives and plasticizers in the terrestrial environment which may have very serious implications on the ecosystem along with the chances of these plasticizers being released into the aquatic and marine ecosystem through runoff and effluents. Halden and coworkers [107] studied the impacts of these plasticizers on the animals and observed that these plasticizers like phthalates may damage the development process of the marine organisms, may have an influence on the reproduction system in the organisms and may lead to the change in genetic behavior of the organisms. Syberg and coworkers [108] stated that the plasticizers may have the capacity to influence the body's endocrine process which in turn will have an impact on the development and reproduction system of the organisms. Rehse and coworkers [109]demonstrated the toxicological effects of plastic additives on the plants. They further stated that the heat resistant plasticizers (like tetrachlorophenol) have very toxic effects on the plankton and these plasticizers are capable of influencing the sedimentation rate diatoms and cryptophytes. Rios and coworkers [110] investigated the presence of these plasticizers and additives in the plastic samples from the north pacific whorl and their results confirmed the presence of PCBs (polychlorinated biphenyls), DDT, aliphatic hydrocarbons and PAHs (polycyclic aromatic hydrocarbons) having their concentration in the range of 27–980 ng per gram, 7–100 ng per gram, 1.1–86,000 μg per gram, 1200 ng per gram, respectively. They further stated that these results are very alarming since the microplastic accumulation in any organism may result in starvation and holes in gastro tracts along with the capacity to transport these microplastic particles into the food chain.

Due to its durability, plastic poses a very serious threat of remaining in the environment from now to the centuries in the future because of its capacity of continuously being increased in the environment from the running waste [41]. Lima and coworkers [111] observed that the buoyant microplastic particles in the oceans have the capacity of getting attached to the microorganisms with the formation of biofilms, this increases the density of the microplastic particles leading to its sinking to the beds of the aqua environment which results in the disturbance of benthic organisms in the water body and its functions. Lusher and coworkers [42] observed that Phthalates are very common additives, generally associated with polyvinyl chloride. In another by Nerland and coworkers [103] observed that the transfer of microplastic additives and plasticizers is highly influenced by its surrounding conditions like water state, concentration gradient, its chemical and physical properties and dominant degradation processes. They further observed that the small plasticizers and chemicals present in the region of non-crystallinity. It has been observed that on the plastic surface, DDT and phenanthrene attain sorption equilibrium within one day [99] whereas it is very difficult for additives like DDE and PCBs to attain sorption equilibrium on the pellets [112].

5 Effects on Human Health

Sources such as primary microplastics in skin care cosmetics, toothpastes, scrubs, and hand washes may expose the general human population to microplastic. Other ill effects of microplastics apart from toxicity, different toxic pollutants like phthalates or PCBs adsorbed on the surface of microplastic may cause dietary exposure of humans [113]. Health of humans that consume seafood is being affected due to the likely buildup of microplastics on the seafood. Microplastics have been carried along the food web as microplastics have been consumed by several sea creatures like mussel, oyster, fish, crab, and sea cucumber. Unfortunately, no information is present that elaborates about the particle size, shape, concentration or composition of microplastic particles that are existing in food items [114]. The German Federal Institute for Risk Assessment has evaluated the health dangers that are been caused by the regular use of hand cleaners, dental care products, face washes and toothpastes which contain PE microplastic particles [114]. The conclusion came out to be that shower products and face pack peeling have microplastic that are greater than 1 μm, these products damage the skin as the PP and PE particles get absorbed in the tissues due to the regular use of these products. The gastrointestinal tract recognized that people without knowing to consume microbeads and microplastic particles while using toothpaste [113]. Obesity, cancer and infertility are major causes of alteration in chromosomes that are the end result of alternate ingestion of microparticles [115]. Estrogenic mimicking chemicals that lead to breast cancer in females. A high risk is created towards the food safety because microplastic pollutants are in a high ratio in seafood and clearly diet consumed by human being exposes them to consume microplastic [116]. Thus, to access the causative risk on human beings' health caused because of contaminated marine food, a comprehensive assessment and analysis of potential health dangers because of microplastics that are present in all the food across the entire diet.

Potential impacts are been caused on human beings' health as the microplastics got into the human food chain because of terrestrial food items and ingestion of seafood, this has been an emerging plane of concern regarding MPs [117]. Lower trophic organisms, that can consume anything of suitable size, also have the accessibility to the MPs as it is small in size. The type of plastic debris that are ingested may differ from organisms to organisms as the bioavailability of MPs in the water column is decided by the density of MPs. Bioavailability of MPs is determined as a rise in the large amount of MPs in marine environment because the probability of organisms to come across MPs in increased [7]. Human beings get directly exposed to MPs that are even $\geq 1\ \mu m$ by consuming gill or skin tissue. A recent study that was executed on soft tissue of mussels firmly proposes that other than ingestion adherence is a novel way for organisms to intake MPs [118].

6 Conclusion

Microplastics was firstly discovered in the marine water and sediment which is generally considered as the sinks of plastic debris. Recently researchers have started to recognize the interaction of Microplastics with human body through inhalation and ingestion. This also explains the relatively smaller number of studies mentioned in this review as compared to the marine microplastic studies. For assessing the occurrence of MPs in human exposure pathways and related health impacts more surveys and studies are required. Drinking water, Marine Food, Table salt and air are the four major human exposure pathways for microplastics. Among these different pathways leading to human body, the inhalation of atmospheric microplastics is estimated to be the most significant (1.9 $\times$ 103 to 1.0 $\times$ 105 items·year^{-1} indoor air; 0 − 3.0 $\times$ 107 items·year^{-1} outdoor air). So far, various related questions on MPs remain unanswered. For example, the exact routes of cellular intake of microplastics, the accumulation of microplastics on tissues, and the potential ill effects of microplastics after a long term exposure in human body are still unknown. The fate and movement of microplastics upon entering a bioorganism through absorption/adsorption and excretion are still unclear. The changes in the body at cellular level or at molecular level and specific mechanisms have not been realized and this knowledge gap is required to be filled.

References

1. Thompson RC, Moore CJ, Saal FSV, Swan SH (2009) Plastics, the environment and human health: Current consensus and future trends. Philos Trans R Soc B Biol Sci 364(1526):2153–2166. https://doi.org/10.1098/rstb.2009.0053
2. Carpenter EJ, Smith KL (1972) "Plastics on the Sargasso sea surface." Science (80-) 175(4027): 1240–1241, doi: https://doi.org/10.1126/science.175.4027.1240
3. Plastics Europe and EPRO (2016) Plastics—the Facts 2016. p 37. [Online]. Available: www.plasticseurope.de/informations
4. Thompson RC et al. (2004) "Lost at sea : Where is all the plastic ?." 304, doi: https://doi.org/10.1126/science.1094559
5. Arthur C, Baker J, Bamford H (2009) "Proceedings of the international research workshop on the occurrence, effects, and fate of microplastic Marine Debris." Group: 530
6. Wagner M et al (2014) Microplastics in freshwater ecosystems: What we know and what we need to know. Environ Sci Eur 26(1):1–9. https://doi.org/10.1186/s12302-014-0012-7
7. Wright SL, Thompson RC, Galloway TS (2013) The physical impacts of microplastics on marine organisms: A review. Environ Pollut 178:483–492. https://doi.org/10.1016/j.envpol.2013.02.031
8. Plastics Europe (2019) "Plastics—the Facts 2019" [Online]. Available: https://www.plasticseurope.org/en/resources/market-data
9. Browne MA, Dissanayake A, Galloway TS, Lowe DM, Thompson RC (Jul.2008) Ingested microscopic plastic translocates to the circulatory system of the mussel, Mytilus edulis (L.). Environ Sci Technol 42(13):5026–5031. https://doi.org/10.1021/es800249a
10. Cole M, Lindeque P, Halsband C, Galloway TS (2011) Microplastics as contaminants in the marine environment: A review. Mar Pollut Bull 62(12):2588–2597. https://doi.org/10.1016/j.marpolbul.2011.09.025
11. Andrady AL (2011) Microplastics in the marine environment. Mar Pollut Bull 62(8):1596–1605. https://doi.org/10.1016/j.marpolbul.2011.05.030
12. Browne MA (2015) "Sources and pathways of microplastics to habitats." In: Bergmann M, Gutow L, Klages M (eds.) Marine anthropogenic litter. Springer International Publishing, Cham, pp 229–244, doi: https://doi.org/10.1007/978-3-319-16510-3_9
13. Gupta DK, Choudhary D, Vishwakarma A, Mudgal M, Srivastava AK, Singh A (2022) Microplastics in freshwater environment: occurrence, analysis, impact, control measures and challenges. Int J Environ Sci Technol. https://doi.org/10.1007/s13762-022-04139-2
14. Van Cauwenberghe L, Devriese L, Galgani F, Robbens J, Janssen CR (2015) Microplastics in sediments: A review of techniques, occurrence and effects. Mar Environ Res 111:5–17. https://doi.org/10.1016/j.marenvres.2015.06.007
15. Lechner A et al (2014) The Danube so colourful: A potpourri of plastic litter outnumbers fish larvae in Europe's second largest river. Environ Pollut 188:177–181. https://doi.org/10.1016/j.envpol.2014.02.006
16. Eriksen M et al (2013) Microplastic pollution in the surface waters of the Laurentian Great Lakes. Mar Pollut Bull 77(1–2):177–182. https://doi.org/10.1016/j.marpolbul.2013.10.007
17. Carr SA, Liu J, Tesoro AG (2016) Transport and fate of microplastic particles in wastewater treatment plants. Water Res 91:174–182. https://doi.org/10.1016/j.watres.2016.01.002
18. Free CM, Jensen OP, Mason SA, Eriksen M, Williamson NJ, Boldgiv B (2014) High-levels of microplastic pollution in a large, remote, mountain lake. Mar Pollut Bull 85(1):156–163. https://doi.org/10.1016/j.marpolbul.2014.06.001

19. Imhof HK, Ivleva NP, Schmid J, Niessner R, Laforsch C (2013) Contamination of beach sediments of a subalpine lake with microplastic particles. Curr Biol 23(19):R867–R868. https://doi.org/10.1016/j.cub.2013.09.001
20. Zbyszewski M, Corcoran PL (2011) Distribution and degradation of fresh water plastic particles along the beaches of Lake Huron, Canada. Water Air Soil Pollut 220(1–4):365–372. https://doi.org/10.1007/s11270-011-0760-6
21. Xiong X, Zhang K, Chen X, Shi H, Luo Z, Wu C (2018) Sources and distribution of microplastics in China's largest inland lake—Qinghai Lake. Environ Pollut 235:899–906. https://doi.org/10.1016/j.envpol.2017.12.081
22. Zbyszewski M, Corcoran PL, Hockin A (2014) Comparison of the distribution and degradation of plastic debris along shorelines of the Great Lakes, North America. J Great Lakes Res 40(2):288–299. https://doi.org/10.1016/j.jglr.2014.02.012
23. Hirai H et al (2011) Organic micropollutants in marine plastics debris from the open ocean and remote and urban beaches. Mar Pollut Bull 62(8):1683–1692. https://doi.org/10.1016/j.marpolbul.2011.06.004
24. Ballent A, Corcoran PL, Madden O, Helm PA, Longstaffe FJ (2016) Sources and sinks of microplastics in Canadian Lake Ontario nearshore, tributary and beach sediments. Mar Pollut Bull 110(1):383–395. https://doi.org/10.1016/j.marpolbul.2016.06.037
25. Shang J, Chai M, Zhu Y (Oct.2003) Photocatalytic degradation of polystyrene plastic under fluorescent light. Environ Sci Technol 37(19):4494–4499. https://doi.org/10.1021/es0209464
26. Feldman D (2002) Polymer weathering: Photo-oxidation. J Polym Environ 10(4):163–173. https://doi.org/10.1023/A:1021148205366
27. Gewert B, Plassmann MM, MacLeod M (2015) Pathways for degradation of plastic polymers floating in the marine environment. Environ Sci Process Impacts 17(9):1513–1521. https://doi.org/10.1039/C5EM00207A
28. Gardette M et al (2013) Photo- and thermal oxidation of polyethylene: comparison of mechanisms and influence of unsaturation content. Polym Degrad Stab 98:2383–2390. https://doi.org/10.1016/j.polymdegradstab.2013.07.017
29. Ng E-L et al (Jun.2018) An overview of microplastic and nanoplastic pollution in agroecosystems. Sci Total Environ 627:1377–1388. https://doi.org/10.1016/j.scitotenv.2018.01.341
30. Tokiwa Y, Calabia BP, Ugwu CU, Aiba S (2009) Biodegradability of plastics. Int J Mol Sci 10(9):3722–3742. https://doi.org/10.3390/ijms10093722
31. Ragusa A et al (2021) Plasticenta: First evidence of microplastics in human placenta. Environ Int 146:106274. https://doi.org/10.1016/j.envint.2020.106274
32. Cózar A et al (2017) The Arctic ocean as a dead end for floating plastics in the North Atlantic branch of the thermohaline circulation. Sci Adv 3(4):1–9. https://doi.org/10.1126/sciadv.1600582
33. Lebreton LCM, Van Der Zwet J, Damsteeg JW, Slat B, Andrady A, Reisser J (2017) River plastic emissions to the world's oceans. Nat Commun 8:1–10. https://doi.org/10.1038/ncomms15611
34. Li WC, Tse HF, Fok L (2016) Plastic waste in the marine environment: A review of sources, occurrence and effects. Sci Total Environ 566–567:333–349. https://doi.org/10.1016/j.scitotenv.2016.05.084
35. Lebreton LC-M, Greer SD, Borrero JC (2012) Numerical modelling of floating debris in the world's oceans. Mar Pollut Bull 64(3):653–661. https://doi.org/10.1016/j.marpolbul.2011.10.027
36. Moore CJ, Moore SL, Leecaster MK, Weisberg SB (2001) A comparison of plastic and plankton in the North Pacific Central Gyre. Mar Pollut Bull 42(12):1297–1300. https://doi.org/10.1016/S0025-326X(01)00114-X
37. Kaiser J (2010) "The dirt on ocean garbage patches." Science (80-) 328(5985): 1506, doi: https://doi.org/10.1126/science.328.5985.1506
38. Law KL et al (2010) "Plastic accumulation in the North Atlantic Subtropical Gyre." Science (80-) 329(5996): 1185–1188, doi: https://doi.org/10.1126/science.1192321

39. Van Sebille E, England MH, Froyland G (2012) "Origin, dynamics and evolution of ocean garbage patches from observed surface drifters." doi: https://doi.org/10.1088/1748-9326/7/4/044040
40. Obbard RW, Sadri S, Wong YQ, Khitun AA, Baker I, Thompson RC (2014) Global warming releases microplastic legacy frozen in Arctic Sea ice. Earth's Futur 2(6):315–320. https://doi.org/10.1002/2014ef000240
41. Barnes DKA, Galgani F, Thompson RC, Barlaz M (2009) Accumulation and fragmentation of plastic debris in global environments. Philos Trans R Soc B Biol Sci 364(1526):1985–1998. https://doi.org/10.1098/rstb.2008.0205
42. Lusher AL, Tirelli V, O'Connor I, Officer R (2015) Microplastics in Arctic polar waters: The first reported values of particles in surface and sub-surface samples. Sci Rep 5(June):1–9. https://doi.org/10.1038/srep14947
43. Desforges JPW, Galbraith M, Dangerfield N, Ross PS (Feb.2014) Widespread distribution of microplastics in subsurface seawater in the NE Pacific Ocean. Mar Pollut Bull 79(1–2):94–99. https://doi.org/10.1016/J.MARPOLBUL.2013.12.035
44. Eerkes-Medrano D, Thompson RC, Aldridge DC (2015) Microplastics in freshwater systems: A review of the emerging threats, identification of knowledge gaps and prioritisation of research needs. Water Res 75:63–82. https://doi.org/10.1016/j.watres.2015.02.012
45. Van Cauwenberghe L, Vanreusel A, Mees J, Janssen CR (2013) Microplastic pollution in deep-sea sediments. Environ Pollut 182:495–499. https://doi.org/10.1016/j.envpol.2013.08.013
46. Su L et al (2016) Microplastics in Taihu Lake, China. Environ Pollut 216:711–719. https://doi.org/10.1016/j.envpol.2016.06.036
47. Zhang K et al (2017) Occurrence and characteristics of microplastic pollution in Xiangxi Bay of Three Gorges Reservoir, China. Environ Sci Technol 51(7):3794–3801. https://doi.org/10.1021/acs.est.7b00369
48. Ivar Do Sul JA, Costa MF (2014) "The present and future of microplastic pollution in the marine environment." Environ Pollut 185: 352–364, doi: https://doi.org/10.1016/j.envpol.2013.10.036
49. Vandermeersch G et al (2015) A critical view on microplastic quantification in aquatic organisms. Environ Res 143(2014):46–55. https://doi.org/10.1016/j.envres.2015.07.016
50. Faure F, Pompini O, de Alencastro LF (2017) "Sources and fate of microplastics in Swiss surface waters." In: Baztan J, Jorgensen B, Pahl S, Thompson RC, J-P B T-F, I of M, Vanderlinden ME (eds) Elsevier, p 5, doi: https://doi.org/10.1016/B978-0-12-812271-6.00004-1
51. Cole M, Webb H, Lindeque PK, Fileman ES, Halsband C, Galloway TS (2014) Isolation of microplastics in biota-rich seawater samples and marine organisms. Sci Rep 4:1–8. https://doi.org/10.1038/srep04528
52. Bergmann M, Gutow L, Klages M (2015) "Marine anthropogenic litter." Mar Anthropog Litter: 1–447, doi: https://doi.org/10.1007/978-3-319-16510-3
53. Maes T, Jessop R, Wellner N, Haupt K, Mayes AG (2017) A rapid-screening approach to detect and quantify microplastics based on fluorescent tagging with Nile Red. Sci Rep 7(March):1–10. https://doi.org/10.1038/srep44501
54. Fok L, Cheung PK, Tang G, Li WC (Jan.2017) Size distribution of stranded small plastic debris on the coast of Guangdong, South China. Environ Pollut 220(Pt A):407–412. https://doi.org/10.1016/j.envpol.2016.09.079
55. Mathalon A, Hill P (Apr.2014) Microplastic fibers in the intertidal ecosystem surrounding Halifax Harbor, Nova Scotia. Mar Pollut Bull 81(1):69–79. https://doi.org/10.1016/J.MARPOLBUL.2014.02.018
56. Retama I et al (2016) Microplastics in tourist beaches of Huatulco Bay, Pacific coast of southern Mexico. Mar Pollut Bull 113(1–2):530–535. https://doi.org/10.1016/j.marpolbul.2016.08.053
57. Laglbauer BJL et al (Dec.2014) Macrodebris and microplastics from beaches in Slovenia. Mar Pollut Bull 89(1–2):356–366. https://doi.org/10.1016/j.marpolbul.2014.09.036

58. Fok L, Cheung PK (Oct.2015) Hong Kong at the Pearl River Estuary: A hotspot of microplastic pollution. Mar Pollut Bull 99(1–2):112–118. https://doi.org/10.1016/j.marpolbul.2015.07.050
59. Claessens M, Van Cauwenberghe L, Vandegehuchte MB, Janssen CR (May2013) New techniques for the detection of microplastics in sediments and field collected organisms. Mar Pollut Bull 70(1–2):227–233. https://doi.org/10.1016/J.MARPOLBUL.2013.03.009
60. Lee J et al (Dec.2013) Relationships among the abundances of plastic debris in different size classes on beaches in South Korea. Mar Pollut Bull 77(1–2):349–354. https://doi.org/10.1016/j.marpolbul.2013.08.013
61. Ivar do Sul JA, Spengler A, Costa MF (2009) "Here, there and everywhere. Small plastic fragments and pellets on beaches of Fernando de Noronha (Equatorial Western Atlantic)." Mar Pollut Bull 58(8): 1236–1238, doi: https://doi.org/10.1016/j.marpolbul.2009.05.004
62. Gregory MR (1999) Plastics and south Pacific Island shores: Environmental implications. Ocean Coast Manag 42(6):603–615. https://doi.org/10.1016/S0964-5691(99)00036-8
63. Corcoran PL, Biesinger MC, Grifi M (Jan.2009) Plastics and beaches: a degrading relationship. Mar Pollut Bull 58(1):80–84. https://doi.org/10.1016/j.marpolbul.2008.08.022
64. Hidalgo-Ruz V, Thiel M (2013) Distribution and abundance of small plastic debris on beaches in the SE Pacific (Chile): a study supported by a citizen science project. Mar Environ Res 87–88:12–18. https://doi.org/10.1016/j.marenvres.2013.02.015
65. Jayasiri HB, Purushothaman CS, Vennila A (Dec.2013) Quantitative analysis of plastic debris on recreational beaches in Mumbai, India. Mar Pollut Bull 77(1–2):107–112. https://doi.org/10.1016/j.marpolbul.2013.10.024
66. Neto AB, De Carvalho DG (2016) "Ocean & coastal management microplastic pollution of the beaches of Guanabara Bay, Southeast Brazil."128: 10–12
67. Turner A, Holmes L (Feb.2011) Occurrence, distribution and characteristics of beached plastic production pellets on the island of Malta (central Mediterranean). Mar Pollut Bull 62(2):377–381. https://doi.org/10.1016/j.marpolbul.2010.09.027
68. Ng KL, Obbard JP (2006) Prevalence of microplastics in Singapore's coastal marine environment. Mar Pollut Bull 52(7):761–767. https://doi.org/10.1016/j.marpolbul.2005.11.017
69. Browne MA et al (Nov.2011) Accumulation of microplastic on shorelines woldwide: sources and sinks. Environ Sci Technol 45(21):9175–9179. https://doi.org/10.1021/es201811s
70. Pham CK et al. (2014) "Marine litter distribution and density in European Seas, from the Shelves to deep basins." PLoS One 9(4): e95839 [Online]. Available: https://doi.org/10.1371/journal.pone.0095839
71. Tubau X, Canals M, Lastras G, Rayo X, Rivera J, Amblas D (2015) Progress in Oceanography Marine litter on the floor of deep submarine canyons of the Northwestern Mediterranean Sea: The role of hydrodynamic processes. Prog Oceanogr 134:379–403. https://doi.org/10.1016/j.pocean.2015.03.013
72. Horton AA, Svendsen C, Williams RJ, Spurgeon DJ, Lahive E (2017) Large microplastic particles in sediments of tributaries of the River Thames, UK—Abundance, sources and methods for effective quantification. Mar Pollut Bull 114(1):218–226. https://doi.org/10.1016/j.marpolbul.2016.09.004
73. Peng J, Wang J, Cai L (2017) Current understanding of microplastics in the environment: Occurrence, fate, risks, and what we should do. Integr Environ Assess Manag 13(3):476–482. https://doi.org/10.1002/ieam.1912
74. Zhang GS, Liu YF (Nov.2018) The distribution of microplastics in soil aggregate fractions in southwestern China. Sci Total Environ 642:12–20. https://doi.org/10.1016/j.scitotenv.2018.06.004
75. Liu M et al (2018) Microplastic and mesoplastic pollution in farmland soils in suburbs of Shanghai, China. Environ Pollut 242:855–862. https://doi.org/10.1016/j.envpol.2018.07.051
76. Albertsson A-C (1980) The shape of the biodegradation curve for low and high density polyethenes in prolonged series of experiments. Eur Polym J 16(7):623–630. https://doi.org/10.1016/0014-3057(80)90100-7

77. Arkatkar A, Arutchelvi J, Bhaduri S, Uppara PV, Doble M (2009) Degradation of unpretreated and thermally pretreated polypropylene by soil consortia. Int Biodeterior Biodegradation 63(1):106–111. https://doi.org/10.1016/j.ibiod.2008.06.005
78. César MEF, Mariani PDSC, Innocentini-Mei LH, Cardoso EJBN (2009) Particle size and concentration of poly(ε-caprolactone) and adipate modified starch blend on mineralization in soils with differing textures. Polym Test 28(7):680–687. https://doi.org/10.1016/j.polymertesting.2009.05.002
79. de Souza Machado AA, Kloas W, Zarfl C, Hempel S, Rillig MC (2018) "Microplastics as an emerging threat to terrestrial ecosystems." Glob Chang Biol 24(4): 1405–1416, doi: https://doi.org/10.1111/gcb.14020
80. Cai L et al (2017) Characteristic of microplastics in the atmospheric fallout from Dongguan city, China: Preliminary research and first evidence. Environ Sci Pollut Res 24(32):24928–24935. https://doi.org/10.1007/s11356-017-0116-x
81. Iwasaki S, Isobe A, Kako S, Uchida K, Tokai T (2017) Fate of microplastics and mesoplastics carried by surface currents and wind waves: A numerical model approach in the Sea of Japan. Mar Pollut Bull 121(1):85–96. https://doi.org/10.1016/j.marpolbul.2017.05.057
82. Dris R et al (2017) A first overview of textile fibers, including microplastics, in indoor and outdoor environments. Environ Pollut 221:453–458. https://doi.org/10.1016/j.envpol.2016.12.013
83. Wright SL, Ulke J, Font A, Chan KLA, Kelly FJ (2020) Atmospheric microplastic deposition in an urban environment and an evaluation of transport. Environ Int 136:105411. https://doi.org/10.1016/j.envint.2019.105411
84. Dris R, Gasperi CJ, Rocher AV, Saad BM, ANRA, BTA (2015) "Microplastic contamination in an urban area : A case study in Greater Paris"
85. Liu C et al (2019) Widespread distribution of PET and PC microplastics in dust in urban China and their estimated human exposure. Environ Int 128:116–124. https://doi.org/10.1016/j.envint.2019.04.024
86. Ambrosini R, Azzoni RS, Pittino F, Diolaiuti G, Franzetti A, Parolini M (Oct.2019) First evidence of microplastic contamination in the supraglacial debris of an alpine glacier. Environ Pollut 253:297–301. https://doi.org/10.1016/j.envpol.2019.07.005
87. Wang X, Li C, Liu K, Zhu L, Song Z, Li D (May2020) Atmospheric microplastic over the South China Sea and East Indian Ocean: abundance, distribution and source. J Hazard Mater 389:121846. https://doi.org/10.1016/j.jhazmat.2019.121846
88. Abbasi S et al (Jan.2019) Distribution and potential health impacts of microplastics and microrubbers in air and street dusts from Asaluyeh County, Iran. Environ Pollut 244:153–164. https://doi.org/10.1016/j.envpol.2018.10.039
89. Liu K, Wang X, Fang T, Xu P, Zhu L, Li D (Jul.2019) Source and potential risk assessment of suspended atmospheric microplastics in Shanghai. Sci Total Environ 675:462–471. https://doi.org/10.1016/j.scitotenv.2019.04.110
90. Dehghani S, Moore F, Akhbarizadeh R (Sep.2017) Microplastic pollution in deposited urban dust, Tehran metropolis, Iran. Environ Sci Pollut Res Int 24(25):20360–20371. https://doi.org/10.1007/s11356-017-9674-1
91. Bergmann M, Mützel S, Primpke S, Tekman MB, Trachsel J, Gerdts G (2022) "White and wonderful? Microplastics prevail in snow from the Alps to the Arctic." Sci Adv 5(8): eaax1157, doi: https://doi.org/10.1126/sciadv.aax1157
92. Li Y et al (Feb.2020) Airborne fiber particles: Types, size and concentration observed in Beijing. Sci Total Environ 705:135967. https://doi.org/10.1016/j.scitotenv.2019.135967
93. Prata JC (Mar.2018) Airborne microplastics: Consequences to human health? Environ Pollut 234:115–126. https://doi.org/10.1016/j.envpol.2017.11.043
94. Browne MA, Galloway TS, Thompson RC (May2010) Spatial patterns of plastic debris along estuarine shorelines. Environ Sci Technol 44(9):3404–3409. https://doi.org/10.1021/es903784e
95. Zhou Q, Tian C, Luo Y (2017) Various forms and deposition fluxes of microplastics identified in the coastal urban atmosphere. Chinese Sci Bull 62:3902–3909

96. Catarino AI, Macchia V, Sanderson WG, Thompson RC, Henry TB (Jun.2018) Low levels of microplastics (MP) in wild mussels indicate that MP ingestion by humans is minimal compared to exposure via household fibres fallout during a meal. Environ Pollut 237:675–684. https://doi.org/10.1016/j.envpol.2018.02.069
97. Tunahan Kaya A, Yurtsever M, Çiftçi Bayraktar S (2018) "Ubiquitous exposure to microfiber pollution in the air." Eur Phys J Plus 133(11) 488, doi: https://doi.org/10.1140/epjp/i2018-12372-7
98. Klein M, Fischer EK (Oct.2019) Microplastic abundance in atmospheric deposition within the Metropolitan area of Hamburg, Germany. Sci Total Environ 685:96–103. https://doi.org/10.1016/j.scitotenv.2019.05.405
99. Bakir A, Rowland SJ, Thompson RC (2014) Transport of persistent organic pollutants by microplastics in estuarine conditions. Estuar Coast Shelf Sci 140:14–21. https://doi.org/10.1016/j.ecss.2014.01.004
100. Derraik JGB (2002) The pollution of the marine environment by plastic debris: A review. Mar Pollut Bull 44(9):842–852. https://doi.org/10.1016/S0025-326X(02)00220-5
101. Fossi MC et al (Nov.2012) Are baleen whales exposed to the threat of microplastics? A case study of the Mediterranean fin whale (Balaenoptera physalus). Mar Pollut Bull 64(11):2374–2379. https://doi.org/10.1016/j.marpolbul.2012.08.013
102. Bravo Rebolledo EL, Van Franeker JA, Jansen OE, Brasseur SMJM (2013) "Plastic ingestion by harbour seals (Phoca vitulina) in The Netherlands.," Mar Pollut Bull 67(1–2): 200–202, doi: https://doi.org/10.1016/j.marpolbul.2012.11.035
103. Nerland IL, Halsband C, Allan IJ, Thomas KV (2014) "Microplastics in marine environments: Occurrence, distribution and effects"
104. Stimmelmayr R, Adams B, Kayotuk C, Pederson M (2019) "Polar bears, plastics, and the pyloric sphincter: A volatile combination." Alaska Mar Sci Symp Anchorage
105. Lithner D, Larsson A, Dave G (Aug.2011) Environmental and health hazard ranking and assessment of plastic polymers based on chemical composition. Sci Total Environ 409(18):3309–3324. https://doi.org/10.1016/j.scitotenv.2011.04.038
106. Wang J, Chen G, Christie P, Zhang M, Luo Y, Teng Y (Aug.2015) Occurrence and risk assessment of phthalate esters (PAEs) in vegetables and soils of suburban plastic film greenhouses. Sci Total Environ 523:129–137. https://doi.org/10.1016/j.scitotenv.2015.02.101
107. Halden RU (2010) Plastics and health risks. Annu Rev Public Health 31:179–194. https://doi.org/10.1146/annurev.publhealth.012809.103714
108. Syberg K et al (May2015) Microplastics: addressing ecological risk through lessons learned. Environ Toxicol Chem 34(5):945–953. https://doi.org/10.1002/etc.2914
109. Rehse S, Kloas W, Zarfl C (Jun.2016) Short-term exposure with high concentrations of pristine microplastic particles leads to immobilisation of Daphnia magna. Chemosphere 153:91–99. https://doi.org/10.1016/j.chemosphere.2016.02.133
110. Rios LM, Moore C, Jones PR (2007) Persistent organic pollutants carried by synthetic polymers in the ocean environment. Mar Pollut Bull 54(8):1230–1237. https://doi.org/10.1016/j.marpolbul.2007.03.022
111. Lima ARA, Costa MF, Barletta M (Jul.2014) Distribution patterns of microplastics within the plankton of a tropical estuary. Environ Res 132:146–155. https://doi.org/10.1016/j.envres.2014.03.031
112. Mato Y, Isobe T, Takada H, Kanehiro H, Ohtake C, Kaminuma T (2001) Plastic resin pellets as a transport medium for toxic chemicals in the marine environment. Environ Sci Technol 35(2):318–324. https://doi.org/10.1021/es0010498
113. Lassen C et al. (2015) "Microplastics," Occurrence, effects and sources of releases to the environment in Denmark. Danish Environmental Protection Agency, Copenhagen K
114. BfR (2015) "Microplastic particles in food."013: 2015
115. GESAMP (2015) "Sources, fate and effects of microplastics in the marine environment: Part 2 of a global assessment"
116. Van Cauwenberghe L, Janssen CR (Oct.2014) Microplastics in bivalves cultured for human consumption. Environ Pollut 193:65–70. https://doi.org/10.1016/j.envpol.2014.06.010

117. Wright SL, Kelly FJ (Jun.2017) Plastic and human health: A micro issue? Environ Sci Technol 51(12):6634–6647. https://doi.org/10.1021/acs.est.7b00423
118. Kolandhasamy P, Su L, Li J, Qu X, Jabeen K, Shi H (Jan.2018) Adherence of microplastics to soft tissue of mussels: A novel way to uptake microplastics beyond ingestion. Sci Total Environ 610–611:635–640. https://doi.org/10.1016/j.scitotenv.2017.08.053

Face Masks: New Source of Microplastic Release in the Environment

Dinesh Kumar Gupta, Amit Vishwakarma, and Archana Singh

Abstract COVID-19 has resulted in the increased use of personal protective equipment (PPEs) all around the globe. These face masks are made from polypropylene and polyurethane material. Now these face masks are becoming the new emerging source of microplastic pollution in the ecosystem. In this study, we have analyzed the release of microplastic particles from different face masks in the aqua medium over a span of time. It has been observed that the surgical face mask releases (3659 particles/ piece) the highest amount of microplastics in the water as compared to other face masks (2300 particles/ piece by KN95, 2908 particles/ piece by branded cloth mask, 3332 particles/ piece by N95) which clearly indicates that it has the highest degradation rate.

Keywords Face mask · COVID-19 · Microplastic · Environmental pollution

1 Introduction

Since the start of the COVID-19 (SARS-CoV-2) pandemic, generation of plastic medical waste has increased upto 410%, which also includes the generated waste due to the rise in demand of packaging plastic [1]. Since the WHO has entrusted that the use of facemasks and maintaining hand hygiene is one of the most effective non-pharmaceutical measures [2] for reducing the contamination of COVID-19, Continuous use of facemasks is expected in the coming days even after the implementation of effective vaccination programs. It has been observed that these used facemasks are not disposed with proper bioremediation, and are being disposed in public places along with municipal waste and in open spaces. This has resulted in the release of plastic particles in soil subgrades and freshwater environment, and finally, to the

D. K. Gupta (✉) · A. Vishwakarma
Department of Civil Engineering, University Institute of Technology RGPV, Bhopal 462033, India
e-mail: gupta.dinesh96@gmail.com

D. K. Gupta · A. Singh
Advanced Materials and Processes Research Institute, Hoshangabad Road, Bhopal 462026, India

K. K. Pathak et al. (eds.), *Latest Developments in Civil Engineering*, Lecture Notes in Civil Engineering 352, https://doi.org/10.1007/978-981-99-2676-3_25

marine environment. It has been recently observed by Torre and Aragaw [3, 4]that these disposable facemasks are made of polypropylene and the same was confirmed by their Fourier transformation spectrometric analysis. They also suggested that due to the mechanical stress from freshwater and marine environment, these plastic polypropylene particles are being released in the ecosystem. Tareq and coworkers [5] stated that the solution to solid waste associated with single use disposable mask is 3D-printing. Since the 3D printed facemasks are made from thermoplastic polymers which are recyclable. However, these 3D printed facemasks might still pose a threat to the environment after continuous recycling and improper disposal, which will eventually lead to solid waste generation releasing the microplastic into the ecosystem. Microplastic fibers and particles can easily infiltrate into our food chains from any point of contact and may accommodate the development of microorganisms on their surface area [6–8]. Therefore, it is high time to quantify the release of microplastics from disposable face masks and its associated risks so that the relevant authorities may initiate a counterplan to avert any possible disaster before it is too late to respond. In this regard, Saliu and coworkers [9] stated that there is a significant effect of weathering on disposable face masks which is resulting in the release of microplastics in the marine environment. They reported that there is a significant amount of fiber released in saline water (around 135,000 particles). In this study, they processed surgical facemasks by inducing artificial weathering. However, they did not study the variation in the release of quantum of particles due to different types of face masks which also includes the different filtering standards.

In this study, four types of disposable masks are taken, which are N95 masks with five layers, KN95 masks with five layers, surgical masks with three layers and branded Cloth masks with three layers. We have quantified the release of MPs from every one of them with reference to time. In this study, the masks are left undisturbed in the water with no other external forces upon them.

2 Materials and Methods

All the masks (N95, KN95, Surgical and Branded Cloth of a single type of manufacturer) were purchased from the local market in the Bhopal city of India. It was made sure that all the masks purchased were of implementing standards for face masks in India. All the masks were fresh and unbroken.

All the types of masks were processed with the conditions as per Table 1 to quantify the release of microplastics from these masks if left in isolated water bodies like lakes, ponds etc. For every condition, the masks were put in 300 ML of tap water in a conical flask and the same were sealed with paraffin foil. After the required duration the masks were removed from the water and the supernatant was vacuum filtered on a 0.50 μm cellulose filter membrane. After the vacuum filtration, the filter membrane was kept in a petri dish and sealed with paraffin foil for further analysis.

Table 1 Different conditions for MPs release

S. no	Water type	Mask type	Duration (in days)	Nomenclature
1	Tap water	Branded cloth	1	B_{1DR}
			7	B_{7DR}
			30	B_{30DR}
			90	B_{90DR}
			180	B_{180DR}
2	Tap water	KN95	1	KN_{1DR}
			7	KN_{7DR}
			30	KN_{30DR}
			90	KN_{90DR2}
			180	KN_{180DR}
3	Tap water	N-95	1	N_{1TR}
			7	N_{7TR}
			30	N_{30TR}
			90	S_{90DR}
			180	S_{180DR}
4	Tap water	Surgical	1	S_{1TR}
			7	S_{7TR}
			30	S_{30TR}
			90	S_{90DR}
			180	S_{180DR}

3 Visual Counting and Qualification of Microplastic

The substrate on the filter paper was analyzed using a metallurgical microscope and images were captured by a 3 Megapixel camera accessory equipped with this metallurgical microscope. Microplastics were examined under the microscope based on their texture, shape and size. A hot needle test technique was carried out for differentiating microplastic and any other organic matter [10] and for the confirmation of the presence of microplastics in the sample.

4 Quality Control and Quality Assurance

All the equipments were prewashed using DI water for avoiding any type of contamination and after prewashing these equipments were kept in a desiccator with the covering of the parafil foil. Cotton masks and laboratory coats were used for avoiding the use of any type of plastic accessories. Proper sanitization process was used for avoiding any type of cross contamination of the sample. One blank sampling was done to avoid any statistical error.

5 Results and Discussions

Microplastics particles have been observed in all types of samples regardless of the type of masks and different conditions. Figures 1, 2 show the microscopic images of the microplastic released from the different face masks.

The average amount of microplasticfibre released from the branded cloth masks varied from 831 particles/piece in one day to 2907 particles/piece in 180 days, from the surgical masks varied from 1471 particles/piece in one day to 3659 particles/piece in 180 days, N95 masks varied from 1010 particles/piece in one day to 3331 particles/piece in 180 days, KN95 masks varied from 739 particles/piece in one day to 2300 particles/piece in 180 days. Highest amount of microplastic particles release was observed from the surgical face masks followed by branded cloth face mask,

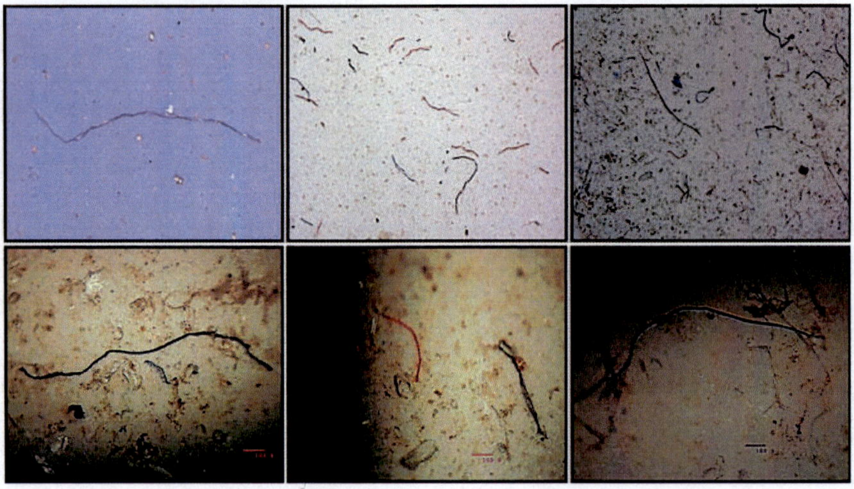

Fig. 1 Microscopic images of MPs fiber released from Branded cloth mask (Upper images) and surgical mask (Lower Images)

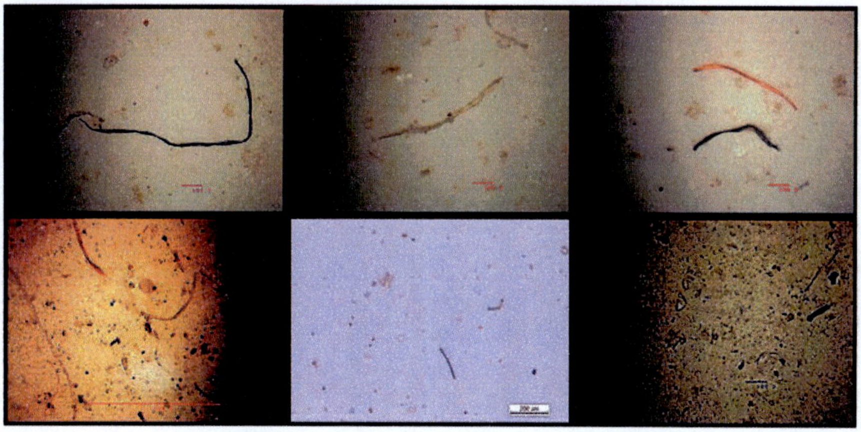

Fig. 2 Microscopic images of MPs fibre released from N95 mask (Upper images), KN95 mask (Lower Images)

N95 masks and least amount was released from KN95 mask. Figures 3, 4 shows the line diagram and bar chart for the microplastic particle abundance over the period of time which clearly indicate that surgical masks releases the highest amount of microplastics.

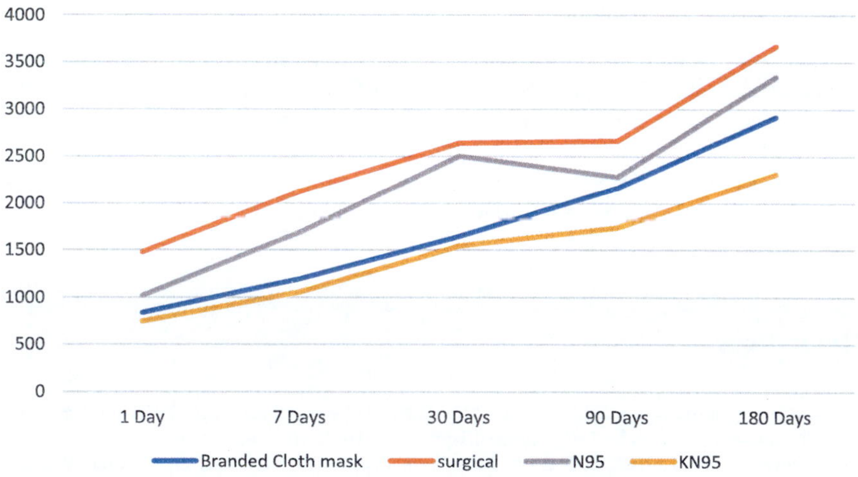

Fig. 3 Microplastic Particle abundance over the period of time (Line diagram)

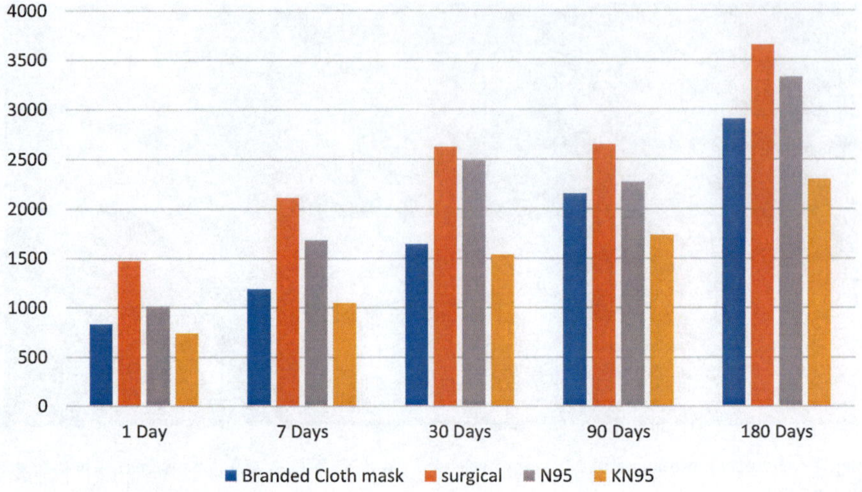

Fig. 4 Microplastic Particle abundance over the period of time (Bar Chart)

6 Conclusions

This study clearly indicates that the face masks have now become the prominent source of microplastic release in the environment be it water air or soil. From this study it can be concluded that:

- Face masks are the emerging source of microplastic release in the environment.
- Surgical masks release highest quantum of microplasticfibre as compared to the other face masks when left undisturbed in the environment for a long period of time.
- There is a need to implement a stringent policy for proper disposal of these face masks after their use to reduce the release of microplastic in the environment.

References

1. Gorrasi G, Sorrentino A, Lichtfouse E (2021) Back to plastic pollution in COVID times. Environ Chem Lett 19(1):1–4. https://doi.org/10.1007/s10311-020-01129-z
2. Davies NG et al (2020) Effects of non-pharmaceutical interventions on COVID-19 cases, deaths, and demand for hospital services in the UK: a modelling study. Lancet Public Heal. 5(7):e375–e385. https://doi.org/10.1016/S2468-2667(20)30133-X
3. De-la-Torre GE, Aragaw TA (2021) What we need to know about PPE associated with the COVID-19 pandemic in the marine environment. Mar Pollut Bull 163:111879. https://doi.org/10.1016/j.marpolbul.2020.111879
4. Aragaw TA (2020) Surgical face masks as a potential source for microplastic pollution in the COVID-19 scenario, Mar. Pollut. Bull., 159. https://doi.org/10.1016/j.marpolbul.2020.111517

5. Tareq MS, Rahman T, Hossain M, Dorrington P (2021) Additive manufacturing and the COVID-19 challenges: An in-depth study. J Manuf Syst 60:787–798. https://doi.org/10.1016/j.jmsy.2020.12.021
6. Galloway TS, Cole M, Lewis C (2017) Interactions of microplastic debris throughout the marine ecosystem.. Nat Ecol Evol, 1(5), p 116, https://doi.org/10.1038/s41559-017-0116
7. Taylor ML, Gwinnett C, Robinson LF, Woodall LC (2016) Plastic microfibre ingestion by deep-sea organisms. Sci Rep 6(1):33997. https://doi.org/10.1038/srep33997
8. Gupta DK, Choudhary D, Vishwakarma A, Mudgal M, Srivastava AK, Singh A (2022) Microplastics in freshwater environment: occurrence, analysis, impact, control measures and challenges. Int J Environ Sci Technol. https://doi.org/10.1007/s13762-022-04139-2
9. Saliu F, Veronelli M, Raguso C, Barana D, Galli P, Lasagni M (2021) The release process of microfibers: from surgical face masks into the marine environment. Environ Adv 4:100042. https://doi.org/10.1016/j.envadv.2021.100042
10. Hidalgo-Ruz V, Gutow L, Thompson RC, Thiel M (2012) Microplastics in the marine environment: A review of the methods used for identification and quantification. Environ Sci Technol 46(6):3060–3075. https://doi.org/10.1021/es2031505

Geotechnical Engineering

Behaviour of Skirt Raft Foundation on Sandy Soil Using Plaxis 2D

Pavan Kumar Elagandular and A. K. Sinha

Abstract Geotechnical engineers are trying to find an alternative and effective method for increasing bearing capacity and reducing foundation settlement. Providing skirts around the foundation system may help to improve the Bearing Capacity of shallow footings on sandy soil by restricting the soil movement beneath them and limiting the plastic flow of the soil. Structural skirts have been utilised to improve the effective depth of foundations in marine and other circumstances where water scour is a serious issue. However; only limited information is available in the literature about the performance of skirted foundations, and therefore in the present study an attempt has been made to analyse the various parameters on raft footings of various widths, skirt angles, and skirt depths. The aim of the study is to enlighten the behaviour of the skirt on sandy soil for the settlement of shallow foundations. For various widths of raft footings namely, 15 m, 20 m, 25 m, and 30 m, and skirt depth ratio of 0B, 1B, 1.5B, 2B, and 2.5B is employed with a skirt angle of 5° and 10°. This study shows the findings of a numerical analysis used to assess the settlement of raft foundations with and without skirts. The results of the skirted foundation show that the percentage settlement is increasing, with a larger skirt depth ratio, less width of footing, and high skirt angle.

Keywords Skirt foundation · Skirt depth ratio · Interior skirt angle · Percentage settlement

P. K. Elagandular (✉) · A. K. Sinha
Post Graduate Student, National Institute of Technology, Jamshedpur, Jharkhand 831014, India
e-mail: 2020pgcege09@nitjsr.ac.in

A. K. Sinha
e-mail: aksinha.ce@nitjsr.ac.in

K. K. Pathak et al. (eds.), *Latest Developments in Civil Engineering*, Lecture Notes in Civil Engineering 352, https://doi.org/10.1007/978-981-99-2676-3_26

1 Introduction

Any construction design must give the guarantee that the design is safe under maximum loading and settlement should be within permissible limits. Now-a-days utilisation of land is more important due to the increase in population and less land area. Engineers are forced to do construction on weak soil like sand, clay, peat, etc. Structures constructed on weak soil should be designed properly by using different improvement techniques. In these modern developing days, there are a lot of improvement techniques used for soil stabilisation but some of them are site restrictions on the field, difficulty in installation, availability of material, and cost expensive. It is also difficult to apply to the existing foundations. In this point of view, a special foundation has its importance and is distinguished based on its performance. The new concept is increasing an optimum alternative for shallow foundations to support structures, especially in available depth. This type of foundation transmits the load from structures to the subsoil and increases the bearing capacity by confinement of soil. This technique has been used for a long time to increase the depth of foundation in offshore areas. This technique does not require any excavation of soil. The presence of a high-water table also will not affect the footing. The skirt restrains the soil beneath laterally and acts as a soil plug to transfer the load from the superstructure to the soil. This foundation is commonly used for wind turbines, offshore structures, storage tanks, windmill towers and oil tanks.

The H shape footing has a slighter higher bearing capacity compared to the square footing. Lower relative density has more effective than higher relative density [1]. The confinement of soil under skirt footing reduces the acceleration foundation and it tends to relieve the transmitted disturbance [2]. The finite element analysis has been done for Mohr–coulomb and hardening soil model, for both models the bearing capacity increases as the skirt depth ratio increases [3]. The skirts are more beneficial on loose sand than the medium and dense sand, which also depends on the relative density of the sand [4]. The bearing capacity and settlement values are very close but not equal when compared between the skirt foundation and pile foundation for same width and depth. Equations has been derived for both the bearing capacity factor and settlement reduction factor which depend on skirt depth to footing width and surface foundation [5]. By practical significance, the reduction of settlement is observed on dense sand using structural skirts. The settlement reduction factor (SRF) is proposed which depends on the stress applied and skirt depth ratio [6]. Experimental study has been carried out for a skirt with open end at the bottom and a closed end at the bottom. Results show that the closed end of skirt foundations has better bearing capacity can the open end of skirt foundation [7]. When skirts are placed inclined for both side and one side, the skirt with both sides inclined with a higher skirt angle gives more improvement than the skirt with one side inclined skirt [8]. The skirt placed on peat soil under the embankment, stability shows good results with skirt footing. The circular footing helps in keeping the load area in stabilisation soil and the settlement has been shown evenly. The effective stress has been increased for skirt footing than the mini pile. The stabilisation of soil under mini piles is not centred

[9]. The bearing capacity is increased for lower relative densities of 30% and 50% compared to relative densities of 70% and 87%. The bearing capacity is increased linearly with footing for both with and without skirts placed to the footing [10]. Using a circular skirt with different parameters the improvement factor and reduction in settlement is found. The higher reduction in settlement is found in a higher skirt depth ratio and improvement factor is increased by 3.25 for higher confinement [11].

2 Numerical Modelling

The Dimensions of soil model used for the finite element analysis is 20B X 20B with different width of raft foundations are 15 m, 20 m, 25 m, and 30 m. The Skirts are placed under the foundation with different skirt depths are 0B, 0.5B, 1B, 1.5B, 2B, and 2.5B. The skirts are placed inclined at the corner with an angle of 5° and 10° with respect to the vertical in clockwise direction and a vertical skirt is placed at the centre of the footing. The load applied on the raft foundation changes from 0 kN/m^2 to 1000 kN/m^2 gradually. Figure 1 show the geometrical model of the unskirt raft foundation and with the skirt placed interior to the footing.

2.1 Geometrical Properties of Sand and Skirt

Table 1 Shows the geometrical properties of soil and skirt used in this analysis.

The model simulation has been done in 12 steps in staged construction mode. In the initial phase, the soil only is activated and in phase 1, the structural steel skirt is

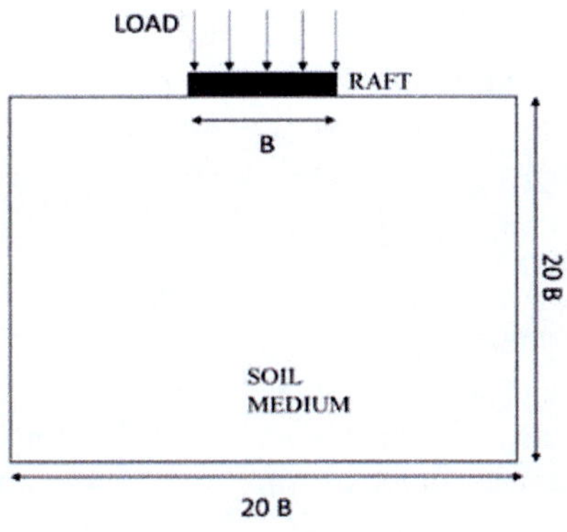

(a) Unskirt Raft Foundation

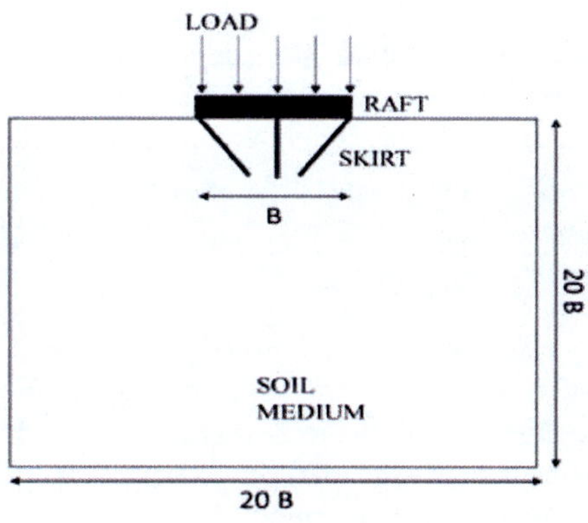

(b) Inclined Skirt Foundation

Fig. 1 Geometrical model of unskirt and skirt raft foundation

Table 1 Geometrical properties

Parameters	Values
Types of material	Sand
Material model	Mohr coulomb
Material behaviour	Drained
E_{50} (kN/m^2)	40,000
γ (kN/m^3)	20
Cohesion(c)	0.1
Poisson's ratio(υ)	0.3
Friction Angle(φ)	35
Angle of dilatancy(ψ)	2
Axial stiffness of steel skirt (kN/m)	31,500

activated. From phase 2 to phase 12 the load is varied from 0 kN/m^2 to 1000 kN/m^2 gradually. In the output for every different load the settlements are noted and the percentage settlement is calculated.

3 Results and Discussion

By using the soil properties and modeling, the analysis is carried out in PLAXIS 2D on a raft foundation placed up on the sandy soil for both unskirt and inclined skirts. The results obtained from PLAXIS 2D has been shown in the Fig. 2 for unskirt raft foundation with a width of 20 m.

Figure 3 shows the Skirt raft foundation for a typical case of width 20 m, skirt depth of 1B, and interior skirt angle placed to the foundation is 5°.

The results have been plotted for different parameters used in the analysis are width of foundation, skirt depth ratio, skirt angle, and different loads. I performed every results in terms of percentage of settlement to know the effect of skirt raft foundation. Here higher percentage of settlement signifies lower value of settlement.

3.1 Effect of Load

For inclined skirts placed interior to the footing shows the settlement is reduced as the skirt depth ratio increases for constant skirt angle and for a given width of footing. Figure 4 shows the load versus percentage settlement graph for different widths with different skirt depth ratio.

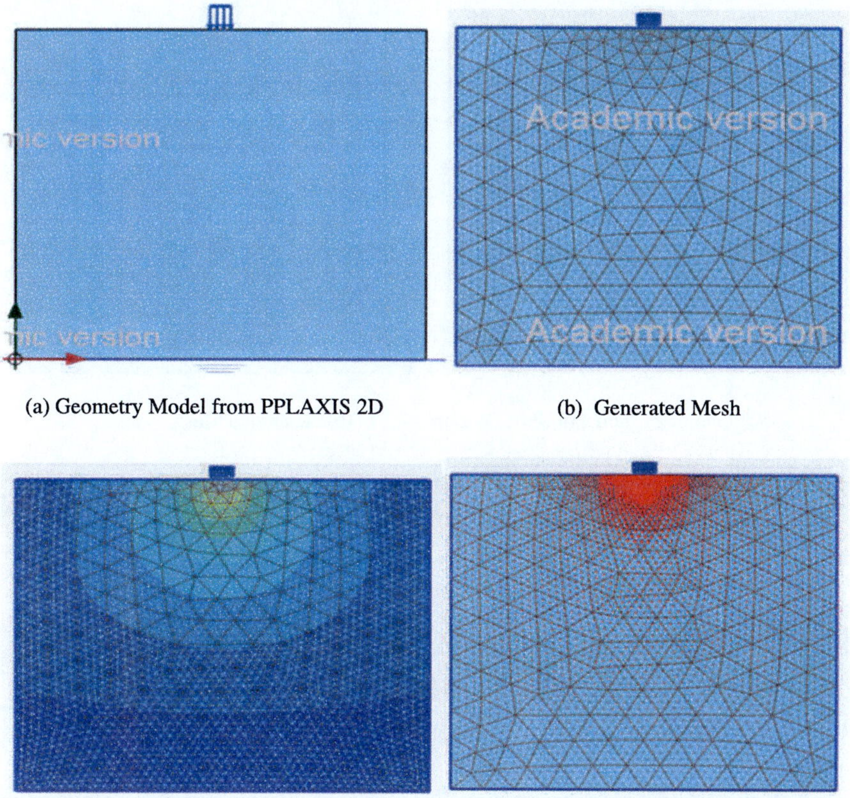

(a) Geometry Model from PPLAXIS 2D

(b) Generated Mesh

(c) Total Displacement Contour

(d) Total Displacement in Arrows

Fig. 2 a–d Raft foundation without skirt

From the above graphs, we can say that the percentage settlement increases with increasing the load for a constant skirt depth ratio. For a given load, the increase in skirt depth ratio has an increase in percentage settlement. We can observe that up to the load of 500 kN/m^2, the percentage settlement has a marginal increment and after 500 kN/m^2 the increase in percentage settlement has significant improvement.

3.2 *Effect of Raft Foundation Width*

From the above Fig. 5, we can observe that the percentage settlement decreases as the width of raft foundation increases for the given skirt depth ratio and the keeping load constant of 1000kN/m^2. For 15 m width, the increases in percentage settlement are more compared to other widths for both skirt angles. For 20 m, 25 m, and 30 m widths the change in percentage settlement is almost same and has a minimal increment.

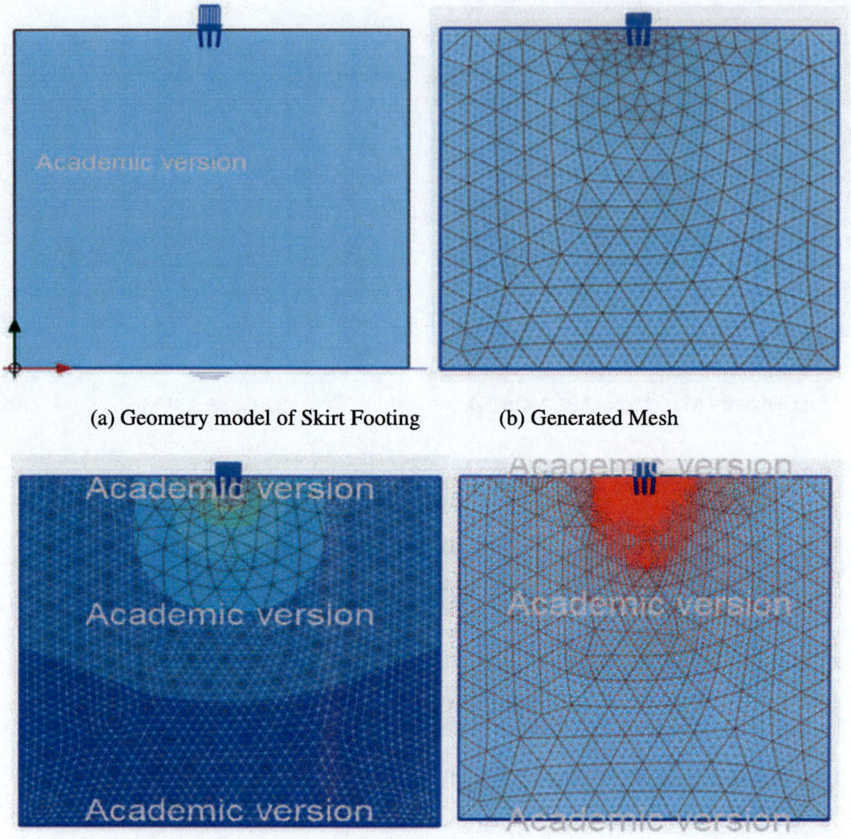

(a) Geometry model of Skirt Footing (b) Generated Mesh

(c) Total Displacement Contour (d) Total Displacement in Arrows

Fig. 3 **a–d** Raft foundation with skirt with interior skirt angle

3.3 *Effect of Skirt Angle*

Figure 6 shows the skirt depth ratio versus percentage settlement for different widths of raft foundation. We can observe that, for any width of foundation the higher skirt angle shows a higher percentage settlement compared to the lower skirt angle.

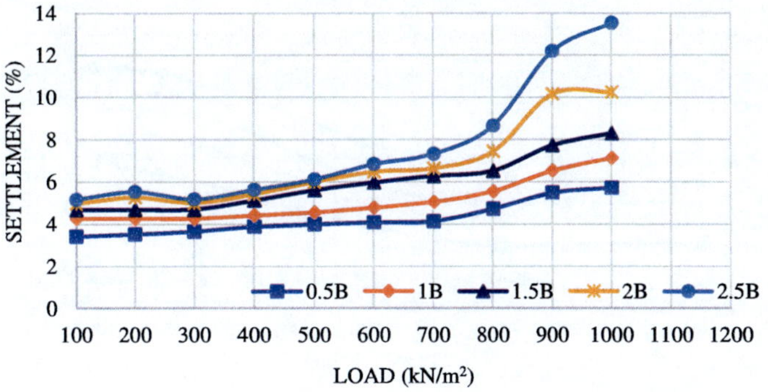

(a) For 15m width, 5° inclination

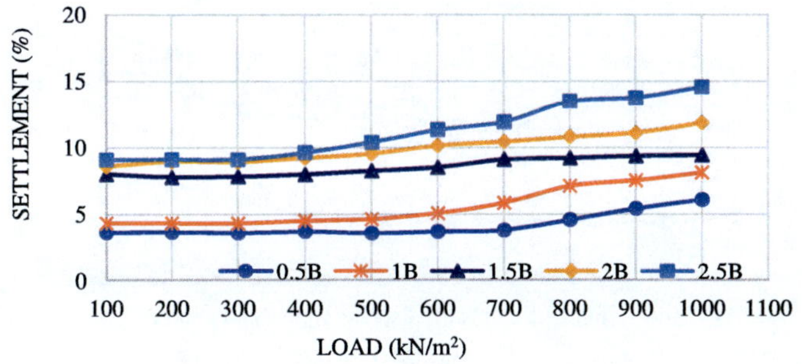

(b) For 15m width, 10° inclination

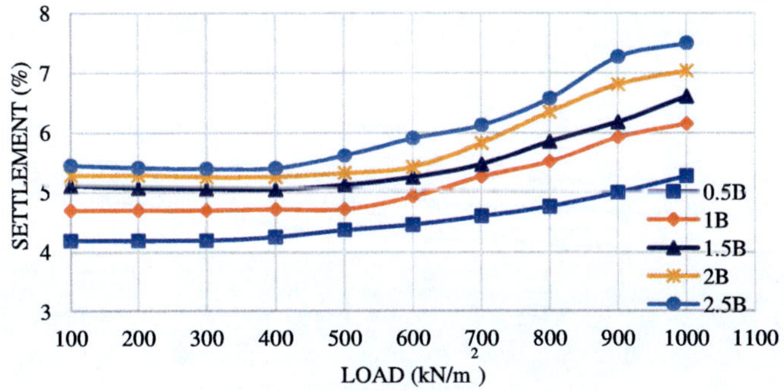

(c) For 20m width, 5° inclination

Fig. 4 a–h Load versus percentage settlement

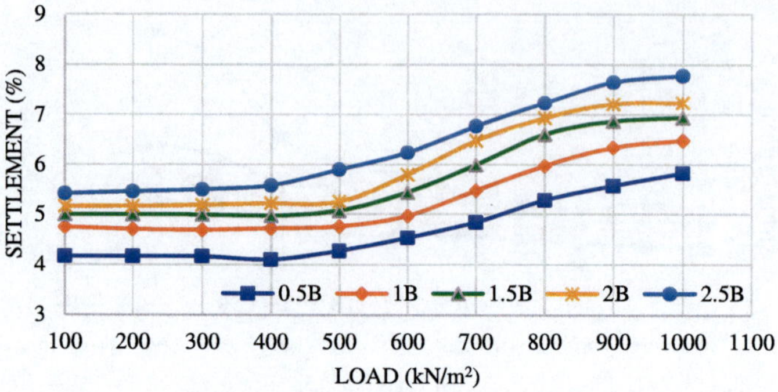

(d) For 20m width, 10° inclination

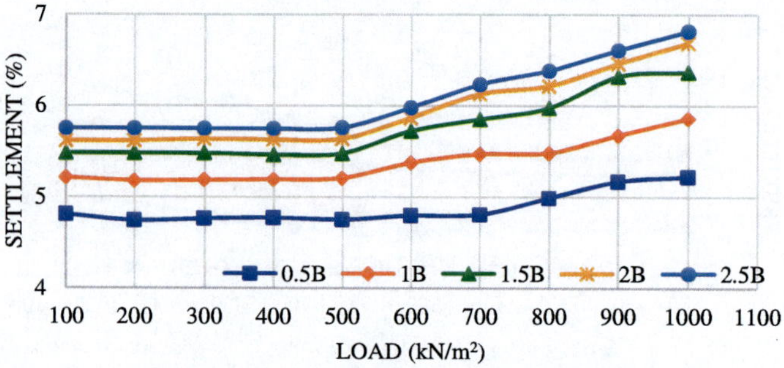

(e) For 25m width, 5° inclination

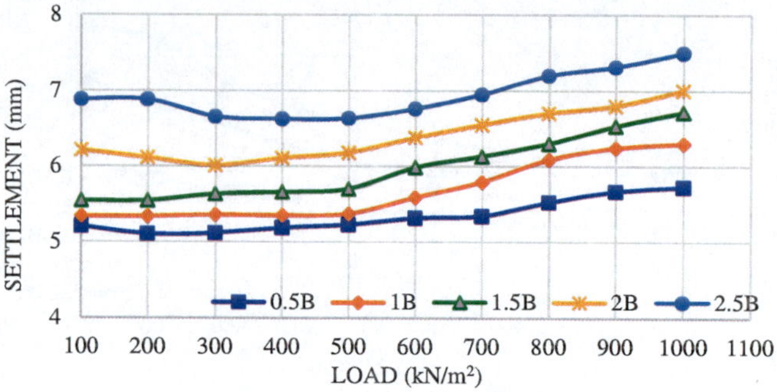

(f) For 25m width, 10° inclination

Fig. 4 (continued)

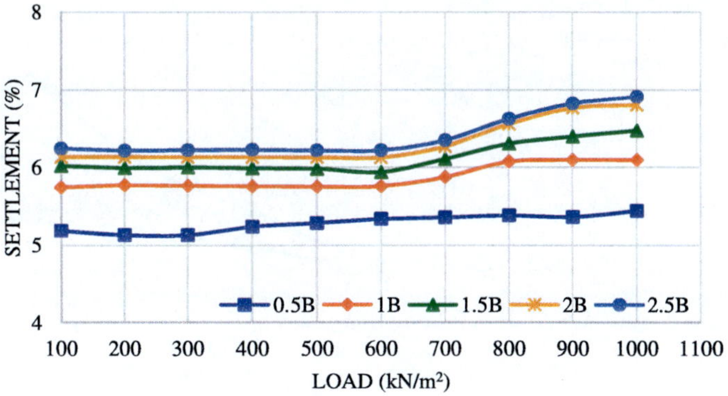

(g) For 30m width, 5° inclination

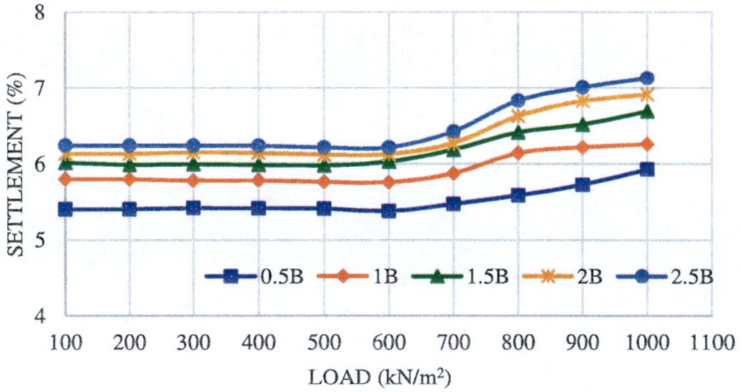

(h) For 30m width, 10° inclination

Fig. 4 (continued)

3.4 Effect of Skirt Depth Ratio

Figure 7 shows the foundation width versus percentage settlement for different skirt depth ratios and both skirt angles. Here we can see that as the width of foundation increases the percentage settlement decreases for a constant skirt depth ratio. Higher the skirt depth ratio has the higher percentage settlement for any foundation widths.

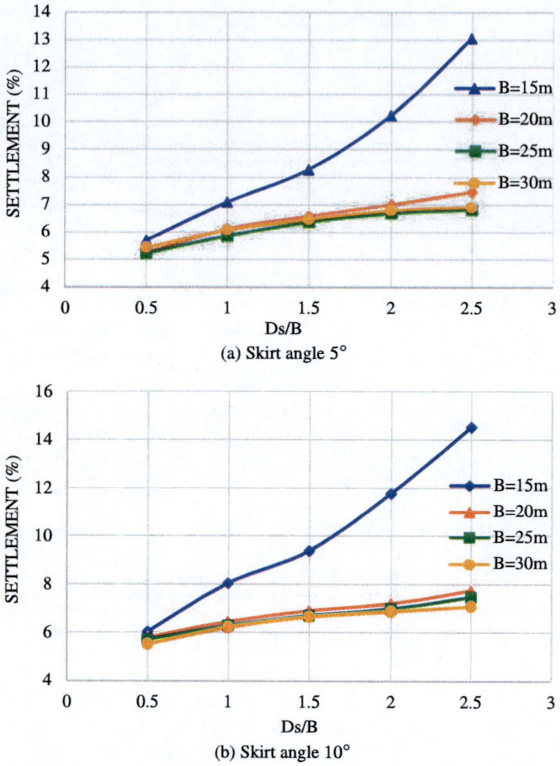

Fig. 5 **a, b** Skirt depth ratio versus percentage settlement for interior skirt angles

4 Conclusion

From all effects of interior skirt placed inclined with different angles for different skirt depths and foundation widths, we can conclude that.

- As the skirt depth ratio increases the percentage settlement also increases for the given width of foundation and Interior Skirt angle.
- The higher degree skirt angle gives more percentage settlement than the lower degree skirt angle for a given skirt depth ratio.
- Lower foundation width has a higher percentage settlement when compared to the higher foundation width for a given skirt angle and skirt depth ratio.
- Using lower width of foundation with higher skirt depth ratio and higher skirt angle will give more percentage settlement.

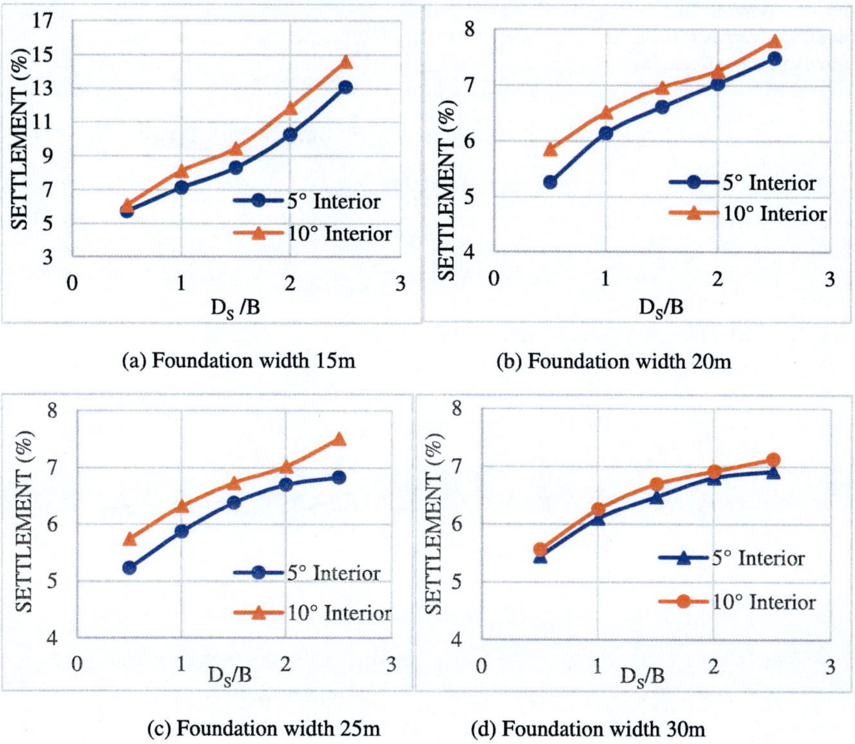

Fig. 6 **a–d** Skirt depth ratios versus percentage settlement for foundation width

Fig. 7 a, b Width of foundation versus percentage settlement for skirt angles

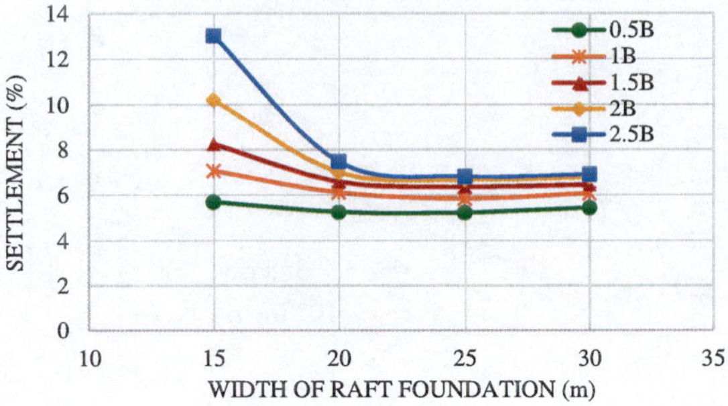

(a) Skirt Angle 5°

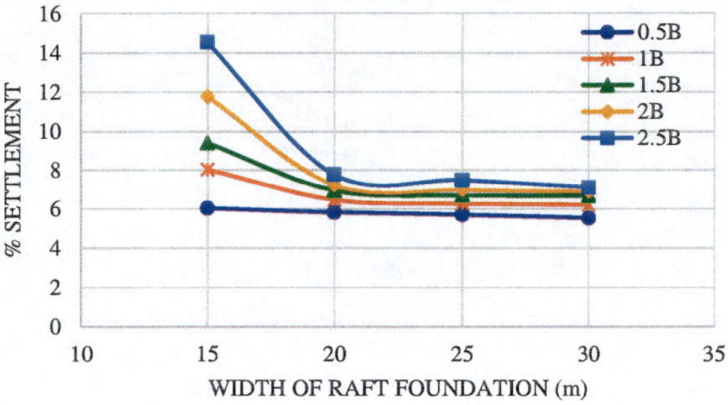

(b) Skirt Angle 10°

References

1. Gnananandarao T, Vishwas NK, Kumar R (2017) Performance of multi-edged skirt footings resting on sand. Indian Geotech J. https://doi.org/10.1007/s40098-017-0270-6
2. Azzam WR (2014) Finite element analysis of skirted foundation adjacent to sand slope under earthquake loading. HBRC J
3. Sarma K, Chetia N (2016) Load displacement behavior of skirted raft foundation sand using Plaxis 2D. Int J Adv Eng Res Sci (IJAERS)
4. Wakil AZ (2013) Bearing capacity of skirt circular footing on the sand. Alex Eng J
5. Hisham T (2013) Bearing capacity and settlement of skirted shallow foundation on sand. Int J Geomech. ASCE/September/October 2013/645
6. Al-Aghbari MY (2006) Settlement of shallow circular foundation with structural skirt resting on the sand. J Eng Res 4(1):11–16
7. Haider L, Haider MM (2018) Response of skirted foundations resting on dry medium dense sand. Civ Eng J, 4(6)
8. Pusadkar SS, Bhatkar T (2013) Behavior of raft foundation with inclined skirt using plaxis 2D. In: Proceedings of Indian Geotechnical Conference December, 22–24, Roorkee
9. Gautama DAW, Supanji BS (2020) The effect of skirt footings for road settlement on peat soil. In: 4th International Conference on Civil Engineering Research
10. Vishwas NK, Debbarma SP, Kumar R, Mohanty B (2017) Pressure settlement behavior of square and rectangle skirted footing resting on sand. Geomech Eng 12(4):689–705
11. Hussian R, Chetia N (2017) A study on the performance of circular skirted foundation on medium dense sand. Indian Geotechnical conference GeoNEst

Use of RBI Grade 81 and Coir Fiber for Stabilization of Expansive Soil

Rajesh Kumar Mahto, Abhishek Dubey, Ramakant Agrawal, and Padma Panday

Abstract Subgrade soil collapse due to inadequate strength, low bearing capacity, cracking and large deformation of problematic soils are commonly seen on the pavement, and this causes huge maintenance costs and repair of highway projects every year. It is essential to decrease these engineering problems and economic losses through economical and eco-friendly methods. Past studies have shown that randomly spread fibers can considerably improve various soil properties. However, there is a lack of complete study on the engineering properties of fiber reinforced high plastic clay. Also, limited mechanical models have been proposed for predicting the shear strength behavior of fiber reinforced clay. In order to investigate these problems, a series of laboratory investigations including compaction, bearing capacity, one-dimensional consolidation, linear shrinkage, desiccation cracking, direct tensile strength, and compression tests should be conducted on unreinforced and Coir fiber reinforced Clay. For this study, the soil samples were prepared with different proportions of RBI grade 81 i.e. (2, 4, 6 and 8% of soil) respectively. After that the coir fibers in different ratios i.e. 0.5, 1, 1.5 and 2% respectively will be added to the sample containing suitable content of RBI grade 81. Then OMC, MDD and CBR values were evaluated for these samples.

Keywords Coir fiber · Highway projects · RBI grade 81 · Tensile strength · OMC

1 Introduction

Expansive soil causes pavement distresses. On the basis of moisture level, expansive soils will experience large changes in volume due to moisture variations from seasonal changes. During the periods of high moisture expansive will "swell" underneath the highway structure. On the other hand during periods of low soil moisture, expansive soil will "shrink" and can result in considerable deformation. These cycles of swell and/or shrinkage may also lead to cracking of pavement. Puppala et al. (2006)

R. K. Mahto (✉) · A. Dubey · R. Agrawal · P. Panday
Department of Civil Engineering, OIST, Bhopal, India
e-mail: errajesh12c12@gmail.com

K. K. Pathak et al. (eds.), *Latest Developments in Civil Engineering*, Lecture Notes in Civil Engineering 352, https://doi.org/10.1007/978-981-99-2676-3_27

shows that expansive soils encountered in many districts primarily in northern Texas are the main causes of pavement failures. Expansive soils situated in regions where cool and wet periods are followed by hot dry periods are more susceptible to such problems.

Subgrade soil collapse due to inadequate strength, low bearing capacity, cracking and large deformation of problematic soils are commonly seen on the pavement, and this causes huge maintenance costs and repair of highway projects every year. It is essential to decrease these engineering problems and economic losses through economical and eco-friendly methods (Fig. 1).

Collapsible soils named as swelling soils or shrink-swell soils are the terms useful to those soils, which have a nature to swell and shrink with the change in moisture content. As a result of which considerable distress in the soil occurs, causing severe damage to the overlying structure. During monsoons, these soils imbibe water, swell, become soft and their capacity to bear water is reduced, while in drier seasons, these soils shrink and become harder due to the evaporation of water. These types of soils are generally found in arid and semiarid regions of the world and are considered as a potential natural hazard, which if not processed well can cause extensive damages to not only to the structures built upon them but also can cause loss of human life. Soils containing the clay minerals montmorillonite generally show these properties. The annual cost of damage to the civil engineering structures caused by these soils is estimated to be £ 150 million in the U.K., $ 1,000 million in the U.S. and many billions of dollars worldwide.

Expansive soils also called as Black soils or Black cotton soils and Regur soils mainly originate over the Deccan lava tract (Deccan Trap) locating Maharashtra, Madhya Pradesh, Gujarat, Andhra Pradesh and areas of Odisha, in the Indian sub-continent. Black cotton soils are which are expansive and collapsible and also found in river valley of Tapi, Krishna, Godavari and Narmada. In the north western part of Deccan Plateau and in the upper parts of Krishna and Godavari, the depth of black soil is very large. Basically, these soils are residual soils left at the place of their formation after chemical decomposition of the rocks such as basalt and trap.

Fig. 1 Cracking appears on the embankment shoulder in road

Fig. 2 Typical dried expansive soil

Also, these types of soils are formed due to the weathering of igneous rocks and the cooling of lava after a volcanic eruption. These soils are rich in lime, iron, magnesia and alumina but lack in the phosphorus, nitrogen and organic matter (Fig. 2).

RBI Grade 81 was developed in 1990s. It was extensively used to construct all manner of roads. RBI Grade 81 was bought by Mr. Josy Cohen from the SA Government, and was further developed and optimized through continuous R&D. After 10 years of R&D South African patent was granted. It took a number of years to introduce soil stabilizers in India through numerous trials using RBI Grade 81. RBI Grade 81 is the pioneer of Soil Stabilization in India. Road Building International (India) Pvt. Ltd. holds the exclusive license to manufacture and market RBI Grade 81 in India. RBI Grade 81 meets the requirement for a well-proven, reliable and very cost-effective method by creating a strong and irreversible impermeable layer resistant to adverse climatic conditions, from very high temperatures to permafrost conditions, and accommodating all vehicular loads. RBI Grade 81 is environmentally friendly and emphasises the use of recycled material, recognizing the lack of readily available resources.

Engineering advantages of RBI GRADE 81

- Increases the California Bearing Ratio (CBR) manifolds.
- Increases Unconfined Compressive Strength (UCS) considerably.
- Increases Modulus of Elasticity value, which results in reduction of pavement crust.
- Reduces Plasticity Index (PI) value.
- Reduces Free Swelling Index (FSI) value.
- By strengthening the existing soil by 12–20 times the initial strength, it helps in replacing the conventional aggregate layers with soil stabilized layers, thus saving aggregate. It also reduces the quantity of bitumen in road construction.
- Since it reengineers any kind of soil and stabilizes it with increased strength, it eliminates the removal and carriage of in-situ soil and replaces it with better soil suitable for construction.

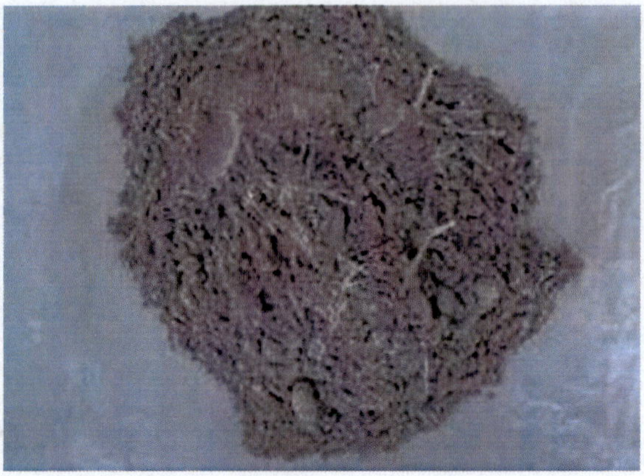

Fig. 3 RBI grade 81 powder

- By using in-situ soil, it reduces the need to transport good soil and aggregate by about 40–60%, thus reducing the carbon emission from the trucks.
- The treated areas are comparatively impermeable to water, thus preventing damage to the road foundation.
- Due to the reduced construction time, air pollution by heavy suspended particles is reduced considerably.
- Durability is increased thus the need for continuous maintenance is reduced.
- RBI Grade 81 technology is very simple and does not require skilled labor.
- RBI Grade 81 can also be used for cold recycling of existing pavement (flexible or rigid) layers, thus saving natural resources, as 90% of road material is being reused (Fig. 3).

Following are the objectives of this work:

- To determine the effect of addition of RBI grade 81 to the soil sample in terms of atterberg limits, Dry density, OMC, and CBR (Unsoaked).
- To determine the optimum percentage quantity of RBI grade 81 to be used as soil stabilizer.
- To determine a suitable percentage of coir fiber for reinforcing a given sample of expansive soil.

2 Literature Review

- **Gangadhara et al. (2021)** In this research, an experiment is made to know the impact on strength characteristics of expansive soil treated with Road Building International Grade 81 (RBI-Grade 81) stabilizer. This stabilizer has a wide response spectrum range. A broad range of studies, such as Atterberg limit, Compaction, UCS, and California bearing ratio (CBR) measures, are confirmed on expansive soil and it is treated with percentages varying between 2 and 10% of

RBI-Grade 81. Due to the formation of new cement materials, which is apparent in the SEM research micrographs, the upgrading in intensity is achieved. The results suggest that the considered stabilizer is efficient in enhancing the engineering properties of expansive soil.

- **Kumar and Khatri (2019)** This study research and analyzes the consequences of a series of Atterberg's limit experiment, Proctor compaction tests, California bearing ratio tests and Unconfined Compressive Strength tests plasticized on Black cotton soil treated with a combination of various Lime and Road Building International (RBI) Grade 81 contents and compacted below the optimum Proctor conditions. These test results illustrate that the geotechnical parameter values are concordant and confirm the bearing capacity improvement of this natural clay, which is translated by a significant increase in soil strength. However, the best performances are obtained for a mix treatment corresponding to 6% Lime and 5% RBI Grade 81 contents. Using 6% Lime we can save up to 25% and using 5% RBI Grade 81 we can save up to 18% on pavement cost.
- **Bernadette et al. (2018)** the present study research the effect of reinforcing the subgrade soils with RBI 81 material. A soil nearby was collected and preliminary tests were performed to classify the soil and it was found from the results that the sample collected was a poorly graded clay. Subsequently Tests such as Proctor Compaction, CBR, and UCC were conducted to study the various engineering properties of the identified soil. In addition to the above tests were also conducted on the soil by reinforcing with varying percentages of RBI 81. From the analysis of test results it was found that this material (RBI 81) will significantly improve the CBR value of the soil. However, it was confirmed that walnut shells can be used as light replacement material instead of fine aggregate up to 30% at 0.38 water/cement ratio without adversely affecting the acceptable compressive strength for structural Portland cement concrete.
- **Shiva and Darga (2017)** the present study deals with the effect of Road Building International Grade 81 (RBI Grade 81) on the strength properties of the clayey subgrade. A wide range of tests was performed on various percentages of RBI Grade 81 subgrade mixtures such as Atterberg limits, compaction characteristics, California Bearing Ratio (CBR) and Unconfined Compression Strength tests (UCS). It may be observed that the CBR of the clayey subgrade has increased signifcantly with the addition of RBI Grade 81. Similarly, the maximum dry density (MDD) has decreased and the optimum moisture content (OMC) has increased with RBI Grade 81 addition to subgrade. The strength of the clay subgrade at 8% of RBI Grade 81 has enhanced 3–3.5 times as compared to the natural clayey subgrade. Overall, it can be concluded that RBI Grade 81 is suggested as a potential stabilizing agent, especially for clayey subgrades.
- **Kumar and Solanki (2017)**, in the present study, a trial is made to improve the engineering properties of the expansive soil available in SVNIT campus of Surat city, Gujarat by stabilizing it with RBI Grade 81. The effect of adding this chemical stabilizer in expansive soil is characterized through various laboratory tests to find out Consistency Limits, Compaction Characteristics, Free Swell Index,

Unconfined Compressive Strength, and California Bearing Ratio. The laboratory investigations indicate that RBI Grade 81 with up to 5% optimizes the engineering properties of expansive soil to be used as subgrade.

3 Material Used

The following materials are used during the research work:

- Raw soil or B.C soil only
- RBI grade 81 powder
- Coir fiber
- Water

The soil used for this investigation is an expansive clay, one type of most problematic soil for subgrade constructions is used for the current study work which is locally found Black Cotton Soil collected from the village Mugaliya chhap near Sehore (Madhya Pradesh) from a depth of 4.5 m from ground level. It contains harmful substances of many sizes. The soil was air dried and pulverized manually. This natural soil is black in colour (Fig. 4 and Table 1).

RBI Grade 81 was purchased from an alchemist dealer in Delhi it was coordinated by Road Building International (Table 2).

Fig. 4 Black cotton soil sample

Table 1 Physical properties of soil

S. No.	Technical properties	Values
1	Specific gravity	2.75
2	Liquid limit (%)	51.90
3	Plastic limit (%)	22.90
4	Plasticity index (%)	29
5	CBR (%)	1.67%

Table 2 Properties of RBI grade 81 taken

S. No.	Technical properties	Values
1	Specific gravity	2.5
2	Appearance	Beige powder
3	PH	12.50
4	Solubility	In water 0.2 pts/100 pts
5	Flammability	Inflammable

Table 3 Physical properties of coir fiber

S. No.	Properties	Value
1	Appearance	Thin like threads
2	Specific gravity	0.128
3	Length	1.5–2 cm

Coconut Fiber Material used is derived from coco fiber obtained by taking the residual yield (waste) of coconut use. Percentage of coconut coir fiber is taken by the weight of soil mixture. The fibers were extracted manually and separated into strands (Table 3).

4 Methodology

Following procedure has been adopted for the work (Figs. 5, 6, 7 and Table 4).

5 Results

Results for B.C soil with altered percentage of RBI grade 81

See Table 5, 6, 7, 8 and Figs. 8, 9.

Results for B.C soil with 8% RBI grade 81 and altered fiber content

See Tables 9, 10 and Figs. 10, 11, 12.

Fig. 5 Liquid limit test

Fig. 6 Compaction test

Fig. 7 CBR testing

Table 4 Cases considered for the study

	Material prepared	Test conducted
Step 1	Raw soil or B.C soil only	• Specific gravity • Consistency indices • Standard proctor's test (light compaction) • CBR test (unsoaked)
Step 2	Black cotton soil with RBI grade 81	
	Black cotton soil with 2% RBI grade 81	• Consistency indices • Standard proctor's test (light compaction) • CBR test (unsoaked)
	Black cotton soil with 4% RBI grade 81	
	Black cotton soil with 6% RBI grade 81	
	Black cotton soil with 8% RBI grade 81	
From the results of above performed tests, suitable RBI grade 81% is chosen to carry out next step of the experiment		
Step 3	Black cotton soil with suitable RBI grade 81 percentage and randomly distributed coir fiber percentage	

(continued)

Table 4 (continued)

	Material prepared	Test conducted
	Black cotton soil with suitable RBI grade 81 and 0.50% coir Fiber	• Standard proctor's test (light compaction) • CBR test (unsoaked)
	Black cotton soil with suitable RBI grade 81 and 1.00% coir Fiber	
	Black cotton soil with suitable RBI grade 81 and 1.50% coir Fiber	
	Black cotton soil with suitable RBI grade 81 and 2.00% coir fiber	

Table 5 Specific gravity of B.C soil and soil with altered RBI grade 81%

RBI grade 81 used	0%	2%	4%	6%	8%
Specific gravity	2.75	2.610	2.45	2.35	2.10

Table 6 Consistency indices of B.C soil with altered percentage of RBI grade 81

Consistency indices	Liquid limit (LL) %	Plastic limit (PL) %	Plasticity index (PI) %
Black cotton soil	51.9	22.90	29.00
2% RBI grade 81 + black cotton soil	47.5	22.51	24.99
4% RBI grade 81 + black cotton soil	46.21	23.22	22.99
6% RBI grade 81 + black cotton soil	44.10	24.11	19.99
8% RBI grade 81 + black cotton soil	41.8	25.21	16.59

Table 7 Results for clayey soil treated with altered percentages of RBI grade 81

Material	MDD (g/cc)	OMC (%)
Black cotton soil	1.75	17.55
2% RBI grade 81 + black cotton soil	1.73	18.38
4% RBI grade 81 + black cotton soil	1.72	19.34
6% RBI grade 81 + black cotton soil	1.71	20.18
8% RBI grade 81 + black cotton soil	1.70	21.15

Table 8 CBR Test results for clayey soil treated with varied percentages of RBI grade 81

Material	CBR (%)
Black cotton soil	2.3
2% RBI grade 81 + black cotton soil	3.30
4% RBI grade 81 + black cotton soil	4.10
6% RBI grade 81 + black cotton soil	6.80
8% RBI grade 81 + black cotton soil	9.50

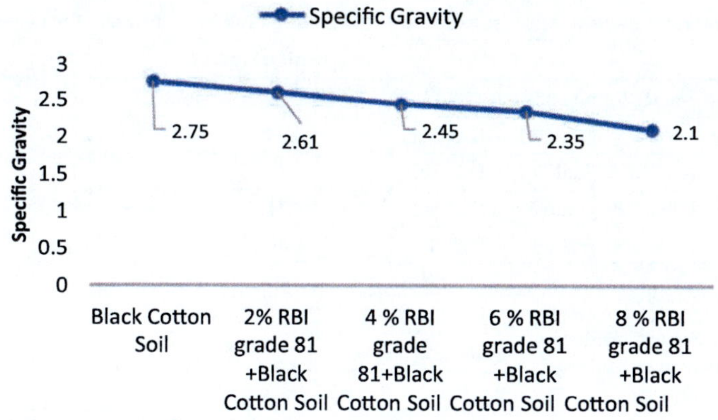

Fig. 8 Specific gravity of soil with altered RBI grade 81%

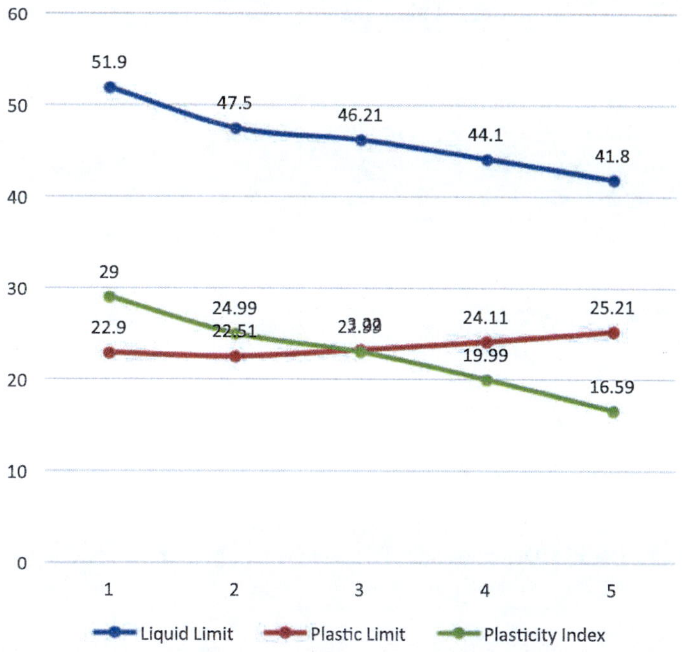

Fig. 9 Slump values observed on coarse aggregate replacement

Table 9 Standard proctor's test results for soil-RBI grade 81 mix with different fiber concentration

Material	MDD (g/cc)	OMC (%)
0% Fiber + 8% RBI grade 81 + B.C Soil	1.70	21.15
0.5% Fiber + 8% RBI grade 81 + B.C Soil	1.73	21.20
1.0% Fiber + 8% RBI grade 81 + B.C Soil	1.75	21.50
1.5% Fiber + 8% RBI grade 81 + B.C Soil	1.71	21.50
2.0% Fiber + 8% RBI grade 81 + B.C Soil	1.69	22.05

Table 10 CBR test results for soil-RBI grade 81 mix with different fiber concentration

Material	CBR %
0% Fiber + 8% RBI grade 81 + B.C soil	9.50
0.5% Fiber + 8% RBI grade 81 + B.C soil	9.70
1.0% Fiber + 8% RBI grade 81 + B.C soil	9.92
5% Fiber + 8% RBI grade 81 + B.C soil	9.90
2.0% Fiber + 8% RBI grade 81 + B.C soil	9.90

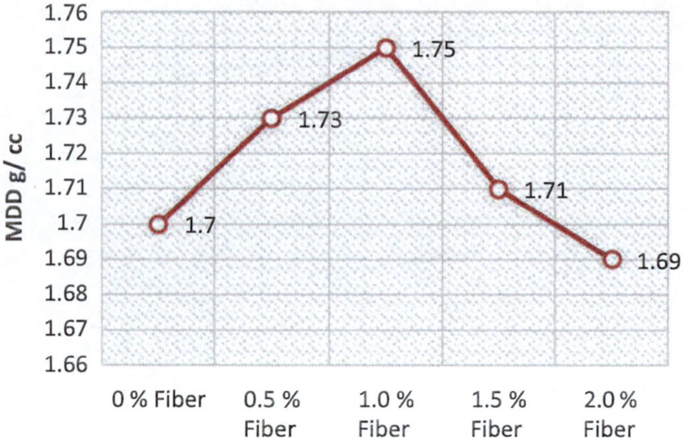

Fig. 10 Comparison of MDD values after fiber addition

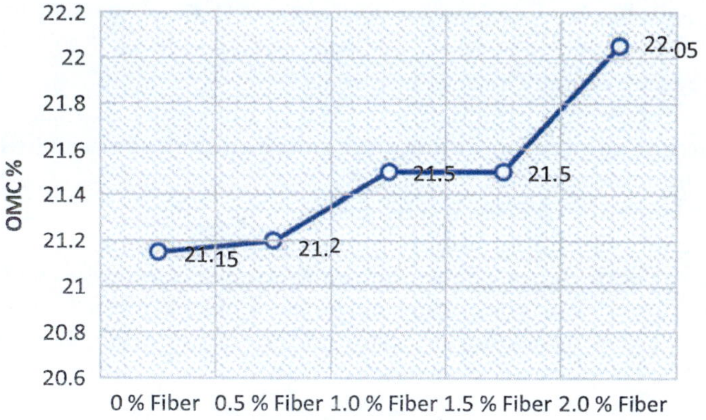

Fig. 11 Comparison of OMC values after fiber addition

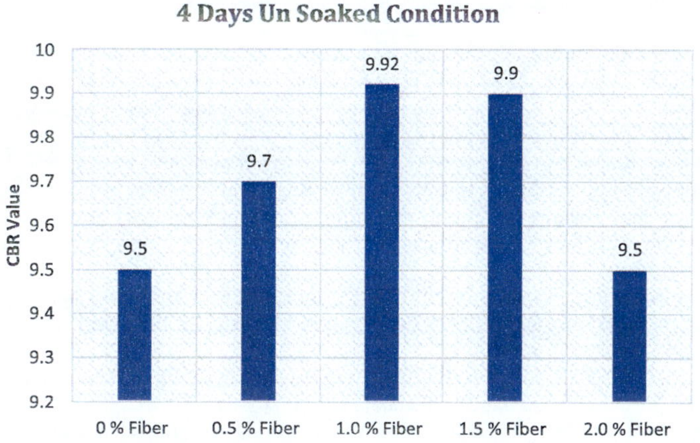

Fig. 12 Comparison of CBR values after fiber addition

6 Conclusion

From the results we can conclude that

- Liquid Limit of the Black cotton expansive and collapsible Soil increases with increase in percent of RBI Grade 81.
- Plastic Limit of Black cotton expansive and collapsible Soil increases with increase in percent of RBI Grade 81.
- Plasticity Index of the Black cotton expansive and collapsible Soil decreases with increase in percent of RBI Grade 81.

- The strength has been compared on the basis of CBR for virgin and RBI grade-81 reinforced soil under un-soaked conditions and it was found to be increasing as the percentage of RBI Grade 81 increased.
- The result shows that when subgrade was reinforced with RBI grade 81 its CBR increases as for virgin soil CBR was 2.3 and it increases to 9.5 with RBI grade 81 under un-soaked condition.
- 8% RBI grade 81 + B.C soil mix provide maximum value of CBR hence 8% RBI grade 81 was selected as a suitable percentage of RBI grade 81.
- On adding fiber to the soil mixture with suitable stabilizing powder (8% RBI grade 81) the CBR value was increased but it became constant after 1% fiber dose.
- Finally it can be concluded that RBI grade 81 along with coir fiber is quite favorable for stabilization of the given expansive soil.

References

1. Nitish et al (2021) Comparative study on soil stabilization using industrial by products and coconut coir. J Phys Conf Ser. In: International conference on physics and energy 2021
2. Bernadette et al (2017) Laboratory study on subgrade soil stabilization using RBI grade 81. In: IOP conf. series: earth and environmental science
3. Kumar KSP, Kumar MA, Kumar ND (2017) Prophecy of plate load test response from theory of elasticity solution and CBR test. Jordan J Civ Eng 11(3):335–345
4. Madurwar KV, Dahale PP, Burile AN (2013) Comparative study of black cotton soil stabilization with RBI Grade 81 and sodium silicate. Int J Innov Res Sci Eng Technol 2(2):493–499
5. Jayatheja M, Kumar ND, Kumar SPK (2016) Influence of zycosoil on consolidation, shear strength characteristics and permeability of clayey subgrade. In: 10th international conference on lowland technology, Manglore
6. Venugopal G, ChetanFakkerappa B (2014) Studies on black cotton soil stabilization using RBI Grade-81. Int J Innov Res Sci Eng Technol 3(12):18205–18210
7. Bujang IZ, Awang MK, Ismail AE (2007) Study on the dynamic characteristic of coconut fibre reinforced composites. In: Regional conference on engineering mathematics, mechanics, manufacturing and architecture, Malaysia
8. Chaple Parag M, Dhatrak AI (2010) Performance of coir fiber reinforced clayey soil. Int J Eng Sci (IJES) 2(4):54–64
9. Mahdi HS, Sheikh ZM, Mahdi AS, Ali Z (2012) A simple review of soil reinforcement by using natural and synthetic fibers. Construct Build Mater 30:100–116
10. Kharade Amit S, Vishal S, Gujar Bhikaji S, Deshmukh Rohankit R (2014) Waste product 'bagasse ash' from sugar industry can be used as stabilizing material for expansive soils. IJRET Int J Res Eng Technol 03(03). EISSN: 2319-1163; PISSN: 2321-7308
11. Shelly M, Singh RR (2014) Improvement of local subgrade soil for road constuction by the use of coconut coir fiber. IJRET Int J Res Eng Technol 03(05)
12. Pillai Sudhakaran M (2003) Ecofriendly plastics/remedial measures for environment sustainability, Coir Board Government of India. In: 4th International R&D conference for water and energy for 21st century, 28–31 January, Aurangabad
13. Karthika R, Amruthalekshmi GR, Peter AK, Mohamed Sajeer M, Raji AK (2011) Study of rut behaviour of coir reinforced black cotton soil using wheel tracking apparatus. In: Proceedings of Indian geotechnical conference, December 15–17, Kochi (Paper no. J-258)

14. Ravi Shankar AU, Chandrasekhar A, Prakash B (2012) Experimental investigation on lithomargic clay stabilized with sand and coir. In: Indian highways: a review of road and road transport development, vol 40, no 2, pp 21–31
15. Reinforced with coir fibres. A thesis submitted in B.Tech Civil Engineering, Malaysia, Phang
16. Singh HP (2013) Strength and stiffness response of itanagar fly ash reinforced with coir fiber. Int J Res Sci Eng Innov Technol 2(9)
17. Amit T, Mahiyar HK (2014) Experimental study on stabilization of black cotton soil by fly ash, coconut coir fiber & crushed glass. Int J Emerg Technol Adv Eng 4(11)
18. Vaidya MK, Chore HS, Kousitha P, Ukrande SK (2013) Geotechnical characterization cement–fly ash–fibers mix. IOSR J Mech Civil Eng (IOSR-JMCE) 60–66. ISSN: 2278-1684

The Adoption of Deep Belief Network Classifier with Shark Smell Optimizer to Predict the Soil Liquefaction

Nerusupalli Dinesh Kumar Reddy, Ashok Kumar Gupta, and Anil Kumar Sahu

Abstract In this paper, post-liquefaction Standard penetration test (SPT) data was collected from the Chi-Chi earthquake and Deep Belief Network (DBN) model using a metaheuristic-based shark smell optimizer (SSO) to improve the predicted accuracy by many iterations. First, the data were normalized and the correlation between variables was determined using chi-square. Next, random sampling was used to select the parameters that were used to train and test the DBN models. Lastly, the SSO was implemented to improve the model's accuracy; it was found that SSO not only improved the model's fitting/accuracy but also sped up the process.

Keywords Soil liquefaction · Standard penetration test (SPT) · Deep belief network (DBN) · Shark smell optimization (SSO)

Artificial neural network (ANN)
Bayesian network (BN)
Support vector machine (SVM)
Genetic Algorithm (GA)
Particle swam optimization (PSO)
Deep Belief Network (DBN)
Shark smell optimization (SSO)

N. D. K. Reddy (✉) · A. K. Gupta · A. K. Sahu
Department of Civil Engineering, Delhi Technological University, Delhi 110042, India
e-mail: dkumarreddy_phd2k18@dtu.ac.in

A. K. Gupta
e-mail: akgupta@dtu.ac.in

A. K. Sahu
e-mail: anilsahu@dtu.ac.in

K. K. Pathak et al. (eds.), *Latest Developments in Civil Engineering*, Lecture Notes in Civil Engineering 352, https://doi.org/10.1007/978-981-99-2676-3_28

1 Introduction

The earthquake-induced soil liquefaction leads to catastrophic disasters, such as landslides, ground cracks, structural settlements, and instability against an unexpected transformation transform in soil stress due to dynamic conditions induced by earthquake leading to soil behaving like fluid, which means the adequate strength of soil is zero [1, 2]. As many soil layers are prone to liquefaction, these layers must be altered to limit the likelihood of liquefaction of disposed soils. In addition to safety considerations, precise predictions of the liquefaction potential may minimize the economic cost of a project due to the high cost of soil modification.

The basic methodologies for determining the liquefaction potential are as follows.

The first method is one that is based on dynamic analysis and a comparison of earthquake-induced soil shear stress. This methodology was developed in the 1970s (laboratory testing cyclic triaxial, cyclic simple shear test, vibration table and centrifuge). Third, statistical procedures employ data from a location where an earthquake has previously occurred and where it is predicted that liquefaction will occur; second, field-based testing techniques such as a standard penetration test; and first, field-based testing approaches such as a standard penetration test.

Each of these methods has benefits and drawbacks that are directly proportional to the usefulness they provide. Both the fabrication of undisturbed samples of half-saturated or loose sand and the high expense of dynamic testing are among the most significant challenges associated with the first technique. During the second round of the evaluation, the region's boundaries will serve as a difficult obstacle. Since Seed and Idriss first proposed the simplified method of analyzing soil liquefaction using relative density and SPT-N values as parameters [3] it has been updated and modified several times without affecting the fundamental equations [4, 5]. These updates and modifications have been made since Seed and Idriss' original proposal. When designing a new and better model, it is important to keep the waveform and length of an earthquake in mind [6].

Due to the unpredictability of field conditions and calculations, conventional methods cannot be investigated precisely despite their popularity and simplicity; therefore, more intelligent methods are required to predict soil liquefaction. The third category is based on contemporary approaches and may compensate for the shortcomings of the first two. Today, machine learning is a widespread computer technique used to develop complicated models. These approaches are used well in geotechnical engineering analysis for non-linear models. In recent years, machine learning techniques like ANN [7], SVM [8], Fuzzy SVM [9], etc. have been applied to soil liquefaction data, proving to be faster than traditional approaches and more accurate in forecasting non-linear consequences.

Nevertheless, the machine learning technique is also concerned with noise and convergence to local minima [10]. Continuously the researchers implemented new techniques to improve the machine learning by using optimizing algorithms such as GA [11, 12] and PSO [9].

1.1 *Contribution Work to This Paper*

Established a new soil liquefaction prediction model by using chi-square for improved correlation features.

Deployed classifier DBN classifier.

Proposes an SSO model for choosing the optimal weights in DBN.

2 Literature Review

Reddy et al. [13] In the year 2022, a one-of-a-kind hybrid-ensemble classifier was developed. Its primary purpose was to enhance generalizability by providing a more precise correlation feature, in addition to the ability to generate chi-square, relief features, and technical signals. The invention was made public in the year 2022. This was successfully completed. In addition to that, itutilizes a variety of different ensemble classifiers, including DBN, LSTM, and SVM, all of which are combined with Bi-GRU. It has been suggested that an AC-SSO model, which is a combination of the Averaged Cat and Salp-Swarm Optimization models, should be used to choose the appropriate weights for the Bi-GRU. This recommendation was made when it was discovered that the AC-SSO model was superior to both of its predecessors. This is due to the fact that an AC-SSO model incorporates both the advantages and the drawbacks of these two models. The strategy that was outlined resulted in an improvement in convergence and addressed difficulties with solutions that were suited for their particular areas.

Zhang et al. [14] In 2021, included SPT data into the GWO model with the intention of enhancing the accuracy of the SVM-based prediction system in the expectation that this would lead to the desired outcome. After determining the optimal value for SVM, a significant number of iterations of GWO were performed in the following steps. It would seem to infer that the performance of SVM-GWO is much superior than that of SVM on its own.

Zhang and Wang [15] proposed a novel hybrid-ensemble classifier in 2021 to increase generalizability by integrating the predictions of seven classifiers using a weighted voting method. The classifiers utilized included BPNN, SVM, DT, KNN, LR, MLR, and NB. Using GA, the weights and hyperparameters of deployed basic classifiers were modified. On three datasets, it demonstrates that the novel classifier ensemble model beat the basic classifiers in terms of various performance metrics, including precision, recall, accuracy, AUC, ROC, and F1-score.

Hu [16] developed two unique Bayesian networks (BN) in 2021 for determining the liquefaction of gravelly soil using a novel hybrid model. The performance of the new hybridized model was validated by comparing it to other existing models, and two innovative BN techniques outperformed traditional methods or models for predicting the liquefaction of gravelly soil.

Alizadeh Mansouri and Dabiri [17] suggested a novel prediction model for predicting the liquefaction potential for a defined range of soil types using ANN in 2021. In Iran, one hundred soil samples were collected utilizing experiential approaches. Subsequently, the outcomes of the experimental schemes were used as training data for ANN, which was evaluated as an alternate method for predicting liquefaction in Iran. The acquired ANN model was used to evaluate the liquefaction prediction.

Cai et al. [18] presented the RBFNN and LSSVM in 2021, and GWO, DE, and GA are the optimization techniques for estimating the possibility for soil liquefaction. Following this, statistical measures such as RMSE were used to assess the model. It was an effective and efficient method for predicting the possibility for liquefaction.

Ghani and Kumari [19] developed a field-based formula in 2021 to assess the influence of soil flexibility on the liquefaction activities of fine-grained soil in Bihar, India. The analytical findings were supported by reliability analysis. The verification of the outcomes with real (Table 1).

Table 1 Review of recent updates in liquefaction studies

Authors	Adopted machine learning model	Features	Challenges
Reddy et al. [13]	Ensembled classifier-AC-SSO	Improved convergence and resolves problems with local optima	Future study must explore greater variable data sets
Zhang et al. [14]	SVM-GWO	Better accuracy and speedy process	Future research must examine bigger variable data sets
Zhang et al. [14]	Ensembled classifier—GA	High recall and Good level of accuracy	Required to work on strong fundamental classifiers
Zhang and Wang [15]	Novel BN	Improved accuracy and Minimal error	Needs to examine cost factors
Hu [16]	ANN	Better reliability	Needs to analyze cost factors
Alizadeh Mansouri and Dabiri [17]	RBFNN and LSSVM-GWO, GA, DE	Improved accuracy and minimal error	Complexity must be examined
Cai et al. [18]	MLR	Less error Improved accuracy	Future research must examine bigger variable data sets

3 Stepwise Interpretation of Soil Liquefaction Prediction Approach

3.1 Deterministic Approach

According to seed and idriss, $CSR_{7.5}$ equation using corrected SPT data expressed as Hanna et al. [7]

$$CSR_{7.5} = 0.65\left(\frac{a_{max}}{g}\right)\left(\frac{\sigma_{v0}}{\sigma'_{v0}}\right)r_d\left(\frac{1}{MSF}\right)\left(\frac{1}{k_\sigma}\right)\left(\frac{1}{k_a}\right) \quad (1)$$

a_{max} = surface peak acceleration,
σ_{v0} = total overburden pressure,
σ'_{v0} = effective overburden pressure,
r_d = reduction in shear stress in depth z,
K_σ, K_a = correction factors,
MSF = magnitude scaling factor,
$CSR_{7.5}$ = equivalent value of CSR for a magnitude of 7.5 corrected by MSF.

CRR is the maximum CSR value that soil can resist before failure which is expressed as Pal [8]

$$CRR = \exp\left\{(N_1)_{60CS}/14.1 + ((N_1)_{60CS}/126)^2 + ((N_1)_{60CS}/23.6)^3 + ((N_1)_{60CS}/25.4)^4 - 2.8\right\} \quad (2)$$

$(N_1)_{60CS}$ = corrected SPT No.

From the above equations, a factor of safety would be

$$F_s = CRR/CSR_{7.5}. \quad (3)$$

3.2 Introduction of a Database of SPT

In the preceding one hundred years, Taiwan has never been struck by an earthquake of the same magnitude as the one that occurred in Chi-Chi [15, 20]. The earthquake was responsible for a significant amount of liquefaction that occurred in Central Taiwan. Following the earthquake, an immediate investigation was carried out. A significant amount of focus has been directed into the investigation and evaluation of liquefaction sites in addition to non-liquefaction places. The simplified SPT-N technique that Seed and Idriss have offered may be used to do an analysis on this inquiry data. When compared to the case histories of other liquefactions that have occurred throughout the globe, the information on liquefaction locations and non-liquefied areas from the Chi-Chi earthquake exhibits the following characteristics.

Due to the fact that an earthquake of magnitude 7.3 on the Richter scale is equivalent to an earthquake of magnitude 7.5 on the typical critical cyclic strength curves of the various liquefaction evaluation approaches, it is possible to circumvent the error that may be caused by the conversion of earthquake.

A number of seismographic stations may be found throughout central Taiwan. A significant amount of information was gathered while the earthquake was occurring. It is not necessary to estimate peak ground acceleration using the attenuation law because one can easily access data on ground acceleration values at both liquefaction sites and non-liquefaction locations by consulting data from a local seismographic station. This removes the need to estimate maximum ground acceleration. As a result, there was an increase in precision. There was a wealth of information gleaned from boreholes and soil particle size distribution curves in the area that was the focus of the study. There are a total of 288 pieces of information, with 164 sets pertaining to liquefaction and 124 sets pertaining to non-liquefaction. The essential data are supplied in Table 2 model data.

Table 2 Model data of post-liquefaction data

L	d (m)	N_m	FC (%)	CC	D50 (mm)	a_{max} (g)	CSR	N_{160cs}
0	9	14	17	4	0.13	0.124	0.14	14.29
0	9	21	14	3	0.23	0.124	0.127	20.7
0	5	16	46	5	0.09	0.124	0.127	20.61
1	7.5	12	55	8	0.08	0.428	0.384	12.14
0	8.2	1	42	6	0.111	0.084	0.069	0.97
1	7.8	7	16	4	0.3	0.42	0.363	6.99
1	1.3	1.5	65	23	0.055	0.789	0.741	3.6
1	4.3	9	26	4	0.14	0.211	0.165	10.65
1	3.6	6	11	3	2	0.42	0.289	7.53
1	4.5	7	26	4	0.135	0.211	0.222	9.85
0	9	19	10	1	0.26	0.124	0.113	17.72
1	6.3	11	30	6	0.11	0.42	0.363	12.13

L = liquefied, d = depth, N_m = SPT no, FC = fines content, CC = clay content, D50 = particle size, CSR = cyclic stress ratio, N_{160cs} = corrected SPT No. $CSR_{7.5}$ = is an equivalent value of CSR for a magnitude 7.5 earthquake

4 Proposed Soil Liquefaction Prediction Model

4.1 *Pre-processing via Data Normalisation*

At the outset, the input data (I_d) are subjected to a pre-processing step known as data normalization, which is a method that organizes the data. The process of arranging data in a methodical fashion and formatting it such that it looks the same across all fields of records is referred to as data normalization [10] in this article. This ultimately leads to the acquisition of data of high quality. The removal of redundant and unstructured data is the primary focus of this method, which is designed to ensure that the logical storage of data is maintained. By removing redundant information, this strategy often results in the database being more consistent. And last, all of the data were updated and checked for errors before being supplied for feature extraction. And finally updated and corrected data $(I_d)_{\text{corrected is}}$ [21] then given for feature extraction.

4.2 *Feature Extraction by Using Chi Square Features*

It [22] is used for selecting the necessary characteristics from a dataset. The Chi-square is applied to the goal and each feature, and as a consequence, the number of features with excellent Chi-square scores stated in Eq. (4) is selected.

$$Fe = \sum \frac{(G - E)^2}{E} \tag{4}$$

In Eq. (4), '*G*' represents the total number of class observations, while '*E*' represents the anticipated number of class observations with no or weak correlation between the class and feature. As the performance of Chi Square improves, the Chi Square ranking system is used, with *Fe* denoting the chosen Chi Square qualities.

4.3 *DBN Classifier*

DBN [23] is a ground-breaking approach to machine learning that consists of a deep neural network that is made up of multiple "Restricted Boltzmann Machines (RBM)" and single-layer classifiers, along with unsupervised greedy layer-by-layer pre-training to give the initial spatial distribution of weights and threshold values. In addition, DBN uses unsupervised greedy layer-by-layer pre-training to give the initial spatial distribution of weights and threshold values. In addition, DBN employs unsupervised greedy layer-by-layer pre-training to offer an initial spatial distribution of weights and threshold values. This is accomplished via the use of the greedy layer-by-layer algorithm. When used for the solution of non-linear problems, this technique brings with it a number of significant advantages that cannot be overlooked. The classic RBM has two layers: one that can be seen, and another that cannot be seen. The layer that can be observed is called the "open" layer. The neurons that are located between the visible and hidden layers have connections to one another; nevertheless, the neurons that are located inside each layer continue to operate independently of one another while having connections to one another.

The learning process is carried out by the use of an energy function. The explanation of energy for visible and hidden neuron (x, y) may be found in Eqs. (5)–(7), where y_l and x_a represent the binary state of hidden unit l and visible unit a, respectively, and where k_a and C_l represent biases.

$$EN(x, y) = -\sum\nolimits_{(a,l)} L_{a,l} x_a y_l - \sum\nolimits_a k_a x_a - \sum\nolimits_l C_l y_l \quad (5)$$

$$\Delta EN(x_a, \overline{y}) = \sum\nolimits_l L_{al} y_l + k_a \quad (6)$$

$$\Delta EN(\overrightarrow{x}, y_l) = \sum\nolimits_a L_{al} x_a + C_l \quad (7)$$

The RBM allots a probability by means of energy function as in Eq. (8), in which PR^F signifies partitioning term as in Eq. (9).

$$c(\overrightarrow{x}, \overrightarrow{y}) = \frac{1}{PR^F} e^{-EN(\overrightarrow{x}, \overrightarrow{y})} \quad (8)$$

$$PR^F = \sum_{\vec{x}, \vec{y}} e^{-EN(\vec{x}, \vec{y})} \quad (9)$$

Figure 1 provides an illustration of how the DBN was put together. In addition to a backpropagation neural network, the top layer is constructed up of a few distinct RBMs. The DBN training method is divided into two separate stages, the first of which is the preliminary training, and the second of which is the fine-tuning. An unsupervised layer-by-layer pre-training technique is used at the beginning of the process to pre-train each RBM layer, beginning at the bottom and working one's way to the top. Because the output of one layer of RBM is used as the input of the next layer, it is possible for each layer to obtain its starting weight and learn the characteristics of the layer that came before it. This is made possible by the fact that the output of one layer of RBM is used as the input of the next layer. Following that, the top-level BP network is put to use in order to carry out top-down supervised learning in order to fine-tune the network in reverse, which eventually leads to incremental improvements in network parameters. The training of the deep belief network is complete after being preceded by pre-training and followed by fine-tuning. The training was completed once both of these steps were completed.

Fig. 1 DBN

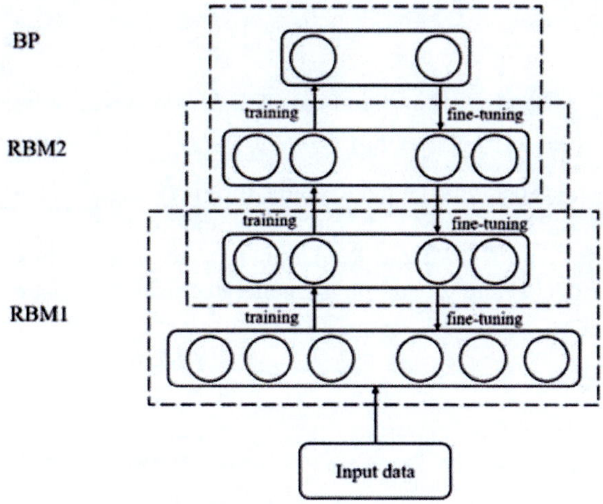

5 Shark Smell Optimization Model

Shark smell optimization model (SSO) [24] model includes many benefits; it controls various limits like premature convergence etc.

SSO includes 4 primary phases namely, "Initialization, Forward movement, Rotational movement, and Position update".

Initialization

In SSO, the initial populations in the searching space are chosen randomly. This odor indicates a possible shark location during the first phase of a search. The starting position vector is represented by Eqs. (10) and (11), where, $N_i^1 = i$ th initial population vector position and np implies the size of the population.

$$N^1 = \left[N_1^1, N_2^1, ...N_{np}^1\right] \quad (10)$$

The optimization problem is represented by Eq. (13), where $N_{i,j}^1 = j$th dimension of the ith location of the shark and nd represents the choice variable count.

$$N_i^1 = \left[N_{i,1}^1, N_{i,2}^1, ...N_{i,nd}^1\right]. \quad (11)$$

Forward Movement

While blood is discharged into the water, all Sharks migrate towards the stronger smell particle with "velocity" V_f in an effort to approach their prey. Consequently, V_f is calculated according to Eq. (12) and each V_f comprises a dimensional element as illustrated in Eq. (13).

$$V_{fi}^1 = \left[V_{f1}^1, V_{f2}^1, ...V_{fnp}^1\right] \quad (12)$$

$$V_{fi}^{1} = \left[V_{f(i,1)}^{1}, V_{f(i,2)}^{1}, \ldots. V_{f(i,nd)}^{1}\right] \tag{13}$$

Therefore, the velocity in all dimensions is calculated using the equation indicated in Eq. (16), wherein, $\upsilon = 1, 2, \ldots.\upsilon \left.\frac{\partial(obj)}{\partial \chi j}\right|_{\chi_{i,j}^{\upsilon}\,max}$ derivative of obj at position $\chi_{i,j}^{\upsilon}$, υ_{max} specifies the cycle count for every shark's forward move, the parameters $R1$, $R2$, $R3$, α_{υ} and β_{υ} are selected at random; however, according to the suggested SSO concept, $R1$, $R2$ and $R3$ parameters are generated using a tent map, as indicated in Eq. (14) and α_{υ} and β_{υ} are generated using a logistic map as shown in Eq. (15).

$$R_{\upsilon+1} = \begin{cases} R_{\upsilon}/0.4, 0 < R_{\upsilon} \leq 0.4 \\ (1 - R_{\upsilon})/0.60.4 < R_{\upsilon} \leq 1 \end{cases} \tag{14}$$

$$R_{\upsilon+1} = 4R_{\upsilon}(1 - R_{\upsilon}) \tag{15}$$

$$V_{f(i,j)}^{\upsilon} = \eta_{\upsilon}.R1.\left.\frac{\partial(obj)}{\partial \chi j}\right|_{\chi_{i,j}^{\upsilon}} \tag{16}$$

In all stages of $V_{f(i,j)}^{\upsilon}$ the velocity limiter is deployed according to Eq. (17), where α_{υ} signifies the inertia coefficient.

$$V_{f(i,j)}^{\upsilon} = \eta_{\upsilon}.R1.\left.\frac{\partial(obj)}{\partial \chi j}\right|_{\chi_{i,j}^{\upsilon}} + \alpha_{\upsilon}.R2.V_{f(i,j)}^{\upsilon-1} \tag{17}$$

The velocity limiter employed for each phase of the SSO model is shown in Eq. (18), where β_{υ} signifies the ratio of the velocity limiter for each phase of υ.

$$\left|V_{f(i,j)}^{k}\right| = \left[\left|\eta_{\upsilon}.R1.\left.\frac{\partial(obj)}{\partial \chi j}\right|_{\chi_{i,j}^{\upsilon}} + \alpha_{\upsilon}.R2.V_{f(i,j)}^{\upsilon-1}\right|, \left|\beta_{\upsilon}.V_{f(i,j)}^{\upsilon-1}\right|\right] \tag{18}$$

The shark's forward movement and its new location $Z_i^{\upsilon+1}$ are computed based on the shark's prior position Z_i^{υ} and velocity (V_i^{υ}), The shark's new position is represented by the equation Eq. (19).

$$Z_i^{\upsilon+1} = N_i^{\upsilon} + V_i^{\upsilon}.\Delta t_{\upsilon} \tag{19}$$

Here, Δt_{υ} represents a time period ($\approx$1).

Rotational Movement

The shark also utilizes rotating movement to locate food with a strong odor. As seen in Eq. (20), this technique is referred to as local search, where $m = 1,2\ldots M$.

$$Y_i^{\upsilon+1,m} = Z_i^{\upsilon+1} + R3.Z_i^{\upsilon+1}. \tag{20}$$

Particle Position Update

As demonstrated in Eq. (21), the shark's seeking pattern combines forward movement with rotating movement as it approaches food with a strong scent; Y_i^{v+1} represents the location of the shark with the highest *Obj* value.

$$N_i^{v+1} = \text{argmax}\left\{obj\left(Z_i^{v+1}\right), Obj\left(Y_i^{v+1,i}\right), ..., Obj\left(Y_i^{v+1,M}\right)\right\} \quad (21)$$

The motions will be repeated till it reaches its optimal value of v.

5.1 The Pseudo-Code of SSO Approach is Shown in Algorithm

Algorithm : SS0
Begain 1. Initializing population Assign mandatory parameters, $np, \alpha_v, v_{max}, \eta_v$ and $v = 1,2,\ldots. v_{max}$ Calculate R1, R2, R3, $\alpha_v$and β_v by using the tent map and logistic map as shown in Eq. (14) and Eq. (15) Generate primary populace with all individuals İnitializing velocity in all dimensions v =1 For v =1: v_{max} 2. Forward movement Compute every element of $V_{f(i,j)}$ Attain novel shark position as revealed in Eq. (19) 3. Rotational movement obtain novel shark position as per rotational movement $Y_i^{v+1,m}$ Choose next shark position depending upon two movements End for v set v =v+1 Choose shark optimal position with higher Obj value End

See Fig. 2.

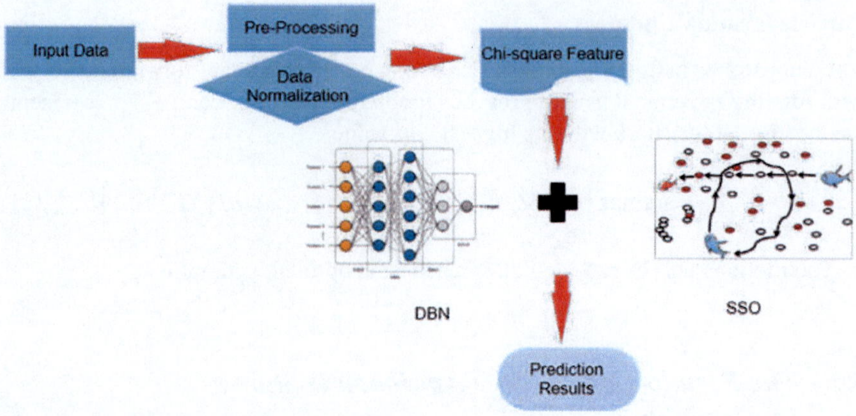

Fig. 2 Developed soil liquefaction prediction model

6 Results and Discussions

Here, the weights $W_1, W_2, \ldots, W_n$ are given to SSO, to determine its optimal weights, this is because the conventional training procedure may get trapped with slower convergence speed stuck in local minima. This can be resolved by minimizing the objective error shown in Eq. (22) (Fig. 3).

$$\text{Objt} = \min(\text{Error}). \tag{22}$$

6.1 Simulation Setup

The proposed setup DBN-SSO based soil liquefaction scheme was executed in Python. And the performance of the adopted models is determined by using various evaluation models [25]. During analysis, the dataset is varied from 60, 70, 80 and 90% for the testing.

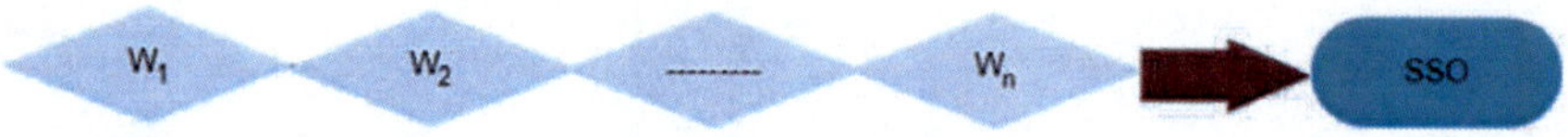

Fig. 3 Weights of DBN

Table 3 Analysis of proposed models

Metrics	DBN	DBN+SSO
Sensitivity (%)	0.0912	0.9231
Recall (%)	0.086944	0.9312
Specificity (%)	0.98572	0.8912
Precision (%)	0.74378	0.91643
F-Measure (%)	0.16592	0.863352
Accuracy (%)	0.732	0.91409
FNR (%)	0.811246	0.0882
NPV (%)	0.709364	0.941766
MCC (%)	0.1842	0.798954
FPR (%)	0.011676	0.101096

6.2 *Performance Analysis*

The corresponding results are shown in a Table 3. The study of the proposed DBN+SSO model is compared to the DBN model alone in terms of the proportion of varied learners (LP). In both optimization and classifier comparisons, the DBN+SSO model has achieved superior results than DBN. All of the created model's outputs for positive measurements grow, while outputs for negative measures decrease. Specifically, the 90th LP yields marvelous results for both current and planned systems. At the 90th LP, the proposed method yields precision values derived mostly from the table (0.91), but other LPs yielded results with very low precision values. In other words, the proposed method achieves accuracy values of 0.81, 0.85, and 0.79 at the 60th, 70th, and 80th LPs, respectively. Consequently, the research demonstrated the enhanced efficacy of DBN+SSO with the incorporation of the optimization approach.

7 Conclusion

This paper has developed a novel soil liquefaction prediction model with DBN+SSO. At first, pre-processing was performed and then varied enhanced features were derived. These features were then given for prediction DBN. The attained waits DBN was given to SSO which offered the final output. At last, the supremacy of offered scheme was established over the conventional schemes concerning diverse measures. In particular, the finest accuracy values (0.91) were accomplished by the developed scheme at 90th LP than the results obtained during other LPs. DBN-SSO not only improved accuracy but also fasten the process. The DBN-SSO model performance is not the best, and it can be improved by hyperparameter optimization of the DBN model.

References

1. Javdanian H (2019) Evaluation of soil liquefaction potential using energy approach: experimental and statistical investigation. Bull Eng Geol Env 78(3):1697–1708. https://doi.org/10.1007/s10064-017-1201-6
2. Zuzulock ML, Taylor ODS, Maerz NH (2020) Soil fatigue hazard screening analyses framework for spacio-temporally clustered induced seismicity with examples of damage potential due to liquefaction. SN Appl Sci 2(6). https://doi.org/10.1007/s42452-020-2878-x
3. Seed HB, Idriss IM (1971) Simplified procedure for evaluating soil liquefaction potential. J Soil Mech Found Div 97(9):1249–1273
4. Hossain MD, Maksud Kamal ASM, Rahman MZ, Farazi AH, Mondal DR, Mahmud T, Ferdous N (2020) Assessment of soil liquefaction potential: a case study for Moulvibazar town, Sylhet, Bangladesh. SN Appl Sci 2:1–2
5. Youd TL, Idriss IM (2001) Liquefaction resistance of soils: summary report from the 1996 NCEER and 1998 NCEER/NSF workshops on evaluation of liquefaction resistance of soils. J Geotech Geoenviron Eng 127(4):297–313
6. Sassa S, Yamazaki H (2017) Simplified liquefaction prediction and assessment method considering waveforms and durations of earthquakes. J Geotech Geoenviron Eng 143(2). https://doi.org/10.1061/(asce)gt.1943-5606.0001597
7. Hanna AM, Ural D, Saygili G (2007) Evaluation of liquefaction potential of soil deposits using artificial neural networks. Eng Comput (Swansea, Wales) 24(1):5–16. https://doi.org/10.1108/02644400710718547
8. Pal M (2006) Support vector machines-based modelling of seismic liquefaction potential. Int J Numer Anal Meth Geomech 30(10):983–996. https://doi.org/10.1002/nag.509
9. Rahbarzare A, Azadi M (2019) Improving prediction of soil liquefaction using hybrid optimization algorithms and a fuzzy support vector machine. Bull Eng Geol Env 78(7):4977–4987. https://doi.org/10.1007/s10064-018-01445-3
10. Gori M, Tesi A (1992) On the problem of local minima in backpropagation. IEEE Trans Pattern Anal Mach Intell 14(1):76–86. https://doi.org/10.1109/34.107014
11. Atangana Njock PG, Shen SL, Zhou A, Lyu HM (2020) Evaluation of soil liquefaction using AI technology incorporating a coupled ENN/t-SNE model. Soil Dyn Earthq Eng 130. https://doi.org/10.1016/j.soildyn.2019.105988
12. Goharzay M, Noorzad A, Ardakani AM, Jalal M (2017) A worldwide SPT-based soil liquefaction triggering analysis utilizing gene expression programming and Bayesian probabilistic method. J Rock Mech Geotech Eng 9(4):683–693. https://doi.org/10.1016/j.jrmge.2017.03.011
13. Reddy NDK, Gupta AK, Sahu AK (2022) A novel soil liquefaction prediction model with intellectual feature extraction and classification. Adv Eng Softw 173:103233
14. Zhang Y, Qiu J, Zhang Y, Xie Y (2021) The adoption of a support vector machine optimized by GWO to the prediction of soil liquefaction. Environ Earth Sci 80(9). https://doi.org/10.1007/s12665-021-09648-w
15. Zhang J, Wang Y (2021) An ensemble method to improve prediction of earthquake-induced soil liquefaction: a multi-dataset study. Neural Comput Appl 33(5):1533–1546. https://doi.org/10.1007/s00521-020-05084-2
16. Hu J (2021) A new approach for constructing two Bayesian network models for predicting the liquefaction of gravelly soil. Compute Geotech 137. https://doi.org/10.1016/j.compgeo.2021.104304
17. Alizadeh Mansouri M, Dabiri R (2021) Predicting the liquefaction potential of soil layers in Tabriz city via artificial neural network analysis. SN Appl Sci 3(7). https://doi.org/10.1007/s42452-021-04704-3
18. Cai M, Hocine O, Mohammed AS, Chen X, Amar MN, Hasanipanah M (2021) Integrating the LSSVM and RBFNN models with three optimization algorithms to predict the soil liquefaction potential. Eng Comput. https://doi.org/10.1007/s00366-021-01392-w

19. Ghani S, Kumari S (2021) Probabilistic study of liquefaction response of fine-grained soil using multi-linear regression model. J Inst Eng (India): Series A 102(3):783–803. https://doi.org/10.1007/s40030-021-00555-8
20. Hwang J-H, Yang C-W. Veri®cation of critical cyclic strength curve by Taiwan Chi-Chi earthquake data. www.elsevier.com/locate/soildyn
21. Jamal P, Ali M, Faraj RH, Ali PJM, Faraj RH (2014) 1–6 data normalization and standardization: a technical repor. https://docs.google.com/document/d/1x0A1nUz1WWtMCZb5oVzF0SVMY7a_58KQulqQVT8LaVA/edit
22. Reddy, N. D. K., Gupta, A. K., & Sahu, A. K. (2023). Optimized ensemble-classification for prediction of soil liquefaction with improved features. Multimed Tools Appl 1–20
23. Wang HZ, Wang GB, Li GQ, Peng JC, Liu YT (2016) Deep belief network based deterministic and probabilistic wind speed forecasting approach. Appl Energy 182:80–93. https://doi.org/10.1016/j.apenergy.2016.08.108
24. Mohammad-Azari S, Bozorg-Haddad O, Chu X (2018) Shark smell optimization (SSO) algorithm. In: Studies in computational intelligence, vol 720. Springer, pp 93–103. doi: https://doi.org/10.1007/978-981-10-5221-7_10
25. Alshammari T, Alshammari N, Sedky M, Howard C (2018) Evaluating machine learning techniques for activity classification in smart home environments

Effect of Geosynthetics Position on Shallow Anchor Uplift Capacity

Buragadda Venkatesh, Orekanti Eswara Reddy, E. Sravani, B. Sai Naresh, and G. Dinesh Kumar

Abstract The plate anchor uplift capacity using geosynthetics was investigated in the present research study. The 1 g laboratory pullout model tests were carried out in the test tank having dimensions of 1000 mm × 1000 mm × 1000 mm using a circular shape anchor size (D) of 100 mm. The embedment depth of plate anchors in a sand bed having a relative density of 70% was 400 mm. Two numbers of polypropylene geotextiles having each size (B_g) of 2D were used to improve plate anchor's uplift capacity. Two different model tests were performed by maintaining the vertical spacing (h) of 0.5D from the anchor plate above placed geotextile and placing the two numbers of geotextiles directly above the anchor plate i.e., h = 0.0D. The present test results reveal that the geotextiles which were placed directly above the anchor plate (0.0D) show a higher improvement in comparison to the spaced geotextiles (0.5D). It could be due to the influence of reinforcement tensile stiffness property which plays a vital role as compared to the reinforcement's interface frictional property.

Keywords Plate anchors · Geosynthetics · Sands · Uplift capacity

1 Introduction

The plate anchors are mostly used in various applications to improve the anchors' uplift capacity such as television towers and transmission towers etc. [1]. However, the plate anchors alone cannot able to resist the uplift forces which were occurred in the field i.e., 10,000 kN [2]. The improvement of anchor uplift capacity can be enhanced by several ways such as by increasing the foundation weight and increasing the foundation embedment depth etc. [3, 4]. However, these methods are not feasible to execute practically in the field for cases of constrained conditions. Hence, geosynthetics have been used by various researchers to improve the anchor uplift capacity [1, 2, 5]. In most of the cases, the geosynthetics are used in various forms to improve

B. Venkatesh (✉) · O. Eswara Reddy · E. Sravani · B. Sai Naresh · G. Dinesh Kumar
Department of Civil Engineering, Sree Vidyanikethan Engineering College, Tirupati 517102, India
e-mail: venkatesh.b@vidyanikethan.edu

K. K. Pathak et al. (eds.), *Latest Developments in Civil Engineering*, Lecture Notes in Civil Engineering 352, https://doi.org/10.1007/978-981-99-2676-3_29

the anchors' uplift capacity i.e., planar form [1, 2] and geocell form [5]. Krishnaswamy and Parashar [1] carried out a series of model pullout tests using plate anchors embedded in the planar form of geosynthetic reinforced soil beds. During the model tests, the researchers varied the several number of reinforcement geometrical parameters such as the number of reinforcements (N), size of reinforcement (B_g), and spacing of reinforcement (h). The test results showed that the immediate placement of geosynthetic reinforcement above the anchor is considered as an optimum position of the first geosynthetic reinforcement. It means that the position of the single number of reinforcements placed immediately above the anchor plate showed a higher amount of improvement in comparison to the placing a greater number of reinforcements (i.e., $N > 1$). Similar tests were also performed by several researchers and reported similar optimum conditional values [2, 5].

Ravichandran [6] carried out model pullout tests to see the effect of reinforcement tensile stiffness by placing double layers (i.e., $N = 2$) of the reinforcement immediately above the anchor i.e., $h = 0$. Also, the tests were performed by placing the same number of reinforcements (i.e., double) with a spacing (h) of 0.5 times the size of the anchor. The test results showed that the reinforcement tensile stiffness does not play a major role in comparison to the interface friction between the soil and reinforcements. However, these test studies were limited to geogrid and geo-composite (geogrid sand witched within two geotextile) reinforcements having dissimilar mechanisms i.e., geogrid (interlocking mechanism) and geo-composite (interface frictional mechanism). However, no further research studies have been carried out to understand the effect of reinforcement tensile stiffness on anchor pullout capacity. Hence, the present research paper describes the effect of reinforcement tensile stiffness in comparison to the reinforcements' spacing (i.e., interface friction) by maintaining the similar properties of geotextile reinforcements.

2 Materials

2.1 *Sand and Reinforcement Material*

Dry river sand was used during the laboratory model pullout tests. The dry sieve analysis was carried out on the collected sand as per ASTM D6913-17 [7], and the respective properties are shown in Table 1. The soil shear strength test results of cohesion (c) and angle of internal friction (ϕ) were determined as 0 kPa and 40°, respectively using small size direct shear test apparatus (60 mm × 60 mm × 25 mm) in accordance with the codal provisions of Bureau of Indian Standards [8].

A white color polypropylene geotextile (PPGT) was used in the present laboratory pullout model tests. The PPGT reinforcement material was collected from Technofab Geosystems Ltd., Mumbai, India. According to ASTM standards, the PPGT reinforcement material properties i.e., Thickness [9], Mass per unit area [10], and Tensile strength [11] were determined and presented in Table 1.

Table 1 Properties of sand and reinforcement material of PPGT

Sand		PPGT	
Property	Values	Property	Values
Specific gravity, G_s	2.69	Thickness (mm)	0.4
D_{10} (mm)	0.33	Mass per unit area (g/m^2)	130
D_{60} (mm)	0.86	Ultimate tensile strength-MD × CMD (kN/m)	29.6 × 23.5
C_u	3.03	Failure strain-MD × CMD (%)	30.2 × 21.4
C_c	1.36	Tensile stiffness at 5% strain (kN/m)	101

MD: Machine direction
CMD: Cross-machine direction

2.2 Test Tank and Anchor Plate

A high rigidity mild steel test tank having a size of 1000 mm × 1000 mm × 1000 mm was used in the present research study. The rigid model anchor plate made with mild steel having a size (D) of 100 mm and thickness of 6 mm was used in the present study. As per the literature of Rahimi et al. [12], the test tank size i.e., respective width and depth should be 7 times and 4 times the size of the anchor plate (D). Hence, the test tank used in the present study satisfied the limits proposed by Rahimi et al. [12].

3 Experimental Test Setup

The experiments were performed in the test tank by maintaining the anchor depth was 400 mm in the sand bed which consists of a desired relative density of 70%. Based on the depth of the anchor maintained in the present study, the anchor was classified as shallow anchor i.e., L/D = 4 [13]. As per the research study of Ilamparuthi and Dickin [14], the sand beds of the present study were prepared at a relative density of 70% by maintaining the tamping technique. A tie wire is one of the main components in the test set setup and it was connected at the base of the anchor plate with nut and bolt. Two pulleys were welded to the reaction frame and tie wire was passed through the pulleys, and a strong mild steel hook was tied at the other end of the tie wire. The loads were applied through the hook system with an incremental rate of 50 N and the applied loads were transferred to the tie wire, as a result, the tie wire along with the anchor moves freely (without any drag) in the upward direction through pulleys. The upward movement of the anchor was measured using two dial gauges that were placed on the light weight steel plate arranged at the surface of the sand bed. A complete photographic view of the test tank is shown in Fig. 1a. As per the earlier test studies performed by Kishore Kumar and Ilamparuthi [2], the two reinforcements (N = 2) having a size (B_g) of 2D were placed immediately above the anchor plate by maintaining a respective reinforcement spacing (h) of 0.0D and 0.5D as shown in Fig. 1b.

a

b

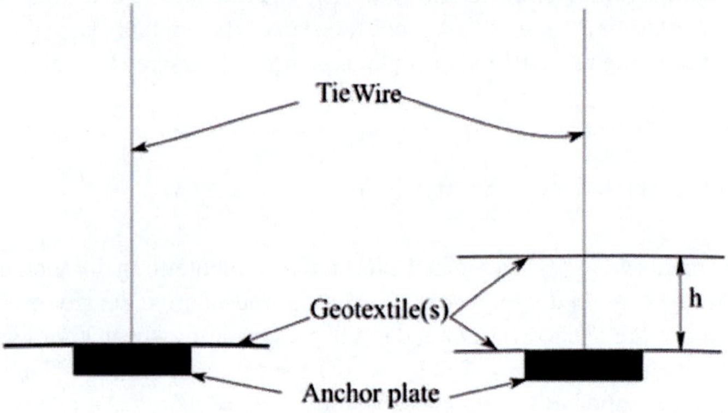

Fig. 1 **a** A Photographic view of complete laboratory test setup. **b** A line diagram of representation of different positions of reinforcements above the anchor plate

4 Results and Discussion

The plate anchor stress-displacement behavior of both unreinforced sand bed and reinforced sand beds is shown in Fig. 2. Figure 2 also shows the test results of the reinforced sand bed of both the cases of reinforcements placed immediately above the anchor plate having a spacing (h) of 0.0D and 0.5D. The unreinforced sand bed failed at an ultimate stress of 152.80 kPa and the test results were identical to the test results of Ilamparuthi et al. [13]. Irrespective of reinforcements' vertical spacing (h), the test results of reinforced sand bed revealed that the anchor uplift capacity of unreinforced

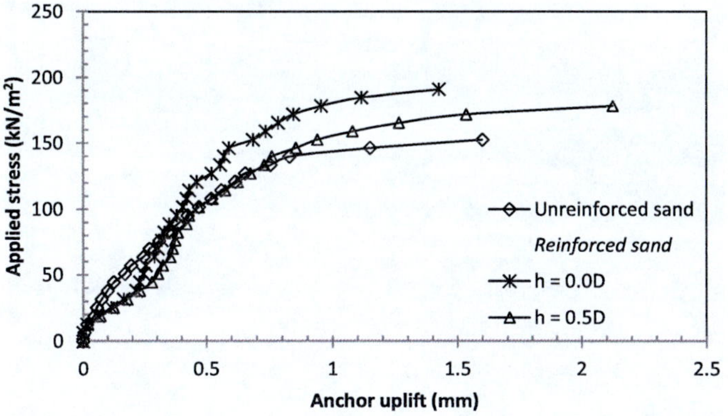

Fig. 2 Variation of stress-displacements of anchor for unreinforced and different reinforcement spacing sands

sand bed is higher than the reinforced sand during the initial stages of uplift. It could be happened due to the lack of proper interaction of soil-geosynthetics interfaces to mobilize sufficient tensile stresses in the reinforcements to resist the anchor uplift during the initial stages.

From Fig. 2, it could be understood that the reinforcements which were having a spacing (h) of 0.0D show a higher uplift capacity (i.e., 191 kPa) in comparison to the reinforcements having h = 0.5D (i.e., 178.27 kPa). It could be due to the enhancing of the stiffness of the reinforcements by placing at h = 0.0D as compared to the h = 0.5D. When the reinforcements are placed at a spacing (h) of more than 0.0D, the improvement of the anchor uplift capacity depends upon the amount of interface frictional forces mobilized along the reinforcement surfaces [2, 6]. The photographic view of the failure of the reinforced sand bed is shown in Fig. 3. During loading, the sand bed undergoes bulging behavior (Fig. 3), and at the end of the test, the corresponding bulging heave was extended up to a distance of 195 mm from the center of the anchor for the cases of unreinforced sand and 265 mm-270 mm from the anchor center, depends on the reinforcements vertical spacing (h) of 0.0D and 0.5D, respectively.

Moreover, the bulging rupture surface for the case of unreinforced sand makes an angle of 22.2° (i.e., $\phi/2$) from the surface edge of the anchor plate to the vertical reference as shown in Fig. 4. The following test result is identical with the findings of literature [13]. However, the following present test results revealed that the reinforcement's tensile stiffness property plays a major role in comparison to the reinforcement interface frictional properties. The present test results are contradicting the test results of Ravichandran [6] and it may be due to the variation of the type of reinforcements (i.e., geogrid and geo-composite) used in the test studies.

Fig. 3 A Photographic view of reinforced sand bed after failure

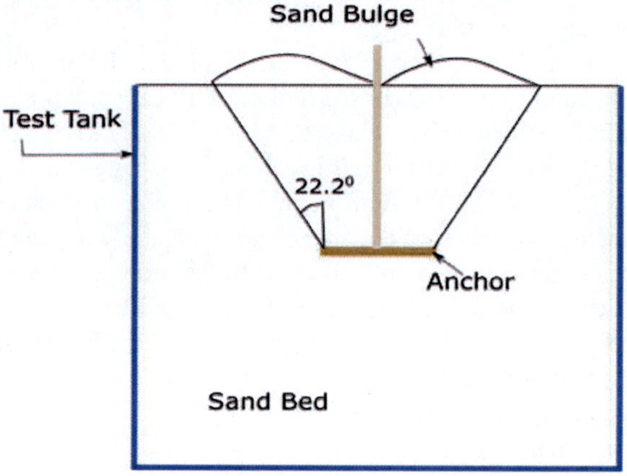

Fig. 4 A schematic view of reinforced sand bed after failure

5 Conclusion

The test results elucidate that the double number of reinforcements placed immediately above the anchor plate (D) with a spacing (h) of 0.5D shows 1.17 times higher uplift capacity as compared to the unreinforced sand. Also, the same type and same number of reinforcements placed at h = 0.0D shows 1.25 times higher compared to the unreinforced and 1.07 times greater than the reinforced sand of h = 0.5D. It could be due to the influence of reinforcement tensile stiffness property which plays a vital role as compared to the reinforcement's interface frictional property.

References

1. Krishnaswamy NR, Parashar SP (1994) Uplift behaviour of plate anchors with geosynthetics. Geotext Geomembr 13(2):67–89
2. Kishor Kumar V, Ilamparuthi K (2020) Performance of anchor in sand with different forms of geosynthetic reinforcement. Geosynth Int 27(5):503–522
3. Giffels WC, Graham RE, Mook JF (1960) Concrete cylinder anchors proved for 345-KV tower line. Electr World 154:46–49
4. Adams JI, Hayes DC (1967) The uplift capacity of shallow foundations. Ontario Hydro Res Q 19(1):1–13
5. Ilamparuthi K, Dickin EA (2001) Predictions of the uplift response of model belled piles in geogrid-cell-reinforced sand. Geotext Geomembr 19(2):89–109
6. Ravichandran PT (2008) Investigations on uplift response of anchors in sand with and without reinforcement. PhD thesis, Anna University, India
7. ASTM D6913 (2017) Standard test methods for particle-size distribution (Gradation) of soils using sieve analysis. ASTM Int, West Conshohocken
8. IS: 2720-13 (1986) Methods of test for soils. Bureau of Indian Standards, New Delhi, India
9. ASTM D5199 (2012) Standard test method for measuring the nominal thickness of geosynthetics. ASTM Int, West Conshohocken
10. ASTM D5261 (2018) Standard test method for measuring mass per unit area of geotextiles. ASTM Int, West Conshohocken
11. ASTM D4595 (2017) Standard test method for tensile properties of geotextiles by the wide-width strip method. ASTM Int, West Conshohocken
12. Rahimi M, Tafreshi SM, Leshchinsky B, Dawson AR (2018) Experimental and numerical investigation of the uplift capacity of plate anchors in geocell-reinforced sand. Geotext Geomembr 46(6):801–816
13. Ilamparuthi K, Dickin EA, Muthukrisnaiah K (2002) Experimental investigation of the uplift behaviour of circular plate anchors embedded in sand. Can Geotech J 39(3):648–664
14. Ilamparuthi K, Dickin EA (2001) The influence of soil reinforcement on the uplift behaviour of belled piles embedded in sand. Geotext Geomembr 19(1):1–22

Use of Polyethylene and Cement Material Is for Strengthening of Soil

Dhanesh Khalotia, Prantik Sen, and Deepak Chaturvedi

Abstract The purpose of this dissertation is to determine the CBR value and the unconfined compressive strength value of soils when they are stabilised using polymer and cement. The stabilization of Soil has frequently been used as an alternative to sub-statute material shortage on site. The usage of non-traditional compounds for the improvement of soil is practised every day. During this investigation, a particular lab try was carried out to access the impact of a polymer which is water borne of an unconfined pressure quality on a sandy soil and California Bearing Ratio (CBR) test on clayey soil. Various tests were performed which includes the size of grain of sandy soil, unit weight of the soil, and unconfined compressive equality test of the soil. The sand along with different quantities of polymer (2, 3, and 4%) and concrete (20, 30, and 40%) were blended together with every one of them into the mixture by hand blending in the research centre conditions. The samples were then exposed to an unconfined pressure test to decide their quality and then following 7 days of restoration. The results of the tests concluded that the waterborne polymer improved the unconfined pressure quality of sandy soils fundamentally which has a tendency of liquefactions. The polymer altered and enhanced the building properties of soil via physical holding. The quantity of polymer which is required to enhance the engineering properties, was directly associated with the soil particle coating thickness and the specific surface. The soils modified using polymer are less strengthened compared to soils modified using cement.

Keywords Soil stabilization · Unconfined compressive strength · CBR test · Polymer

D. Khalotia (✉) · P. Sen · D. Chaturvedi
IES College of Technology, Bhopal 462044, India
e-mail: Dhaneshkhalotia@gmail.com

K. K. Pathak et al. (eds.), *Latest Developments in Civil Engineering*, Lecture Notes in Civil Engineering 352, https://doi.org/10.1007/978-981-99-2676-3_30

1 Introduction

Stabilization of soil can be referred to as a regulated process to enhance the soil and its properties by using additives so it can be used in base or sub-base courses to carry the pavement loads and expected traffic. Several methods can be used by which soils are often stabilised [1].

There are two ways to strengthen the characteristics of sandy soils: mechanical stabilization, which involves mixing natural soil and stabilizing material to create a homogeneous mixture; and adding stabilizing material to undisturbed soils to create interaction by allowing it to permeate through soil voids. In chemical stabilization, the qualities of the soil that is readily available locally are altered to strengthen its features. Lime stabilization and cement stabilization are the two chemical stabilization techniques that are most frequently utilised [2–5].

In the past, various laboratory tests, such as the California bearing proportion (CBR), Hveem stabilometer and cohesiometer test (The Hveem technique), and R-esteem tests to describe the asphalt materials, had been used to determine the quality of the subgrade soil used in the development of asphalt. In any event, due to the static nature of their stacking circumstances, neither of these methodologies takes into account the effect of cyclic stacking of the vehicle strain on the asphalt.

The continuous development of asphalt configuration includes the introduction of a solidity-based modulus, referred to as the tough modulus, which controls the ongoing stacking situation on the materials to be tested, thereby simulating real-world vehicle stacking. To determine the versatile modulus, the continuing stacking triaxial test is carried out inside the dirt's flexible range [6]. However, the everlasting distortion controls the periodic stacking of materials past the examples as far as feasible or occasionally up to their disappointment in order to evaluate the materials' rutting performance (single-stage tests) and various inquiry stages (or cut-off points) (multi-stage tests).

1.1 Objectives

- To investigate the negative effects of an increase in soil moisture content and the corresponding strength characteristics. The strength and stiffness of soils decrease significantly as the moisture level rises over the optimum; yet, soil performs admirably when it is at or below the optimal moisture content (dry side of optimum) (wet side of optimum).
- To investigate the soil's ability to contract or expand when there is a high water table. The in situ moisture content affects the soils' ability to shrink and swell [7, 8].
- To investigate how soils stabilize when combined with cement and polymer.
- To contrast the various stabilising methods according to the kind of soil. The strength/stiffness of the subgrade soil is simply insufficient to support the loads

of the pavement due to the soft nature of the soil in some areas and the presence of high water levels [9]. Engineers have utilised a variety of strategies based on the soil types in order to address this issue. For instance, fine-grained soils are preferred over mechanical stabilisation. However, it is typical in some areas with soft clay subgrade and high water level to treat the soils with chemical stabilisers or stabilisers high in calcium[10].

- To forecast how the subgrade soil will behave at the ideal moisture level. Because the majority of soils have in situ moisture contents that are greater than the ideal, it is insufficient to forecast subgrade behaviour based solely on properties that are in or close to the optimum on either side. Pavement construction on subgrades with high moisture contents and weak soil has become simpler thanks to the employment of various stabilisers that reinforced the qualities of the raw to treated/ stabilised subgrade soil [11, 12].

2 Literature Review

General

There has been a great deal of research done on the use of common stabilisers, particularly lime and concrete. The adjustment tools for lime and concrete have long since been abandoned, and various applications have demonstrated the effectiveness of these traditional stabilisers. However, there has been relatively little research on the use of non-traditional stabilisers, such as synthetic polymers and magnesium chloride, and their present history has evolved.

According to **Sharifah Zaliha et al.** [13], finding the best soil stabilisers to combat problems caused by delicate soils is still a top priority. This includes looking at both the cost and the impact on the environment in addition to achieving the necessary soil designing properties. This paper's goal was to examine the soil adjustment techniques that had been developed based on exploratory analyses. Examining several materials to determine their suitability as a soil stabiliser had been done which used cementitious binders, fly debris geo-polymeric folio, additional sodium hydroxide additive, and other remnants. These materials were discussed in this research, and their suitability for balancing out fragile soils could be shown from the obtained findings, simply in terms of solidarity, based on the directed California Bearing Ratio (CBR) test and unconfined compressive quality (UCS) test. With the use of these compounds, the quality of fragile soils was substantially improved, and it was expected that they would be effective soil stabilisers in field applications.

Yıldız and Soğancı [14] examined small patches of sediment residue that had crystallised and found flat ice focal points perpendicular to the freezing direction as well as vertical ice-filled shrinkage splits that were joined to shape segments by polygonal cross segments. On the pressure-driven conductivity of three compacted dirts with different properties, Othman and Benson investigated the effects of compaction conditions (shaping water content and compactive exertion) and outside conditions (temperature slope, extreme temperature, dimensionality of freezing, number of freeze-defrost cycles, and condition of weight). According to research facility analysis, the number of freeze-defrost cycles, freezing rates, and pressure circumstances have the most effects on changes in water driven conductivity. As the rate of freezing, the number of freeze-defrost cycles, and the overburden pressure increase, so does the water driven conductivity. Different factors, such as a specific temperature, the dimension of freezing, and the accessibility of an exterior water gracefully, don't seem to have a significant effect.

Cabalar et al. [15], metropolitan zones are presently battling with the high volume of strong squanders particularly the development and destruction of materials. In this examination, the squashed waste cement (CWC), which is viewed as perhaps the greatest segment of strong waste, was utilised to improve some geotechnical properties of natural soil. The CWC at the proportions of 5, 10, 15, and 20% were added to natural so as to direct an escalated arrangement of tests. The lab tests were incorporated as far as possible by fall cone, adjusted compaction, unconfined compressive quality (UCS), and growth rate. The outcomes show that when the CWC rates were expanded to half, there were diminishes of about 30 and 60% in fluid breaking point and pliancy record of dirt, individually. An expansion of about 35% in γ_{drymax} for the natural soil was seen when the CWC content was expanded from 10% to half. A decrease of about half of the weight for natural soil appeared by expanding the CWC rate to half. The UCS estimations of the natural residue increment by around 25% by expanding the CWC rate upto half. The expanding rate expanded by adding CWC upto 30%, and afterward diminished with the option of CWC upto half.

Tang et al. [16] investigated how to fly debris containing heavy metals may be adjusted or hardened using Portland concrete as a cover. It is found that the mechanical (i.e., compressive quality) and drainage practises of the settled fly debris blends are significantly influenced by the concrete/fly debris proportion and restoring time. The expansion of compressive quality percentage rises from 42.24 to 80.36% when the concrete/fly debris ratio increases from 4:6 to 8:2, while the draining amount of heavy metals decreases by 2.33 to 85.23%. The compressive quality percentage of blends increases from 240.00 to 414.29% as the restoring time increases from 3 to 56 days, while the draining percentage of heavy metals decreases by 16.49–88.70%. The expansion of fly debris stacking, which prevents the development of ettringite and destroys the structure of hydration items, can be blamed for the lessening of compressive quality with the lower concrete/fly debris proportions and less restoring time. This results in the pozzolanic response and obsession with water atoms. Additionally, the presence of concrete reduces drainage because of the placement of ettringite and the restriction of heavy metal particle movement in many structures, such as C–S–H gel and adsorption.

2.1 The Value of Polymers

1. Polypropylene (PP) fibre: It is the most frequently used integration in soil adjustment lab tests. PP filaments are used today to enhance the qualities of dirt quality, reduce the properties of shrinkage, and combat concoction and biological corruption. Additionally, PP fibre fortification reduces the volumetric shrinkage stresses and swell weights of the broad mud and enhances the dirt's unconfined compressive quality (UCS).

It was determined from examinations of field test segments that a 203-mm thick coating of sand fibres was sufficient to support substantial amounts of military truck traffic and that the method displayed amazing potential for military runway and street applications. Field investigations also showed that, in order to prevent fibre pullout under traffic, the surface needed to be fixed with an emulsion folio [17].

Maheshwari blended polyester strands of 12 mm length alter from 0 to 1% in extremely compressible clayey soil. The results demonstrated that adding arbitrarily dispersed filaments to profoundly compressible clayey soil increased the definitive bearing limit and decreased settlement at the definitive burden. They assumed that the safe bearing weight (SBP) and the dirt bearing limit (DBL) both increase with an increase in fibre content up to 0.50% before decreasing as more strands are added [18].

2. Polyethylene (PE) filaments: It has also been partially investigated whether it is possible to strengthen soil using polyethylene (PE) strips and strands. It has been determined that a small amount of high thickness PE (HDPE) filaments can increase the soil's ability to break down. These days, GEOFIBRES are typically blended or mixed into sand or earth soils in the form of one to two long distinct PP as well as PE fibrillated or tape strands. But be aware that some researchers have referred to the PP filaments used in soil support as "geofibre" [19–21].

Sobhan and Mashand showed the significance of utilizing strength as a proportion of execution. These investigations indicated that increments in rigidity with included HDPE strips were not understood yet huge increments in sturdiness coming about because of expanded strain limit was watched. With expanding sturdiness, a great part of the normal presentation benefits because of fibre incorporation is in the post-top burden bit of the pressure strain conduct. Hence, as the filaments create pressure, an improved pressure strain reaction is the outcome. In any case, enhancements in weariness conduct were not noted.

3. Glassstrands: Consoli et al. show that taking into account glass threads in silty sand improves top quality. The effect of PP, PET, and glass filaments on the mechanical conduct of fibre-fortified solidified soils was examined by Consoli et al. in a different study. Their findings showed that although the deviatoric fears of disappointment somewhat decreased, the consideration of PP fibre significantly enhanced the delicate conduct of solidified soils [22]. In contrast to the case of PP fibre, the consideration of PET and glass strands slightly decreased the fragility and slightly increased the deviatoric anxieties about disappointment. Maher and Ho studied the conduct of kaolinite-fibre (PP and glass filaments) composites and discovered that the instances with glass fibre reinforcement had more defined UCS expansion.

4. Nylon fibre: Kumar and Tabor contemplated the quality conduct of nylon fibre strengthened silty earth with various levels of compaction. The examination shows that the pinnacle and leftover quality of the examples for 93% compaction are essentially more than the examples compacted at the higher densities.

3 Experimental Investigations

3.1 Methodology

Cement and polymer are employed in this dissertation to stabilise the soil. Sandy and clayey soils both contain different amounts of cement and polymers. In an experiment with varied cement and polymer compositions, the impact of unconfined compression strength and CBR values is investigated. First, two distinct types of soil are mechanically analysed. Next, the soil is combined with various amounts of cement and polymer, and any variations are investigated.

3.2 Material Used

(1) **Sandy Soil**
(2) **Clay Soil**

Table 1 Engineering properties of sandy soil

Property	Sample
Specific gravity	2.75
Grain size	
(4.75–20) mm (%)	2.5
(<4.75 mm) (%)	97.5
Max. void ratio (e_{max})	0.8
Min. void ratio (e_{min})	0.42
Void ratio (%)	1.024
Optimum moisture content (%)	15
Maximum dry unit weight (g/cm^3)	1.86
Soil classification (USCS)	S

Table 2 Index properties

Description of index properties	Experimental value (%)
Liquid limit	30
Plastic limit	18.50
Plastic index	11.50
Shrinkage limit	14.65

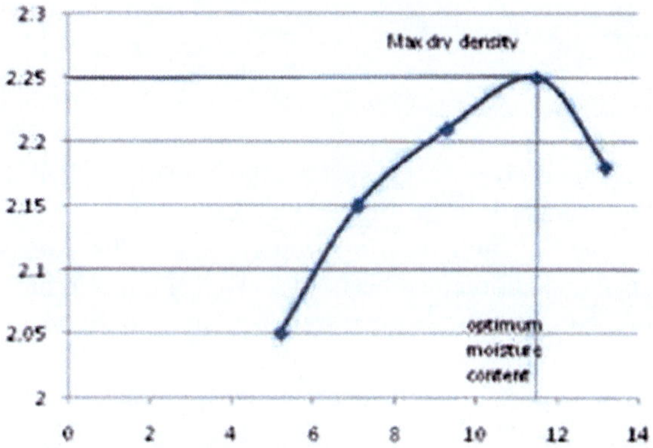

Fig. 1 Result of modified proctor compaction

Table 3 Cement properties

Properties of OPC43 grade cement								
Fineness (sq. m/kg) min	Soundness (mm) Max	Setting time		Compressive strength				
		Initial (mts) min	Final (mts) Max	Day Mpa	3 Day Mpa	7 Day Mpa	28 Day Mpa
225	10	30	600	N S	23	23	43

(3) **Cement**
(4) **Polymers**

3.3 Tests Conducted

(1) **California Bearing Ratio (CBR) Testing**

The CBR is the sole test that can determine a subgrade's strength. We shall compare the strength of several subgrade materials using this test. The CBR test is carried out in accordance with a standard procedure through which one can ascertain or plan the thickness or strength of the subgrade layer. The CBR value is inversely related to

Table 4 Physicochemical properties of received emulsion

Name	Acrylic-copolymer watered solut ion
Physical state	Liquid-white colour
Solvability in water	Solution
Boiling point	100°
Water absorption	1% max
Non-self-burning	Non-explosive
Applicable temperature	Not less than 10°
Density (g/cm^3)	1.11 (20°)
Toxicity	Non toxic

pavement layer thickness. Less thickness is needed if the subgrade is stronger since the stronger it is, the higher the CBR value.

The California bearing proportion, or CBR, is expressed as the rate at which the heap opposition (test heap) of a given soil test to the standard burden at an entrance of 2.5 or five mm is expressed in relation to the standard burden;

$$\text{CBR} = (\text{Test load}/\text{Standard load}) * 100$$

The standard loads are 1370 and 2055 kg for two penetrations of 0.5 and 5 mm, respectively. The CBR test is performed using probing ring divisions to penetrate the dial reading on a tiny scale. At first, research was done to identify several soil characteristics, such as grain size distribution and index qualities. In the future, rigorous compaction tests were carried out to determine the ideal moisture content and matching maximum dry density [23].

Then CBR tests were performed at OMC, and analysis was done to look into how CBR varied over the course of multiple soaking days, from unsoaked (day 0) to soaked (day 1) to (day 4). Calculations were made on the quantity of soil utilised for CBR using the moisture content and associated dry density. Each soil sample was soaked for one day, two days, three days, and four days before the sample was examined using CBR instruments to determine the relevant CBR values.

The soil needed to fill a typical CBR mould into a 4 kg bag is the initial step in the preparation of all CBR samples. The dirt is blended and evenly distributed with the amount of water needed to get it to the ideal moisture level. To moisten the soil mixture, a few chemical additions are additionally combined at the necessary concentration. The soil–water combination is then combined with the chemical that has been introduced.

(2) **Unconfined Compression Testing**

The polymer emulsion-treated fine-grained material was tested using unconfined compression (UC). By substituting a 4-inch plate for the infiltration cylinder, the soil test stacking template used for CBR testing was altered.

Tests were compacted using adjustable delegate compaction in a 4 inch by 8-inch form. Tests for compaction were removed from the form once they were completed and wrapped in a rubber film to prevent excess air from reviving the samples. Estimates were obtained every 0.01 crawls up to an absolute strain of 15% using a strain rate of 0.6 inches per second (up to 1.23 inches). To study the compressive quality of balanced out samples, sandy soils were used with a varying rate of concrete blending and different amounts of waterborne polymer added to the soils. Before using the dirt in the blends, it was dried. The required amounts of polymer as a level of the test's dry load and concrete were first combined, and then they were applied to the dry soils. The weight of the dry example was used to determine the amounts of fluid polymer, which were chosen to be 2, 3, and 4%, and concrete, which were chosen to be 20, 30, and 40%, respectively. The form contained the blending test. Following a 24-h waiting period, the samples were taken out of the moulds, stored in a room with a soothing temperature range of 21–25 °C, and then tested after seven days. The polymer mixture was properly hand-kneaded to transform it into dough. The uniformly combined dough was then added to a steel mould that was 300 mm in diameter and 150 mm tall.

4 Results and Discussions

4.1 Results of the Unconfined Compression Test

Figures 2 and 3 display the results of 7 days of cure on unconfined compression strength.

With more curing time, stabilised samples' unconfined compression strength gets stronger. Both specimens, which contained 20–40% by weight of cement and 2–4% by weight of polymer, underwent a 7-day air curing process. Therefore, by increasing the polymer content, the soil's strength rose along with the cross-linking between polymer networks. Figures show that the stabilised soils' compressive strength rose as the air curing conditions' curing time increased (Table 5).

4.2 Unconfined Compression Test Result

The results of the CBR test (Tables 6, 7, 8, 9, 10 and 11) performed on a soil sample taken at 2% polymer content and 20% cement content under various soaking durations are shown in Figs. 4, 5, 6, 7 and 8.

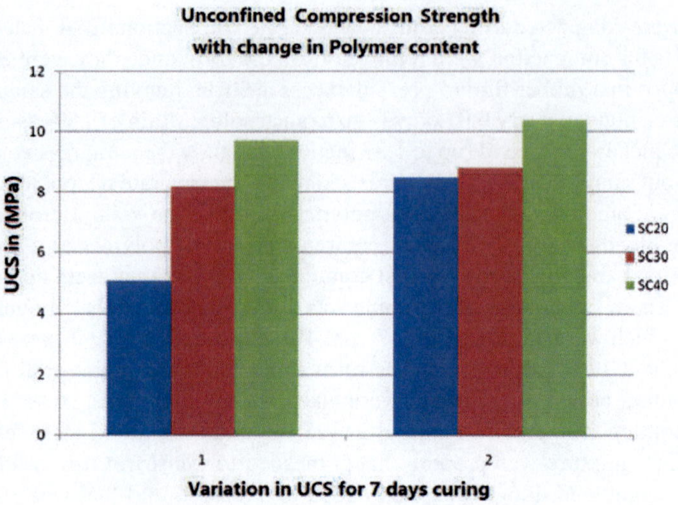

Fig. 2 SC20 indicates soil plus 20% cement. Soil + 30% cement = SC30; SC40 = 40% cement + soil

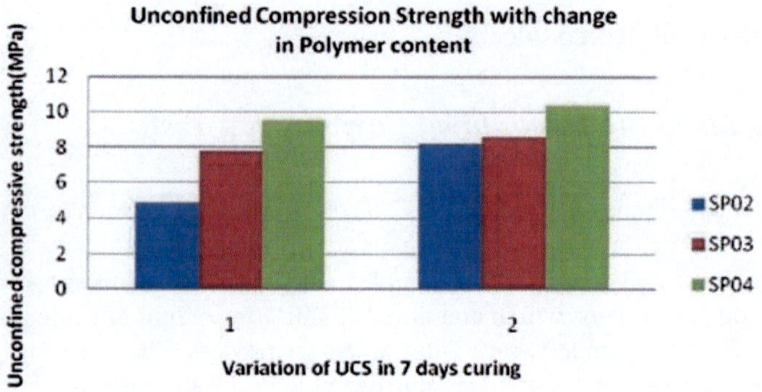

Fig. 3 Mix-SP02 = soil + 2% polymer SP03 = soil + 3% polymer SP04 = soil + 4% polymer

Table 5 Variation of cement versus polymer content

	Variation in cement content			Variation in polymer content		
Unconfined compression strength (MPa)	20%	30%	40%	2%	3%	4%
	5.1	8.2	9.7	4.9	7.8	9.56
	8.5	8.8	10.4	8.2	8.6	10.35

Table 6 CBR values for unsoaked 0 day

Soil with different stabiliser s	Penetration at	
	2.5 mm	5 mm
Without stabilisers	45.40	44.91
Polymer 3%	52.55	54.55
Cement 20%	57.66	59.56

Table 7 CBR Values for soaked (1 day)

Soil with different stabilisers	Penetration at	
	2.5 mm	5 mm
Without stabilisers	10.66	13.24
Polymer 3%	13.87	15.18
Cement 20%	15.18	16.30

Table 8 CBR values for soaked (2 days)

Soil with different stabilisers	Penetration at	
	2.5 mm	5 mm
Without stabilisers	8.83	10.51
Polymer 3%	13.14	16.55
Cement 20%	14.67	18.83

Table 9 CBR values for soaked (3 days)

Soil with different stabilisers	Penetration at	
	2.5 mm	5 mm
Without stabilisers	5.11	5.64
Polymer 3%	7.15	6.33
Cement 20%	8.98	7.49

Table 10 CBR values for soaked 4 day

Soil with different stabilisers	Penetration at	
	2.5 mm	5 mm
Without stabilisers	4.74	5.35
Polymer 3%	5.25	6.03
Cement 20%	5.77	6.28

Table 11 Increase in CBR values

Days of soaking	0	1	2	3	4
S	44.91	13.24	10.51	5.64	5.35
SP02	54.55	15.18	16.55	6.33	6.03
SC20	59.56	16.30	18.83	7.49	6.28

Fig. 4 Un-soaked (0 day)

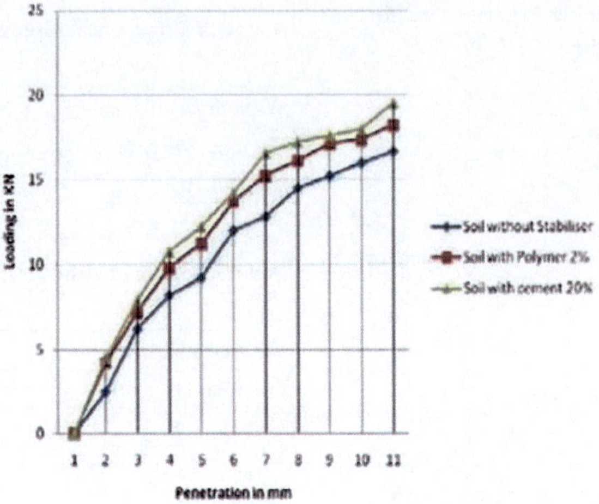

Fig. 5 Soaked (1 day)

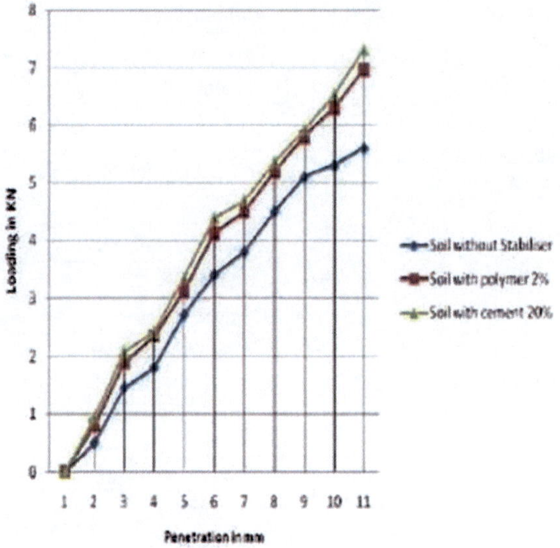

4.3 Flexible Pavement Design (As Per IRC: 37-2012)

Assume N = 100 msa and Traffic Volume = 100 msa

From IRC: 37-2012,

Pavement thickness at CBR = 5.35 is 714 mm.

Pavement thickness at CBR = 6.03 is 684 mm.

Pavement thickness at CBR = 6.28 is 675 mm.

Pavement thickness is decreasing as follows:

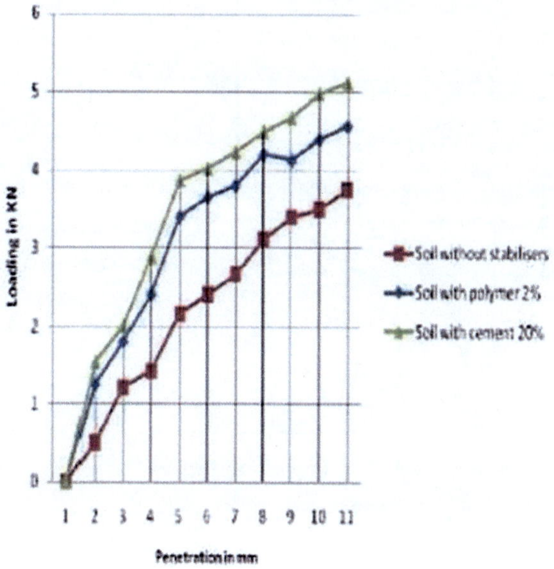

Fig. 6 Soaked (2 days)

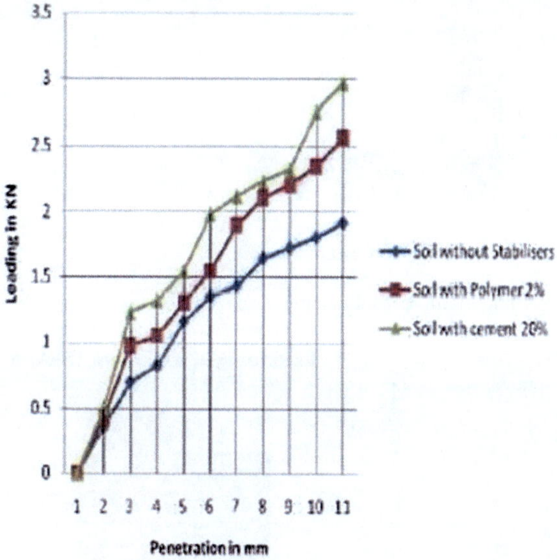

Fig. 7 Soaked (3 days)

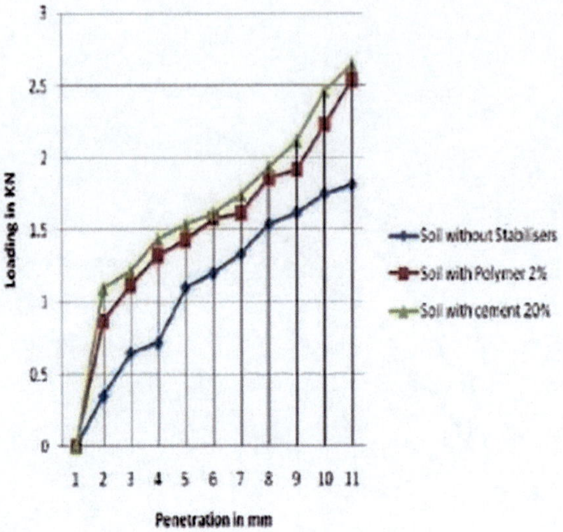

Fig. 8 Soaked (4 days)

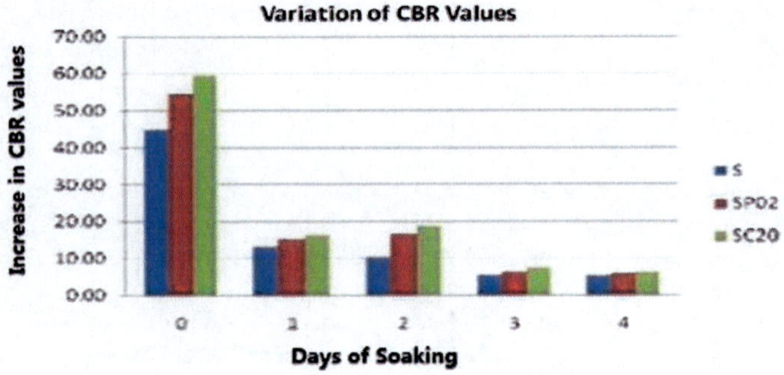

Fig. 9 CBR value variation with respect to days of soaking

Fig. 10 % Decrease in the thickness of the pavement

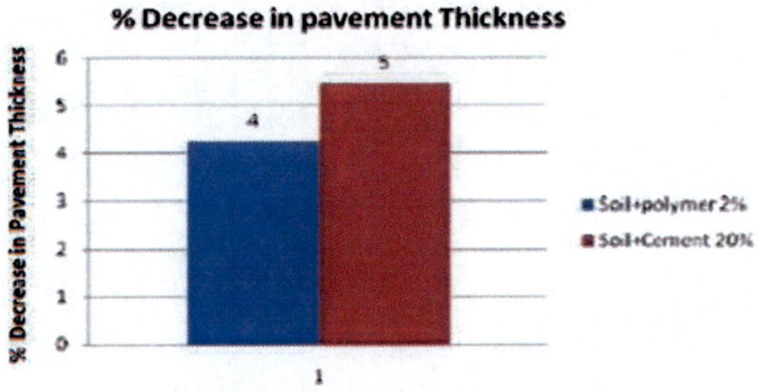

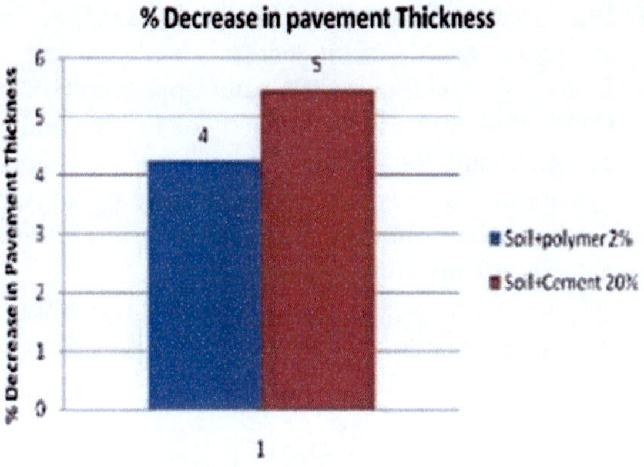

Fig. 11 % Decrease in cost of pavement

When using 2% polymer = 30 mm
When using 20% cement = 39 mm.

4.4 *Cost Analysis for Pavement Per Kilometer of Road Length*

See Fig. 11.

5 Conclusion

The tests conducted for this dissertation's findings are summarised in the conclusions that follow.

- From the initial time of curing evaluation, the addition of polymer to the natural soil resulted in an improvement in its mechanical capacities as measured by unconfined compression tests. The ideal polymer content for liquefiable sandy soils was calculated at 2% based on strength considerations.
- Above a cement level of about 30%, the strength of sandy soil mixtures almost becomes constant. Cement content increases the strength of sandy soil mixtures. The sheer fact that the tiny cement granules were positioned and covered in the sand grains helps to explain this phenomenon.
- If the soil's polymer content is maintained at less than 4%, it is evident from Fig. 1 that the soil's unconfined compressive strength will increase as well. This phenomenon is explained by the fact that the polymer content is increased, it covers the entire sample area, and it increases cross-links. Additionally, there isn't much of an impact on strength when cement and polymer contents vary.

- When 20% cement is first added to the sand, the gain in unconfined compressive strength is greater, but it decreases when cement concentration is raised.
- It has been noted that CBR values rise with rising cement and polymer content; CBR values also rise significantly on the first and second days of soaking, but not as significantly on subsequent days.
- When soil is stabilised with cement and polymer, CBR values are greatly impacted, yet when their impact is properly scrutinised, cement and polymer content does not vary significantly.
- Adding polymer and cement to the clayey soil reduces the thickness of the pavement and, thus, the cost of paving to a good extent.

6 Future Research Potential

Fly ash is frequently combined with another addition, such as lime, murrum, cement, and other similar materials, and the quantity should be changed to create the simplest stabilising composition feasible.

References

1. Jones D (2007) Development of performance-based tests for nontraditional road additives. Transp Res Rec J Transp Res Board 2(1989):142–153
2. McGown A, Andrawes Z, A1-Hasani M (1978) Effect of inclusion properties on the behavior of sand. Geotechnique 28:327–346
3. Murray J, Frost D, Wang Y (2000) The behavior of sandy soil reinforced with discontinuous fiber inclusions. Trans Res Rec 1714:9–17
4. Attoh-Okine NO (1995) Lime treatment of laterite soil. Constr Build Mater 9(5):283–287, from Butterworth Heinmann
5. Newman JK, Tingle JS (2004) Emulsion polymers for soil stabilization. US Army Engineer Research and Development Center, Vicksburg, MS
6. IRC-37 (2012) Guidelines for the design of flexible pavements. IRC, New Delhi
7. Kaniraj R, Gayathri V (2003) Geotechnical behavior of fly ash mixed with randomly oriented fiber inclusions. Geotext Geomembr 21:123–149
8. Kumar A, Walia BA (2007) Influence of flyash, lime, and polyester fibers on compaction and strength properties of expansive soil. J Mater Civ Eng ASCE 19:242–248
9. Hejazi SM et al (2011) A simple review of soil reinforcement by using natural and synthetic fibers. Constr Build Mater 30:100–116
10. Kolias et al (2005) Stabilisation of clayey soils with high calcium fly ash and cement. Cem Concr Compos 27:301–313. Sciencedirect.com
11. Kumar S, Tabor E (2003) Strength characteristics of silty clay reinforced with randomly oriented nylon fibers. EJGE 127:774–782
12. Arora KR Soil mechanics and Justo and Khanna Highway engineering
13. Sharifah Zaliha SZ, Kamarudin H, Mustafa Al Bakri AM, Binhussain M, Siti Salwa MS (2013) Review on soil stabilization techniques. Aust J Basic Appl Sci 7(5): 258–265
14. Yıldız M, Soğancı AS (2012) Effect of freezing and thawing on strength and permeability of lime-stabilized clays. Scientia Iranica 19(4):1013–1017

15. Cabalar AF, Abdulnafaa MT, Karabash Z (2016) Influences of various construction and demolition materials on the behavior of a clay. Environ Earth Sci 75(841):1–9. https://doi.org/10.1007/s12665-016-5631-4
16. Tang Q, Liu Y, Gu F, Zhou T (2016) Solidification/stabilization of fly ash from a municipal solid waste incineration facility using Portland cement. Adv Mater Sci Eng 2016. Article ID 7101243. https://doi.org/10.1155/2016/7101243
17. Terzaghi K, Peck R, Mesri G (1996) Soil mechanics in engineering practice, 3rd edn. Wiley-Interscience, p 592
18. Maheshwari V (2011) Performance of fiber reinforced clayey soil. EJGE 16:1067–1087
19. Biswajeet S, Devadatta N (2009) A study of subgrade strength related to moisture
20. Sobhan K, Mashnad M (2002) Tensile strength and toughness of soil–cement–flyash composite reinforced with recycled high density polyethylene strips. J Mater Civ Eng ASCE 14:177–184
21. Tang C, Shi B, Chen W (2006) Strength and mechanical behavior of short polypropylene fiber reinforced and cement stabilized clayey soil. Geotext Geomembr 24:1–9
22. Santoni RL, Tingle JS, Webster SL (2002) Stabilization of silty sand with nontraditional additives. Transp Res Rec J Transp Res Board 1787:61–70
23. Maher H, Gray H (1990) Static response of sand reinforced with randomly distributed fibers. J Geotech Eng ASCE 116:1661–1677

"Use of Fly Ash in Stabilization of Soil with Shrinking and Swelling Properties"

Dhanesh Khalotia, Sarwar Imam, and Rahul Saini

Abstract More than 51.8 million hectares of land are covered by the extensive soil in India (mainly Black Cotton soil). Soil stabilization, which is widely used in constructions and street asphalt developments, is one of the most important angles for development objectives; this is on the grounds that such an adjustment system improves designing properties of the soil, for example, volume soundness, quality and strength. In this cycle, evacuation or supplanting of the dangerous soil is done; substitution is finished by a superior quality material, or the soil is treated with an added substance. In the current study, stabilization of black cotton soil from Nagpur utilizing fly ash from Sesa Sterlite, Jharsuguda, Odisha is attempted. Expansive soils are stabilized using different ratios of this addition, such as 10, 20, 30, 40, and 50%. Fly ash has no plasticity, hence the plasticity index (P.I.) of clay-fly ash mixtures shows a decrease in the value when the fly ash percentage is increased. In conclusion, the addition of fly ash brings about abatement in pliancy of the extensive soil, and increment in usefulness by changing its grain size and colloidal response. Tried under both splashed and un-doused conditions, the CBR estimations of mud with fly ash blends were watched. Examination of the previously discovered outcome uncovered the potential of fly ash as an additive that could be used for improving the engineering properties of expansive soils.

Keywords Black cotton soil · Expansiveness · Swelling · Shrinkage · Stabilization

D. Khalotia (✉) · R. Saini
IES University, Bhopal 462044, India
e-mail: Dhaneshkhalotia@gmail.com

S. Imam
Department of Mechanical Engineering, RKDF University , Bhopal, India

K. K. Pathak et al. (eds.), *Latest Developments in Civil Engineering*, Lecture Notes in Civil Engineering 352, https://doi.org/10.1007/978-981-99-2676-3_31

1 Introduction

About 3% of the world's land is covered by the black cotton soil (i.e., about 340 million hectares). They are primarily found in Africa, specifically in South Africa, Ethiopia, Tanzania, and the Gezira cotton fields of the southern black cotton plains of Sudan. They are widely distributed in Asia, particularly on the Deccan Plateau of India. Additionally, they were found in large portions of Australia, the West Indies, and Russia which take up nearly 20% of its land area in India. In India these, soils are predominant in the states of Gujarat, Maharashtra, Madhya Pradesh, Andhra Pradesh, Karnataka and Tamilnadu [1].

Black cotton soil or clayey soil, is an expansive soil that expands or contracts excessively in response to changes in moisture content. Depending on the stress level and the pressure of the soil swelling, an engineering structure connected to this kind of soil experiences either settlement or heave. Design and construction of civil engineering structures on and with expansive soils for geotechnical engineers, is a difficult task [2].

The engineering behavior of fine-grained soils depends on their water content. Liquid limit and plastic limit are important water contents as well as two important parameters of plasticity index, This is the primary index parameter used for classification of fine-grained soils. Plasticity index has also been used in correlation with many other engineering properties like internal friction angle, undrained shear strength, lateral earth pressure over consolidation ratio etc. Another crucial characteristic is the shrinkage limit (SL), where soils tend to contract when their moisture content decreases. Particularly vulnerable to shrinkage and the ensuing volume change are fine-grained soils. The behavior of the soils may be negatively impacted by shrinkage, which might result in soil cracking. That is generally acknowledged that the liquid limit test effectively evaluates the shear strength of soil that is so soft it is on the verge of liquidity. Heave is swell that occurs vertically. Among the illite, kaolinite and montmorillinite clay minerals, the montmorillinite possesses the greatest ability to swell by illite. Kaolinite doesn't swell up. In a dry state, black cotton soils are exceptionally hard and have a high bearing capacity. Black cotton soils in India exhibit shrinkage limits of 9–14%, plasticity indices of 20–65%, and liquid limit values of 50–100%. As the tides rise, the quantity of swell typically rises the measure of plasticity. The crystal lattice structure, cation exchange capability, water absorption capacity, density, and water content all affect the swelling potential of clay minerals [3].

2 Fly Ash

Fly ash is a waste product that is taken out of the gases that come from coal-fired furnaces, usually in a thermal power plant. Fly ash is quite similar to the volcanic ashes that were used as hydraulic cements in the ancient times, which was one of the main uses for volcanic ashes. These remains were accepted to be outstanding amongst other pozzolanas (restricting specialist) widely used throughout the world.

The interest of intensity flexibly has exponentially elevated these days because of expanding urbanization and industrialization wonders. Therefore, this development has brought about the expansion in the number of intensity providing warm force plants that utilizes coal as a consuming fuel to create power. The fly ash is the mineral accumulation left over after burning coal. The force plants' Electro Static Precipitator (ESP) collects these fly remnants. After being pulverised, fly ash is the ash that was recovered from the flue gases using an electrostatic precipitator. Among bottom ash, pond ash, and fly ash, it is the smallest particle. The primary component of fly ash, with a little amount of unburned carbon, is the non-combustible particulate matter. Typically, these are made up of silt-sized particles. Fly ashes have been divided into four categories based on a test for lime reactivity, as follows:

- Cementitious fly ash
- Cementitious and pozzolanic fly ash
- Pozzolanic fly ash
- Non-pozzolanic fly ash

The use of fly ash can be broadly classified into three categories:

The Low Value Utilizations, such as backfilling, structural filling, construction of roads, stabilizing soil, construction of embankments and dams, building ash dykes, etc. The Medium Value Uses, such as grouting, cellular cement, pozzolana cement, bricks, blocks and soil amendment agents, prefabricated building blocks, fly ash concrete, weight aggregate, etc.

High-Value Applications, such as fly ash paints, ceramic industry, magnetite extraction, distempers, metal recovery, acid refractory bricks, floor and wall tiles, and so on.

3 Objcctivc

To determine the extent to which adding additives can reduce expansiveness and increase bearing capacity value.

Also to establish the use of fly ash as an additive, so allowing for its use as a fine waste product from thermal power plants.

4 Literature Review

Chijioke Christophor [4], extensive soils are hazardous because of the exhibitions of their earth mineral constituent, which makes them display the therapist swell qualities. The psychologist swell practices make broad soils wrong for direct designing application in their characteristic structure. While trying to make them more possible for development purposes, various materials and procedures have been utilized to balance out the dirt. In this examination, the added substances and strategies applied for balancing out sweeping soils will be centered around, as for their proficiency in improving the designing properties of the dirt. At that point we examined the micro-structural communication, concoction measure, monetary ramifications, nanotechnology application, just as waste reuse and maintainability. A few issues in regards to the viable use of the rising patterns in sweeping soil adjustment were given three classifications, in particular geo-environmental, normalization and advancement issues.

Udayashankar D. Hakari and S. C. Puranik [5], urbanization and development in the economy of level 2 urban areas of India have prompted the lofty increment in the structure development exercises and has required the execution of framework undertakings, for example, expressways, railroads, landing strips, water tanks, recovery and so on. These undertakings perpetually require quality earth in gigantic amount.

In urban territories, acquire earth isn't effectively accessible which must be pulled from a significant distance. Frequently, huge regions are secured with profoundly plastic and far reaching soil, which isn't reasonable for such reason. The twin city of Hubballi–Dharwad is a quickest developing level 2 city of Karnataka state and is the second biggest city of the state only close to Bangalore.

The wide spread of the dark cotton soil in the twin city of Hubballi–Dharwad has presented difficulties and issues to the development exercises. An errand was in this manner attempted to explore and improve the designing properties of the dark cotton soils of Hubballi–Dharwad.

Indrani Parimala and B. Ganesh [6] have studied that historical changes are evidences occurring in the world, from Egyptian pyramids to the today's world twin tower, and any change in structural development depends on the deep study of soil existing by civil engineers. Extensive soil, which is one among the hazardous soils, is a term utilized for any dirt that has a high potential for contracting or expanding because of any difference in dampness content.

Akshaya Kumar Sabat [7], explain that expansive soil is a tricky soil for structural specialists in light of its low quality and cyclic swell-contract conduct. Adjustment utilizing strong squanders is one of the various strategies for treatment, to improve the building properties and make it appropriate for development. The gainful impacts of some conspicuous strong squanders as acquired in research facility examines in adjustment of extensive soil have been explained.

Yongzhen Cheng and Xiaoming Huang [8], black cotton soil (BCS) structures a significant soil bunch in Kenya and is portrayed by high therapist/swell potential when presented to water. An extensive arrangement of lab tests were performed on BCS treated with lime (0–9%), volcanic ash (VA, 0–25%), and their blends so as to consider the physical–mechanical properties and mineralogical changes of the settled BCS. In addition, a test street which supplanted the BCS with the lime–VA-settled BCS was developed to explore the dampness change and soil development in the BCS establishment. The outcomes uncovered that BCS settled with mixes of lime and VA shows bigger California bearing proportion (CBR) and unconfined compressive quality (UCS) values when contrasted and a solitary stabilizer.

Shi He [9], traditional soil stabilizers, for example, lime and concrete are generally used to diminish swell and shrinkage conduct and improve quality properties of broad soils through the development of cementitious items. However, the manufacturing process of these calcium-based stabilizers, such as lime and cement, need large amounts of water and emit gases such as CO, CO_2, NO_x, and SO_2 that are harmful to the environment. Hence, environmentally-friendly techniques are often sought out by the civil infrastructure industry [10, 11].

5 Methodology and Materials

5.1 Materials

Shrink and Swell Soil—The extensive black cotton soil was obtained for this investigation, from the Khairi site in Nagpur, Maharashtra. The obtained black cotton soil was taken to the laboratory in bags.

To find the soil's natural moisture content, a tiny sample of the material was sieved through a 4.75 mm sieve, weighed, and then dried by air before being reweighed. The following are a few of the soil's different geotechnical characteristics Shown in below Table 1.

Fly ash—Fly ash was procured from Sesa Sterlite in Jharsuguda, Odisha, for the purpose of the investigations in this study. This fly ash was sieved using a 2 mm sieve to remove the plant and foreign particles. Before further use, the samples were in the oven to dry for exactly 24 h.

Table 1 Geotechnical properties of the procured soil

S. No.	Properties	Code referred	Value
1	Specific gravity	IS 2720 (Part 3/Sec1)-1980	2.44
2	Maximum dry density (MDD)	IS 2720 (Part 7)-1980	1.52 g/cc
3	Optimum moisture content (OMC)	IS 2720 (Part 7)-1980	22.65%
4	Natural moisture content	IS 2720 (Part 2)-1973	7.28%
5	Free swell index	IS 2720 (Part 40)-1977	105%
6	Liquid limit	IS 2720 (Part 5)-1985	65%
7	Plastic limit	IS 2720 (Part 5)-1985	37.08%
8	Shrinkage limit	IS 2720 (Part 6)-:1972	17.37%

Table 2 IS test specifications code

S. No.	Test	IS code
1	Grain size analysis	IS2720 (part IV) 1985
2	Atterberg's limits	IS2720 (Part V) 1985
3	Modified proctor test	IS2720 (Part VII) 1983
4	Linear shrinkage	IS2720 (Part20)-1992
5	Free swell	IS2720 (Part40)-1977

5.2 *Methodology Adopted*

The effectiveness of fly ash as a stabilizing ingredient in expansive soils was investigated through a number of research. Fly ash content in the expanded soil ranged from 10 to 50% (multiples of 10) by weight of the entire amount taken.

The below Table 2 lists the IS code specifications that are followed for all tests.

6 Results and Discussion

6.1 *Standard Proctor Test for Soil Fly Ash Mixture*

The test is carried out in 6 different samples, in which the fly ash is mixed in the variation of 0, 10, 20, 30, 40 and 50%. The maximum value is discovered after these samples are evaluated using the standard proctor test (Figs. 1, 2 and Tables 3, 4).

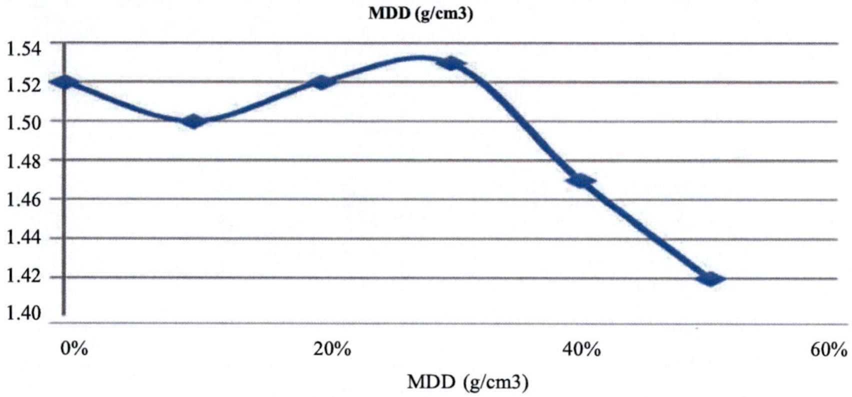

Fig. 1 Variation of MDD for different fly ash mixtures

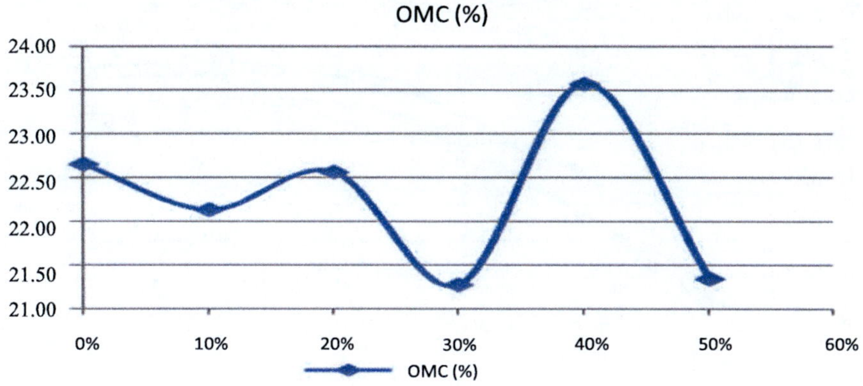

Fig. 2 Variation of OMC for different fly ash mixtures

Table 3 Variation of MDD for different fly ash mixtures

Sample No.	% Fly ash (%)	MDD (g/cm^3)
1	0	1.52
2	10	1.50
3	20	1.52
4	30	1.53
5	40	1.47
6	50	1.42

Table 4 Variation of OMC for different fly ash mixtures

Sample No.	% Fly ash (%)	OMC (%)
1	0	22.65
2	10	22.13
3	20	22.56
4	30	21.27
5	40	23.57
6	50	21.34

6.2 Unconfined Compressive Strength (UCS) Test

See Figs. 3, 4 and Table 5.

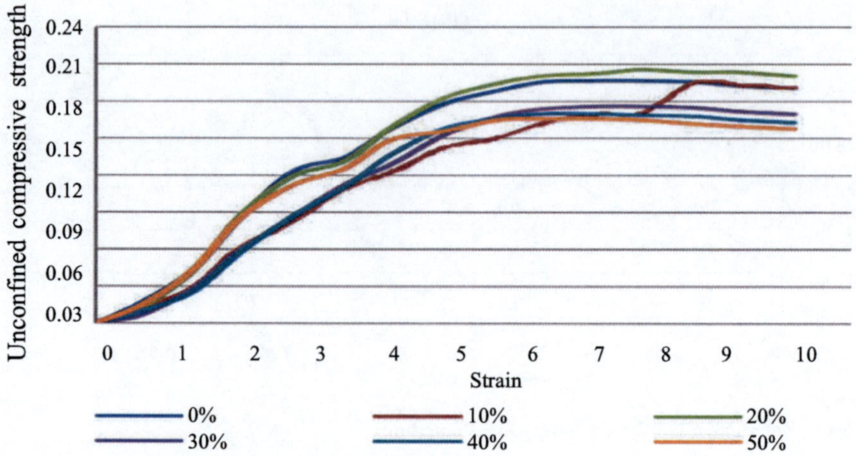

Fig. 3 Graph showing comparison of UCS test readings

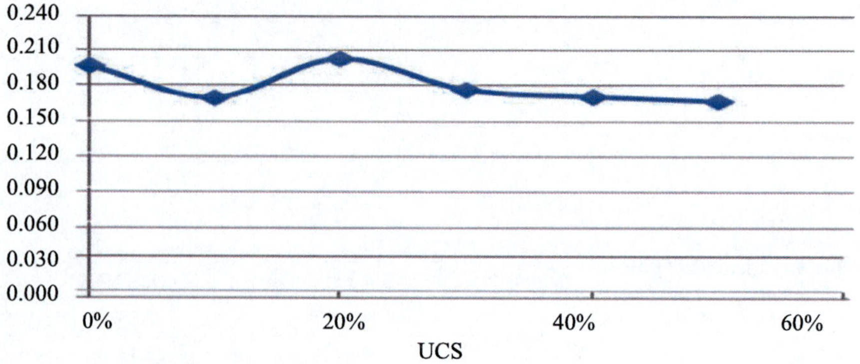

Fig. 4 Graph showing variation of maximum UCS test

6.3 *California Bearing Ratio (CBR) Test*

See Figs. 5, 6 and Tables 6, 7.

6.4 *Free Swell Ratio and Plasticity Index*

See Figs. 7, 8 and Table 8.

Table 5 Variation of UCS for different fly ash mixtures

S. No.	Strain	UCS for different % fly ash					
		0%	10%	20%	30%	40%	50%
1	0	0.000	0.000	0.000	0.000	0.000	0.000
2	0.6	0.017	0.012	0.013	0.007	0.010	0.015
3	1.3	0.044	0.029	0.043	0.029	0.026	0.044
4	1.9	0.085	0.060	0.087	0.059	0.057	0.085
5	2.6	0.123	0.082	0.119	0.086	0.087	0.112
6	3.3	0.135	0.109	0.132	0.113	0.111	0.126
7	3.9	0.158	0.122	0.159	0.129	0.137	0.149
8	4.6	0.178	0.142	0.183	0.153	0.158	0.156
9	5.3	0.189	0.150	0.195	0.168	0.166	0.165
10	5.9	0.196	0.162	0.201	0.174	0.170	0.166
11	6.6	0.197	0.169	0.203	0.176	0.170	0.166
12	7.2	0.197	0.170	0.206	0.176	0.169	0.165
13	7.9	0.196	0.195	0.204	0.175	0.168	0.162
14	8.5	0.193	0.193	0.204	0.172	0.165	0.160
15	9.2	0.191	0.191	0.201	0.170	0.163	0.158

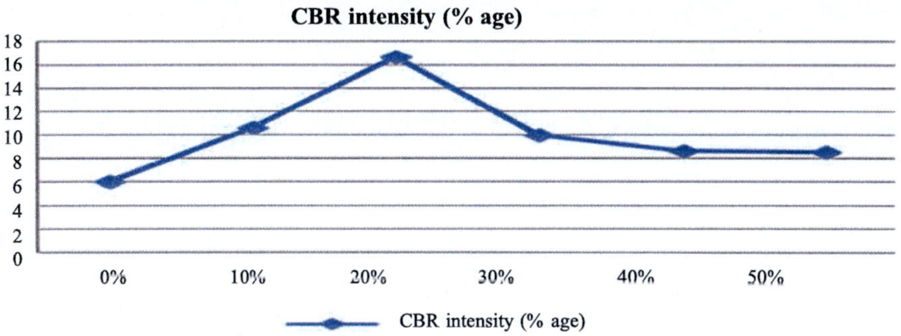

Fig. 5 Graph showing variation of maximum unsoaked CBR intensity

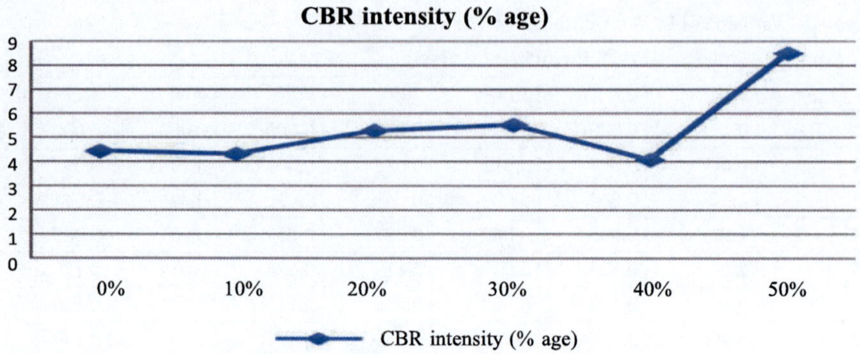

Fig. 6 Graph showing variation of maximum soaked CBR intensity

Table 6 Variation of unsoaked CBR intensity for different fly ash mixtures

Sample No.	% Fly ash (%)	CBR intensity (% age)
1	0	5.99
2	10	10.58
3	20	16.67
4	30	9.95
5	40	8.61
6	50	8.51

Table 7 Variation of soaked CBR intensity for different fly ash mixtures

Sample No.	% fly ash (%)	CBR intensity (% age)
1	0	5.99
2	10	10.58
3	20	16.67
4	30	9.95
5	40	8.61
6	50	8.51

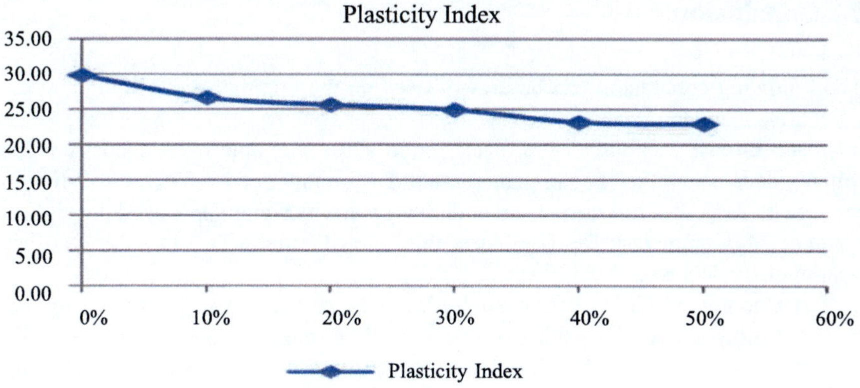

Fig. 7 Graph showing variation of plasticity index

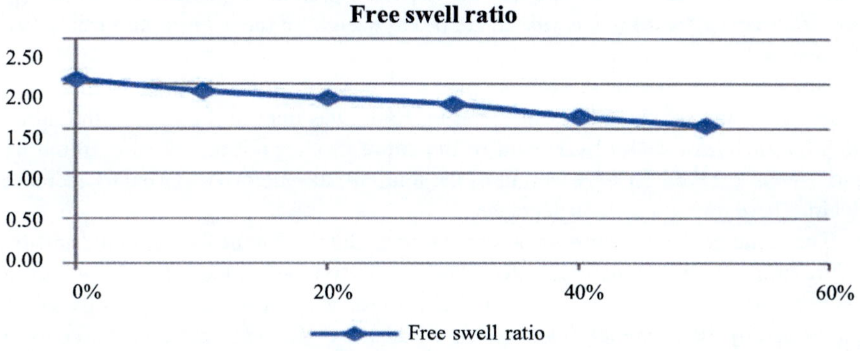

Fig. 8 Graph showing variation of free swell ratio

Table 8 Variation of the plasticity index and the free swell ratio in expansive soil with fly ash content in expansive soil

Mixture	Liquid limit (LL)	Plastic limit (PL)	Plasticity index (PI)	Free swell ratio
Only soil	65.6	35.8	29.8	2.05
Soil + 10% fly ash	61.2	34.6	26.6	1.92
Soil + 20% fly ash	58.8	33.2	25.6	1.84
Soil + 30% fly ash	56.4	31.5	24.9	1.77
Soil + 40% fly ash	51.8	28.67	23.13	1.63
Soil + 50% fly ash	49.2	26.3	22.9	1.53

7 Conclusion

The following conclusion can be drawn based on the findings of the study and the comparisons made:

When fly ash was added, the black cotton soil's Maximum Dry Density value initially decrease. It then began to increase as the amount of fly ash in the soil-fly ash mixture increased. A mixture of soil and 30% fly ash by weight produced the highest value of Maximum Dry Density. After that, the Maximum Dry Density readings continuously decrease.

The Unconfined Compressive Strength (UCS) of the soil with varying fly ash content followed a similar pattern to that of the Maximum Dry Density values, with the exception that the peak value was recorded with a fly ash content of 20% by weight.

In soil experiments using the un-soaked California Bearing Ratio (CBR) that were conducted with varied fly ash contents, the CBR gradually increase as the fly ash content increased until it was valued at 20% by weight of the overall mixture; it then decreases.

However, the change in soil samples with variable fly ash contents that were tested using the soaked California Bearing Ratio was uneven. Following the initial addition of fly ash (10% by weight of the entire mixture), it increases until the fly ash content reached 30% by weight of the total mixture before it started to decrease again. Thereafter, the values decrease.

The value of the free swell ratio decreased as the amount of fly ash in the soil-fly ash mixture increased, as was noted. The plasticity index values also decreased in response to this decline. The swelling behavior of the mixture of soil and fly ash is influenced by the plasticity index values, which are directly related to the percentage of swell in an expansive soil.

Therefore, adding fly ash to the black cotton soil reduces swelling and improves its strength.

References

1. Public Works Department Pune/Integrated Public Works Division, Pune. M/s H.J. Tekawade, Taluka-Purandhar, District-Pune "Woven geo textile TFI 5300 for subgrade stabilization"
2. Gourley et al (1993) Problems related to black cotton soil
3. Arora KR Soil mechanics
4. Ikeagwuani CC, Nwonu DC (2019) Emerging trends in expansive soil stabilisation: A review. J Rock Mech Geotech Eng 11(2):423–440
5. Udayashankar DH, Puranik SC (2012) Stabilisation of black cotton soils using fly ash, Hubballi Dharwad Municipal Corporation Area, Karnataka, India. Glob J Res Eng 12(2):21–29
6. Indrani P, Ganesh R, Chitra S (2017) Stabilization of expansive soil using fly ash and lime. Int J Current Eng Sci Res 4(2):5–10
7. Sabat AK, Mohanta S (2016) Performance of limestone dust stabilized expansive soil-fly ash mixes as construction material. Int J Civil Eng Technol (IJCIET) 7(6):482–488

8. Cheng Y, Wang S, Li J, Huang X, Li C, Wu J (2018) Engineering and mineralogical properties of stabilized expansive soil compositing lime and natural pozzolans. Constr Build Mater 187:1031–1038
9. He SJ, Wang JW, Li Y, Xu ZY, Wang XX, Lu X, Fu Y (2019) Nickel-catalyzed enantioconvergent reductive hydroalkylation of olefins with α-heteroatom phosphorus or sulfur alkyl electrophiles. J Am Chem Soc 142(1):214–221
10. Puppala AJ, Congress SS, Bheemasetti TV, Caballero SR (2018) Visualization of civil infrastructure emphasizing geomaterial characterization and performance. J Mater Civil Eng 30(10):04018236
11. George AL (2019) Case studies and theory development: the method of structured, focused comparison. In: Caldwell D, George AL (eds) A pioneer in political and social sciences. Pioneers in arts, humanities, science, engineering, practice, vol 15. Springer, Cham. https://doi.org/10.1007/978-3-319-90772-7_10
12. Brooks (2009) Effect on swelling of black cotton soil by stabilizing rise husk ash-fly ash
13. Ramesh HN, Krishnaiah AJ, Shilpashet S (2013) Effect of lime on the index properties of black cotton soil and mine tailings mixtures. UVCE, Bangalore University, Bangalore, India
14. Agrawal V, Gupta M (2011) Expansive soil stabilization using marble dust. M. Tech, Structural Engg. Department, MNIT, Jaipur
15. US Army, US Navy, US AirForce (2005) Soil stabilization for pavements. University Press of the Pacific, Honolulu, Hawaii
16. Mehta A, Parate K, Ruprai BS (2013) Stabilization of black cotton soil by Fly Ash. GNIEM, Nagpur
17. IS 2720 (Part 4) (1985) Method of test for soil (Grain size analysis)
18. IS 2720 (Part 5) (1985) Method of test for soil (Determination of liquid and plastic limit)
19. IS 2720 (Part 20) (1992) Method of test for soil (Determination of linear shrinkage)
20. IS 2720 (Part 40) (1977) Method of test for soil (Determination of free swell index of soil)
21. IS 2720 (Part 8) (1983) Method of test for soil (Determination of optimum moisture content and maximum dry density for modified proctor test)
22. IS 1498 (1970) Classification and identification of soils for general engineering purposes

Impact of Soil Property in Selection of Stabilisation Technique—A Review

Himanshu Jangde, Farhan Khan, Mohammad Irshad Ansari, and Kaushal Prajapati

Abstract The selection of stabilisation technology is heavily influenced by the intrinsic qualities of the soil. Various soil characteristics affect the application intensity of the stabilising procedure. Consequently, a variety of stabilising measures should be used such that they complement soil qualities and aid to enhance various soil metrics. Lime and cement are a common alternative for binding, however reusing the same procedure might pose environmental and economic issues. They may cause unintended responses, which merely exacerbates the situation. The authors gave case studies in which the character of the soil strongly influenced the selection of binders, the availability of other foreign substances, and a number of external circumstances. In this paper, there are three main determining factors for the selection of chemicals: (i) soil properties in the selected area, (ii) the behaviour of mixed external matter in soil, and (iii) external factors such as temperature and compaction, which are discussed through the inclusion of case studies from previous papers. This review attempted to cover the selection of novel chemical binders such as FARmLG (Fly ash, Red Mud, Lime, Gypsum) for marine dredge soil, a combination of fly ash and lime for long-term stabilisation of allophanic soil, and EICP (enzyme-induced carbonate precipitate) slurry (made from urea, calcium chloride dehydrates, jack bean, nonfat milk) for stabilising the various Ottawa. The influence of organic matter and 10% Ca-Bentonite with Pulp Fly ash (PFA) as a stabilising agent for this kind of soil is also discussed. Also covered is the adaptation of the solidification/stabilisation (s/s) technique for the treatment of heavy metals. The impact of the use of these binders on various strength parameters and compaction investigations, such as the enhancement of Yield strength and compression parameter using alkali-activated MgO for marine dredge soil, is reviewed. Also, the scientists discovered that adding Ca-Bentonite and PFA increases the compressibility of allophanic soil. By analysing previous research, a brief summary of the positive and negative effects of using different binders on key soil properties is provided. According to the results, allophanic soil may be stabilised using a mixture of 3% lime and 3% fly ash. Sulfate-containing soils will benefit from C3A-containing binders; otherwise, heavy metal mobilisation

H. Jangde · F. Khan (✉) · M. I. Ansari · K. Prajapati
Department of Civil Engineering, Rungta College of Engineering and Technology, Bhilai, Chhattisgarh, India
e-mail: Farhan1@rungta.ac.in

K. K. Pathak et al. (eds.), *Latest Developments in Civil Engineering*, Lecture Notes in Civil Engineering 352, https://doi.org/10.1007/978-981-99-2676-3_32

may occur. The hydraulic resistivity of marine dredge soil is enhanced by alkali-activated MgO, which also provides compressibility. Although the scope of this work is confined to 50 past research publications, the topic's usefulness is extensive. The authors endeavoured to provide a concise overview of the issue so that the next generation of the scientific community would be motivated to investigate this area.

Keywords Stabilisation techniques · FARmLG · Allophanic soil · Ca-bentonite · Pulp Fly Ash · Solidification/stabilisation (s/s) method

1 Introduction

The word "soil" refers to the uppermost layer of the earth, which often has a porous, unconsolidated surface and promotes plant development. According to one definition used in engineering, "soil is the substance formed by disintegration of rocks containing unconsolidated solid particles" [1]. Soil engineering qualities include characteristics for strength, modulus, physical parameter, consolidation properties, index properties, permeability, and shear strength [2].

The availability of a suitable soil for engineering applications is one of the basic needs of humans. Continuous development created a challenge in the fulfilment of this rudimentary requirement. The soil on earth is divided into different categories, which possess different properties. This difference in properties compels engineers to limit their areas of exploration. To increase the uses of different soils for engineering purposes, soil improvement has become obligatory [3–8].

Soil stabilisation is the process of enhancing soil qualities via various techniques. The most significant parts of soil improvement are raising its strength (bearing capacity and shear strength), its stiffness, and managing its hydraulic conductivity and seepage [9]. Mechanical and chemical procedures are equally legitimate means of achieving this objective. The chemical method includes the use of several chemicals known as soil binders [10, 11]. The process of chemically interacting with soil particles to insert additives in the soil's pores and bind soil particles together increases the soil's strength and stability [9, 10]. Different additives have qualities that react chemically with soil particles to put the additives in the soil pores. Binders such as cement, fly ash and lime are often used [12], but the selection of the binder relies greatly on a number of characteristics. Binders have the potential to cause serious issues without it.

Numerous chemical binders are often used for soil improvement, stabilising the soil via chemical connections between soil grains [3, 13–15]. The micro-level cooperation between water and fine content governs the chemical approach to binding employed for soil growth [16]. The soil's mineralogical composition and plasticity index also affect its mechanical characteristics [17]. Since soil does not alter on its own, chemicals are required to preserve soil stability. Depending on the kind of soil, the impact of different chemical binders will vary. Globally, lime and cement are two of the most often used binders. The interaction between lime and the hydration process gives sufficient stability and the calcium component of cement assists in the binding of soil particles. However, the scientific community debated these binders because of several environmental and economic problems. An adverse reaction between cement and some kinds of soil may be since lime's strength-building process is substantially slower, requiring more compaction work and increasing the project's total cost.

Furthermore, we can say that soil binders interact differently with various kinds of soil. For example, various soils containing sulphate and other heavy metals react together [18–22]. Consequently, it is difficult to ensure that the same old chemical binders will operate in all soil types. Diverse soil qualities impact the selection of additives and should be done on the basis so it can (i) complement the different soil properties (ii) leverage the soil properties to enhance the stabilisation process (iii) should not start mobilisation of hazardous process, (iv) conveniently accessible, exhibit excellent workability and should be cost-efficient. From this research, the authors attempted to deliberate on distinct soil qualities and select stabilising strategies according to them. It should be verified that the essential needs listed above will be satisfied.

2 Discussion

2.1 Factors Affecting Selection of Soil Binders

Stabilisation does not magically enhance the characteristics of all soils. The selection of binders is determined by the soil qualities that must be altered. Durability, strength, permeability, compressibility, and volume stability are the most sought-after attributes by engineers [23, 24]. Various binders are incompatible with different soil properties. Listed below are characteristics that do influence the choosing of binders.

2.1.1 Available Soil Type

The chemical composition of different binders affects the soil's characteristics in various ways. They chemically react and produce specific results. For one set of enhancements, we need a chemical that increases just specific soil parameters. Therefore, we must choose binders based on the soil to complement each other and enhance the desired soil characteristic. The author discovered a few instances in which the kind of soil affected the choice of the binder by examining several historical articles.

Marine Dredge Soil. This maritime soil includes a high concentration of salts, which might react with the component of the binder. Therefore, compounds that do not participate in hazardous reactions during the stabilising process are essential. The physical parameters of marine dredging soil are shown in * MERGEFORMAT

Table 1 Properties of marine dredge soil. *Source* [25]

Properties	Marine dredged sediment (MDS)
Maximum dry densities (g/cm^3)	1.502
Optimum moisture content (OMC) (%)	17.55
Liquid limit LL (%)	29.6
Plastic limit PL (%)	N.P.
Specific gravity	2.69
D50 (mm)	0.1
Particles less than 75 μm (%)	16.34
Color	Dark green

Table 1. FARmLG is a unique stabilising technique that combines fly ash (Fa), lime (L), gypsum (G), and red mud (R) (Rm). The combination of the waste products offers us a more effective stabilising strategy that actively enhances the many geotechnical characteristics of Maryland (strength, stiffness, and hydraulic resistance capacity). Another advantage is a non-hazardous MD product is stabilised with a new binder mixture [25]. Occasionally, heavy metals that are soluble in coastal soil have the potential to produce an undesirable scenario since typical binders may cause these metals to migrate. To prevent this, magnesium oxide (MgO) is used as an absorbent to allow the application of binders without the risk of releasing heavy metals. It absorbs heavy metals and facilitates the operation of the binder. As it absorbs the heavy metal, MgO has the additional effect of increasing yield stress and curing time while decreasing the initial void ratio. MgO's reactivity with water influences the mechanical properties as well. $Mg(OH)_2$ production generates an alkaline environment [26, 27].

Authors found that use of new cementless binders provides protection against heavy metal and also excess of water in the marine dredge soil does not impact the stabilising rate which helps in gaining strength over a period of time.

Allophanic soil. In New Zealand Allophone mineral is termed as problematic soil, due to which pavement construction is difficult there. The physical properties of Allophanic soil are described in * MERGEFORMAT Table 2.

When this soil is treated with lime, it shows improvement in strength for some time but reverts to its original state making it uneconomical due to repetition of reverting. The past study describes the different combinations of binders to stabilise allophanic soil. Some combinations performed well some performed poorly. From different combinations used in the past paper, it is observed that the combination of cement and lime prevents a reduction of strength. However, samples with fly ash and lime perform well in every stage of testing. It is recommended that 3% lime + 3% of fly ash should be considered when allophanic soil is encountered [22, 28, 29].

Authors found in the study that different combinations of lime and fly ash in the stabilisation process do not provide satisfactory results with optimum amount.

Table 2 Properties of allophanic soil. *Source* [22]

Properties	Westley road soil	Rangitoto road soil
Colour with sodium fluoride	Red	Red
Percentages of allophane	14%	36%
Aluminium: silicon ratio	1.67	1.59
Liquid limit	65	243
Plastic limit	56	151
Plasticity index	9	92
Clay content	46%	40%
Silt content	40%	42%
Natural water content	67.9%	215.6%

Also, higher percentage of these materials only contribute in increase of project cost. Higher percentage of combination of lime and fly ash only helps in reducing the strength of the soil with time.

Ottawa sand. In Saudi Arabia there are 4 different types of soils which possess different properties which are described in * MERGEFORMAT Table 3. Also SEM imaging is shown in Fig. 1 For this type of soil enzyme induced carbonate precipitate (EICP) is applied in-situ to stabilise it. Slurry of urea, calcium chloride dehydrate ($CaCl_2 \cdot 2H_2O$), jack bean meal urease enzyme, and non-fat dry milk into deionised (DI) water is made. Three different concentrations of EICP were investigated. The first EICP cementing solution's concentration (E1) was taken as (0.5 M urea, 0.335 M $CaCl_2 \cdot 2H_2O$, 1.5 g/L of urease enzyme, and 4 g/L of non-fat dry milk). The second EICP cementing solution's concentration (E2) was taken as (1 M urea, 0.67 M $CaCl_2 \cdot 2H_2O$, 3 g/L of urease enzyme, and 4 g/L of non-fat dry milk). A higher concentration solution (E3) is prepared for the third type of solution (2 M urea, 1.34 M $CaCl_2 \cdot 2H_2O$, 6 g/L of urease enzyme, and 4 g/L of non-fat dry milk). Although attempt is made to improve the soil, the solution is found to be ineffective for Al-Nafud sand due to the presence of organic matter.

Authors found that effect of strength improvement by slurry outperformed cement binders in Al-Rasheed sand. But hybrid mix of slurry and cement negatively impact the stabilisation process [3, 30–35]. Hybrid mix provides reduction in strength and makes the stabilisation process ineffective.

2.1.2 Presence of Different Foreign Matters

Impurities in soil not only cause a poor base for engineering application but also react with the chemical used for the stabilization process and minimise improvement in soil parameters. Assortment of chemicals for binding should be done while keeping

Table 3 Properties of different Ottawa sand [3]

Parameter	Ottawa sand	Al-Nafud sand	Al-Rasheed sand	Filtration sand
Maximum void ratio, e_{max}	0.67	0.66	0.84	0.83
Minimum void ratio, e_{min}	0.47	0.45	0.61	0.61
Specific gravity of soil, Gs	2.65	2.64	2.65	2.64
Maximum dry density, γ_{dmax} (g/cm^3)	1.80	1.83	1.65	1.65
Minimum dry density, γ_{dmin} (g/cm^3)	1.59	1.60	1.44	1.45
D_{10} (mm)	0.45	0.14	0.46	0.45
D_{60} (mm)	0.62	0.25	0.65	0.62
Uniformity coefficient, C_u	1.38	1.79	1.41	1.38

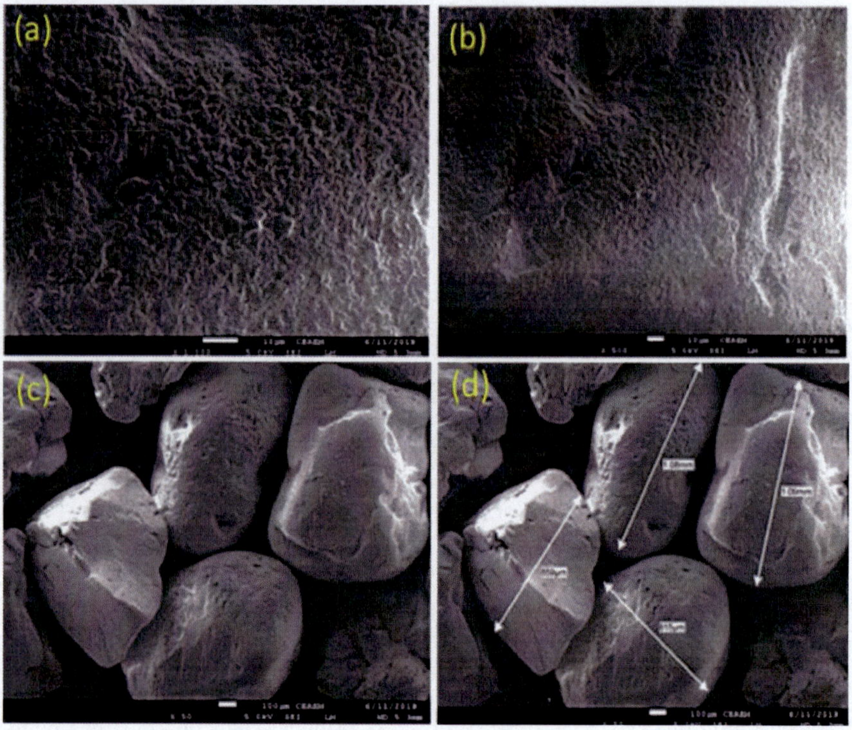

Fig. 1 SEM images of Ottawa sand showing particle size, shapes, and surface texture at magnification of **a** 1000, **b** 500, and **c**, **d** 50 [33]

in mind various foreign matters and their effect on soil as well as reaction with chemicals too.

Presence of Organic Matter in soil. About 436 million hectares of land are covered with organic soil, of which 110 million are situated in Canada only [4, 5]. The presence of organic sand is surveyed through GIS methods which save time and create an efficient way to survey [36–40]. The organic soil reacts with chemicals and creates obstacles to the binding process. To treat organic soil, a mix of 10% Ca-Bentonite with 30% pulp fly ash is used. Properties of Ca-bentonite pulp fly ash (PFA) and organic soil properties are discussed in Table 4. These three SEM imaging and X-ray diffractions are also shown in Figs. 2 and 3, respectively. Results suggest that undrained shear strength increased remarkably (5.8 times), and Cc and Cu decreased significantly (almost 55%). Also, the study suggests that the formation of CSH gel increases the compressibility of the treated soil [6].

Presence of Heavy Metals. Different industries release different kinds of byproducts, and untreated chemicals are harmful to the soil. Some industries inject heavy metals like suphfate carbonate, lead, and sulphide directly into the soil, which deteriorates the natural balance of the soil. These heavy metals are so unstable that using

Table 4 Geotechnical properties of soil, PFA and Ca-Bentonite. *Source* [41]

Parameters	Average value		
	Soil	PFA	Ca-Bentonite
Field density (kg/m^3)	1213 ± 22	–	–
Natural moisture content (%)	206 ± 26	–	–
Organic content (%)	26 ± 2.1	17.42	–
Swelling index (mL/2 g)	–	–	12
Loss on ignition (LOI %)	28.8 ± 1.6	20.6	–
Specific gravity	2.17 ± 0.04	2.7	2.49–2.72
Ash content (%)	74 ± 2.1	–	–
Fiber content (%)	23 ± 2	–	–
Shrinkage limit (%)	35 ± 0.8	–	–
Undrained shear strength (kPa)	2.53		
pH	7.19 ± 0.1	13	8.26
Liquid limit (%)	104 ± 5	106.8	209
Plastic limit (%)	84 ± 4	66	21
Plasticity index	20 ± 4	40.8	188
Particle size distribution (%)			97.2
>0.075 mm	18	20	–
0.002–0.075 mm	60	63	–
0.002 mm	22	17	–

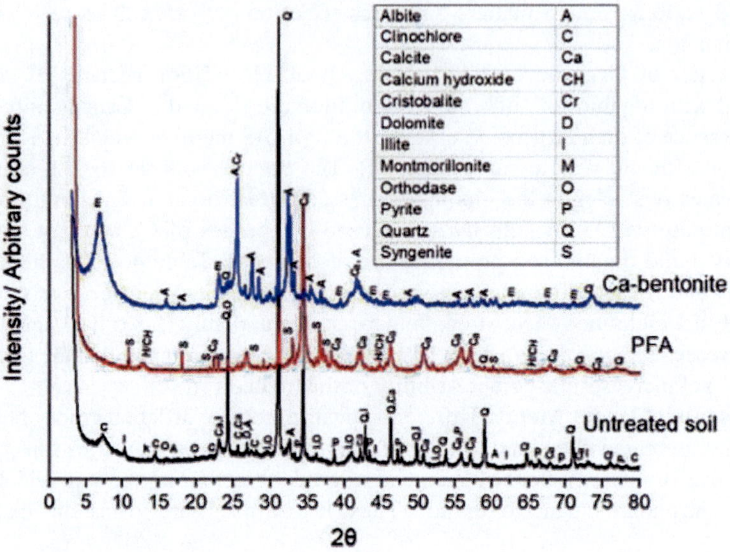

Fig. 2 X-ray diffraction pattern for untreated soil, PFA, and Ca-bentonite [41]

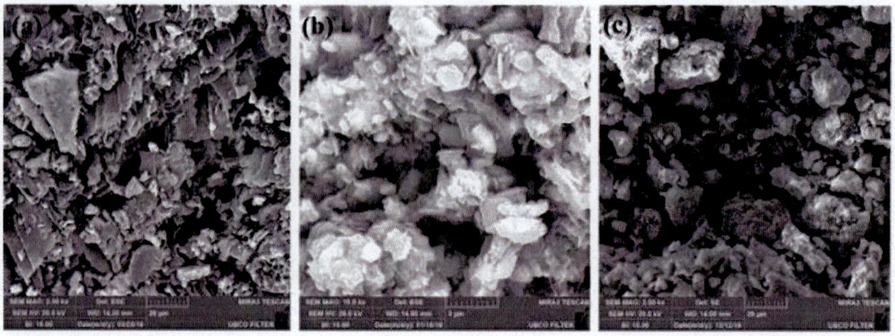

Fig. 3 SEM image of **a** Organic soil, **b** Pulp mill fly ash, and **c** Ca-bentonite [41]

the old stabilisation method like cement lime only causes these heavy metals to mobilise, which leads to the formation of expansion cracks causing damage to the structure [18, 42–44]. To prevent this phenomenon, some novel adaptations should be considered. For the stabilisation of sulphate-containing soil, high C3A content is required. Normal OPC reacts with sulphate and causes it to mobilise. Hence Ground granulated blast furnace slag is adopted, which has a high per cent of C3A, and it regulates mobilisation of sulphate by 89% [7, 45–49]. Lead also affects the traditional soil binders. SEM imaging of lead-contaminated soil is represented in Fig. 4. For Pb contaminated soil solidification/stabilisation (S/S) is adopted [13]. The solidification/stabilisation process combines both chemical immobilisation and physical

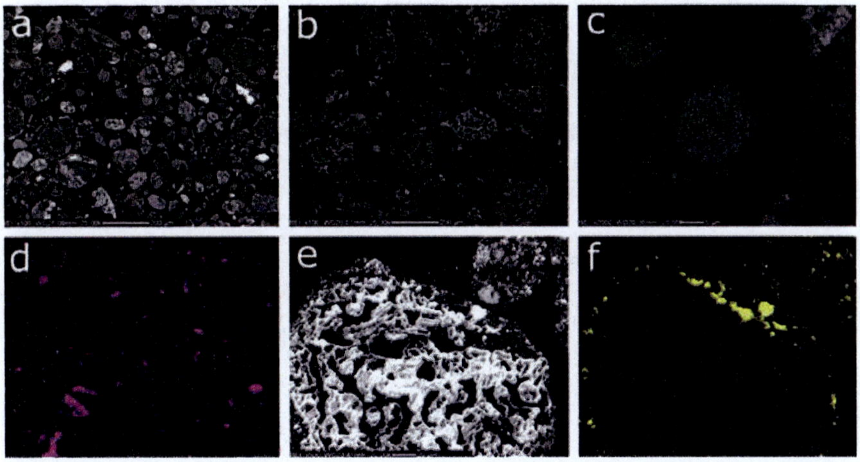

Fig. 4 SEM micrographs of the contaminated soil. **a** Back-scattered electron image of the bulk soil; **b** back-scattered electron image of the material sampled from the purple layer, showing a large amount of iron oxides; **c** back-scattered electron image of framboidal pyrite found in the light-yellow layer; **d** map of the Pb distribution relative to image **a**; **e** iron oxide particle; and **f** Pb mapping showing the metal adsorption on the particle in image **e** [44]

stabilisation of the soil. Hydraulic binders are reacted with contaminated content (typically Ordinary Portland Cement, OPC is used) [14].

Authors suggests that despite of OPC's extensive use, occasionally formation of toxic substances from lead reacting with OPC is reported in previous work [15]. Hence novel calcium aluminate cement (CAC) and an alkali-activated metakaolin binder are used, which results in a reduction in the mobilisation of Pb. Associating CSH gel alkali solution of NaOH and metakaolin helps in the complete retention of Pb [16].

2.1.3 Different Factors Influencing Selection of Stabilisation Methods

Compaction

The soil density is a major factor affecting the intensity of the soil binding process. Higher dry density is achieved in the stabilised soil compared to untreated soil, and optimum moisture content rises as binder per cent increases. It is required to compact the soil as soon as binding agents are mixed with soil, as it starts to harden the very soil moment of addition [23, 50].

Studies reveal that delays in compaction cause more effort to achieve the same level of strength improvement. It also results in bond breakage and strength loss. Despite all the negative impacts, lime addition gains an advantage by delaying compaction as it requires some time to complete hydrolysis. For reduced plasticity in lime-mixed soil, re-compaction should be adopted.

Water Content

In soil stabilisation, sufficient water is necessary not only for the hydration of binders but also for effective compaction. Cement and lime take 20 and 32% of water to hydrate of their own weight, respectively. Insufficient water content results in the incomplete chemical reaction of binders and affects the final strength [17].

Temperature

Pozzolanic reactions show sensitivity towards temperature changes. In low temperatures, the reaction is slow and thus results in low strength; hence in cold regions, stabilisation is advised to be done in the warmer season [18].

3 Conclusion

As technological advancements progress, more chemicals and stabilising agents will be tested. Compatibility, durability, strength parameters, etc., are going to be improved by these novel methods. By going through 50 past works of researchers, the author collected some of the works in the area of selection for stabilisation techniques in accordance with the soil type, their properties, and inherent materials. Some factors which affect the choice of stabilisation are also discussed in this paper. Different papers suggest that,

1. The selection of binders should be based on compatibility between the soil and its stabilising agents. For example, allophanic soil requires binders that match soil properties. For this, a 3% lime + 3% fly ash mix (dry weight of soil) is found to be suitable. Only lime will provide strength but will revert to its original strength after some time, making it uneconomical.
2. When a wrong selection is made, the properties of binding agents can start unwanted reactions when chemically reacted with soil and inherent matters in soil. This is why the treatment of soil containing heavy metals is done through the solidification/stabilisation process, in which prevention of the mobilisation of heavy metals is the primary objective.
3. The proper selection of binders improves the shear strength parameter, compressibility, and solidification, but the wrong choice can lead to a reduction in structure stability, waste of resources, and risk of overbudgeting. For example, a sulphate-containing soil having a higher C3A-containing binder should be adopted. Standard OPC can lead to the mobilisation of heavy metals, thus resulting in poor strength parameters of the building. The effects of global warming and the changing environment can also affect the binder's efficiency. The impact of temperature decreases pozzolanic reaction and reduces the chances of binders reacting appropriately with the soil.
4. Different soil parameters are also found to be affected by applying different soil binders. Yield strength parameters are improved using alkali-activated MgO. Using NaOH as a precursor created an alkaline environment in the soil and

improved hydraulic resistivity for marine dredged soil. Compression strength is also enhanced by using MgO in aquatic dredged soil, Ca-Bentonite, and PFA improves allophanic soil's compressibility.

4 Future Scope

Going through a lot of previous research, it is highlighted that not enough studies are available on selection for stabilisation methods of different Indian soil having various inherent properties. This is a promising area which researchers should explore. Also, there is an immense opportunity to study the bitumen and other synthetic binding agents' behaviour under the temperature variation issue. So, the authors suggested intense in-situ research for all these novel binding agents. Also, the area for binder application in future works can be maximised so that economic feasibility can be studied and relevance in normal life can be justified.

Acknowledgements and Consent for Publication Not applicable.

Author's Contribution Each author contributed to the conceptualisation and design of the research. Himanshu Jangde and Farhan Khan handled the material preparation, data gathering, and analytic tasks. Himanshu drafted the first version of the text, while the second author, Farhan Khan, provided feedback on earlier draughts. All authors reviewed and approved the final version of the article.

Funding The authors state that they did not receive any funding, grants, or other support during the writing of this paper.

Availability of Data and Material Not Applicable.

Statement and Declaration All Authors have read, comprehended, and cooperated with the declaration on "Ethical Responsibilities of Authors" given in the instructions for authors, and are aware that, with limited exceptions, no changes to authorship may be made after the article has been submitted.

References

1. Source https://yamannvinci069.blogspot.com/2016/04/what-is-soil-in-geotechnical-engineering.html
2. Source https://lisbdnet.com/how-do-engineers-define-soil/
3. Bagriacik B (2021) Utilization of alkali-activated construction demolition waste for sandy soil improvement with large-scale laboratory experiments. Constr Build Mater 302. https://doi.org/10.1016/j.conbuildmat.2021.124173
4. Yong RN, Ouhadi VR (2007) Experimental study on instability of bases on natural and lime/cement-stabilized clayey soils. Appl Clay Sci 35(3):238–249. https://doi.org/10.1016/j.clay.2006.08.009

5. Liu P (2007) Polymer modified clay minerals: a review. Appl Clay Sci 38(1):64–76. https://doi.org/10.1016/j.clay.2007.01.004
6. Latifi N, Eisazadeh A, Marto A (2014) Strength behavior and microstructural characteristics of tropical laterite soil treated with sodium silicate-based liquid stabilizer. Environ Earth Sci 72(1):91–98. https://doi.org/10.1007/s12665-013-2939-1
7. Al-Mukhtar M, Lasledj A, Alcover J-F (2010) Behaviour and mineralogy changes in lime-treated expansive soil at 20°C. Appl Clay Sci 50(2):191–198. https://doi.org/10.1016/j.clay.2010.07.023
8. Khan F, Bhave HD, Rao T (2018) Classification of foundation soil: using geoinformatics (Gis). Int J Civ Eng Technol 9:1199–1207
9. Han J (2015) Principles and practices of ground improvement. Wiley, Hoboken, New Jersey
10. Makusa GP (2013) Soil stabilization methods and materials in engineering practice: state of the art review
11. Chang I et al (2020) Review on biopolymer-based soil treatment (BPST) technology in geotechnical engineering practices. Transp Geotech 24. https://doi.org/10.1016/j.trgeo.2020.100385
12. Karol RH (2003) Chemical grouting and soil stabilization, 1st edn. CRC Press. https://doi.org/10.1201/9780203911815
13. Blanck G, Cuisinier O, Masrouri F (2014) Soil treatment with organic non-traditional additives for the improvement of earthworks. Acta Geotech 9(6):1111–1122. https://doi.org/10.1007/s11440-013-0251-6
14. Horpibulsuk S et al (2010) Analysis of strength development in cement-stabilized silty clay from microstructural considerations. Constr Build Mater 24(10):2011–2021. https://doi.org/10.1016/j.conbuildmat.2010.03.011
15. Latifi N et al (2016) Strengthening montmorillonitic and kaolinitic clays using a calcium-based non-traditional additive: a micro-level study. Appl Clay Sci 132–133:182–193. https://doi.org/10.1016/j.clay.2016.06.004
16. Sharma NK, Swain SK, Sahoo UC (2012) Stabilization of a clayey soil with fly ash and lime: a micro level investigation. Geotech Geol Eng 30(5):1197–1205. https://doi.org/10.1007/s10706-012-9532-3
17. Naeini SA, Ziaie_Moayed R (2009) Effect of plasticity index and reinforcement on the CBR value of soft clay. Int J Civ Eng 7(2):124–130
18. Diaz Caselles L et al (2020) Stabilization of soils containing sulfates by using alternative hydraulic binders. Appl Geochem 113. https://doi.org/10.1016/j.apgeochem.2019.104494
19. Li S et al (2020) Evolution of heavy metals during thermal treatment of manure: a critical review and outlooks. Chemosphere 247:125962. https://doi.org/10.1016/j.chemosphere.2020.125962
20. Chen W et al (2020) A comprehensive evaluation of the treatment of lead in MSWI fly ash by the combined cement solidification and phosphate stabilization process. Waste Manage 114:107–114. https://doi.org/10.1016/j.wasman.2020.06.041
21. Wang J et al (2020) Comparison of cadmium uptake and transcriptional responses in roots reveal key transcripts from high and low-cadmium tolerance ryegrass cultivars. Ecotoxicol Environ Saf 203:110961. https://doi.org/10.1016/j.ecoenv.2020.110961
22. Kett IJ, Ingham J, Evans J (2010) Identifying an effective binder for the stabilisation of allophanic soils. Int J Pavement Eng 11(3):223–236. https://doi.org/10.1080/10298430903033347
23. Sherwood P (1993) Soil stabilization with cement and lime. State of the art review. Transport Research Laboratory, HMSO, London
24. Al Tabbaa A, Stegemann JA (eds) Stabilisation/solidification treatment and remediation. In: Proceedings of the international conference on stabilisation/solidification treatment and remediation, 1st edn, 12–13 April 2005, CRC Press, Cambridge, UK
25. Do TM et al (2018) Development of a new cementless binder for marine dredged soil stabilization: strength behavior, hydraulic resistance capacity, microstructural analysis, and environmental impact. Constr Build Mater 186:263–275. https://doi.org/10.1016/j.conbuildmat.2018.07.130

26. Wang D, Wang H, Wang X (2016) Compressibility and strength behavior of marine soils solidified with MgO—a green and low carbon binder. Mar Georesour Geotechnol 35(6):878–886. https://doi.org/10.1080/1064119x.2016.1258095
27. Gabr M, Bowders JJ (2000) Controlled low-strength material using fly ash and AMD sludge. J Hazard Mater 76(2–3):251–263. https://doi.org/10.1016/S0304-3894(00)00202-8
28. White G, Gnanendran C (2003) Variability in unconfined compression strength of pavement materials stabilised with cementitious binders. In: Transport: our highway to a sustainable future: proceedings of the 21st ARRB and 11th REAAA conference, May 2003, Cairns
29. Kett IJ (2005) Identifying an effective binder for the stabilisation of allophanic soils. Thesis (Master of Engineering (Civil)), University of Auckland
30. Almajed A, Khodadadi Tirkolaei H, Kavazanjian E Jr (2018) Baseline investigation on enzyme-induced calcium carbonate precipitation. J Geotech Geoenviron Eng 144. https://doi.org/10.1061/(ASCE)GT.1943-5606.0001973
31. Kayastha AM, Das N (1999) A simple laboratory experiment for teaching enzyme immobilization with urease and its application in blood urea estimation. Biochem Educ 27(2):114–117. https://doi.org/10.1016/S0307-4412(98)00231-3
32. Das N, Kayastha AM, Srivastava PK (2002) Purification and characterization of urease from dehusked pigeonpea (Cajanus cajan L.) seeds. Phytochemistry 61(5):513–521. https://doi.org/10.1016/S0031-9422(02)00270-4
33. Almajed A et al (2020) Enzyme-induced carbonate precipitation (EICP)-based methods for ecofriendly stabilization of different types of natural sands. J Clean Prod 274. https://doi.org/10.1016/j.jclepro.2020.122627
34. Andriesse J (1988) Nature and management of tropical peat soils. Food and Agriculture Organization (FAO) of United Nations, Rome
35. Harper KA, Kershaw GP (1997) Soil characteristics of 48-year-old borrow pits and vehicle tracks in shrub tundra along the CANOL No. 1 Pipeline Corridor, Northwest Territories, Canada. Arctic Alpine Res 29(1):105–111. https://doi.org/10.2307/1551840
36. Khan F, Das D, Dewangan N (2021) Determination of geotechnical properties and stability of expansive soil using fly ash. Walailak J Sci Technol (WJST) 18. https://doi.org/10.48048/wjst.2021.22783
37. Hebib S, Farrell ER (1999) Some experiences of stabilizing Irish organic soils. In: Proceeding of dry mix methods for deep soil stabilization. Balkema, Stockholm, pp 81–84
38. Hicks R (2002) Alaska soil stabilization design guide
39. Khan F et al (2021) A review on the feasibility and application of geospatial techniques in geotechnical engineering field. https://doi.org/10.1016/j.matpr.2021.02.108
40. Khan F, Das D (2021) Geospatial approach to determine Soil bearing capacity of Nagpur city, Maharashtra India.https://doi.org/10.1088/1755-1315/796/1/012069
41. Pokharel B, Siddiqua S (2021) Effect of calcium bentonite clay and fly ash on the stabilization of organic soil from Alberta, Canada. Eng Geol 293. https://doi.org/10.1016/j.enggeo.2021.106291
42. Saussaye L et al (2013) Influence of chloride and sulfate ions on the geotechnical properties of soils treated with hydraulic binders. Road Mater Pavement Des 14(3):551–569. https://doi.org/10.1080/14680629.2013.779303
43. Bernal SA et al (2014) Role of carbonates in the chemical evolution of sodium carbonate-activated slag binders. Mater Struct 48(3):517–529. https://doi.org/10.1617/s11527-014-0412-6
44. Contessi S et al (2020) Stabilization of lead contaminated soil with traditional and alternative binders. J Hazard Mater 382:120990. https://doi.org/10.1016/j.jhazmat.2019.120990
45. Mulligan CN, Yong RN, Gibbs BF (2001) Remediation technologies for metal-contaminated soils and groundwater: an evaluation. Eng Geol 60(1):193–207. https://doi.org/10.1016/S0013-7952(00)00101-0
46. Batchelor B (2006) Overview of waste stabilization with cement. Waste Manage 26(7):689–698. https://doi.org/10.1016/j.wasman.2006.01.020

47. Wang Y-S et al (2018) Influence of lead on stabilization/solidification by ordinary Portland cement and magnesium phosphate cement. Chemosphere 190:90–96. https://doi.org/10.1016/j.chemosphere.2017.09.114
48. Wang L et al (2018) Low-carbon and low-alkalinity stabilization/solidification of high-Pb contaminated soil. Chem Eng J 351:418–427. https://doi.org/10.1016/j.cej.2018.06.118
49. Guo B et al (2017) The mechanisms of heavy metal immobilization by cementitious material treatments and thermal treatments: a review. J Environ Manage 193:410–422. https://doi.org/10.1016/j.jenvman.2017.02.026
50. Afrin H (2017) A review on different types soil stabilization techniques. Int J Transp Eng Technol 3(2). https://doi.org/10.11648/j.ijtet.20170302.12

Analysis of Alternative Soil Binders and Their Effect on Soil: A Review

Himanshu Jangde, Farhan Khan, Mohammed Irshad Ansari, and Kaushal Prajapati

Abstract The traditional binders of soil have a negative influence on the environment, and scientists need to develop eco-friendly alternatives. In this research, we have chosen few suitable novel binders from prior studies and offer a comprehensive evaluation of them. Newly adopted binders, their composition, features, and potential to influence soil parameters have been reviewed in five major regions in this research. Bitumen with crumb rubber and the possible rejuvenation of bitumen-aged pavement with vegetable oil and waste frying oil are only a few of the strategies proposed within the study. Soil qualities may be affected by the origin and properties of Biopolymer based treatment (BPST) like guar gum, agar gum, xanthan, and Gellen gum, as well as enzyme-based carbonate precipitation (EICP). In addition, a strategy for enzyme stabilization caused by bacteria is described in this chapter. Stabilizing agents included construction demolition waste (CDW) and slag generated during steel making from iron in a basic oxygen furnace (BOFS), agricultural waste ash, bone ash, and lime, which are discussed in this article. Various soil properties such as plasticity, unconfined compressive strength,, shearing strength, stability of heavy metal-containing soil, bearing capacity, permeability, and CBR are examined and reported in this article concerning the study of combined effect of these innovative binders on soil. In general, the authors claim that most binders provide favorable outcomes and improve a variety of soil qualities. Authors found reduction in cracks, enhanced viscosity and cutting in project cost by use of vegetable

---

Statement of novelty

The objective of this paper was to collect data on a variety of innovative soil binders. The 96 previously published articles were read by the authors, and various binders were discussed. The amount and application of the soil binder are addressed. These binders' impact on soil quality is also explored.

The authors observed that most of the binder improves the soil's compressive strength, but not all binders increase other soil attributes. Also, some novel binders are associated with problems to apply. This research also discusses the impact of a particular soil feature on the behavior of the soil binder.

H. Jangde · F. Khan (✉) · M. I. Ansari · K. Prajapati
Department of Civil Engineering, Rungta College of Engineering and Technology, Bhilai, Chhattisgarh, India
e-mail: farhan1@rungta.ac.in

K. K. Pathak et al. (eds.), *Latest Developments in Civil Engineering*, Lecture Notes in Civil Engineering 352, https://doi.org/10.1007/978-981-99-2676-3_33

oil in pavement construction. EKO soil reduces plasticity on Ottawa soil. The study provides reduction in plasticity using different biopolymers. Agricultural-waste ash imparts compressive strength to the soil. Bone ash with lime (5% + 9%) provide better heavy metal stabilization. As a consequence of the usage of various novel binders, the authors attempt to offer a study on the change of various soil functions, which will give a better knowledge of these novel binders and are helpful in evaluating the influence of additional novel binders on soil functions.

Keywords Alternate binders · Stabilizing agent · Biopolymer based treatment (BPST) · Construction demolish waste (CDW) · Enzyme · Sustainable binder

1 Introduction

Stabilization process can be expressed as enhancing specific soil properties. For all subgrade projects, stabilization of the soil is a standard procedure [1, 2]. For a very long period, stabilizing procedures in poor soils have been practiced. Numerous ancient civilizations, including the Chinese, Greeks, and Romans, used diverse ways of stabilizing the soil [3]. For example, Before the Christian period, additives were used for route enhancement [4].

In 1920, when various limits were put on enterprises, paper mills used their waste as a dirt road. This paved the way for a new age of soil stabilization. The present period of soil stabilization began in the 1960s and 1970s when research revealed that the hardness of dirt roads was caused by the chemical interaction of the soil with waste materials [5]. Post 70s, there were crises for aggregate supply which forced engineers to deem unconventional options other than aggregate to achieve the required engineering characteristics [3]. Every engineering undertaking requires a solid foundation. Stabilization creates a strong, stable soil surface. The structure's weight is readily transferred to the ground through solid surfaces. Weak soil cannot do this. If soil is not stable, it will compress and swell when exposed to water, which will ultimately cause the whole structure to shift. Through pozzolanic reaction, the weak soil may be transformed into sufficiently solid and robust soil to support structural loads. Reducing plasticity with a stabilizing mechanism also gives oedema protection. Stabilization may also prevent wall buckles. On the other hand, numerous route subgrades are also stabilized to prevent future cracking and pavement deterioration [6, 7].

Current stabilization practices encompass a various of ways to enhance soil characteristics. The first technique to enhance soil qualities is pulverization, which produces a uniform environment without adding any additives. Adaptation in situ is preferable for pulverization. When FDR (full depth reclamation) is available, and it is guaranteed that the soil surface can be made solid enough to withstand traffic loads, pulverization is applied to the soil. The second strategy for enhancing the soil strength characteristics is the use of an additive. The most common additives are cement and lime, and the usage of cement has expanded substantially since its introduction. Additive stabilization is necessary, which aids in the enhancement of the required soil qualities. The use of additives was advantageous for reducing the moisture content of soil and particle cohesiveness [8]. Today Cement as well as lime are the two most common binders in use. Bitumen is also utilized to enhance the subgrade of paths. As a general rule, a plasticity index (PI) of 10 serves as the threshold for selection criteria, and this threshold limit is attained by cement, making it the most advantageous stabilizing component [4, 9]. Due to the production of $Ca(OH)_2$ upon contact with water, the calcium component of lime and cement combines with soil (five percent of the dry weight of soil) to boost its capacity to bear. The selection criteria for cement and lime are plasticity and CBR (California bearing ratio). When the CBR value is below six, lime is used as an additive. The slow rate of strength growth owing to carbonation provides long-term stability against leaching, swelling, and a weak subgrade. Cement, on the other hand, provides rapid stabilization. The heat of hydration created by the chemical interaction between the calcium content of cement and water forms potent, complex binding agents that adhere to the soil particles, providing considerable stability to the weak soil. In the past, only dry mix was used; however, now slurry mix with water is also favored. The third stabilization method is the employment of numerous chemicals to get superior outcomes. In this method, it is essential to put one layer of additive (either dry or wet) on the soil's surface prior to adding another addition in the slurry form [2, 5].

There are numerous factors that alter the charecteristics of a soil stabilizing agent and its ability to bind soil particles. To reduce these factors and maximize the utility of the binding agent, it is necessary to experiment with various stabilization agents that complement the soil's properties and enhance the intended purpose, i.e., stabilizing the soil [10–12]. Classical binding methods employ cement and lime, but cement production increases greenhouse gas emissions; in fact, cement production contributes 7% of CO_2 emissions. The rapid increase in demand made cement a more profitable business, causing industries to produce more and more cement without having regard for the environment. However, the cement's high strength is not always required in the project, also the carbonation of lime produces CO_2 emissions. Similarly, its slow rate of strength gain encourages individuals to seek alternatives to these traditional binders [13–15]. However, there is no need for the great strength provided by these agents in every situation. This prompted researchers to broaden their knowledge and hunt for unorthodox binding agents and techniques of soil stabilization.

This research aimed to provide a detailed evaluation of the unconventional materials employed for soil stabilization. In addition, a secondary objective is to describe the effects of various soil binders on key soil strength metrics.

2 Advancement on The Alternate Novel Binding Agent

People today use a variety of novel binders to strengthen the weak soil. This paper focuses primarily on the relatively novel third type of binding technique, i.e., mixing different admixtures. Here, some of the binders are categorized.

2.1 Alternate Binders for Which Bitumen is the Main Content

Bitumen is the primary component in this category of adhesives. Using bitumen as the foundation material, additives are blended with it to enhance its qualities and are then thoroughly combined with bitumen. Here, one of the bitumen-based new binders is described.

Asphalt amended with rubber grindings. The disposal of used tires is a major global issue. They are non-degradable, and chemical reactions involving these scraps may pose grave dangers. Instead of just disposing of them in a landfill, the rubber may be mixed with asphalt to increase the quality of the pavement [16–20]. This section aimed to shed light on the rheological characteristics of the asphalt binder that has been rubberized. CR (crumb rubber) diameters vary from 4.75 to 0.0075 mm and are used to construct roads. Tires are ground using both ambient and cryogenic processes. However, ambient grinding produces a rough surface on CR, which improves oil and resin absorption from asphalt owing to the increased surface area [19, 21–23]. Particle swelling and chemical degradation are the two primary mixing mechanisms between CR and asphalt [24, 25]. The inclusion of crumb rubber altered the viscosity of bitumen, which improved its workability. Additionally, it improves moisture resistance, thermal cracking, and compressibility.

Alternate binder by blending vegetable oil in bitumen. Previously, vegetable oil has been used as a construction material in Japan and various countries. However, a new research introduced a novel method by blending vegetable oil with bitumen. Basically, the authors rejuvenate the bitumen pavement by mixing the rejuvenating material (in this case, vegetable oil) through the hot mixing and grinding process in scrap collected from the old pavement. Recycling provides a cheap and better alternative to entirely rejuvenating the pavement layer. The deciding factor for this experiment is the viscosity and quality of aged bitumen. The effect of the blend of oil in bitumen is shown in Table 1 by comparing different test results on rejuvenated bitumen. Authors claimed that 4–12% of plant-based oil is sufficient for rejuvenating bitumen [26]. One of the concerning factors in using plant-based oil is the price. High oil prices will increase the project's cost, making it not feasible to adopt.

Alternative binders using bio-asphalt made from waste cooking oil. In addition to using vegetable oil as a rejuvenating agent, the cost of virgin plant oil is a major factor that plays a vital role. To mitigate this situation, a different research paper suggested a system of use of waste cooking oil (WCO) to rejuvenate bitumen. After frying food, industries dispose waste oil which causes problems to the environment.

Table 1 The impact of mixing virgin oil and bitumen on penetrability, softening threshold, and viscosity [26]

Oil content %	Penetration (mm)	Softening point (°C)	Viscosity			Temperature (°C)
			120 °C	150 °C	180 °C	
0	56	50.4	1.074	0.231	0.074	160
2	89	46.6	0.844	0.200	0.064	156
4	121	44.3	0.723	0.172	0.06	152
6	155	41.1	0.607	0.151	0.053	149
8	222	37.2	0.504	0.131	0.046	146
10	285	3.7	0.429	0.117	0.044	143

Table 2 Chemical charecteristics of waste cooking oil [27]

Type of free fatty acid	% in waste cooking oil
Oleic acid	43.67
Palmitic acid	38.35
Linoleic acid	11.39
Stearic acid	4.33
Myristic acid	1.03
ɣ- Linolenic acid	0.37
Lauric acid	0.34
Linolenic acid	0.29
Cis-11-Eicosenoic acid	0.16
Heneicosanoic acid	0.08
Total	100

However, the author suggested an alternate way to rejuvenate bitumen by applying this waste oil (properties of WCO are given in Table 2). The viscosity change is shown in Table 3. The application of WCO in this fashion shows positive results in managing it and provides a cheap alternative for rejuvenating bitumen pavement [27].

Table 3 Viscosity value based on different temperature [27]

Oil content (%)	Viscosity (Pa s)		
	120 °C	150 °C	180 °C
0	1.074	0.231	0.074
2	0.844	0.200	0.064
4	0.723	0.172	0.060
6	0.607	0.151	0.053
8	0.504	0.131	0.046
10	0.429	0.117	0.044

2.2 Enzyme-Based Alternative Binders

Enzymes as binders are a novel approach to stabilizing the soil. Bio-Enzymes are biological, organic, and liquid-concentrated chemicals that are utilized to improve the subbase integrity of flexible pavements. The clay particles are surrounded by water, bearing a positive charge (cations), and this water layer gives flexibility to the soil by employing the fermentation of certain microorganisms that offer a huge quantity of stabilizing enzymes. These enzymes help speed up reactions of cations exchange to stabilize the soil without being a part of the final product [28]. Some enzyme-based binders are given below.

Use of Eko soil for stabilizing red mud. The aluminium industry creates red mud as a waste product (SEM image is represented in Fig. 1). Red mud has a bright cherry red color with a high dry density and specific gravity. The chemical structure of red mud is indicated in Table 4. Additional features of Red mud are stated in Table 6 [29]. Eko soil employed in this work is an imitation of a natural binding agent, "termite saliva". It speeds the cations to react with clay particles to stabilize them. Eko soil properties are summarized in Table 5 [29]. The outcome of Eko soil stabilizing Red mud soil is discussed further in this work.

Fig. 1 SEM image of red mud [29]

Table 4 Chemical composition of red mud [29]

S.N	Constituents	Composition
1	Al_2O_3	20.10
2	$CaCO_3$	3.99
3	Fe_2O_3	34.21
4	Na_2O	5.0
5	SiO_2	7.8
6	TiO_2	15.5
7	Loss of ignition	11.0
8	Others	2.4

Table 5 Geotechnical property of Eko soil [29]

Properties	Value
Specific gravity	1.05
Boiling point	212 °F
Evaporation rate and vapor pressure	Same as water
Appearance and odor	Liquid, brown color, Slight ferment
Solubility in water	Infinite
pH	4–5.5

Table 6 Geotechnical properties of red mud [29]

Laboratory tests	Numeric value
Specific gravity	3.02
Gravel (%)	0
Sand (%)	8
Silt (%)	75
Clay (%)	17
Color	Red
Liquid limit (%)	45.5
Plastic limit (%)	32.4
Plasticity index (%)	13.01
Shrinkage limit (%)	2.78
Indian standard soil Classification	MI
Maximum dry density (%)	1.59
Optimum moisture Content (%)	33
CBR, soaked (%)	1.422
CBR, unsoaked (%)	6.219
UCS, (MPa)	0.143
Permeability	5.786
pH	11.3

Enzyme-induced carbonate precipitation (EICP). EICP is a urea-induced enzyme that precipitates carbonate inside soil pores by ureolyzing urea in the presence of salt. EICP is an eco-friendly method of soil stabilization. Free urea is present in agricultural resources such as jack beans, soybeans, pea plants, and watermelon seeds [30, 31]. Using this technique, a slurry was created by calcium chloride dehydrates ($CaCl_2.2H_2O$), dissolving urea, urease enzyme, jack bean meal, and nonfat dry milk in deionized (DI) water to stabilize the soil in Ottawa, which is then combined with sand. The outcomes of the mixing are elaborated over later in this study [32–34]. However, this single-phase method is high priced. Another study regarding

EICP proposed a method of multiphase EICP treatment for granular soil treatment. In this method cost driven urea is substituted with soyabean crude urea to catalyze the $CaCO_3$ via EICP to stabilize the soil. In one cycle of bio cementation, the percolation of soya crude urea is followed by the percolation of the cementing solution, which is repeated several times. This method improves production of carbonate precipitate by almost four times and also clogging of carbonate precipitation is limited in multiphase stabilization. The Unconfined compressive strength of the soil reached 10 MPa with the carbonate precipitation amount being 20% [35].

Bacterial-induced mineralization for soil solidification. The solidification and stabilization (s/s) method mainly adopt cement as its chief constituent. But the negative impact of cement (discussed earlier) made researchers find a novel cementless method called bacterial-induced mineralization (BIM). The basic concept of the method lies in the fact that microorganisms generate acid ions (CO_3^{2-}, PO_4^{3-}, SO_4^{2-}) [36]. These acid ions react with heavy metals within soil mass and help stabilize these [37, 38]. 10% of the binder content improves the UCS (unconfined compressive strength) of treated soil, but no significant improvement is noticed for a higher percentage (i.e., 10–30%). BIM reduces soil permeability, which helps increase effective stress [39]. Table 7 shows the use of multiple EPS-producing bacterial species against heavy metal contamination [40].

2.3 *Soil Remediation Based on Biopolymer*

Recently, the usage of biopolymer in soil stabilization has increased [53–55]. Biopolymers are polymers generated from natural assets including silk, marine bacteria, and Gellan gum [53, 56]. It is environmentally safe and extensively utilized in culinary and medicinal fields [57, 58]. Recently, the notion of reinforcing soil using biopolymer has gained widespread acceptance. It also aids in controlling permeability, preventing erosion, and mitigating dust [59, 60]. Common biopolymers include the following.

Agar gum. Agar gum is a linearly connected polysaccharide biopolymer composed of galactose molecules [61]. The chain structure of Agar gum is shown in Fig. 2. Agar gum's primary feature is the creation of reversible gels by chilling hot aqueous solutions without requiring any further chemical reaction [62]. Agar gum is often used as a thickening agent and food stabilizer. It is also used in medicine [63]. The dehydration of agar gum strengthens the soil and increases its compressive strength [64, 65].

Guar gum. Guar gum derived from the seeds of the leguminous plant Cyamopsis tetragonoloba is a member of the galactomannan family. In cold water systems, guar gum hydrates rapidly, resulting in high viscosity at low concentrations [67]. The chemical structure of guar gum is represented in Fig. 3. It is recommended to supervise the application of guar gum for 2 h since uncontrolled application might actually reduce the viscosity [57]. This phenomenon happened due to guar gum's

Table 7 Utilization of EPS containing diverse bacterial species against heavy metal contamination [40]

Species	Heavy metals	Initial concentration (in ppm)	PH	Temperature (°C)	Contact time (in hours)	Remediation efficiency (in percentage)	References
Paenibacillus jamilae	Pb	303.03	–	25	24	70	[41]
Azotobacter chroococcum XU1	Pb	33.5 of Pb	4.4	25	–	40.48	[42]
	Hg	38.9 of Hg	4	25		47.87	
Enterobacter cloacae	Cr	8.3	–	–	80	60–70	[43]
Paenibacillus polymyxa	Cu, Pb	111.11	5	30	2	90	[44]
Ochrobactrum sp. HG16	Hg	5	–	30	48	30	[45]
Serratia marcescens HG19	Hg	2	–	30	48	90	[46]
Alteromonas macleodii subsp. fijiensis	Pb	75	5	37	3	55	[47]
Alcaligenes feacalis	Cd	10 of Cd	7	30	72 of Cd	70 of Cd	[48]
	Pb	10 of Pb			98 of Pb	98 of Pb	
Burkholderia cenocepalia	Cd	100	-	30	48	60	[49]
Bacillus cereus KMS3-1	Cd, Cu, Pb, Zn	5	7	27	48	–	[50]
Bacillus licheniformes	Cr	200	7.4	37	24	–	[51]
Bacillus sp. S3	Cd	20	5	28	7	–	[52]
	Cr	150					
	Cu, Pb	150					
	Sb	150					

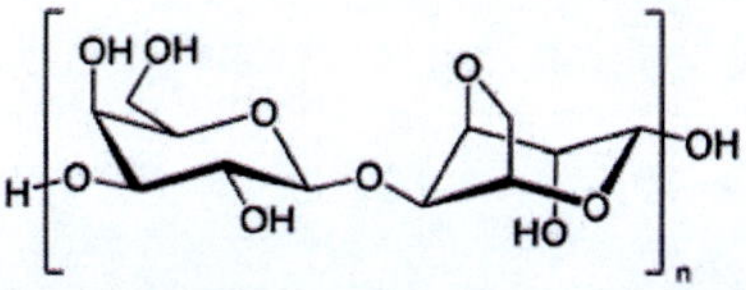

Fig. 2 Agarose chain structure of agar gum [66]

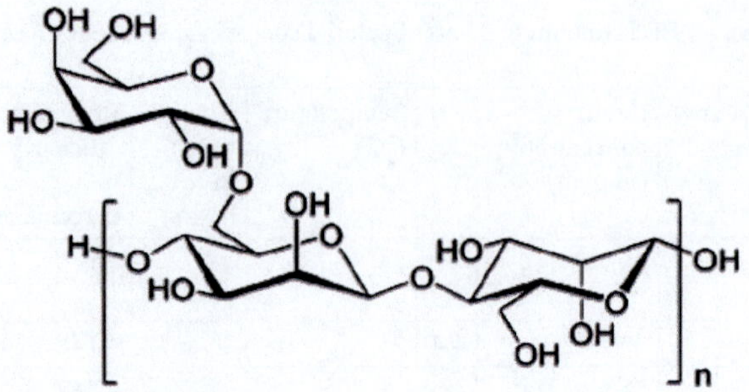

Fig. 3 Chemical structure of guar gum

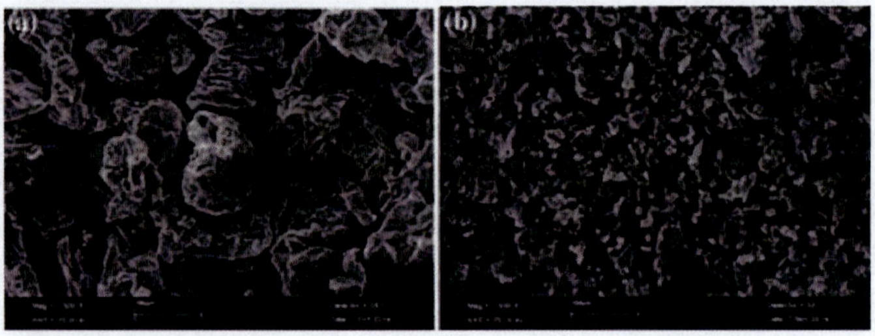

Fig. 4 SEM images of guar gum in **a** Gelation process **b** Gel breaking process [57]

gelation and gel-breaking processes. Figure 4 shows SEM pictures of both processes, respectively.

Gellen Gum. Gellen gum has a viscoelastic, cation-dependent gelation behavior. It is made from Sphigomonas elodea through microbial fermentation. Temperature variation changes its structure from double-helical strands to single helical strands due to low to high temperatures, respectively [68, 69]. Application of gellen gum in soil stabilization improves the soil strength significantly due to the formation of firm hydrogels, providing clogging effects in cohesion less soil, and helping in water prevention. Other than that, the application of gellen gum shows greater resistance against wetting and drying cycles [53].

Xanthan gum. This has a helical structure made of Xanthomonas campestries shown in Fig. 5, Xantham gum shows a linear increase in viscosity with Xanthan gum content. Xanthan gum solution provides high stability against temperature, and pH. From its rheological characteristics, Xanthan gum application improves UCS and provides economic feasibility [70–80].

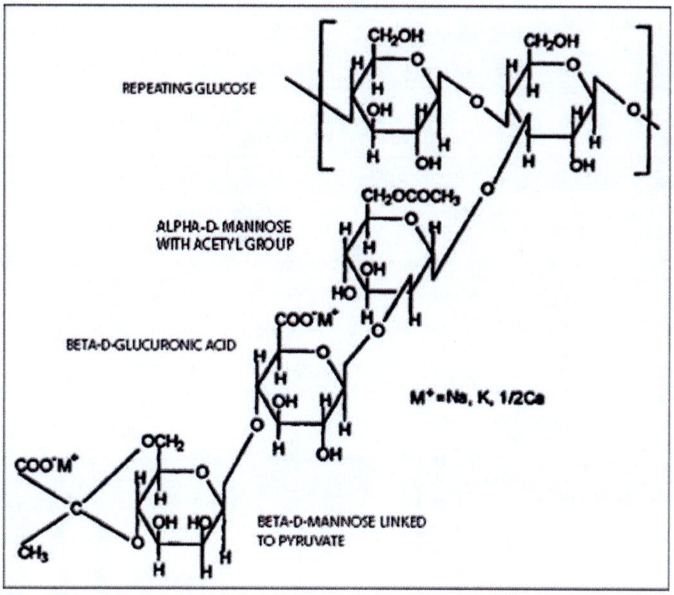

Fig. 5 Chain structure of Xanthan gum [35]

2.4 *Creating Alternatives from Waste Materials*

Various processes generate waste, which is harmful to the environment. Some biodegradable garbage only causes short-term annoyance, but industrial waste has severe effects on the environment. Mixed with the soil's water, trash containing heavy metals creates unfavorable conditions. It is crucial to managing garbage from many sources effectively. Utilization of industrial waste is a strategy for waste management, although chemical neutralization is required before use. Listed below is one instance in which garbage has been employed for soil stabilization.

Alkali-activated demolition waste (CDW) as a soil binder. Different types of civil engineering applications generate demolition trash in the form of bricks, tiles, steel components, concrete fragments, ceramic glass, wood, plastic, etc. Approximately 27% of Europe's landfill garbage is produced by CDW [81, 82]. The sand used in this instance (properties given in Table 8 and sieve analysis distribution displayed in Fig. 6) is stabilized with alkaline activated CDW. For alkaline media, the molarities of NaOH, KOH, and CaO are 10, 10, and 6 M, respectively, and 16% CDW was determined to be ideal for stabilizing sand in that study [83]. Figure 7 depicts the CDW ratio with optimum cohesiveness.

Different Agro Waste Binders. Agro waste ashes (AWA) are used to form alkali-activated concrete, which have (SiO2) and Al_2O_3 in sufficient amount. bagasse (BA),Rice husk ash (RHA), palm oil fuel ash (POFA), corn cob ash (CCA), wheat straw ash (WSA), and sugarcane straw ash (SSA) are used to analyze the potential

Table 8 Chemical analysis result for sand [83]

Component	SS (%wt)	CDW (%wt)
SiO_2	96.15	19.2
Al_2O_3	2.10	8.7
Fe_2O	0.6	–
Fe_2O_3	–	5.6
MgO	0.04	4.3
Cao	0.22	58.4
K_2O	0.39	0.8
Na_2	0.07	1.2
TiO_2	–	0.4
P_2O_5	–	0.3

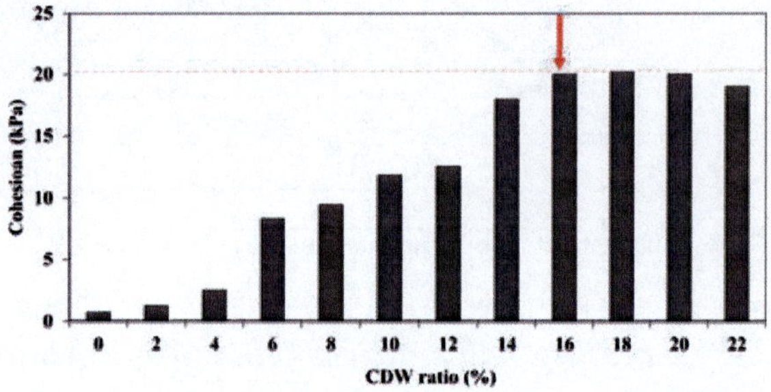

Fig. 6 Cohesions for various mixing proportions [83]

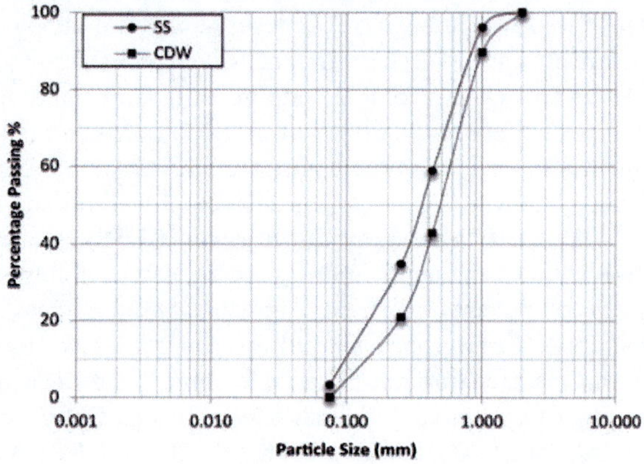

Fig. 7 Sieve analysis for sandy soil (SS) and construction demolition waste (CDW) [83]

of AWA as a binder. Nowadays, fly ash (FA) is considered very useful as it has a high silica content and is also cheap compared to all ashes. All ashes have the same amount of silica as fly ash except rice husk ash. Bagasse ash shows the least specific gravity, whereas all other ashes have a range of specific gravity situated between 2.1

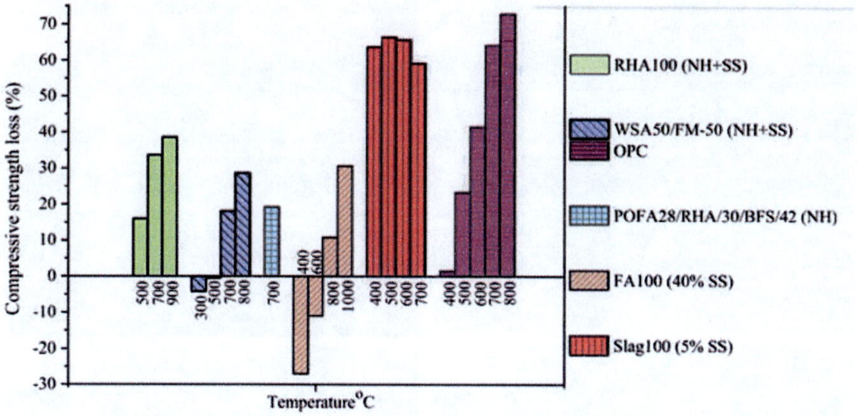

Fig. 8 Preformance of agro-waste ashes-based AAB specimens exposed to elevated temperature [66]

and 2.7. Different microstructures are found for BA, WSA, and SSA (SEM images are shown in Fig. 9). For higher molarity, efflorescence is observed. However, it is found that the higher molarity of BA, RHA, FLY ash, and slag, i.e., 8 and 14 M, increases the compressive strength. For a temperature of more than 600 °C, silica fuses and causes a swell of the specimen also causing a reduction in strength [66, 84–89]. Performance against compression of agro-waste under temperature increment is shown in Fig. 8.

2.5 *Sustainable Alternative to Soil Binders*

The solidification of soil polluted with heavy metals is an essential technique, and the use of OPC is quite widespread [90–93]. However, the continuation of this practice results in releasing a significant quantity of greenhouse gases using a significant quantity of energy [93]. As a result, a sustainable binder agent is necessary, one that not only plays a critical role as a stabilizing agent but also prevents any malfunctions that might result in the production of greenhouse gases.

Basic oxygen furnace slag (BOFS). During the transformation of iron into steel, a byproduct known as BOFS is created in the basic oxygen furnace. The quantity of BOFS accounts for about 10–15 percent of basic oxygen steel production [94]. The fundamental mineral that may be found in BOFS is di-calcium silicate and tri-calcium silicate, which is one of the primary constituents in the Portland cement that increases strength. The use of BOFS as a stabilizing material may thus help alleviate the issue of where to dispose off the 400 million tons of steel slag that are produced annually [93, 95]. In actual fact, it has been discovered that the operation of BOFS in high temperatures throughout the steel-making process results in a gradual

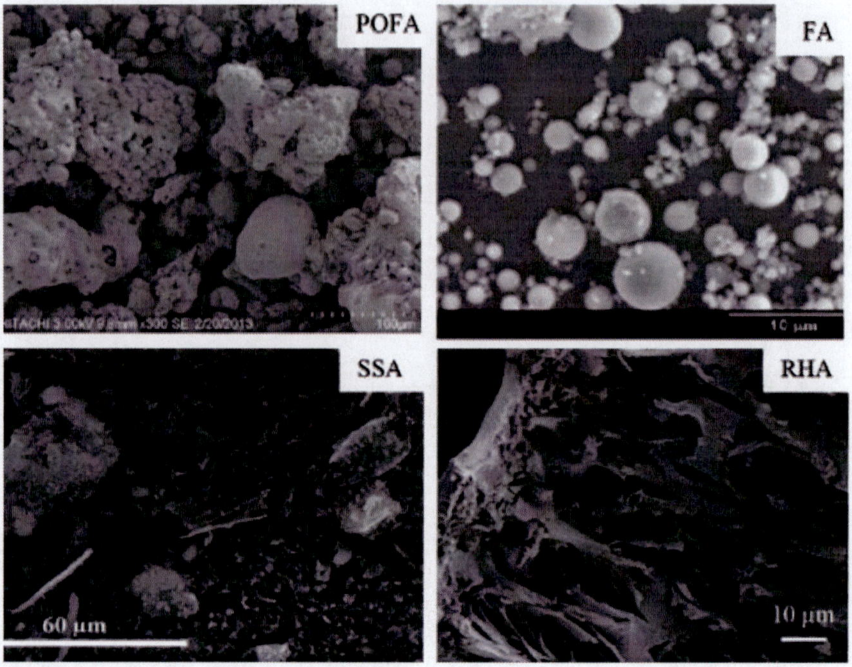

Fig. 9 SEM images of different binder used to stabilize [66]

increase in the material's strength during its early stages. Small quantities of calcium carbonate ($CaCO_3$) and silicon dioxide (SiO_2) are combined with calcium carbide residue that contains calcium hydroxide ($Ca(OH)_2$) in order to accelerate the reaction [91, 93, 95]. The hydration product C–S–H gel ettringite (Aft) portlandite improves the soil strength. Additionally, metal-bearing hydroxide, the absorptivity of C–S–H, the encapsulation of heavy metals by C–S–H, and the ion exchange of heavy metals with AFt can result in the immobilization of heavy metals in the solidified/stabilized soil matrix [90, 91, 93]. It can be seen in the SEM picture of the BOFS binder on how the BOFS binder absorbs various heavy metals. Figure 10 demonstrates the different types of absorbed heavy metal, which ultimately led to heavy metal stabilization and immobilization [93]. In addition, Fig. 11 displays X-ray diffractograms of BCP binder pastes that have been spiked with various ions that are found in contaminated soil.

Use of lateritic soil as building material by improving its characteristics. Different industries producing different building materials cause greenhouse gas emissions. To avoid this situation, one can find an alternative by improving the characteristics of locally available building materials and preferring that material. Obianyo et al. tried to improve the characteristics of lateritic soil by using various percentages of lime and bone ash (the chemical composition is shown in Table 9). The author also studied the effect of different curing methods on the soil sample. The

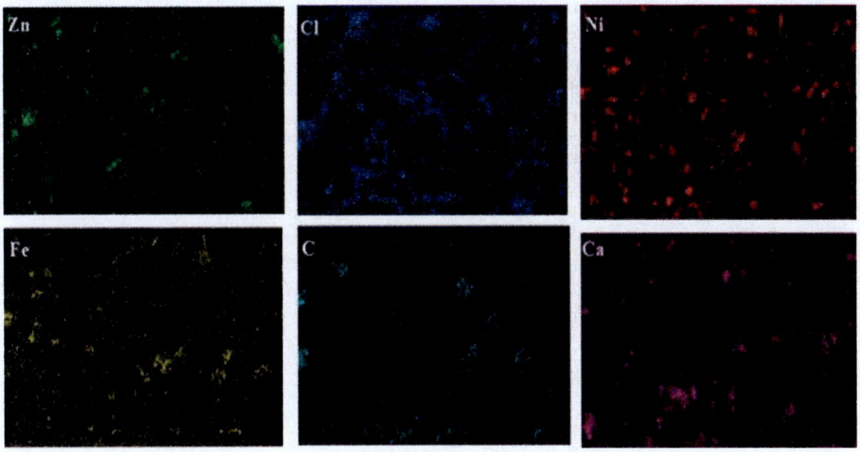

Fig. 10 SEM images of various heavy metals like Zn, Cl, Ni, Fe, C, Ca, absorbed through BCP binder [93]

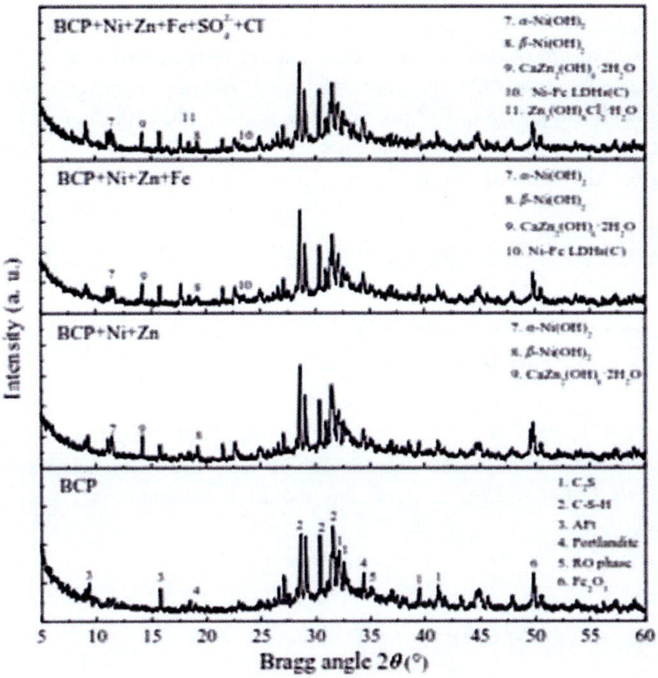

Fig. 11 X ray diffractograms of BOFS binder pastes spiked with different ions contained in contaminated soil [93]

Table 9 Chemical composition of bone ash used for the study [35]

Chemical compositions (%)	Percentages
Calcium oxide (CaO)	52.2020
Magnesium oxide (MgO)	2.0770
Aluminium Oxide (Al_2O_3)	0.6130
Phosphorus Oxide (P_2O_5)	48.0770
Sodium Oxide (Na_2O)	1.3290
Potassium Oxide (K_2O)	0.0907
Strontium oxide (SrO)	0.8670
Iron oxide (Fe_2O3)	0.0303
Zinc oxide (ZnO)	0.0069
Sulphur (S)	0.2124
Chloride (Cl)	0.3400
Chromium oxide (Cr_2O_3)	0.0011
Nickel oxide (Ni_2O)	0.0003
Manganese oxide (MnO)	0.0002

research provided the outcome of that per cent combination of (9% + 5%) of lime and bone glue respectively and was found to be very useful for strength improvement. Also, the morphology of reduced porosity is achieved after 28 days of curing [35]. In this study, the authors observed that only cattle bone ash was used. Therefore, the authors suggest that a set of different animal bones would be used, and a study can be performed on whether different bone structures impact the strength improvements.

3 Changes In Soil Parameters Resulting From the Use of Different Soil Binders

The authors in this section tried to compile the effects of using these binders under different soil parameters as subcategories which were influenced. Different binders influence the different soil parameters, and under these categories, collective effects are reported.

3.1 Compressive Strength

The majority of the binders discussed in this article improve the unconfined compressive strength of the different soils. The unconfined shear strength of sandy soil may be increased by using NaOH as an activator to break down particles [83]. When a BCP binder is used to stabilize soil, the compressive strength of the soil which is being stabilized is significantly improved [93]. The rise in UCS may be attributed to the electrostatic contact that occurs between the soil and the biopolymer. The improvement by guar gum was twelve percent, whereas the improvement by agar gum was just three percent [6] When applied to fine soils, BPST demonstrates greater efficacy. The increase in compressive strength that may be attributed to EICP for the sand subjected can be as high as 700 kPa [33]. BIM in little portion improves the UCS, but a higher percentage of BIM does not impact UCS [39]. Using agro-waste also increases the shear compressive strength of soil when used in high molarity i.e., 8–10 M [66, 85–89].

3.2 *Impact on Plasticity*

When using the hybrid technique of EICP, adding 8%, lime will increase the plasticity index; however, adding more will have a detrimental effect on Ottawa sand [33]. When added to the sample soil, i.e., red mud, Eko soil has the effect of raising the plastic limit while simultaneously lowering the liquid limit [29]. Additionally, BPST causes a reduction in the soil plasticity [6].

3.3 *Capacity to Support Loads*

Because of the CSH gel and ettringite mineral in the BCP binder, the geoenvironmental properties of the soil are altered, which ultimately leads to an increase in the bearing capacity of the subjected soil [93]. Also, demolition debris that has been treated with alkaline solutions has a greater capacity for bearing loads. NaOH initiates the hydration process that provides CDW with its higher bearing capacity at the optimal 16% of the dry weight of sand [83]. When applied on Ottawa soil at a concentration of one percent, BPST results in a bearing capacity that is ten percent higher than that of the application of cement as a binder [6]. In addition, the use of Gellen gum imparts bearing strength to the soil mass [53]. Also, the study indicates that a dry mix of lime with bone ash (9% + 5%) is said to be optimum for increasing bearing strength [35].

3.4 *Shear Strength*

Shear strength of soil mass refers to the maximum amount of load that a soil mass can safely bear before it experiences a failure. It is an essential component in the planning and design of any building. Clay and sand both benefit from better apparent cohesiveness when BPST is added, and this benefit is independent of the water content. The shear resistance of sand may be improved by using agar gum and guar gum [6]. Also, in the instance of Eko soil, the addition of four percent of it and three to four percent of lime increases the shear strength of the soil that has been exposed to stress [29].

3.5 *Heavy Metal Stability*

The relative binding intensity index (IR) is used to determine the amount of BCP that must be used in order to make heavy metals more chemically stable. In addition, infrared may be used to measure the mobility of heavy metals [96, 97]. The content of the heavy metals Ni and Zn is reduced when the curing period of the BCP is increased. On a scale that is semi-logarithmic, IR reveals a linear relationship with heavy metals. The equations that describe the relationship are as follows:

$$\mathrm{logCNi(n)} = \mathrm{aIR(Ni)} + \mathrm{b} \tag{1}$$

$$\mathrm{logCZn(n)} = \mathrm{cIR(Zn)} + \mathrm{d} \tag{2}$$

where the content of nickel or zinc is denoted by the symbol CNi or CZn, respectively. n may be either 1 or 2, and relative binding is denoted by IR (Ni) and IR(Zn) [93]. The stabilization of the red mud by Eko soil (ES) lowers the concentration of heavy metals, bringing them to a level that is only slightly higher than the one recommended by the World Health Organization (WHO) for the use of water. The quantity of heavy metals in red mud that contains 4 percent ES is very low [29]. BIM reacts with heavy metal presented in contaminated soil and reduces the concentration of contamination of heavy metal in soil [37, 38].

3.6 *An Increase in Both CBR and Permeability*

Adding 4% Eko soil, along with gypsum and lime, may raise the CBR value of red mud by 274.87% when it is unsoaked, and it can rise to 854.20 percent when it is soaked. An increase in the ES content will have the effect of decreasing both the permeability and the leachability. However, adding BIM to soil mass reduces permeability and increases effective stress [39].

4 Conclusion

The use of conventional binders enables us to achieve appropriate stability for problematic soil. However, the adverse consequences brought about by these binders drive scientists to continue their research and make efforts to minimize and lessen these impacts. For this reason, research into new methods that stabilize problematic soil is taking a considerable amount of time at this point. These binders' most important characteristics should be their efficacy, low cost, and the fact that they should not harm the environment. The authors present a review on new binding agents by examining and evaluating 96 previous articles. These novel binding agents have been evaluated for some time and are being used in various parts of the globe. Authors provide a summary of a variety of alternative binders, including their characteristics, composition, and the influence that they have on various soil parameters. The following is a list of this review's findings:

- An increase in unconfined strength is often associated with the use of almost any innovative binder.
- Asphalt mixed with crumb rubber modifies the property of the pavement. A blending of vegetable oil helps decreases cracking and provides enhanced viscosity. Also, a study on waste oil is conducted, which helps reduce the project cost and helps rejuvenate bitumen.
- For Ottawa sand, reducing the plasticity of the EICP with 8% lime will result in an increased plastic index. In addition, Eko soil and BPST contribute to the reduction of plasticity.
- Not only does adding crumb rubber to asphalt make it more resistant to water damage, but it also makes it less prone to cracking from compression, heat, and other types of damage.
- Eko soil and EICP are both enzyme-based stabilizers that help increase compressive strength, decrease flexibility, and impart cohesion. Both of these properties are desirable in a stabilizer. Heavy metals may be more easily immobilized with the aid of Eko soil. Additionally, bearing strength, shear strength, and CBR improved significantly due to the usage of Eko soil.
- use of bacterial-induced mineralization solidification and stabilization method increase compressive strength and a reduction in plasticity achieved. Also, heavy metal ions react with these microbes, reduce the concentration of heavy metal and help in the immobilization of the heavy metal. Permeability is reduced using these binders, which eventually helps in an increase in effective stress.
- use of different biopolymers helps in enhancing soil characteristics. For example, Xanthan gum, Gallen gum, and gaur gum enhance the bearing strength of the soil when used as a binder. They reduce plasticity and increase shear resistance.
- use of agro waste ash in higher molarity cause efflorescence. However, the higher molarity of these agro-waste ash imparts compressive strength to the soil mass. However, a higher temperature (600 °C) reduces the strength.
- The use of waste from building and demolition at a rate of around 16 percent (the ideal dosage) may increase bearing capacity; however, this waste must be activated first by any alkaline medium.
- bone ash with lime in (5% + 9%) respectively provide the better heavy metal stabilization using solidification and stabilization (s/s) technique. It reacts with heavy metal ions, traps them, and does not allow them to mobilize.

5 Future Scope

Although the use of new binders provides access to a vast array of opportunities, some topics for further research are outlined below.

5.1 *Shoulder Strengthening*

The road pavement construction not only comprises the route subgrade which holds the weight of cars but the shoulder alongside road pavement should be robust enough to bear traffic load also. Scope for a new stabilizing agent may be taken in mind so naturally accessible boulders can be employed successfully for the building of pavement shoulders with better bearing strength.

5.2 *Effect of These Binders on the Agricultural Sector*

Increased use of conventional chemical binders enters into the environmental cycle. Chemicals may flow and can migrate into agricultural soil which finally enters our food chain. Hence the influence of these innovative binders on agricultural output and their effect on the food chain should be researched.

5.3 *Effect of Different Oil Types on Rejuvenation of Asphalt*

In the paper of Wan Nur Aifa Wan Azahara et al. [27] the potential of waste oil is described, but the authors suggest that the effect of different oil, i.e., palm oil, groundnut oil, olive oil, mustard oil, etc., should be taken into consideration, and the effect on rejuvenation of bitumen pavement using all these waste fried oils also be analyzed.

5.4 *Different Types of Bone from Different Animal*

In the paper of Obianyo et al. [35], cattle bone ash is used for stabilization. The authors suggest the use of different bone sets for stabilization. Different animal bones can be used for stabilization, and results can be compared.

Acknowledgements and consent for publication: Not applicable.

Statement and Declaration All Authors have read, comprehended, and cooperated with the declaration on "Ethical Responsibilities of Authors" given in the instructions for authors, and are aware that, with limited exceptions, no changes to authorship may be made after the article has been submitted.

AuthorContribution Author's contribution Each author contributed to the conceptualization and design of the research. Himanshu Jangde and Farhan Khan handled the material preparation, data gathering, and analytic tasks. Himanshu drafted the first version of the text, while the second author, Farhan Khan, provided feedback on earlier draughts. All authors reviewed and approved the final version of the article.

Availability of data and material Not Applicable.

Funding The authors state that they did not receive any funding, grants, or other support during the writing of this paper.

References

1. Anjan P (2019) 3–soil stabilization, in geotechnical investigations and improvement of ground conditions. In: Anjan P (ed). Woodhead Publishing, pp 19–27
2. Torralvo FA, Pereira CF, Piqueras OF (2017) By-products from the integrated gas combined cycle in IGCC systems. https://doi.org/10.1016/B978-0-08-100167-7.00014-7
3. Source: https://www.trstabilisation.co.uk/History-of-Soil-Stabilisation.html
4. Kowalski TE, Starry DW, America JW (2007) Modern soil stabilization techniques
5. Source: https://roadpackersolutions.com/the-history-of-modern-soil-stabilization/#:~:text=The%20beginning%20of%20modern%20(chemical,during%20the%20expanding%20industrial%20era
6. Chang I et al (2020) Review on biopolymer-based soil treatment (BPST) technology in geotechnical engineering practices. Transp Geotech 24. https://doi.org/10.1016/j.trgeo.2020.100385
7. Source: https://mintekresources.com/the-importance-of-soil-stabilization/.
8. Addo JQ, Thomas PMC, Sanders G (2004) Road dust suppression: effect on maintenance stability. Saf Environ Phases 1–3
9. Afrin H (2017) A review on different types soil stabilization techniques. Int J Transp Eng Technol 3(2). https://doi.org/10.11648/j.ijtet.20170302.12
10. source: https://www.earthlok.com/blog/the-importance-of-soil-stabilization/
11. Beckham TL, Hopkins TC (1997) Stabilization of subgrade soil using hydrated lime product
12. Khan F, Das D (2021) Geospatial approach to determine Soil bearing capacity of Nagpur city, Maharashtra India
13. Jafer HM et al (2018) Development of a new ternary blended cementitious binder produced from waste materials for use in soft soil stabilisation. J Clean Prod 172:516–528. https://doi.org/10.1016/j.jclepro.2017.10.233
14. Karim MR et al (2013) Fabrication of a non-cement binder using slag, palm oil fuel ash and rice husk ash with sodium hydroxide. Constr Build Mater 49:894–902. https://doi.org/10.1016/j.conbuildmat.2013.08.077
15. Specht E, Redemann T, Lorenz N (2016) Simplified mathematical model for calculating global warming through anthropogenic CO_2. Int J Therm Sci 102:1–8. https://doi.org/10.1016/j.ijthermalsci.2015.10.039
16. Karger-Kocsis J, Mészáros L, Bárány T (2013) Ground tyre rubber (GTR) in thermoplastics, thermosets, and rubbers. J Mater Sci 48(1):1–38. https://doi.org/10.1007/s10853-012-6564-2

17. Uriarte-Miranda M-L et al (2018) Reverse logistic strategy for the management of tire waste in Mexico and Russia: review and conceptual model 10(10):3398
18. Asaro L et al (2018) Recycling of rubber wastes by devulcanization. Resour Conserv Recycl 133:250–262. https://doi.org/10.1016/j.resconrec.2018.02.016
19. Sienkiewicz M et al (2012) Progress in used tyres management in the European Union: a review. Waste Manage 32(10):1742–1751. https://doi.org/10.1016/j.wasman.2012.05.010
20. Zheng W et al (2021) A review on compatibility between crumb rubber and asphalt binder. Constr Build Mater 297. https://doi.org/10.1016/j.conbuildmat.2021.123820
21. Hassan NA et al (2014) A review of crumb rubber modification in dry mixed rubberised asphalt mixtures 70(4)
22. Blumenthal M, Producing ground scrap tire rubber: a comparison between ambient and cryogenic technologies. (1996) American society of mechanical engineers. NY (United States), New York
23. Xiang Y et al (2019) Ultraviolet irradiation of crumb rubber on mechanical performance and mechanism of rubberised asphalt 20(7):1624–1637
24. Jamrah A, Kutay ME, Varma SJTRR (2015) Backcalculation of swollen crumb rubber modulus in asphalt rubber binder and its relation to performance 2505(1):99–107
25. Wang S, Cheng D, Xiao F (2017) Recent developments in the application of chemical approaches to rubberized asphalt. Constr Build Mater 131:101–113. https://doi.org/10.1016/j.conbuildmat.2016.11.077
26. Helen K, Bailey, PP (2009) ASPHALT REJUVENATION, U.S.P.A. Publication, Editor: United States
27. Azahara WNAW, Bujang M, Jaya RP, Hainin MR, Mohamed A, Ngadi N, Jayanti DS (2016) The potential of waste cooking oil as bio-asphalt for alternative binder—an overview. Penerbit UTM Press, pp 111–116
28. Ravi Shankar AU, Rai HK, Mithanthaya R (2009) Bio-enzyme stabilized lateritic soil as a highway material
29. Kushwaha SS et al (2018) Stabilization of Red mud using eko soil enzyme for highway embankment. Mater Today: Proceed 5(9):20500–20512. https://doi.org/10.1016/j.matpr.2018.06.427
30. Das N, Kayastha AM, Srivastava PK (2002) Purification and characterization of urease from dehusked pigeonpea (Cajanus cajan L.) seeds. Phytochemistry 61(5):513–521. https://doi.org/10.1016/S0031-9422(02)00270-4
31. Kayastha AM, Das N (1999) A simple laboratory experiment for teaching enzyme immobilization with urease and its application in blood urea estimation. Biochem Educ 27(2):114–117. https://doi.org/10.1016/S0307-4412(98)00231-3
32. Pooni J et al (2021) Mechanism of enzyme stabilization for expansive soils using mechanical and microstructural investigation 21(10):04021191. https://doi.org/10.1061/(ASCE)GM.1943-5622.0002164
33. Almajed A et al (2020) Enzyme-induced carbonate precipitation (EICP)-based methods for ecofriendly stabilization of different types of natural sands. J Clean Prod 274. https://doi.org/10.1016/j.jclepro.2020.122627
34. Almajed A, Khodadadi Tirkolaei H, Baseline K (2018) Investigation on enzyme-induced calcium carbonate precipitation. J Geotech Geoenviron Eng 144. https://doi.org/10.1061/(ASCE)GT.1943-5606.0001973
35. Hao M et al (2021) Multiple-phase enzyme-induced carbonate precipitation (EICP) method for soil improvement. Eng Geol 294:106374. https://doi.org/10.1016/j.enggeo.2021.106374
36. Mohapatra S et al (2011) Development and evaluation of enzymatically triggered multiparticulate colon targeted drug delivery system. Int J Res Ayu Pharm 2:211–215
37. Anbu P, Kang CH, Shin YJ, So JS (2016) Formations of calcium carbonate minerals by bacteria and its multiple applications. Springer Plus 5(1):1–26. https://doi.org/10.1186/s40064-016-1869-2
38. Choi SG, Chang IH, Lee MH, Lee JH, Han JT, Kwon TH (2020) Review on geotechnical engineering properties of sands treated by microbially induced calcium carbonate precipitation

(MICP) and biopolymers. Constr Build Mater 246:118415. https://doi.org/10.1016/j.conbuildmat.2020.118415
39. Han L, Li J, Xue Q et al (2020) Bacterial-induced mineralization (BIM) for soil solidification and heavy metal stabilization: a critical review. Sci Total Environ. https://doi.org/10.1016/j.scitotenv.2020.140967
40. Mahapatra B et al (2020) Application of bacterial extracellular polymeric substances for detoxification of heavy metals from contaminated environment: a mini-review. Mater Today: Proceed 30:283–288. https://doi.org/10.1016/j.matpr.2020.01.490
41. Morillo JA, Aguilera M, Ramos-Cormenzana A, Monteoliva-Sánchez M (2006) Curr Microbiol 53:189–193
42. Rasulov BA, Yili A, Aisa HA (2013) J Environ Prot (Irvine Calif) 04:989–993
43. Iyer A, Mody K, Jha B (2004) Mar Pollut Bull 49:974–977
44. Mokaddem H, Azouaou N, Kaci Y, Sadaoui Z (2014) Chem Eng Trans 38:31–36
45. François F, Lombard C, Guigner J-M, Soreau P, Brian-Jaisson F, Martino G, Vandervennet M, Garcia D, Molinier A-L, Pignol D, Peduzzi J, Zirah S, Rebuffat S (2012) Appl Environ Microbiol 78:1097–1106
46. Loaëc M, Olier R, Guezennec J (1997) Water Res 31:1171–1179
47. De J, Ramaiah N, Vardanyan L (2008) Mar Biotechnol 10:471–477
48. Wang X, Zhang X, Liu X, Huang Z, Niu S, Xu T, Zeng J, Li H, Wang T, Gao Y, Huang M, Cao L, Zhu Y (2019) Metallomics 11:1252–1264
49. Krishnamurthy M, Jayaraman Uthaya C, Thangavel M, Annadurai V, Rajendran R, Gurusamy A (2020) Carbohydr Polym 227:115369
50. Kavitha V, Radhakrishnan N, Gnanamani A, Mandal AB (2011) Biol Med 3:16–26
51. Zeng W, Li F, Wu C, Yu R, Wu X, Shen L, Liu Y, Qiu G, Li J (2020) Bioprocess Biosyst Eng 43:153–167
52. Zeng W, Li F, Wu C, Yu R, Wu X, Shen L, Liu Y, Qiu G, Li J (2020) Role of extracellular polymeric substance (EPS) in toxicity response of soil bacteria Bacillus sp. S3 to multiple heavy metals. Bioprocess Biosyst Eng 43:153–167. https://doi.org/10.1007/s00449-019-02213-7
53. Chang I, Im J, Cho G-C (2016) Introduction of microbial biopolymers in soil treatment for future environmentally-friendly and sustainable geotechnical engineering 8(3):251
54. Lentz RD (2015) Polyacrylamide and biopolymer effects on flocculation, aggregate stability, and water seepage in a silt loam. Geoderma 241–242:289–294. https://doi.org/10.1016/j.geoderma.2014.11.019
55. Cabalar AF, Awraheem MH, Khalaf MM (2018) Geotechnical properties of a low-plasticity clay with biopolymer 30(8):04018170. https://doi.org/10.1061/(ASCE)MT.1943-5533.0002380
56. Jamshidian M et al (2010) Poly-lactic acid: production, applications, nanocomposites, and release. Studies 9(5):552–571. https://doi.org/10.1111/j.1541-4337.2010.00126.x
57. Mudgil D, Barak S, Khatkar BS (2014) Guar gum: processing, properties and food applications—a review. J Food Sci Technol 51(3):409–418. https://doi.org/10.1007/s13197-011-0522-x
58. Osmałek T, Froelich A, Tasarek S (2014) Application of gellan gum in pharmacy and medicine. Int J Pharm 466(1):328–340. https://doi.org/10.1016/j.ijpharm.2014.03.038
59. Latifi N et al (2017) Improvement of problematic soils with biopolymer. Environ Friend Soil Stab 29(2):04016204. https://doi.org/10.1061/(ASCE)MT.1943-5533.0001706
60. Cabalar AF, Canakci H (2011) Direct shear tests on sand treated with xanthan gum 164(2):57–64. https://doi.org/10.1680/grim.800041
61. Araki C (1966) Some recent studies on the polysaccharides of agarophytes
62. Imeson A (2010) Food stabilisers, thickeners and gelling agents. Chichester. John Wiley & Sons, U.K
63. Laurienzo P (2010) Marine polysaccharides in pharmaceutical applications: an overview 8(9):2435–2465
64. Smitha S, Sachan A (2016) Use of agar biopolymer to improve the shear strength behavior of sabarmati sand. Int J Geotech Eng 10(4):387–400. https://doi.org/10.1080/19386362.2016.1152674

65. Chang I et al Soil strengthening using thermo-gelation biopolymers. Constr Build
66. SOURCE: https://www.altrafine.com/blog/chemical-properties-of-guar-gum-that-supports-biological-applications/
67. Risica D, Dentini M, Crescenzi V (2005) Guar gum methyl ethers. Part I. Synthesis and macromolecular characterization. Polymer 46(26):12247–12255. https://doi.org/10.1016/j.polymer.2005.10.083
68. Giavasis I, Harvey LM, McNeil B (2000) Gellan gum. Crit Rev Biotechnol 20:177–211. https://doi.org/10.1080/07388550008984169
69. Chandrasekaran R, Radha A (1995) Molecular architectures and functional properties of gellan gum and related polysaccharides. Trends Food Sci Technol 6:143–148. https://doi.org/10.1016/S09242244(00)89022-6
70. Chang I, Lee M, Thi Phuong Tran A, Lee S, Kwon Y-M, Im J, Cho G-C (2020) Review on biopolymer-based soil treatment (BPST) technology in geotechnical engineering practices. Transp Geotech. https://doi.org/10.1016/j.trgeo.2020.100385
71. Chang I, Im J, Prasidhi AK, Cho G-C (2015) Effects of xanthan gum biopolymer on soil strengthening. Constr Build Mater 74:65–72. https://doi.org/10.1016/j.conbuildmat.2014.10.026
72. Kwon Y-M, Ham S-M, Kwon T-H, Cho G-C, Chang I (2020) Surface-erosion behaviour of biopolymertreated soils assessed by EFA. Géotech Lett 10:1–7. https://doi.org/10.1680/jgele.19.00106
73. Chang I, Kwon Y-M, Im J, Cho G-C (2019) Soil consistency and interparticle characteristics of xanthan gum biopolymer–containing soils with pore-fluid variation. Can Geotech J 56:1206–1213. https://doi.org/10.1139/cgj-2018-0254
74. Qureshi MU, Chang I, Al-Sadarani K (2017) Strength and durability characteristics of biopolymer-treated desert sand. Geomech Eng 12:785–801. https://doi.org/10.12989/gae.2017.12.5.785
75. Cabalar A, Wiszniewski M, Skutnik Z (2017) Effects of Xanthan Gum biopolymer on the permeability, odometer, unconfined compressive and triaxial shear behavior of a sand. Soil Mech Found Eng 54:356–361. https://doi.org/10.1007/s11204-017-9481-1
76. Lee S, Chang I, Chung M-K, Kim Y, Kee J (2017) Geotechnical shear behavior of xanthan gum biopolymer treated sand from direct shear testing. Geomech Eng 12:831–47. https://doi.org/10.12989/gae.2017.12.5.831
77. Im J, Tran ATP, Chang I, Cho G-C (2017) Dynamic properties of gel-type biopolymer-treated sands evaluated by Resonant Column (RC) tests. Geomech Eng 12:815–830. https://doi.org/10.12989/gae.2017.12.5.815
78. Lee S, Im J, Cho G-C, Chang I (2019) Laboratory triaxial test behavior of xanthan gum biopolymer-treated sands. Geomech Eng 17:445–52. https://doi.org/10.12989/gae.2019.17.5.445
79. Ayeldeen MK, Negm AM, El Sawwaf MA (2016) Evaluating the physical characteristics of biopolymer/soil mixtures. Arab J Geosci 9:1–13. https://doi.org/10.1007/s12517-016-2366-1
80. Bouazza A, Gates W, Ranjith P (2009) Hydraulic conductivity of biopolymer-treated silty sand. Géotechnique 59:71–72. https://doi.org/10.1680/geot.2007.00137
81. Vieira CS, Pereira PM (2015) Use of recycled construction and demolition materials in geotechnical applications: a review. Resour Conserv Recycl 103:192–204. https://doi.org/10.1016/j.resconrec.2015.07.023
82. Gálvez-Martos J-L et al (2018) Construction and demolition waste best management practice in Europe. Resour Conserv Recycl 136:166–178. https://doi.org/10.1016/j.resconrec.2018.04.016
83. Bagriacik B (2021) Utilization of alkali-activated construction demolition waste for sandy soil improvement with large-scale laboratory experiments. Constr Build Mater 302. https://doi.org/10.1016/j.conbuildmat.2021.124173
84. Khan F, Das D, Dewangan N (2021) Determination of geotechnical properties and stability of expansive soil using fly ash. Walailak J Sci Technol (WJST) 18. https://doi.org/10.48048/wjst.2021.22783

85. Karim MR, Zain MFM, Jamil M, Lai FC (2015) Development of a zero-cement binder using slag, fly ash, and rice husk ash with chemical activator. Ann Mater Sci Eng. https://doi.org/10.1155/2015/247065
86. Salih MA, Farzadnia N, Abang Ali AA, Demirboga R (2015a) Effect of different curing temperatures on alkali activated palm oil fuel ash paste. Construct Build Mater 94:116e125. https://doi.org/10.1016/j.conbuildmat.2015.06.052
87. Memon SA, Khan MK (2018) Ash blended cement composites: eco-friendly and sustainable option for utilization of corncob ash. J Clean Prod 175:442e455. https://doi.org/10.1016/j.jclepro.2017.12.050
88. Moraes JCB, Tashima MM, Akasaki JL, Melges JLP, Monzo J, Borrachero MV, Soriano L, Paya J (2016) Increasing the sustainability of alkali-activated binders: the use of sugar cane straw ash (SCSA). Construct Build Mater 124:148e154. https://doi.org/10.1016/j.conbuildmat.2016.07.090
89. Guo X, Shi H, Dick WA (2010) Compressive strength and microstructural characteristics of class C fly ash geopolymer. Cement Concr Compos 32:142e147. https://doi.org/10.1016/j.cemconcomp.2009.11.003
90. Sharma HD, Reddy KR (2004) Geoenvironmental engineering: site remediation, waste containment, and emerging waste management technologies. John Wiley & Sons, New York
91. Spence RD, Shi C (2004) Stabilization and solidification of hazardous, radioactive, and mixed wastes. CRC Press, Boca Raton
92. Du Y-J et al (2014) New phosphate-based binder for stabilization of soils contaminated with heavy metals: Leaching, strength and microstructure characterization. J Environ Manage 146:179–188. https://doi.org/10.1016/j.jenvman.2014.07.035
93. Feng YS et al (2021) Geoenvironmental properties of industrially contaminated site soil solidified/stabilized with a sustainable by-product-based binder. Sci Total Environ 765:142778. https://doi.org/10.1016/j.scitotenv.2020.142778
94. Shi C, Qu B, Provis JL (2019) Recent progress in low-carbon binders. Cem Concr Res 122:227–250. https://doi.org/10.1016/j.cemconres.2019.05.009
95. Zhang T et al (2011) Preparation of high performance blended cements and reclamation of iron concentrate from basic oxygen furnace steel slag 56(1):48–55
96. Han FX, Banin A, Kingery WL, Triplett GB, Ding WX (2003) New approach to studies of heavy metal redistribution in soil. Adv Environ Res 8(1):113–120. https://doi.org/10.1016/S1093-0191(02)00142-9
97. Gusiatin ZM, Kulikowska D (2014) The usability of the IR, RAC and MRI indices of heavy metal distribution to assess the environmental quality of sewage sludge composts. WasteManage 34(7):1227–1236. https://doi.org/10.1016/j.wasman.2014.04.005

Ground Improvement of Cohesive Soil with Demolished Construction Waste—An Experimental Study

Aisha Khan and Ramakant Agrawal

Abstract A clayey soil consists of clay particles (size less than 2 microns) possessing plasticity and cohesion. Water is an influential parameter for the strength characteristics of clays. Clay endures as the heaviest and densest soil type when compacted. Most of the construction activities are encountered with clayey soil during the initial stages of construction. The clays due to their high activity, expand and contract beyond the permissible limits of construction To arrest such volumetric changes and ground improvement in the soil, in this work the clay soil was added with demolished construction waste (basically concrete) fines at varying percentages 10, 20, 30, 40, 50, 60% for examining the soil in compression and shear loads. The soil on addition with demolished particles of concrete was tested to observe dry densities, specific gravity, permeability, and shear strength. The dry density, specific gravity and permeability values saw a significant rise as the content of construction demolished waste was raised in the clays. While studying shear parameters the cohesion increased continuously however the value of internal friction angle rises upon addition of 30% demolished waste after which it declines continuously.

Keywords Compaction · Dry density · Specific gravity · Permeability · Shear · Demolished concrete waste · Ground improvement · Cohesion

1 Introduction

Cohesive soils are particularly fine particles which are seived through the 75 micron I.S. sieve. The fraction passing through this sieve exceeds more than 50%. However this fine soil comes under the category of clays when greater than 50% fraction of it passes through a 2 micron I.S. Sieve [6]. Cohesive soil exhibits high plasticity and can be remolded in any shape when water is added to it.

A. Khan (✉) · R. Agrawal
Department of Civil Engineering, OIST, Bhopal, M.P 462022, India
e-mail: civil.aisha84@gmail.com

K. K. Pathak et al. (eds.), *Latest Developments in Civil Engineering*, Lecture Notes in Civil Engineering 352, https://doi.org/10.1007/978-981-99-2676-3_34

Clay particles are strongly adhered to each other using electrostatic forces among them. The clay particles are positively and negatively charged which is responsible for their particulate arrangement to be either flocculated or dispersed. The past history of the loads acted on the soil is also responsible for the particulate arrangement of soil, whether the clays are under, normal or over consolidated clay. Since the particles of clay are very fine, having very low void ratio makes it less porous bringing the permeability to very lower values. However with the existence of active minerals like montmorillonite in the soil, the clay expands and shrinks to a much larger values on addition or reduction of water in its voids. This water inside the voids of the soil causes large volumetric changes, which is a major demerit of the soil containing cohesion which is to be used as a soil for structures carrying heavy loads or for flexible pavements where chances of uneven settlements are high.

In this study a replacement material named Demolished Construction Waste (basically concrete) was utilized in which the particulate matter of size was ranging from 19 mm to 425 microns was infused into the soil and was tested (I.S. 2720) for compaction properties along with permeability and specific gravity with variations in the shear parameters.

Previous studies for improving clays using altering values of demolished materials or with the varying values of moisture content were performed. However, the patterns of the results received were similar with the study. But previous studies did not cover the potentials of demolished materials at a constant moisture content for studying different engineering properties.

The experimental study performed in this research sphere explores the possibilities of large range of percentage of demolished material to be infused into clay at a particular moisture content. This study also addresses the ill effect of optimum moisture in line with the content of the demolished material added while performing tests for observing the deviations in engineering properties of the clays.

2 Related Works

The authors worked on using sustainable material for improving the geotechnical properties of soft clays. According to the work done by authors 5–30% of crushed concrete was added to the soft clays and concluded with 16% OMC and 1.71 g/cc of dry density [1]. They also found the rise in cohesion values as the percentage of demolished material in the soil was increased.

While performing compaction test for Stabilization of soil for sub grade using aggregates of demolished concrete, as stated by the authors, they used demolished aggregate waste for stabilizing weak sub grade soil, and found 40% of the waste material to be an optimum mix, with the dry density raised as the percentage of the demolished material was increased [2] in the mix.

In 2016 work was done by blending Demolished Construction waste and recycled clay. According to authors on addition of the demolished material ranging from 0 to 100%, the value of internal friction angle first saw a decline of upto 70% demolished waste [3], after which on addition of more demolished waste in the clay soil the friction angle increased.

According to the authors the study was done on Demolished Building Waste to stabilize expansive soil to reduce the excessive settlement in soil, where the demolished material was used for stabilizing the expansive soil in varying proportions from 5 to 25%. The dry density along with cohesion & internal friction angle saw a continuous rising curve with increment in the percentage of the demolished material [4].

3 Methodology

3.1 Compaction

This section deals with the compaction as an engineering property of the soil when loaded dynamically. The aim of the test was to define the relation between the dry density of soil and the moisture content added to it. The Maximum value of Dry Density of the clay at which it can be compacted corresponding to the Optimum Moisture Content (OMC) was identified [5] on addition of the demolished material.

About 5 kg of air dried sample was taken for the test and added with moisture content from 8% to 12% for different samples and mixed thoroughly.

↓

The compaction mould of volume 998.2 cm^3 was attached to the base plate and weighed.

↓

The soil was compacted with the rammer of 2.59 kg falling from the height of 31 cm with 25 blows in three layers.

↓

The compacted soil along with the mould was weighed and the bulk density of soil was calculated.

↓

From the calculated bulk density and known moisture content, the dry density of soil was evaluated and the compaction curves were plotted.

↓

Similarly this test procedure stated above is repeated on the clays mixed with varying amount of demolished material upto 60% and the variation in dry density was observed and compared.

3.2 *Specific Gravity*

The aim to perform the Pycnometer test for the specific gravity was to observe and compare the variations in specific gravity of clays [5] when mixed with varying proportions of the demolished material.

A clean empty Pycnometer of capacity 900 ml was taken and weighed.

↓

Approximately 200 grams of cohesive soil comprising of desired amount of demolished material, fraction of both passing 4.75 mm I.S. sieve was taken in the Pycnometer and weighed (W_2).

↓

The water was now added in the Pycnometer completely full along with the soil sample with entrapped air removed and weighed (W_3).

↓

The soil was now removed from the Pycnometer and was filled only with water completely and weighed (W_4).

↓

The following relation is used in this test to determine the specific gravity of soil.

$$G= [(W_2-W_1) / (W_2-W_1) - (W_3-W_4)]$$

↓

The above steps were repeated for three consecutive samples on same percentage of demolished material to obtain an average value of specific gravity.

↓

The test was repeated for different content of demolished material in the clay varying upto 60% and the variation in specific gravity was observed and compared.

3.3 Permeability

The Falling Head Permeability test was performed to compute the permeability of fine grained clay [9] when the demolished material was infused into it in certain proportions ranging upto 60% and to observe the variations in the permeability value.

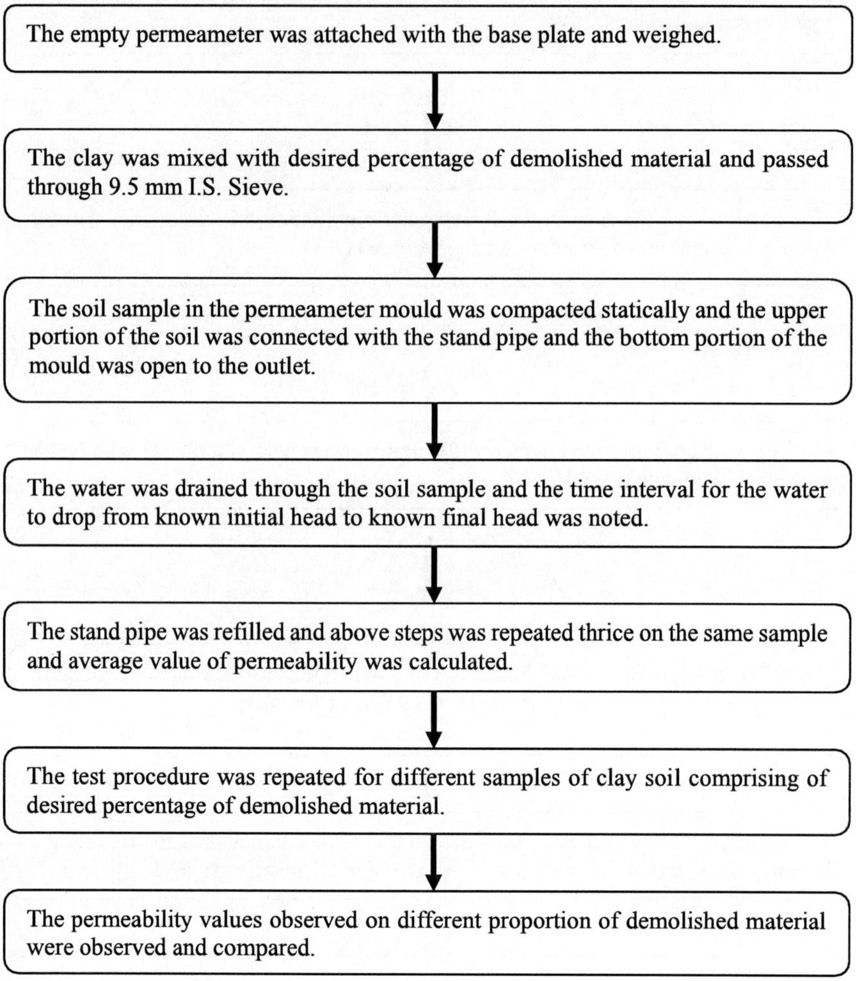

3.4 Shear

On addition of the demolished material in varying proportions, the variation in parameters for shear, namely, cohesion and internal friction angle of particles were observed and compared through Direct Shear Test [8].

The clay soil sample infused with desired amount of demolished material was passed through 4.75 mm I.S. Sieve and 12% moisture was added to it.

↓

Sample of dimension 60 mm and thickness 25 mm was prepared and trimmed and inserted in the shear box.

↓

The two halves of shear box were connected with the connecting pins and the serration plates were placed on the either side of the sample perpendicular to the line of force.

↓

The shear box along with sample was placed in the loading frame of the direct shear testing machine & the connector pins were removed.

↓

The normal load was applied and the horizontal seating load was applied to establish contact between shear box and the proving ring and the dial gauges were set to zero.

↓

The readings on the proving ring were noted corresponding to the shear displacement of the sample and the loads were evaluated.

↓

The test was performed for same value of demolished material on 0.5, 1.0 & 1.5 kg/cm^2 normal stress.

↓

The test procedure in above steps was repeated for different samples of clay soil comprising of varying proportions of demolished material and cohesion values along with values of internal friction angle were observed and compared.

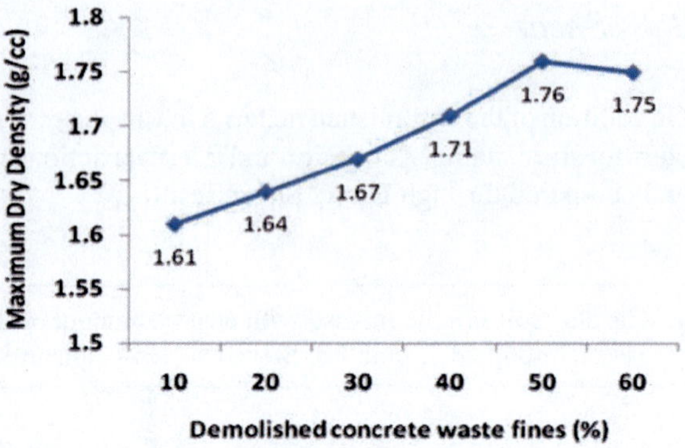

Fig. 1 Displays the variation of Maximum values of dry density with respect to percentage of demolished material at the Optimum water content of 12%

4 Results and Discussions

4.1 Compaction

The I.S. Light Compaction test or SPT was performed on the clay on addition with varying proportion of the demolished material of concrete ranging upto 60% and the relation between dry densities and moisture contents was observed [7]. A steep rise in the Maximum values of dry densities was observed when the percentage of the demolished waste of concrete in the clays increased from 10% to 50% [2]. The Maximum values of Dry Density (MDD) obtained at 10% was 1.61 g/cc which raises upto 1.76 g/cc [1] when 50% demolished waste was added into the soil by weight (Fig. 1).

Since, clay has the tendency to expand when its particles comes in contact with water due to the high activity value resulting in volumetric enlargement of voids in the soil. In this condition the finer particles of the demolished material seems to occupy the voids in the soil thus enhancing the compaction process. The Curve ascends steeply upto 50%. However on further addition of the demolished material beyond 50% in the soil, the curve starts descending as upto 60%.

4.2 Specific Gravity

The specific gravity of the clay soil sample infused with the demolished material was found using the Pycnometer test [5]. Specific gravity witnesses a consistent rise when the infusion in clay was raised upto 60%.

The finer particles of the demolished material occupy the voids between the particulates making it less porous, thus susceptible to the denser nature [4]. No fall in the curve is observed anywhere which indicates that specific gravity is increased with the addition of the demolished material in the soil [1].

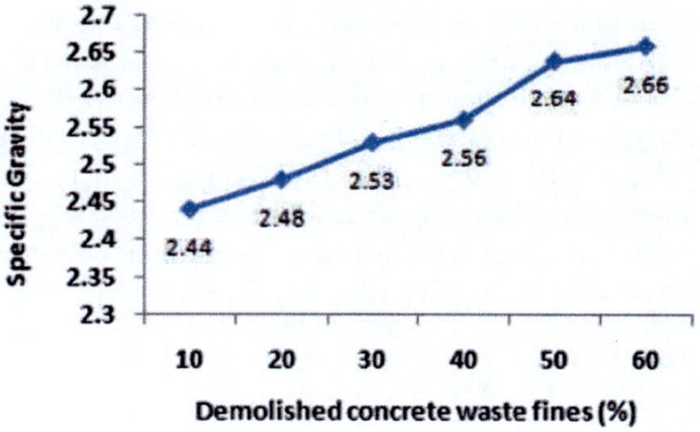

Fig. 2 Exhibits the variation in specific gravity of clay with increasing proportion of demolished material

However a sharp rise is observed between 40% and 50% addition of demolished waste. As the curve moves further beyond 50% it flattens with a gradual rise of upto 60% (Fig. 2).

4.3 *Permeability*

For Vertical Permeability in the soil the falling head test was performed [9]. The permeability value from 10% to 60% cohesion ranges from 0.00979 cm/s to 0.0196 cm/s (Fig. 3). These voids enhance the permeability values of the soil, however this drainage is necessary in the absence of which the voids may get clogged with water. This water in the voids referred to as the pore water increases the pore water pressure in the soil after load applications thereby reducing the effective stress in the soil [3]. The load bearing capacity in that case may get drastically reduced and another factor will be the volumetric expansion in the soil caused by water absorption.

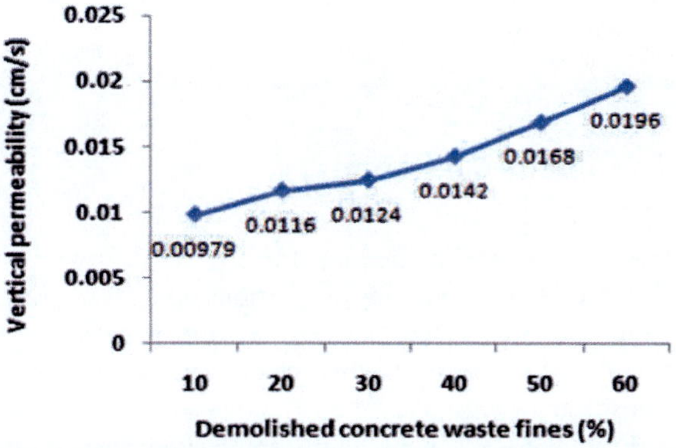

Fig. 3 displays the rise of vertical permeability in soil with altering proportions of demolished material

In order to overcome these factors, here the finer demolished concrete particles had occupied the majority of voids between the particulates. The cohesive soil particles absorbed some moisture due to high activity of clays and expanded volumetrically occupying a major volume of voids by demolished fines. However the vertical permeability values of the soil rises due to increase in void ratio. But majority of voids filled with demolished waste fines flows away with the water through the interconnected voids. Demolished waste particles however have larger permeability than clays, hence enhancing the vertical permeability of the soil mix.

4.4 Shear

Shear Test in Unconsolidated Undrained condition on the remolded soil specimens of clays after addition with varying percentage of demolition waste upto 60% [8], to test its stability against shear force. As the content of demolition waste is increased in the clay soil with every successive sample to be tested, the cohesion value is likely to increase which can be observed (Fig. 4). The values of cohesion into the soil ranged from 0.652 to 1.784 kg/cm^2 with an appreciable rise of almost 173% corresponding to the demolish material percentage increasing till 60% [1]. This rise in the demolished material along with cohesion into the soil mix is the governing shear parameter to resist the shear loads in extreme situations of slope stabilities and retaining walls. The gradual rise in the cohesion curve can be observed when the percentage of the demolished material is increased. However the curve climbs rapidly when the share of the demolished material in the clay goes beyond 40% and continues to rise at the same rate upto 60%. The ϕ-value witnesses a substantial rise from 21.8° to 30.9° when the demolished material content increases and saw a peak value at 30% [4]. However, as the percentage of infusion increases beyond 30% the ϕ-value starts declining from 30.9° to 19.3°. Since there is no drainage for the pore water to dissipate, which in the above case resulted in reduced internal friction among the granular particles of demolished material, thus, reducing the ϕ-value among the soil particles.

5 Conclusion Summary

The clay soil was studied for improving the ground on infusing it with Demolished Concrete Waste in varying proportions upto 60%. The following clay and demolished waste mix was tested and assessed under certain parameters governing the engineering properties of the mix which were compaction, specific gravity, permeability and shear.

From the I.S. Light Compaction test performed on the clays mixed with varying percentage of demolished materials of concrete upto 60% where the Maximum values of Dry Density which is the major parameter for compaction saw a significant rise

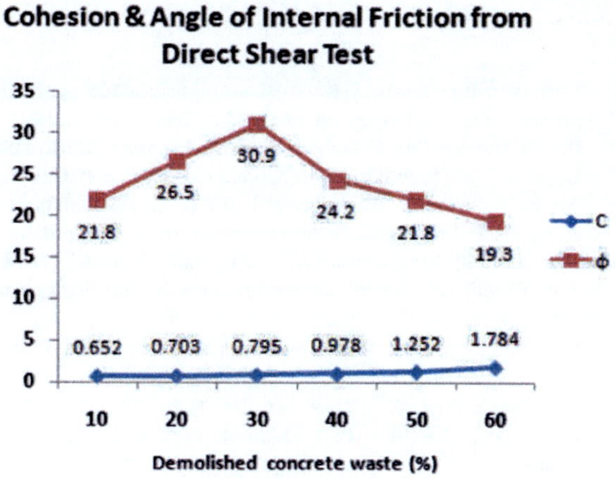

Fig. 4 Depicts the Cohesion with Internal friction angle at different proportions of demolished material in clay

upto 50% demolished waste, after which the curve falls. The highest value of MDD was detected to exist when the demolished material in the soil lies from 40% to 50%.

Another result obtained from Light Compaction test was the percentage of Optimum Moisture for conducting further tests. However 12% moisture content was formalized to be the OMC.

For specific gravity using the Pycnometer test, a continuous rise in specific gravity was found as the percentage of the demolished material in the soil increased. Specific gravity was observed to be maximum when the demolished material was 60% but due to some unknown anomalies in the test, demolished material content from 40% to 50% can be considered as an optimum mix for ground improvement.

The results from Falling Head test for permeability saw a continuous rise in the vertical permeability value of the cohesive soil when infused with the demolished material. Therefore due to volumetric changes in the cohesive soil, the demolished material percentage can be kept above 30% for ground improvement.

The results from the Direct Shear Test on cohesive soil when mixed with the demolished material in varying proportions, the value of internal friction angle saw a peak value when the demolished material was 30% beyond which it decreases continuously, however, the cohesion increases gradually without any decrement as the percentage of the demolished material raises. In order to maintain an adequate balance of both parameters for shear, the demolished material percentage in the soil can exist between 30 and 50% as an optimum mix for ground improvement.

As per the results retrieved on performing above tests, the best possible mix to be selected for improving the clays with construction demolished concrete particulate waste can range from 30 to 40%.

References

1. Karkush MO, Yassin S (2020) Using sustainable material in improvement the geotechnical properties of soft clayey soil. J Eng Sci Tech 15(4) 2208–2222
2. Paul H, Cyrus S (2016) Stabilization of weak sub grade soil using demolished concrete aggregate. December 2016, Indian Geotechnical Conference IGC2016, IIT Madras, Chennai, India
3. Arisha A, Gabr A, El-Badawy S (2016) Using blends of construction & demolished waste materials and recycled clayhttps://doi.org/10.1016/j.proeng.2016.06.14
4. Ganesh KM, Vara Prasad ASS, Jagapathi Raju M (2018) Experimental study on building demolished waste stabilized expansive soil with potassium chloride, Int J Eng Tech 7(3.31): 214–218
5. I.S. Code 2720-1980 (Part 3) for the procedure adopted for determining the Specific Gravity of soil
6. I.S. Code 2720-1985 (Part 4) for classifying the coarse grained and fine grained soil
7. I.S. Code 2720-1980 (Part 7) for defining the relation between maximum dry density and optimum moisture content using light compaction test
8. I.S. Code 2720-1986 (Part 13) for the methodology to determine the shear parameters using Direct Shear
9. I.S. Code 2720-1986 (Part 17) for the methodology for determining the Vertical Permeability of soil mix

Structural Engineering

Optimization of Intze Type Water Tank Using Machine Learning

S. Ganesh Reddy, Aditya Komaravolu, Kamalini Devi, Shrihari Saduwale, and A. Obulesh

Abstract India has an urban population that makes up more than 34% of the country's total population. Water, petroleum, oil products, and other liquids are kept in storage tanks and overhead tanks. These tanks have been employed in several different fields, including agriculture, the food industry, paper mills, firefighting, and water distribution. It is well known in the current situation that nearly every industry is heading toward machine-based automation, starting with the most basic systems up to master-level ones. Among them, Machine Learning (ML) is one of the crucial tools that resembles Artificial Intelligence (AI) the most. It does this by allowing some well-known data or experience to automatically enhance or estimate the behaviour or condition of the provided data using a variety of methods. To avoid leaks, every tank is designed with a framework free of cracks. The objective of the structural design process is to provide a safe design that complies with all design code criteria while keeping design expenses to a minimum. A recent idea, structural design optimization is based on random information interchange, determining the best design using more precise mathematical algorithms, and methodologies as opposed to human judgment. This study uses machine learning (ML) to bring down the price of planning and designing a tank to a specific water tank capacity. In this

S. Ganesh Reddy (✉) · K. Devi · S. Saduwale
Department of Civil Engineering, Vidya Jyothi Institute of Technology (A), Hyderabad, Telangana, India
e-mail: ganeshreddyseelam252@gmail.com

K. Devi
e-mail: kamalinidevi1@gmail.com; kamalinidevi_civil@cbit.ac.in

S. Saduwale
e-mail: shriharistructure@gmail.com

A. Komaravolu · A. Obulesh
Department of Artificial Intelligence, Vidya Jyothi Institute of Technology (A), Hyderabad, Telangana, India
e-mail: avuku06@gmail.com

Present Address:
K. Devi
Department of Civil Engineering, Chaitanya Bharathi Institute of Technology (A), Hyderabad, Telangana, India

K. K. Pathak et al. (eds.), *Latest Developments in Civil Engineering*, Lecture Notes in Civil Engineering 352, https://doi.org/10.1007/978-981-99-2676-3_35

study, programs are created to promote the Intze type reinforced concrete water tank design with the lowest possible cost.

Keywords Optimization · Machine learning · Intze type water tank

1 Introduction

Liquid storage is provided by overhead tanks and reservoirs. Irrigation work, human consumption, fires, production units, rainwater collection, and water storage are all frequent uses for these tanks. The capacity of tanks with zone ii, zone iii, zone iv, and zone v in the empty scenario, as well as the condition with the full water load, were evaluated using a nonlinear static analysis by Gondalia et al. [1], Rao et al. [2]. Using plots of the whole base shear rather than the top displacement curve, it was determined that base shear would rise as the tank's peak and filling level in the storage tank increased. SVM could be a regulated machine learning calculation that may be utilized for characterization or relapse issues. It utilizes a procedure known as the bit lure to alter your data and subsequently taking into consideration these changes it finds a perfect limit between the conceivable yields. It will come to a good degree complicated information changes, then is sensible in a way to isolate your information taking into account the names or yields you've characterized.

For the most recent trend in building construction, Jintao Wei et al. (2017) suggested a novel method for form optimization or geometrical optimization for the nonlinear analysis of structures. It was discovered that this strategy was more recent than the traditional optimization techniques.

The cost of reinforced concrete structures with medium ductile shear walls that were identical in plan and height and had been improved by using a genetic algorithm and artificial neural network was studied by Nazary and Moosa Mazlom [3], Carstensen [4], Davies et al. [5], Danhaive and Mueller [6]. Machine learning is considered a subset of artificial intelligence and a branch of computational science that analyses and interprets data (which can take the form of anything like pattern reorganization, tracing, or tracking of data) to improve a system and assist in making decisions with little to no human interference [7–9].

A manual design of a water tank with working stress was proposed by Adilakshmi et al. [10]. Staging factors for the Intze tank were designed with the aid of the limit state method for a specific capability. For a specific water maintaining ability, this project provides the most accurate estimations of the required amount of concrete and steel [11, 12].

The major reason for the layout of tanks are economical, strength, provider life, and to offer secure transportable consuming water after storing for a long term additionally faces up to unique situations like wind and earthquakes. The Design of the tanks relies upon the location of the tank i.e., above or below, on the floor degree. The overhead tanks are usually built at a positive peak from the floor degree for the usage of columns and braces, and for direct distribution of water via way of means of gravity. In any case, the underground tanks are relaxed below the floor degree.

1.1 Intze Tank

This is a particularly rare form of the elevated tank that can hold a lot of water. If flat-bottom plates are provided, very big capacity round tanks are uneconomical. An upper dome is supported by a ring beam installed on a cylinder wall in Intze-type tanks. The ring beam and the conical ceiling are above the walls. There is also a lower dome, which is supported by a ring beam. The horizontal thrust from the cone base and the horizontal thrust from the lower dome is balanced in the cone and lower dome. A circular girder, which is supported by several pillars, supports the cone and lower dome. A partition divides the tank into two parts for big capacities.

In the field of structural optimization, a considerable amount of research was performed. Most of these studies deal with academic or minor issues. For decades, this technique has been avoided by practicing structural engineers as esoteric with little practical value. However, over the last decade, remarkable efforts have been made to incorporate structural optimization techniques into the practice of structural engineering.

Support Vector Machine (SVM) is a classification and regression prediction tool that automatically prevents data from over-fitting while maximizing predicted accuracy. Support Vector Machines are systems that use the hypothesis space of linear functions in a high-dimensional feature space and are trained using an optimization theory-based learning method that incorporates a learning bias. SVM gains notoriety and when given pixel maps as input, it performs handwriting recognition tasks with accuracy that rivals that of highly developed neural networks with extensive features. Numerous more uses include face analysis, handwriting analysis, and other areas, particularly those based on pattern categorization and regression.

After defining the objective function, the standard optimization process begins by creating a design population by converting the actual design parameters into a binary string using binary encoding. The objective functional score for each string is a measure of survival potential called fitness. The higher the fitness score, the more likely the string will be selected for mating and breeding. Since each chromosome or chain in the solution space is independent of other chromosomes or chains, the search method is a multipath search that searches for many peaks in parallel. This helps to reduce the chances of being trapped.

The objectives are to investigate the design and analysis of water tanks, and calculate the amount of steel reinforcement in concrete while manually analyzing and designing the water tank by IS Codes. To comprehend the most crucial elements that affect the cost-effective and secure design of water tanks to satisfy demand. Using a support vector machine to do parametric studies on the design of water tanks while effectively leveraging several design restrictions and variables and gathering a sizable amount of data from prior studies. By using this approach, it will be possible to evaluate the study and design of a reinforced concrete water tank of the Intze type while saving money and reducing the reliance on manual processes.

Artificial creativity, genetic programming, and binding graphs are applications of machine learning that are utilized for mechanized automatic control system design, water distribution system design, distributed computer network architecture, and cellular infrastructure optimization.

2 Methodology

1. For this investigation, a typical Intze tank with the appropriate capacity is chosen.
2. The tank's preliminary dimensions are estimated for the necessary capacity.
3. Next, the water tank construction is manually created using the relevant Indian Code provisions, working stress technique, the draught of the crack-free concrete section, and total cost assessed by reference to regularly scheduled rates.
4. STAADPRO software is used to model and analyse the water tank's structural components.

Figure 1 displays the water tank's structural components.

The following elements will be included in the tank's design:

- The meridian and latitude are strengthened along the thickness of the upper dome, which is typically 100–150 mm. Typically, the pitch is one-fifth of the span.
- Ring beam supporting the dome: To counteract the horizontal component of the dome thrust, a ring beam is required. For induced hoop stress, the ring beam is constructed.
- The cylindrical walls need to be built to withstand the hoop stress brought on by horizontal water pressure.

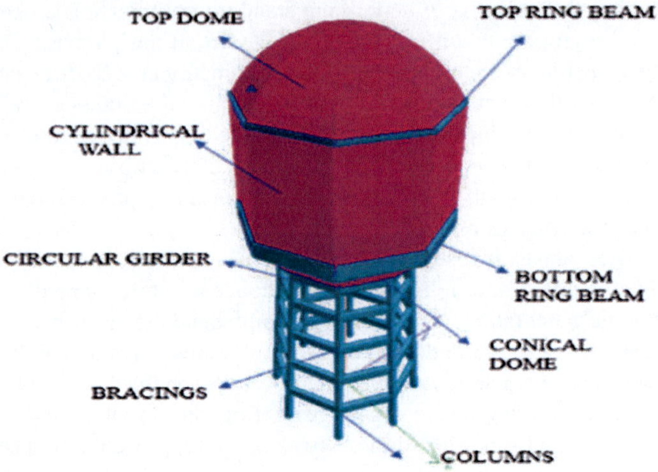

Fig. 1 Intze tank's parts

- To resist the horizontal component of the reaction of a conical wall to a cylindrical wall, there is a ring beam at the intersection of the two types of walls. For induced hoop stress, the ring beam is constructed.
- Conical plate Tire stress caused by water pressure was taken into account when designing. The plate can also be made to span between the lower ring bearing and the top ring beam.
- Tank bottom: The base might have an arched or round shape. A ring carrier supports this plate.
- Ring Bearer: It's made to hold up the tank and everything inside of it. The beams must be built to withstand the bending moments and torsions that will come from the columns supporting them.
- Columns: They ought to be constructed to support the entire load. The pillars, which are fixed at regular intervals, must be built to withstand seismic and wind loads.
- Foundation: In general terms, all columns should be supported by the combined foundation. A ring bearer and a circular plate are typically used to create a foundation.

2.1 *Manual Design*

The design is done as per the I.S-3370 (Part II-1967), I.S-3370 (Part IV-1967) and for the water tank which is situated at location: Vivekananda Nagar, Kukatpally Hyderabad.

Population = 7500

Per capita demand wlmultiplied by population gives the Intze water tank's capacity.

$= 195 * 7500 = 1462500$ L

$= 1500\ m^3$

Taking the volume as $V = \pi/4\ D^2 * 0.75\ D$

$$1500 = \pi/4\ D^2 * 0.75$$

$D = 13.68$ m ≈ 14 m.

Design of Tank

Design of a 15,00,000-L intze tank.

Soil has a 200 kN/m^2 safe bearing capability

Wind pressure according to IS875 is 1230.8 N/m^2. Assuming M30 concrete

For which $\sigma_{cbc} = 10$ N/mm^2,

$\sigma_{cc} = 8$ N/mm^2

Direct tension $\sigma_t = 5$ N/mm^2

Tension in bending = 2 N/mm^2

Modular ratio m = 9.33

For Steel stress,

Tensile stress in direct tension = 115 N/mm^2. Tensile stress in bending on liquid face = 115 N/mm^2 for t < 225 mm.

Dimensions

Dia of tank (D) = 14 m

Dia of Bottom Dome (D_O) = 5/8D = 8.75 m

Dia of Top Dome (h_1) = 1/6D = 2.33 m

Rise of Bottom Dome (h_2) = 1/8D = 1.75 m

Height of Conical Dome (h_0) = 3/16D = 2.6 m

Height of Cylindrical Portion (h):

$$V = (\pi/4 * D^2 * h) + \pi/12 * h_o(D^2 + D_O^2 + D * D_O) - \pi/3 * h_2^2(3R_2 - h_2)$$

Here

$$\begin{aligned} R2 &= (D_0/2)^2 + h_2^2/2 * h_2 \\ &= [(8.75/2)^2 + 1.75^2/2 * 1.75] \\ &= 6.34 \text{ m} \\ h &= 7.53 \text{ m} \end{aligned}$$

Height of Cylindrical Portion (h): 7.53.

Design of Top Dome

If R = Radius of the dome

D = Dia at base = 14 m

r = central rise (1/6 * 14) = 2.33 m

$$\begin{aligned} R &= (D/2)^2 + r^2/2 * r \\ &= [(14/2)^2 + 2.33^2/2 * 2.33] \\ &= 11.68 \text{ m} \end{aligned}$$

Sinθ = D/2/R = 7/11.68 = 0.59

$$\begin{aligned} \text{Cos}\theta &= \text{Dia of bottom dome/Radius of dome} \\ &= 8.75/11.68 = 0.74 \end{aligned}$$

Meridional force (T_1) = WR/1 + Cosθ

Here w = Total load

Self weight of dome = 0.1 * 25 = 2.5 kN/m^2

Thickness of dome slab(t) = 100 mm

Live load = 1.2 kN/m^2

Finishes = 0.1 kN/m^2

$$\begin{aligned}\text{Total load (w)} &= 2.5 + 1.2 + 0.1 = 3.8\ \text{kN/m}^2 \\ &= \text{WR}/1 + \text{Cos}\varphi \\ &= 3.8 * 11.68/1 + 0.74 \\ &= 25.50\ \text{kN/m}\end{aligned}$$

$$\begin{aligned}\text{Meridional stress} &= 25.50 * 10^3/1000 * 10 \\ &= 0.25\ \text{N/mm}^2\end{aligned}$$

$$\begin{aligned}\text{Hoop force } (\text{T}_2) &= \text{WR}_1(\text{Cos}\theta - 1/1 + \text{Cos}\theta) \\ &= 3.8\ *\ 11.68(0.74{-}1/1 + 0.74) \\ &= 7.33\ \text{kN/m}\end{aligned}$$

$$\begin{aligned}\text{Hoop stress} &= 7.33 * 10^3/1000 * 100 \\ &= 0.0733\ \text{N/mm}^2 < 8\ \text{N/mm}^2\end{aligned}$$

Hence ok

Provide Nominal Reinforcement (0.24%)

A_{st} = 0.24/100 * 1000*100 = 240 mm^2

Provide 6 bars of 8 mm dia @200 mm c/c spacing

Design of Top Ring Beam

Horizontal component of meridional force (T_1)

W = T_1 * Cosθ

25.50 * 0.74 = 18.87 kN/m

Total Hoop tension in beam

= W * D/2

= 18.87 * 14/2

= 132.09 kN

$$\begin{aligned}\text{A}_{\text{st}} \text{ for Hoop Tension} &= \text{T}/\sigma_{\text{st}} \\ &= 132.09\ *\ 10^3/115 \\ &= 1148.60\ \text{mm}^2\end{aligned}$$

Provide 12 mm dia 12 bars (Ast = 1356 mm^2)

$\sigma_{ct} = T/A_{gt}(m - 1)A_{st}$ (IS456-2000Pg-80)

$= 132.09 * 10^3/bd + (m - 1)A_{st} \leq 1.5\ b = 300$ mm

$= 132.09 * 10^3/300 * D + (9.33-1) * 1356$

$= 255.9 \leq D$

Provide D = 300 mm

Size of beam = 300 * 300 mm

Provide 8 mm dia 2 legged stirrups

$S_v = 0.87 * fy * A_{sv}/0.4b$

$A_{sv} = 100\ mm^2$

$S_v = 0.87 * 415 * 100/0.4 * 300$

$= 300$ mm

Provide 8 mm dia 2 legged stirrups @225 mm c/c

Provide 12 mm dia 8 bars as main Reinforcement.

Design of Cylindrical Tank Wall

Maximum hoop tension at the base of the wall

$F_t = (whD/2)$

Where w = Density of water = 10 kN/m^3

h = depth of water

Therefore $F_t = 10 * 7.5 * 4/2$

$= 525$ kN/m

Tension Reinforcement per meter height

$A_{st} = 525 * 10^3/115$

$= 4565.2\ mm^2$

Provide 20 mm dia 16 bars @130 mmc/c on each face ($A_{st} = 5026.4\ mm^2$)

For Thickness of Wall

$\Sigma_{ct} = T/A_{gt}(m - 1)A_{st}$

$T = 525$ KN/m

$A_g = 1000t$

$= 525 * 103/1000t + (9.33 - 1)5026.4 \leq 1.5$

$= 308.13$ mm

Adopt 350 mm thick walls at bottom Gradually reducing 200 mm at top.

Distribution Steel

Provide min steel of 0.24%

$A_{st} = 0.24/100 * 1000 * 350 = 840\ mm^2$

Provide 12 mm dia 8 bars @120 mmc/c ($A_{st} = 904.36mm^2$)

$A_{st} = 0.24/100 * 1000 * 200 = 480\ mm^2$ (top)

Provide 8 mm dia 10 bars @100mmc/c.

Bottom Ring Beam

$$\begin{aligned}\text{Load due to top dome} &= T_1 \sin\theta \\ &= 25.5 * 0.59 \\ &= 15.04\ \text{KN/m}\end{aligned}$$

$$\begin{aligned}\text{Load due to top ring beam} &= 0.3 * 0.3 * 25 \\ &= 2.25\ \text{KN/m}\end{aligned}$$

$$\begin{aligned}\text{Load due to wall} &= 0.2 * 6.7 * 25 \\ &= 33.5\ \text{kN/m}\end{aligned}$$

Here

$h = (7.5 - 0.3 - 0.5) = 6.7$ m

Assume size of beam = 1000 * 500 mm

$$\begin{aligned}\text{Load due to self weight} &= 1 * 0.5 * 25 \\ &= 12.5\ \text{kN/m}\end{aligned}$$

Total load on Beam (w_1) = 63.29 kN/m

$$\begin{aligned}\text{Horizontal component } (H_1) &= w_1 * \tan\beta \\ &= 63.29 * \tan 45 \\ &= 63.29\ \text{kN/m}\end{aligned}$$

$$\begin{aligned}\text{Water pressure } (H_2) &= r_w * h * d \\ &= 10 * (7.5 - 0.3) * 0.5 \\ &= 36\ \text{kN/m}\end{aligned}$$

$H = H1 + H2 = 63.29 + 36 = 99.29$ kN/m

$$\begin{aligned}\text{Hoop Tension} = H * D/2 &= 99.29 * 14/2 \\ &= 528\ \text{kN}\end{aligned}$$

$A_{st} = 528 * 10^3/115 = 4591.30\ mm^2$

Provide 20 mm dia 16 bars ($A_{st} = 5026\ mm^2$)

Provide a ring beam of 1000 * 500 mm with 16 bars of 20 mm dia and

10 mm dia stirrups @180 mm c/c spacing.

Design of Conical Dome

$$\text{Avgdia of conical dome} = 14 + 8.75/2$$
$$= 11.3\ \text{m}$$

Avg depth of water = (7.5 + 2.6)/2 + 7.5 = 8.8 m

$$\text{Weight of water} = (\pi * d * \text{depth} * \text{width}) * 10$$
$$= 3.14 * 11.3 * 8.8 * 2 * 10$$
$$= 6247.99\ \text{kN}$$

Width of slab = 2.83 m

Weight of slab = (π * 11.3 * 2.83 * 0.60 * 25)

Assume 600 mm thick slab

$$= 1506\ \text{kN}$$

Weight from top dome, wall, beam

$$= (\pi * D) * W_1 = (\pi * 14) * 63.29$$
$$= 2783.63\ \text{kN}$$

$$\text{Total load} = 6247.99 + 1506 + 2783.63$$
$$= 10537.62\ \text{kN}$$

Load per m (w2) at the base of slab

$$W2 = 10537/\pi * 8.75$$
$$= 383.31\ \text{kN/m}$$

Meridional force (T_3):

$$T_3 = W_2/\text{Cos}\beta$$
$$= 383.31/\text{Cos}45$$
$$= 547\ \text{kN/m}$$

Meridional stress:

$$= 547.14 * 10^3/1000 * 500$$
$$= 1.09\ \text{N/mm}^2 < 8\ \text{N/mm}^2$$

Therefore the stress is within safe limits.

Hoop Tension:

$$P = \text{water pressure} = r_w * h$$
$$= 10 * 7.5$$
$$= 75\ \text{kN/m}^2(\text{top})$$

$$q = \text{selfweight of slab}$$
$$= 1 \ * \ 0.50 \ * \ 25/\text{Area}$$
$$= 12.5 \text{ kN/m}^2$$
$$\theta = 45$$

H_t at top of conical dome

$$= [(10 \ * \ 7.5)\text{cosec}45 + 12.5\text{cot}45] \ * \ 14/2$$
$$= 829.92 \text{ kN}$$

At centre

$h = 7.5 + 2.6/2$ and $D = 14$ m

$$H_t = [(10 \ * \ 8.8)\text{cosec}45 + 12.5\text{cot}45] \ * \ 14/2$$
$$= 958.65 \text{ kN}$$

At Bottom

$h = 7.5 + 2.6$ and $D = 8.75$ m

$$H_t = [(10 \ * \ 10.1)\text{cosec}45 + 12.5\text{cot}45] \ * \ 8.75/2$$
$$= 679 \text{ kN}$$

Max Hoop tension is at top = 829.92 kN

$$A_{st} = 829.92 \ * \ 10^3/115$$
$$= 7216.69 \text{ mm}^2$$

$$A_{st} \text{ on each face} = 7216.69/2$$
$$= 3608.34 \text{ mm}^2$$

Provide 25 mm dia bars of 8 nos @ 150 mm c/c ($A_{st} = 3926.9\text{mm}^2$)

Distribution Steel

Provide min steel of 0.24%

$A_{st} = 0.24/100 * 1000 * 500 = 1200 \text{ mm}^2$

A_{st} on each face = 600 mm^2

Provide 10 mm dia 8 bars @130 mmc/c ($A_{st} = 628 \text{ mm}^2$).

Design of Bottom Spherical Dome

Meridional force (T_4) = WR/1 + Cosθ

Assume thickness of dome slab = 300 mm

Dia at base (D) = 8.75 m

r = central rise (1/5 * 8.75) = 1.75 m.

If r = radius of the dome

$(2R_2 - r)r = (D/2)^2$

$(2R_2 - 1.75)1.75 = (8.75/2)^2$

$3.5R - 3.06 = 19.14$

$R_2 = 6.34$ m

$$\text{Weight of water} = 1 * [(7.5 + 2.6) - 1.75] * 10$$
$$= \text{width} * \text{height} * \text{Density}$$
$$= 83.55\ \text{kN/mm}^2$$

Self weight = 0.3 * 25 = 7.55 kN/mm^2

$$\text{Total load} = 83.5 + 7.5$$
$$= 91\ \text{kN/mm}^2$$

$$\text{Sin}\theta_2 = D_o/2/R_2$$
$$= 8.75/2/6.34$$
$$= 0.690$$

$\theta_2 = 45.41$

$$T4 = 91 * 6.34/1 + \text{Cos}45.41$$
$$= 325.72\ \text{kN/m}$$

$$\text{Meridional stress} = 325.72 * 10^3/1000 * 300$$
$$= 1.08\ \text{N/mm}^2 < 8\text{N/mm}^2$$

Provide min steel of 0.24%

$A_{st} = 0.24/100 * 1000 * 300 = 720\ \text{mm}^2$

Provide 12 mm dia bars @150mmc/c.

Design of Circular Girder

$$\text{Inward thrust from conical dome} = T_3\cos\theta$$
$$= 547.14 * \cos 45$$
$$= 386.88\ \text{KN/m}$$

$$\text{Outward thrust from spherical dome} = T_4\cos\theta$$
$$= 325.72 * \cos 45.41$$
$$= 251.21\ \text{KN/m}$$

$$\text{Hoop compression on beam} = wd/2$$
$$= 135.67 * 8.75/2$$
$$= 593.55\ \text{KN}$$

Hoop stress = F/A = $593.55 * 10^3/500 * 1000$

$1.18\ \text{N/mm}^2 < 8\ \text{N/mm}^2$ safe

Vertical loads on beam

$T_3\ \text{Sin}\theta_1 + T_4\ \text{Sin}\theta_2$

547.14sin45 + 325.72sin45.41

617.19 kN/m

Self weight of beam = 0.5 * 1.0 * 25 = 12.5 kN/m

Total UDL = 617.19 + 12.5 = 630 kN/m

Total load = $(\pi * D) * W_1 = (\pi * 8.75) * 630$

= 1378.02 kN

Max −ve BM at support

= $C1WR^2(2\theta)$

= $0.066 * 630 * 4.34^2 * \pi/4$

= $615.111\ \text{kN} \cdot \text{m}$

Max +ve BM at midspan

= $C1WR^2(2\theta)$

= $0.030 * 630 * 4.34^2 * \pi/4$

= $279.59\ \text{kN} \cdot \text{m}$

Max Torsional moment

= C3WR2(2θ)

= $0.055 * 630 * 4.34^2 * \pi/4$

= $46.59\ \text{kN} \cdot \text{m}$

Max shear force at support = W/2 * no. of columns

= 17318.02/2 * 8

= 1082.37 kN

$M = Qbd^2$

$615.11 * 10^6 = 1.80 * 500 * d^2$

d = 826.71 mm
Provide d = 950 mm, cover 50 mm.

$A_{st} = M/\sigma_{st} * j * d$

Here $j = 0.86$

$A_{st} = 6546.85\ mm^2$

Provide 25 mm dia 14 nos ($A_{st} = 7389 mm^2$)

σ_{sv} = permissible tensile stress in shear reinforcement which shall not be taken >230 N/mm^2

$S_v = 100.56$ mm

Provide 12 mm dia 4 legged vertical stirrups @100 mm c/c.

Columns

Vertical load on one column at top = $17318.02*10^3/8$

= 2164752 N

Let α be the inclination of the column with vertical

$\tan\alpha = 1/12\ \alpha = 4.76$

$\sin\alpha = 0.83, \cos\alpha = 0.99$

$$\text{Actual length of column} = 1\sec\alpha$$
$$= 12.04 \text{ m}$$

Providing 500 mm dia of column

Weight of column = $\pi/4 * 0.5^2 * 12.04 * 25$

= 59102 N

$$\text{Total vertical load} = 2164752 + 59102$$
$$= 2223854 \text{ N}$$

$$\text{Corresponding axial load} = 2223854/0.995$$
$$= 2235029.14 \text{ N}$$

Weight of water resting on the conical slab

$$W_w = r_w * A * 2 * (8.75/2 + x)$$
$$= 1000 * A * 2 * (8.75/2 + x)$$
$$A = 3/2(7.5 + 10.1)$$
$$= 26.4 \text{ m}^2$$
$$x = 0.8 \text{ m}$$
$$= 858408.77 \text{ N}$$

Weight of water resting on dome

$= r_w(\pi/4 * d^2 * h - \pi * h_c/3)(3R - h_c)$

$= 10.45 * 10^6$ N

Weight of water = 858408.77 + 10.45 * 10^6

= 11.30 * 10^6 N

Weight of water transmitted on one column

= 11.30 * 10^6/8

= 1.413 *10^6 N

Vertical load on one column when the tank is empty

= 2223854 – 1.413 * 10^6

= 810854 N

Corresponding Axial load = 810854/0.99

= 819044 N

$c(A - A_{sc}) + A_{sc} = 2235029$

$8(\pi/4 * A_{sc}) + 190 * A_{sc} = 2235029$

= 1739 mm^2

Min Reinforcement of steel = 0.8%

= 0.8/100 * π/4 * 500^2

= 1571 mm^2

Provide 8 bars of 16 mm dia (A_{st} = 1608.48 mm^2)

More steel has been subjected since the column is subjected to bending moment caused by wind load.

Design of Braces

Moment in brace = 2 * Moment for the column * sec45

Moment for the column = Horizontal shear column * Distance/2

Total weight of staging = weight of column + weight of braces

Weight of column = 8 * π/4 * 0.5^2 * 12.04 * 25

= 472809.7 N

Height of column = 12.04 m

No. of columns = 8

Dia = 500 mm

No. of braces = 4

Size of braces = 300 * 500 mm

Weight of braces = clear length of braces b/w two column * size * density * No. of braces * No. of columns

= 3.4 * 0.3 * 0.5 * 25 * 4 * 8

= 408000 N

Total weight = 472809.7 + 408000

= 880809 N

Wf = 17318.02 * 10^3 + 880809/3
(When tank is full)

= 17611623 N

Sesmic force = 0.081 * 17611623

= 1426541.46 N

Shear force for column = 1426541.46/8

= 178317.68 N

Max B.m for the column

= 178317.68 * 3/2

= 267476.52 N-m

Moment in brace = moment for the column * sec45

= 267476.52 * sec45

= 378268.92 N-m

Provide 300 * 500 mm section and designing as doubly reinforcement beam with equal steel at top and bottom

$A_{sc} = A_{st}$ = 378268.92 * 1000/220 * 420

= 4023 mm^2

Provide 14 bars of 20 mm dia at top and equal amount of steel at bottom

$$\begin{aligned} \text{Shear force for brace} &= \text{B.m for brace}/0.5 * \text{span} \\ &= 378268/0.5 * 3.4 \\ &= 222510 \text{ N} \end{aligned}$$

$$\begin{aligned} \text{Therefore Nominal shear stress} &= \tau v = \text{s/bd} \\ &= 222510/30 * 460 \\ &= 1.61 \text{ N/mm}^2 < 8 \text{ N/mm}^2 \end{aligned}$$

Hence the section is adequate with shear reinforcement

Therefore provide nominal stirrups say 4 legged 8 mm dia stirrups@160 mm c/c.

Foundation

Load carried by column wc = 2164.75 kN

Safe bearing capacity of soil = 200 kN/m^2

Grade of concrete = M20

Grade of steel = Fe415

Side of square column:

$$
\begin{aligned}
b &= 0.717D \\
&= 0.717 * 500 \\
&= 358.5 \\
&= 360 \text{ mm}
\end{aligned}
$$

$$
\begin{aligned}
\text{Area of footing} &= \text{Load/sbs} \\
&= 2381.225/200 \\
&= 11.906 \text{ m}^2
\end{aligned}
$$

Size of square footing = $(3.5)^{1/2}$

Hence size of square footing for circular column = 3.5 * 3.5 m

Factored soil pressure on footing (q_u):

$$
\begin{aligned}
q_u &= \text{factored load/actual area} \\
&= 1.5 * 2164.75/3.5 * 3.5 \\
&= 265.071 \text{ kN/m}^2
\end{aligned}
$$

Depth of footing by bending moment criteria.
The critical section for Bm is taken as face of column

$Mu = q_u * B * (B - b/2) * (B - b/2)/2$

$q_u * B = 265.071 * 3.5$

= 927.75 kN/m

$Mu = 1143.405 \text{ kN} \cdot \text{m}$ (Bm at critical section)
In Equilibrium conditions:

$1143.405 * 10^6 = 0.138 * 30 * 3500 * d^2$

d = 280.90 mm.

Table 1 Estimation and Schedule of rates for finished items of work

S. no.	Quantity	Description	Amount
1	52.49 (ft)	Earthwork excavation	68241.472
2	660.16 (m^3)	Concrete	3201736
3	34709.5 (kg)	Steel	3211318.5
4	1608.04 (m^2)	Plastering	578894.4
5	15356.35 (ft^2)	Painting	460690
6	10%	Contractors profit	752627.9

2.2 Estimation

The cost of each item of work is calculated and the rates of different items of work are taken as per the schedule of rates or current workable rates for finished items of work are tabulated in Table 1.

For H/D = 0.91081, TOTAL COST = 82,78,906/-

2.3 Machine Learning Procedure

There are a number of feature factors that must be taken into account when constructing any tank. These are a few of the parameters:

1. Tank volume, first
2. Dimensions of the tank
3. The tank's total height
4. H1
5. H2
6. H3
7. H/D Ratio

These factors taken together provide a conclusive result on a tank's cost. A good collection of data with the mentioned parameters can be used and integrated with Machine Learning for a better, faster, and more reliable cost prediction.

Machine Learning

The goal of machine learning, a subfield of artificial intelligence, is to gradually increase accuracy by simulating human learning through the use of data and algorithms. The same considerations that have increased the appeal of statistical methods are also the reasons why machine learning is regaining prominence, such as rising data volumes and diversity, less expensive and more potent computational processing, and affordable data storage. With all of this in mind, it is possible to quickly and automatically develop models that can analyse more, more complex data and provide faster, more accurate responses—even on a gigantic scale. An organization's chances of seeing significant possibilities or averting unforeseen threats are increased by creating detailed models.

Table 2 shows a small portion of the generated dataset along with the values for several feature parameters.

If the tank's volume and diameter (which can be pre-fixed for calculation of the rest of the values) were given as input, the model should forecast both the tank's cost and the H/D ratio. The H/D ratio is enough to derive the total height of the tank, h1, h2, and h3.

The data can be combined with Supervised Machine Learning methods to produce the best results because the model is given both corresponding output parameters as well as a number of input parameters.

There are other controlled methods of training algorithms that serve the purpose but it is equally important to choose the right one. Heatmaps or correlation matrices

Table 2 Sample of feature parameters from the dataset

Volume	Diameter	H1	H2	H3	Height	H/D	Cost (lakh)
1000	11.95	1.99	7.08	2.24	11.31	0.95	59.06
990	11.91	1.98	6.95	2.23	11.17	0.94	58.79
980	11.87	1.98	6.89	2.23	11.09	0.93	58.42
970	11.83	1.97	6.83	2.22	11.02	0.93	58.12
960	11.79	1.96	6.78	2.21	10.95	0.93	57.69
950	11.75	1.96	6.72	2.20	10.88	0.93	57.34
940	11.71	1.95	6.68	2.20	10.83	0.92	56.89
930	11.67	1.94	6.62	2.19	10.75	0.92	56.24
920	11.62	1.94	6.55	2.18	10.66	0.92	55.86
910	11.58	1.93	6.50	2.17	10.60	0.92	55.25
900	11.54	1.92	6.45	2.16	10.54	0.91	54.92
890	11.5	1.91	6.39	2.15	10.46	0.90	54.65
880	11.45	1.90	6.34	2.14	10.39	0.90	53.98
870	11.41	1.90	6.3	2.13	10.34	0.90	53.29
860	11.37	1.89	6.25	2.13	10.27	0.90	52.88
850	11.32	1.88	6.21	2.12	10.21	0.90	52.02
840	11.28	1.87	6.18	2.11	10.17	0.90	51.49
830	11.23	1.87	6.13	2.10	10.10	0.89	50.98
820	11.19	1.86	6.08	2.09	10.04	0.89	50.26
810	11.14	1.85	6.04	2.08	9.98	0.89	49.56
800	11.09	1.84	6	2.07	9.92	0.89	48.95
790	11.05	1.84	5.94	2.07	9.85	0.89	47.69
780	11.00	1.83	5.9	2.06	9.79	0.89	46.98
770	10.95	1.82	5.86	2.05	9.73	0.88	46.36

that show the quantitative relationship between these factors can be used to determine the best model.

The cost and h/d ratio can be accurately predicted by a well-chosen model.

3 Results and Discussions

Figure 2 shows the linear relation between two feature parameters Volume and Cost.

Figure 3 shows the non-linear relation between diameter and cost.

Figure 4 shows data points and SVM model prediction with a polynomial kernel.

Figure 5 shows the data points and Linear Regression model prediction.

Figure 6 shows the data points and Decision Tree model prediction.

Figure 7 shows the data points and Polynomial Regression model prediction.

Figure 8 shows the data points and SVM model with linear kernel prediction.

Figure 9 shows the data points and Lasso model prediction (Fig. 10).

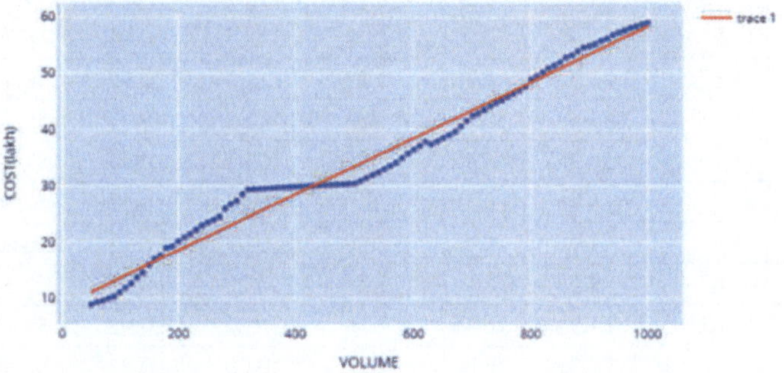

Fig. 2 Volume and cost relation

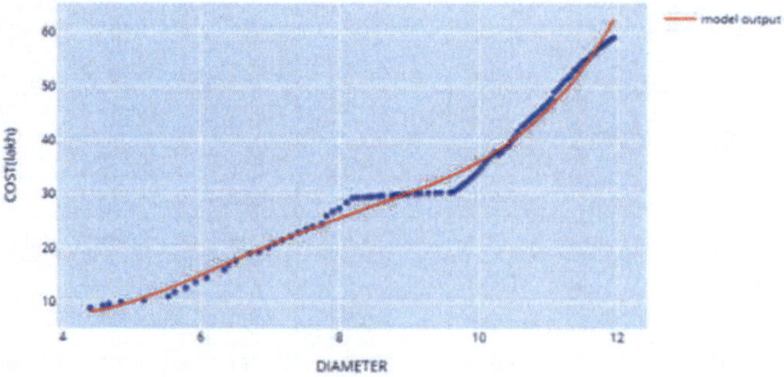

Fig. 3 Diameter and cost relation

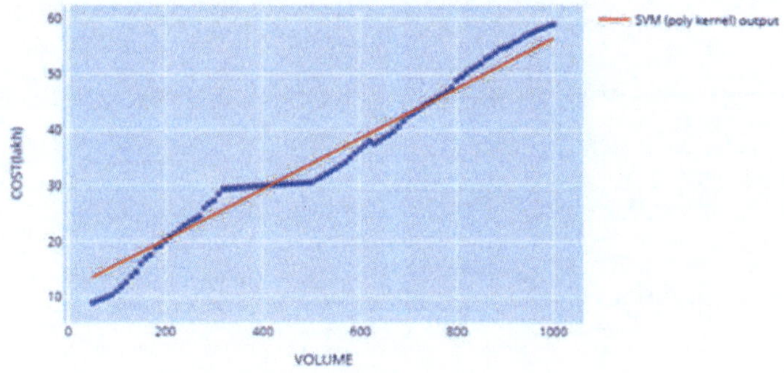

Fig. 4 SVM (kernel = 'poly') plot diagram (training)

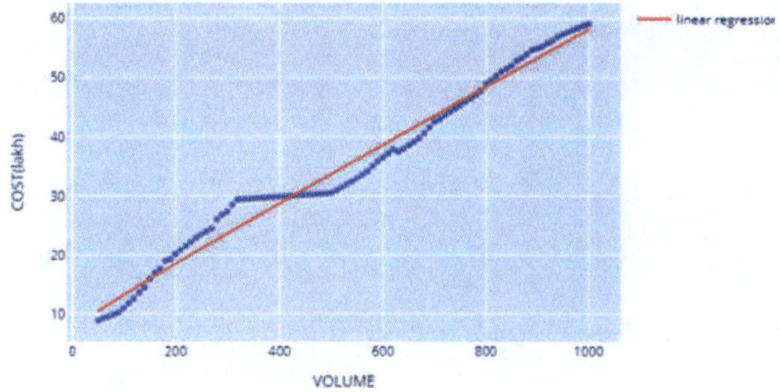

Fig. 5 Linear regression plot diagram (training)

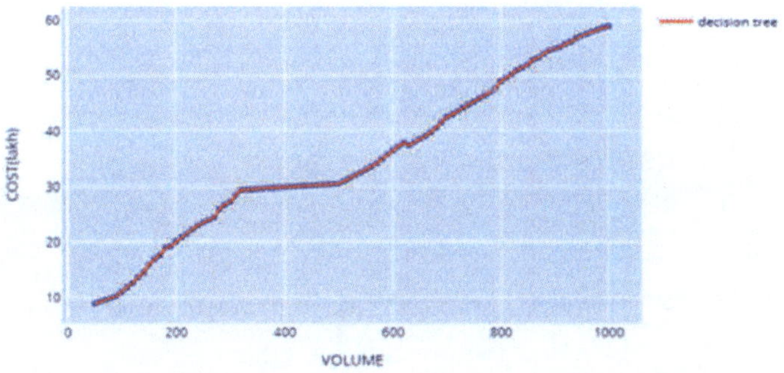

Fig. 6 Decision Tree plot diagram (training)

Figure 11 shows the programmatic results generated by the SVM model when volume = 1500 m^3 and diameter = 13.68 m are given as input parameters.

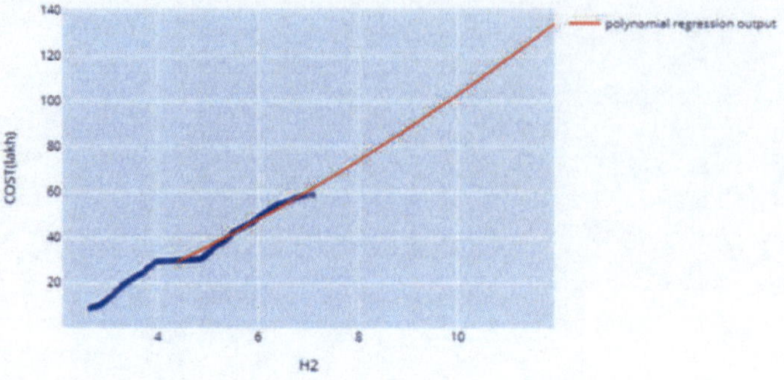

Fig. 7 Polynomial regression plot diagram (training)

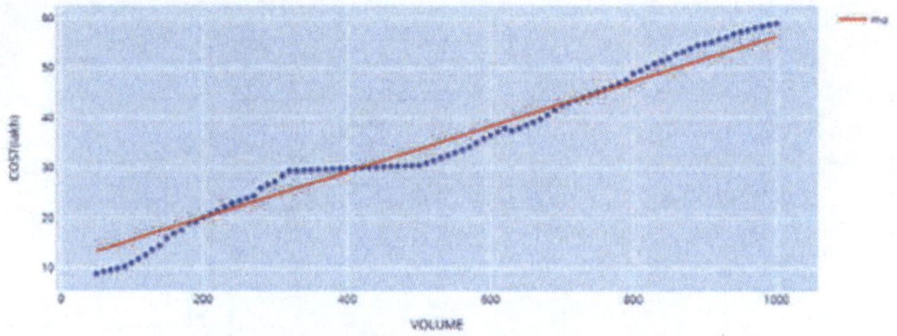

Fig. 8 SVM (kernel = 'linear') plot diagram (training)

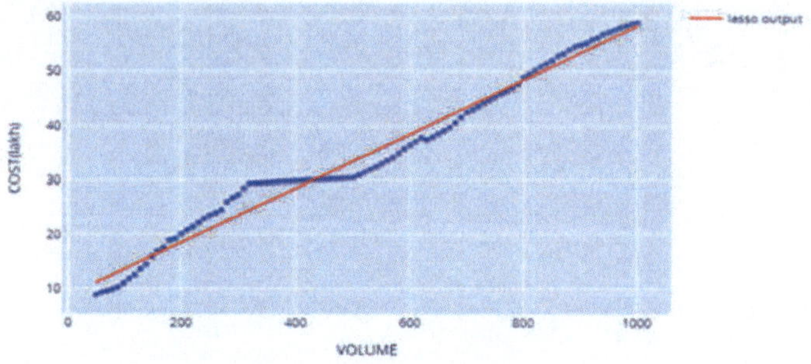

Fig. 9 Lasso plot diagram (training)

The following individualities were observed:

- The association involving cost and volume is linear, but the association involving cost and diameter is polynomial.
- Since Volume and Diameter are both considered as input factors, holding separate mathematical equations describing the costing of the tank, linear models like

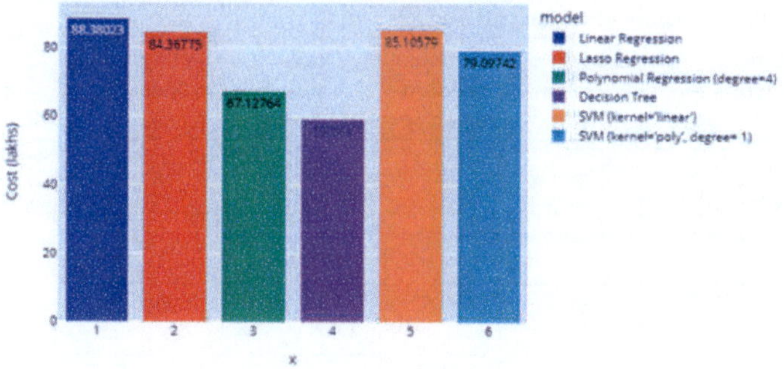

Fig. 10 Bar graph depicting the cost predictions of various supervised ML models

```
Enter volume of tank:
1500
Enter diameter of tank:
13.680



*******************************************************
                     MODEL OUTPUT
*******************************************************
Total Height of tank required: 12.20019709090909
H|D ratio: 0.89183
Cost of the tank: 79.09742 Lakhs
*******************************************************
```

Fig. 11 SVM (kernel = 'poly') model output

Linear Regression, and Lasso Regression give out of bound values in the resulting costs.

- The necessary link between the parameters is established through nonlinear models.
- SVM had different outputs for different kernels
  - Linear Kernel had similar results to Linear Regression and Lasso Regression.
  - Radial Basis Function Kernel created excessive higher dimensional data than required, which caused too much deviation from accurate values.
  - Polynomial kernel gave the most accurate training and testing results out of all models.

Table 3 Prediction values of various ML models

	Linear regression	Lasso regression	Polynomial regression (degree = 4)	Decision Tree	SVM (kernel = 'linear')	SVM (kernel = 'poly', degree = 1)
Cost (lakhs)	88.38023	84.36775	67.12764	59.06400	85.10579	79.09742
H/D ratio	1.06053	0.92250	0.92249	0.94657	0.89183	0.89183
Height (m)	14.50810	12.61986	11.53467	12.94905	12.20019	12.20019

Table 4 Comparison between SVM and manual calculation

	Manual calculation	SVM (kernel = 'poly', degree = 1) Prediction
Cost (lakhs)	82.78906	79.09742
H/D ratio	0.91081	0.89183
Height (m)	12.4300	12.20019

3.1 Comparison of Results

For *Volume*: 1500 m^3 *Diameter*: 13.68 m.

Table 3 shows the predicted values of various supervised ML models for a given volume of 1500 m^3 and diameter of 13.68 m.

Table 4 shows SVM model prediction in comparison to manually calculated values.

4 Conclusion

- Both manually and using STAAD Pro, the tank is to be designed; the program's results show that the design is safe.
- Through design is safe but we observe that reinforcement is less when compared with manual calculations.
- A responsible and accurate model should give values less or equal to manual results (with the least variance).
- In a comparison of values from Tables 3 and 4, the following conclusions are made:
- As stated in the results, computationally SVM (kernel = 'linear') and Lasso have better variance values (1.87 and 2.72%) than SVM (kernel = 'poly') which produced (4.34%). But the cost required by the former models is more than the manual calculations. Hence, they cannot be taken into consideration.

- Polynomial Regression had a good trace of the data points in the training phase but gave more than 20% more false positives during testing.
- Decision Tree and Random Forest had the highest training accuracy of 98% but gave the least accuracy in testing due to the overfitting of the data.
- From Fig. 6, the variation in output of Decision Tree compared to manual output is approximately 23.72 Lakhs which isn't the best model to use.
- Likewise, from Fig. 7 the variation in output of Polynomial Regression compared to manual output is approximately 15.66 Lakhs which isn't the best model to use.
- Out of all the models, SVM with the polynomial kernel (degree = 1) gave the best result with 95% test accuracy.
- SVM with polynomial kernel stands out with the least variance of 4% and cost difference of 3.69 Lakhs when compared with the manual calculations.

References

1. Gondalia R, Patel D (2017) Non-linear static pushover analysis on elevated storage reservoir. Int J Adv Eng Res Dev 4(4):1–6.
2. Rao UR, Goverdhan P, Divyavani B (2018) Analysis and design of RCC overhead water tank for seismic and wind loads. Indian J Sci Res 17(2):425–428
3. Nazary M, Mazloom M (2016) Optimization of shear walls with the combination of genetic algorithm and artificial neural networks. Indian J Sci Technol 9:43
4. Carstensen JV (2020) Topology optimization with nozzle size restrictions for material extrusion-type additive manufacturing. Struct Multidisc Optim 62:2481–2497. https://doi.org/10.1007/s00158-020-02620-5
5. Davies A, Veličković P, Buesing L, Blackwell S, Zheng D, Tomašev N et al (2021) Advancing mathematics by guiding human intuition with AI. Nature 600:70–74. https://doi.org/10.1038/s41586-021-04086-x
6. Danhaive R, Mueller CT (2021) Design subspace learning: structural design space exploration using performance-conditioned generative modeling. Autom Constr 127:103664. https://doi.org/10.1016/j.autcon.2021.103664
7. Chaillou S (2020) Archigan: artificial intelligence X architecture. In: Architectural intelligence. Springer, Singapore, pp 117–127. https://doi.org/10.1007/978-981-15-6568-7_8
8. Chang K-H, Cheng C-Y (2020) Learning to simulate and design for structural engineering. In: International conference on machine learning (PMLR), pp 1426–1436
9. D'Amico B, Myers RJ, Sykes J, Voss E, Cousins-Jenvey B, Fawcett W et al (2019) Machine learning for sustainable structures: a call for data. Structures 19:1–4. https://doi.org/10.1016/j.istruc.2018.11.013
10. Adilakshmi B, Suribabu P (2016) Design, analysis and optimization of intze type water tank for different parameters as per Indian codes. Int J Adv Res Sci Eng 5(1):1–7
11. Soni NK, Singh DP, Varma G (2018) Cost analysis of overhead tank foundation with varying depth of soil above footing. Int J Res Appl Sci Eng Technol 6(3):2435–2439
12. Chandana T, Surendhar SV (2019) Comparative seismic and cost analysis of RCC circular, rectangular and intze elevated water tank. Int J Innov Technol Explore Eng 8(8):1–8
13. Ramamrutham S (2010) Textbook: design of reinforced concrete structures, Dhanpat Rai Publishing Co Pvt Ltd, New Delhi, India
14. I.S-3370 (Part IV-1967) Code of practice for concrete structures for the storage of liquids
15. I.S-3370 (Part II-1967) Code of practice for concrete structures for the storage of liquids

Seismic Analysis of Building on Sloping Ground Including Soil-Structure Interaction

Dulganti Dheeraj Reddy and Pallavi Badry

Abstract An earthquake causes the loss of lives, property and damage to buildings and even collapse. Buildings that are irregular and unsymmetrical are more liable to earthquakes as they suffer large torsional and shear forces. Depending on the topography, buildings constructed on the hilly areas and sloping grounds are constructed in different configurations. The buildings are prone to damage when subjected to seismic forces. Buildings that are too flat and too steep experience increased story stiffness and increased displacements. The stepback-setback building and step-back building with induced bracings perform better and are recommended for the construction on the sloping ground. In this attempt, the behavior of the buildings is evaluated by performing static and dynamic analysis using the numerical model technique. A comparative analysis and parametric observations have been noted the structures considering fixed base and soil-structure interaction. To attain this objective, G + 4 step back building with induced inverted V bracings with fixed base and soil-structure interaction are considered. The buildings on various inclinations and soil types are modeled that was carried out for seismic analysis. The building with a suitable inclination of slope on various soil types has been evaluated. From the analysis, it has been found that the static analysi sand a dynamic analysis are essentially required to describe the seismic response of the building through various responses. The numerical analysis tool used in this study gives a rough idea on the behavior of the structures founded on the inclined ground. From the parametric study, it has been concluded that the buildings on inclinations ranging from 25° to 30° show better performance.

Keywords Soil-structure interaction · Step back building · Seismic analysis

D. D. Reddy (✉)
Utracon Structural Systems Pvt. Ltd., Panjagutta, Hyderabad, Telangana, India
e-mail: dheerudheeraj2296@gmail.com

P. Badry
Civil Engineering, Vidya Jyothi Institute of Technology, Aziz Nagar Gate, Hyderabad, Telangana, India

K. K. Pathak et al. (eds.), *Latest Developments in Civil Engineering*, Lecture Notes in Civil Engineering 352, https://doi.org/10.1007/978-981-99-2676-3_36

1 Introduction

1.1 Buildings on Sloping Ground

In present times, the demand for construction of the buildings in the hilly regions with sloping ground has increased with the speedy growth in the economy and urbanization in the hilly areas. Structures on the slopes are prone to failures due to the actions of earthquake and earth pressures imposed due to the slopy ground. Certain areas are grouped into significant earthquake zones on the basis of the past records of earthquake data considering the intensity and magnitude of the earthquake. Any structure to be constructed in a specific zone, must be ensured that the structure should have the possible seismic intensity. The topographical characteristics of the earth and its configuration affect the seismic vulnerability as well as performance of any structure built on slopes.

The earthquakes occur on account of the adjustment or sliding of the tectonic plates with respect to the fault plane. Usually, the region of hard rock strata with the trap rock or basaltic rock is more prone to earthquakes. As per the past earthquake data in India, most of the north-eastern parts and Himalayan regions categorized under the seismic zones IV and V are more prone to severe earthquakes of magnitude 8 and above since the Indian plate moves at the rate of 50 mm every year towards the Eurasian plate (web source).

To investigate the seismic response of a building of an adopted configuration, the structures are carried out for the numerical analysis through various engineering analytical tools as per the codal recommendations of IS 1893:2002. The suitable building-configurations possible on the sloping ground are Set-back buildings, Step back build- ings and a fusion of step-back and set-back as shown in Fig. 1.

As per the literature, the step back-setback building performs well. With the incorporation of bracings, building with the step back building configuration showed better performance in comparison to the step back-setback building configuration.

Generally, buildings on the sloping ground are with hard rock or soft soil confirming to analysis where the base is considered as fixed or flexible base analysis. Considering the interaction of foundation system (Soil and footing) and structure different types of soil (C, Φ, C-Φ) effect the efficiency of the building is due to the seismic activity. In fixed-base analysis, disregarding the soil type does not affect

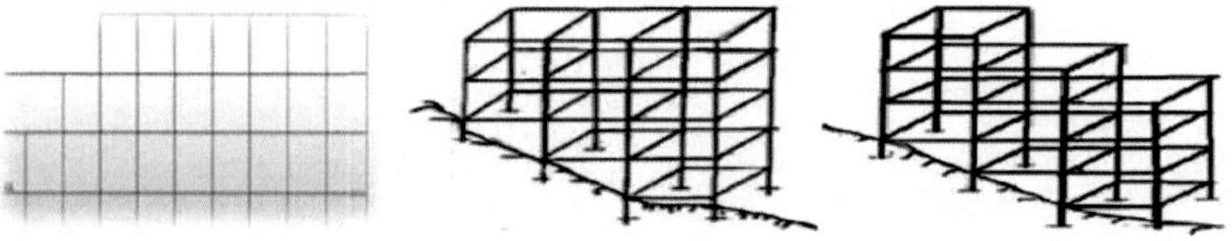

Fig. 1 Building configurations on the sloping ground (Ref. Birajdar, Nalwande). **a** Set-back building **b** Step-back building **c** Stepback-setback building

the response and the building performance whereas soil structure interaction shows significant behaviour. With the performance of seismic analysis, it has been found that the buildings without the consideration of soil-structure interaction system consider the heavy stiffness of columns and forces acting on the building get overestimated than actual. Besides the seismic responses of the building for various parameters gets underestimated. Hence by considering the soil-structure interaction system the lesser forces imposed on the building and the displacement responses of greater magnitude are observed.

2 Literature Review

The dynamic analysis has been performed on 24 numerical building models in various building configurations considering step-back, set-back, stepback-setback buildings. The models were analyzed and their responses such as the time period, story displacement and the base shear were analyzed. They summarized that Stepback-Setback buildings are more suitable for efficient construction on the inclined ground [1]. In 2013, Pradeep Kumar Ramancharla and Ajay Kumar Sreerama studied the earthquake behavior of G + 3 building at inclinations ranging from 0° to 45°. To perform the seismic analysis they applied the Northridge ground motion (N90E component of 0.565 g PGA and magnitude M6.7) with numerical tool using lateral load analysis and incremental dynamic analysis. From the fragility curves they concluded that the hinge mechanism developed with an increase in the inclination of ground at the slender columns.

Suresh and Arunakanthi [2] worked on the three dimensional seismic responses of the structure incorporated with Bracings on the sloping ground. The dynamic behavior of the building was analyzed through the response spectrum analysis and concluded that step back building with induced bracings in step back building performed better than the stepback-setback building of the respective configuration.

Halemani and Sreenivasa worked on the influence of the bracing systems in the buildings on sloping ground under Wind Loads and revealed that the X and inverted V bracing performed better [3].

In 2016, Karthik and Manjunath investigated the earthquake analysis of multi story Reinforced concrete buildings on slopes with various bracing systems. They performed Pushover analysis and the building response has been evaluated. They concluded that buildings with inverted V bracings performed well and the use of eccentric bracings is a successful way for the building safety.

Sharma and Pandey, 2011 worked out the seismic behavior of all possible building configurations on the sloping ground with various soil conditions and discovered that with the consideration of soil-structure interaction, the building showed drastic change in the responses with the height of the building. Beer, Kan and Clayton [4] studied the soil-foundation-structure interaction The building with step back-set back configuration recommended for the buildings on slopes in all the soils limiting the maximum ht to 30m Raghuvanshi, Sakalle and Arya [5]. Soil stiffness is an important parameter for controlling the time period, displacement, bending moments, story drifts and story shear of the building. The effect of soil-structure interaction varies as per the soil slope incli- nation. It was observed that the seismic response of the building was large with the consideration of the interaction system compared to the effect of inclination [6].

3 Methodology

3.1 General

The present study has been aimed to reduce the forces in the buildings on sloping ground where it is vulnerable to failures. As per the codal provisions in IS 1893, there are various methods to analyze the seismic performance of any structure statically and dynamically those deliver the responses of the proposed structure when exposed to seismic forces of certain magnitude and acceleration. The rigorousness of the effect depends on the earthquake zone, the magnitude of the earthquake, the configuration of the building and the type of soil on which the structure has been resting.

3.2 Modeling

For the evaluation of the seismic response of the buildings on the inclined ground, a five-story rectangular step-back building of height 15 m with 4 bays (each of 3 m) in X-Horizontal and Z-Vertical directions respectively has been considered for the analysis. The ground with inclinations of 20°, 25°, 30° and 35° are considered for the analysis. With an increase in the slope, the length of columns varies largely in the ground story. Table 1 depicts the lengths of the columns in the ground story for the buildings with the respective inclination of the ground.

For the steel bracing members, the steel tube sections are modeled as truss members in inverted V form and attached to the external columns in the efficient

Table 1 Length of the columns in the ground story of the building models considered

Building	Number of bays (in X & Z directions)	Length of the bottom column (m)				
		Col 1	Col 2	Col 3	Col 4	Col 5
0° slope	4	3	3	3	3	3
20° slope	4	3	4.1	5.18	6.27	7.36
25° slope	4	3	4.4	5.8	7.2	8.6
30° slope	4	3	4.73	6.46	8.2	9.93
35° slope	4	3	5.1	7.2	9.3	11.4

Table 2 Geometrical properties of the structural elements

Geometrical Property	Dimensions
Inter story height	3 m
Length of the building along the slope	12 m
Width of the building across the slope	12 m
Thickness of the slab	150 mm
Beam cross-section	230 × 300 mm
External column cross-section	230 × 450 mm
Internal column cross-section	300 × 450 mm
Section of steel bracing	TUBE 122 × 61 × 5.4
Number of bays along the slope	4
Number of bays across the slope	4

configuration as per Ghosh and Debbarma [7]. The connections between the steel bracings with the frame are made rigid. Table 2 shows the geometrical properties of the structural elements incorporated in the building models.

From Pandey et al. [8] for three different kinds of soil, Table 3 shows the elastic properties of the foundation soil and Table 4 shows the spring constants for the isolated footing of the respective soil types.

The building structural system for all the considered models consists of beams, columns, slabs, walls, and foundations. Horizontal (beams) and vertical (columns) members are modeled as two noded beam elements with six degrees of freedom at every node. All the vertical load resisting elements are considered under diaphragms category and are modeled as four noded plate elements with six degrees of freedom at every node. The foundation for the building models with a fixed base is assumed to be fixed and for the buildings with the foundation system and structure interaction

Table 3 Elastic properties of foundation soil (Ref., Birajdar B. G & S. S. Nalwande)

Type of soil	Shear modulus G (KN/m^2)	Elastic modulus E (KN/m^2)	Poisons ratio μ
Hard	2700	6750	0.25
Medium	450	1200	0.33
Soft	85	255	0.5

Table 4 Spring constants for isolated footing (Ref., Birajdar B. G & S. S. Nalwande)

Type of soil	K_x (kN/m)	K_y (kN/m)	K_z (kN/m)	KR_x (kN/rad)	KR_y (kN/rad)	KR_z (kN/rad)
Hard	7310	7310	8122	1780	1778	2670
Medium	1252	1252	1520	335	335	445
Soft	250	250	367	80	80	84

is considered as flexible and the fixed base is replaced with equivalent linear springs. The material property is assumed to be isotropic. The buildings are modeled in the step- back building configuration. To improve the performance of the building as discussed earlier inverted V bracings of tube cross-section are incorporated into the building.

4 Seismic Analysis

For the determination of seismic responses of the building, static and dynamic analyses were performed by equivalent static analysis (static analysis) and response spectrum analysis (dynamic analysis) respectively for the fixed base and soil-structure interaction system (flexible base) conditions. The analyses were performed using the structural analysis tool STAAD Pro v8i software. The seismic parameters for the present study are listed in Table 5.

The buildings of the present consideration are assumed to be located in north-eastern India under the Indian seismic zone IV and intended for residential use. A live load of 3 kN/m^2 is considered for the analyses. The Indian codes are 456:2000, IS 875 (part-I), IS 875 (part-II), and IS 1893:2016 (part-1) are referred for the analysis of RCC frame (i.e., dimensions of beams, columns and slab), dead load, live load and earthquake load respectively. The parameters to be studied are base shear, spectral acceleration, time period, frequency, story drift, story shear, story displacement, absolute stresses, bending moments and shear forces.

Table 5 Seismic parameters considered for the analysis as per IS:1893:2016 (Part 1)

Title	Description
Seismic zone	IV
Zone factor, z	0.24 from (Table 2)
Type of building	SMRF from Table 7
Occupancy	Residential
Response Reduction Factor, R	5 from Table 7
Importance factor, I	1.0 (residential building) from Table 6
Type of structure	General building
Type of soil	Hard, medium and soft
Damping ratio	5%

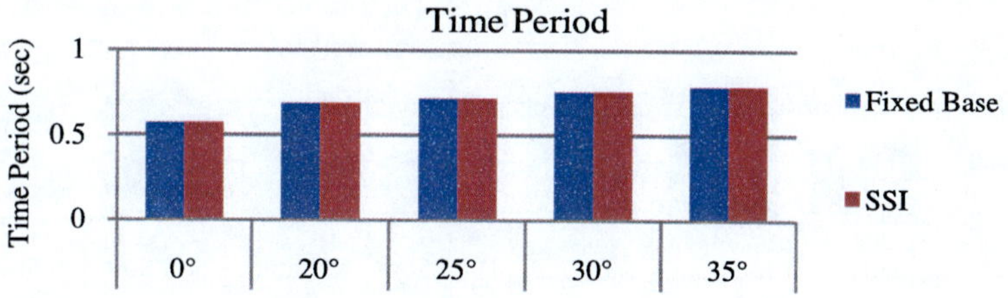

Fig. 2 Time period at varying slopes for fixed base and interaction system

5 Results and Discussion

All the building models were analyzed using the STAAD Pro software for linear static and dynamic behaviors as per IS 1893:2016 (part-1) and their responses are computed and compared.

5.1 *Time Period*

The variation of time period with slope is shown in Fig. 2.

From the bar graph profile Fig. 2, the time period in fixed base analysis and interaction analysis remained unchanged. It is noted from the observation that with an in- crease in the slope of the building time period is increased at the rate of 4%. The magnitudes of the time period in fixed base analysis and interaction analysis remained unchanged by performing the equivalent static analysis on the building models of various inclinations.

5.2 *Base Shear, V_B (kN)*

The variation of base shear with slope and type of soil is shown in Fig. 3.

From Fig. 3, the magnitudes of the base shear in fixed base analysis and interaction analysis remained unchanged. The results show that with an increase in the slope there is a decreasing trend in base shear with variations upto 4% and base shear of 25° and 30° is almost the same. The base shear of hard soil is 36% less than that of medium soil and it is 22% less than that of soft soil. The magnitudes of the base shear in fixed base analysis and interaction analysis remained unchanged by performing the equivalent static analysis on the building models of various inclinations.

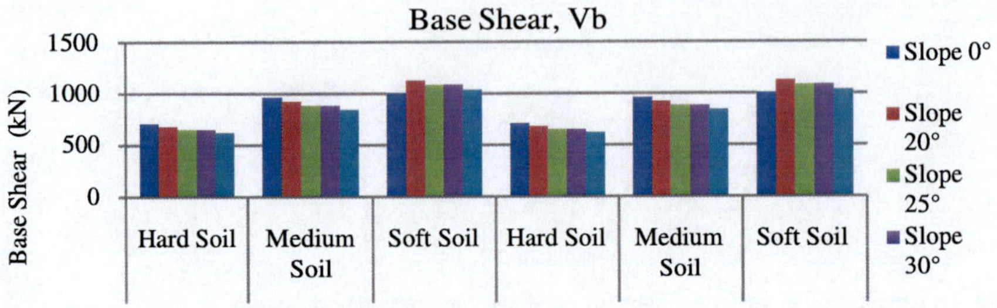

Fig. 3 Base Shear for different soils at varying slopes

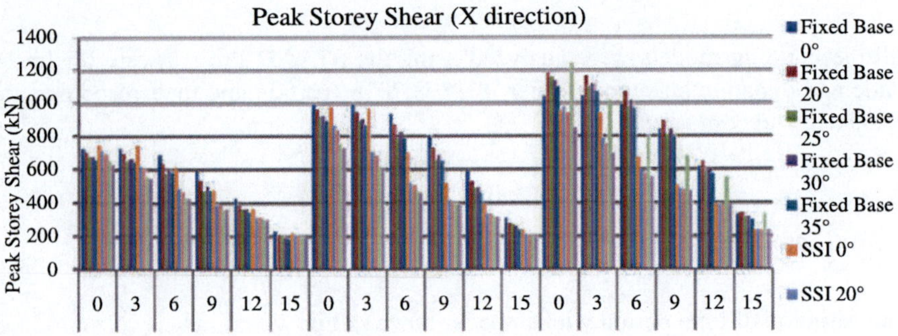

Fig. 4 Peak story shear in X-direction for different soils at varying slopes

5.3 Peak Story Shear in X Direction

The variation of time period with slope is shown in Fig. 4.

From Fig. 4, the peak story shear in the X direction had a decrease of 5% in hard soil, a decrease of 10% in medium soil and around 20% decrease in soft soil at all the inclinations of the ground with SSI.So with the consideration of the soil structure interaction in the buildings, the peak story shear had a drastic decrease in the magnitude that describes the decreased shear capacity of the building elements at each floor level.

5.4 Peak Story Shear in Z Direction

The variation of peak story shear in the z-direction is shown in Fig. 5.

From Fig. 5, the peak story shear in Z direction had a decrease of 5% in hard soil, a decrease of 10% in medium soil and around 20% decrease in soft soil at all the inclinations of the ground with the consideration of SSI. So with the consideration of the soil structure interaction in the buildings, the peak story shear had a drastic decrease in the magnitude that describes the decreased shear capacity of the building elements at each floor level.

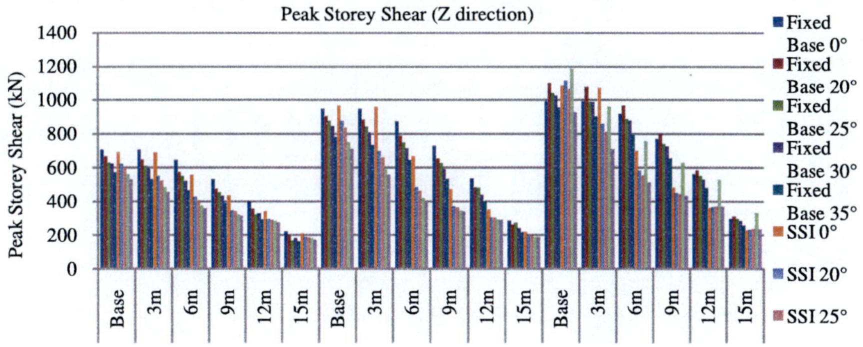

Fig. 5 Peak story shear in Z-direction for different soils at varying slopes

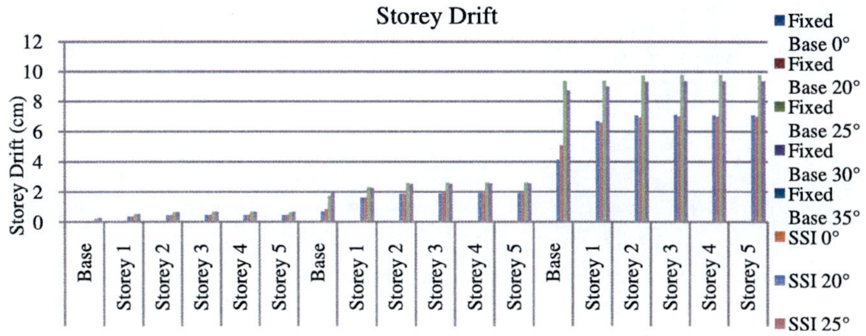

Fig. 6 Story drifts for different soils at varying slopes

5.5 Story Drift

The variation of story drift with slope and type of soil is shown in Fig. 6.

From Fig. 6, the story drift of the buildings had an increase of 8–10 times in hard soil, an increase of 50–60 times in medium soil and an increase of 400–500 times in soft soil at all the inclinations of the ground with the consideration of soil-structure interaction. So with the consideration of the soil structure interaction in the buildings, the story drift had a drastic increase in the magnitude that describes the increased deflection of the building elements at each floor level.

5.6 Maximum Bending Moments

The variation of maximum bending moments is shown in Fig. 7

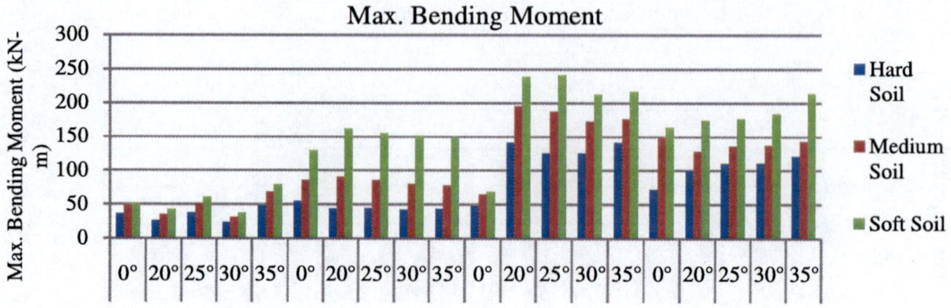

Fig. 7 Maximum bending moments for different soils at varying slopes

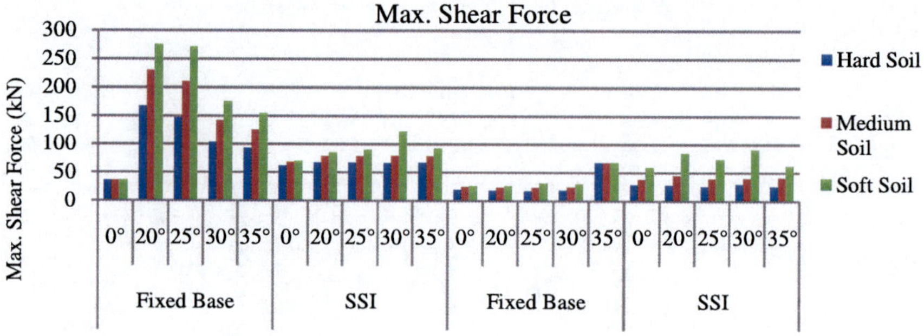

Fig. 8 Maximum shear forces for different soils at varying slopes

From Fig. 7, the longitudinal bending moment, My in all soils and at all the inclinations had an increase in the value from a fixed base to the interaction system about 40–50% whereas, the transverse bending moment, Mz had a decrease in the value from fixed base to the interaction system by about 5–20%.

5.7 *Maximum Shear Forces*

The variation of maximum shear forces is shown in Fig. 8.

From Fig. 8, the shear force, Fy in all soils and at all the inclinations had a decrease in the value from fixed base to the interaction system by about 40–50% whereas Fz had an increase in the value from fixed base to the interaction system by about 5–20%.

5.8 *Maximum Absolute Stress (kN/m^2)*

The variation of maximum absolute stress is shown in Fig. 9.

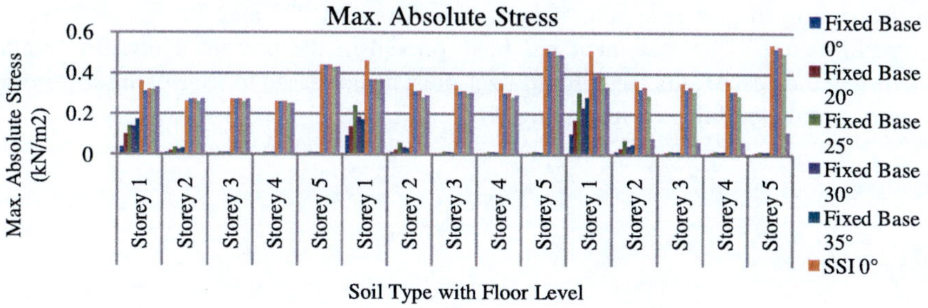

Fig. 9 Maximum absolute stress for different soils at varying slopes

From Fig. 9, the maximum absolute stresses at all the inclinations in all the soils in the buildings with interaction system had a drastic increase of 50% in the first story and at all other storys, the stresses are about 500 times with respect to the buildings on the fixed ground.

6 Conclusions

- The deviation in the results from the fixed base analysis and including SSI are negligible for the parameters such as time period and base shear.
- The peak story shears in X and Z directions are much reduced in the case of SSI condition than in fixed base condition and decreased with story level for the respective soil types in both the conditions. From the fixed base anal- ysis, higher slopes offer lesser peak story shear and in SSI, the construction in hard and medium soils in slopes ranging from 25° to 30° but in soft soil, the buildings on the 30° slope showed higher story shears. So the recom- mended slope is 25°.
- Story drifts are constant and are within the permissible limits (i.e., inter story drift is not more than 0.004 times the interstorey height) and in- creased with an increase in the slope in fixed base analysis. In the buildings with SSI consideration, story drifts increased with the change in the soil type. The drifts in the medium and soft soils exceeded the permissible limits of inter-story drift and they lead to the soft story condition. This can be minimized by increasing the column sizes.
- The maximum bending moments and shear forces are less for 25° and 30° slopes in hard and medium soils whereas, in soft soil the 30° slope showed more stresses. Hence the slope 25° to 30° are recommended for the construction in hard and medium soils and 25° is recommended for soft soil.
- The absolute stresses in the buildings in hard and medium soils have almost the same and lesser magnitude for buildings on the sloping ground than on the plain ground. In soft soil, the buildings on 35° slope showed reduced stresses.

- According to past research, step back building on 27° and 30° showed better performance. This statement has been proved in the present study. To obtain further the best results the column sizes must be increased to control the soft story effect in the buildings considering SSI conditions.

References

1. Birajdar BG, Nalawade SS (2004) Seismic analysis of buildings resting on sloping ground. In: 13th world conference on earthquake engineering, Vancouver, B.C., Canada, Paper No. 1472
2. Suresh G, Arunakanthi E (2014) Seismic analysis of buildings resting on sloping ground and considering bracing system. Int J Eng Res Tech (IJERT)3(9). ISSN: 2278-0181 IJERTV3IS090871
3. Halemani S, Sreenivasa MB (2015) Influences of bracing system in RC structures on sloping ground under wind loads. Int J Eng Sci Res Tech (IJESRT) Halemani 4(11). ISSN: 2277-9655
4. Clayton P, Kam WY, Beer A (2014) Interaction of geotechnical and structural engineering in the seismic assessment of existing buildings. NZSEE Conf
5. Raghuvanshi DS, Sakalle R, Arya R (2017) Analysis of amultistorey building frame for lateral forces at sloping strata under the effect of seismic forces using STAAD Pro. Int J Eng Sci Res Technol (IJESRT) 6(9):September. ISSN: 2277-9655
6. Suryawanshi AC, Bogar VM (2019) Seismic analysis of building resting on sloping ground with soil-structure interaction. Int Res J Eng Technol (IRJET) 06(07):2395–0072. e-ISSN: 2395–0056. p-ISSN: 2395–0072
7. Ghosh R, Debbarma R (2019) Effect of slope angle variation on the structures resting on hilly region considering soil-structure interaction. Int J Adv Struct Eng 11:67–77
8. Pandey AD, Kumar P, Sharma S (2011) Seismic soil structure interaction of buildings on hill slopes. Int J Civ Struct Eng (IJCSE) 2(2)
9. Sreerama AK, Ramancharla PK (2013) Earthquake behavior of reinforced concrete framed buildings on hill slopes. In: International conference paper, new technologies for Urban safety of mega cities in Asia
10. Manjunath CS, Siddu Karthik CS (2016) Seismic performance of RC buildings on sloping grounds with different types of bracing systems. Int J Res Eng Tech (IRJET) 05(02). eISSN: 2319-1163, pISSN: 2321-7308

Optimization of Reinforced Concrete Shear Wall by Machine Learning

N. B. S. Priyadarshini, M. Ashritha, Kamalini Devi, and A. Obulesh

Abstract In the existing work, cost optimization for an R/C structure system was performed by using machine learning. In fetch optimization, the aspects of the shear wall were contemplated as design variables, and the objective was to find the optimal shear wall aspects that keep down the overall cost of the material. The limitations of the structural inflation problem are designed according to the exigencies of the R/C spec stipulation named "IS 456:2000", IS 13920:1993, and INDIAS Seismic Code, which came into force as IS 1893:1975. Support elements are typically based on the designer's technical knowledge, experience, and intuition. In empirical design implementations, the eventual aspects are usually chosen as one of the bulk appropriate among numerous standard-compliant design choices. Though solely these blueprint alternatives may not be cheap, and the cheapest blueprint could only be apportioned through a more laborious optimization process. An application of the software is also being developed to determine the optimum shear wall dimensions for designing structural systems at a minimal cost. The recommended algorithm trivializes structural costs, including concrete and reinforcement costs, where fetch associated with transport, labor, and shuttering prices are excluded. An 18-story shear wall system in Sikkim, INDIA is a numerical example.

Keywords Machine learning · Structural optimization · Cost optimization · Shear wall

N. B. S. Priyadarshini (✉) · K. Devi
Department of Civil Engineering, Vidya Jyothi Institute of Technology (A), Hyderabad, Telangana, India
e-mail: priyaneerudu144@gmail.com

M. Ashritha · A. Obulesh
Department of Artificial Intelligence, Vidya Jyothi Institute of Technology (A), Hyderabad, Telangana, India
e-mail: avuku06@gmail.com

K. Devi
Department of Civil Engineering, Chaitanya Bharathi Institute of Technology (A), Hyderabad, Telangana, India
e-mail: kamalinidevi1@gmail.com; kamalinidevi_civil@cbit.ac.in

K. K. Pathak et al. (eds.), *Latest Developments in Civil Engineering*, Lecture Notes in Civil Engineering 352, https://doi.org/10.1007/978-981-99-2676-3_37

1 Introduction

Today, the need for housing increases with population and urbanization; Therefore, the construction sector has become more and more important. Although, the fact that buildable land is limited and expensive, particularly in thickly populated areas, it is observed that an optimal valuation of this land is required. To a great extent, as prices continue to rise, construction costs rise, making both aspects and fetch optimization inevitable, as well as crucial. When designing an erection, the geometric aspects of the constituent that make up the structural system of the structure are generally determined by engineering skills and experience accumulated over time. When sizing, the tensile forces to which the material is to be subjected must be within safe limits and the sizes selected must conform to specifications (IS 456:2000).

In building design, the pre-design details provided do not generally change significantly; the quantities acquired in the next or at least in the tertian explications are taken as quantities of the conveyance system.

The transportation system can be dimensioned in infinite ways to provided all inevitable circumstances, and the costs of individual transportation system alternatives may differ. The essential objective in engineering is to find an aspect with inexpensive fetch and guarantees the expected restraint. Optimization is widely used in numerous engineering streams today, and profuse arithmetical data processing approaches are used to provide optimal explication.

The bulk of these procedures acknowledges aspect variables as "continuous" and this feature may not apply to all technical issues. Especially within the construction industry, the employment of "discrete" style values is usually necessary for addition to mathematical programming techniques. The machine learning approach is one of the approaches using "distinct" aspects fluctuating.

1.1 Shear Wall

It is an upright component of a system designed to withstand in-plane lateral impacts, most commonly wind and tectonic loads. There exists excellent performance of buildings with the shear wall even under seismic force. Shear walls are now extensively used for all earthquake-resistance designs.

It should have good ductility under reversible and repeated loads. They provide lateral stiffness to the system as well as carry the gravity load.

2 Methodology

- An18 story commercial building is chosen for this study and designed layout by using auto cad isituated in Sikkim, India zone 4.

- The preliminary dimensions of the shear walls are taken according to the layout of the modeling.
- The Height and width of the building have not changed in modeling and only the length of shear walls ranges from 12 to 14 m.
- The thicknesses of shear walls are constant and equal to 300 mm. All building components such as beams and columns are the same in all models and have not changed.
- The shear wall structure is then designed manually by using the relevant provisions of the Indian seismic Code, and the Total cost is estimated by the reference to standard scheduled rates.
- The shape and position of the shear walls influence the behavior of the structure, the best position for shear walls is in the center of the building
- By manual design take design variables as dimensions of the shear wall (length and height), objective function as the total cost of steel, and total cost of reinforced concrete for a shear wall
- Optimize the data taken from manual designing by applying machine- learning techniques

The vital principle of machine learning is to escalate the standard of fitness of the objective function by accumulating new independents from the precursory generation at each step and to acquire the bulk advisable value as a result by providing the objective function in a way that is justified by some restrictions. In machine learning, the functioning first commences by ascertaining an initial generation. Each independent that makes up this generation is erect by being encoded in a binary number system. These independents are called artificial chromosomes.

- These chromosomes contain a finite number of successions and each 0 or 1 integer in the succession constitutes an artificial gene. Binary encoding is performed for each design variable that constitutes its sequence number in the set. This is called a sequence.
- In the optimization problem, a sequence whose number is identical to.

The number of design variables is written next to each other, thus forming an individual. A simple machine learning that works well for numerous complications entails three principal activities: breeding, mating, and gene crossing. Playback is the step in which independent successions are copied (played) and situated by the object's function values. In an efficient breeding operation, indistinguishable successions are higher likely to be chosen from independents with good fitness in the next step, proportional to the level of fitness.

Living individuals are combined into a mating group and randomly mated using a crossover operator to form a new generation.

The intersection methodology is split into three stages: initially, the generation's independents are chosen in sets. Furthermore, a random location is determined to perform a transverse operation among the genes of the selected individuals. Eventually, the crossing procedure is concluded by modifying the M.L relevant data at the specified location. Through implementing such an operation for every centrist of the preceding generation, the immediately preceding production is procured.

2.1 Optimization

The intended function of the frame in the optimization of steel structure systems is typically assumed to have been maintaining the structure weight low. Since steel systems have been made of a single component, limiting the mass of the material usually implies dropping the value of the material.

When it comes to optimizing reinforced concrete systems, however, the situation is quite different. Because reinforced concrete structures are made of two different materials, the structure's minimal mass does not always imply its low cost. Different materials, the minimal mass of the structure do not always mean that its cost is minimal.

In the optimal control enigma for reinforced concrete systems, the objective function should be an expense depreciation rather than mass simplification. The criterion f(x) in this scenario could be composed:

- (fx) = Cost of steel + Cost of reinforced concrete
- Design variable = Dimensions of the shear wall (length (lw) and height of shear wall)

The length values of the width section of the wall panels—lw are assumed as design values.

Because it has a different value for each changing design variable lw, the degree of reinforcement in the shear wall constituent appears as a reliant variable in the complication.

2.2 Machine Learning

- Machine learning is a sub division of Artificial Intelligence that focuses on data analysis. The methods weres the first proposed by Goldberg (1989).
- One of the first applications of genetic algorithm to civil engineering problems was done by Rajeev and Krishnamoorty (1992), and the method was applied as a three-bar truss system problem, in detail.
- The cost of reinforced concrete structures with medium ductile shear walls that were identical in plan and height had been improved by using a genetic algorithm and the artificial neural network was studied by Nazary and Mazlom (2016).
- In the study of Saka (1998), a genetic algori-thm was presented for the optimum design of grillage systems to decide the cross-sectional properties of members from a standard set of universal beam sections.

- In Sahab (2005) paper, cost optimization of reinforced concrete slab buildings according to the British Code of Practice (BS8110) is presented. The objective function is the total cost of the building including the cost of floors, columns and foundation.
- The paper prepared by Castilho (2007) describes the use of a modified GA as an optimization method in structural engineering for minimizing the production costs of slabs using precast prestressed concrete joists.
- Govindaraj and Ramasamy (2005) paper present the application of genetic algorithm for the optimum detailed of reinforced concrete continuous beams based on Indian Standard specifications.

3 Modelling

In this research, an 18-story building with the same height and AutoCAD floor plan was used for modeling, Fig. 1. During the modeling, only the length of the wall panels, varying from 12 to 1 m, was changed to maintain the height and width of the building.. Unchanged, the thickness of the shear wall is constant and corresponds to 300 mm. For each model, the beams and supports that make up the structure are the same and remain unchanged.

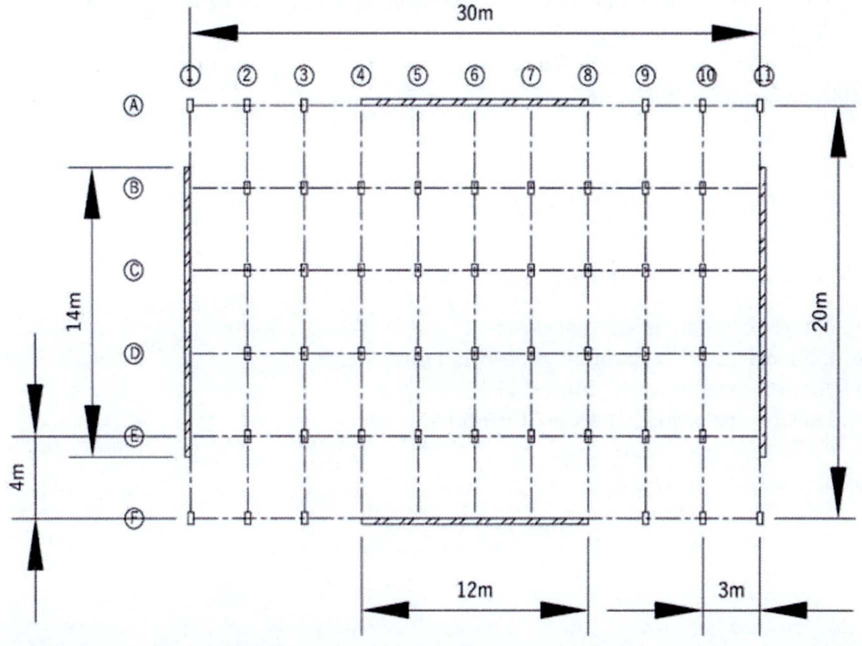

Fig. 1 *Note* Case study plan in Fig. 1

3.1 Floor Planning Details

- Floor plan of an 18-floored structure =30*20 m
- No of Columns for each floor - 66
  Dimensions of the column = 300*600 mm
- No of Beams for each floor = 115 Dimensions of the beams = 300*600 mm 60 Beams, 3 m (span) 55 Beams, 4 m (span).
- 4 Shear walls for a floor
- 2 shear walls with 12 m length (lw_1)
- 2 shear walls with 14 m length (lw_2)

3.2 Manual Design

Floor plan of an18 story building of plan dimensions 30*20 m.

shear walls are to be provided on each story of the building to resist the tectonic forces

The axial load of each wall is 1500 KN due to both dead and live loads.

The height b/w the floors is 3.0 m Dead load per unit area of the story: Floor slab + finishes etc. = 4.5 kN/m^2

W.t of partitions of the story = 3 kN/M ^{2}

As a result of the concentration of live load on a single story = 3.5 KN/m^2

And roof live load = 1.5 KN/m^2

The soil surface beneath the substructure is Hard and the building is situated in Sikkim, INDIA (ZONE 4).

3.3 Floor Planning Details

Story plan of an 18-floored structure = 30*20 m

- No of Columns for each floor—66
- Dimensions of the column = 300*600 mm
- No of Beams for each floor = 115
- Dimensions of the beams = 300*600 mm
- 60 Beams,3 m (span)
- 55 Beams,4 m (span)

3.4 Material Properties

Material properties of reinforced concrete shear wall assumed as

- Concrete grade—M_{30}
- F_{CK} (characteristics of compressive strength of concrete)—30 MPa
- Reinforcement grade HYSD Fe 415

3.5 Seismic Weight of the Building

According to the regulations, the percentage of calculated live load that must be taken into account when calculating the seismic resistance is 25% of the floor slab, and the deck live load. So the effective weight of each layer is : 4.5 + 3.0 + 0.25*4 = 8.5 KN/m^2

And that at the roof= 4.5 KN /m^2

- Weight of 115 Beams:
- 60 Beams with a 3 m span, at each floorandroof0.3*0.6*(3*60)

  *25=810KN
  55 Beams with 4 m span, at each floor and roof = 0.3*0.6*(4*55)
  *25=990 KN

- weight of 66 columns on each floor: (66–18) = 48
- 0.3*0.6*2.4*48*25 = 518.4 KN
- Weight of columns at roof = ½*518

  = 259 KN

- Modelled area the of the story is 30*20
  = 600 mm^2
- Equivalent Load at Roof Level = 4.5*600
  +(810+990) +259 = 4759 KN
- Equivalent load at each floor = 8.5*600 + (810 + 990) + 518.4 = 7418.4 KN
  Seismic weight of the building (w) = Equivalent wt @each floor + Equivalent wt @ roof
  w= 7418.4 +4759 * 17 = 88321.4 KN

3.6 Base Shear

Base shear (V_b) = I/R *Z/2 *Sa/G * w (where W = 88,321.4 KN).

Where (h) = height of the building (here18 floors,3 m each = 54 m).

d = plan dimension (20 m)

T= 0.09*54/$\sqrt{20}$ = 1.0062

Building situated at Sikkim –zone 4
Coefficient of acceleration of the average response (Sa/g) = 1.81.

3.7 Lateral Loads and Shear Forces at Different Floor Levels Are Given

Equivalent load at roof level = 4759 KN

The basic natural vibration period T of a building with shear walls is given by (T) = 0.09 h/$\sqrt{d}$.

Design horizontal coefficient (Ah) = I/R

*Z/2 * Sa/g , where

I = importance factor – 1 R = response reduction factor – 4 Z = zone factor - 0.24.

(Ah) = 1/4 * 0.24/2 *1.81 = 0.0543

Base shear (V_b) = I/R *Z/2 *Sa/G *w = 0.0543*88,321.4 = 4795.85202 KN.

Equivalent load at each floor = 7418.8 KN Base shear (V_b) = 4795. 85,202 KN (Table 1).

Table 1 Calculation of lateral loads and shear

Mass No	We (KN)	Hi(m)	Wi (Hi)2	Wi(Hi)2/$\sum Wi*(Hi)$ 2	Qi (KN)	Vi (KN)
1	4759	54	13,877,244	0.1042929	500.1733	500.1733
2	7418.8	51	19,296,298.8	0.145019	695.489	1195.662
3	7418.8	48	17,092,915.2	0.128459	616.0703	1811.7323
4	7418.8	45	15,023,070	0.11290425	541.472	2353.2046
5	7418.8	42	13,086,763.2	0.0983521	471.6821	2824.886
6	7418.8	39	11,283,994.8	0.0848036	406.702	3231.588
7	7418.8	36	9,614,764.8	0.0722587	346.542	3578.1307
8	7418.8	33	8,079,073.2	0.06071739	291.191	3869.3217
9	7418.8	30	6,676,920	0.05017966	240.654	4109.9757
10	7418.8	27	5,408,305.2	0.0406455	194.929	4304.9047
11	7418.8	24	4,273,228.8	0.03211498	154.018	4458.922
12	7418.8	21	3,271,690.8	0.02458803	117.920	4576.842
13	7418.8	18	2,403,691.2	0.01806468	86.635	4663.477
14	7418.8	15	1,669,230	0.0125449	60.163	4723.64
15	7418.8	12	1,068,307.2	0.0080207	38.466	4762.106
16	7418.8	9	600,922.8	0.0045161	21.658	4783.764
17	7418.8	6	267,076.8	0.002007186	9.626	4793.39
18	7418.8	3	66,769.2	0.0005017	2.406	4795.796

$$\sum Wi * (Hi)2 = 133060266$$

3.8 Moment and Shear Force

Provides two shear walls in each direction to resist seismic forces. The lateral force acting on the shear wall will be half the calculated displacement shown in Fig. 2

The shear wall will be aspected as a cantilever beam, installed at the base, and braced at the top (Height of shear wall 54 m in Fig. 2).

Maximum shear force at base V = V_b /2

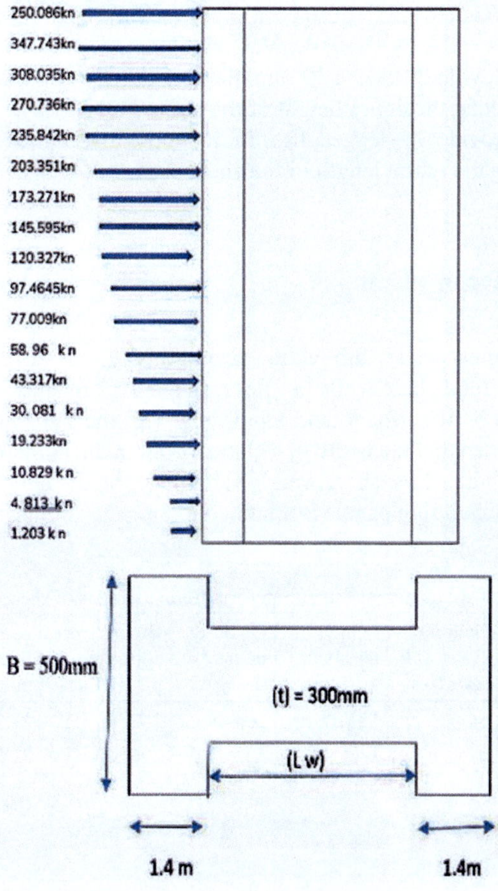

Fig. 2 Elevation of shear wall

= 4795.85202 / 2 = 2397.926 KN **MAXIMUM BENDING AT BASE (M)** (M) = (1.203*3) +(4.813*6) +(10.829*9) + (19.233*12) + (30.081*15) + (43.317*18) + (58.96*21) + (77.009*24) + (97.464*27) + (120.327*30) + (145.595*33) + (173.271*36) +(203.351*39) +(235.842*42)+ (270.736*45)+(308.035*48)+(347.743*51)+(250.086*54) (M) = 98006.16 KN m

3.9 Stresses and Material Properties

Typical stress results for a shear wall from the analysis section above are:

Let's take A partial safety factor = 1.5

Factored shear force (Vu) = 1.5*V 1.5* 2397.926 = 3596.889 KN

Factored bending moment (Mu) = 1.5*M 1.5* 98,006.16 = 147,009.24 KN.

Axial force on boundary element

=1.5*1500 = 22500 KN

SHEAR REINFORCEMENT (AS perIS 9.2.5 of IS:13920–1993)

Sv = 0.87*fy*Ah*dw/Vus

Provide 2- legged 12 mm dia horizontally aligned closed stirrups at 100 c/c.

Along the entire height of the shear walls

Provide 2 – legged 12 mm dia vertically aligned closed stirrups at 100 mm c/c along the entire length of the shear wall.

4 Estimation

The shear wall width section size of this system varies in horizontal section, so there are 4 total shear walls per floor, two are 12 m shear walls, and two are 14 m shear walls on the ground and a building with the same size and shear wall height. It is assumed that the width of the shear wall in the horizontal section is constant and that the width of the shear wall is variable along the length of the section. The thickness of the building panels is assumed to be constant throughout the building (Table 2).

Table 2 Manual cost estimation for different lengths of shear wall

S. No	Height of shear wall (m)	Length of the shear wall (m)	Quantity of steel (in kg)	Quantity of concrete (in m3)	Amount of steel (in lakhs)	Amount of concrete mix (in lakhs)	Total amount (in lakhs)
1	54	12	36,849.4	257.92	27.268	15.884	43.152
2	54	14	42,997.1	296.52	31.817	18.221	50.038

Table 3 Specimen designs for ML data

S.No	Height (in m)	Length (in m)	Concrete cost (in lakhs)	Steel cost (in lakhs)	Total cost (concrete + steel cost in lakhs)
1	54	8	11.21	18.17	28.381
2	54	9	12.14	20.44	32.824
3	54	10	13.54	22.71	36.267
4	54	11	14.71	24.99	39.710
5	54	12	15.88	27.26	43.152
6	54	13	17.05	29.54	46.595
7	54	14	18.22	31.81	50.038
8	54	15	19.38	34.09	53.481
9	54	16	20.55	36.36	56.924
10	54	17	21.72	38.64	60.366
11	54	18	22.89	40.91	63.809
12	54	19	24.06	43.18	67.252
13	54	20	25.23	45.46	70.695
14	54	21	26.39	47.73	74.138
15	54	22	27.56	50.01	77.580

5 Utilizing the Specimens Designed for Machine Learning

To acquire the data need for the machine learning techniques to link the specimens, 100 shear wall specimens of various lengths were used and their fetch was estimated separately. Machine learning consists of a set of (processing units) the connections between them, and adjustable weights (depending on the problem conditions) (Table 3).

6 Results and Discussion

Here following results were obtained by considering the design variable as dimensions of the shear wall (length and height) and objective function as the cost of steel and concrete of a shear wall by applying various supervised machine learning models for different lengths (Figs. 3, 4, 5 and 6) (Table 4).

→ The complete implementation of the model is done in google collaboratory and the dataset is acquired from the internet.

Data Set Link:-

→ Import the required packages such as NumPy, pandas, and sci-kit learn. NumPy is a Python library used for working with arrays. It also has functions for working in

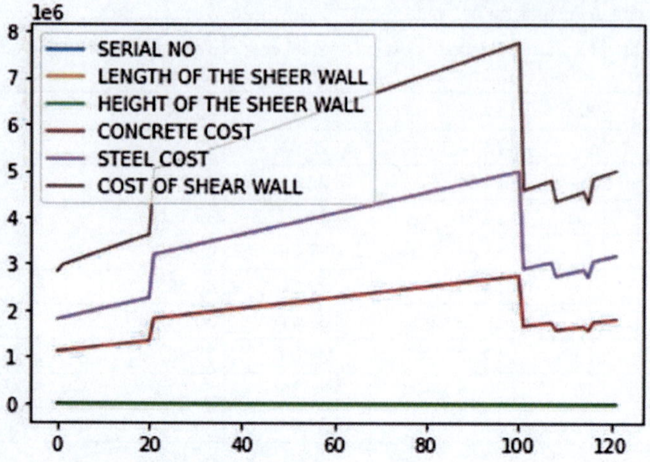

Fig. 3 Scattered plot of design variables

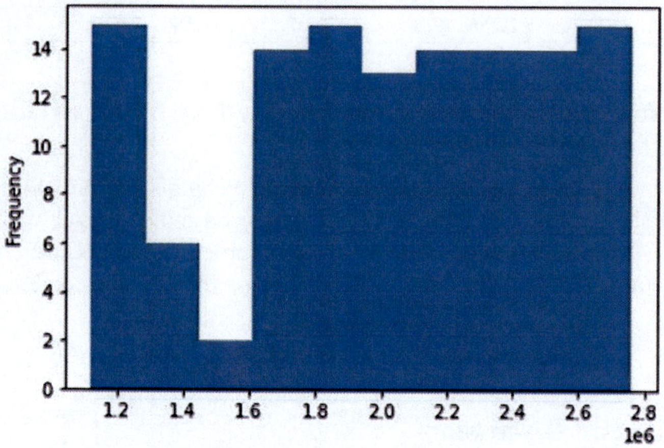

Fig. 4 Concrete cost frequency histogram

the domain of linear algebra, Fourier transform, and matrices. Pandas are used for playing with the data such as manipulating the data, analyzing the data, storing the data, and many more. Scikit learn is a library that stores all the machine learning algorithms.

→ Read the data set which is in CSV format with the help of pandas. The panda's package here converts the dataset into the data frame.

→ Now, go on exploring the dataset with the help of small functions such as Shape, which gives the dimensions of the dataset

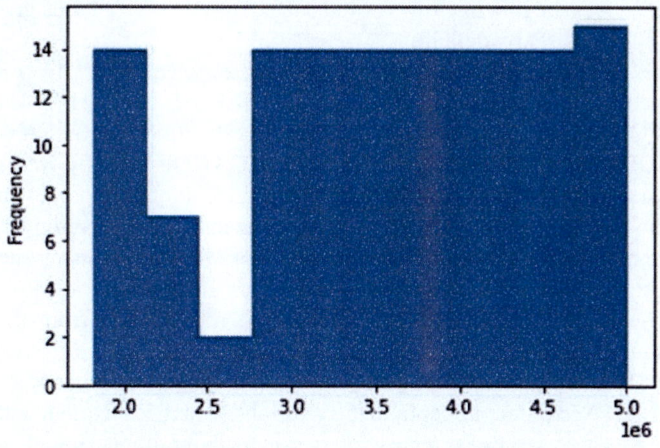

Fig. 5 Steel cost frequency histogram

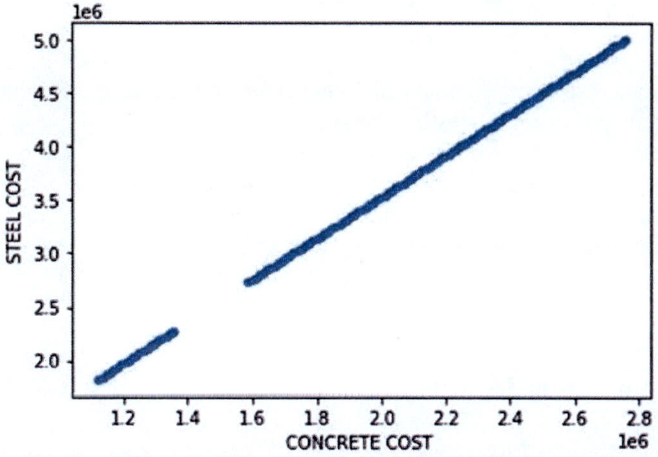

Fig. 6 Concrete and steel graph from data

Table 4 Cost predicted values of various supervised ML models

S.No	Length of the shear wall (m)	Linear regression (cost in lakhs)	Lasso regression (cost in lakhs)	Decision tree (cost in lakhs)
1	12	43.023	38.229	35.381
2	14	50.017	47.810	46.973

Head, this gives the first five rows of the dataset by default Tail, this gives the last five rows of the dataset by default

Columns, this gives all the column names in the dataset.

→ The columns in the provided dataset are 'SERIAL NO', 'HEIGHT in (m)', 'LENGTH OF SHEER WALL in (m)', 'CONCRETE COST OF SHEAR WALL ', 'STEEL COST OF A SHEAR WALL,' TOTAL COST OF SHEAR WALL.

→ **Preprocessing:-** This is one of the important techniques in model building. In this step, the data is cleaned. The major thing done in this step is removing the null values. The empty values are directly dropped from the dataset using the function dropna().

→ Now, the data is error-free and insightful. refined data from the original columns, so the original columns can be dropped.

→ Now all the columns except the scour depth are considered in the × 1, × 2, × 3….. Variables. And the scour depth in the y variable which is the predicted variable.

→ Here we have used linear regression whose math intuition is based on y=mx+c the x parameters keep increasing based on the columns available. The equation might seem this way.

$$y = m(x1 + x2 + x3 + x4 + x5 \ldots\ldots) + c$$

→ The data is split into 80% and 20%, 80 percent of the data is used in the training phase and 20 percent for the testing phase.

→ Linear regression is imported from the sci-kit learn library and the model is built, and also the accuracy of the model is also measured using the metrics such as r2, root mean squared error, and mean squared error.

→ The individual values of the r2 = 0.999

Mean squared error = 4597.74 Root mean squared error = 67.80

7 Comparison of Results

Here considered 2 types of shear walls with different lengths $(l_w)_1 = 12$ m and $(l_w)_2 = 14$ m Height of shear wall = 54 m lasso regression model prediction is compared to manually calculated values (Table 5).

Table 5 Comparison between the lasso regression model and manual estimation

S. No	Length (m)	Manual (cost in lakhs)	Lasso regression (cost in lakhs)
1	12	43.152	38.229
2	14	50.038	47.810

8 Conclusion

- The Total cost is estimated by reference to standard scheduled rates.
- Manual calculations are optimized with various supervised ML models
- Decision tree and linear regression had the highest training accuracy of 95% but gave the least accuracy in testing due to overfitting of the data
- Out of all the models, lasso regression gave the best result
- An 18-story commercial building is chosen for this study and designed manually and layout by using auto cad and it is situated in Sikkim, India zone 4.
- Through design is safe and observe that reinforcement is less when compared with manual calculations.

References

1. IS 13920 (1993) Ductile detailing of reinforced concrete structures subjected to seismic forces - Code of practice [CED 39: Earthquake Engineering]
2. IS 456 (2000) Plain and Reinforced Concrete - Code of Practice [CED 2: Cement and Concrete]
3. IS:4326 (1993) Earthquake resistant design and construction of buildings
4. SP:22(S&T) (1982) Indian standard explanatory handbook on codes for earthquake engineering - IS:1893–1975 & IS:4326–1976
5. Goldberg DE (1989) Genetic algorithms in search, optimization and machine learning. 13th ed. Addison- Wesley
6. Specifications for structures to be built in the disaster areas (1997) Available from: https://www.yumpu.com/en/document/view/23906558/specification-for-structures-to-be-built-in- disaster-areas
7. Rajeev S, Krishnamoorty CS (1992) Discrete optimization of structures using genetic algorithm. J Struct Eng 118(5):1233–1250
8. Feng C, Liu L, Burns SC (1997) Using genetic algorithm to solve construction time-cost trade off problems. J Comp Civil Eng 11(3):184–189
9. Stanley KO, Miikkulainen R (2002) Evolving neural networks through augmenting topologies. Evolut Comput 10(2):99–127
10. Saka MP (1998) Optimum design of grillage systems using genetic algorithm. Comp Aided Civil Infrast Eng 13(4):297–302
11. Sahab MG, Ashour AF, Toropov VV (2005) Cost optimization of reinforced concrete flat slab buildings. Eng Struct 27(3):313–322
12. Saini B, Sehgal VK, Gambhir ML (2007) Least-cost design of singly and doubly reinforced concrete rebar using genetic algorithm optimized artificial neural network based on levenberg-marquardt and quasi-newton back propagation learning techniques. Struct Multidiscip Optim 34(3):243–260
13. Ramasamy JV, Rajasekaran S (1996) Artificial neural network and genetic algorithm for the design optimization of industrial roofs - a comparison. Comput Struct 58(4):747–755
14. Pezeshk S, Camp CV, Chen D (2000) Design of nonlinear framed structures using genetic optimization. J Struct Eng 126(3):382–388
15. Keyhani A, Shafiee A (2016) Effectiveness of Seismic retrofit for existing concrete buildings using nonlinear static-anal- ysis. Indian J Sci Technol 9(2):1–8
16. Gen M, Cheng R (1997) Genetic algorithms and engineering design. John Wiley and Sons, Inc., New York

17. Paykani A, Akbarzadeh A, Shervanitabar MT (2011) Experimental investigation of the effect of exhaust gas recirculation on performance and emissions characteristics of a diesel engine fueled with biodiesel. Intern J Eng Technol 3(3):239–243
18. IITK-BMTPC Earthquake Tips: Learning Seismic Designand Construction

Computation of Stress Block Parameters of Rectangular RC Column at Elevated Temperatures

Chaitanya Akkannavar, M. H. Prashanth, and Sayed Abrar Korpalli

Abstract This paper proposes an approach to determine the load carrying capacity of Reinforced Concrete (RC) columns at elevated temperatures using Stress block parameters. The method uses a Strain distribution diagram for the RC column with various cases of eccentricity and utilizes maximum strain and minimum strain mentioned as per Eurocode. A modified strain distribution diagram is plotted using temperature-dependent strain properties, and a modified distribution diagram calculates stress at that level. Stress block parameters (C1, C2) are computed using these values, which can be used to determine Pu and Mu for various cases of eccentricities. Further, these values can be used to plot interaction curves for multiple temperatures following the standard time–temperature curve.

Keywords RC column · Fire resistance · Stress block parameters · Interaction curve

1 Introduction

When subjected to accidental fire, the structure's behaviour varies compared to the standard load (Static, Dynamic) condition for which they are analyzed. Hence, capturing the structure's response when subjected to high temperatures becomes essential. Column plays a significant role in any structure, and damage can cause progressive structure collapse. When fire accidents happen in a structure column may be subjected to 1/2/3/4 face heating depending upon the position of the column in the structure [1, 2].

Many codes still give prescriptive approaches for the design of columns at elevated temperatures [3–5]. After major fire accidents like the WTC collapse, Broadgate

C. Akkannavar (✉) · S. A. Korpalli
School of Civil Engineering, KLE Technological University, Hubballi, Karnataka, India
e-mail: chaitanya.d.405@gmail.com

C. Akkannavar · M. H. Prashanth
Department of Civil Engineering, NITK, Surathkal, Karnataka, India

K. K. Pathak et al. (eds.), *Latest Developments in Civil Engineering*, Lecture Notes in Civil Engineering 352, https://doi.org/10.1007/978-981-99-2676-3_38

fire, Fire at Ronan apartment in east London, and Windsor tower fire at Madrid [6–8], studies to capture the behaviour of structural elements have taken pace and researchers are coming up with empirical relations/experimental data which can help design the columns at elevated temperature [9–13].

Mohamed Bikhiet et al. [14] inspected the behaviour of RC columns of 15 × 15 × 100 (cm) size, performed non-linear FE analysis and compared results with experimental results. Studies indicate that smaller diameter bars and mild steel performed better than high strength larger diameter bars. Franssen [9] investigated different parameters influencing the capacity of the column in fire conditions. In the experimental study, parameters affecting the column's load-carrying capacity are analyzed. Roudsari et al. [15] conducted experimental studies on short axially loaded columns with fire load; studies indicate that column load carrying capacity decreases as fire duration increases. Balaji et al. [16] investigate the axial capacity of RC columns exposed to fire using the 500° isotherm method to evaluate column capacity. Further finite element analysis is performed to measure the effect of different parameters on column performance at elevated temperatures. Jaszczak et al. [17] studied on comparison of various simplified methods mentioned in Eurocode for calculating the load-bearing capacity of RC columns exposed to fire. Outcomes were compared against each method and experimental results. Barreto et al. [18] proposed an approach for the evaluation of axial force and moment carrying capacity for RC sections using modified stress–strain relationships.

The current design guidelines don't provide an approach to designing columns at elevated temperatures. In the design of short columns, stress block parameters play an important role in constructing interaction curves. These stress block parameters, in turn, depend upon strain distribution across the cross-section. An attempt is made here to revise those stress block parameters when subjected to elevated temperature. This can be further used to produce an interaction curve for different temperatures.

2 Methodology

The design approach involves

- To generate the time–temperature curve as per the standard fire curve.
- Temperature-dependent strain values from Eurocode EN 1992-1-2-2004 are tabulated.
- To generate a modified stress–strain variation diagram for the RC column considering the position of the Neutral Axis (NA) and temperature.
- To determine the depth of Fulcrum for the considered temperature.
- To calculate stress block parameters for the particular temperature.

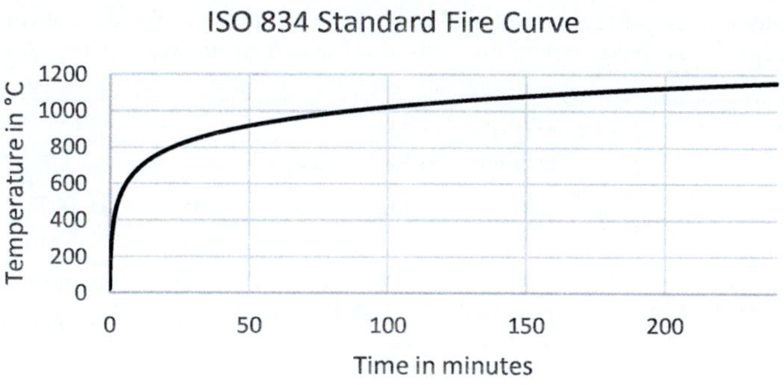

Fig. 1 ISO-834 standard fire curve

3 ISO-834 Standard Fire Curve

Based on the building type and purpose, the temperature build-up inside the structure changes. Hence it is essential to capture fire behaviour once fire sets. There are many standard time–temperature curves, out of which ISO 834 standard time temperature curve is used in the study (Fig. 1).

The following formula describes the typical temperature–time curve:

$$T = T_0 + 345, \log_{10}(8t + 1) \tag{1}$$

where

t time (min)
T heating temperature (°C) at time t
T_0 initial heating temperature (°C).

4 Temperature-Dependent Strain Values in Concrete for Siliceous and Calcareous Type Aggregates

Temperature-dependent strain in concrete for siliceous and calcareous aggregates is tabulated as shown in Table 1.

5 Calculation of Depth of Fulcrum in Strain Distribution

It will be observed that, as the eccentricity goes on increasing from zero to e_1 corresponding to $X_u = D$ the maximum strain ε_{max} along the highly compressed edge also goes on increasing from 0.0055 to 0.025 as per 200 °C temperature strain value

Table 1 Values for the primary variables affecting the stress–strain relationships of normal weight concrete with siliceous or calcareous materials at elevated temperatures. (Source: *EN 1992-1-2–2004*)

Concrete temperature θ (°C)	Siliceous aggregates			Calcareous aggregates		
	fc,θ/ fck	εc1,θ	εcu1,θ	fc,θ/ fck	εc1,θ	εcu1,θ
20	1.00	0.0025	0.0200	1.00	0.0025	0.0200
100	1.00	0.0040	0.0225	1.00	0.0040	0.0225
200	0.95	0.0055	0.0250	0.97	0.0055	0.0250
300	0.85	0.0070	0.0275	0.91	0.0070	0.0275
400	0.75	0.0100	0.0300	0.85	0.0100	0.0300
500	0.60	0.0150	0.0325	0.74	0.0150	0.0325
600	0.45	0.0250	0.0350	0.60	0.0250	0.0350
700	0.30	0.0250	0.0375	0.43	0.0250	0.0375
800	0.15	0.0250	0.0400	0.27	0.0250	0.0400
900	0.08	0.0250	0.0425	0.15	0.0250	0.0425
1000	0.04	0.0250	0.0450	0.06	0.0250	0.0450
1100	0.01	0.0250	0.0475	0.02	0.0250	0.0475

taken from Eurocode 1992-1-2-2004. The relation between ε_{max} and the minimum strain ε_{min} at the opposite edge for the Case of $X_u = D$ is obtained as under.

On superimposing strain diagrams of Case I ($X_u = \infty$) and Case III ($X_u = D$) the two diagrams intersect each other at point' O', which by properties of similar triangles, happens to 7D/9 with X_u lying between ∞ and D, the strain diagram is assumed to pass through fulcrum O for all strain distribution lines with the depth of the neutral axis X_u greater than D as per Indian Standards.

In Fig. 2 line A_1OB_1 is for Case-I ($Xu = \infty$), A_2OB_2 is for Case-II ($X_u > D$) and A_3OB_3 for Case-III ($X_u = D$).

Since point O lies on A_1OB_1, the strain at 'O' is 0.0055, which is taken as per Indian Standard diagrams.

From similar triangles OA_1A_3 and OB_1B_3,

$$\frac{\text{OA1}}{\text{OB1}} = \frac{\text{A1A3}}{\text{B1B3}} = \frac{\text{AA3} - \text{AA1}}{\text{B1B3}} = \frac{0.025 - 0.0055}{0.0055} = \frac{0.0195}{0.0055} = \frac{32}{9}$$

$$\frac{\text{0A1}}{\text{0A1} + \text{OB1}} = \frac{32}{32 + 9} = \frac{32}{41} = \frac{7}{9}$$

$$\text{But OA1} + \text{OB1} = \text{D Therefore OA1} = \frac{7D}{9}$$

$$OB1 = 1 - OB1 = 1 - \frac{7D}{9} = \frac{2D}{9}$$

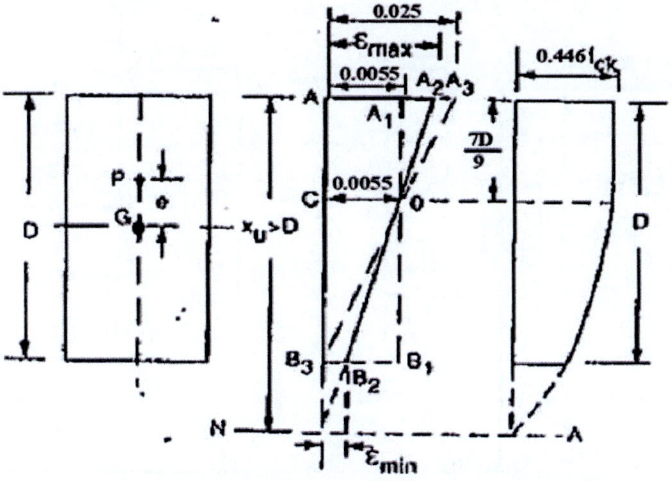

Fig. 2 Column under combined axial compression and bending for $x_u > D$

6 Calculation of Stress Block Parameters

According to the Indian Standard assumption for combined bending and axial compression, the strain ε is 0.0055 at a distance of $\frac{7D}{9}$ from the highly compressed edge. Therefore, from the properties of the stress–strain curve of concrete, the stress is constant and equal to f_{max} and hence, the stressblock is rectangular over a section where ε > 0.0055,i.e. up to a point 'O'. The stress distribution is parabolic on the remaining part where ε < 0.0055 with zero at N. A and maximum value at point 'O', The stress is, thus, ABCDF, as shown in Fig. 3. The properties of the stress block will now be derived from the above figure.

$$Letx_u = k_u D, g = EF = (f_{\max} - f_{\min})$$

where, $f_{\min}$ = stress at least compressed edge.

Now the equation of parabola DF, with vertex at D and y axis as an axis of symmetry for the parabolic curve, is given by: $y = ax^2$

Where x is a distance measured from vertex D at which the parabola is tangential to CD, and for obtaining the value of the constant 'a', we use the boundary conditions.

$$y = f_{\max}\ at\ x = \left(x_u - \frac{7D}{9}\right) \text{i.e.} \left(k_u D - \frac{7D}{9}\right)$$

Therefore

$$f_{\max} = a\left(k_u D - \frac{7D}{9}\right)^2$$

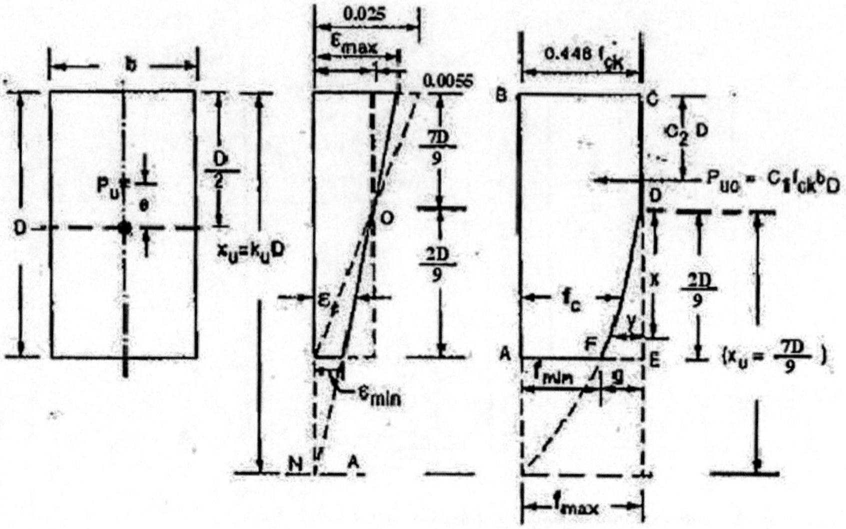

Fig. 3 Stress—blocked parameters for combined axial compression and bending for $x_u > D$

Therefore

$$a = \frac{f_{\max}}{(kuD - \frac{7D}{9})^2}$$

Therefore

$$y = \frac{f_{\max}}{\left(kuD - \frac{7D}{9}\right)^2} x^2$$

$$\text{Therefore } at\ x = \frac{2D}{9}, y = g$$

$$g = f_{\max}\left[\frac{\frac{2D}{9}}{kuD - \frac{7D}{9}}\right]^2 = f_{\max}\left[\frac{2}{9ku - 7}\right]^2 = f_{\max}\left[\frac{2}{(9 * 1.1 - 7)}\right]^2$$

$$g = f_{\max} * 0.47562 \tag{2}$$

Substituting $f_{\max} = 0.446 f_{ck}$ prescribed by IS code

$$g = 0.446 f_{ck}\left[\frac{2}{9k_u - 7}\right]^2$$

Area of stressblock = Area ABCE − Area DEF

$$A = f_{\max} D - \left(\frac{1}{3}\right) g \left(\frac{2D}{9}\right) = f_{\max} D - 0.07 g D$$

$$A = 0.446 f_{\max} D - 0.446 f_{\max} \left(\frac{2}{9k_u - 7}\right)^2 * 0.07 D$$

$$A = 0.446 f_{ck} D \left[1 - 0.07 * \left(\frac{2}{9k_u - 7}\right)^2\right]$$

$$\text{Putting}\, C_3 = \frac{4}{9}\left[\frac{2}{9k_u - 7}\right]^2 = \frac{4}{9} * 0.47562 \tag{3}$$

$$\boldsymbol{C_3} = 0.209$$

$$\text{Therefore } A = 0.446\left(1 - \frac{C_3}{6}\right\}.f_{ck} D = C_1 fckD$$

$$\text{Where } C1 = 0.446(1 - \frac{C3}{6} \tag{4}$$

$$C_1 = 0.446\left(1 - \frac{0.209}{6}\right)$$

$$\boldsymbol{C_1} = 0.430$$

$$A\overline{X} = Area\, of\, ABCDE * \frac{D}{2} - Area\, DEF * \left(D - \frac{1}{4} * \frac{2D}{9}\right)$$

$$= (0.446 f_{ck} D) * \left(\frac{D}{2}\right) - (\frac{1}{3} * g * \frac{2D}{9})(D - \frac{2D}{36})$$

$$= 0.446 f_{ck} \frac{D^2}{2} - 0.06 * g D^2$$

$$= 0.446 f_{CK} D^2 [0.5 - 0.06\left(\frac{2}{9k_u - 7}\right)^2$$

Therefore

$$A\overline{X} = 0.446 f_{ck} D^2 \left(0.5 - \frac{C_3}{7}\right)$$

$$\overline{X} = \frac{A\overline{X}}{A} = \frac{0.446 f_{ck} D^2 (0.5 - \frac{C_3}{7})}{0.446 f_{ck} D (1.0 - \frac{C_3}{6})}$$

$$\overline{X} = \left[\frac{0.5 - \frac{C_3}{7}}{1.0 - \frac{C_3}{6}}\right] * D\text{Putting } \overline{X} = C_2 * D$$

Therefore

$$C_2 = \frac{\left(0.5 - \frac{C_3}{7}\right)}{\left(1.0 - \frac{C_3}{6}\right)} \tag{5}$$

$$C_2 = \frac{0.50 - 0.06 * 0.47562}{1 - 0.07 * 0.4756}$$

$$C_2 = 0.487$$

From referring to Eqs. 3–5, the modified stress block parameters for NA outside the section can be determined by considering strain values for different temperatures. This process can be continued for various temperatures, as shown in Table 2.

From referring to Eqs. 3–5, the modified stress block parameters for NA at the edge of the section can be determined by considering strain values for different temperatures. This process can be continued for various other temperatures, as shown in Table 3.

Table 2 Stress block parameters when NA falls outside the section

Code	K_u	ε_{max}	ε_{min}	C_1	C_2	X_1	X_2
20 °C IS 456 2000	1.1	0.0035	0.0020	0.3845	0.4428	3/7	4/7
Eurocode 20 °C	1.1	0.0200	0.0025	0.4403	0.4953	7/8	1/8
Eurocode 100 °C	1.1	0.0225	0.0040	0.4357	0.4915	5/6	1/6
Eurocode 200 °C	1.1	0.0250	0.0055	0.4305	0.4871	7/9	2/9
Eurocode 300 °C	1.1	0.0275	0.0070	0.4270	0.4841	3/4	1/4
Eurocode 400 °C	1.1	0.0300	0.0100	0.4167	0.4749	2/3	1/3
Eurocode 500 °C	1.1	0.0325	0.0150	0.3984	0.4573	1/2	1/2
Eurocode 600 °C	1.1	0.0350	0.0250	0.3643	0.4199	2/7	5/7
Eurocode 700 °C	1.1	0.0375	0.0250	0.3711	0.4279	1/3	2/3
Eurocode 800 °C	1.1	0.0400	0.0250	0.3769	0.4346	3/8	5/8
Eurocode 900 °C	1.1	0.0425	0.0250	0.3832	0.4415	2/5	3/5
Eurocode 1000 °C	1.1	0.0450	0.0250	0.3867	0.4452	4/9	5/9
Eurocode 1100 °C	1.1	0.0475	0.0250	0.3913	0.4501	1/2	1/2

Table 3 Stress block parameters when NA at the edge of the section

Code	K_u	ε_{max}	ε_{min}	C_1	C_2
20 °C IS 456 2000	1	0.0035	0.0020	0.3610	0.4160
Eurocode 20 °C	1	0.0200	0.0025	0.4274	0.4845
Eurocode 100 °C	1	0.0225	0.0040	0.4196	0.4775
Eurocode 200 °C	1	0.0250	0.0055	0.4134	0.4718
Eurocode 300 °C	1	0.0275	0.0070	0.4088	0.4675
Eurocode 400 °C	1	0.0300	0.0100	0.3964	0.4554
Eurocode 500 °C	1	0.0325	0.0150	0.3774	0.4351
Eurocode 600 °C	1	0.0350	0.0250	0.3398	0.3884
Eurocode 700 °C	1	0.0375	0.0250	0.3469	0.3980
Eurocode 800 °C	1	0.0400	0.0250	0.3531	0.4060
Eurocode 900 °C	1	0.0425	0.0250	0.3593	0.4138
Eurocode 1000 °C	1	0.0450	0.0250	0.3634	0.4188
Eurocode 1100 °C	1	0.0475	0.0250	0.3673	0.4235

7 Conclusion

In the present study, a method for designing RC columns under fire-conditions have been proposed. The Indian Standard Method of designing the column serves as the foundation for the proposed approach to design at elevated temperatures. Using Strain values at different elevated temperatures provided in Eurocode 1992-1-2, the distance of the Fulcrum is calculated using these values; further, the Stress Block parameters at different elevated temperatures are calculated. For further investigation, these values can be used to determine Pu and Mu for various eccentricities, and the Interaction diagram can be plotted at different temperatures following the standard time–temperature curve. The strength of column at any particular time of standard time temperature curve can be found using the interaction curve.

References

1. Tan HK, Yao Y (2003) Fire resistance of four-face heated reinforced concrete columns. Struct Competency Archit 262–271. https://doi.org/10.4324/9780203583159-39
2. Tan KH, Yao Y (2004) Fire resistance of reinforced concrete columns subjected to 1-, 2-, and 3-face heating. J Struct Eng 130:1820–1828. https://doi.org/10.1061/(ASCE)0733-9445(2004)130:11(1820)
3. ASTM International—ASTM E119-10 (n.d.) Standard test methods for fire tests of building construction and materialslengineering 360. https://standards.globalspec.com/std/1273811/astm-e119 Accessed 14 July 2022
4. Bureau of Indian Standards (BIS), IS 456 (2000) Reinforced concrete—code of practice. Ind Stand 1–106

5. EN 1992-1-2 (2011) Eurocode 2: design of concrete structures—Part 1–2: General rules—structural fire design 1
6. Bailey C (2015) The windsor tower fire, madrid, one stop shop in structural fire engineering. University of Manchester, pp 1–5
7. Pearson C, Delatte N (2005) Ronan point apartment tower collapse and its effect on building codes. J Perform Constr Facil 19:172–177. https://doi.org/10.1061/(ASCE)0887-3828(2005)19:2(172)
8. Ahrens M, Evarts B (2020) NFPA: fire loss in the United States during 2019. NFPA Res Rep 1–11. https://www.nfpa.org/News-and-Research/Data-research-and-tools/US-Fire-Problem/Fire-loss-in-the-United-States
9. Dotreppe JC, Franssen JM, Bruls A, Baus R, Vandevelde P, Minne R, van Nieuwenburg D, Lambotte H (1997) Experimental research on the determination of the main parameters affecting the behaviour of reinforced concrete columns under fire conditions. Mag Concr Res 48:117–127. https://doi.org/10.1680/macr.1997.49.179.117
10. Shah AH, Sharma UK, Kamath P, Bhargava P, Reddy GR, Singh T (2016) Fire performance of earthquake-damaged reinforced-concrete structures, Materials and Structures/Materiaux et. Mat Struct/Mater et Constr 49:2971–2989. https://doi.org/10.1617/s11527-015-0699-y
11. Shah AH, Sharma UK (2017) Fire resistance and spalling performance of confined concrete columns. Constr Build Mater 156:161–174. https://doi.org/10.1016/j.conbuildmat.2017.08.167
12. Lie TT (1989) Fire resistance of reinforced concrete columns: a parametric study. J Fire Prot Eng 1:121–129. https://doi.org/10.1177/104239158900100402
13. Chinthapalli HK, Agarwal A (2020) Effect of confining reinforcement on fire behavior of reinforced concrete columns: experimental and numerical study. J Struct Eng 146:04020084. https://doi.org/10.1061/(ASCE)st.1943-541x.0002617
14. Mohamed Bikhiet M, El-Shafey NF, El-Hashimy HM (2014) Behavior of reinforced concrete short columns exposed to fire. Alexandria Eng J 53:643–653. https://doi.org/10.1016/j.aej.2014.03.011
15. Roudsari SS, Abu-Lebdeh TM (2019) Evaluation of fire effects on reinforced concrete columns using finite element method. Am J Eng Appl Sci 12:227–235. https://doi.org/10.3844/ajeassp.2019.227.235
16. Balaji A, Luquman MK, Nagarajan P, Madhavan Pillai TM (2016) Studies on the behavior of reinforced concrete short column subjected to fire. Alexandria Eng J 55:475–486. https://doi.org/10.1016/j.aej.2015.12.022
17. Jaszczak B, Kuczma M, Szymkuć W (2021) Comparison of the load-bearing capacity of reinforced concrete columns under fire conditions using the method A, zone method and isotherm 500 method. Fire Saf J 124 https://doi.org/10.1016/j.firesaf.2021.103396
18. Caldas RB, Sousa JBM, Fakury RH (2010) Interaction diagrams for reinforced concrete sections subjected to fire. Eng Struct 32:2832–2838. https://doi.org/10.1016/j.engstruct.2010.05.002

Innovative Sustainable Design and Techniques: A Review of Literature

Rajesh Kumar, Vanita Aggarwal, and S. M. Gupta

Abstract Imperishable development is the most trending issue all over the globe which is proposed to decrease the significant influence of the construction industry on the surroundings, society and economy. Several developed nations have embraced the Green Building Construction considering the most affirmative solution to the conservation of their natural resources, less consumption of energy and minimize the construction negative impacts on the climate and environment. A critical review of the literature has been conducted exploring the implementation of the Green Building Construction (GBC) in India which consists of relevant works of sustainability in construction materials, building rating tools and the comparative studies, with the goal to optimize design effort. This study also takes considerations of the green building evaluation tools for rating the performance of different buildings include Comprehensive Evaluation Procedure for Built Environment Efficiency (CASBEE), Green Building Tool (GBTool), Leadership in Energy and Environmental Development (LEED), GRIHA, Building Research Establishment Environmental Assessment Method (BREEAM). This rating will persuade the constructors to work more in the direction of constructing more efficient green buildings and convert the old building to a green building.

Keywords Sustainable design · Construction materials · Environmental assessment

R. Kumar (✉)
M.M. Engineering College, M.M. (Deemed to Be University), Mullana, Ambala 133207, Haryana, India
e-mail: Raj1rathee@gmail.com

V. Aggarwal
Civil Engineering Department, M.M. Engineering College, M.M. (Deemed to Be University), Mullana, Ambala 133207, Haryana, India

S. M. Gupta
Department of Civil Engineering, NIT Kurukshetra, Kurukshetra 136119, Haryana, India

K. K. Pathak et al. (eds.), *Latest Developments in Civil Engineering*, Lecture Notes in Civil Engineering 352, https://doi.org/10.1007/978-981-99-2676-3_39

1 Introduction

A Sustainable/livable Building is a structure that is planned in such a manner to manufacture remodeled worked or reused in a biological and asset effective way. Green buildings are designed in such a way to connect determined objectives like saving occupant well-being; using water, energy and further more efficient resources and decreasing the influence on environment. Energy and substance depletion in buildings can put up noteworthy to worldwide change in weather conditions. The residential sector has the highest energy consumption amongst all sectors. Consequently, addressing energy efficiency in housing by providing comfortable environments that reduce the need for mechanical cooling and heating is essential. The cost of energy has risen as an outcome of the depletion of fossil fuel distribution in the whole globe. Due to this, nations around the globe have established sustainable plans between the formations of policy apparatus. Practically, all the fields including manufacturing, business, transportation, construction, have involved sustainable practices into their ongoing business plans to ensure surroundings safety [1]. According to the scientists and researchers, the way to diminish the hazards to the environment is to construct the structure imperishable and energy effective.

At design stage, the architect blueprint the structure through advanced software and techniques which evaluate, estimate and calculate the surroundings performance attributes of a building [2]. The environmental evaluation tools for buildings structure have been invented to furnish an objective assessment of quality of indoor environmental, assets use, and sustainable loadings, etc. [3]. These tools provides several methods to determine the criteria of sustainable buildings. The primary blueprint behind green building embraces efficient abate in consumption of energy, water and other resources, recycling of wastes and fortifying health of occupants. Green building uses alternate energy sources like solar energy that is renewal of electrical energy during daytime. Rain water harvesting systems are used in green building in such a way that water is efficiently reused before final disposal. A Green building is an energy efficient building which optimizes water use, and partially relies on natural sources of energy, generates less waste and provides a better indoor air quality for its occupants. It results in enhanced comfort level and productivity of its occupants.

2 Literature Survey

2.1 Sustainable Building Design: A Review

Wasilah et al. in 2019 studied the design process methods for green buildings and design with the principles of implementation in the structural design, the planning of the site and the natural energy for the conservation of the building's natural and thermal light. The study also explained that Matano Lake's potential views and sustainability are improved by the lake resort which is based on green buildings and design with natural principles [4]. Hammond et al. in 2019 studies the gap between acceptance and slow adoption and implementation of the Green Building Agenda that has been studied. From an economic point of view we use game theory insights, especially the game model "StagHunt" as an explanation of the decision-making procedure for building buildings stakeholders in respect to understand how risks, cost, system of the market and by-laws can generate obstacles to the overall implementation and adoption of sustainable buildings. In addition, we look at how the asymmetry of awareness and information complicates such issues further. This model was concluded to be more effective [5].

Huang et al. in 2018 discussed the limits of existing building practices to the future energy conservation index for climate envelopes in Taiwan. Second, a new index, OTTV, is based entirely on the simulation results, and has attempted to convert existing baseline criteria for the new OTTV. Studies have shown that the OTTV equation has to change its coefficient over time due to warming climates, so that the changing climate reacts accurately. The analysis of this paper provides a way for the formulation of a climate-responsive index to pursue a future more energy-efficient building [5, 6].

Richard N. Lacroix et al. in 2007 overviewed the concept of "Green Building" or "Green Architecture". They examined the relative terms, the efficiency and the functionality of their application through references to international practices (Asia, Europe, and USA) and techniques of eminent green builders and green designers [7]. N. Huberman et al. in 2007 studied those existing structures are completely answerable for plentiful vitality utilization and resultant environmental emanations in numerous nations. Investigation into vitality effectiveness of structures has boss spotlight on the necessity of vitality for a structure's progressing utilization, even the vitality epitomized in its creation is regularly neglected. This type of perspective has initiated in last few years to strategies which strengthen the thermal performance of a structure yet which depends on high encapsulated vitality (EE) materials and items. Despite the fact that database and appraisal techniques have created in most recent couple of years, the genuine EE force for a given substance might be exceptionally reliant on neighborhood innovations and transportation separations. The primary motive is to distinguish building materials which may improve the vitality necessity of a structure over as long as it can remember cycle. It was discovered that the exemplified vitality of the structure represents some 60% of the general life-cycle vitality utilization [8].

Rebecca C. Retzlaff 2009 studied that this article was motivated to provide organizers with a preface to the idea of green structures and frameworks for building assessments and to identify and examine important themes in writing, as well as arrangements. Six subjects are recognized: scope, weighting, subjectiveness, precision, adaptation, and examination of the life cycle. Despite the fact that there is plenty of room for developing work in the field of green structure problems, organizers have started to create and actualize green structure arrangements and projects. The article ends with an expansive motivation of research, which can help to advance the exchange of work in the field of green structure issues. In order to add more weight to the problem, arrangements must progressively take place in green structures [9].

Kevern, J. T. 2011 discussed a system to consolidate the practical structure and think as another course of structural construction and meetings between the pilot's offers. Important territories are designed to help all designers understand maintenance in accordance with conventional building standards. In order to introduce ideas of structural support and basis support, introduction of green-building experts, green-building rating frameworks were used. Structural specialists can provide proactive solutions to a developing worldwide foundation by improving their understanding of support through training [10].

2.2 Innovative Sustainable Ratings

Sunita Bansal et al. in 2018 studied that green construction trends now give a lot of attention in the world today. In order to determine whether or not the specific building is green, the governing body develops many criteria for examining the growing practice of green building. The different criteria called the rating system are owned by each country, which is different according to the category, as in existing buildings and new buildings. The parameters of energy efficiency, materials, water efficiency, health etc. can be estimated. GRIHA and GBCI are essentially followed in India, ESGB in China and BREEAM in UK. In this paper, the various aspects of the existing green rating systems used in different countries were also overviewed [11]. Chandra Shekhar Singh et al. suggested in 2018 a green building concept involving a building designed, constructed, managed and maintained to improve the health of occupants, increase productivity, use natural resources wise and reduce the impact on the environment. This process focuses on the general environmental impact of the building including the design, construction, operation and maintenance phases, efficiency of development, resource efficiency, etc. The green buildings design should therefore start with the choice of environmentally friendly materials and the use of environmentally friendly materials. The fastest way to initiate the coordination of manageable plan ideas can be new application materials, reuse and, maintainable items and the careful selection of eco-logical feasible structural substances and the use of green assets. This paper emphasizes the contribution of sustainable construction materials to the production of healthy buildings that are environmentally sustainable. A small concept for saving the environment and saving energy: This paper presents GREEN CONSTRUCTION CONCEPT, a major step in eco-friendly energy and environment design in the building industry [9].

In 2015, Bhagyashri Kshirsagar et al. talked about the novelty and importance of the concept of sustainability in the construction sector worldwide. Today's market has innovative green building materials and technologies as green buildings are becoming increasingly aware. This paper examines the comparative analysis of BREEAM, LEED, Green Globes, IGBC and Eco Housing green building rating systems. The purpose of this paper is to propose amendments to Indian green building classification systems in the current scenarios in order to deliver the most efficient and healthy green construction [12]. Cesar A. Poveda et al. in 2014 explored the prospective advantages of execution ERS in industries apart from the construction sector have been explored. The growing demand for the exploration and exploitation of natural resources has given greater attention both to the organization's impact and its stakeholders. It must be regulated with the intervention of government department and legal needs as single solution to diminish the negative influence. These procedures often give small practical help to companies in meeting triple end-point sustainability objectives while providing general guidance. More recently, a variety of tools have been developed to help businesses take decisions that best fit these objectives. This paper identifies potential benefits for industries to choose to use these tools to evaluate their sustainable performance [13].

Kushagra Varma et al. in 2014 proposed that buildings are one of the largest energy consumers and greenhouse gas producers. Buildings produce 35% of CO_2, 49% of SO_2, and 25% of NO found in the air, according to the NIBS (USA). In addition to thinking and implementing sustainable development in every new building, it has imposed immediate requirements. Green buildings agree to the principle of careful handling of the natural resources, which means that the use of surroundings friendly substances and the utilization of renewable energy assets cause as little environmental interference as possible. This paper concentrates on green design as a vital change in the architecture of contemporary development countries [6]. Farzad Jalaei et al. in 2014 studied that the building energy consumption is quite high as compared to normal; a major concern must thus be demonstrated with regard to sustainable construction projects and energy performance. Energetic analytical tools (EAT) have been used for years to design energy-efficient buildings. (BIM) Building Information Modeling has the potentiality to assist users evaluate different design substitutes, choose key energy strategies and systems at the imaginary design phase. With BIM tools, designers can choose the exact material type during the blueprint phase. A process that links EAT and BIM with a green building certification system is the principal goal of this study. It helps designers identify and measure potential energy losses or gains in various design substitutes and carried out possible LEED points, can pile up and get and choose the best one accordingly [13].

Peng Wu et al. in 2010 studied that the building sector is the greatest reason of greenhouse gas emissions in the world. One problem that is begging to be addressed originating from both interior and outer drivers for development and building organizations is being manageable or green. Several green rating systems have been developed to evaluate how sustainable or green the building is. This research focuses on comparing the Green Globes, LEED and the BCA Green Mark for comprehension present practices and addressing the importance of project management in sustainable and green construction. The discoveries recommend that the designing and development organizations take venture of the executives regarding both the training and the procedure into thought while satisfying the prerequisites of being green [7].

2.3 Green Building Challenge

Sabnis and Pranesh et al. in 2016 explained the impact of energy of generally adopted three formwork techniques in India and also briefed their inductions as far as collaboration esteems. They consider three diverse connection conditions I1, I2 and I3 as association between encapsulated vitality and materials and epitomized vitality. All around association was taken as logarithmic total of three affiliation. It was found that the total correspondence estimate for each square meter of the progress zone was based on ordinary formworks with steel floor plates, which differed from the estimated 31,781 and 30,490 domain improvements for coated systems with squeezed aluminum board and wood floor plates. In this respect, the supreme impact on the substance can be reduced by picking aluminum coating instead of customary coating by approximately 33% [14]. Joel Ann Todd et al. in 2015 stated that Green Building Challenge (GBC) was designed for the test, the improvement and the discussion of an assessment structure, criteria and instrument to push the flowering edge of the project execution examination. GBC was not designed for this application on particular commercial markets, contrary to other proprietary assessment systems. The job of GBC over the last five years was to give reference technology, structure and tools to promote new frameworks [12]. Tayyab Ahmad et al. declared in 2015 that the sustainable use of sustainable building systems and technology is subject to a sensible construction. Long-lasting economic benefits, health and environmental protection and designers need to optimize combinations of specific techniques and systems to provide user comfort. This research focuses on suggesting various approaches to implement multiple sustainable building technology systems and techniques into the blueprint process in order to optimize the aim of the design effort. Interviews of building practitioners and researchers who help to express design priorities focused on semi-arid climatic conditions. At the same time, a model is proposed that incorporates all three dimensions of sustainable development into building counterfeit programs, ease the implementation of green design [15].

Ahmed et al. in 2014 studied critical literary review in implementation of the GBC in Ghana was investigated. The study examined the current situation in Ghana and the drivers who are responsible for the current state. The implications of these drivers in the construction of Ghana are also discussed. The main step towards increased awareness and publicity campaigns, national education, the enforcement of binding public policies and regulations and financial incentives and market-based incentives, among others, should be taken to strengthen the concept in Ghana and Sub-Saharan Area (SSA) [16]. Niklaus Kohler et al. in 2014 explained the achievements and objectives of the international Green Building Challenge project that are analyzed. GBC is situated within the criteria of other international environmental methods. The differences between data problems outcome by these are discussed. The circumstance of a possible headway for widening GBC into Life Cycle Assessment philosophy, into different periods of life cycle and adjusting GBC for use with the current structure stocks are proposed. The 'green' structures idea is supplanted by a bigger idea of supportable improvement. The subject of how significant the GBC proposed targets are in connection to the long-haul manageable improvement of structures stock, structures and urban situations is checked [17].

Huang et al. in 2013 assessed the environmental impact of China on the basis of material flow analysis, taking up to 2050, the consumption of the material in the construction industry. The results showed that natural extraction of materials and the effect of GHG in China will decrease in the years to come. They emphasized the reuse and recyclability of materials, in any case, to maintain practical progress. In the 2141 MJ/cum request, block stone work expends high rates of vitality among a few sub-systems comprised of a structural framework. SMBs use 715 MJ/cum compared to brick masonry. Around 38–45% of the brick block work vital is consumed by hollow concrete block. The most vital productive elective material in SMB manufacturing is closed [5]. Santamouris and Kolokotsa et al. discussed the application of passive cooling strategies in 2013, with the result that superior quality of indoor air and thermal solace with very low energy exhaustion significantly decreased their thermal impact. The three main methodologies for passive cooling are: the use of soil as a heat sink, evaporative cooling and, finally, the night cooling method. Passive strategies as an alternative to existing air conditioning at the building design stage are therefore of paramount importance [15].

2.4 *Innovative Sustainable Materials*

Sagheb et al. in 2011 studied that conventional material can be replaced by alternative materials with low incarnated energy in a building construction. Also admitted to the importance of low energy alternative materials. Based on their case study, they have found that a third decrease in carbon dioxide emissions is feasible by using low energy construction materials. The affordable housing policy and practice is important for energy conservation or usage of the recycling material. Constructing sustainable housing on a big quantum is the major challenge, committing a solution to the slum's density, unplanned increase, travelling congestion, and uncontrolled development of real estate. An academic view on policy for affordable housing and practiced that universally, affordable housing implementations are bifurcated into two wide strategic perspectives. Denmark, the Netherlands, Singapore and Sweden and other nations follow the global approach, where the whole population is furnished with decent and sustainable housing. On the other side, like the Canada, USA, and Malaysia, most of the European Unions, wherein poor sections aims, they would not get the same opportunities of the housing market.

Shahriar Shams et al. in 2011 declared that the building makes a large-scale contribution to greenhouse gas (GHG) emissions by emitting nearly 40% of carbon dioxide. Different types of construction materials emit CO_2 at different magnitudes. The choice of suitable building substances can considerably reduce CO_2 emissions and increase the sustainability and energy efficiency of our buildings. The research is a try to deal with the issues of imperishable construction of buildings and how building materials can be selected to reduce CO_2 emissions. In order to see CO_2

variation, a case study for a typical 90 m^2 home was analyzed. The case research showed that building substances like steel and aluminum should be encouraged less than glass and wood [18].

Sinem Korkmaz et al. in 2010 stated that green buildings of high performance require close building systems integration with an extraordinary spotlight on day lighting, vitality and material investigation during their plan forms. The present paper shows a case study of the Early Childhood Learning Centre's integrated design process carried out by a team of university students and faculty. During the blueprint development phases of the case research, the key decision on procedure modeling approach, necessary consultants, and virtual building prototypes was made. This case demonstrates that process modelling and visualization tools were used to give a precise information procedure for collective teams. From this experience, visualization tools and process modelling were seen as valuable components to accomplish superior structure objectives and diminish configuration process squander [1]. Daniel Castro-Lacouture et al. in 2008 explained that buildings cause a remarkable impact on the environment which is continuously increasing, firstly, due to the emission of major proportion of carbon and secondly, because of using a noteworthy number of energy resources. The development of green structure rose to limit these impacts and improved the structure development process. This effect ought to bring huge monetary, ecological, money related, and social advantages. So as to help the determination of the correct materials, the study proposes a multi-model number upgrade combining expenditure plans and structure limits while increasing the amount of loans provided to the rating system of LEED [19]. The renewable energy application for an existing building and reducing existing building energy depletion. To achieve this, two main perspectives are considered: One is to minimize the energy needs by implementing energy efficiency computations, and the other one is to offset the unused needs of building energy by optimizing renewable energy measures.

Huberman and Pearlmutter et al. in 2008 demonstrated that the use of life by the substances used in the designs of a casing absorb about half of the absolute vitality consumed by the structure and by the use of the elective low-exemplified building materials such as soil concrete squares, empty solid squares, fly debris blocks... The views correspond to studies conducted by [20, 21]. Thormark 2001, 2002 and Blengini 2009 note that the use of recycled materials can reduce energy consumption [22]. This is in line with Reddy and Jagdish's 2003 conclusions. In 2008, Dimoudi and Tompa added above to the situation by showing that 11 and 62% of steel and cement, about 8% of brick envelopes and about 4 per cent of aluminum composites contribute to GHG outflow. As detailed by Baird and Chan, Buchanan and Honey used vitality coefficients in the 1994 decision-making processes for all the carbon encapsulation. They have strongly suggested a move from solid steel and 45 aluminums to increased use of wood, which would cut CO_2 emissions significantly [20].

Reddy and Jagadish et al. In 2003 suggested alternative materials to enable reduction in the consumption of vitality saw that brick work comprises a significant segment in a structure development. In the Indian setting, they did ponder on five innovations of the building and looked at the utilization of vitality for each situation. They concluded that, stones use up least energy in the manufacturing process however; some energy is used during transportation of stones to site. They also observed that Cement is also one of the high energy consuming construction materials with a minimum consumption of energy 5.85 MJ/kg [23]. Catarina Thormark et al. 2001 state that during the existence cycle of a structure, vitality utilization is a fast-developing examination field. In structures, where vitality is low, the epitomized vitality makes up a rising piece of the utilization of vitality. Reusing of materials likewise also gives the chances to reduce the encapsulated vitality by considering reusable recyclable materials parts and reused materials. This paper tells esteems on required vitality for any activity, encapsulated vitality, and the capability of reuse of the most vitality e1cient condo lodging in Sweden (45 kWh = m^2). In 50 years of life expectancy, epitomized vitality represented 45% of the complete vitality need. The reusing potential was somewhere in the range of 35 and 40% of the encapsulated vitality [24].

Shabha et al. in 1997 evaluated a maintenance cost relevant study in high-rise buildings situated in Birmingham [25]. The authors highlight a low-energy renovation scheme. The structure cost and sustainable cost aspect of high-rise structures were described in the research. They described that an approximate cost of building construction at the element stage can be create by taking into consideration the component cost and then evaluating how much of the component is needed to complete the building.

2.5 Innovative Value Engineering

Pratheeban et al. in 2020 studied sustainable and low budgeted housing value engineering in India [26]. As there is a shortfall of requirement-based, i.e., affordable housing due to fast track increasing population, the aim must be on value engineering to achieve the gap/shortfall of housing requirements. The researchers focused at considering the value engineering process for cost optimization, quality and time. Also, they focused at increasing the total project value. They resulted that alternating a material with same motives aids in minimizing the project cost, by adding its value. They evaluated that the study would highlight the benefits of value engineering in affordable and sustainable construction and the use of substitute substances to enhance the output and success of a project. They also establish that low budgeted housing has to be a decision for equalizing the requirements of the concerned people. PPP's motive is to aim on the housing shortfall problem for the low-income section. The policy requires an aggressive plan of action having capacity of changing the situations. PPP model, can resolve the smart housing shortfall problem in India at a certain level.

V. Aggarwal et al. studied the substantial solicitation of sustainable prefabricated pre-engineered wall technology for low budgeted and energy-efficient homes in India [27]. The team evaluated that affordable buildings in India have an essential increase in execution and blueprint aspect. It is proposed that the enhancement of the bio-based non-rural infrastructure is to adopt the climatic circumstances by using sustainable substances in the construction. The sustainable substances including agro-industrial products, insulating substances, precast elements, prefab techniques, non-burnt bricks, and other materials. A simulation assessment was evaluated with respect to cost and energy analysis for 18 cases included in the research. As the results come out, the researchers endorse that the practicability for affordable large scale housing that is obtained for the dwelling units above 100, each considering 25 m^2 of area. This is because the single-unit construction is not viable, and prefab techniques of construction is more expensive in terms of all types of cost like erection cost, material cost, transportation cost.

The latest outlook on more imperishable and low-budgeted housing for lower-income segments in Turkey was taken in to consideration by researchers. As per the researchers, sustainability and affordability in the housing sector are the toughest concepts to conserve the environment universally and to avail social equity. It can be measured by optimization of life cycle measures. The cost analysis of the life cycle has been utilized in a sustainability concern for buildings. Construction industry used LCC to measure the number of entire buildings, their components and systems, substances costs and analyzing was ensued through the life cycle. Every sector of the society must be reached sustainably to conserve the surroundings completely.

3 Conclusion Summary

Energy usage, environment, mindful building design have become high priorities. The merits of green design are to provide buildings sustainable environment to society in general and building owners or users in particular. The development of such green buildings results in minimized devastation of natural habitats as well as biodiversity, decreased air pollution and water pollution, lower water consumption, and enhanced user productivity. It is certainly relevant that all our future buildings should be designed to function as "Green buildings". The necessity of these buildings is rising on a daily basis stimulating the need for novel and better ways of construction. This includes the use of new and sustainably improved materials of construction and application of new and developed techniques for construction.

References

1. Kibert CJ (2007) The next generation of sustainable construction. Build Res Inf 35(6):595–601
2. Morledge R, Jackson F (2001) Reducing environmental pollution caused by construction plant. Environ Manag Health 12(2):191–206
3. Cole RJ (2000) Editorial: cost and value in building green. Build Res Inf 28(5–6):304–309
4. Todd JA, Crawley D, Geissler S, Lindsey G (2001) Comparative assessment of environmental performance tools and the role of the green building challenge. Build Res Inf 29(5):324–335. https://doi.org/10.1080/09613210110064268

5. Thormark C (2002) A low energy building in a life cycle—its embodied energy, energy need for operation and recycling potential. Build Environ 37(4):429–435
6. Debnath A, Singh SV, Singh YP (1995) Comparative assessment of energy requirements for different types of residential buildings in India. Energy Build 23:141–146
7. Wu P, Low SP (2010) Project management and green buildings: lessons from the rating systems. J Prof Iss Eng Educ Pract 136(2), 64–70
8. Wasilah W, Hildayanti A, Hamzah H (2019) Green building with nature concept on lakeside resort design. Int J Environ Sci Sustain Dev 4(1):31–43
9. Retzlaff RC (2009) Green buildings and building assessment systems. J Plan Lit 24(1):3–21
10. Kevern JT (2011) Green building and sustainable infrastructure: sustainability education for civil engineers. J Prof Iss Eng Educ Pract 137:107–112
11. Bansal S, Grover T, Saini N, Sharma S. A review on existing green building rating system between different countries. Eng Technol. ISSN:0976-3104
12. Blengini GA (2009) Life cycle of buildings, demolition and recycling potential: a case study in Turin, Italy. Build Environ 44:319–330
13. Poveda CA, Young R (2015) Potential benefits of developing and implementing environmental and sustainability rating systems: making the case for the need of diversification. Int J Sustain Built Environ 4(1):1–11
14. Singh CS. Green construction: analysis on green and sustainable building techniques. ISSN:2575-8950
15. Ahmed T (2015) Selection of building technology based on sustainability requirements—Brazilian context. Archit Eng Des Manag 11(5):1–15
16. Ahmed K, Hatira L, Valva P (2014) How can the construction industry in Ghana become sustainable, pp 1–86
17. Buchanan AH, Honey BG (1994) Energy and carbon dioxide implications of building construction. Energy Build 20:205–217
18. Korkmaz S, Messner JI, Riley DR, Magent C (2010) High-performance green building design process modeling and integrated use of visualization tools. J Archit Eng 16(1):37–45
19. Castro-Lacouture D, Sefair JA, Florez L, Medaglia AL (2009) Optimization model for the selection of materials using a LEED-based green building rating system in Colombia. J Plan Lit 24(3–21)
20. Asif M, Muneer T, Kelley R (2007) Life cycle assessment: a case study of a dwelling home in Scotland. Build Environ 42:1391–1394
21. Chen TY, Burnett J, Chau CK (2001) Analysis of embodied energy use in the residential building of Hong Kong. Energy 26:323–340
22. Huberman N, Pearlmutter D (2008) A life-cycle energy analysis of building materials in the Negev desert. Energy Build 40:837–848. www.elsevier.com
23. Lacroix RN (2007) Green architecture and sustainable development: applications and perspectives, July 24–26
24. Dimoudi A, Tompa C (2008) Energy and environmental indicators related to construction of office buildings. Resour Conserv Recycl 53:86–95
25. Alkass S, Alhussein M, Moselhi O (1997) Computerized crane selection for construction projects. Assoc Res Constr Manag 2:427–436
26. Partheeban P, Dinakaran V, Raguraman D (2020) Powering sustainable development through the integration of teaching and research in engineering education. In: 6th International conference on advanced computing and communication system 2020. IEEE, Coimbatore
27. Aggarwal V, Gupta S, Sachdeva S (2012) Investigation on fly ash concrete for pavements. 3(3)
28. Hammond S (2019) Stakeholders embrace green construction as the right direction: but as individuals they make self-interested decisions. CIB World Building Congress
29. Tao H, Shi F, Tanikawa H, Fei J, Han J (2013) Materials demand and impact of buildings construction and demolition in China based on dynamic material flow analysis. Resour Conserv Recycl 72:91–101
30. Huberman N, Pearlmutter D (2008) A life-cycle energy analysis of building materials in the Negev desert. Energy Build 40:837–848

31. Reddy BVV, Jagadish KS (2003) Embodied energy of common and alternative building materials and technologies. Energy Build 35:129–137
32. Sabnis A, Pranesh MR (2016) Sustainability development index (SDI) for highrise buildings with concept of figure of merit. Int J Res Eng Technol 5(9)
33. Azadeh S, Vafaeihosseini E, Kumar P (2011) The role of building construction materials on global warming: lessons for architects. In: National conference on recent trends in civil mechanical engineering, CEEIT Hyderabad, India, March
34. Mattheos S, Kolokotsa D (2013) Passive cooling dissipation techniques for buildings and other structures: the state of the art. Energy Build 57:74–94
35. Shams S, Mahmud K, Al-Amin Md (2011) A comparative analysis of building materials for sustainable construction with emphasis on CO_2 reduction. Int J Environ Sustain Dev 10(4)
36. Jalaei F, Jrade A (2014) Integrating building information modeling (BIM) and energy analysis tools with green building certification system to conceptually design sustainable buildings. J Inf Technol Constr (ITcon) 19:494–519
37. Shams S, Mahmud K, Al-Amin M (2011) A comparative analysis of building materials for sustainable construction with emphasis on CO_2 reduction. Int J Environ Sustain Dev 10(4):364–374
38. Thormark C (2006) The effect of material choice on the total energy need and recycling potential of a building. Build Environ 41(8):1019–1026
39. Kohler N (1999) The relevance of green building challenge: an observer's perspective. Build Res Inf 27(4/5):309–320
40. Varma K, Chaurasia M (2014) Green building architecture: a literature review on designing techniques. 4(2):1. ISSN:2250-3153
41. Kshirsagar B, Mane V. Comparative analysis of green building rating systems. 3(2). ISSN:2349-4476

Analysis of a High Rise Building Frame Under Wind Pressure Considering Steel RCC Composite Structure

Rahul Prajapati and Sanjeev Kumer Verma

Abstract Tall structures and innovative architectural designing are now a days on trend. Every one wants beautifwith safety parameters in a high rise building. For creating such beautiful structures architects and structural engineers always keep safety and structural stability first to avoid any hazardous situation or collapse of structure. Thus in this study we are presenting a technique of structure where we can consider both safety as well as innovative design.As we all known that composite structures are now a days in trend to enhance structural strength and stability, therefore, in this present work we are considering a symmetrical building frame of G + 10 and assigning lateral load i.e. wind load for a selected region of wind speed 39 m/s.I this study we are considering two cases—RCC building and Composite structure—and comparing both in terms of forces, displacement, moment and cost effectiveness using analysis tool ETABS. Since composite structures withstand 30% greater stresses and displacement than RCC structures, the authors of this research conclude that they are more stable and cost-effective.

Keywords Lateral load · Structure · Wind pressure · Composite · Moments · Displacement · Optimization

1 Introduction

In this world, due to drastic increase in population there's a need for multi storey structures so as to settle the rise in population. In terms of civil engineering designers need to plan such tall structures keeping in mind parameters such as structural safety, appropriate ventilation and load resisting structure.

Designing tall structures which are capable enough to resist lateral forces namely seismic and wind pressure, there's a need for utilizing materials and techniques which are capable enough to provide enough resistance and stability to the structure.

R. Prajapati (✉) · S. K. Verma
Department of Civil Engineering, School of Engineering and Technology, Sanjeev Agrawal Global Educational (SAGE) University Bhopal, Bhopal, MP, India
e-mail: rahulprajapati251297@gmail.com

K. K. Pathak et al. (eds.), *Latest Developments in Civil Engineering*, Lecture Notes in Civil Engineering 352, https://doi.org/10.1007/978-981-99-2676-3_40

Composite structure are those materials which are generally made of two different combination of materials which are able to provide compression and tensile strength along with stability.

Lateral forcesare considered as a threat to structures and civilization from the first day of its existence, devastating lives, property and structures. The very recent hazards due to lateral forces was faced in the neighboring country Nepal, causing such a massive destruction to the country and its people. It is such an unpredictable hazard that for survival it is must to ensure that the strength of the building should be capable of withstanding seismic forces. Therefore research studies are going on all over the world to develop a technique to encounter such hazardous impacts and for stable performance of the structure. Obviously, structures designed with lateral load resisting techniques to resist such hazardous damage during earthquake activity causes higher cost of construction than general structures, but it is needed for safety against such harmful impacts under lateral forces.

In this dissertation work we are analyzing a tall structure of G + 10 with considering Wind pressure as per I.S. 875-III. In this study we are utilizing the ETABS analysis tool to perform dynamic effect of lateral forces. In this study we are considering the lateral load resisting technique i.e. composite structural members (Steel and RCC) to resist forces and moment and to minimize the cost using thin structural members.

1.1 Composite Structures

The increased population density came about into developing interest for tall structures. In light of the steel's drawbacks compared to RCC buildings and the unexpectedly high cost of steel, composite structures are increasingly being used as a viable alternative.

Most analyses of structures use the simplifying assumption that the loads acting on them are static. Since the structure does not respond strongly to dynamic loads, increasing the number of examples might make the analysis more difficult and time-consuming, hence it has not yet been examined. This situation, where the dynamic load isn't taken into account, it may be dangerous, especially in case of seismic loads, in particular.

The massive loss of life and property caused by earthquakes is unprecedented when compared to other natural catastrophes such as floods and other comparable events. Therefore, the only practical solution for resisting seismic loads is to design and analyse a structure taking into account lateral load resisting elements. Each incident that might have been catastrophic has yielded useful data that has been included into the design process in an effort to preserve the lives of building inhabitants. Composite structures can be said as those structures which are created or formed using two or more different aspects materials to enhance the performance and stability of the structural members for strengthening the structural stability (Fig. 1).

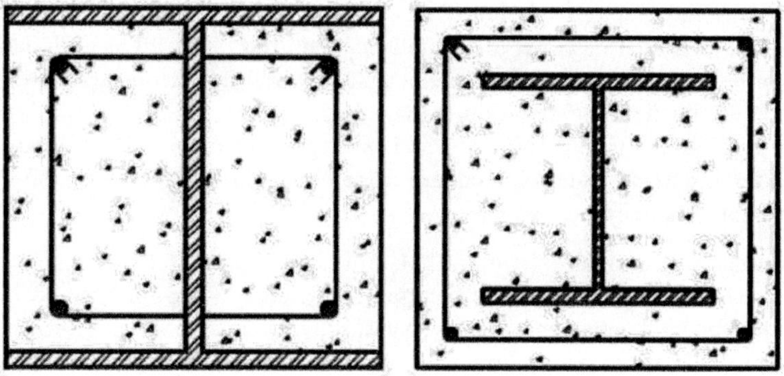

Fig. 1 Cross section of composite structure

2 Literature Review

Chao Gong et al. (2022). An academic work investigated the seismic response of buildings during earthquake mainshock-aftershock sequences. To illustrate the impact of utilising BRB and SCBRB components, we compare two varieties of the extra BRB elements and also compare them to the original steel frame. Here, we apply the identical mainshock-aftershock sequences as before to evaluate the seismic performance of BRBF and SCBRBF and draw comparisons between the two. A great deal of energy was lost during motion because of the flag-shaped hysteretic curve of the SCBRB parts.

The findings indicated that the reactions of buildings equipped with SCBRB components to earthquakes were less severe than those of conventional buildings. Exhausted energy is increased. The structure's frequency is increased by including SCBRB and BRB components. The SCBRBF's residual deformations are over 6.6 times less than SFs, while the BRBFs are over 2.11 times smaller. According to the findings, including SCBRB components into a building in a seismic zone might not only lessen the amount of damage caused by earthquakes, but also enhance the building's other qualities, such its maintainability and its service life.

Moushtakim Billah et al. (2019). This study provided a description of cold formed steel by discussing its characteristics, qualities, manufacturing methods, and element classifications. Guidelines and rules for cold-formed steel buildings, including the significance of its design requirements, connecting membranes, and durability difficulties, which were also given.

Because of its many useful qualities, cold-formed steel is becoming more popular as a structural material. Besides these positive aspects, cold-shaped steel also has a few features that affects its structural implementation. Engineers have challenges in making sure a cold-shaped steel structure is properly planned and built due to factors including the thickness of the regions involved, the framing process, and the complexity of the structure's design. This study has looked at the evolution, qualities, codes, and determinations of cold-formed steel buildings, as well as a fundamental design principle, energy use, fire safety, and research advancements. In order to overcome the challenging situation, enhance its display, and update the laws and rules, new research on cold-formed steel is strongly encouraged. Due to the results of these tests, experts and architects are more confident in the use of cold-shaped steel to enhance a building's appearance.

Sureshbabu and SenthilSelvan (2019). This study presented the results of an experimental investigation into the flexural behaviour of Cold-formed-Steel (CFS) members' lipped-channel corrugated sections, with consideration given to three distinct sets of corrugated sections: I horizontal corrugated back-to-back lipped-channel sections without gap; (ii) horizontal corrugated back-to-back lipped-channel sections with gap; and (iii) vertical corrugated lipped-channel sections.

The analytical findings showed that the load bearing capability of vertical corrugated sections was increased by as much as 9.7%.

The corrugation serves as a web stiffener for the shaft and provides excellent protection for the vertically creased pillar against the two-point loading conditions that are imposed. For the vertically creased region, the highest loads obtained were 41KN from the Abaqus result and 31KN from the test, with respective avoidance values of 3.63 and 5.53 mm. Horizontally corrugated specimens with and without spacing were resisting a load that was about equal to the range of estimates for loads that would be resisted by a corrugated beam.

Raffaele Landolfo (2019). This study described in detail the several Cold formed steel research applications that have been conducted at the University of Naples Federico II over the last two years, focusing on the seismic behaviour of "Stick-Built" buildings and new structural advancements.

CFS structural system's widely acknowledged fundamental execution in addition to its increased degrees of building, security, strength, and supportability, are rapidly propagating this development framework over the globe. At the same time, advancements in fresh research are needed to meet the actual need for explicit construction codes, particularly for the applications in a seismic zone. The article described a few studies that have been conducted recently at the University of Naples to prepare for the expected growth.

Padmanaban and Suresh Babu (2019). Using cold-formed steel (CFS) edge segments for its construction, this article demonstrated the results of its testing on a flat pratt truss. Each model of the 1 m cold-framed Flat pratt support has three points spaced evenly apart and measuring $50 \times 50 \times 3.0$ mm. Five connections were prepared at the points where the top chord people and the vertical chord met, and five connections were constructed at the points where the bottom chord individuals met. Harmony and vertical people converged at the junction, where they were linked by jolts to the gusset plates. Five aggregated loads were applied at each junction throughout the experimental study until frame deformation was observed. The truss was also subjected to a numerical study using finite element analysis, and the findings were compared to experimental data. The trial and investigational results show that the bracket fails more often at its top corner junction, where there is just a shear strain, and at its midpoint, where there is a drooping or twisting failure in the base harmony portion. The test findings showed that the shear quality of the jolts in the bracket and the proximity of the gusset plates increased a definite load on the truss by 33%. Definitive load removals were also recorded at the crossroads and their behaviour was investigated in great detail via a comparison of experimental and analytical findings.

The results showed that the averaged ultimate load obtained experimentally by the flat pratt truss junction was 14,744 KN, that the failure was seen in the model noted at the top corner joints (which was a shear failure) and base centre joint (which was a twisting or sagging failure), and that the yielding property of steel has been noted through the bends demonstrated where the expansion in loads was not significantly different from the initial value. As a result, we determined that the addition of the gusset plate and bolts enhanced the truss's ultimate load bearing capability by 33%. This work paved the way for the widespread use of cold-formed-steel (CFS) sections as the only material for roof trusses.

Fayaz et al. (2019). This research paper was concerned with investigation of the strength of members by undertaking the buckling profile responsible to commemorate application of the load on the entire structure. When considering the load specifics, it is helpful to think about the pointed ends that reimburse a perfect parabolic curve. The pre-stressing of the tendon was the determining factor in the examination's principal carrying load. The compression part's distortional buckling characteristics were reported by the numerical study, which continued the simulation procedure. Harmonic sine(or)cosine wave application in design science is a vital aspect that has recently been brought to light, however when compared to parametric analysis, the results were disappointing.

The entire examination presented that the variation in ultimate load from the parametric study along with numerical analysis was in between 30 and 40%.

Sattainathan Sharma et al. (2018). In addition to presenting a non-linear finite element analysis of steel beam column connections built of Cold Formed Steel Sections, this study illustrated the different qualities and behaviour of a CFSS when employed as principal structural membranes in the mode to use it for a futuristic approach.

Cold-formed steel "I" and "Channel" section bolted joint beam-column connections were studied, with the connection occurring at the flange and web of the column, respectively, to determine structural performance.

The results presented that I section behaved dominantly in comparison to Channel Section when subjected to axial loads. There has been a 34.78 rate ascend for I segment than of Channel segment by methods for the diagnostic outcome esteems. Also, the data obtained from experiments have shown that the I section improves by 50% more than the Channel section does using the same approaches. This rate worth demonstrates that I segments are efficient for the lightweight cold-formed steel structures and furthermore it presented web association that gives preferred outcomes over flange proportion associations of the pillar. This suggests that having welded joints in lightweight steel constructions is beneficial. In addition, the experimental study shows that the I section is far more effective than the Channel section in withstanding pivotal stresses, and it also hints to the benefits of using a darted joint in a lightweight construction. The investigation additionally presented that flexural part being associated with the joints at the web segment of the pressure part greatly affecting the compression member under loading condition.

Loragayle Doctolero and Mustafa Batikha (2018). This research paper presented a comparative analysis of a four storey commercial structure with the use of three different materials namely Hot Rolled Steel Sections (HRSS), Reinforced Concrete (RC) and Cold Formed Steel Sections (CFSS).The primary structural members of the entire building was designed as per BS standards along with Eurocodes with the use of linear elastic analysis. Design results were extracted on the basis of Building Mass, cost of the material, cost of construction and the overall cost (Material Cost + Construction Cost) and the time durations of the construction. Compared to Reinforced Concrete (RCC) or Hot Rolled Steel, Cold Formed Steel's (CFS) proven results show that it is relatively cost-effective in terms of materials and total cost of the project, even lowering the duration of the construction (HRS). It was also discovered that the cost of materials has a significant influence on the total cost of an HRS building, while the cost of construction has a substantial impact on the cost of an RCC structure.

The conclusion derived from the research paper stated that the structures of Cold formed steel members are 67% lighter in comparison to Reinforced sections, even 5% more lighter than Hot Rolled Steel Members. CFS was found to be 34% cheap in comparison to RC members and 89% cheap then Hot Rolled Steel Members. In terms of construction time period, CFS was found to be 164% less time consuming than RC structures and 38% less than HRS. The consolidated index of the cost and development time terms that CFS has 325% numerous advantages concerning the expense than RC and 86% more advantages than HRS. Then again, HRS registers 239% advantageous development list than RC.

Sudarshan Bhutekar et al. (2018). Here, the author evaluates the performance of steel and a composite (steel–concrete) building under a G + 15 story earthquake load. While steel has a high strength to weight ratio, composite building incorporates the advantages of both steel and concrete, including lower cost, faster construction, fire protection, etc.

The steel structure performed better than the steel concrete composite frame. The purpose of this research was to examine how steel frame structures may outperform RCC in terms of seismic performance while also being more cost-effective. The results of the pushover study showed that the steel frame structure was one of the safest options for building in a seismic area.

Sanjay Kulkarni et al. (2017). Here, the authors conducted an in-depth analysis of RCC and composite structures in seismic zone III, and their research article is based on those findings. Response spectrum and nonlinear time-history analysis and were used to assess the seismic behaviour of the study frames developed using the suggested technique. The modelling and analysis were done using ETABS.

The results presented that from the analysis of the model it was seen that the hinges are formed in beam element first rather than the column and conclusions stated that the composite frame follows a strong column weak beam concept or capacity based design concept.

Objectives

The main objectives of this study are:

1. Comparison of structural performance of conventional structure with composite structure.
2. To determine the variation in forces, moment and displacement in both the cases considering wind pressure.
3. To determine the cost analysis of both the structures to find out the economic structure as per S.O.R.
4. To determine the formation of composite structure in analysis tool ETABS.
5. To justify the positive effect of using composite members to resist forces and displacement of a tall structure.

3 Methodology

In this study we are considering a symmetrical building frame of 24 m × 30 m and height of the building frame is considered as G + 10.

Wind load analysis is performed as per I.S. 875-III considering Bhopal region where the basic wind speed is 39 m/s as per appendix-A of the code.

Cases considered for study are:

Case 1: RCC structure (Fig. 2).

Case 2: Composite structure (Fig. 3 and Tables 1 and 2).

The analytical work is performed on a symmetrical building structure with a 24 m × 30 m frame and a G + 10 frame height. Following steps are considered as follows:

Step-1: First step is to prepare a literature survey of composite structures, analysis tool, lateral forces and tall structures by reviewing journals, books and blogs.

Step-2: To understand the work progress till date and decide aims and objectives of our study and its scope in future.

Step-3: To select geometry, height and material to be use in this study (Fig. 4).

Step-4: To create and assign sectional properties (Fig. 5).

Step-5: Create sections for RCC and Composite Structural Members (Fig. 6):

Step-6: To Assign Support conditions and release moment in secondary members.

Step-7: To Assign loading conditions as per I.S. 875-I & II for dead and live load (Fig. 7).

Step-8: Assign Lateral load (wind load) as per I.S. 875-III as a non-linear static load (Fig. 8).

Fig. 2 RCC conventional structure

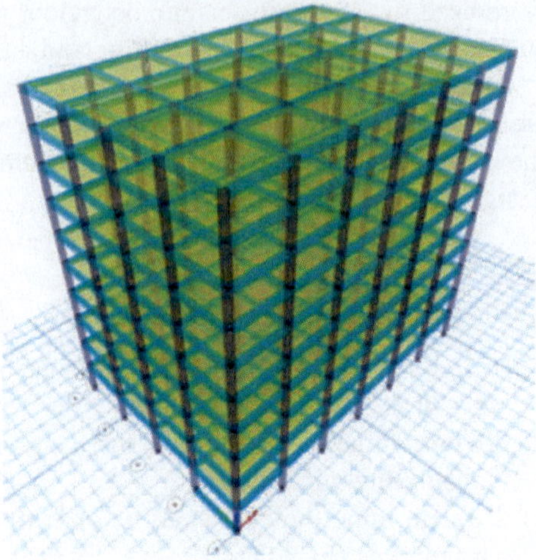

Fig. 3 Composite structure

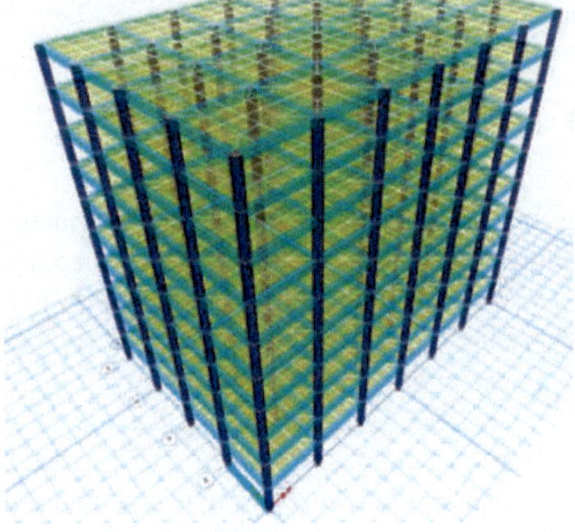

Step-9: To analyze both the cases to determine forces, displacement, drift, base reaction and quantity of materials (Fig. 9).

Step-10: To conclude the outcome of the study (Fig. 10).

Table 1 Geometrical description of the structure

1	Number of stories	Ground + 10 storey
2	Height of stilt floor	3.3 m
3	Height of upper stories	3.3 m
4	Depth of foundation	−1.5.0 m
5	Grade of concrete for RCC structure	M 25
6	Grade of concrete for composite structure	M 25
7	Steel used for longitudinal reinforcement	HYSD 500
8	Steel used for lateral reinforcement	HYSD 415
9	Steel sections	Fe 345
10	Masonry	Infill brick
11	Wind region (basic wind speed)	39 m/s

Table 2 Sectional design

Conventional reinforced concrete frame		
1	Column	450 mm × 450 mm
2	Beam	300 mm × 250 mm
3	Slab	125 mm thick
4	Masonry	130 mm thick
Composite structure		
1	Columns (steel + RCC)	450 mm × 450 mm
2	Beam	300 mm × 250 mm
3	Slab	125 mm thick
4	Masonry	130 thick

4 Analysis Results

4.1 The Greatest Possible Torque of Bending in KN-M

Inferences

As shown in the Table 3 it can be observed that with steel casing composite members one can minimize the bending moment by 30% which means reinforcement requirement will be decreasing, hence the structure becomes more economical (Fig. 11).

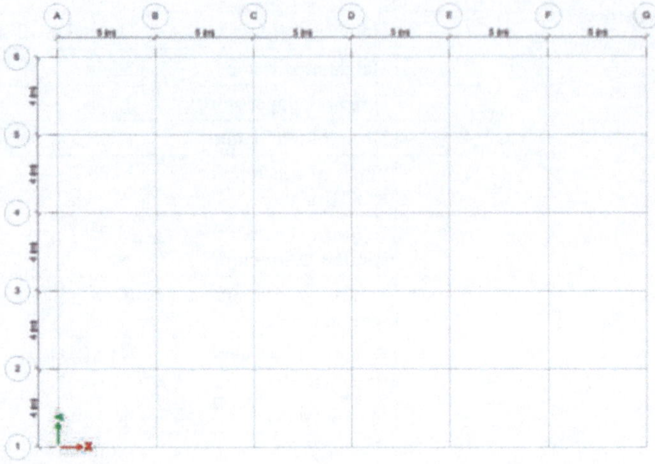

Fig. 4 Selected geometry

Shear Force

See Table 4.

Inferences

In the Fig. 12 it is clearly observed that unbalanced forces are less in composite structure in comparison to RCC structure due to proper distribution of forces in compression and tensile zone.

Axial Force

See Table 5.

Inferences

Axial forces are the vertical forces distribution beam load from column to end condition. In the Table 5 it is observed that the composite structure is distributing it properly confining proper settlement of load to the foundation level (Fig. 13).

Storey Stiffness

See Table 6.

Inferences

It is the ratio of drift to shear of the structure due to lateral forces. From the results of this research, we may draw the conclusion that composite structures are more robust and sturdy (Fig. 14).

Storey Displacement

See Table 7.

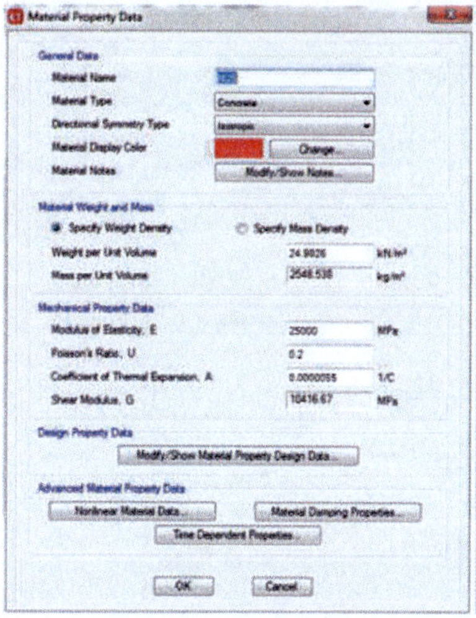

a. Creating concrete

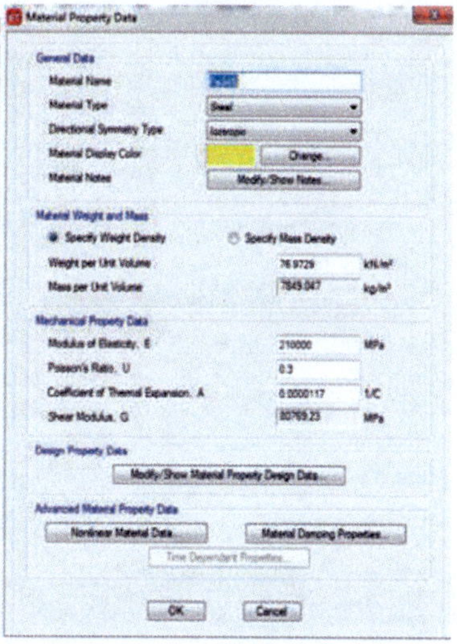

b. Create Steel Property

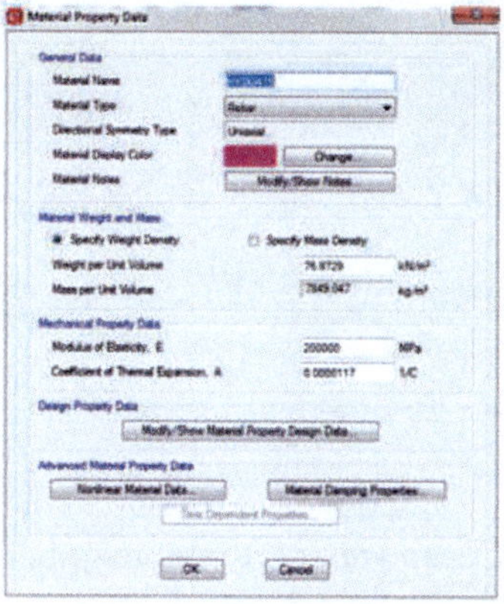

c. Create Fe 415 rebar property

Fig. 5 Material properties

Inferences

As shown in Fig. 15 it can be said that the composite structure is comparatively more stable than the RCC structure as it is more resistable to lateral forces in comparison to the general RCC structure.

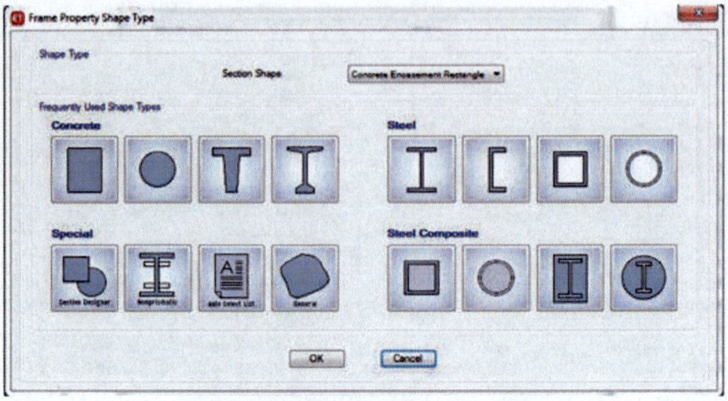

a. RCC sectional properties

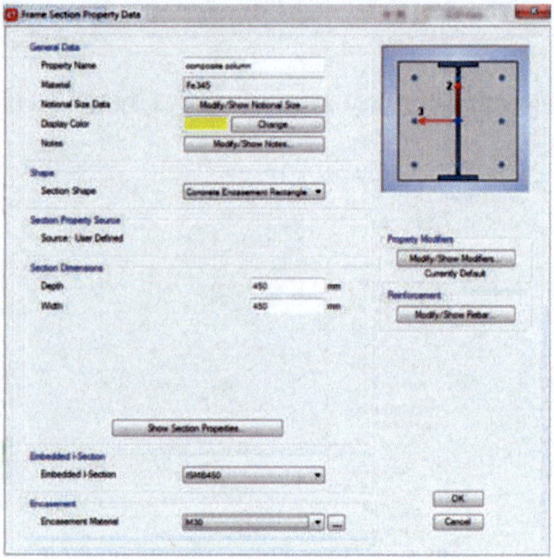

b. Composite sectional members

Fig. 6 Assigning structural Members

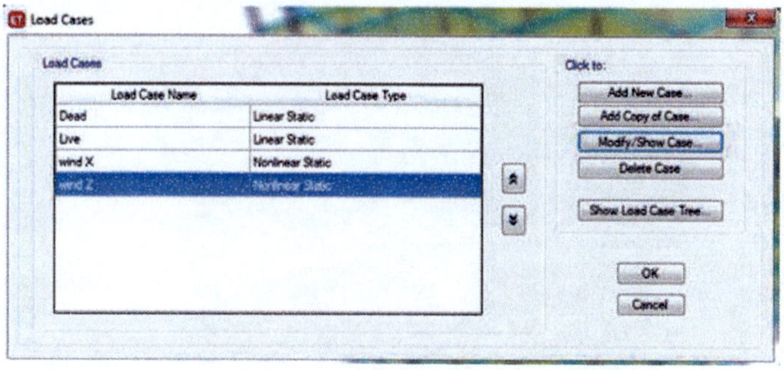

Fig. 7 Load conditions

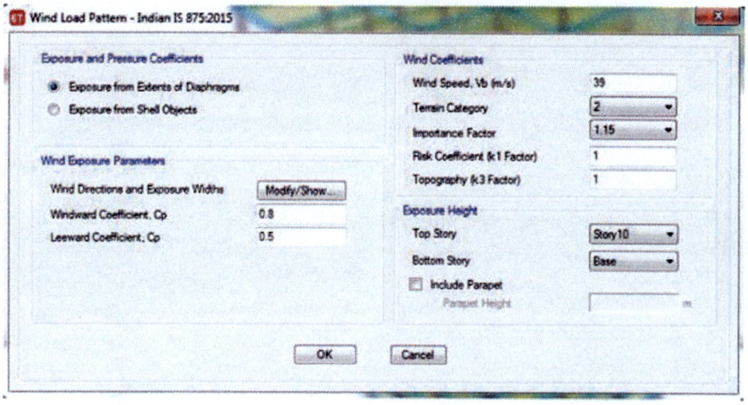

Fig. 8 Wind pressure

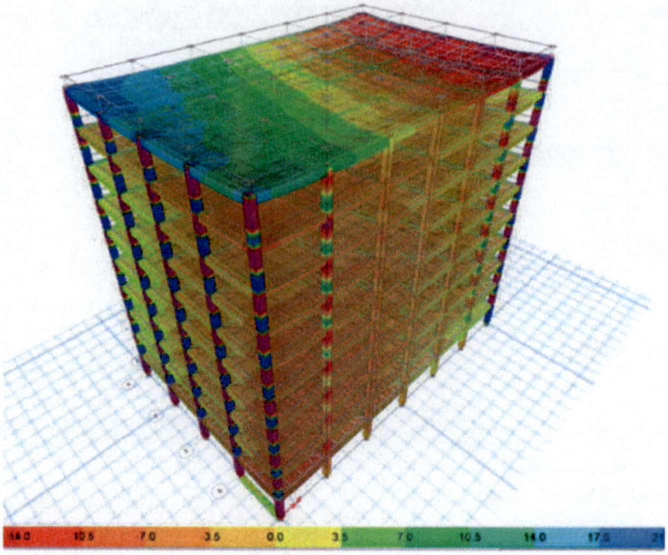

Fig. 9 Analysis outcome

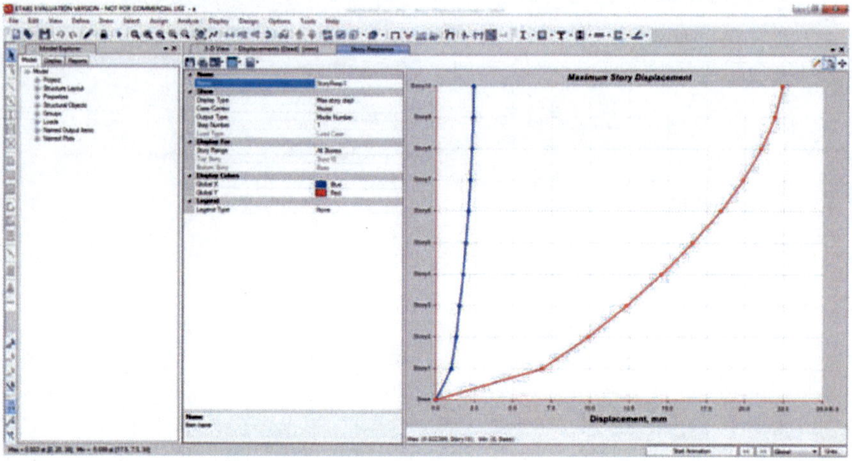

Fig. 10 Comparative results

Table 3 The greatest possible torque of bending in KN-m

Maximum bending moment KN-m			
Beam	RCC structure	Composite structure	Ratio
10th	347.982	336.04	1.0355374
9th	345.011	328.98	1.0487294
8th	342.998	321.92	1.0654759
7th	327.476	314.86	1.0400686
6th	326.578	307.8	1.0610072
5th	296.68	285.98	1.0374152
4th	280.393	264.16	1.0614514
3rd	271.916	242.34	1.1220434
2nd	259.046	220.52	1.1747052
1st	258.108	198.7	1.2989834
Base	230.919	176.88	1.3055122

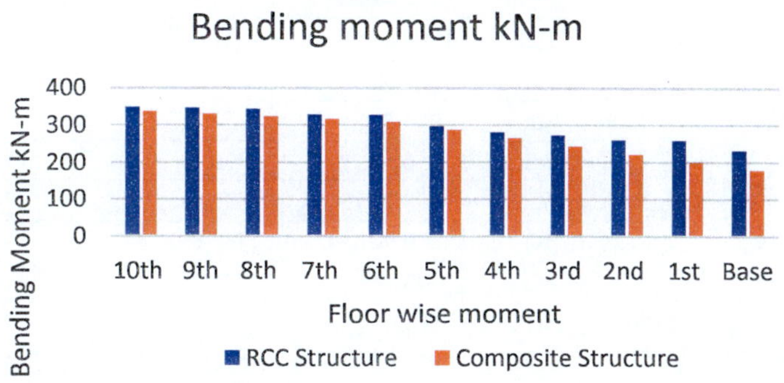

Fig. 11 Bending moment

Table 4 Shear force

Shear force in KN			
Beam	RCC structure	Composite structure	Ratio
10th	558.395	380.65	1.46695
9th	549.39	368.76	1.48983
8th	594.162	356.87	1.66493
7th	595.803	344.98	1.72707
6th	593.089	333.09	1.78057
5th	584.084	321.2	1.81844
4th	507.999	309.31	1.64236
3rd	444.339	297.42	1.49398
2nd	506.358	285.53	1.7734
1st	437.812	273.64	1.59996
Base	370.371	261.75	1.41498

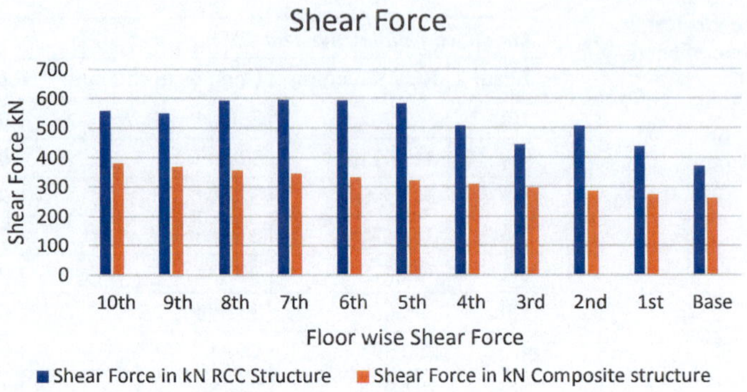

Fig. 12 Shear force

Table 5 Axial force KN

Axial force in KN			
Beam	RCC structure	Composite structure	Ratio
10th	1089.86	1021.09	1.06735
9th	1032.08	986.9	1.04578
8th	974.3	952.71	1.02266
7th	916.52	896.09	1.0228
6th	858.74	839.47	1.02296
5th	800.96	782.85	1.02313
4th	743.18	726.23	1.02334
3rd	685.4	669.61	1.02358
2nd	627.62	612.99	1.02387
1st	569.84	556.37	1.02421
Base	512.06	499.75	1.02463

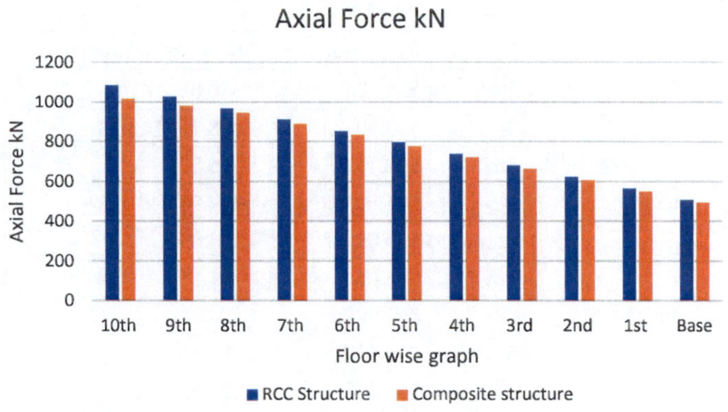

Fig. 13 Axial force

Table 6 Storey stiffness

Storey stiffness			
Beam	RCC structure	Composite structure	Ratio
10th	1,190,652	1,856,412	0.64137271
9th	2,072,134	2,198,054	0.94271296
8th	2,953,616	2,539,696	1.16298014
7th	3,835,098	2,881,338	1.33101288
6th	4,716,580	3,222,980	1.46342205
5th	5,598,062	3,564,622	1.57045039
4th	6,479,544	3,906,264	1.65875732
3rd	7,361,06	4,247,906	1.73285991
2nd	8,242,508	4,589,548	1.79593023
1st	9,123,990	4,931,190	1.8502613
Base	10,005,472	5,272,832	1.89755183

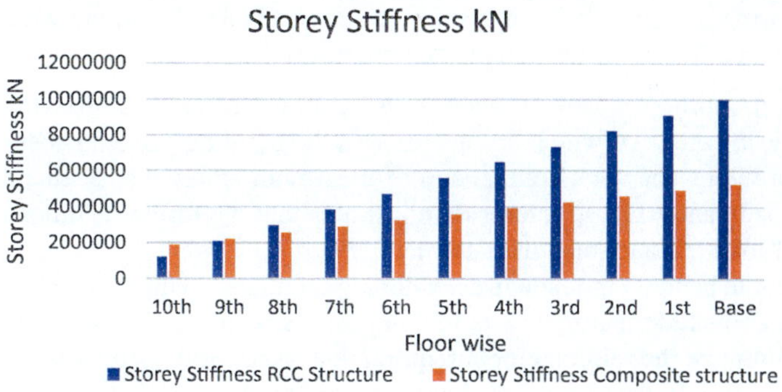

Fig. 14 Storey Stiffness in KN

Table 7 Storey displacement

Storey displacement (mm)		
Beam	RCC structure	Composite structure
10th	342.34	265.98
9th	327.65	253.65
8th	312.96	241.32
7th	298.27	228.99
6th	283.58	216.66
5th	268.89	204.33
4th	254.2	192
3rd	239.51	179.67
2nd	224.82	167.34
1st	210.13	155.01
Base	195.44	142.68

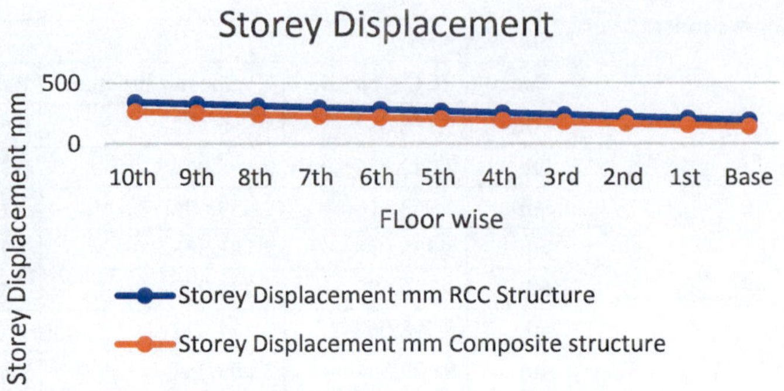

Fig. 15 Storey displacement

5 Conclusion

The following are the results of a study on the analysis of a frame with the RCC structure and a frame with a Composite structure:

1. As the proportion of cement in composite sections decreases, so does the available supply, the effect of which is a decrease in the building's overall cost.
2. The previous chapter's comparison of maximum storey displacement between composite and RCC structures demonstrates that the former is more stable and lateral load resistant than the latter by 32%.
3. Maximum bending moment in each floor is calculated in the above chapter where we can conclude that moments in composite structures are less by 30% which can minimize the reinforcement requirement (Ast.) of the structure.
4. It is clearly observed that unbalanced forces are less in composite structure in comparison to the RCC structure due to proper distribution of forces in the compression and tensile zone.
5. Axial forces are the vertical forces distribution beam load from column to end condition. It is observed that the composite structure is distributing it properly confining proper settlement of the load to the foundation level.
6. Composite constructions should be increasingly chosen for buildings meant to withstand seismic loads and wind speeds since damping in RCC structures is seen to be 3–17% greater.

References

1. Folić R, Radonjanin V et al (2016) Design and analysis of steel-concrete composite structure. University of Novi Sad, Faculty of Technical Sciences, Department of Civil Engineering Novi Sad, Serbia
2. Uy B (2012) Recent innovations for steel and composite steel-concrete structures in Australia. In: CIES, vol II

3. Damam V (2016) Design of steel concrete composite structure as comparative with reinforced concrete structure by adapting STAAD. PROV8I. IJESMR [Damam*, 3(1):January 2016]
4. Datta D (2016) Steel-concrete composite construction—new trend in India. IOSR J Mech Civil Eng. e-ISSN: 2278-1684, p-ISSN: 2320-334X
5. Zona A, Barbato M et al (2008) Nonlinear seismic response analysis of steel–concrete composite frames, 0.1061/_ASCE_0733-9445_2008_134:6_986
6. Liu J, Liu Y (2008) Seismic behavior analysis of steel-concrete composite frame structure systems. In: The 14th world conference on earthquake engineering, October 12–17, 2008, Beijing, China
7. Zaveri BH, Gadhiya JA et al (2016) A review on the comparative study of steel, RCC and composite building. IJIRSET 5(1):Januray 2016
8. Pandey R (2014) Comparative seismic analysis of RCC, steel & steel-concrete composite frame. National Institute of Technology, Rourkela
9. Gupta T, Sharma RK (2015) Steel concrete composite structures: state of art. Int J Latest Res Sci Technol 4(2):149–153, March–April 2015
10. Muhammed Sabith K, Sabeena MV (2017) Seismic analysis of irregular composite structures with shear connectors using ETABS. IJSRSET 3(3). Print ISSN: 2395-1990, Online ISSN: 2394-4099

Parametric Study of Castellated Steel Beams

A. J. Mehetre, N. U. Mate, S. B. Kandekar, and R. S. Ingole

Abstract In this study, a detailed parametric investigation of the castellated beams (CBs) is done by analytical and experimental methods. The usage of CBs or perforated web beams has become very widespread nowadays as a result of its valuable structural applications. A lot of research work has been supported for improving sizes of castellated steel beams (CSBs) and henceforth there is a requirement to optimize the steel beams with varying depths of hexagonal opening. The objective of this research is to do a comparative study of the parent beam that is I-beam and CBs with changing depth. The study is a performance-based investigation on bending strength, deflection, lateral-torsional buckling moment, bending compressive stress, and shear capacity of I-beam and also investigates the moment of resistance, deflection of CBs with standard hexagonal shape of the opening. These are done on an analytical and experimental basis.

Keywords Castellated beam · I-beam · Parametric study · Analytical study · Experimental analysis

1 Introduction

A steel beam having a perforated web is termed a (CBs) castellated beam [1]. It is steel beam having opening in web which is prepared from a single rolled steel wide flange beam segment. The desired shape is prepared using gas flame, cutting

A. J. Mehetre (✉) · N. U. Mate · S. B. Kandekar · R. S. Ingole
Department of Civil Engineering, AVCOE, Sangamner, MS, India
e-mail: mehetreaj@rediffmail.com

N. U. Mate
e-mail: nilesh.u.mate@gmail.com

S. B. Kandekar
e-mail: sbkandekar@gmail.com

R. S. Ingole
e-mail: ramakant.ingole@avcoe.org

K. K. Pathak et al. (eds.), *Latest Developments in Civil Engineering*, Lecture Notes in Civil Engineering 352, https://doi.org/10.1007/978-981-99-2676-3_41

the web portion in a specific pattern and rejoining the sections by using electric arc welding to shaped a regular hole pattern in the web such as hexagonal, diamond, oval, rectangular, circular, and sinusoidal or in combination [2]. The first recognized use of CBs in the United State (US) was by Horton H. E. of the Bridge in Chicago. G. M. Boyed established the awareness of CBs in 1935, who in Argentina worked as a steel fabricator. His first idea was to stack two beams of a minimum depth on top of one another to increase the depth. Boyed then decided to cut a section of hot rolled steel beams along with the web in a saw tooth pattern, separate the two halves, and weld the web posts back together at the high points. By this process, the depth of the beam increases by approximately 50%, therefore increasing the strength and stiffness by about 20–30% without increasing the beam weight [3, 4].

Most of the existing research studies were carried out for hexagonal, rectangle, circular and sinusoidal shape of the opening of CBs; however, the very little study was reported on the parametric study with varying depth of castellated beams. In fact, this varying depth is very useful for improving moment resistance capacity of the section and control the deflection. So therefore, in this paper, the emphasis is given on varying depth of hexgonal opening and its effect over the CBs in terms of the ultimate moment carrying capacity, deflection, bending moment, and shear stress as depicted in Tables 1 and 2.

Table 1 Details of I-beam geometry used for parametric study

Beams → Properties↓	ISMB150	ISMB200	ISMB300
W N/m	146.902	248.528	451.58
A mm^2	1910	3230	5870
h mm	150	200	300
b_f mm	75	100	140
t_f mm	8	10.8	13.1
t_w mm	5	5.7	7.7
I_{xx} mm^4	$717.6 * 10^4$	$2235.4 * 10^4$	$8985.7 * 10^4$
I_{yy} mm^4	$46.8 * 10^4$	$150 * 10^4$	$486.3 * 10^4$
r_{yy} mm	15.7	21.5	28.6
Z_{ex} mm^3	$95.7 * 10^3$	$224 * 10^3$	$599 * 10^3$
Z_{px} mm^3	$107.65 * 10^3$	$249.69 * 10^3$	$670.484 * 10^3$
d mm	113.4	152.7	240.2
I_t mm^4	$3.15 * 10^4$	$9.57 * 10^4$	$25.348 * 10^4$
I_w mm^6	$23.59 * 10^8$	$1.34 * 10^{10}$	$7 * 10^{10}$
λLT	1.84	1.57	1.13
φLT	2.38	1.88	1.23
χLT	0.257	0.34	0.58

Table 2 Design parameters of I-beam

I-beam	Depth (mm)	Load (kN)	Deflection (mm)	fbd (N/mm^2)	Md (kNm)
ISMB150	150	130	2.830	58.57	6.466
ISMB200	200	130	0.9	77.27	19.73
ISMB300	300	130	0.226	93.18	62.47

2 Advantages of CBs

1. As the depth of a beam increases which enhances the strength without changing its weight.
2. The CBs is highly steel efficient and it has an optimal load carrying capacity for the same cross-section area of the I-beam.
3. To attain good strength against web crippling, web buckling and local failure the physical properties of the beam can be varied.
4. It is mostly lighter in weight. Fewer support columns with longer spans are characteristic of a CBs project.
5. It can also utilize for the placement of secondary installation, like electrical, mechanical and plumbing which is impossible by considering a wide flange beam having a solid web portion.

2.1 Disadvantages of CBs/Limitations

1. Stress concentration—It has high stress concentration in and around the opening.
2. Shear failure—It has a chance to fail in shear.
3. Potential mode of failure—It has six failure modes [5–7].
4. The market price of steel is high and also requires highly skilled workers.
5. It has more fabrication cost due to cutting and welding.

3 Methodology

The recent study objectives is to compare the result of analytical [8] and experimental outcomes of plain webbed beams (I-beam) and castellated steel beams [9, 10]. Three types of I beams and CBs are considered for the present study. The structural results are measured in terms of bending strength, lateral-torsional buckling moment, bending comparative stress, shear capacity, deflection and moment of resistance [11–13]. The point load system is formed by simple logic that is one movable C-sections that are attached with a steel rod welded to them. To adjust the application of load on L/2 distance from both faces of the CB, the C-section is moving parallel along the length of the steel I-section. This detachable point load assembly is positioned on a parent beam which applies the single-point load. A wide flange I-steel beam with

Fig. 1 Method of fabrication for castellated beams. **a** Web cut **b** completed beam

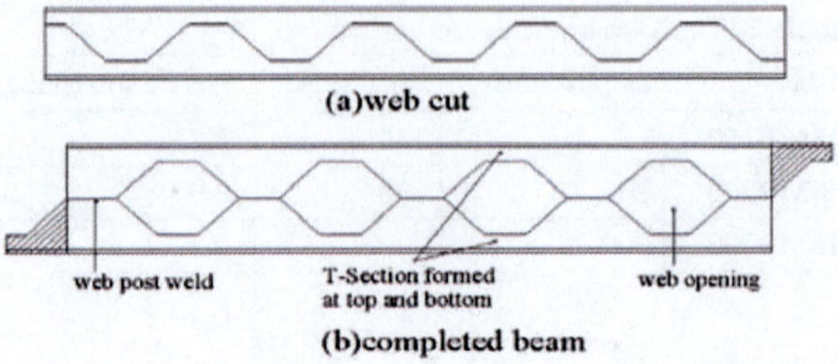

point loads was parallelly resting on the parent beam. By applying a single point vertical load, the central part of the CB is undergoing change in flexure and nominal shear. UTM of 1000 kN capacity is used to apply the load. The bending moment is the workout from the applied load during testing. The corresponding deflection is measured with center at every increment in load with the help of a dial gauge having 0.01 mm least count. Details of the fabrication program are given in Fig. 1 and Table 2.

3.1 Procedure of Study

The analysis is supported by adopting the procedure as given below [14–16].

1. Three I-Beams were carefully chosen for the study.
2. Numerical analysis is done for I-beam and castellated beam as per IS 800:2007.
3. The experimental study is performed over castellated beams using U.T.M. (Universal Testing Machine).
4. A dial gauge is used to measure a deflection throughout the experimental work.
5. Numerical comparison of I-beam and castellated beam using deflection criteria.
6. Experimental comparison of the CBs with changing depth versus deflection.

3.2 Parametric Study

The parametric investigation is done for three types of I-beams as mentioned below [17]. The geometrical properties of the I-beam are worked out manually using IS 800: 2007 standard code procedure as depicted in Table 1. The design parameter as listed in Table 2 is also found out manually.

4 Results and Discussion

Figure 2 and Table 3 show the analytical deflection of all three I-beams and castellated beams respectively. The result indicates that as the beam depth increases, the bending strength, shear capacity, lateral-torsional buckling and moment of resistance of the beams increases. At the same time, for increases in depth, deflection decreases due to more lateral stiffness.

4.1 Comparison of Deflection Between I-beams and Castellated Beams [18–20]

The analytical deflection is calculated for the I-beams and it is compared with the experimental deflection of castellated beams, as depicted in Fig. 3.

From Fig. 3, there is no such a big difference in deflection in both the beams that is I-Beam and CBs. But it is noted that the CBs fail in shear however I-beams fail in buckling. The aesthetic view of castellated beams is much superior to I-beams.

Table 4 and Fig. 4 show the load against deflection outcomes of all three castellated beams. Figure 5 shows the experimental setup for testing of castellated beams.

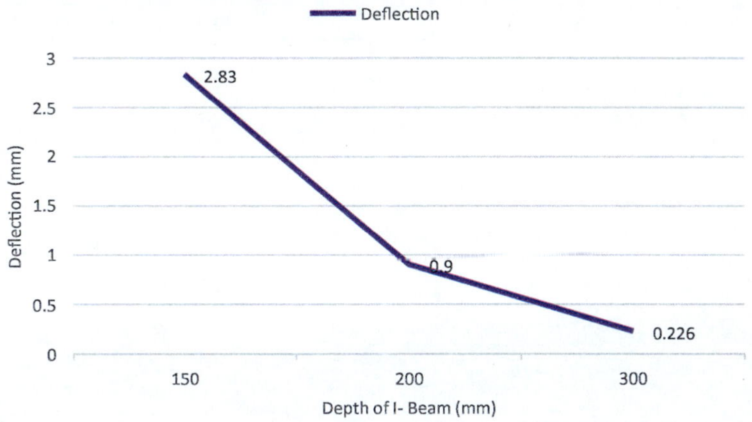

Fig. 2 Depth of beams versus deflection (I-beam)

Table 3 Analytical deflection of CBs

Castellated beam	Depth (mm)	Deflection (mm)
IC225	225	2.139
IC300	300	0.66
IC450	450	0.163

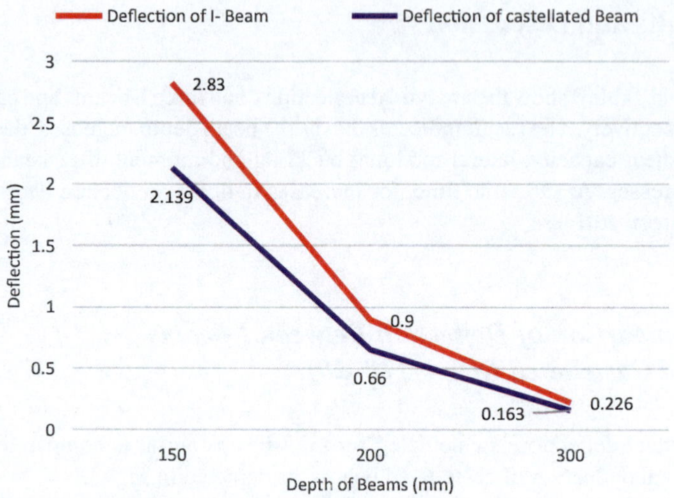

Fig. 3 Depth versus deflection of I-beam and castellated beam

Table 4 Experimental deflection of CBs

Beam type	Depth (mm)	Load (KN)	Deflection (mm)
IC225	225	130	3.1
IC300	300	130	2.2
IC450	450	130	1

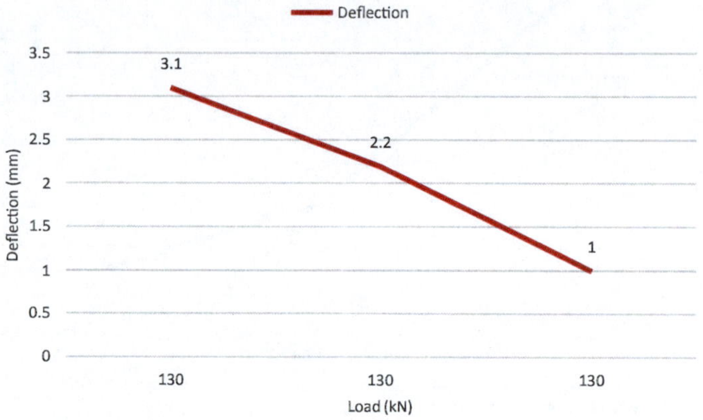

Fig. 4 Load versus deflection of CBs tested experimentally

Fig. 5 A figure shows experimental test set up for testing of castellated beams

4.2 *Results and Discussion Summary*

The CBs is fabricated from the I-Beam. So, it is recommended to check the deflection of ordinary I-Beam. From Table 1, it is concluded that as the section increases the depth of the standard beam increases but decreases the deflection. The same thing happens for Castellated Beam. By comparing both the beams, it is seen that I-beam deflects more than the CB. Keeping all the parameters same only the depth varies and the height of the CB increases by 1.5 times of I-beam as per the strength and serviceability consideration. So, it is found that a CB is more effective and efficient than a solid I-beam. Analytically the beam is designed considering depth criteria. Experimentally load–deflection criteria are satisfied. So, there was very little difference in deflection.

5 Conclusions

1. It concluded that the deflection of CB reduces by 0.25% to that of the I-beam. The depth of the CB improved by 1.5 of the I-beam. The I-beam has a higher section than the CB for the same geometric and loading condition.
2. On the basis of analysis, the permissible deflection is normally on the higher side so, there is no need to provide a stiffener in castellated beams.
3. In real practice during the entire life of beams, it can carry less load than the design load, so here for experimental investigation low value of the load is considered.
4. As far as deflection is concerned, there find a very marginal differences in deflection of CBs analytically and experimentally.
5. It is noticed that the failure in CBs is mostly due to shear failure, however, the failure in solid I-beam is a local buckling failure.

6 Scope of the Future Work

1. This work can be extended for different loading conditions considering different patterns of web opening or in combination.
2. The tapered CBs can also be considered for further studies.
3. Under combined loading of flexure, shear, torsion, or axial forces the behaviour of castellated steel beams can be investigated in the future.

Acknowledgements The work presented in this research is the work done by the authors at Amrutvahini College of Engineering, Sangamner.

References

1. Amayreh L, Saka M (2005) Failure load prediction of castellated beams using artificial neural networks. Asian J Civ Eng (Build Hous) 6(1–2):35–54
2. Boyer JP (1964) Castellated beams-new developments. In: AISC National engineering, conference, Omaha, Nebr, pp 104–108
3. Durif S, Bouchair A (2016) Analytical model to predict the resistance of cellular beams with sinusoidal openings. J Constr Steel Res 121:80–96
4. Djebli B, Kerdal DE, Abidelah A (2014) The total deflection of composite cellular beams under transverse loading. Arab J Sci Eng. https://doi.org/10.1007/s13369-014-1227-z
5. Ellobody E (2011) Interaction of buckling modes in castellated steel beams. J Constr Steel Res 67:814–825
6. Kerdal D, Nethercott DA (1984) Failure modes for castellated beams. J Constr Steel Res 4:295–315
7. Kawani S, Wijaya PK (2017) Lateral torsional buckling of castellated beams analysed using the collapse analysis. Procedia Eng 171:813–830
8. Gandomia AH, Tabatabaei SM, Moradian MH, Radfar A, Alavi A (2011) A new prediction model for the load capacity of castellated steel beams. J Constr Steel Res 67:1096–1105
9. Morkhade SG, Gupta L (2015) An experimental and parametric study on steel beams with web openings. Int J Adv Struct Eng 7:249–260
10. Morkhade SG, Kshirsagar M, Dange R, Patil A (2019) Analytical study of effect of web opening on flexural behaviour of hybrid beams. Asian J Civ Eng. https://doi.org/10.1007/s42107-019-00122-4
11. Kaveh A, Shokohi F (2015) Optimum design of laterally-supported castellated beams using CBO algorithm. Steel Compos Struct 18(2):305–324
12. Bedi KS, Pachpor PD (2011) Moment and shear analysis of beam with different web openings. Int J Eng Res Appl
13. Wang P, Wang X, Ma N (2014) Vertical shear buckling capacity of web-posts in castellated steel beams with fillet corner hexagonal web openings. Eng Struct 75:315–326
14. Duggal SK. Limit state design of steel structure. Tata McGraw Hill Education
15. IS 800 General constructions in steel-code of practice (2007)
16. IS 2062 Specification of steel for general structural purposes (2006)
17. Ellobody E (2012) Nonlinear analysis of cellular steel beams under combined buckling modes. Thin-Walled Struct 52(1):66–79
18. Mehetre AJ, Talikoti RS (2020) Effect of fillet radii on moment carrying capacity of sinusoidal web opening castellated steel beams in comparison with hexagonal web openings. Iranian J Sci Technol Trans Civ Eng. https://doi.org/10.1007/s40996-020-00378-w

19. Mehetre AJ, Talikoti RS (2020) Prediction of ultimate load carrying capacity of castellated beams by the experimental and analytical investigation. Int J Struct Eng 11(2):107–126
20. Mehetre AJ, Talikoti RS. The effect of angle of web opening for prediction of ultimate failure load of castellated beams by experimental investigation. In: Recent trends in civil engineering. Lecture notes in civil engineering, Chapter 9, vol 77, pp 121–134

Effect of Incremental Loading with Construction Sequence Analysis of an RCC Floating Column Building Located in High Seismic Prone Areas

Aleti Ganesh, C. Vivek Kumar, G. V. V. Satyanarayana, and Joga Himavarsha

Abstract In this paper, the effects of the floating column (FC), as well as the transfer beam (TB), were studied to know the behavior of the structure under seismic conditions in zone IV whereas analysis was done using ETABS 2018. In a nonlinear analytic approach, the structure is analyzed at different phases in accordance with the order of construction, and the partial necessary forces are applied sequentially at each phase. The current study includes Equivalent static analysis (ESA), Response spectrum analysis (RSA), and Construction sequence analysis (CSA) and the final output is the behavior of structure with base shear, story drift, story displacement, bending moment, shear force, and deflection that were represented as graphs and tables. From this, it can be concluded that the CSA is a suitable analysis as it gives more accurate values for the floating column structure compared to ESA and RSA.

Keywords Transfer beam · Floating column · ESA · RSA · CSA

1 Introduction

CSA is a non-linear analysis method that involves the structure being assessed at various phases in accordance with the order of construction, as well as a partial requisite load which is successively applied at each phase [1]. Typically, structures are studied and planned in a single-phase using earthquake or gravity analyses under

A. Ganesh (✉)
Structural Engineering, Gokaraju Rangaraju Institute of Engineering and Technology, Hyderabad, India
e-mail: aletiganesh246@gmail.com

C. Vivek Kumar · G. V. V. Satyanarayana
Civil Engineering, Gokaraju Rangaraju Institute of Engineering and Technology, Hyderabad, India

J. Himavarsha
VNR Vignana Jyothi Institute of Engineering and Technology, Hyderabad, India

K. K. Pathak et al. (eds.), *Latest Developments in Civil Engineering*, Lecture Notes in Civil Engineering 352, https://doi.org/10.1007/978-981-99-2676-3_42

the underlying presumption that they would be completely loaded at the same time. Key facts in tall structures which have a major impact on analytical reliability are infrequently considered in practice. The impact of sequentially applying loads during development of structural elements in terms of responses. the result of diverse tributary areas' effects on column supports' unequal shortening of the exterior and interior supports [2].

The structural members were added incrementally as when the structure is constructed, and the portion of the building that has reached that level of completion is responsible and carrying the dead load [3]. As a result, the stresses as well as displacements dispersion in the portion of the completed structure at every stage caused by the partial dead load of members implemented at that stage may not rely on the geometry, properties, or existence of the members making up the remainder of the overall structure and can be accurately obtained by summing the results of analysis at each stage. Ignoring this influence could lead to inaccurate results from the examination of the overall structures. As a result, each stage construction should incorporate the structure to be analyzed.

The principle of incremental loading is taken into consideration by construction sequence analysis, commonly referred to it as staged construction analysis. As the structure frame has been constructed, load is gradually placed to it. When analyzing for strength, stability, including deflection just at conclusion of each stage, CSA is a more realistic and accurate form of analysis because it takes into consideration the numerous ways in which load is delivered to the frame [4]. The positioning of the building's numerous parts in the right sequence is also crucial.

2 Concept of Floating Column

In addition to being vertical elements, floating columns are often referred to as hanging columns [5] play area, parking, and function rooms can be located on the lower story because the hanging columns are typically constructed above the bottom floor. Structures with these FC have an irregular distribution of loads, which causes them to be more flexible and, as a result, less resilient to earthquakes. The floating columns-supported building is seismically unstable due to abnormalities in strength and stiffness both in-plane and out-of-plane. Under vertical loading conditions, this particular type of structure poses no risk [6]. Due to irregularities in strength and stiffness both in-plane and out-of-plane, the building with floating columns is seismically vulnerable. This type of structure has no issues when subjected to vertical loading. High shear capacity beams and deep beams are used to support the floating column [7]. A transfer beam (TB) is built to carry the entire load of the column as a single-point load.

3 Method of Analysis

"The effects of design earthquake load on structures can be analyzed using three methods". The following are the different methods of analysis herewith:

Equivalent static method. Most seismic assessments continue to be done with the premise that the lateral (horizontal) force is the same as the actual (dynamic) loading. The base shear, which is the overall horizontal force on the structure, is calculated using the structural mass, fundamental period of vibration, and associated mode shape. "According to the Code formula, the base shear is distributed in terms of lateral forces together with the height of buildings." This method is often conservative for low to medium-height buildings with regular configurations [8].

Response spectrum method. "A response spectrum is a graph of the peak or steady-state response of a sequence of oscillators with different natural frequencies that are all triggered by the same base vibration or shock". The resulting graphic can then be used to isolate the response of any linear system based on its natural oscillation frequency.

Construction sequence analysis. "A construction sequence load case automates the inclusion of stories in a model to account for the sequence effects of construction [9]. The sequential application of structure and load as a building is built may result in a significantly different distribution of forces (i.e., dead load) than would occur where the load is to be applied only after the building is complete" [10].

The auto construction sequence case performs a multi-step analysis. That tracks the building's sequential construction. On applications of CSA, you will get as shown in Fig. 1.

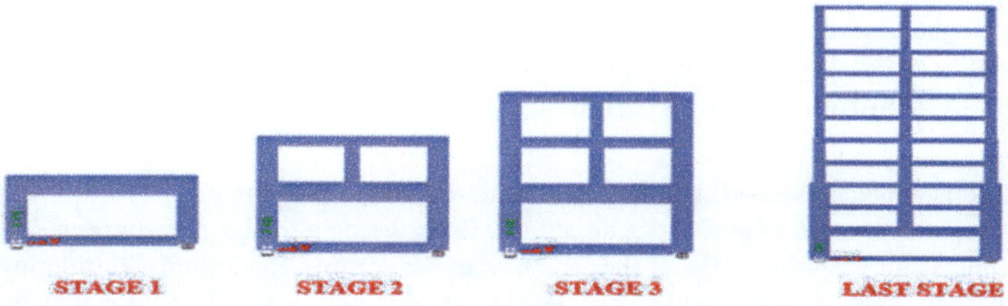

Fig. 1 Sequence analysis

4 Significance of Study

The software ETABS 2018 was used to construct the design phase. For this study, authors in [11] modelled several situations that have investigated to foresee how the structure will behave in zone IV under various combinations or seismic circumstances [12]. These buildings were built in accordance with the Indian Code of Practice for Seismic Resistant Building Design [13]. For this investigation, OMRF (ordinary moment resisting frame), which is situated in zone IV and has medium soil conditions, and the SMRF (special moment-resisting frame), which is positioned around the perimeter of the G+10 RCC building [14] are used. The base shear and maximum bending moment at each level for both frames were compared using FC positioned on various levels [15].

These failure spots are identified by this investigation. The use of safety measures to prevent structural damage can be beneficial to engineers. If FC are present on the outer perimeter of both regular and irregular structures, the displacement will be at its maximum [16]. Thus, in earthquake-prone regions, FC-containing irregular constructions have suffered significant damage [17]. In seismic zones, it is therefore advisable to avoid this type of situation as much as possible [18]. The risk component of providing FC is that it improves the structure's damage when an earthquake occurs. According to this study's findings, maximum displacement and inter-story drift both increase as the bottom floor column height increases. Both overturning as well as base shear moments differ as the column size varies [19]. The weight reduction upon these floors would result in a smaller story shear on the top floors than on the bottom stories. It was found that the drift and displacement in RC structures were decreased by shear wall, infill wall, and steel bracing [20] along with increased pillar dimensions to enhance the stiffness of the structure [21]. It indicates that the stiffness is rising while the drift values are declining.

5 Methodology and Modelling

In this analysis, a G+10 building is considered with a plan area of 1225 m^2 with a span of 35 m in both x and y directions. Considering a total 6 different cases to understand the importance of the structure under seismic conditions. Considering that the building is located in Zone IV type II soil conditions, and SMRF with an importance factor of 1 (Fig. 2).

Model 1 (M1). This is a normal G+10 building without any floating columns.

Model 2 (M2). In this model, the floating columns are located on the ground floor (GF) in FCB2, FCB9, FCI2, and FCI9. Here the building does not sustain the seismic load and failed.

Model 3 (M3). As the M_2 failed to keep the building plan as same as M_2 changed the dimensions of over-stressed columns and beams. Designed RCC transfer beam

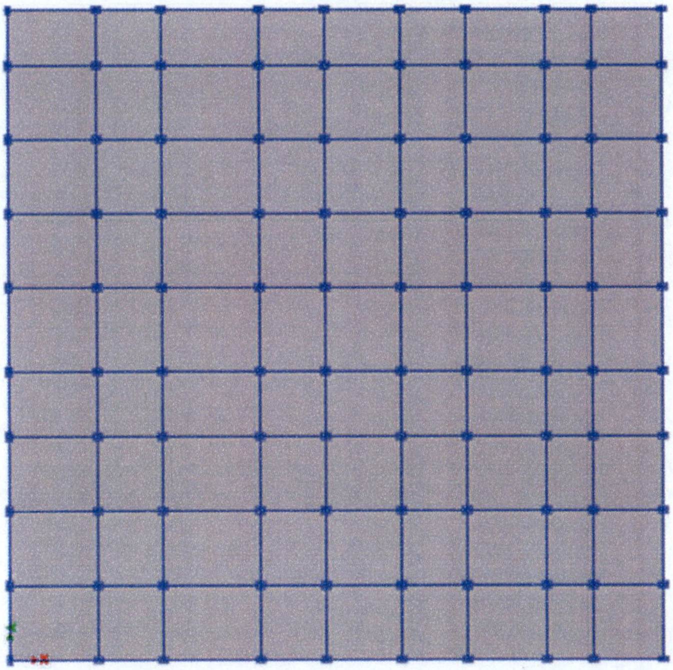

Fig. 2 Plan of G+10 RCC framed structure

(TB1) with 600 * 1200 M_{50} grade concrete. The transfer columns dimensions are as TC_1 600 * 900, and TC_2 750 * 900.

Model 4 (M4). The building was modeled the same as M_3 with the addition of ISMB 450 (TB_2) bracing at GF. To distribute the lateral loads.

Model 5 (M5). The building was designed as M_3 with a Composite transfer beam (CTB_1) located at GF.

Model 6 (M6). Building same as M_5 with the addition of ISMB 450 (CTB_2) bracing at GF.

Note. All the Beam and Column dimensions are in mm (Figs. 3, 4 and 5; Table 1).

6 Results and Discussion

Story drift. "Displacement of one-story relative to other story is called story drift. The story shall not exceed 0.004 times the story height", according to IS 1893(part 1): 2016 (cl 7.11.1) (Tables 2, 3, 4, 5, 6, 7 and 8).

Story displacement. The building's lateral displacement, also known as joint displacement, is an important factor to consider while constructing a multi-story

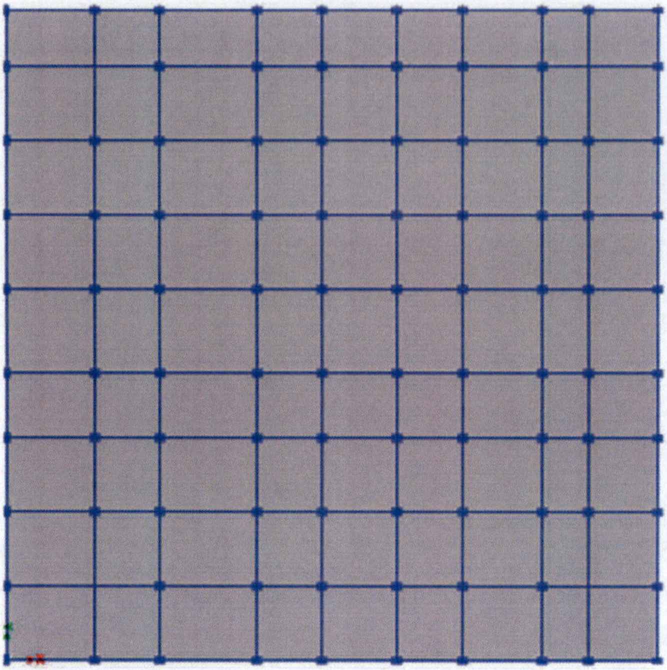

Fig. 3 Model 2 (structure with FC)

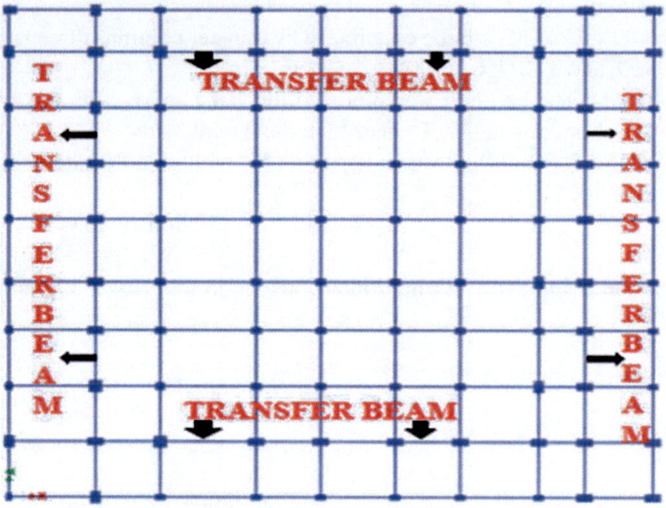

Fig. 4 Model 3 (FC with the inclusion of TB)

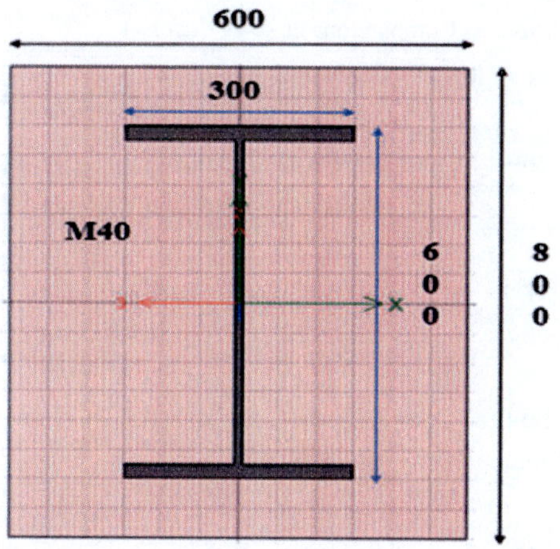

Fig. 5 Cross section showing the CTB

Table 1 Properties of building model

Type of structure	SMRF	*Type of soil*	Medium soil (II)
No of story's	G+10	*Zone factor*	0.24 (Zone IV)
Height of each story	Bottom story 4 m	*Grade of steel*	Fe 415
	Remaining story's 3 m		Fe 500
Dimension of plan	35 m × 35 m	*Grade of concrete*	Columns M_{40}
Response reduction factor (R)	5		Beams M_{30}, M_{50}
Damping ratio (ζ)	5%		Slab M_{30}
Thickness of slab	150 mm	*Thickness of parapet wall*	100 mm
Thickness of wall	230 mm		

Table 2 Comparisons of story drift EQ_x

Story level	Story drift EQ_x					
	M_1	M_2	M_3	M_4	M_5	M_6
10F	0.000835	0.000834	0.000894	0.000894	0.000885	0.000885
8F	0.001276	0.001272	0.001372	0.001373	0.001357	0.001358
5F	0.001361	0.001357	0.001464	0.001464	0.001448	0.001449
4F	0.001667	0.001662	0.001795	0.001796	0.001775	0.001776
GF	0.001893	0.001887	0.002038	0.002039	0.002015	0.002017

construction According to Indian Codes, the maximum displacement of the building in the case of wind load should be within 1/500 times the building height, and in the case of earthquake load, it should be 1/250 times the building height. The displacement of the structure in the X and Y directions for various models is estimated for

Table 3 Comparisons of story drift EQ_y

Story level	Story drift EQ_y					
	M_1	M_2	M_3	M_4	M_5	M_6
10F	0.001336	0.001331	0.001427	0.001461	0.001414	0.001453
8F	0.001784	0.001777	0.001909	0.001955	0.00189	0.001943
5F	0.001529	0.001524	0.001635	0.001674	0.001619	0.001664
4F	0.001859	0.001852	0.001989	0.002037	0.00197	0.002024
GF	0.002116	0.002108	0.002263	0.002316	0.002241	0.002302

Table 4 Comparisons of story drift RS_x

Story level	Story drift RS_x					
	M_1	M_2	M_3	M_4	M_5	M_6
10F	0.000795	0.00079	0.000907	0.000916	0.000889	0.000898
8F	0.001167	0.00116	0.001308	0.001321	0.001287	0.001299
5F	0.001172	0.001166	0.001278	0.001289	0.001263	0.001274
4F	0.001393	0.001387	0.001491	0.001503	0.001476	0.001488
GF	0.001555	0.001548	0.001638	0.001651	0.001625	0.001637

Table 5 Comparisons of story drift RS_y

Story level	Story drift RS_y					
	M_1	M_2	M_3	M_4	M_5	M_6
10F	0.00139	0.001382	0.001555	0.001615	0.001535	0.001605
8F	0.001719	0.001711	0.001882	0.00194	0.001865	0.001933
5F	0.001337	0.001333	0.001416	0.001448	0.001411	0.001447
4F	0.001576	0.001572	0.00164	0.001669	0.001638	0.001669
GF	0.001776	0.001772	0.001826	0.001853	0.001827	0.001854

Table 6 Comparisons of story drift CSA_x

Story level	Story drift CSA_x					
	M_1	M_2	M_3	M_4	M_5	M_6
10F	0.003734	0.003738	0.003645	0.003644	0.003646	0.003637
8F	0.002111	0.002119	0.001933	0.001935	0.001953	0.001942
5F	0.001002	0.001012	0.0008	0.000798	0.000822	0.000815
4F	0.000161	0.000153	0.000413	0.000407	0.000366	0.000365
GF	0.000991	0.000982	0.001248	0.001246	0.001203	0.001205

Table 7 Comparisons of story drift CSA_y

Story level	Story drift CSA_y					
	M_1	M_2	M_3	M_4	M_5	M_6
10F	0.004051	0.004036	0.003938	0.003907	0.003923	0.003885
8F	0.001989	0.001981	0.001767	0.0017	0.001772	0.001688
5F	0.001111	0.001107	0.000894	0.000823	0.000899	0.000818
4F	0.000177	0.00018	0.000442	0.000515	0.000412	0.000497
GF	0.001114	0.001112	0.001384	0.001465	0.001355	0.001449

Table 8 Comparisons of base shear

Models	Base shear (kN)					
	ESA		RSA		CSA	
	EQ_x	EQ_y	RS_x	RS_y	CSA_x	CSA_y
M_1	8841.622	8019.201	8870.438	8075.34	9886.9016	9061.706
M_2	8796.499	7990.164	8843.13	8062.804	9841.5533	9028.812
M_3	9500.922	8537.013	9348.117	8392.643	10,544.632	9576.585
M_4	9509.361	8715.168	9351.162	8437.041	10,554.008	9755.128
M_5	9383.263	8453.93	9283.273	8339.167	10,428.174	9490.387
M_6	9399.483	8661.495	9294.221	8397.31	10,444.421	9698.878

comparative ESA, RSA, and CSA as indicated in the table below (Tables 9, 10, 11, 12, 13, 14 and 15; Figs. 6 and 7).

The decrement value of the base shear is observed in M_2. This trend is followed for all types of analysis (i.e., ESA, RSA, CSA) in both directions. The decrement in value is observed as there are floating columns in M_2. The Base shear (V_b) values are observed to be increasing in CSA compared to ESA and RSA. But these values are more in x-direction than y-direction. In M_5 these values are less as compared to M_3 due to Composite Transfer Beam (CTB_1). Base shear values as seen as 8.94%, 9.7% higher in CSA as compared to ESA and RSA in M_1. Similarly, 9.04%, 8.88% in M_3 and 8.9%, 7.9% in M_5 (Figs. 8, 9, 10, 11, 12 and 13).

Table 9 Comparisons of story displacement EQ_x

Story level	Story displacement EQ_x (mm)					
	M_1	M_2	M_3	M_4	M_5	M_6
10F	51.849	51.917	50.044	50.061	50.12	50.098
8F	45.517	45.599	43.247	43.262	43.394	43.369
5F	30.755	30.884	27.358	27.364	27.682	27.644
4F	25.137	25.284	21.32	21.322	21.71	21.668
GF	5.892	6.069	3.483	3.481	3.805	3.791

Table 10 Comparisons of story displacement EQ_y

Story level	Story displacement EQ_y (mm)					
	M_1	M_2	M_3	M_4	M_5	M_6
10F	60.138	60.159	58.119	57.853	58.277	57.917
8F	50.781	50.837	48.111	47.604	48.365	47.728
5F	34.267	34.385	30.45	29.523	30.877	29.756
4F	27.982	28.123	23.741	22.657	24.231	22.931
GF	6.564	6.757	4.206	2.971	4.474	3.113

Table 11 Comparisons of story displacement RS_x

Story level	Story displacement RS_x					
	M_1	M_2	M_3	M_4	M_5	M_6
10F	41.578	41.625	38.613	38.886	38.855	39.07
8F	37.173	37.239	33.788	34.018	34.089	34.265
5F	26.749	26.859	22.756	22.902	23.151	23.249
4F	22.593	22.72	18.401	18.514	18.831	18.899
GF	5.875	6.052	3.381	3.389	3.696	3.695

Table 12 Comparisons of story displacement RS_y

Story level	Story displacement RS_y					
	M_1	M_2	M_3	M_4	M_5	M_6
10F	49.546	49.69	45.2	44.48	45.889	44.854
8F	42.826	42.995	37.977	37.061	38.703	37.454
5F	30.781	30.978	25.591	24.463	26.302	24.854
4F	25.933	26.139	20.66	19.452	21.359	19.841
GF	6.71	6.928	3.992	2.781	4.243	2.972

Table 13 Comparisons of story displacement CSA_x

Story level	Story displacement CSA_x					
	M_1	M_2	M_3	M_4	M_5	M_6
10F	12.19	12.188	12.151	12.181	12.168	12.193
8F	29.63	29.661	28.859	28.845	28.869	28.797
5F	29.232	29.371	26.44	26.394	26.689	26.574
4F	25.15	25.328	21.671	21.61	22.006	21.886
GF	6.591	6.81	4.014	3.928	4.357	4.26

Table 14 Comparisons of story displacement CSA_y

Story level	Story displacement CSA_y					
	M_1	M_2	M_3	M_4	M_5	M_6
10F	15.239	15.191	15.281	15.378	15.249	15.361
8F	33.263	33.145	32.371	32.127	32.24	31.948
5F	32.773	32.667	29.741	28.772	29.696	28.607
4F	28.166	28.089	24.413	23.191	24.407	23.052
GF	7.5	7.68	4.738	3.333	4.973	3.448

Table 15 Comparisons of bending moment, shear force and deflection

Beam type	Bending moment (kN-m)			Shear force (kN)			Deflection (mm)		
	ESA	RSA	CSA	ESA	ESA	CSA	ESA	RSA	CSA
TB1	134.8003	111.971	1991.65	63.214	54.2094	1125.156	0.499	0.4	6.66
TB2	89.6384	76.9396	1160.244	42.154	38.0069	701.7319	0.352	0.283	3.838
CTB1	126.6041	108.4417	1138.579	61.2876	52.859	664.7585	1.104	0.925	8.314
CTB2	74.9082	66.1083	708.534	36.1924	32.3934	462.1914	0.681	0.573	5.613

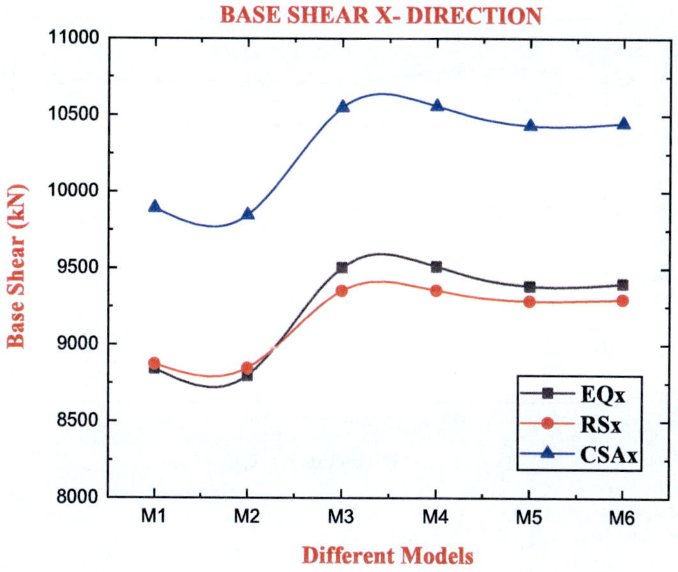

Fig. 6 Base shear EQ_x

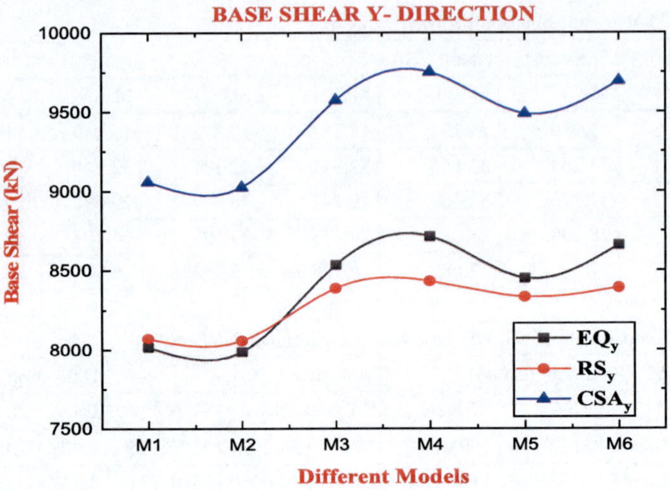

Fig. 7 Base shear EQ_y

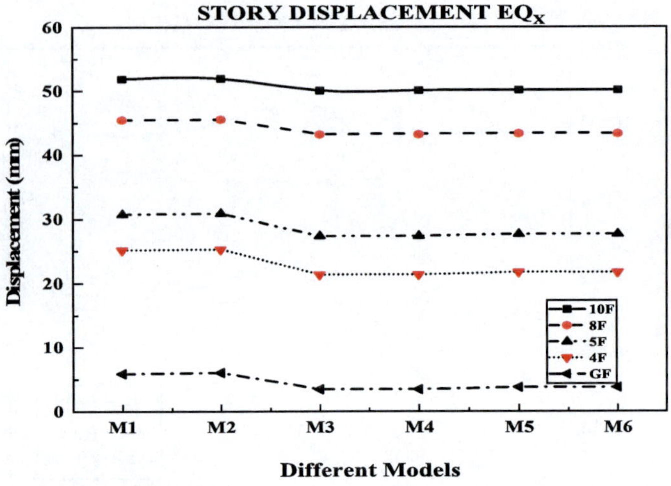

Fig. 8 Story displacement EQ_x

- From bottom story to top story, the story displacement value increases by 3 times in M_2 in EQ_x direction because of FC. In M_3, M_4, M_5 and M_6 there is a decrement of 0.35 times, 0.43 times, 0.28 times and 0.29 times respectively in the Ground floor when compared to M_1. At the Top floor values of displacement increased as 6.625 times, 11.33 times, 12.35 times, 10.17 times and 11.21 times in M_2, M_3, M_4, M_5 and M_6 respectively when compared to GF. Now in EQ_y direction there

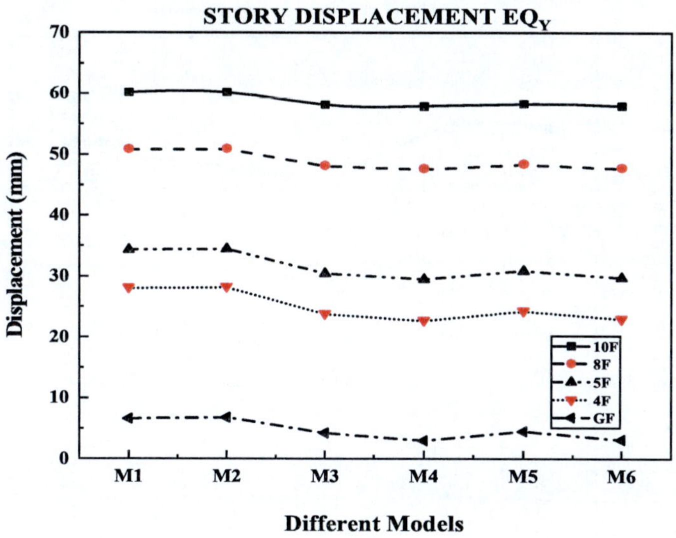

Fig. 9 Story displacement EQ_y

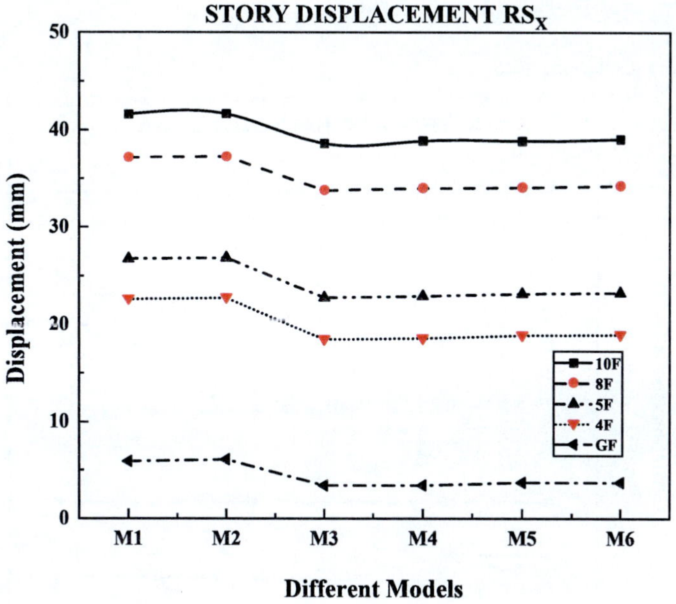

Fig. 10 Story displacement RS_x

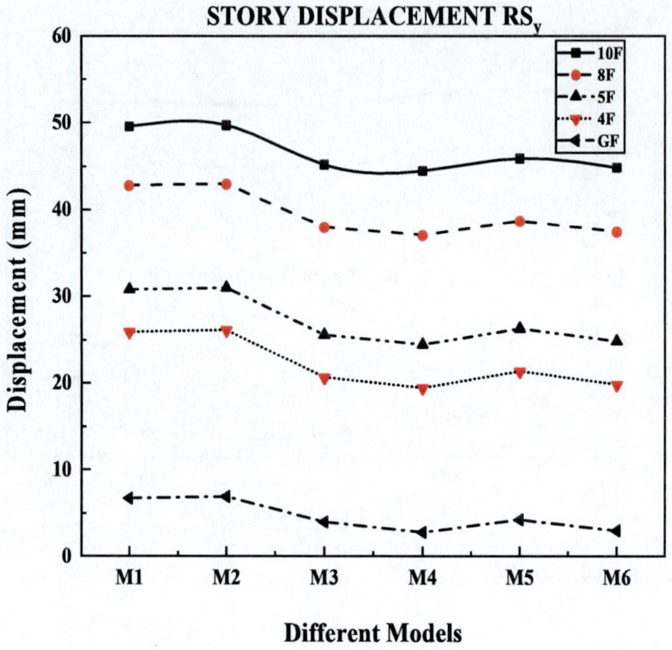

Fig. 11 Story displacement RS_y

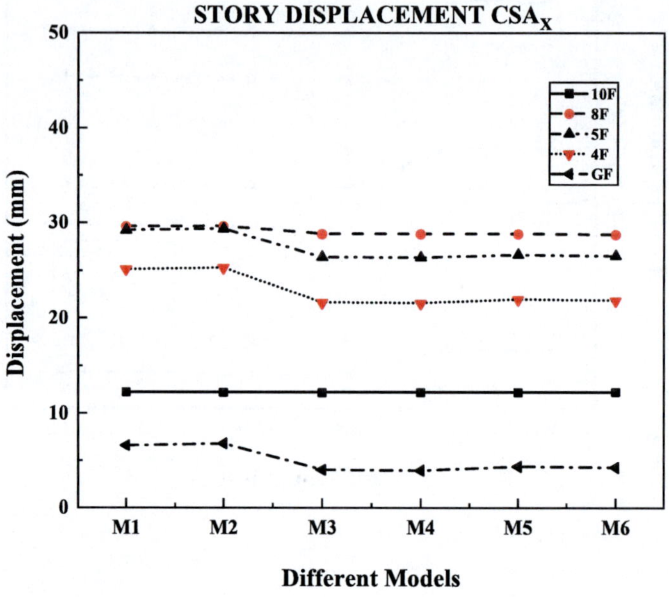

Fig. 12 Story displacement CSA_x

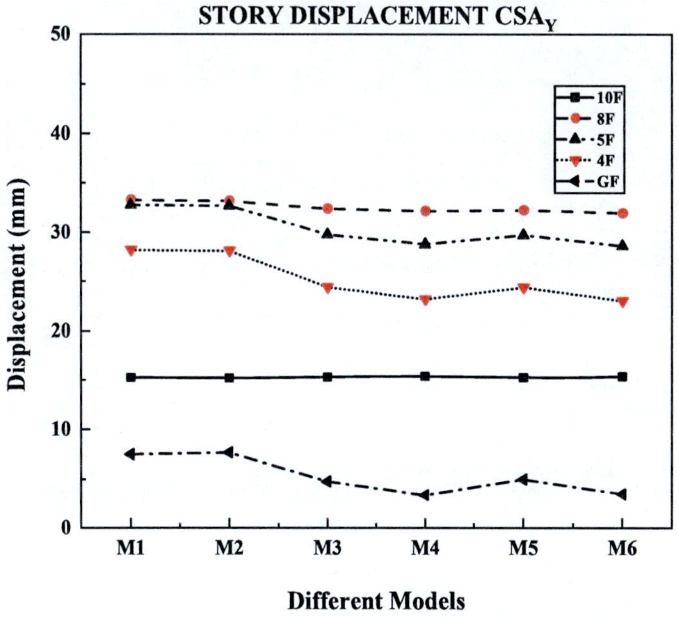

Fig. 13 Story displacement CSA_y

is a rise of 0.1 times in M_2, M_4 and M_6 and 0.12 times in M_3 and M_5 as compared to EQ_x in the top story.

- In RS_x direction the displacement decreased as 6.30%, 5.78%, 5.5%, 4.93% for M_3, M_4, M_5 and M_6 respectively as compared to M_2. And for RS_y the displacement values decreased to 9.82%, 5.56%, 7.6% in M_3, M_4 M_5 and M_6 respectively compared to M_2.
- In CSA there is an increase in displacement values up to 8F in both x and y directions. And there is a sudden fall down at 9F and 10F because of the variation in dimensions of the beams and columns (Figs. 14, 15, 16, 17, 18 and 19).
- Story drift increases up to 5F and then falls down in 6F to 10F. The Drift values are more effective in y-direction than x-direction in all analyses. From the graph it can be seen that the drift values follow a similar pattern in 5F. Static analysis gives more drift values than Staged construction analysis and Response spectrum.
- The drift values are more in M_2 compared to remaining models, and less in M_3 compared to all other models (Figs. 20 and 21).
- The BM values are high in CSA compared to ESA and RSA. There is 6.76 times and 5.622 times increase in TB_1 in ESA and RSA respectively. But for CTB_1 there is only 6.35 times, and 5.44 times increase. For TB_2 4.5 times and 3.86 times more in ESA and RSA respectively this is because of the inclusion of bracings and it is 3.76 times and 3.31 times in CTB_2. From the above, we can observe that CBT_2 is the best section.

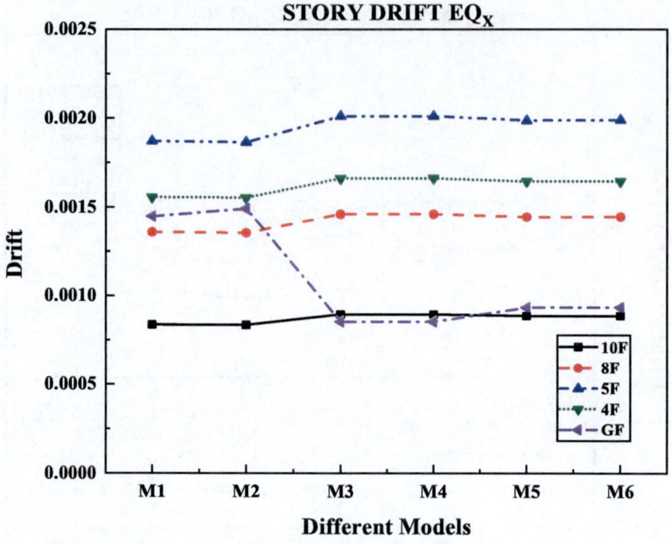

Fig. 14 Story drift EQ_x

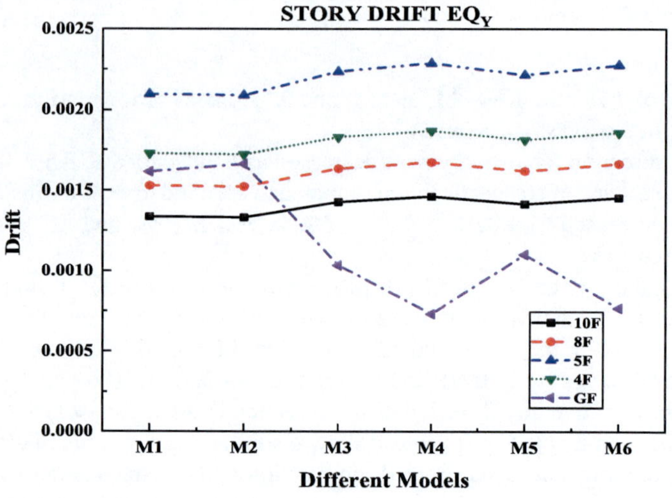

Fig. 15 Story drift EQ_y

- The (SF) Shear force values are higher in CSA compared to ESA and RSA. There is 5.61 times and 4.81 times rise for ESA and RSA respectively in TB_1. But for CTB_1 there is only 5.41 times and 4.69 times rise. For TB_2 it is 3.74 times and 3.37 times more in ESA and RSA respectively this is because of the provision of bracings. And 3.21 times and 2.89 times in CTB_2 (Fig. 22).

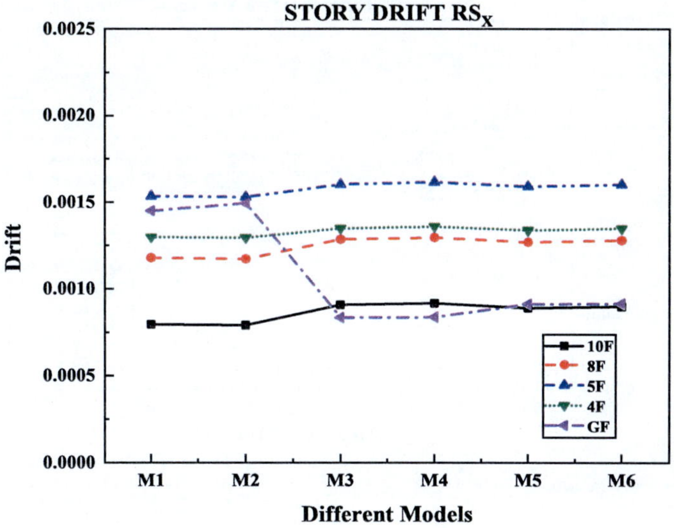

Fig. 16 Story drift RS_x

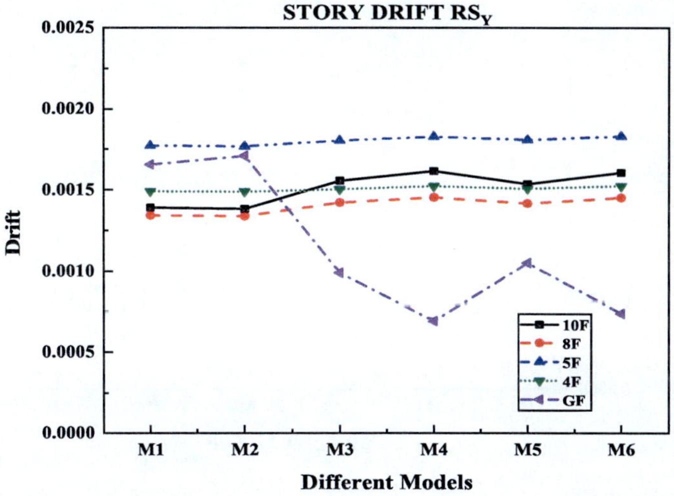

Fig. 17 Story drift RS_y

- Deflection is high in CSA compared to ESA and RSA. They are 6.66 mm, 3.838 mm, 8.314 mm and 5.613 mm in TB1, TB2, CTB1 and CTB2 respectively. The deflection is reduced in TB2 because of more depth of the beam with bracings. If the same design sections are used for CTB2 the deflection may even be very less as it seen from the BM and SF values which are less in CTB2.

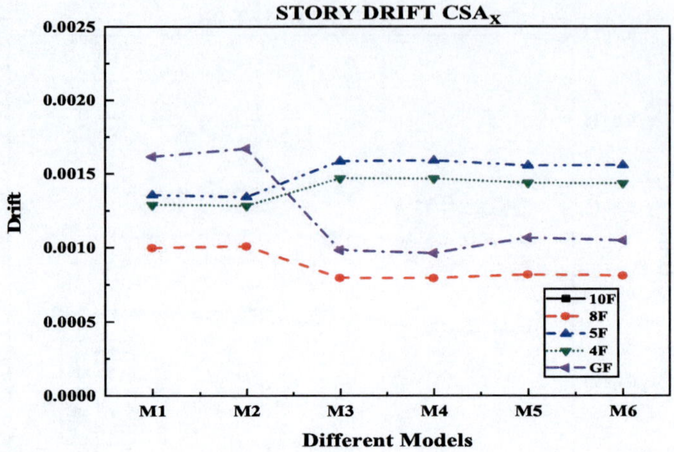

Fig. 18 Story drift CSA$_x$

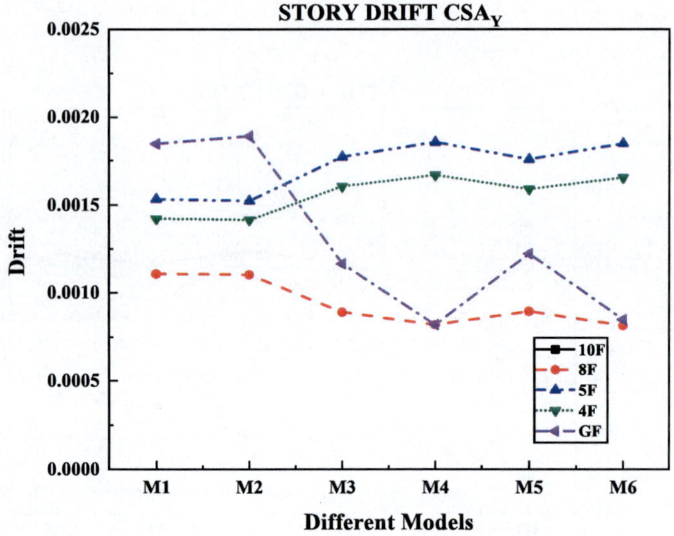

Fig. 19 Story drift CSA$_y$

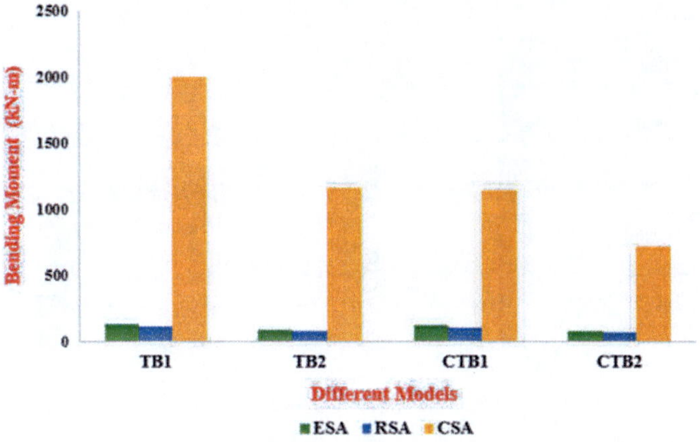

Fig. 20 Bending moment

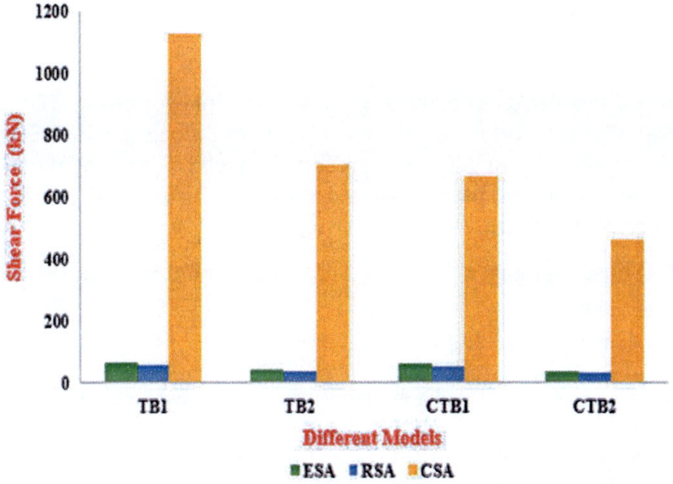

Fig. 21 Shear force

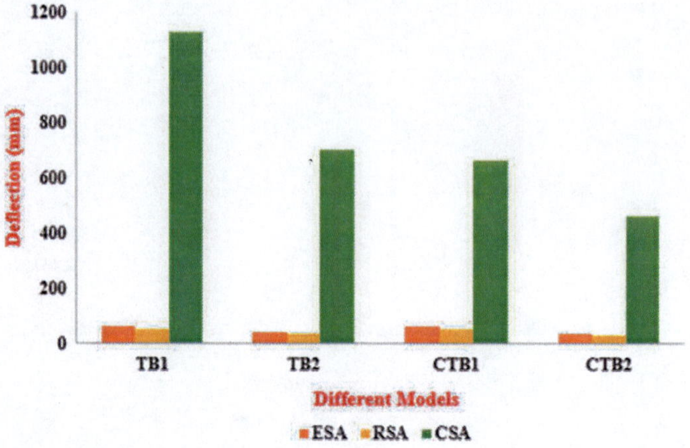

Fig. 22 Comparison of deflection for different analysis

7 Conclusion

It is concluded that using different models with the transfer beam (TB1 and TB2) and composite transfer beam (CTB1 and CTB2) in a floating column building in high seismic prone areas to derive the parameters such has bending moment, shear force, story displacements, base shear, and story drift for zone IV, seismic analysis of G+10 story buildings are investigated using ESA, RSA and CSA.

- The (BM) bending moment and (SF) Shear force of a Composite Transfer Beam (CTB2) is reduced.
- CSA has a higher bending moment, shear force, and Deflection than ESA and RSA.
- The presence of bracings reduces base shear, story displacement, BM, SF, and Deflections.
- CSA is the best analysis as it gives realistic values for (FC) hanging column structures.
- CTB is stiffer than TB. So, the way of acting of CTB is more effective than TB.
- The FC and Transfer Beams are designed for higher values compared to all other methods i.e., ESA, RSA, CSA
- There was significant variation in BM, SF, and Deflection values compared to ESA, RSA, and CSA.
- Transfer beam plays a major role in stub column (FC) building.

References

1. Jain N, Ahirwar S (2022) Comparative study of seismic analysis of multi-storey building with and without floating column: a review. IJRPR Int J Res Publ Rev 3(8):7–11
2. Uplenchwar AR (2022) A review: seismic analysis of a structure with soft storey and floating column. IJPRSE 3(3):59–62
3. Gaidhankar DG, Pitamberwale YV, Kulkarni MS, Shinde SN, Karad V (2021) A comparative study of construction sequence with conventional analysis. Turk J Comput Math Educ 12(13):6482–6496
4. Abbas DHS, Kharnoob DMM, Atia DNS, Bassam DBF (2022) Analysis or design of composite column by using ETABS software. Webology 19(1):5289–5301. https://doi.org/10.14704/web/v19i1/web19355
5. Mahaveer DN, Ravindra PM (2019) Analysis of an RC building incorporating the effects of sequential construction. Int J Res Eng Sci Manag 2(6):576–581
6. Elansary AA, Metwally MI, El-Attar A (2021) Staged construction analysis of reinforced concrete buildings with different lateral load resisting systems. Eng Struct 242(March 2020):1–12. https://doi.org/10.1016/j.engstruct.2021.112535
7. Sreenivasulu K, Sowjanya K (2019) Seismic analysis and designing of multi story building with floating columns by construction sequence analysis by using ETABS. IAETSD J Adv Res Appl Sci VI(244):244–248
8. Thant NN, Kyaw TY (2019) Study on the effect of response spectrum analysis and construction sequence analysis on setback steel structure. Publ Int J Trend Sci Res Dev 3(4):1349–1355. https://doi.org/10.31142/ijtsrd25142
9. Pranay R, Sreevalli IY, Kumar S (2021) Study and comparison of construction sequence analysis with conventional lumped analysis using ETABS. Int J Sci Eng Technol Res 10:16–27
10. Lanjewar B, Khedikar PA (2020) Seismic response of RC building under main shock-aftershock sequence. J Interdiscip Cycle Res 12(1409):1409–1419
11. Sawai G, Atif M, Khan Y (2021) Seismic analysis of multi-storey building with floating and non floating column. IJECS 4(7):16–19
12. Banugariya ND, Solanke SS (2018) Analysis of multi-storey structures using sequential analysis—a review. IJIES 3(3):4–7
13. Gupta A, Verma VK, Agari C (2019) Comparative seismic study of G+10 building with and without floating column. Int J Res Eng Sci Manag 2(8):323–327
14. Udhav SB, Shaikh AN, Ravi GM (2015) Analysis of multistorey building with floating column. Int J Eng Res 4(9):475–478. https://doi.org/10.17950/ijer/v4s9/902
15. Pandey S, Jamle G (2018) Optimum location of floating column in multistorey building with seismic loading optimum location of floating column in multistorey. Int Res J Eng Technol 5(10):971–976
16. Patil S, Patil PVB (2017) Seismic analysis of bottom rigid beam storey and intermediate soft storey having moment transfer beams. Int Res J Eng Technol 4(4):840–845 [Online]. https://www.irjet.net/archives/V4/i4/IRJET-V4I4215.pdf
17. Goud R (2017) Study of floating and non-floating columns with and without earthquake. Int J Sci Technol Eng 4(1):152–157
18. Sharma S, Pastariya S (2020) Effect of floating columns on seismic response of multistory building. Int J Adv Sci Res Eng Trends 5(12):795–800
19. Bhensdadia H, Shah S (2015) Pushover analysis of RC frame structure with floating column and soft story in different earthquake zones. Int J Res Eng Technol 4(4):114–121. https://doi.org/10.15623/ijret.2015.0404020
20. Kumbhar A, Banhatti G (2016) Seismic retrofitting of building with so storey and floating column. IRJET 3(7):1917–1921
21. Purushothaman V, Sukumaran A (2017) Comparative study on seismic analysis of multi storied buildings with composite columns. Int J Eng Res V6(06):555–561. https://doi.org/10.17577/ijertv6is060248

Analysis and Design of Irregular Multi-story Building Having Visco-elastic Dampers

Jagannadham Rohith Kumar and G. V. V. Satyanarayana

Abstract In the paper we discuss about analysis and design of G+15 plan irregular buildings under seismic loads, there are 4 different models. Model 1 has no dampers placed, Model 2 has dampers placed around the building placed at alternate floors starting from the 6th floor till the 15th floor, Model 3 has dampers placed around the building placed at alternate floors starting from the ground floor till the 14th floor, and Model 4 has dampers placed around the building starting from the 6th floor till the 15th floor. We can conclude that model 3 is effective compared to remaining models and there is a significant reduction in story drift, story displacement. This can also be the use of dampers.

Keywords Story displacement · Story drift · Base shear · Visco-elastic dampers

1 Introduction

As a result of the rapid amount of population growth there is increase in demand for high rise buildings. People are opting for apartments in metropolitan cities. In today's cities, a great number of residential and commercial structures have multiple floors. There hasn't been much research into the effects of floors with dampers [1]. The breakdown of these structures during earthquakes serves as a reminder of the necessity for energy dissipation systems. Using energy dissipation devices and comparing different models, an effort has been made to determine the variance in time period, story displacement, base shear (V_b), and drift of structure [2]. Recently, passive dampers have been employed in both the design of new buildings and the strengthening of existing ones. Without isolating the structure, energy dissipating

J. R. Kumar (✉)
Structural Engineering, Gokaraju Rangaraju Institute of Engineering and Technology, Hyderabad, India
e-mail: jagannadhamrohithkumar@gmail.com

G. V. V. Satyanarayana
Civil Engineering, Gokaraju Rangaraju Institute of Engineering and Technology, Hyderabad, India

K. K. Pathak et al. (eds.), *Latest Developments in Civil Engineering*, Lecture Notes in Civil Engineering 352, https://doi.org/10.1007/978-981-99-2676-3_43

viscous dampers can now be used. Although both systems have the same goal of limiting earthquake damage, while their implementation methods are different [3]. Dampers can be employed all across the building. Visco-elastic dampers can provide up to 30% or even greater dampening. These dampers can be utilized in both new and old buildings [4]. VED can be installed normal along the story height of the buildings above the cellar level in a building with multiple basements, so that it is simple to set up and manage. Visco-elastic dampers will increase resistance in the building's lateral system [5]. The efficiency of horizontal dampers for structures with multiple stories of basements is investigated in this study.

The VE dampers are used in this investigation among several types of dampers which are most effective [6].

1.1 Visco Elastic Dampers

A passive energy dissipation device is a viscoelastic (VE) damper. This type of damper dissipates the structure's mechanical energy and turns it to heat [7]. The efficiency of the damper system is influenced by a number of factors, including ambient temperature and loading frequency. Visco-elastic dampers are shown to improve the whole dynamically responsive performers' performance buildings by greatly increasing the total dampening of the structure. Furthermore, Visco-elastic dampers (VED) are the most auspicious gadgets and have been put in a number of structures throughout the world [8]. It is made up of steel plates attached to layers of VE material (copolymers or glassy solids). Shear deformation of the VE material placed between steel plates dissipates vibration energy. Figure 1: A viscoelastic damper is installed horizontally [9]. A highly dissipative polymeric substance is included in the structure of a viscoelastic damper. These materials' elastic stiffness generates a lateral acceleration force, while the viscous component creates a "velocity-dependent force". For the past 20 years, viscoelastic dampers have been utilized as wind vibration absorbers in the World Trade Center raised towers in NYC, and they have recently been added to a number of additional structures [10].

Please note that the first paragraph of a section or subsection is not indented [11].

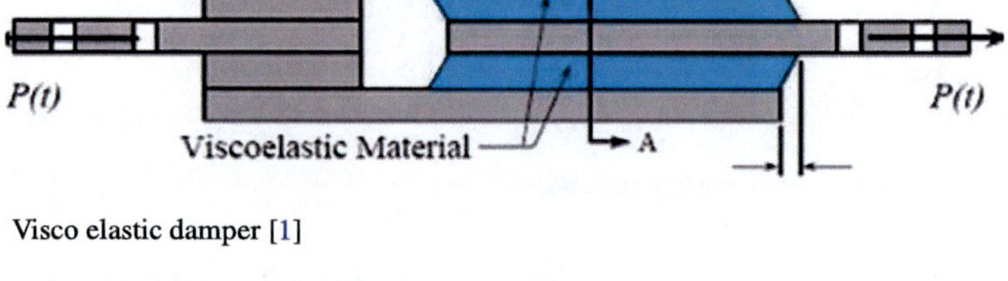

Fig. 1 Visco elastic damper [1]

2 Analysis

"The effects of design earthquake load on structures can be analyzed using two methods".

(a) Equivalent static approach. (b) Dynamic analysis method.

Equivalent static method: The majority of seismic assessments still assume that the lateral (horizontal) force is the same as the real (dynamic) loading. "The base shear, which is the overall horizontal force on the structure, is calculated using structural mass, fundamental period of vibration, and associated mode shape". In terms of lateral forces, the base shear is distributed along with the height of structures according to the Code formula. "This method is usually conservative for low-to medium-rise buildings with a regular shape" [12].

Dynamic analysis: On the other hand, dynamic analysis can be done in three different methods. The time history method, the modal time history method, and the response spectrum approach are all examples of time history methods (IS 1893 (Part 1c)).

Response spectrum method: "A response spectrum is a graph depicting the peak or steady-state response of a series of oscillators with varied natural frequencies, all triggered by the same base vibration or shock". The generated image can then be used to isolate any linear system's response using its natural oscillation frequency [13].

3 Methodology

- We use ETABS 2018 for modeling and applying the lateral loads. For modelling purposes, frame elements for columns and beams were chosen, whereas membrane elements were used for slabs.
- The link property had been used to model the "VED".
- "Damper properties K of 20,000 KN/m and the damping coefficient C of 10,000 KNs/m" and Modelling.

4 Modelling

A plan of 28 m × 25 m multi-story (G+15) building is modelled with SMRF (special moment resting frame) for this study. It was considered to be located in zone iv type 2 soil.

Buildings were modified by adding visco-elastic dampers at various locations and changing the number of dampers.

M1: Building without Visco-elastic dampers located in ZONE IV (Fig. 2).

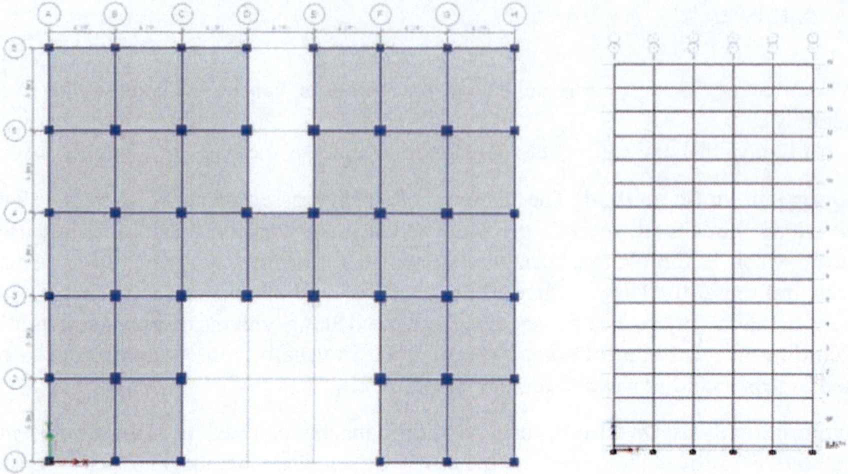

Fig. 2 Normal building

M2: Building with Visco-elastic dampers at ZONE IV @ 6, 8, 10, 12, 14 at each of the respective floors we have provided dampers around the face of building (Fig. 3).

M3: Building Visco-elastic dampers at ZONE IV @ GF, 2, 4, 6, 8, 10, 12, 14 at each of the respective floors we have provided dampers around the floor (Fig. 4).

M4: Building Visco-elastic dampers at ZONE IV @ 6, 7, 8, 9, 10, 11, 12, 13, 14, 15 at each of the respective floors we have provided dampers around the floor (Fig. 5; Table 1).

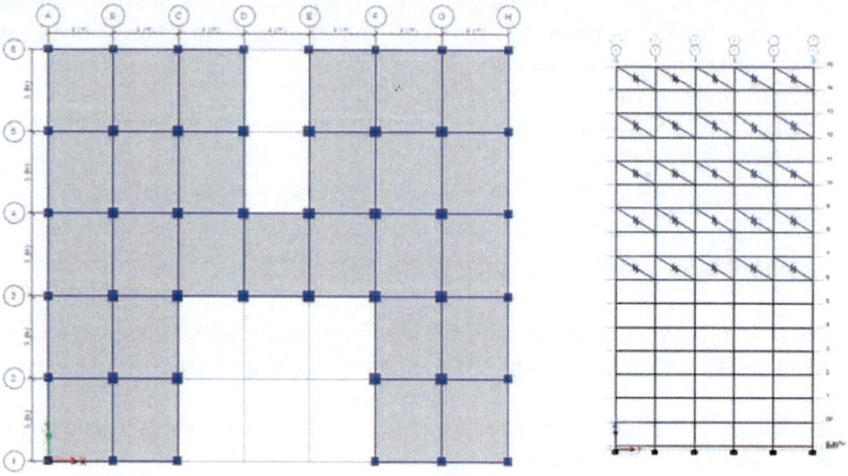

Fig. 3 Model 2

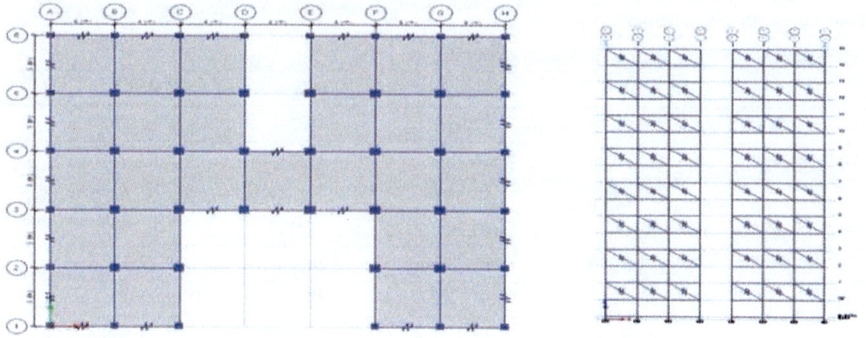

Fig. 4 Model 3

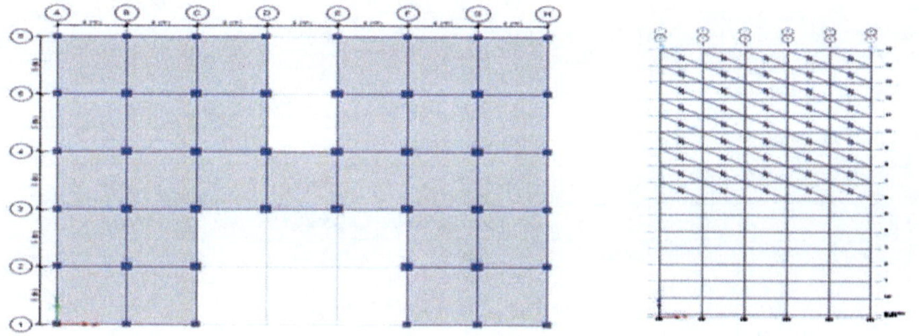

Fig. 5 Model 4

Table 1 Properties of building

Type of building	SMRF	No of story's	G+15
Height of story	3 m	Grade of concrete	M30, M40
Grade of steel	Fe415, Fe500	R	5
I	1.5	Damping ratio	5%
Soil type	2	Slab	150 mm

5 Results and Discussion

- The story displacement in model 3 is 46.43% less compared to conventional structure in EQ_X (Table 2; Figs. 6, 7, 8 and 9)
- The story displacement in model 3 is 46.32% less compared to conventional structure in EQ_Y

Table 2 Story displacement in M3A

Story level	Story displacement (mm)			
	EQ_X	EQ_Y	RS_X	RS_Y
15	44.287	47.416	32.741	19.912
14	41.958	46.139	31.054	19.638
13	39.115	43.592	29.047	18.894
12	36.76	42.26	27.359	18.619
11	33.258	38.458	24.937	17.333
10	30.92	37.112	23.288	17.066
9	26.988	32.436	20.606	15.309
8	24.745	31.13	19.053	15.06
7	20.612	25.911	16.22	12.893
6	18.563	24.708	14.825	12.669
5	14.451	19.23	11.891	10.15
4	12.714	18.192	10.72	9.982
3	8.828	12.685	7.712	7.3
2	7.531	11.864	6.833	7.167
1	4.055	6.512	3.813	4.081
GF	3.311	5.927	3.288	3.972

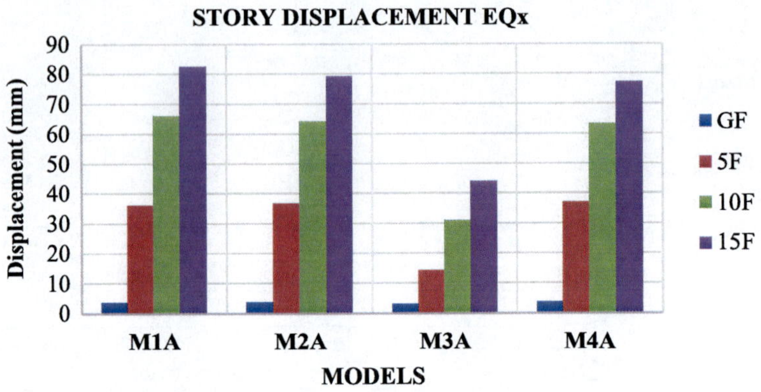

Fig. 6 Story displacement EQ_X

- The story displacement in model 3 is 47.60% less compared to conventional structure in RS_X
- The story displacement in model 3 is 70.34% less compared to conventional structure in RS_Y (Table 3; Figs. 10, 11 and 12)
- The story drift in model 3 is 29.46% less compared to the conventional structure in EQ_X

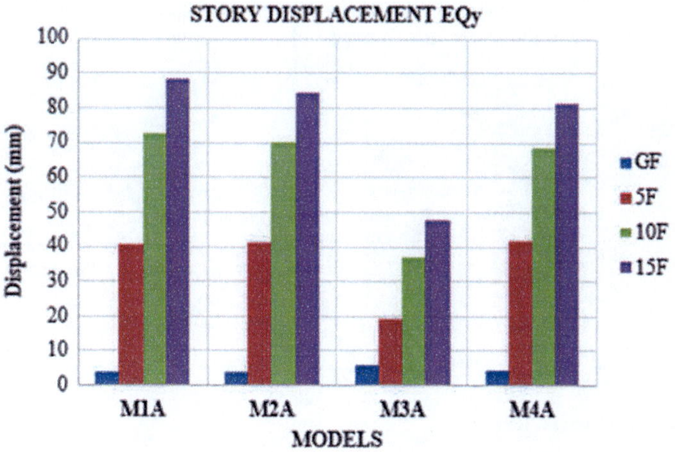

Fig. 7 Story displacement EQ_Y

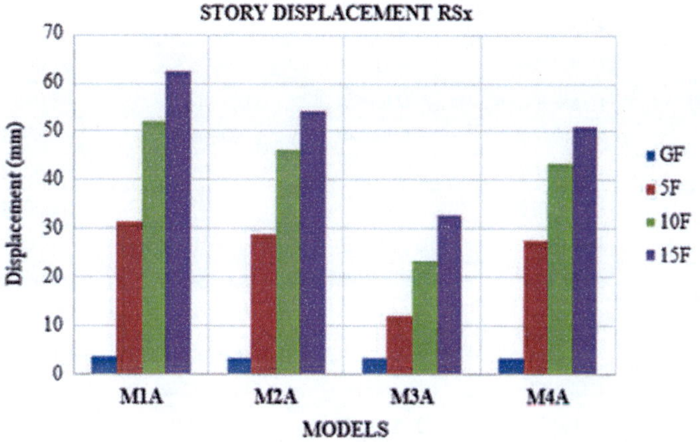

Fig. 8 Story displacement RS_X

- The story drift in model 3 is 59.25% less compared to the conventional structure in EQ_Y
- The story drift in model 3 is 40.86% less compared to the conventional structure in RS_X
- The story drift in model 3 is 85.26% less compared to the conventional structure in RS_Y
- We can observe that model 3 is effective in bring the structure to its original position (Figs. 13, 14 and 15; Table 4)
- We can observe that base shear is more in model 3 compared to other models

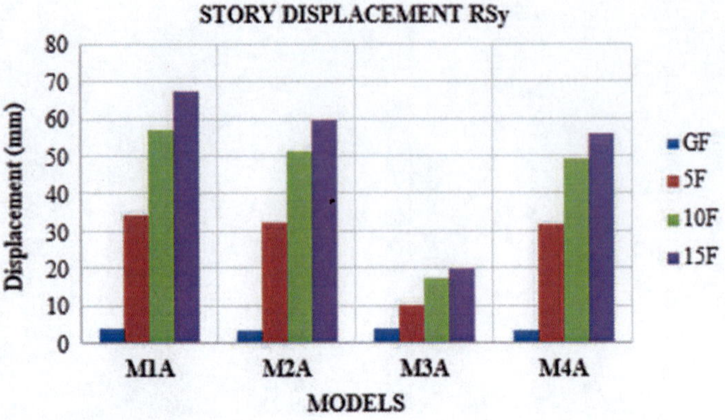

Fig. 9 Story displacement RS_Y

Table 3 Story drift ratio in M3A

Story level	Story drift ratio			
	EQ_X	EQ_Y	RS_X	RS_Y
15	0.000776	0.000426	0.000577	0.000133
14	0.000948	0.000849	0.000732	0.000374
13	0.000785	0.000444	0.000584	0.000139
12	0.001168	0.001267	0.000918	0.000628
11	0.000779	0.000449	0.000578	0.000141
10	0.001311	0.001559	0.001012	0.000778
9	0.000748	0.000435	0.000553	0.000135
8	0.001377	0.00174	0.00104	0.000833
7	0.000683	0.000401	0.000504	0.000123
6	0.001371	0.001826	0.001037	0.000864
5	0.000579	0.000346	0.000428	0.000106
4	0.001295	0.001835	0.001028	0.000951
3	0.000432	0.000274	0.000323	8.40E-05
2	0.001158	0.001784	0.001013	0.001037
1	0.000248	0.000195	0.000191	6.40E-05
GF	0.001066	0.001904	0.001058	0.001269

- The base shear (V_b) value in X-direction, when compared EQ_X, is 1.53% more in RS_X
- The base shear value in Y-direction when compared EQ_Y is 58.48% less in RS_Y.

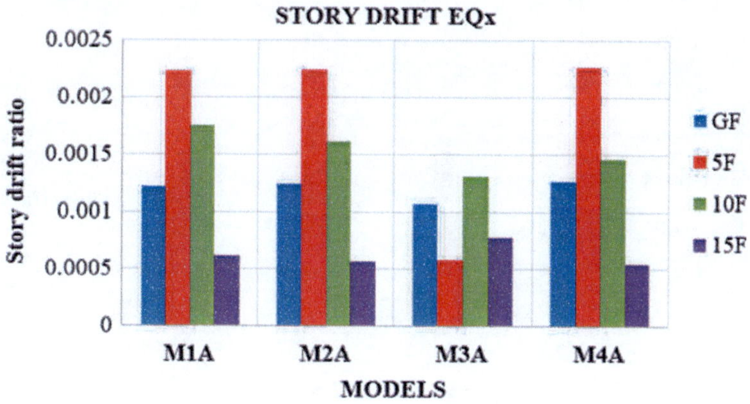

Fig. 10 Story drift EQ_X

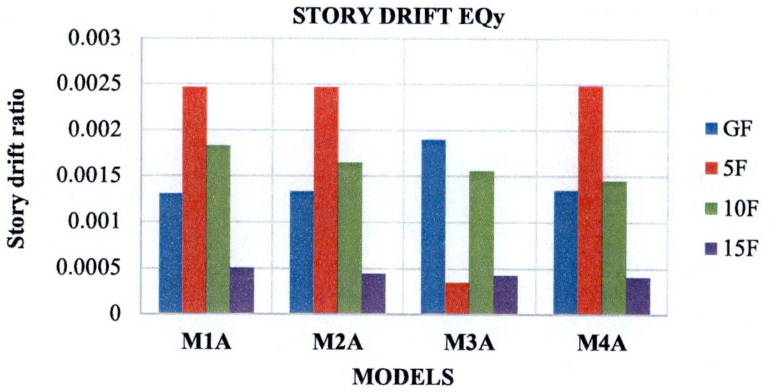

Fig. 11 Story drift EQ_Y

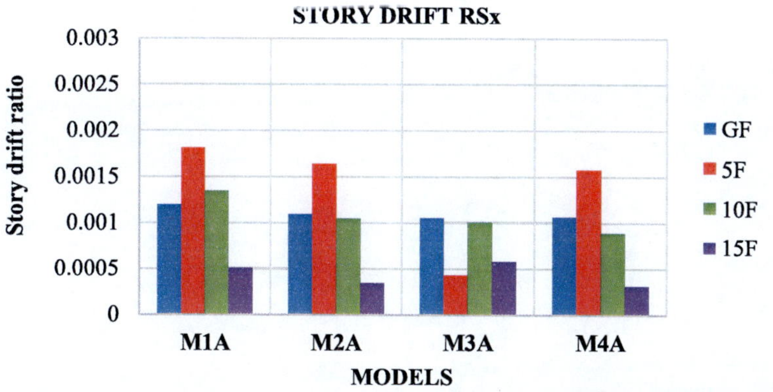

Fig. 12 Story drift RS_X

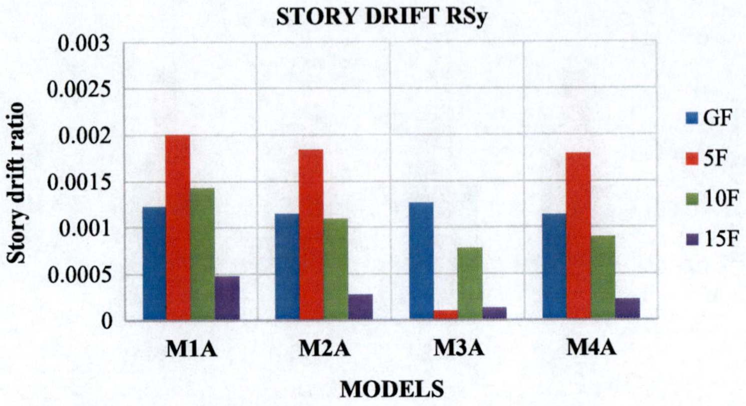

Fig. 13 Story drift RS_Y

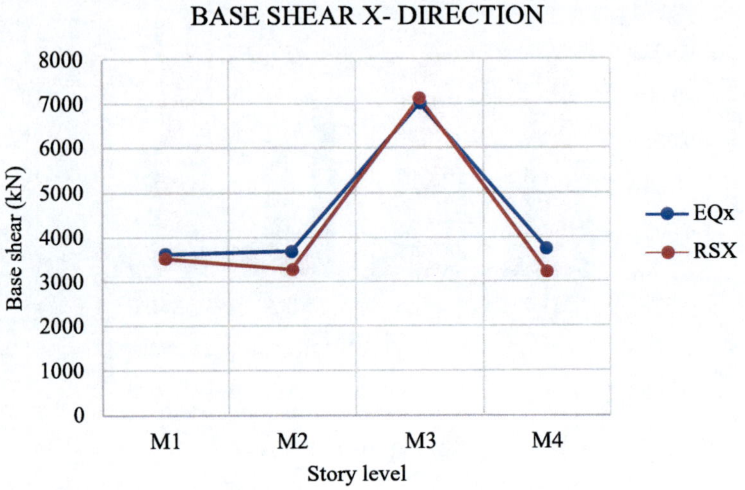

Fig. 14 Base shear X-direction

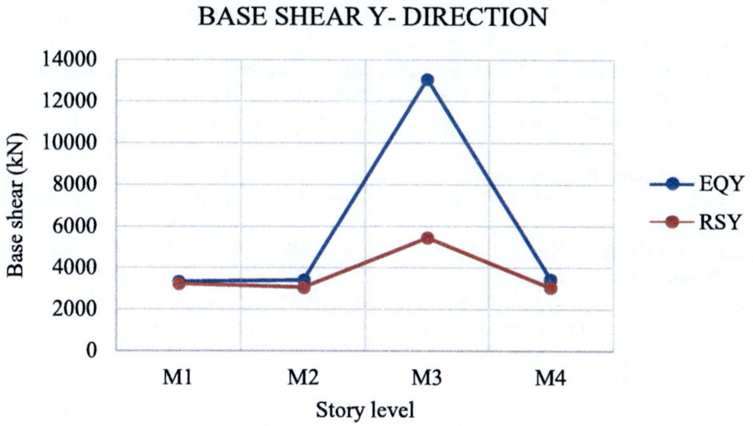

Fig. 15 Base shear Y-direction

Table 4 Base shear

Models	Base shear (KN)			
	ESA		RSA	
	EQ_X	RS_X	EQ_Y	RS_Y
M1A	3628.335	3532.094	3329.2269	3221.438
M2A	3704.853	3283.814	3398.8663	3040.315
M3A	7017.382	7125.156	13,065.308	5423.669
M4A	3756.165	3252.208	3440.3123	3024.012

6 Conclusion

- The study of the investigation shows that model 3 is effective compared to other models which have dampers
- The placement of the dampers in model 3 is effective when compared to other models having dampers
- Model 3 has less story drift, and story displacement compared to the conventional model and models having dampers

References

1. Patowary MZ (2017) Seismic analysis for multi-story building horizontally damped above basement level. Doctoral dissertation, Universiti Teknologi Malaysia
2. Heysami A (2015) Types of dampers and their seismic performance during an earthquake. Curr World Environ 10(Special-Issue1):1002–1015

3. Ghanchi N, Kewate S (2015) Dynamic analysis of 25 storey RCC building with and without viscous dampers. Int J Sci Eng Res 6:63–68
4. Constantinou MC, Symans MD (1993) Experimental study of seismic response of buildings with supplemental fluid dampers. Struct Des Tall Build 2(2):93–132
5. Fu Y, Kasai K (1998) Comparative study of frames using viscoelastic and viscous dampers. J Struct Eng 124(5):513–522
6. Dadkhah E, Shiri B, Ghaffarzadeh H, Baleanu D (2020) Visco-elastic dampers in structural buildings and numerical solution with spline collocation methods. J Appl Math Comput 63(1):29–57
7. Xu ZD, Liao YX, Ge T, Xu C (2016) Experimental and theoretical study of viscoelastic dampers with different matrix rubbers. J Eng Mech 142(8):04016051
8. Lewandowski R, Bartkowiak A, Maciejewski H (2012) Dynamic analysis of frames with viscoelastic dampers: a comparison of damper models. Struct Eng Mech 41(1):113–137
9. Fujita K, Moustafa A, Takewaki I (2010) Optimal placement of viscoelastic dampers and supporting members under variable critical excitations. Earthq Struct 1(1): 43–67
10. Kang JD, Tagawa H (2013) Seismic response of steel structures with seesaw systems using viscoelastic dampers. Earthq Eng Struct Dyn 42(5):779–794
11. Banisheikholeslami A, Behnamfar F, Ghandil M (2016) A beam-to-column connection with visco-elastic and hysteretic dampers for seismic damage control. J Constr Steel Res 117:185–195
12. Bhatti AQ (2013) Performance of viscoelastic dampers (VED) under various temperatures and application of magnetorheological dampers (MRD) for seismic control of structures. Mech Time Dep Mater 17(3):275–284
13. Mazza F, Vulcano A (2011) Control of the earthquake and wind dynamic response of steel-framed buildings by using additional braces and/or viscoelastic dampers. Earthq Eng Struct Dyn 40(2):155–174

Frequency Analysis of Self-supported Steel Chimneys

Ashish Kumar Gupta, Sudhir Singh Bhadauria, and Aruna Rawat

Abstract Chimneys are the most significant industrial components used to release toxic and undesirable fumes into the airspace. Various kinds of chimneys are used in industries, such as RCC chimneys, masonry chimneys, steel chimneys, etc. The work focuses on the frequency analysis of self-supported steel chimneys using single-degree-of-freedom (SDOF) system, multi-degree-of-freedom (MDOF) system, and Dunkerley's method. The two chimneys of heights 275 and 120 m having varying diameter and thickness, varying diameter and uniform thickness, and uniform diameter and thickness are considered. The frequencies are evaluated and compared with the above methods for all the chimney models. It is observed from the results that all three methods considered in the current study approximately give the same values of fundamental natural frequency. The chimney model with uniform diameter and thickness and the fundamental frequency were observed to be more as compared with other two models.

Keywords Dunkerley's method · Natural frequency · Steel chimney · Time period

1 Introduction

Tall structures are needed in the present world. These tall chimneys are primarily used in industries to release toxic and undesirable gases into the air. The dominant loads on chimneys are self-weight of the chimney, wind load, earthquake load, and temperature loads. Steel chimneys are manufactured in steel plants and are supported by large base foundations for their stability. The self-supporting chimney is modeled as a cantilever beam.

A. K. Gupta (✉) · S. S. Bhadauria · A. Rawat
Department of Civil Engineering, University Institute of Technology (UIT), RGPV, Bhopal 462033, India
e-mail: ag057511@gmail.com

K. K. Pathak et al. (eds.), *Latest Developments in Civil Engineering*, Lecture Notes in Civil Engineering 352, https://doi.org/10.1007/978-981-99-2676-3_44

2 Literature Review

The steel chimney has been the subject of prior studies that include: Pritchard [1] compared the performance of 65 full-scale chimneys. It reported the performance of chimneys and also examined the safe case of the isolated chimneys. It was observed that the correlation with the observation was not very well. It was also understood that it is over-conservative. When the Zorilla method was used then this method gives a slightly better correlation which was based on the mass and damping. This method was unsafe in three cases-marginal safety, stiffness, and damages. The correlation was further improved by using British code BS4076 when stiffness was an addition to mass damping, but we found that this method was not safe for the case of marginal stability. Simonović et al. [2] studied 60 m tall steel chimney for its starting root section failure. It was observed that the cracks developed in the steel walls of the windshield affected the integrity of the chimney. Analytical and numerical analyses were carried out to study the failure of the root section. The ultimate stress value has been identified numerically by the finite element method in the steel structure. From the results, it was concluded that the industrial chimneys were damaged due to mechanical-chemical-thermal action. Deng et al. [3] examined the chimney with spiral guide plate installed on its outer wall. From the analyses, it was verified that self-vibration, wind vibration frequencies for this chimney reduce wind load effect on it and also on the flange connection. It was concluded that the steel chimney with a spiral guide plate can attain a more stable outer structure.

Kuras et al. [4] carried out a dynamic analysis of steel chimneys with the help of an interferometric radar and robotic total station. The experimental analysis of freely supported chimneys confirmed the results of amplitude, frequency, and damping of vibration. The instruments were employed to study a chimney in the tripod, which was used to observe movement in the chimney and damper. The synchronization of the clocks in both instruments proceeded with the measurements. Kawecki and Zuranski [5] studied cross-wind vibration of 100 m steel chimneys. This case history provides detailed information regarding cross-wind vibration of chimneys caused by vortex excitation, and useful guide for wind actions on structures.

Recently, Ellingsen et al. [6] studied the field test slender light and low damped chimneys for super-critical cross-wind vortex-induced vibration (VIV) at medium wind velocity for various field tests. An analysis of the dominating frequency and amplitude responses was done statistically to determine the probability distribution as a function of wind speed and direction.

In the current work frequency of steel chimneys using the single-degree-of-freedom (SDOF) system, multi-degree-of-freedom (MDOF) system, and Dunkerley's method for evaluation of lumped masses was performed. The two chimneys of heights 275 and 120 m having varying diameter and thickness, varying diameter and uniform thickness, and uniform diameter and thickness are considered.

3 Methodology

In the present methodology the self-supported chimney is modeled as a cantilever beam having a number of elements with lumped masses and stiffness. Initially, the single-degree-of-freedom system of the chimney as shown in Fig. 1a is modeled. For cantilever beam the exact formulae for calculating the first three frequencies is as follows:

$$\omega_1 = 3.516\sqrt{E \times I/\rho \times A \times L^4} \quad (1)$$

$$\omega_2 = 22.03\sqrt{E \times I/\rho \times A \times L^4} \quad (2)$$

$$\omega_3 = 61.70\sqrt{E \times I/\rho \times A \times L^4} \quad (3)$$

where ρ is density of the chimney material, A is the cross-sectional area, L is the length of the chimney, E is modulus of elasticity of the chimney material and I is the moment of inertia.

In the case of the multi-degree-of-freedom (MDOF) system, the frequency equation of motion is as follows:

$$[M]\{\ddot{x}\} + [K]\{x\} = 0 \quad (4)$$

where $[M]$, $[K]$, and $\{x\}$ are mass, stiffness, and displacement matrices, respectively. The eigenvalue equation of motion is given as:

$$[K]\{\phi_i\} = \{\omega_i^2\}[M]\{\phi_i\} \quad (5)$$

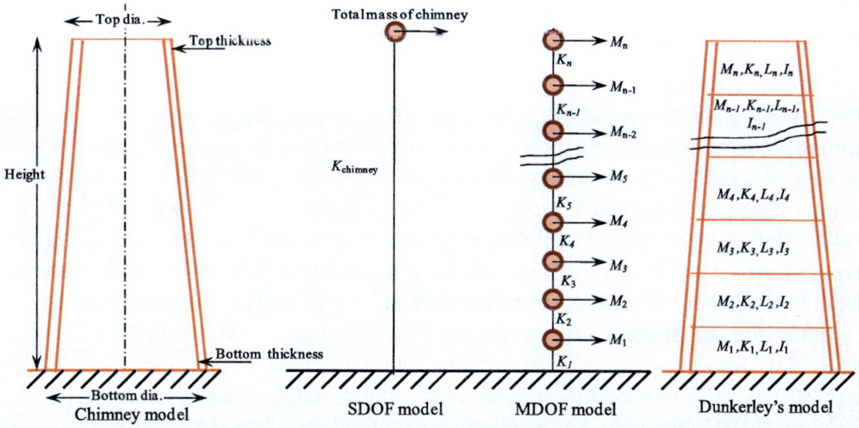

Fig. 1 Steel chimney models

where the natural frequency with a natural mode shape is ω_i ($i = 1, 2 \ldots N$).

$$[M] = \begin{bmatrix} M_1 & 0 & \cdots & 0 \\ 0 & M_2 & 0 & \vdots \\ \vdots & \vdots & \ddots & 0 \\ 0 & \cdots & 0 & M_n \end{bmatrix} \quad (6)$$

Here, M_i is the lumped mass which is considered at each node of the chimney (for $i = 1$ to n), (K_n) is the stiffness matrix and damping coefficient of elemental member ($i = 1$ to n).

$$[K] = \begin{bmatrix} K_1 + K_2 & -K_2 & \cdots & 0 \\ -K_2 & K_2 + K_3 & -K_3 & \vdots \\ 0 & -K_3 & \ddots & -K_n \\ 0 & \cdots & -K_n & K_n \end{bmatrix} \quad (7)$$

Dunkerley's method for lumped masses method is used for calculating the frequencies for all chimney models' idea. The cantilever chimney is divided into a number of elements and for each element the specific mass, heights, and properties of the chimneys are determined, and the natural frequency for each element is calculated and the whole chimney's natural frequency is given as:

$$\frac{1}{\omega_{\text{chimney}}^2} = \frac{1}{\omega_1^2} + \frac{1}{\omega_2^2} + \frac{1}{\omega_3^2} + \cdots + \frac{1}{\omega_n^2} \quad (8)$$

4 Numerical Study

Two chimneys of 275 m and 120 m heights are considered, and the mass is distributed among the various nodes, and bottom to top has been taken at equal intervals of height of 12.5 m and 10 m, respectively. Three types of models have been taken for calculation of natural frequencies. These are of varied diameter and thickness (Model 1), varied diameter and uniform thickness (Model 2), and uniform diameter and thickness (Model 3) as given in Table 1. Steel has a density of 7850 kg/m^3 and an elastic modulus of 2.0×10^8 kN/m^2. For the study of beam elements, on each node of the chimney, the lateral degree-of-freedom is considered, and eliminating the rotational degree-of-freedom with the help of condensation methods is done. The frequency of steel chimneys using the single-degree-of-freedom (SDOF) system, multi-degree-of-freedom (MDOF) system, and Dunkerley's method for lumped masses is evaluated as shown in Fig. 1.

Table 1 Dimensions of steel chimney models

Dimension (m)	Model 1	Model 2	Model 3
275 m height			
Bottom diameter	25	25	25
Top diameter	12	12	25
Bottom thickness	0.795	0.795	0.795
Top thickness	0.237	0.795	0.795
120 m height			
Bottom diameter	5	5	5
Top diameter	2.5	2.5	5
Bottom thickness	0.096	0.096	0.318
Top thickness	0.050	0.096	0.318

5 Results

The natural frequencies, time period, and mode shapes are examined for all chimney models using all the three methods.

5.1 Frequencies and Time Period

Tables 2 and 3 give the natural frequencies and time periods for 275 m and 120 m heights of chimneys for all the models, respectively. All three approaches utilized in the present investigation are compared to the natural frequencies. Figures 2 and 3 show fundamental frequencies and time periods for all chimney models. It can be observed that fundamental frequency and time period values are approximately the same for all three methods. As in Model 3 uniform diameter and thickness is having more mass as compared with the other two models, and fundamental frequency for Model 3 is more as compared to other two models 1 and 2.

Figure 4 shows the variation of the first five frequencies for all three models. It can be observed that as the mode number increases the value of frequency also increases. The frequency values for Model 3 for both height of chimney are more as compared to other two models for all the mode numbers.

5.2 Mode Shapes

The first three mode shapes values are extracted and plotted in Fig. 5. The normalized mode shapes values are plotted along the height of the chimney. The mode shapes gives an idea about the deflected shapes of the chimney for the first three modes. The mode shapes of the chimneys construct a unique deflected pattern for a particular

Table 2 Fundamental natural frequencies

	SDOF (rad/s)	MDOF (rad/s)	Dunkerley's method (rad/s)
275 m height			
Model 1	1.91	1.83	2.00
Model 2	1.78	1.72	1.67
Model 3	2.00	1.92	1.89
120 m height			
Model 1	1.96	1.94	1.99
Model 2	1.85	1.89	1.87
Model 3	2.14	2.10	1.98

Table 3 Time period

	SDOF (rad/s)	MDOF (rad/s)	Dunkerley's method (rad/s)
275 m height			
Model 1	3.29	3.43	2.71
Model 2	3.53	3.65	3.76
Model 3	3.14	3.27	3.32
120 m height			
Model 1	3.21	3.24	2.83
Model 2	3.39	3.32	3.36
Model 3	2.94	2.99	3.24

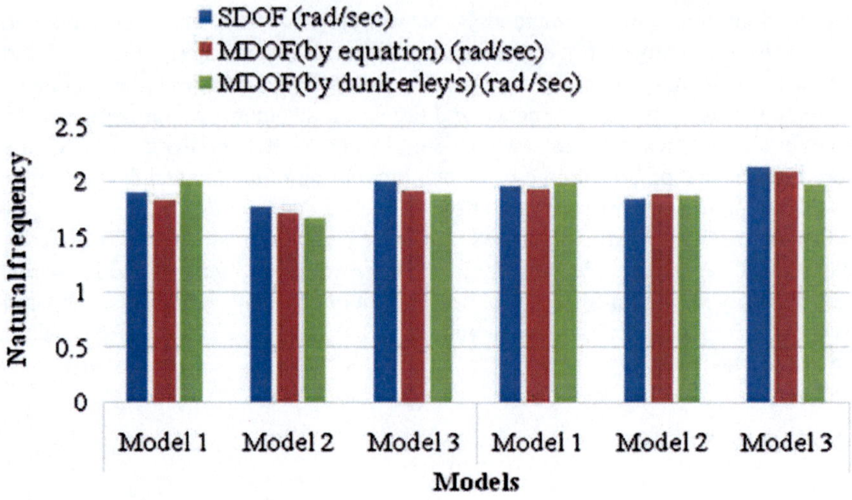

Fig. 2 Fundamental natural frequency for all models

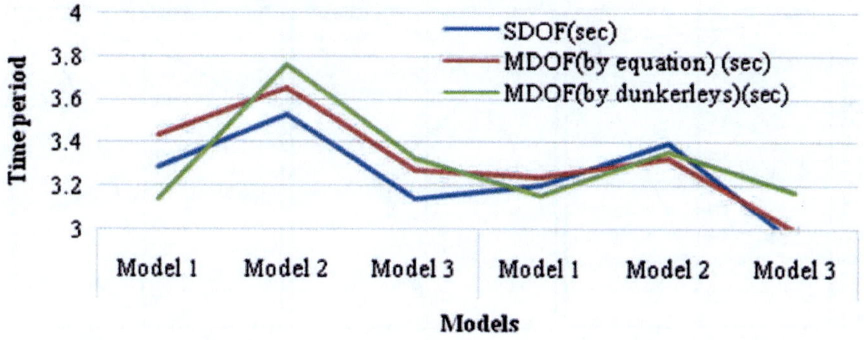

Fig. 3 Fundamental time period for all models

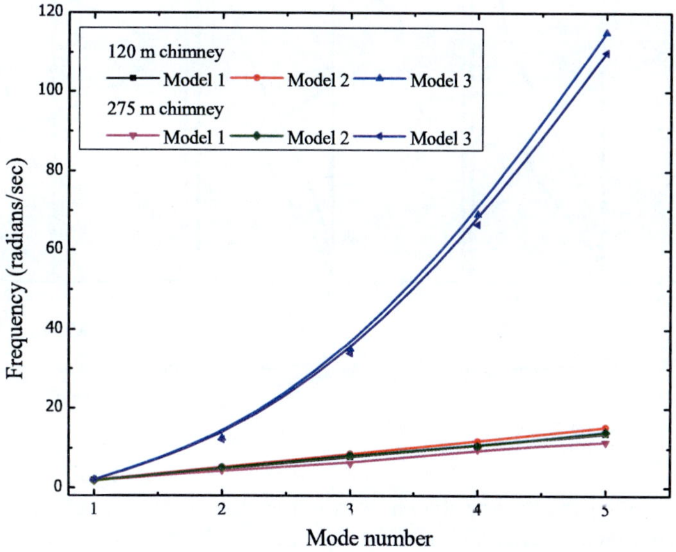

Fig. 4 Variation between frequency and mode number for all models

natural frequency. All mode shapes are combined to give the original deflection shape of the chimneys.

6 Conclusions

In the present work the frequency of steel chimneys using the single-degree-of-freedom (SDOF) system, multi-degree-of-freedom (MDOF) system and Dunkerley's method for lumped masses is evaluated. The two chimneys of heights 275 and 120 m

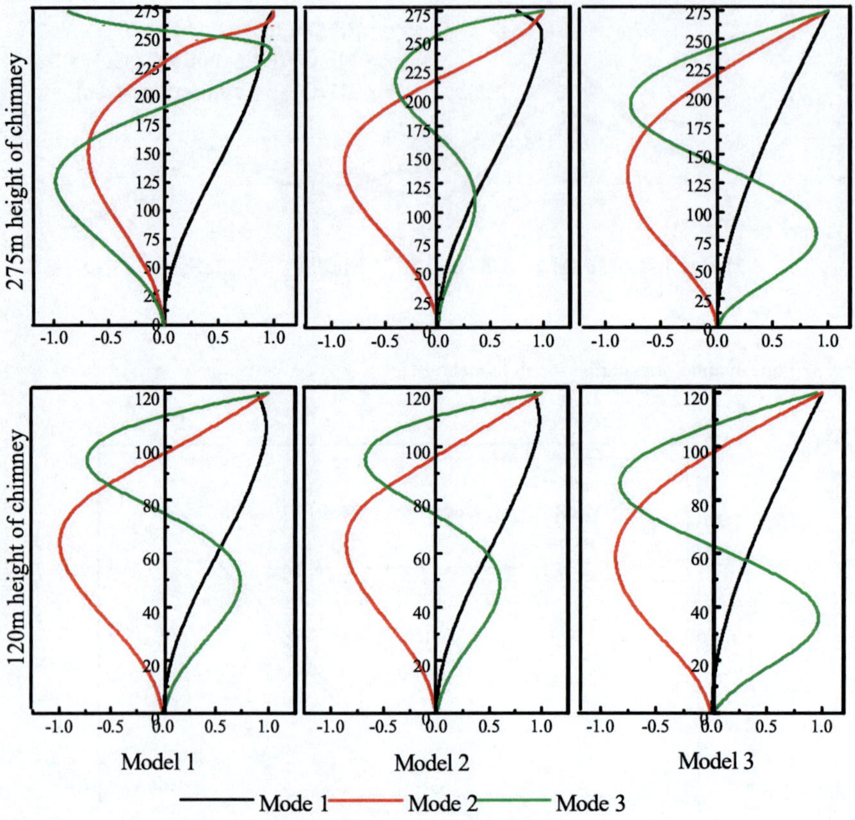

Fig. 5 Variation mode shapes for all models

having varying diameter and thickness, varying diameter and uniform thickness, and uniform diameter and thickness are considered. The following conclusions are observed from the present study:

1. All three methods considered in the present study approximately give same values of of the fundamental natural frequency.
2. The multi-degree-of-freedom (MDOF) system results are more accurate and are nearly closer to the accurate values of fundamental natural frequency of the chimney. The accuracy also depends on the number of beam elements considered for analyses. The mode shapes can also be evaluated and gives a unique deflected pattern for a particular natural frequency.
3. For Model 3 having uniform diameter and thickness is having more mass as compared to the other two models, the fundamental frequency for Model 3 is more as compared to the other two models 1 and 2. The mode number increases as the value of the frequency also increases. The frequency values for Model 3 for both height of chimney are more as compared to other two models for all the mode numbers.

References

1. Pritchard BN (1984) Steel chimney oscillations: a comparative study of their reported performance verses predictions using existing techniques. Eng Struct 6:315–323
2. Simonović AM, Stupar SN, Pekovic OM (2008) Stress distribution as a cause of industrial steel chimney root section failure. FME Trans 36:119–125
3. Deng X, He J, Zhang C, Li J, Li Y (2014) Modal and wind analysis about steel chimney outer wall with a spiral guide plate based on workbench. Appl Mech Mater 508:215–218
4. Kuras P, Ortyl TO, Kocierz R (2016) Analysis of effectiveness of steel chimneys vibration damper using surveying methods. Eng Mater Sci 1–8
5. Kawecki J, Zuranski, JA (2007) Cross-wind vibration of steel chimneys—a new case history. J Wind Ind Aerodyn 95:1166–1175
6. Ellingsen ØM, Flamand O, Amandolese X, Coiffet F, Hémon P (2022) Field tests on a full-scale steel chimney subjected to vortex-induced vibrations. Struct Eng Int 32(1):55–61

Effect of Lateral Stiffness on Propped Embedded Retaining Walls' Structure

Toshi Bhavsar, Suresh Singh Kushwah, and Aruna Rawat

Abstract The present work investigates the effects of lateral stiffness on propped embedded retaining walls. The deflection of a retaining wall is affected by many aspects such as loading on the outside of the excavation, strength, type and stiffness of the soil, embedded depth, retained height, type and stiffness of the retaining wall, and the water table. This work focuses on impact of different types of stiffness of the retaining wall and excavation depth. The comparative finite element analysis (FEA) of 2-D and 3-D embedded retaining walls was studied considering the effects of lateral stiffness. The total displacement of soil and wall displacement was evaluated. The results show that with an increase in the wall stiffness the displacement reduces and the lateral stiffness reduces the central maximum displacement of soil.

Keywords Propped retaining wall · Lateral stiffness · Embedded depth · Stiff clay · Plaxis 2-D and 3-D

1 Introduction

The lateral (i.e., horizontal) stiffness of a retaining wall has the effect of reducing the displacements of both retaining wall and supported ground. Thus, the lateral stiffness reduces the displacements due to corner effect as exhibited in Fig. 1. The corner effect is where the displacement of a retaining wall is at or near zero. This is due to the greater stiffness of the section in the corner. This section of higher stiffness has a 3-D effect of reducing the displacements along the wall [1]. The extent of this effect along

T. Bhavsar (✉) · S. S. Kushwah · A. Rawat
Department of Civil Engineering, University Institute of Technology (UIT), RGPV, Bhopal 462033, Madhya Pradesh, India
e-mail: toshibhavsar@gmail.com

S. S. Kushwah
e-mail: skushwah@rgtu.net

A. Rawat
e-mail: arunarawat@rgtu.net

K. K. Pathak et al. (eds.), *Latest Developments in Civil Engineering*, Lecture Notes in Civil Engineering 352, https://doi.org/10.1007/978-981-99-2676-3_45

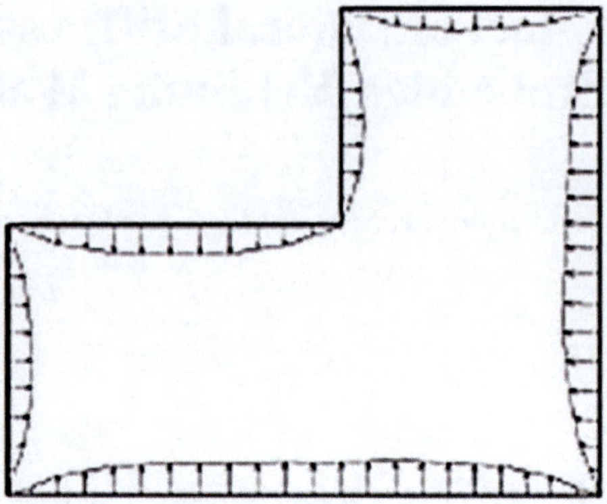

Fig. 1 Corner effect (plan view) of the horizontal displacements in retaining wall [3]

the wall depends on the stiffness of the wall and joints in the wall. To incorporate this effect into the prediction of the behavior of the wall, 3-D Finite Element Analysis (FEA) or empirical calculations are the most commonly used methods [2].

The lateral stiffness effect can also be seen in the difference between 2-D and 3-D FEA, this is because a 2-D FEA doesn't take into account the horizontal dimension of the retaining wall. The increased wall displacements were found by Schweiger et al. [4] for a small strutted deep excavation in problematic soil in Saltsburg, Austria when they compared 3-D and 2-D FEA. The results showed that the 2-D analysis could model the displacements (presumed the maximum displacements) in the wall correctly, however, a 3-D FEA is more accurate for looking at the displacements of the rest of the wall, this means that in some cases the effect of lateral stiffness has little/no impact on the maximum displacement of the structure [4]. This view is somewhat supported when looking at global databases of cantilever walls, which show that there is a lack of correlation between wall stiffness and maximum horizontal displacements [5, 6]. These authors argued that this was due to the arching effects of the soils and also that displacements are mainly controlled by base heave. This indicates that both base heave (the upward movement of soil due to the reduction in vertical pressure) and arching play a dominant role in the displacement behavior of a cantilever retaining wall to such an extent that the stiffness of the wall is irrelevant when looking at the maximum displacements of a retaining wall.

The diaphragm walls are generally constructed using stiff concrete of the same stiffness throughout. Previous studies have shown that with increase in wall flexibility, the stress imposed by the soil on the wall redistributes and structural forces on the wall decrease.

Gauging the extent and distribution of ground movements, adjoining to an excavated wall is a crucial part of the analysis and design process when excavating in an urban environment. Three-dimensional effects caused by the greater stiffness at the corners of an excavation lead to minor ground movements near the corners and major ground movements toward the center of the excavation wall.

Bakker, Finno et al. and Zhou [3, 7, 8] found that 2-D FEA overpredicted the movements in the wall at a rate in proportion to the length of the wall as can be seen in Fig. 2. With Bakker found that the maximum displacement of a 3-D FEA was around 60% of the 2-D FEA for a small "L" shaped propped excavation (it was thought that this was due to arching and corner effects which require lateral stiffness to be transferred) [3]. Finno et al. found that a 2-D FEA overpredicts the displacements for stiff walls more than flexible walls [7]. The authors were of the opinion that it was because 2-D FEA doesn't take the effect of the corners of the retaining walls into account, which will affect a stiff wall more than a flexible wall. These differences in the modeling approaches were also shown by Zhou who modeled a rectangular propped excavation in Amsterdam using both PLAXIS 2-D and PLAXIS 3-D. The author also found that the differences in the maximum displacements between 2-D and 3-D FEA increase when the plan of the excavation decreases. It shows that the smaller the width of the excavation, the more effect the lateral stiffness has on the maximum displacements [8].

The empirical equation presented by Finno and Roboski predicted the maximum lateral movement along a tied-back retaining wall in clay using an empirical equation based on the height and length of the excavation, this showed that there is a "hockey stick" relationship between the distance from the corner of the excavation and the

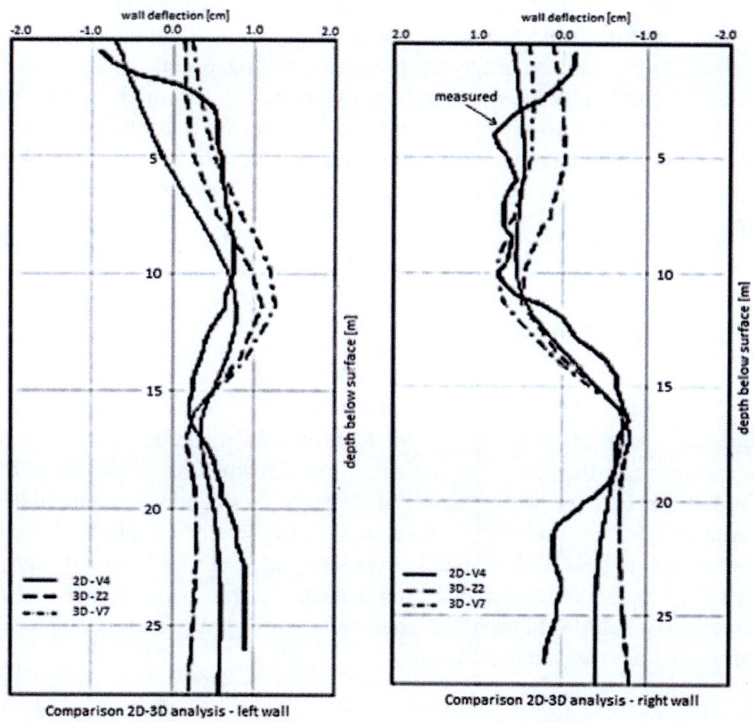

Fig. 2 Comparing 2-DFEM, 3-D FEM and 3-D FEM with a "gap" (named 3-D-V7) with the measured results [4]

horizontal displacement in the wall, it is thought that the same relationship occurs in cantilever embedded retaining walls [1]. The methodology adopted by Finno and Roboski [1] of calculating the lateral displacement is based only on the height and length. It is different as compared to the deductions offered by other authors [4, 9–12], which all state that the lateral displacement is based on the stiffness or rigidity of the structure as well as the depth and length of the excavation [1, 4, 9–12]. It is thought that the stiffness of the structure is included in the equation by the numerical factors which could possibly mean the Equation will be inaccurate for walls with different stiffnesses [1]. There has also been research to show that effect of the stiffness of the corner of the excavation can have a large effect on the displacements of the retaining wall [11].

2 Finite-Element Model and Procedures

The commercially available PLAXIS 3-D V21 and 2-D V21.01 [13, 14], three-dimensional and plane strain geotechnical finite element software packages, respectively, were used to conduct the parametric study. Structural elements were modeled with anisotropic linear and nonlinear elastic elements. Soil elements are 15-node wedge elements that are created by the projection of 2-D, 6-node triangle elements. Variations in stratigraphy interfaces, i.e., non-horizontal were modeled by 13-node pyramid elements and 10-node tetrahedral elements. Support structure elements consist of 3-node line elements for beams, and 6-node and 8-node plate elements for walls. Soil-structure interaction is simulated by 12-node and 16-node interface elements.

2.1 Parametric Variables

A total of 18 models of 2-D and 3-D finite-element analyses were made to evaluate the influence of excavation depth and structural parameters on horizontal soil deformation and compare the computed horizontal soil deformation distributions for the 20 m by 20 m excavation at an excavation depth of 8, 10 and 12 m and embedded depth ratio of 0.5. The sides of the excavation are supported by embedded retaining walls, which are braced by horizontal struts at an interval of 5 m in stiff clay. The values of 32, 320 and 3,200 were used to represent flexible, medium and stiff walls, respectively. The 3,200 value is higher than most "stiff" wall systems and represents an upper bound on the system stiffness [7].

2.2 Soil and Structural Parameters

The soil layer was modeled using the HS model. This effective stress model is formulated within the framework of elasto-plasticity. Table 1 summarizes the wall stiffness parameters. The horizontal bending stiffness is computed assuming that the wall is 20 times more flexible in the horizontal direction. The two directions as given in Table 2 indicate the direction along the horizontal direction, to account for the rotations in the connections of a sheet pile wall and the lack of continuity in stiffer wall systems in this direction.

Table 1 Soil parameters used in parametric study [7]

Parameter	Stiff clay
$E^{ref}{}_{50}$ (kPa)	17,723
$E^{ref}{}_{oed}$ (kPa)	12,406
c^{ref} (kPa)	1
Φ (°)	32
Ψ (°)	0
M	0.85

Table 2 Wall stiffness parameters [7]

Parameter	Wall		
	Flexible	Medium	Stiff
Plane strain FE parameters			
System stiffness, S	32	320	3,200
Bending stiffness, EI (kN m^2/m)	50,400	504,000	5,040,000
Axial stiffness, EA (kN/m)	3,427,000	34,270,000	342,700,000
Element thickness (m)	0.42	0.42	0.42
Poisson's ratio	0	0	0
Three-dimensional FE parameters			
Young's modulus E_1 (kPa)	8,160,000	81,600,000	816,000,000
Young's modulus E_2 (kPa)	408,000	4,080,000	40,800,000
Young's modulus E_3 (kPa)	200,000,000	2,000,000,000	20,000,000,000
Shear modulus G_{12} (kPa)	408,000	4,080,000	40,800,000
Shear modulus G_{13} (kPa)	400,000	4,000,000	40,000,000
Shear modulus G_{23} (kPa)	1,330,000	13,300,000	133,000,000
Element thickness (m)	0.42	0.42	0.42
Poisson's ratio	0	0	0

3 Results and Discussions

See Figs. 3, 4, 5 and 6.

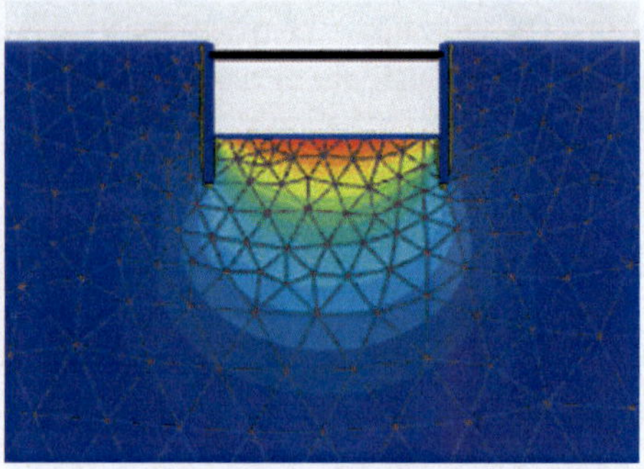

Fig. 3 Total deformation: 2-D FEA stiff wall

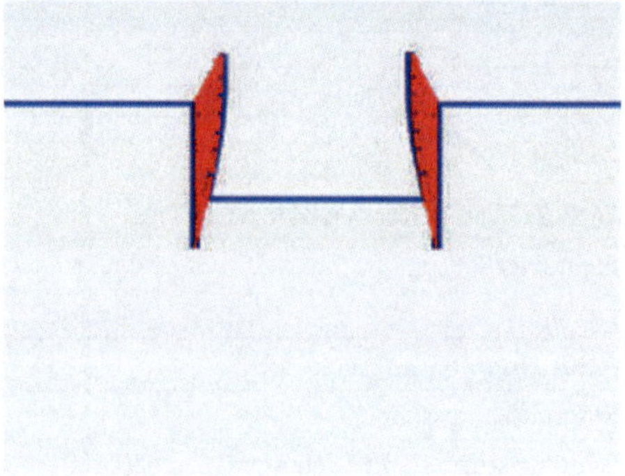

Fig. 4 Wall displacement: 2-D FEA stiff wall

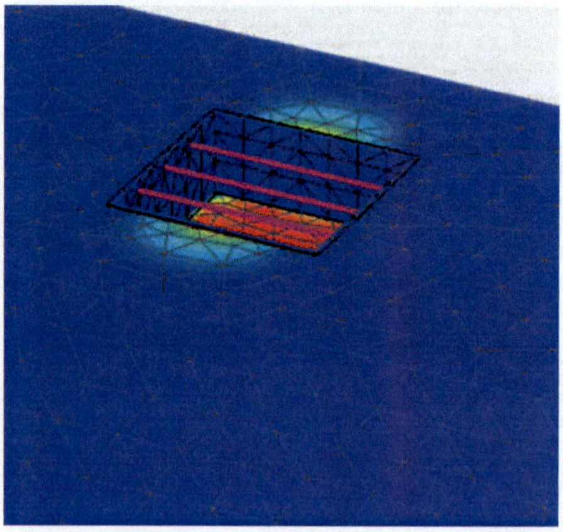

Fig. 5 Total deformation: 3-D FEA stiff wall

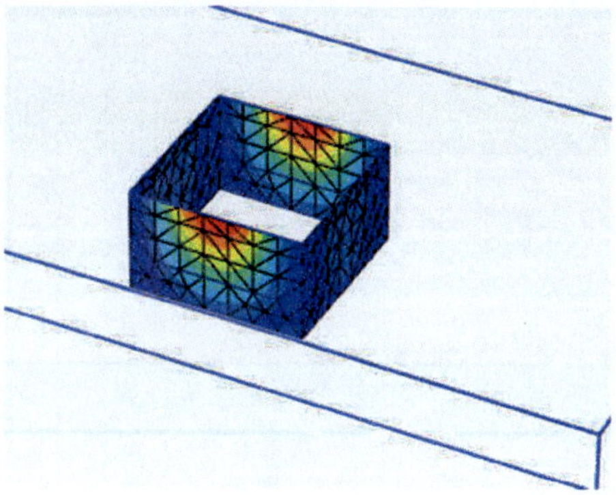

Fig. 6 Wall displacement: 3-D FEA stiff wall

3.1 Effect of Wall Stiffness

See Table 3 and Figs. 7, 8 and 9.

3.2 Effect of Lateral Stiffness

The graphs show that the wall with higher stiffness shows lesser displacement at the center of the wall for any given excavation dimensions. Also, increase in excavation depth, the stiffer wall shows the lesser displacement (Tables 4 and 5; Figs. 10, 11, 12 and 13).

4 Limitations and Further Scope

4.1 Limitations

The present work is based on software analysis. It may be verified by experimental results also.

4.2 Further Scope

The present work may be extended to include the following studies:

i. The study may be extended to include different wall types.
ii. The study can also be carried out on different soil profiles.
iii. Also, the effects of water table can be studied.

Table 3 Wall displacement at center (m)–stiffness

Finite element method (FEM)	2-D			3-D—at center			3-D—at corner		
Wall	Excavation depth (m)								
	8	10	12	8	10	12	8	10	12
Stiff	0.0096	0.01256	0.01641	0.00576	0.00828	0.01157	0.0032	0.0045	0.004
Medium stiff	0.0116	0.02173	0.03836	0.01008	0.0199	0.03158	0.003	0.004	0.004
Flexible	0.0311	0.06904	0.1437	0.02783	0.06088	0.1099	0.006	0.016	0.02

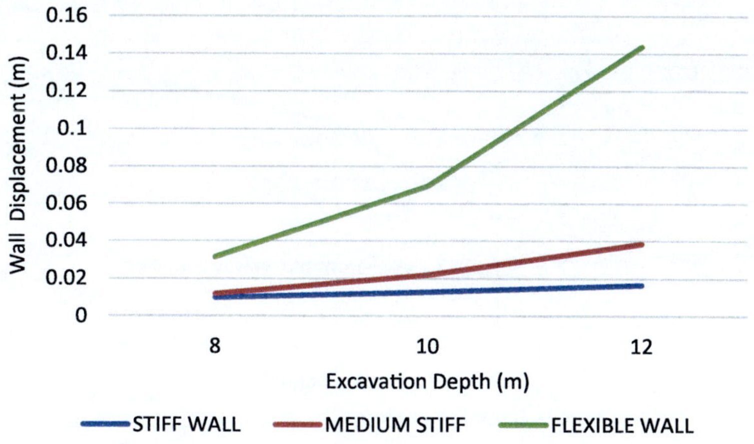

Fig. 7 Wall displacement at center—2-DFEA model

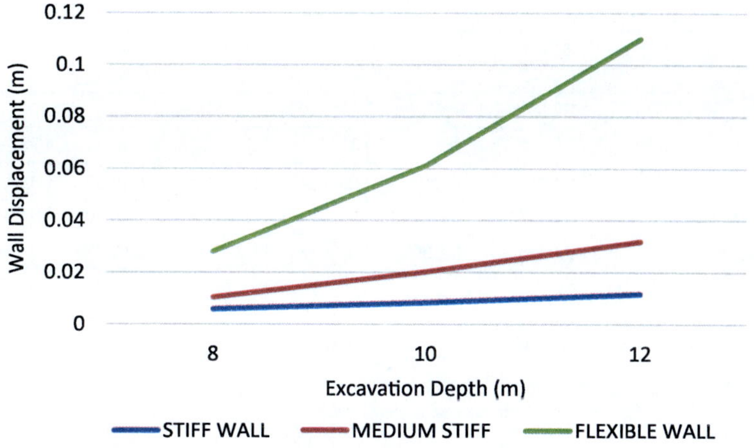

Fig. 8 Wall displacement at center—3-D FEA model

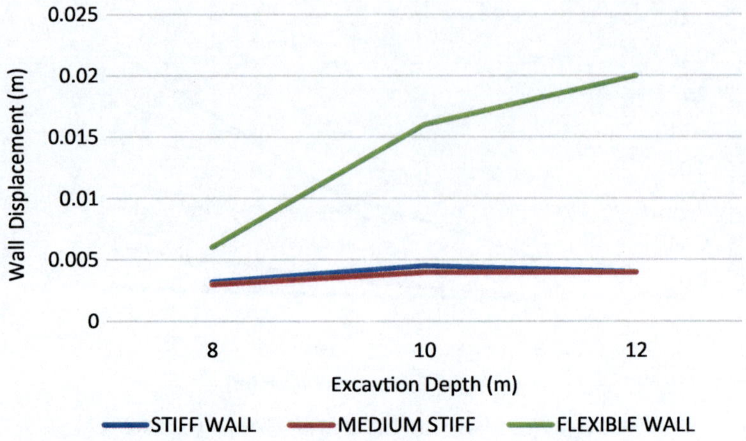

Fig. 9 Wall displacement at corner—3-D FEA model

Table 4 Wall displacement at center (m)–lateral stiffness

Finite element method (FEM)	2-D			3-D		
Wall	Excavation depth (m)					
	8	10	12	8	10	12
Stiff	0.009615	0.01256	0.01641	0.005758	0.00828	0.01157
Medium stiff	0.01167	0.02173	0.03836	0.01008	0.0199	0.03158
Flexible	0.0311	0.06904	0.1437	0.02783	0.06088	0.1099

Table 5 Total displacement (m)

Finite element method (FEM)	2-D			3-D		
Wall	Excavation depth (m)					
	8	10	12	8	10	12
Stiff	0.04558	0.0519	0.0576	0.03525	0.0478	0.0511

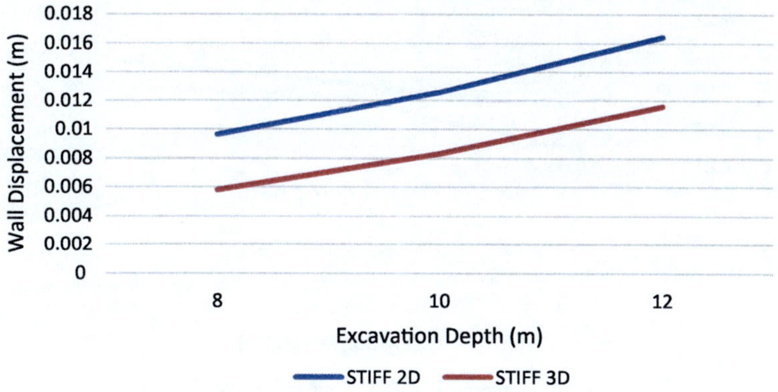

Fig. 10 Wall displacement at center—stiff wall

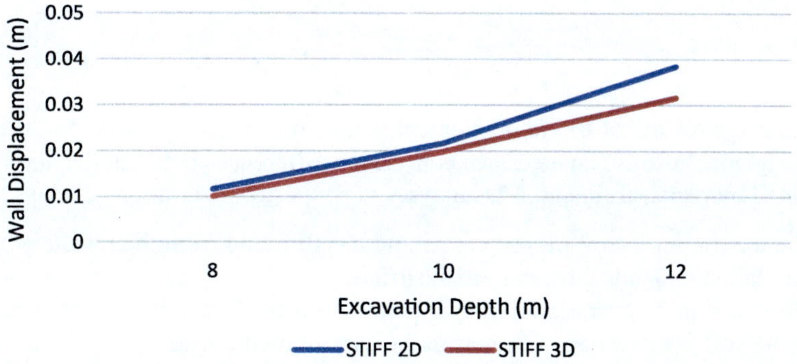

Fig. 11 Wall displacement at center—medium stiff wall

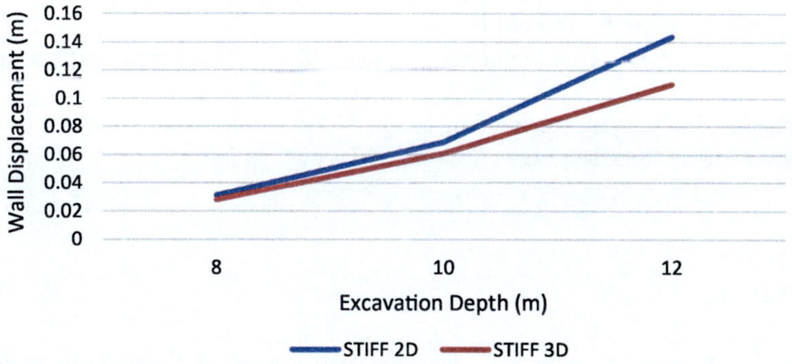

Fig. 12 Wall displacement at center—flexible wall

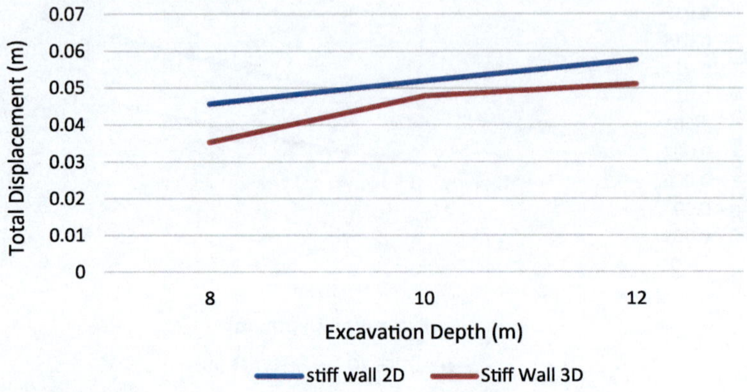

Fig. 13 Total displacement compression 2-D FEA and 3-D FEA

5 Conclusion

Based on the results of the finite-element parametric studies presented herein, the following conclusions can be drawn concerning differences between displacements computed by plane strain and 3-D analyses of supported excavations in clays:

1. The maximum lateral movement behind a wall found from the results of a 3-D simulation depends on wall system stiffness.
2. There is significant variation in 2-D plane strain and 3-D FEA results in the case of the stiff wall system. Whereas in the case of medium stiff and flexible walls, the results vary relatively less.
3. For any given wall the displacement at the wall center obtained in 3-D modeling is less than 2-D modeling. Hence, the lateral stiffness plays a significant role in reducing wall displacement [3].
4. The displacement at the wall center reduces with increase in the stiffness of the wall system.
5. Total displacement of surrounding soil in 2-D plane strain has significantly more displacement then 3-D FEA.

References

1. Finno RJ, Roboski SM (2005) Three-dimensional responses of a tied-back excavation through clay. J Geotech Geoenviron Eng 131(3):273–282
2. Fuentes R, Devriendt M (2010) Ground movements around corners of excavations: empirical calculation method. J Geotech Geoenviron Eng 136(10):1414–1424
3. Bakker KJ (2006) A 3-D FE model for excavation analysis. In: Kwast EA, Bakker KJ, Broere W, Brezuijen A (eds) Geotechnical aspects of underground construction in soft ground, Amsterdam, The Netherlands, 15–17 June 2005. Taylor & Francis Group plc, London

4. Schweiger HF, Vermeer PA, Wehnert M (2009) On the design of deep excavations based on finite element analysis. Geomech Tunn 2(4):333–344
5. Long M (2001) A case history of a deep basement in London clay. Comput Geotech 28(2):397–423
6. Moormann C (2004) Analysis of wall and ground movements due to deep excavations in soft soil based on a new worldwide database. Soils Found 44(1):87–98
7. Finno RJ, Blackburn JT, Roboski JF (2007) Three-dimensional effects for supported excavations in clay. J Geotech Geoenviron Eng 133(1):30–36
8. Zhou T (2015) 3-D FEM analysis for sequential excavation. Master's thesis, Delft University of Technology
9. Zdravkovic L, Potts DM, St John HD (2005) Modelling of a 3-D excavation in finite element analysis. Géotechnique 55(7):497–513
10. Chua HY, Bolton MD (2005) Modelling of horizontal arching on retaining walls. In: The 16th international conference on soil mechanics and geotechnical engineering (16ICSMGE), 2005-9-12 to 2005-9-16, Osaka, Japan, pp 1455–1459
11. Tan Y, Wei B, Diao Y, Zhou X (2014) Spatial corner effects of long and narrow multipropped deep excavations in Shanghai soft clay. J Perform Constr Facil 28(4):4014015, 1–17
12. Dong YP, Burd HJ, Houlsby GT (2016) Finite-element analysis of a deep excavation case history. Géotechnique 66(1):1–15
13. PLAXIS 2D CONNECT edition V21.01
14. PLAXIS 3D tutorial manual CONNECT edition V21

Pushover Analysis Method Applied to a Building with and Without a Shear Wall and Analyzing Placement of Shear Walls

Ashala Sharath Kumar, C. Vanadeep, and Joga Himavarsha

Abstract Today's earthquakes have a greater impact on concrete buildings, which have either collapsed or suffered significant damage. Therefore, it is necessary to assess seismic adequacy. Future earthquakes cannot be prevented, but we may manage or prepare the buildings to ensure their structure is safe and to lessen the level of loss and damage. Although Pushover analysis was originally used in the 1980s, its potential has been understood for the past 20 years. Estimate the base shear and the related displacement of the given structure in this process. A particularly helpful method for assessing both new and old buildings is pushover analysis. In this paper, we discuss about the pushover analysis applied on a building. ETABS software is used to simulate a G + 10 multi-storied building and pushover analysis (PA) is performed utilizing shear wall conditions. The four model cases—a building without a shear wall, one with a corner shear wall, one with a center shear wall, and one with a shear wall with an alternate shear wall—are used.

Keywords Earthquakes · Shear wall · Pushover analysis · ETABS software

1 Introduction

Pushover evaluation, a sort of nonlinear static evaluation, has been utilized in training for the reason that 1970s, even though its capability has simplest simply come to light. This technique is broadly speaking used for calculating the seismic call for a present

A. S. Kumar (✉)
Structural Engineering, Gokaraju Rangaraju Institute of Engineering and Technology, Hyderabad, India
e-mail: sharathashala@gmail.com

C. Vanadeep
Civil Engineering, Gokaraju Rangaraju Institute of Engineering and Technology, Hyderabad, India

J. Himavarsha
VNR Vignana Jyothi Institute of Engineering and Technology, Hyderabad, India

K. K. Pathak et al. (eds.), *Latest Developments in Civil Engineering*, Lecture Notes in Civil Engineering 352, https://doi.org/10.1007/978-981-99-2676-3_46

shape subjected to a delegated earthquake in addition to the energy and waft functionality of the shape [5]. This technique could also be used to assess the suitability of sparkling structural designs. Pushover evaluation has been covered in several earthquake recommendations in current years because of the method performance and computational ease [2].

1.1 Reinforced Concrete (RC) Shear Walls

Reinforced concrete shear partitions are structural partitions blanketed in systems, which are especially constructed to face up to horizontal stresses which are created inside the wall's aircraft through wind, earthquakes, and different factors [7]. Shear partitions can resist heavy horizontal mass forces and maintain gravity masses on the identical time due to their extraordinarily excessive in-aircraft stiffness and strength. The thickness of the strengthened concrete partitions can vary from 140 to 500 mm, relying at the building's age, the want for thermal insulation, and horizontal stresses from wind, earthquakes, etc. [4]. Although maximum of those partitions is non-stop all through the peak of the structure, a few partitions are damaged up at the road degree or basement degree to offer room for parking or business areas. When as compared to at least one axis of symmetry inside the plan, the wall association is regularly symmetrical [5]. Shear walls can resist lateral loads by shifting the weight of an earthquake or wind to the ground. Additionally, they bear gravity stresses and give the system lateral rigidity. From the views of financial system and manipulation of horizontal displacement, shear partitions might also additionally emerge as critical. The shape continues to be broken thanks to a few or the alternative purpose in the course of earthquakes[2]. The weight distribution, lateral stiffness, and energy of the building's creation all have an effect on the way it responds to seismic motion [6]. The shape makes use of bolstered concrete shear partitions for the reduction of the effect of earthquakes.

2 Analysis

"The effects of design earthquake load on structures can be analyzed using two methods".

(a) Static analysis approach. (b) Dynamic analysis method.

Equivalent Static Method: This technique is the simplest, requiring the least quantity of computing attempt and counting on equations discovered inside the code of practice. The shape is seemed as a discrete gadget with focused hundreds at ground degrees in all the code's counseled methods of assessing multi-tale structures.

Pushover Analysis Method: Pushover evaluation is the technique of subjecting a mathematical version that immediately carries the nonlinear load-deformation traits of person constructing additives and factors to gradually better lateral hundreds that constitute earthquake inertia forces till a predetermined "goal displacement" is exceeded.

Dynamic analysis: On the contrary, dynamic analysis can be done in three different methods. The time history method, the modal time history method, and the response spectrum approach are all examples of time history methods. (IS 1893 (Part 1c).

Response spectrum method: In engineering, the term spectrum refers to the behavior of structures across a wide variety of time periods that may be described in a particular single graph. When it involves predicting drifts and member forces in structural systems, the reaction spectra technique of seismic evaluation gives considerable computing benefits.

Time History method: This approach must be used on a suitable ground motion and must be carried out according to known dynamics principles.

3 Methodology

- Model generation using ETABS software.
- Applying loads and seismic parameters to the models.
- The results of the models are compared.

4 Modeling

A plan of 15 × 13 m multi-story (G + 10) building is modeled with SMRF (special moment resting frame) for this study. It was considered to be located in the zone V, type 2 soil. Buildings were modeled by placing the shear wall portions at various locations.

M1: Building without shear walls located in ZONE-V (Fig. 1).

M2: Model with the shear walls located at the corners of the building in ZONE-V (Fig. 2).

M3: Model with the center shear walls located in building in ZONE-V (Fig. 3) (Table 1).

M4: Model with the shear walls located at alternative positions of the building in ZONE-V (Fig. 4).

5 Results and Discussion

See Figs. 5, 6, 7, 8, 9, 10, 11, 12, 13, 14, 15, 16, 17 and 18.

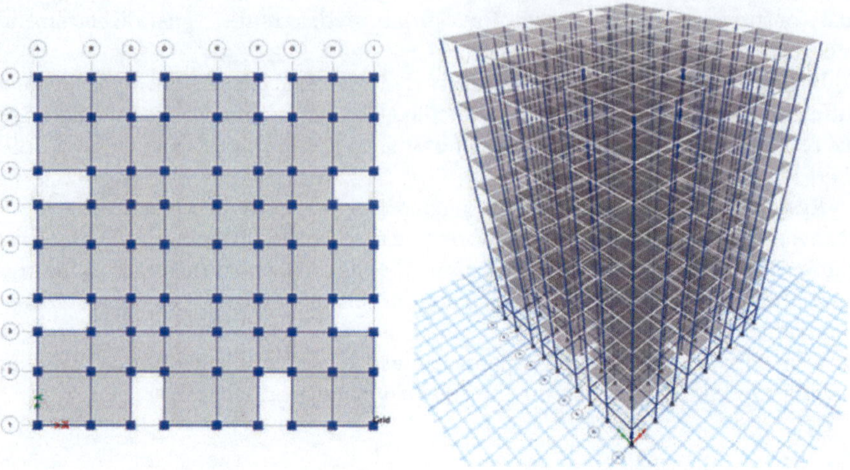

Fig. 1 Building without shear wall partitions

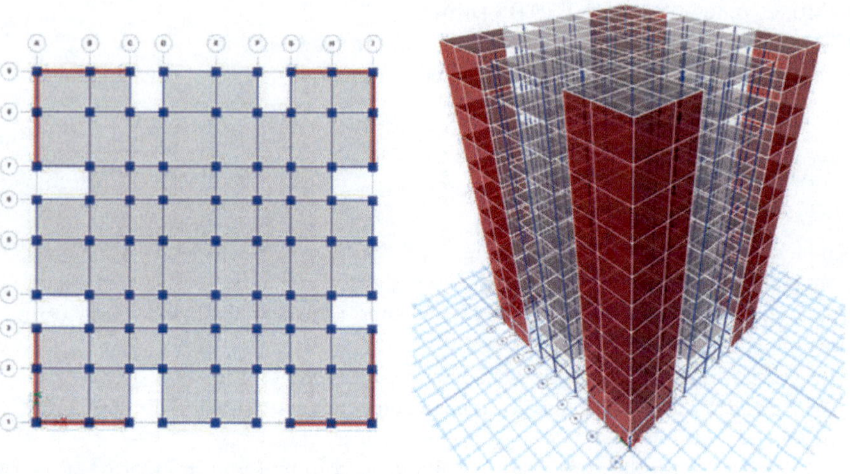

Fig. 2 Building with shear walls placed at the corner

6 Conclusion

After supplying the shear wall portion in a shorter direction, the top deflection was lowered and attained within the permitted deflection.

In comparison to alternative sites, the shear walls' placement was shown to be more efficient for shorter columns.

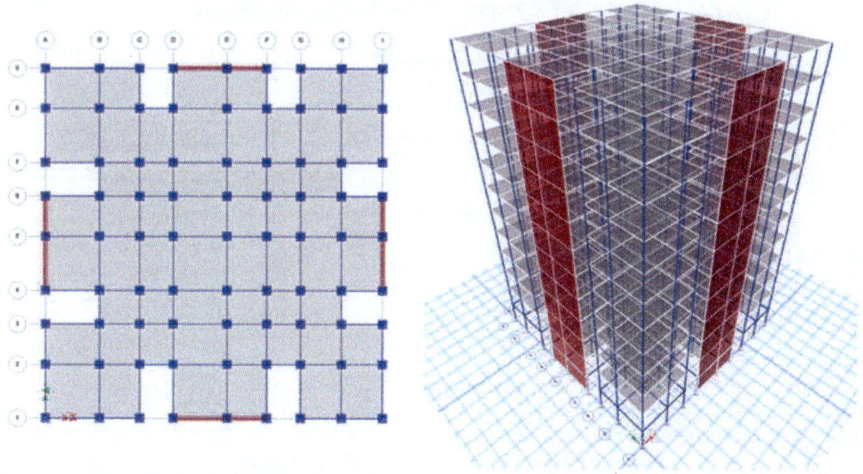

Fig. 3 Building with center shear walls

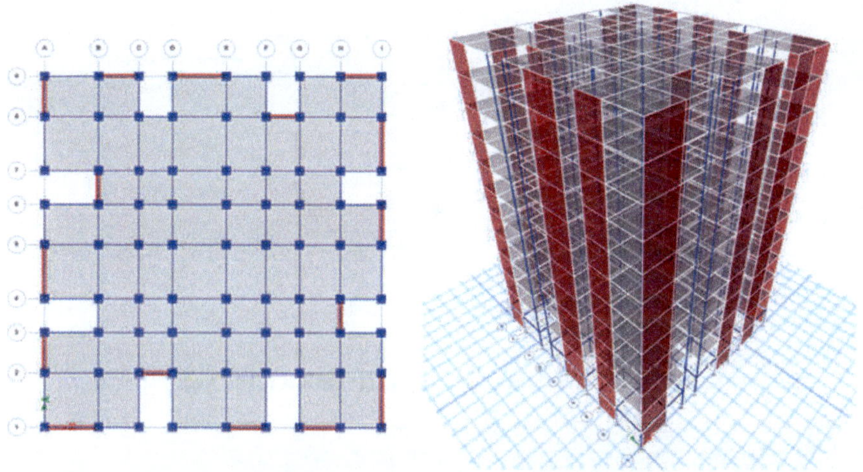

Fig. 4 Building with alternative shear walls

Symmetrically placed shear walls in the extremely outermost moment-resisting frames provide superior results for producing regular shapes.

Compared to the other situations, the values of drift are determined to be smaller for buildings having shear walls at the center.

Structures with shear walls in corner positions had lower values for shear in X and Y directions than other buildings.

Compared to the other structures, the structure with the shear wall arrangement at the corner location had a lower value for bending.

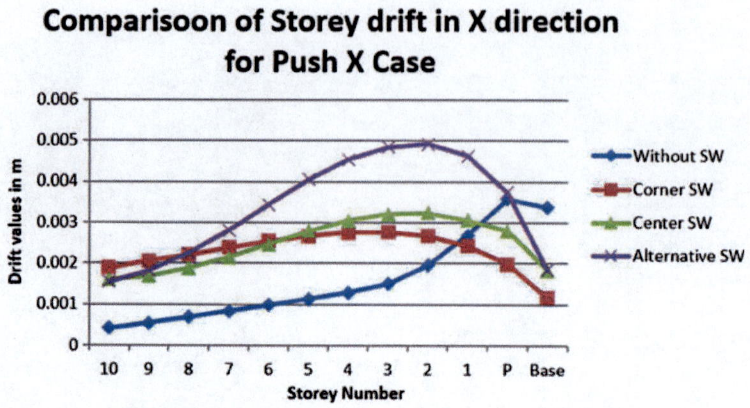

Fig. 5 Story drift in meter Push X in X direction

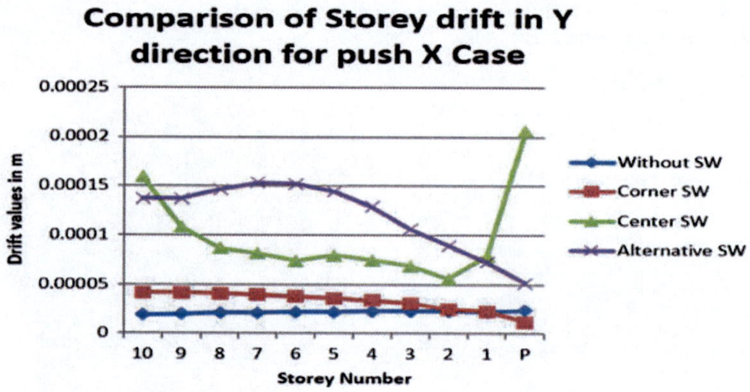

Fig. 6 Story drift in meter Push X in Y direction

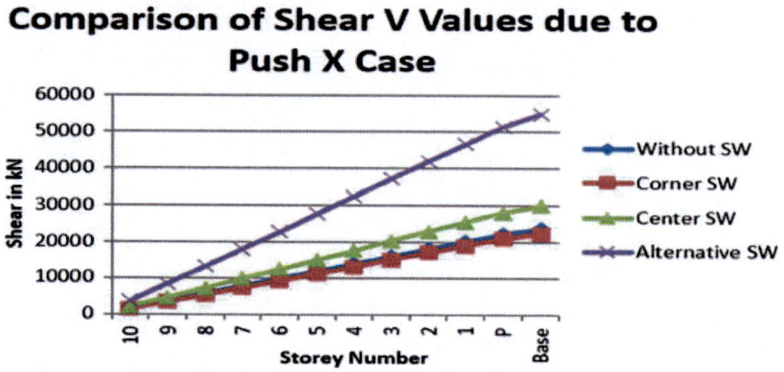

Fig. 7 Story shear in kN Push X

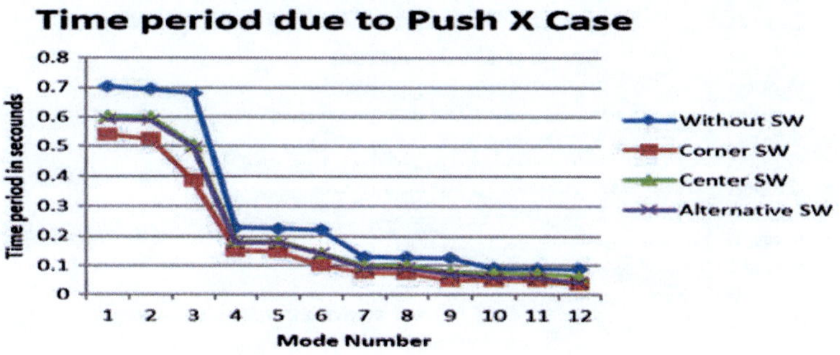

Fig. 8 Time Period in seconds Push X

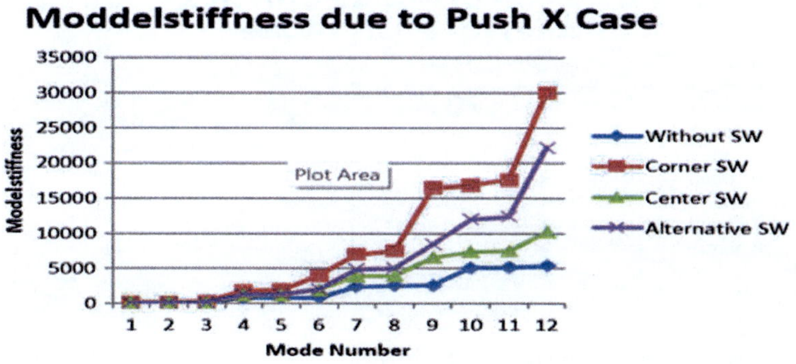

Fig. 9 Model Stiffness Push X

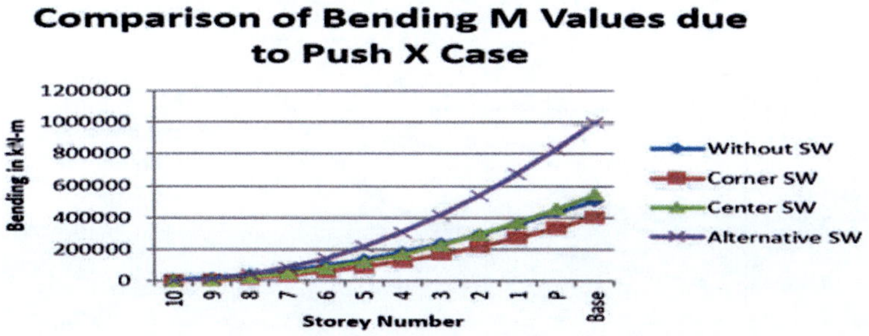

Fig. 10 Bending Moment in Kn-m Push X

Compared to other examples, the values of the building's deflection are determined to be smaller for buildings with shear wall arrangement at the center.

When the opening position is changed from one position to another position, it has been seen for a specific opening in a wall.

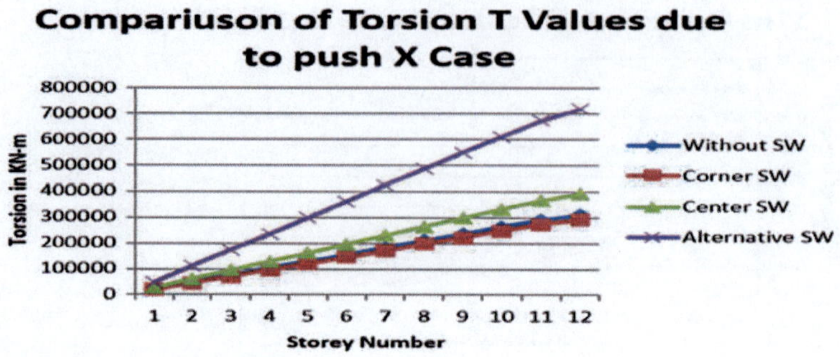

Fig. 11 Torsion in Kn-m Push X

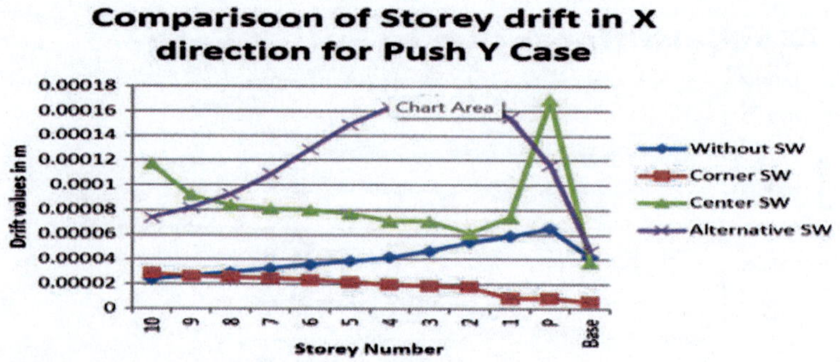

Fig. 12 Story drift in meters of Push Y in X direction

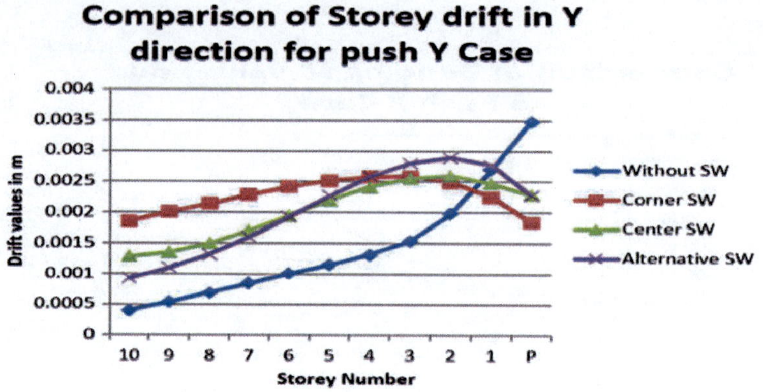

Fig. 13 Story drift in meters of Push Y in Y direction

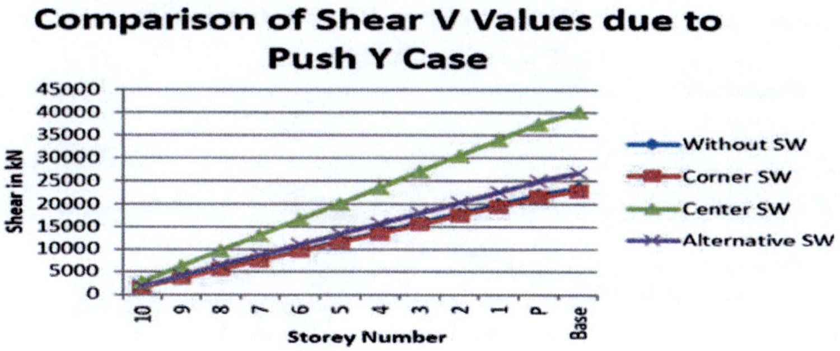

Fig. 14 Story shear in kN of Push Y

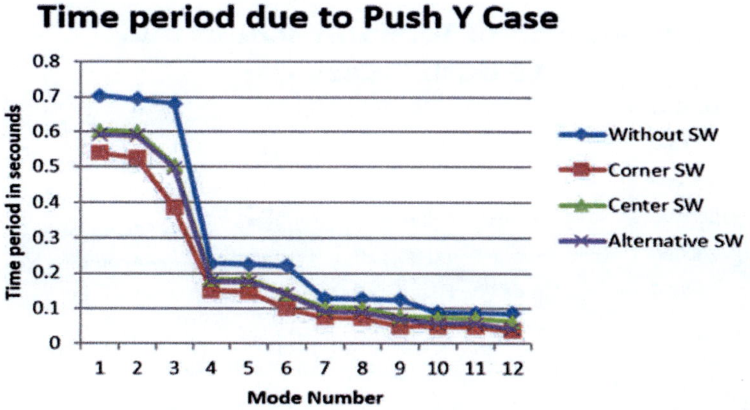

Fig. 15 Time period in seconds of Push Y

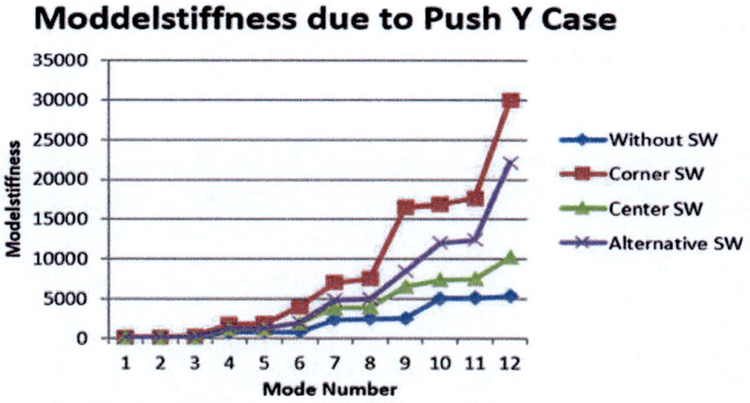

Fig. 16 Model Stiffness of Push Y

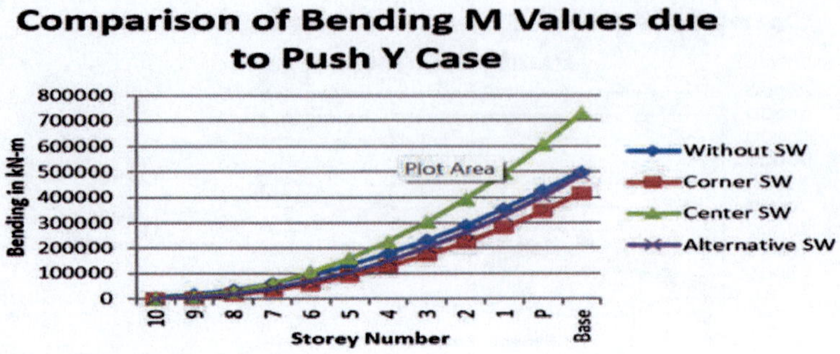

Fig. 17 Bending Moment in Kn-m of Push Y

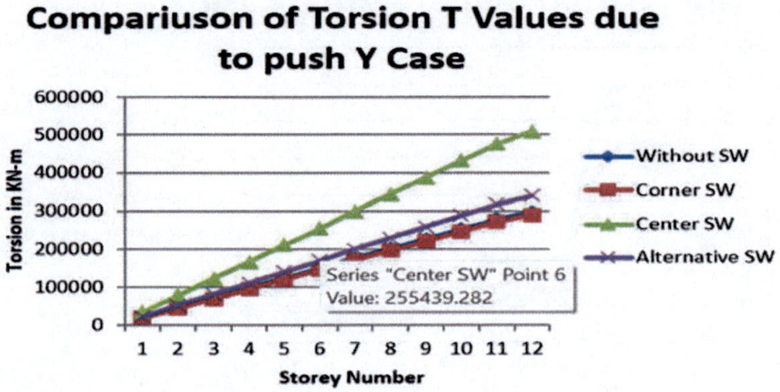

Fig. 18 Torsion in Kn-m of Push Y

It Was found from this investigation that when the fraction of shear walls rises, drift and deflection decrease while shear force (SF) and bending moment (BM) increase.

Table 1 Properties of building

Type of Building	SMRF	No of story's	G + 10
Height of story	3 m	Grade of concrete	M40
Grade of steel	Fe415,Fe500	R	5
I	1.5	Damping ratio	5%
Soil type	2	Slab	150 mm

References

1. Asnhuman S, Dipendu Bhunia, Bhavin Ranjiyani (2011) Solution of shear wall location in multi-storey building. Intern J Civil Struct Eng Res
2. Shaik Kamal Mohammed Azam, Vinod Hosur (2013) Seismic performance evaluation of multistoried RC framed buildings with shear wall. J Sci Eng Res 4(1)
3. Chandurkar PP, Dr Pajgade PS (2013) Seismic analysis of RCC building with and without shear wall Intern J Mod Eng Res 3
4. Chaitanya Kumar JD, Lute Venkat (2013) Analysis of multi storey building with precast load bearing walls. Intern J Civil Struc Eng 4
5. Lakshmi KO, Prof Jayasree Ramanujan, Mrs. Bindu Sunil, Dr. Laju Kottallil, Prof. Mercy Joseph Poweth (2014) Effect of shear wall location in buildings subjected to seismic loads. ISOI J Eng Comp Sci
6. Aainawala MS, Dr. Pajgade PS (2014) Design of multistoried R.C.C. buildings with and without shear walls. Intern J Eng Sci Res Techn
7. Tarun shrivastava, Prof. Anubhav Rai, Prof. Yogesh Kumar Bajpai (2015) Effectiveness of shear wall-frame structure subjected to wind loading inmulti-storey building. Intern J Comput Eng Res 5
8. IS I893(Part I) (2002) Criteria for earthquake resistant design of structures- general provisions and buildings, fifth revision. Bureau of Indian Standards, New Delhi
9. IS: 875 (Part 1) (1987) Code of practice for design loads (Other than Earthquake) for buildings and structures dead loads. Bureau of Indian Standards, New Delhi
10. IS: 875 (Part 2) (1987) Code of practice for design loads (Other than Earthquake) for buildings and structures -imposed loads. Bureau of Indian Standards, New Delhi

Flexure Behavior of Hybrid-Reinforced Concrete Beams Consisting of PVA Fibers and GFRP Bars

Kotte Sai Krishna, K. Hemalatha, and V. Srinivasa Reddy

Abstract In the current study, PVA fibers were examined at doses of 0.125, 0.250, 0.375, and 0.50% in relation to concrete volume. Two different types of plasticizers, SNF-based and PCE-based, were employed in the preliminary study to understand the various effects on concrete and to use polyvinyl alcohol fibers to improve the plasticizer dosage in fiber-reinforced concrete. To determine which plasticizer is more useful and practical to employ, the compressive and split tensile strengths are examined. We have discovered that the best dosage of PVA fibers for efficacy with PCE-based plasticizer is 0.9%. The primary study has then been preceded. For the purposes of our experiment, we cast the beams using pure FRP, which contains 0% PVA and 0.25% PVA, and HFRP, which contains 0% PVA and 0.25% PVA, with GFRP bars installed in the tension face of the beam at the corners. The flexure capacity of pure FRP and HFRP was then contrasted. Then, based on the results, we discovered that HFRP with 0.25% PVA can be used to increase flexural capacity. Since there is no yielding point, the main disadvantage of FRP bars is their brittleness. Consequently, numerous studies have been done to improve their flexure capacity by introducing hybrid reinforcement, where both steel bars and FRP bars are placed in tensile reinforcement so that they can meet minimum flexure capacity.

Keywords Hybrid fiber reinforced concrete · Polyvinyl alcohol · Flexural load capacity

K. Sai Krishna (✉) · K. Hemalatha
Structural Engineering, Gokaraju Rangaraju Institute of Engineering and Technology, Hyderabad, India
e-mail: saikrishnakotte1515@gmail.com

V. Srinivasa Reddy
Civil Engineering, Gokaraju Rangaraju Institute of Engineering and Technology, Hyderabad, India

K. K. Pathak et al. (eds.), *Latest Developments in Civil Engineering*, Lecture Notes in Civil Engineering 352, https://doi.org/10.1007/978-981-99-2676-3_47

1 Introduction

The PVA fiber improves ductility and capability for releasing energy. Additional studies were conducted to examine the impact and fracture characteristics of hybrid fiber-reinforced concrete. According to research, adding hybrid fibers to pumice lightweight aggregate concrete improves its compressive strength, modulus of elasticity, split tensile strength, and modulus of rupture and allows it to satisfy structural requirements [1]. Other studies using steel and polyvinyl alcohol reveal that they improve ductility, flexure, rupture modulus, and compressive strength. In the current study, beam specimens were subjected to an experimental inquiry on curing times of 7, 14, and 28 days in M30 grade concrete [2].

1.1 Glass Fiber Reinforced Polymer (GFRP)

Glass Fiber Reinforced Polymer (GFRP) substances are being hired all around the global to restore and adapt old infrastructures, together with bridges and homes. These homes have visible considerable strength and stiffness losses through the years due to harsh climatic elements like dampness, salinity, and alkali solutions. The corrosion problem may be solved and the strength and stiffness of the internally bolstered with GFRP bars may be notably improved via way of means of the usage of superior fiber composite substances like GFRP [3]. However, the hyperlink among the GFRP plate and floor of the RC beam substantially affects the strength of externally bolstered RC beams, which might be bolstered with GFRP plates and textiles and subjected to harsh environmental conditions. Therefore, its miles critical to examine the general overall performance of the RC beams which have been externally bolstered with GFRP plates and textiles and subjected to numerous environmental elements. The technological sectors have grown notably over the preceding numerous a long time and are anticipated to keep developing further. This is in most cases on account of those substances' excessive stiffness and strength [4].

1.2 Advantages of GFRP Rebars

Rebar made of glass fiber reinforced polymer is a highly valuable building material. Governments and other large-scale infrastructure providers increasingly recognize that GFRP is an affordable building material with the ability to increase the lifespan of public facilities where corrosion can have a significant negative impact on the economy and the environment. The use of fiberglass reinforcing material has grown significantly because of the growth in corrosion caused by climate change. Future versions of these sophisticated composite materials would be able to exhibit their strengths and characteristics more clearly [5]. The use of GFRP rebar has a number of benefits, some of which are listed below.

- A high-quality vinyl ester resin that is corrosion-resistant and extends the life of a concrete structure is one of the components of GFRP [6].
- GFRP rebar has double the tensile strength of steel while weighing a quarter as much as typical steel rebar.

1.3 Poly Vinyl Alcohol Fibers (PVA)

Vinyl alcohol (PVA) makes up at least 50% of the fiber-forming substance in vinylon fibers, which also contain at least 85% of the fiber's total weight in vinyl alcohol units as well as one or more acetal units. Another name for the fiber is vinal or vinylal.

Vinyl monomer is not stable; hence vinyl resin cannot be produced from it. In its place, polyvinyl acetate is hydrolyzed to yield polyvinyl alcohol (PVOH) [8]. This fiber can replace cotton in many applications since the fabrics manufactured from it feel like cotton, have an excellent affinity for water, don't release harmful fumes when burned, and are simple to wash and dry.

PVA fiber is an appropriate reinforcement material for cementitious composites due to its properties. An excellent environmentally friendly cement reinforcement material is polyvinyl alcohol fiber. It, because to its distinctive molecular structure, has alkali and weather resistance, therefore it enhances the concrete's resistance to frost. improving bending strength, impact strength, and fracture strength; improving permeability; and improving impact and seismic resistance of concrete by forming a strong affinity with cement. Increased toughness is a result of concrete's improved brittleness, impact resistance, and bending strength. Strong bonding with cement matrix is one of PVA fiber's notable properties.

1.4 Applications of PVA Fibers

- PVA Fibers' "stealth" quality is their main advantage. They can be cast all over the concrete and hardly or not at all be visible on the completed surfaces. This is true whether you are producing troweled finished concrete or concrete cast in a mould.
- The ductility or flexibility of PVA-reinforced concrete is exceptional. Instead of having a quick catastrophic failure, the concrete section will flex substantially before it fractures. It only fails in a different way, which does not imply that it is more durable than AR Glass fiber in use [10]. In comparison tests, AR Glass performs better, although it isn't always feasible to utilize it for a particular application.

- Concrete and PVA are said to "chemically bind" together. This assertion looks doubtful to ordinary observation. We won't give that assertion much credence and will just leave it alone. If it's accurate, we prevail. If not, we won't worry about it.

1.5 Plasticizer

These are chemicals that are added to the concrete mix whenever it is necessary to speed up the setting process or use less water. Because the lack of nearby room for development, they are employed in areas where natural resources are few or where concrete must be transported over great distances.

Sulphonated Naphthalene Formaldehyde (SNF)

By lowering the amount of water in the mixture, we have employed this plasticizer to hasten the strength increase at young ages. Additionally, it makes concrete more fluid without using any additional water, which lessens the labor-intensive process of polishing and smoothing the concrete. Additionally, it enhances pumpability, lessens the effects of segregation, and bleeding, and revamp the workability of concrete. For high strength, low water absorption concrete, the recommended range of plasticizer is 1–3 L per 100 kg of cementitious ingredients. For 100 kg of cementitious ingredients, 0.70–2 L of concrete can be worked (Table 1).

Polycarboxy Ether-Based (PCE)

This plasticizer is utilized when the project or building demands exceptional durability and performance. It is primarily utilized in pre-cast, pre-stressed buildings with high levels of early strength and longevity. High strength may be attained without bleeding or segregation. For a standard grade of concrete, an appropriate dose of 500–1500 mL per 100 kg of cementitious ingredients should be maintained. Plastic shrinkage cracks, bleeding, segregation, and increased initial and final setting times are all consequences of using too much plasticizer (Table 2).

Table 1 Typical properties of sulphonated naphthalene formaldehyde

Aspect	Brown liquid
Specific gravity	1.18 @ 25 °C
Chloride content	Nil to BS 5075/BS: EN934
Air entrainment	Less than 2% additional air entrained at normal dosages

Table 2 Technical data of polycarboxy Ether-based

Aspect	Reddish-Brown liquid
Relative density	1.08 ± 0.20 at 25 °C
pH	Greater than or equal to 6
Chloride ion content	Less than 0.2%

2 Mix Design of Concrete

Concrete mix design is a process that involves choosing the right elements for concrete and their appropriate ratios with the goal of preparing concrete as affordably (value engineered) as feasible while maintaining a specific minimum strength, desirable workability, and durability.

Gather the following information before choosing a concrete mix design, as some design requirements are predetermined based on this information.

2.1 Mix Proportions

Cement = 394 kg/m^3
Water = 197 kg/m^3
Fine aggregates = 732 kg/m^3
Coarse aggregate = 1139 kg/m^3
Water-cement ratio = 0.50
mix for M30 grade concrete is 1:1.86:2.89 at w/c of 0.50

3 Beam Detailing

See Figs. 1 and 2.

4 Test Results and Discussions

In this project, we have tested the specimen of M-30 grade concrete for 7 and 28 days of curing are tested in the laboratory. Results are tabulated and graphs are presented.

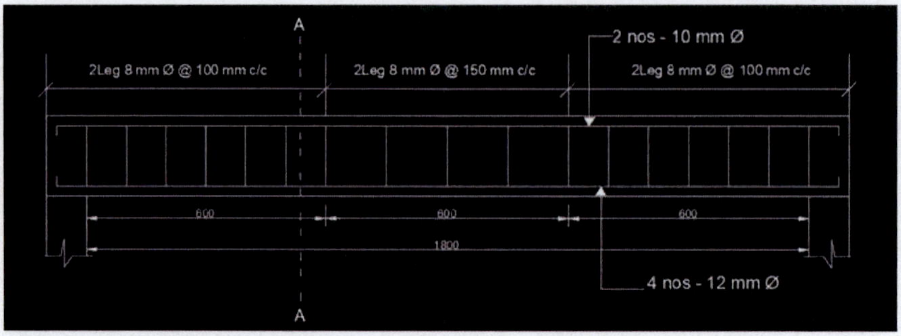

Fig. 1 Longitudinal section of the beam

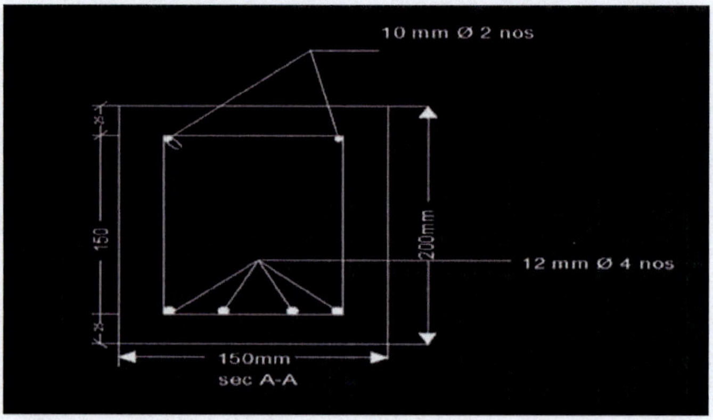

Fig. 2 Cross-section of the beam

4.1 Workability

The test is conducted on different types of concrete mixes. We tried to maintain the slump value between 50 and 100 mm. So as to maintain the slump value for different mixes, we added different percentages of plasticizer (Table 3).

4.2 Compressive Strength Test

The specimen experiences the force from two sides like the specimen squeezed in a test. Generally, the cubes are placed in between two plates, the bottom plate is fixed, and the top plate is moveable which exerts pressure on the cube, which tends to flatten the specimen (Tables 4 and 5).

Table 3 Workability of different mixes

Serial number	Sample	Plasticizer used	Plasticizer (%)	Slump value (mm)
1	CP-0	SNF-based	0.3	84
2	CP-0.125	SNF-based	0.4	80
3	CP-0.250	SNF-based	0.6	72
4	CP-0.375	SNF-based	1.0	65
5	CP-0.500	SNF-based	1.5	63
6	PP-0	PCE-based	0.3	90
7	PP-0.125	PCE-based	0.7	87
8	PP-0.250	PCE-based	0.9	85
9	PP-0.375	PCE-based	1.2	73
10	PP-0.500	PCE-based	1.6	69

Table 4 Compressive strength test results for SNF plasticizer of 28 days

Sample	Compressive strength of fiber concrete	% change
CP-0	35.4	–
CP-0.125	36.3	+2.5
CP-0.250	39.4	+11.3
CP-0.375	32.6	−7.9
CP-0.500	29.5	−16.7

Table 5 Compressive strength test results for PCE-based plasticizer of 28 days

Sample	Compressive strength of fiber concrete	% change
PP-0	35.4	–
PP-0.125	36.2	+2.25
PP-0.250	40.4	+14.12
PP-0.375	39.2	+10.73
PP-0.500	38.6	+9.03

$$\text{Compressive strength} = \frac{\text{Compressive load (kN)}}{\text{Crosssectional Area of Cube (mm}^2)}$$

4.3 Split Tensile Strength Test

In this Test, we use cylinders for testing we place them in the longitudinal direction as shown in the Fig. 3. The load applied in a longitudinal direction due to which the cracks formed. We use the same testing machine as the compressive strength testing machine.

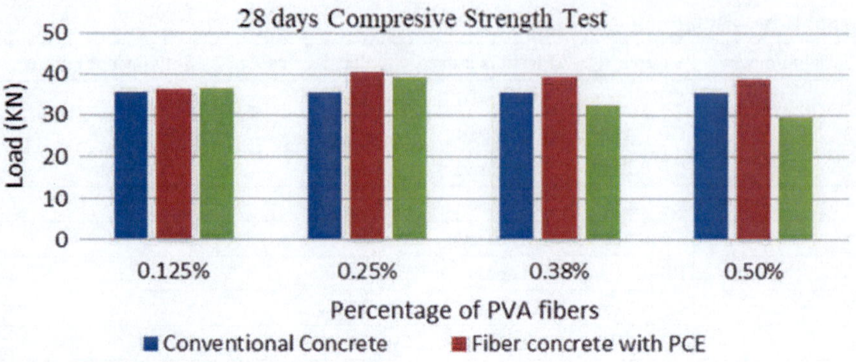

Fig. 3 Compression of compressive strength

$$\text{Split Tensile Strength} = \frac{2 \times \text{LOAD}}{\pi \times \text{Diameter} \times \text{Length}}.$$

4.4 *Split Tensile Strength of 28 days*

The split tensile strength test was carried out for 28 days on cylinders with a diameter of 150 mm and a length of 300 mm. The results are listed below for the various blends (Tables 6, 7 and Fig. 4).

Table 6 Split strength test results for SNF-based plasticizer of 28 Days

Sample	Split tensile strength	% change
CP-0	2.83	–
CP-0.125	3.12	+12.37
CP-0.250	3.50	+26.37
CP-0.375	3.40	+20.14
CP-0.500	2.72	+18.02

Table 7 Split strength test results for PCE-based plasticizer of 28 days

Sample	Split tensile strength	% change
PP-0	2.83	–
PP-0.125	3.18	+10.24
PP-0.250	3.64	+28.42
PP-0.375	3.43	+21.20
PP-0.500	3.34	−3.89

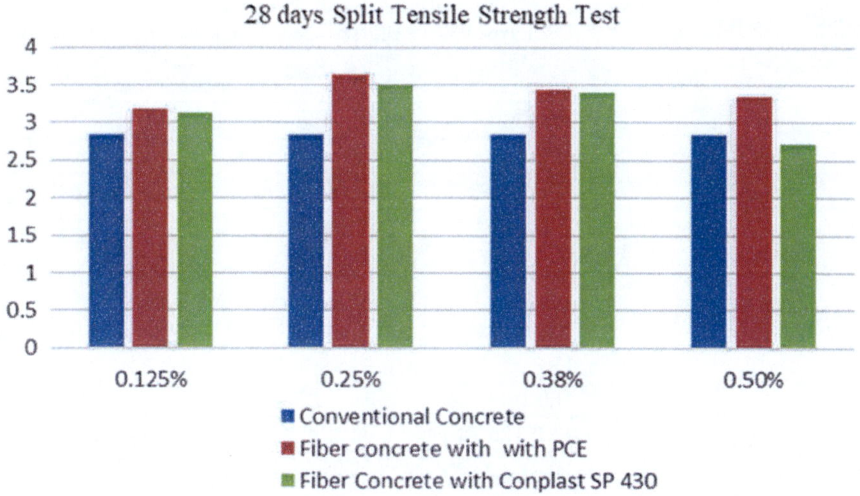

Fig. 4 Compression of split tensile strength

- From the above tests, we have understood that PCE has a good effect on the PVA fiber at a dosage of 0.9% compared to 0.3, 0.7, 1.2, and 1.6%, respectively. From that, we can observe the compression strength and split tensile strength is maximum at a dosage of 0.9% PVA, i.e., 40.4 and 3.64 N/mm^2, respectively, compared to SNF at 0.9% dosage of PVA fibers in fibers reinforced concrete. So, we have finalized the percentage to use for different proportions of PVA fibers in concrete. Then we tested beams for flexural behavior.

4.5 Flexure Behavior of Reinforced Concrete Beams

The HFRP beam is attached to the testing equipment and the load is applied to it. The ultimate load and ultimate moment were used to investigate the flexural behavior.

4.6 Four-Point Bending Test Results

We casted eight beams of different dosage of PVA fibers with 0% and 0.25%, respectively, with respect to FRP and HFRP. We have provided beam dimensions of Length (L) = 2000 mm of width (b) 150 mm and with a depth (d) of 200 mm and the top and bottom covers are provided with 25 mm, respectively. We have used reinforcement bars of 12 mm GFRP which are placed at the bottom corners of the beam and steel bars of 12 mm which are placed in the tension face of the beam and hanger bars of 10 mm which are placed at top of the beam in compression face of the beam & 8 mm

Table 8 Test results from the experiment

Beam type PVA	Ultimate load (kN)	Ultimate moment (kN/mm)	% change
FRP 0%	257.8	64.25	–
HFRP 0%	304	76	+18.28
FRP 0.25%	379	94.75	+47.47
HFRP 0.25%	445	111.25	+73.151

Table 9 Ultimate deflection

Beam type	Ultimate deflection (mm)	% change
FRP 0% PVA	22	–
HFRP 0% PVA	18	18.18
FRP 0.25% PVA	42.26	92.090
HFRP 0.25% PVA	21.05	4.31

Table 10 First crack load

Beam type	First crack load (kN)	% change
FRP 0% PVA	46.9	–
HFRP 0% PVA	266.4	4.68
FRP 0.25% PVA	198.4	3.23
HFRP 0.25% PVA	389.5	7.30

dia two legged stirrups @ 150 mm c/c at mid span at supports 100 mm c/c from L/4 distance. To know the flexure capacity of beams, we used four-point loading test, the beam is over hanged from the both supports at a distance of 600 mm for both the ends and the load is applied between the supports. From the results, we have taken average value of the beams. From the testing results, we can conclude that the HFRP beams give better resistance when compared to FRP beams. HFRP is a combination of FRP and steel as we know comparatively HFRP beams having more yielding than FRP beams, so the load bearing capacity is greater (Tables 8, 9, 10 and Figs. 5, 6, 7, 8).

5 Conclusions

The findings that were drawn from research on beams were as follows:

- Poly Carboxylic Ether (PCE)-based mixtures are stronger and easier to deal with than Sulphonated Naphthalene Formaldehyde (SNF)-based mixes.
- Based on the data, the PVA Fiber dosage that results in the maximum compressive strength and split tensile strength is 0.25%.

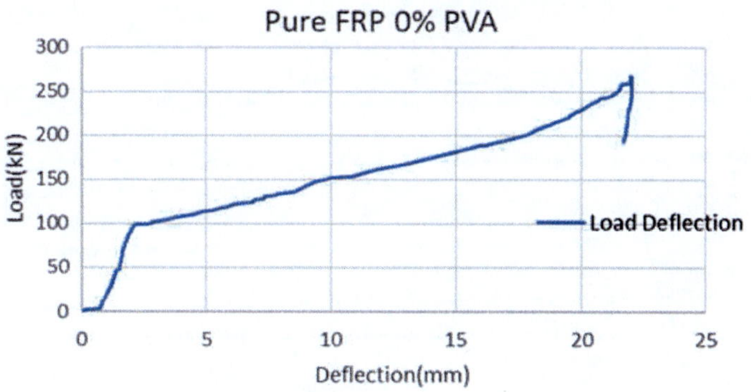

Fig. 5 Load deflection curve of pure FRP 0% PVA

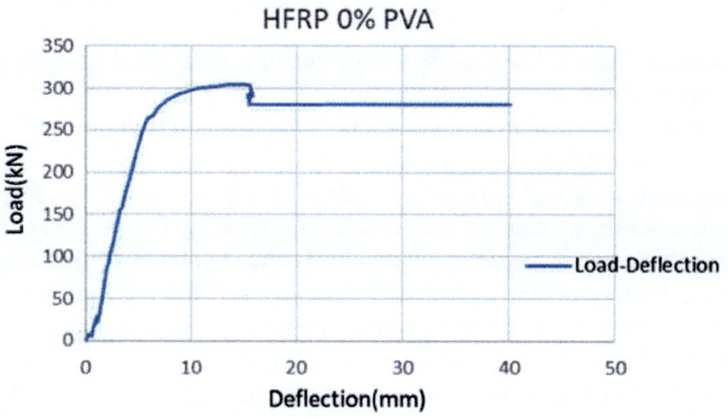

Fig. 6 Load deflection curve of HFRP 0% PVA

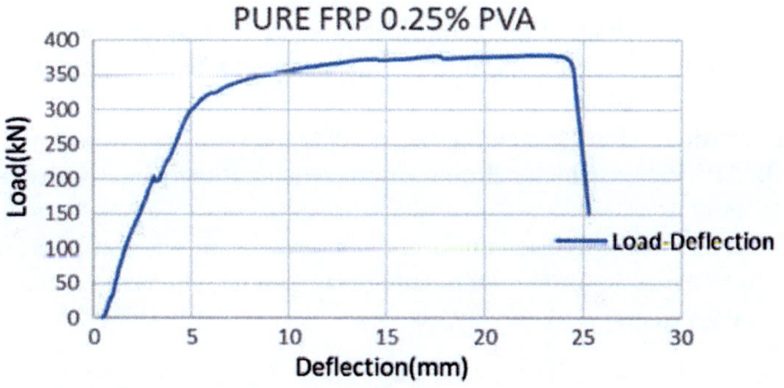

Fig. 7 Load deflection curve of pure FRP 0.25% PVA

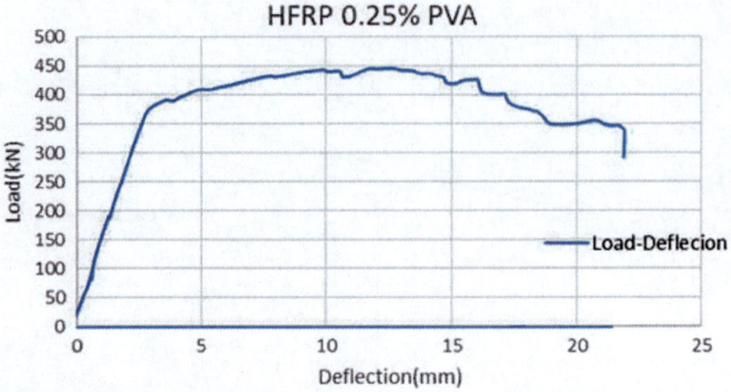

Fig. 8 Load deflection curve of HFRP 0.25% PVA

- Due to this, we have chosen to use this percentage as the ideal percentage for the remaining mixtures.
- When compared to combinations based on "Sulphonated Naphthalene Formaldehyde (SNF)" and Poly Carboxylic Ether (PCE), PCE had a positive impact on the PVA Fiber at a dosage of 0.9% compared to 0.3, 0.7, 1.2, and 1.6%, respectively. As a consequence, we can see that fiber-reinforced concrete has higher compression strength and split tensile strength with a dosage of 0.9% PVA (40.4 and 3.64 N/mm^2, respectively) than SNF does at a dosage of 1% PVA.
- When PVA fibers are added to concrete, the strength effectiveness is 14.13 and 11.25% for PCE-based and SNF-based, respectively.
- Strength effectiveness of concrete in split tensile strength with the addition of PVA fibers is 28.42 and 26.37%, respectively, by PCE-based and SNF-based.
- Split tensile strength of concrete with the inclusion of PVA fibers is 28.42 and 26.37% stronger by PCE-based and SNF-based, respectively.
- Observations show that, in terms of ultimate moment and load-bearing capacity, the HFRP beam with 0% PVA fiber is 18.28% more effective than the Pure FRP beam with 0% PVA fiber.
- When compared to Pure FRP with 0.25% PVA, the HFRP beam with 0.25% PVA has a 73.151% increase in the ultimate moment and ultimate load with a deflection of 21.05 mm.
- When compared to a hybrid reinforced concrete beam without fibers and pure GFRP, the increase in ultimate moment capacity is 73.151 and 47.47% at a dose of 0.25% PVA fibers, respectively.

References

1. Dasgupta A et al (2018) Retrofitting of concrete structure with fiber rein- forced polymer. IJIRST 4(9):February 2018, ISSN (online) 2349-6010
2. Yang E, Li VC. Rate dependence in engineered cementitious composites. Department of Civil and Environmental Engineering, University of Michigan, USA
3. Sarker P, Begum M, Nasrin S et al (2011) Fiber reinforced polymers for structural retrofitting: a review. J Civil Eng (IEB) 39(1):49–57
4. Devi M, Kannan L, Ganeshkumar M et al (2017) Flexural behavior of polyvinyl alcohol fiber reinforced concrete. SSRG Int J Civil Eng (SSRG–IJCE) 4(6):June 2017
5. Yalburgimath C, Rathod A, Bhavanishankar S et al (2018) Retrofitting of reinforced concrete beam using carbon fiber reinforced polymer (CFRP) fabric. Int Res J Eng Technol (IRJET) 5(10):October 2018, e-ISSN: 2395-0056
6. Alexander A, Nimesh Mohan M et al (2015) Retrofitting of reinforced concrete beam with glass fiber reinforced polymer strips and sheet. Int J Eng Res Technol (IJERT). ISSN: 2278-0181. Published by www.ijert.org
7. Nagarathinam N, Vijayalakshmi R, Sabaritha P (2019) Experimental study on mechanical properties of PVA fiber reinforced concrete using m-sand and flyash. Int Res J Eng Technol (IRJET) 6(4):April 2019
8. Pammar RP, Ramesh V (2015) Experimental study on combined effect of steel and glass fibers on compressive strength, flexural strength and durability of concrete and comparison with conventional concrete. Int J Civil Struct Eng Res 3(1):146–150
9. Guru Prasad MC, Tanuja K, Vasu Deva Naidu N (2016) Experimental study & strength of concrete by using steel & glass fibers. Int Res J Eng Technol (IRJET) 3(9)
10. Patil TR, Burile AN (2015) Comparative study of steel and glass fiber rein- forced concrete composites. Int J Res (IJSR). ISSN (Online): 2319-7064

Performance of a High-Rise Building Frame with Composite Columns in Non-linear Analysis Using ETABS

Dhanesh Khalotia, Dharmendra Singh Meena, and Gulshan Kumar Meena

Abstract The compression element used in composite columns is made up of steel tubes or concrete-filled steel tubes. Concrete steel composite columns combine concrete and steel, so they take advantage of each material's benefits. In this research, the software programmed ETABS is used to simulate a Ground + 10 story framed structure exposed to nonlinear dynamic stress of Zone V in accordance with IS 1893–2016. Comparable loading conditions have been applied to two models that were created with various types of columns, Reinforced Cement Concrete Column and CFST Column. These two models were broken down, and the results were examined in terms of how the structures would perform given the following parameters: maximum story displacement, story shear, story drift, story overturning moment, and section size reduction.

Keywords Analysis · ETABS · Column · Composite material · Displacement · Moment · Forces

1 Introduction

India is a developing country, yet comparison to other countries; it consumes far less steel in the construction field. The density of population in cities is growing everyday as a result of the enormous growth in the population, the intensity of development within and around metropolitan areas, and the scarcity of large tracts of land. The need for high-rise structures has arisen as a consequence of the higher population size. Because of the buildup of loads from across all story's in high-rise buildings, vertical gravitational loads upon columns predominate in the planning of the structure of the building. Because of their benefits over reinforced concrete columns as well as the exorbitant cost of steel structures, composite structures are being adopted as replacements to steel ones [1–5, 7, 20].

D. Khalotia (✉) · D. S. Meena · G. K. Meena
IES University, Bhopal 462044, India
e-mail: Dhaneshkhalotia@gmail.com

K. K. Pathak et al. (eds.), *Latest Developments in Civil Engineering*, Lecture Notes in Civil Engineering 352, https://doi.org/10.1007/978-981-99-2676-3_48

Generally speaking, the majority of conventional structures are designed under the assumption that any associated load is steady. Because the structure is occasionally subjected to dynamic loads, the effect of dynamic load [18] isn't taken into account. Instead, this causes the arrangement to become more complicated and laborious. This aspect of ignoring special abilities could occasionally lead to disaster. Especially if an earthquake disruption [13, 20] was about to occur [2, 4–6, 8, 11].

Considering the following benefits, steel concert hybrid column components are now being marketed as an alternative to bare steel/reinforced cement concrete columns (Fig. 1).

- Significant load-bearing capacity.
- Strong fire resistance.
- Materials that are inexpensive [8].
- Quick to construct.

Although composite columns can support more weight than R.C.C. columns, structural designers can create thinner columns with greater resistance to bending moment and shear stress [8, 10, 11, 20].

This investigational evaluation hopes to learn more about the behavior of an enclosed structure made up of composite segments and equipped with standard R.C.C. slab and beam [4, 5, 9]. For the research and design of the construction,

Fig. 1 Specific parts of composite columns

the structural examination and plan programming ETABS [17, 25] will be used, this study's primary goal is to ascertain how composite columns affect a high-rise structure when subjected to non-linear dynamic stress [2, 5, 6, 8, 11, 18].

2 Objectives

The primary goals of this study are–

1. To evaluate the high-rise building's structural performance:
   A. Considering broad concepts R.C.C. Building [1, 7].
   B. Considering a construction with RCC Beam and Slab and CFST Composite Columns. The structural parameters that are taken into consideration when comparing the structural performance [3, 4] include story stiffness, story shear, maximum story displacement, modal period, overturning moment, story drift, and frequency.
2. CFST composite column size optimization [3, 5, 8].
3. A column comparison of the two constructions' costs is done [8].

3 Literature Review

Campian et al. (2015): A composite made of steel and solid signifies that perhaps the steel area's components are encapsulated in cement while the concrete slab as well as profiled deck slab is mechanically attached to the steel beam such that it work like one unit. A Corresponding Static Method of Assessment takes into account the IS:-1893-(Part-1):2002 provisions for seismic stacking for the G+15 story located in quake zone IV. Reinforced bond solid and steel-solid composition alternatives are indeed taken into account. The Composites and Reinforced Cement Concrete structure presentation utilizes STAAD genius programming. This research primarily discusses on seismic construction and performance of compound steel-solid edges. Steel and reinforced bond solid constructions are used to complete the examination of characteristics such as timing, relocation, minutes, and burden transporting limit. Analysis of the results reveals that composite structures are even better from some angles.

In tests, it was discovered that although composite segments built with low-quality cement bombed gradually and their bearing capacity steadily dropped, High-quality cement-built segments collapsed suddenly due to breakage [30].

Due to their superior performance, this exploratory research shows that totally enclosed compound sections can be utilized in seismic and non-seismic zones as an alternative. During tests, composite sections using HSC showed a weak, disappointing design; hence, further exploratory study is being done in this area.

Netravathi et al. (2017): In this research, the performance of a standard Reinforced cement on Crete column and a composite column was examined using study of one tab by response spectrum. Either regular or irregular architectures were compared to traditional Reinforced cement concrete columns for composite columns. For rectangular or circular composite column sections in regular structures, displacement was reduced by 40–50%, though shear was raised by 60–70%, and drift was improved by 35–40%. Displacement decreased by 40–50% in irregular structures as well, but shear raised by 60–70% and drift increased by 35–40%. This means that using composite columns does not depend on the geometry of the building. Since all of the elements used in this research were composite sections, there is still room to examine how well each particular composite element performs in comparison to other R.C.C. structural elements [31].

Renavikar et al. (July 2016): They conducted a comparison of the cost and analysis of Reinforced cement concrete and steel-composite structures. The study analyzes a residential building that is made of steel-concrete composite and reinforced cement concrete. The planned structure consists of four multi-story structures with floor heights of 3.0 m for floors Ground+9, Ground+12, Ground+15, and Ground+18 (plan dimension 15m x 9m). STAAD-Pro 2007, a 2D modeling program, was used for the analysis, with load combinations chosen in accordance with the IS Code. In order to compare the costs of a composite construction to an equivalent RCC structure, the project requires the analysis of an analogous RCC structure. Composite structures outperform traditional RCC structures due to their inherent ductility qualities. The RCC structure has higher axial forces, seismic forces, bending moments, and deflections than the composite structure. As a result of smaller element size, steel structures are less expensive than RCC structures. For high-rise buildings, the composite option is preferable to RCC because the weight of composite structures is lower than that of RCC structures, which lowers the foundation costs and subjects them to less force from earthquakes [32].

Desai et al. (2013): The seismic response of soft story composite columns is investigated in this research. STAAD pro software was used to assess the functionality of four different models. In model-01, the building's story elevation was kept constant, however, in model-02, and only the ground-level columns were substituted with composite columns. In model-03, the ground floor and first floor columns were changed with composite columns, whereas in model-04, the ground floor heights were increased to 4 m. It was revealed from above that the stiffness of the ground floor for composite columns is greater compared to R.C.C. columns in model-01. Models-02 and 04, which include a composite column solely on the ground floor, generate additional stiffness than models-01 and 03. According to the aforementioned study, a shear wall should be constructed with composite columns to improve the stiffness and ductility of the soft story, and first floor columns ought to be stiffer than ground floor columns [30].

4 Methodology

Steps:

I. Assess the geometry that will be used in the study.
II. Choosing the boundary conditions, such as the sections, materials, and loading standards.
III. Prepare the structure modeling using the 339 analysis tool.
IV. Assign the chosen column type, loading, and composite sections.
V. Perform a nonlinear dynamic analysis in accordance with Indian Standards.
VI. Evaluating the outcome in terms of forces, moments, and displacement (Tables 1, 2 and 3).

Table 1 Selected geometry

1	Quantity of stories	G + 10&mumpty
2	Floor stilt height	3.2mtr
3	Upper story height	3.2mtr
4	Foundation depth	−2.0mtr
5	Concrete grade of RCC beam and slab	M25
6	Concrete grade of columns	M25
7	Longitudinal reinforcement utilizing steel	HYSD500
8	Lateral reinforcement utilizing steel	HYSD415
9	Sections of steel	Fe345
10	Masonry	Brick filler
11	Zone of seismicity	Zone:V

Table 2 Data by section

Conventional Frame of Reinforce Concrete (Unit—mm)		
1	Column	650 × 650
2	Beam	300 × 400
3	Slab	150 (thick)
4	Masonry	130 (thick)
R.C.C. Slab and Beam Composite Column (Unit—mm)		
1	CFST composite Column	450 × 450
2	Beam	300 × 400
3	Slab	150 (thick)
4	Masonry	130 (thick)

Table 3 Loading scenario

Name	Type of analysis	Self wt. multiplier
Dead	Linear static	1 (Dead)
Live	Linear static	0 (Live)
Super Dead	Linear static	0 (Super imposed)
EQ + X	Nonlinear dynamic	0 (Seismic)
EQ + Y	Nonlinear dynamic	0 (Seismic)

5 Analysis of Results

See Figs. 2, 3, 4, 5, and 6.

5.1 Cost Analysis

Hence, the total cost of RCC Column building is Rs. 1.25 crores, while the cost of composite Column construction is Rs. 1.20 crores. Adopting composite columns saves 4% of the cost, while lighter foundation designs save even more money due to reduced dead weight of the building.

See Table 4.

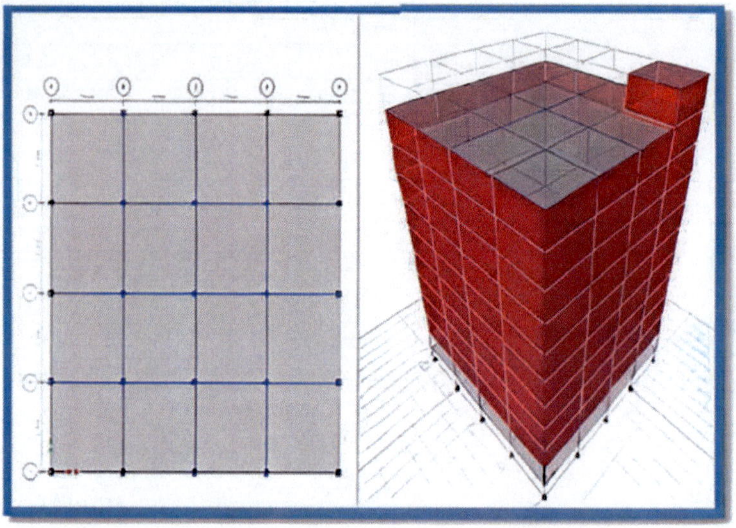

Fig. 2 Structural modelling in ETAB

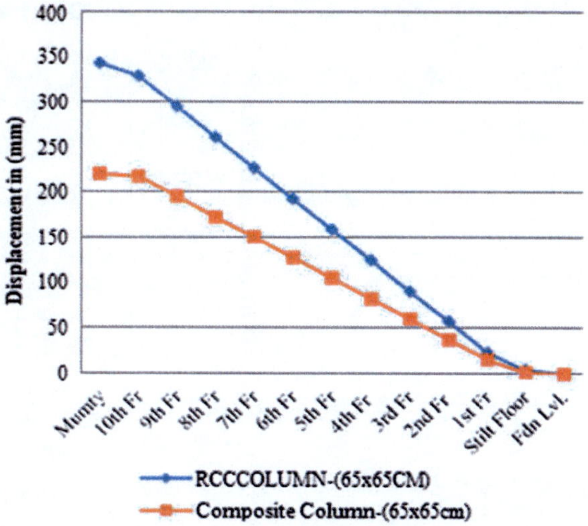

Fig. 3 Maximum story displacement

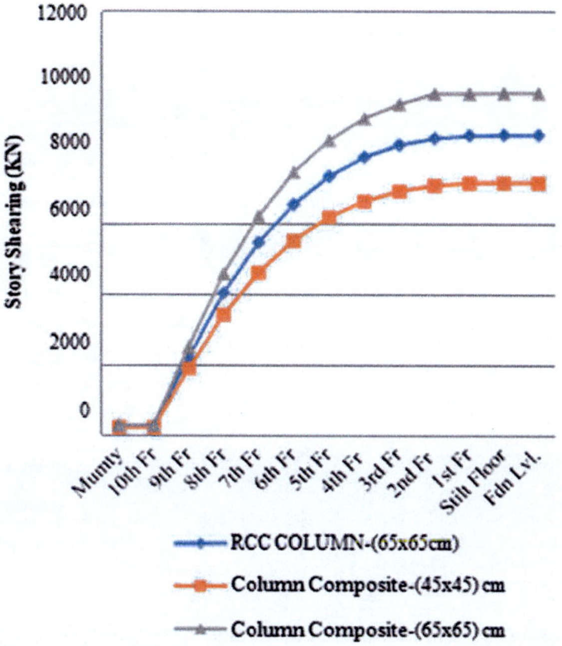

Fig. 4 Story shear

5.2 Minimum Section Size of Columns

The column section size required in RCC for a similar structure with identical loading is 650 × 650 mm, while when the exact same structure is planned utilizing composite

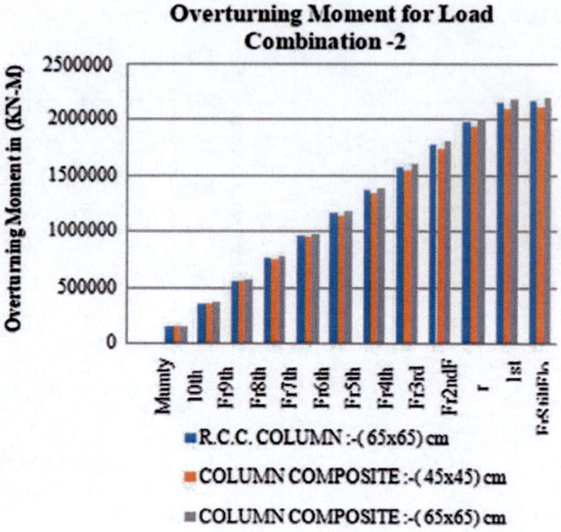

Fig. 5 Overturning moment

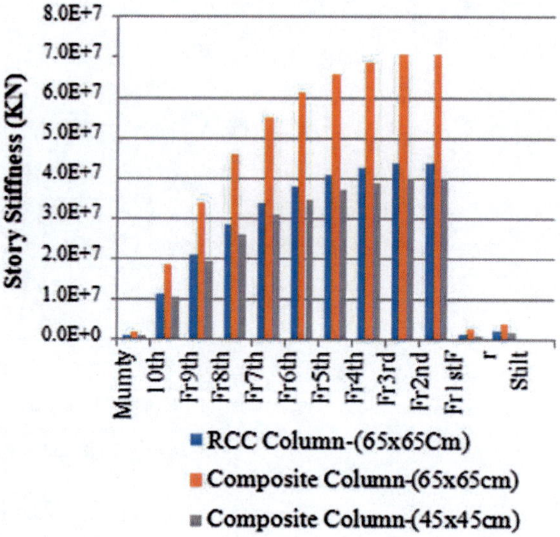

Fig. 6 Story stiffness

columns; the required section size is 450 × 450 mm. The 52% reduction in section size will not only allow for more room and better aesthetics but it will also save a significant amount of money and resources (Fig. 7).

Table 4 Analysis of cost

S. No	Details	Units	Quantity	Rate (Rs.)	Amount (Rs.)
The R.C.C. column					
1	M-25 Grade of R.C.C	Cu.m	433.69	7749.00	3,360,797.04
2	Shuttering P/D	Sq.m	2566.3	467.90	1,200,620.06
3	Steel Cutting and Binding	Kg	128,662	58.00	7,462,376.28
4	12 mm thick Plaster	Sq.m	2566.2	186.90	479,503.80
R.C.C. column total costing (Rs.)					**12,503,297.20**
Columns composite					
1	M-25 Grade of R.C.C	Cu.m	189.79	7749	1,470,790.411
2	Steel Cutting and Binding	Kg	13,468	58	781,163.72
3	Steel structure	Kg	145,044	67.2	9,746,987.04
Composite column total costing (Rs.)					**11,998,941.10**

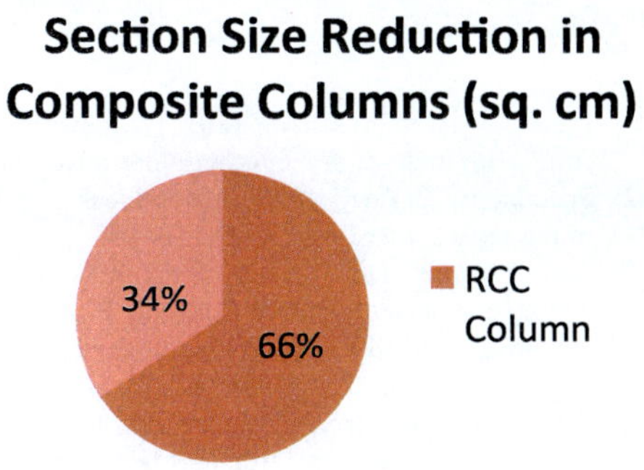

Fig. 7 Composite column section size reduction

6 Conclusion

1. The CFST sections' waste conveying limit has been increased as a result of cement repression. For this model, the segment area needed in RCC is 650 × 650 mm; however, the segment size was reduced to 450 × 450 mm while planning the identical model using composite segments.
2. When compared to composite segments of the same area size, in R.C.C. segments, the maximum story displacement is 49–55% greater. The area size needed for composite segments is smaller as segment sizes are smaller; however, the most severe narrative displacement for composite segments is 6–12% more than for R.C.C. segments.
3. The maximum story shear for an edge comprised of R.C.C. sections is 17–19% more than for a casing made of composite components (65 × 65 cm) (45 × 45 cm). Because of the stress reduction on the structure when utilizing composite sections, story shear in combination segments is minimized.

4. Overturning mins in (45 × 45) cm combination segments are similar to those in (65 × 65)cm

R.C.C. segments.

5. Story stiffness in R.C.C. sections of size 65 × 65 cm is 8–26% more than in composite segments of size 45 × 45 cm.
6. Although R.C.C. columns provide 4–18% better damping, composite sections should be slowly adopted for constructions built to resist seismic and wind stresses.
7. The value of composite segments is 4% more expensive than the value of R.C.C. segment as a result of the section size reduction. The construction dimension and concept for composite sections are likewise light as a result of the structure's reduced dead weight.

References

1. Preetha V, Arun Prasad MC (2017) Comparative study on behaviour of RCC and steel-concrete composite multi-storey building. Intern Res J Eng Techn 4.2
2. Spacone E, El-Tawil S (2004) Nonlinear analysis of steel-concrete composite structures: state of the art. J Struct Eng 130(2):159–168
3. Azna PA, Abraham R (2021) Structural performance of multi-sectional CFST columns with double corrugated plate. In Dasgupta K, Sudheesh TK, Praseeda KI, Unni Kartha G, Kavitha PE, Jawahar Saud S (eds) Proceedings of SECON 2020. SECON 2020. Lecture Notes in Civil Engineering, vol 97. Springer, Cham. https://doi.org/10.1007/978-3-030-55115-5_1
4. Ömer Faruk Kültür, Ahmet Taşkıran, Baris Sayin (2022) Determination of the effectivity of composite members on structural behavior through different combinations, ISSN 2214-5095, https://doi.org/10.1016/j.cscm.2022.e01436
5. Liang, Qing Quan (2018) Analysis and design of steel and composite structures. CRC Press
6. Abbas, Rafaa, Dheeb, Ahmed (2018) Linear and nonlinear static analysis of high-rise buildings under wind load using direct analysis method. Intern J Sci Res (IJSR), 7:848- 854. https://doi.org/10.21275/ART201870
7. Comparative study on the behavior of R.C.C, Steel & Composite Structures (B+G+20Storey) (2015) Sattainathan .A, Nagarajan .N. Intern J Applic Civil Environ Eng 1(3):21–26
8. Cost, Analysis and Design of Steel-Concrete Composite Structure RCC Structure (2014) Anamika, Dr. Savita maru. IOSR J Mech Civil Eng (IOSR-JMCE), e-ISSN:2278–1684, ISSN:2320- 334X, 11(1) Ver.II:54–59
9. Comparative Study of RCC and Composite Multi-storey Buildings (2013) Shashikala. Koppad, Dr. S . V.Itti. Intern J Eng Innov Techn (IJEIT) 3(5)
10. AISC 360–05 (2005) Specification of structural steel building, An American national Standard, American Institute of Steel Construction, Inc.
11. Otani S (1980) Nonlinear dynamic analysis of reinforced concrete building structures. Can J Civ Eng 7(2):333–344
12. Aydınoğlu MN (2004) An improved pushover procedure for engineering practice: incremental response spectrum analysis (IRSA). In International Workshop on Performance-based Seismic Design: Concepts and Implementation (pp 345–356)
13. De Stefano M, Pintucchi B (2008) A review of research on seismic behaviour of irregular building structures since 2002. Bull Earthq Eng 6(2):285–308

14. Herrera RG, Soberon CG (2008) Influence of plan irregularity of buildings. In The 14th World Conference on Earthquake Engineering
15. Poonam AK, Gupta AK (2012) Study of response of structurally irregular building frames to seismic excitations. Intern J Civil Struct Environ Infrastruc Eng Res Devel 2(2):25–31
16. Pirizadeh M, Shakib H (2013) Probabilistic seismic performance evaluation of non-geometric vertically irregular steel buildings. J Constr Steel Res 82:88–98
17. Guleria A (2014) Structural analysis of a multi-storeyed building using ETABS for different plan configurations. Intern J Eng Res (IJERTI) ISSN, pp 2278–0181
18. Sharma M, Maru DS (2014) Dynamic analysis of multistoried regular building. IOSR J Mech Civil Eng (IOSR-JMCE) e-ISSN, pp 2278–1684
19. IS2000:456 (2000) Code of practice for plain and reinforced concrete code of practice. Bureau of Indian Standards, NewDelhi
20. IS: 11384 (1985) Code of practice for composite construction in structural steel and concrete. Bureau Of Indian Standards, New Delhi
21. IS:1893 (2002) Criteria for earthquake resistant design of structures-general provisions for buildings. Part1, Bureau of Indian Standards, New Delhi
22. IS: 875 (2002) Code of practice for design load (other than earthquake) for buildings and structures. Bureau of Indian Standards, New Delhi
23. IS: 800 (2007) Code of practice for general construction in steel, Bureau of Indian Standards, New Delhi
24. Naeim F (1989) The seismic design handbook. Springer Science & Business Media
25. ETABS (2016) Software verification examples. Computer and Structure, Inc., 2017
26. Indian Standard, IS (2002) 1893: 2002 (Part-III). Earthquake resistant design of structure
27. Indian Standard, IS (2000) 456: 2000. Plain and reinforced concrete code of practice
28. Indian Standard, IS (1987) 875: 1987 (Part-I). Dead loads—unit weights of building materials and stored materials
29. Indian Standard, IS (1987) 875: 1987 (Part-II). Imposed load
30. Qureshi UM, Hussain A (2019) A review paper on composite structure used in column for various analysis. Int J Res Eng Appl Manag 5(4)
31. Nethravathi SM, Thouseef T (2017) Performance analysis of regular and irregular structure under seismic effect for RCC and steel composite column using response spectrum. Int J Eng Res Sci 3(8):2–6
32. Renavikar AV, Suryawanshi Y (2016) Comparative study on analysis and cost of RCC and steel composite structure. Int J Sci Res (IJSR) 5(7)

Maximal Support Reaction Analysis of the Impact of Infill Masonry and Floating Columns on a Seismic Zone-II High-Rise Building

Dhanesh Khalotia, Pranita Murkute, and Bhanu Pratap Singh Sikarwar

Abstract Our country's population is quickly expanding, so we need to build towering structures that can withstand high drift and displacement as a result of lateral forces to accommodate this massive increase in people. The failure of tall structures can also be caused by earthquakes or other seismic events. This issue should be solved before towering structures are built. Floating columns, which are vertical elements given over beams and do not rely on the ground, can be used to make these tall structures safe. This strategy is used in earthquake-prone areas to keep loads coming down on the ground consistent so that the path of transfer does not alter. 13 different loading combinations are used in this study for each scenario, they are all based on Indian Standards, with column and beam measurements of 450 mm × 450 mm and 400 mm × 300 mm, respectively. We discovered that infill walls may distribute stresses evenly; the position of the floating column affects the structure's forces and bending moment in any instance when the number of story is varied. This causes the forces and bending moment to drop, creating an affordable section.

Keywords Seismic · Lateral pressures · Tall structures · Floating columns · STAAD Pro

1 Introduction

An earthquake is a sudden movement or shaking of the Earth's crust that usually starts at or below the surface. The term "Normal" is essential since it disallows shock waves produced by atomic bombs, artificial explosions, etc. About 90% of all earthquakes have structural causes, which are mostly associated with volcanism's breakdown of impossibly large objects or human-made influences [1–3].

D. Khalotia (✉) · P. Murkute · B. P. S. Sikarwar
IES University, Bhopal 462044, India
e-mail: Dhaneshkhalotia@gmail.com

K. K. Pathak et al. (eds.), *Latest Developments in Civil Engineering*, Lecture Notes in Civil Engineering 352, https://doi.org/10.1007/978-981-99-2676-3_49

Modern multi-story buildings are built with a delicate story on the bottom level for purposes such as residential, commercial, industrial, etc. The ground level is maintained clear of all developments, with the exception of those that transfer the structure's burden to the earth's surface for commercial buildings, and is used for halting areas most frequently. Guard room campaigns typically take place in the ground floor [4, 5].

1.1 Movable Column

A part is a section that protrudes from the surface and descends to pile on the floor. The phrase "drifting section" also refers to a perpendicular portion that, due to the structural layout or site conditions, closes at its ground floor (last level) and rests on a support that is an even component [4, 5].

The term "coasting section" refers to a segment that is working with gliding and has a space outline with a few segments that are not contacting the surface segments and bars supporting with greater limits would be thoroughly designed and constructed Skimming sections, also known as floating segments, are those that float (hang) or move in storylines over others to the point that they take up more space [6] (Fig. 1).

The majority of the buildings in Gandhi Dham and Ahmadabad are safeguarding the most possible area on a plot under the established standing guidelines [7]. Since galleries are not checked in the floor space inventory (FSI), the building has exhibitions that are undulate in the higher stories past the fictitious territory at the base story [8, 9]. Overhangs up to 1.2–1.5 m in course of motion are frequently provided on each side of the structure. The border parts from the ground level are stopped in the higher story, and coasting portions are provided along the building's overhanging edge [7, 10, 11].

Fig. 1 Structure having movable column

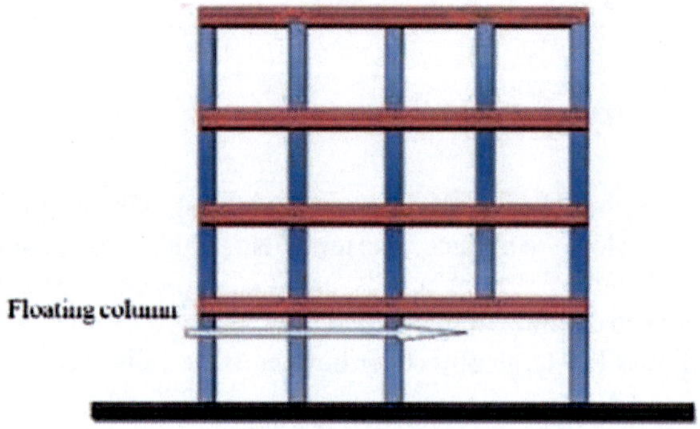

2 Literature Review

In Arlekar et al. [12], such features, are highly bad in buildings constructed in seismically active areas; this has been confirmed in several contacts with substantial shaking during previous earthquakes. They discussed the need for explicitly recognizing the proximity of the open first story during the building's inspection. Additionally, they recommended reducing the stiffness of the tale above and the open first story to decrease the oddity the open first story displayed.

Balsamoa, Colombo et al. [13] conducted pseudo dynamic tests on a CFRP overlay-repaired RC structure. The potential for seismic repair of reinforced cement (RC) structures opened by the use of Carbon Fiber Reinforced Polymer (CFRP) composites was assessed on a full-scale double framework put through pseudo dynamic testing in the ELSA research facility. The goal of the CFRP repair was to restore the fundamental characteristics that the edge had prior to the seismic activity by adding greater distortion limit to the two sections and joints. A variety of fiber surfaces were used to describe the repair, all of which relied on the central system that governs each segment. The article presents the exploratory testing results as well as the guiding criteria in the CFRP repair's plan.

In Natarajan [14], structures' seismic response was taken into account, and an assessment was carried out using the reaction range technique and ETABS' programming with finite components. For the investigation, consideration is given to a private functioning of the G+ 15 erratic structures. It is crucial to accurately measure the seismic response of the dividers since the qualities of these seismic shear dividers control how the structures react. The findings show that shear dividers in an unexpected structure are more resistant to parallel stresses. When partitions are placed in advantageous locations within a structure, they may be quite efficient at resisting sidelong stresses caused by earthquakes or wind.

3 Experimental Study

3.1 Objectives

(1) To ascertain how seismic forces may the effect of floating columns on a structure [15].
(2) In order to compare axial force, SF, BM, and displacement of story in the X and Z axes [15].
(3) Assessment of the building using vertical and horizontal placement of floating columns [16, 17].
(4) STAAD Pro v8i software is used to do the analysis [15].

3.2 Structure Model

- The following is the technique used to achieve the above-mentioned targets:
- For the context analysis, choose two unique edge structure models.
- These models selected gliding segments for the structure, both with and without an infill structure.
- Shear constraint, bowing minute, removal, hub power, and narrative relocation are all compared.
- Conclusion and Discussion (Fig. 2).

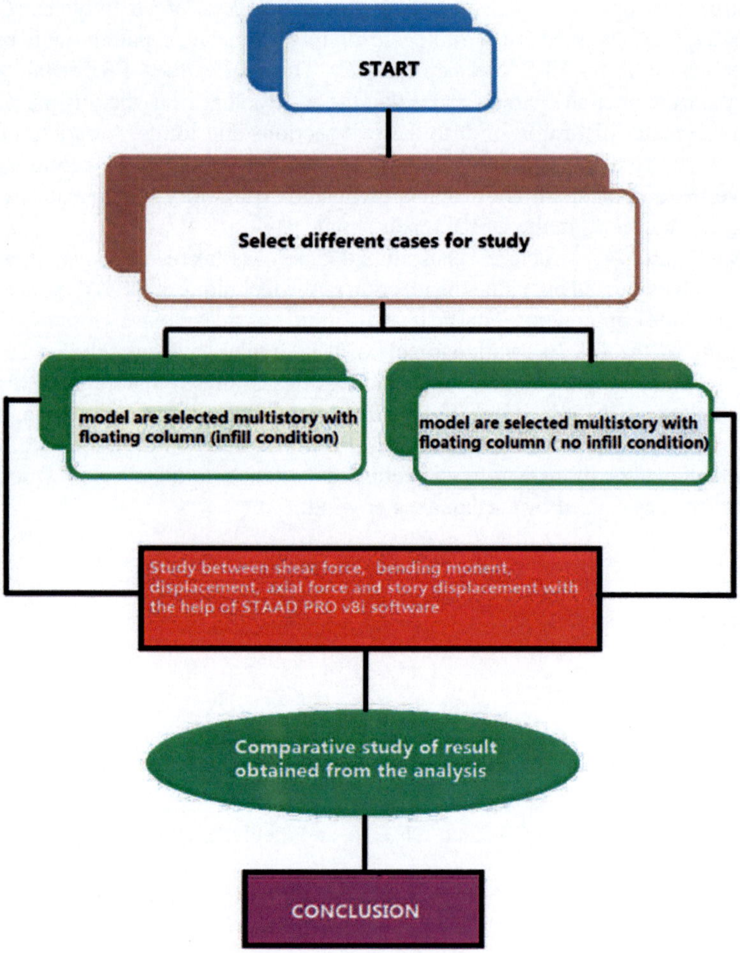

Fig. 2 Methodology steps

3.3 Structure Description

For the purposes of analysis, two types of models are taken into account: (Geometry: 16 m × 16 m with 4 m-long bays in the x and z directions).

- Model A: Special Moment Resisting Frame with five stories (G+ 4) and no infill condition.
- Model B: Special moment-resisting frame with an infill condition that is five stories tall (G+ 4) (Fig. 3).

Model A

Nine scenarios have been taken into account for model A.

- In Case i, a floating column exists because the outer column at the bottom floor is discontinuous (model pht1).
- In Case ii, the ground-floor position-2's horizontal orientation has been disrupted, resulting in a floating column (model pht2).
- In Case iii, the ground floor's center column has disrupted, resulting in a floating column (model pht3).
- In Case iv, a floating column exists because there has been a disruption to the ground floor's perpendicular position (pht4 model).

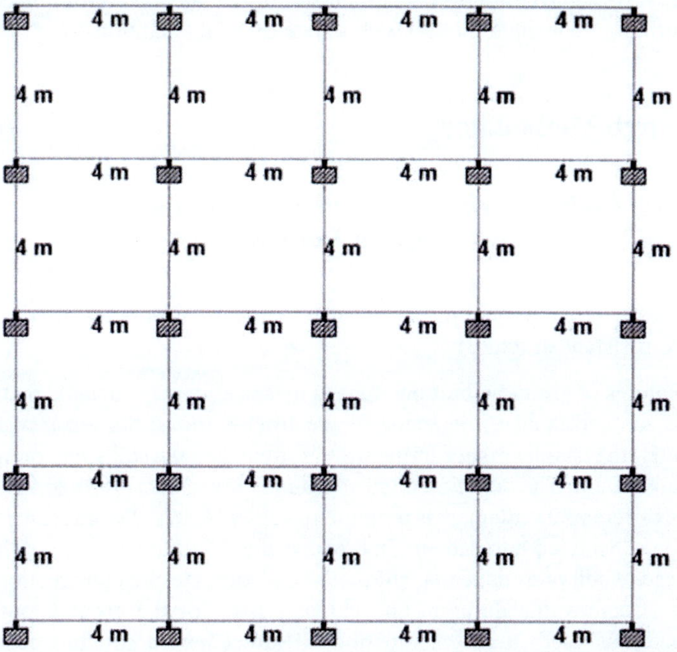

Fig. 3 TopView

- In Case v, the ground-floor side column has disrupted owing to a floating column (model pht5).
- In Case vi, a floating column exists because the outside column at the bottom floor has disrupted (PVT1 model).
- In Case vii, a floating column exists at the first floor because of the break in the side column (PVT2 model).
- In Case viii, a Movable column exists at the second floor as an outcome of the side column discontinuity (PVT3 model).
- In Case ix, a Movable column exists because the lateral column at the 3rd floor has disintegrated (PVT4 model).

Model B

Nine scenarios have been taken into account for model B.

- Case 1: The exterior column at the ground level (model pht1) has been discontinued, leaving a floating column with the remaining piece in an infill state.
- Case 2: The Longitudinal position-2 discontinuity at the ground level (pht2 model) has resulted in a floating column with an infill condition.
- Case 3: The ground floor's center column (Model pht3) has a discontinuity, resulting in a floating column with an infill condition.
- Case 4: A floating column exists as a result of a horizontal alignment discontinuity at the ground level (pht4 model) with a state of infill.
- Case 5: A movable column exists as a result of the lateral columns discontinuity at the ground floor (pht5 model) with a state of infill condition.

4 Research Methodology

4.1 Using Codal Provisions to Guide the Design of an Earthquake-Resistant Building

The following are the guiding ideas and design philosophies used to create earthquake-resistant structures:

1. The features of ground vibrations caused by seismic forces depend on the intensity of the earthquake, the scope of the force's focus, the distance from the epicenter, the characteristics of the wave's route, and the soil strata on which the building is built. The direction of most ground movements is horizontal [18, 19].
2. Where differential settlement is prohibited, vertical factors that cause earthquakes must be considered into account in broad spans.
3. The kind of soil present near a building's base, the size and construction method of the structure, the duration and characteristics of the ground motion, and the materials used in its construction all affect how a structure responds to earthquakes.

4. The design strategy aspires for structures to resist big earthquakes (MCE) without collapsing, ensuring that they have a minimum strength to sustain frequent small earthquakes (DBE) without damage, moderate earthquakes without substantial damage, while some non-structural damage may occur. Actual forces operating on the buildings are significantly larger than the forces of design given here. but the difference between actual and design lateral forces is believed to be explained by ductility, which is caused by inelastic material behavior and design, as well as overstrength, which is caused by extra strength that has been reserved in the constructions over the design strength [8, 9, 20].
5. According to IS: 456 and IS: 1343, members with reinforcement and pre-stress are appropriately built to guarantee that there is no early failure induced by shear or bond.
6. High ductility is achieved in steel constructions by proportioning the members and their connections.
7. The influence of a structure's motion inside the foundational medium is illustrated by the influence of soil structure. The seismic evaluation of buildings supported by rocks often does not take the structural interaction into account.
8. The structures' two orthogonal horizontal directions must be taken into account while designing lateral forces. Only two orthogonal directions must be taken into account when designing lateral force for constructions with parts resisting lateral force. In accordance with paragraph 2.3.2 of IS 1893 (part 1): 2002, structures with lateral resistive components in two directions other than orthogonal must be assessed. When considering both horizontal and vertical forces, load combinations must follow clause 2.3.3 IS1893 (part 1): 2002.

4.2 Analyzing Linearly

- Two main types of loads are often applied to structures built on the earth:
- Static
- Dynamic
  - The term "static load" refers to loads that remain constant over time, whereas "dynamic load" refers to loads that change over time.
  - The assumption that all applied loads are static is often used in the design of the majority of civil engineering structures. Because the structure is rarely subjected to any dynamic load, the influence of dynamic load is frequently ignored. Additionally, taking into account dynamic loads during analysis makes the solution more complicated and time-consuming.

4.3 Basic Provisions of the Code

For the following buildings, dynamic analysis should be carried out to determine the design seismic force and its distribution to various lateral load-resisting elements and levels along the height of the building [21, 22]:

- Regular buildings—In zones IV and V, individuals who are longer than 40 min and those who are taller than 90 m. in zones II, III.
- Irregular buildings—Zones IV and V are all structures with framed heights larger than 12 min; zone II and III are hose with heights greater than 40 min.

4.4 Techniques for Response Spectrum Analysis in STAAD Pro

The Indian IS: 1893 (Part 1)–2002 Eqs. 7.8.4.5c and 7.8.4.5d are used by STAAD to calculate the intended lateral shear force for every floor and mode.

$$W_b = Ah \times \text{ Weight of building}$$

5 Result Analysis

For various circumstances, the behavior of a building's frame with and without a floating column is investigated. Varied models for the current study, various methodologies and strategies, code suggestions, etc., have been discussed in earlier chapters. The STAAD Pro V8i Software's Linear, or Response Spectrum Method, was used to examine the many models that were taken into consideration for this study.

5.1 Load Up the Roof

5.1.1 Dead Load

a. Slab-Weight of the slab = $(0.12 \times 25 \times 1 \times 1) = 3$ KN/Sq m.
b. Finishing load equals to $1 \times 1 \times 0.06 \times 24$ equals to 1.44 KN/Sq m.
   Total equals to $3 + 1.44$ KN/Sq m equals to 4.44 KN/Sq m.
c. Infill Brick Wall
   Weight of wall = $20 \times 0.23 \times (1.0) \times 1$ (@20KN/cum) = 4.6 KN/m.
   Imposed Load/Live Load-Live load at roof level = 1.5 KN/Sq m.

5.2 *Loading for STAAD Pro*

a. Member Load equals to 4.6 KN/m.
b. Floor Load (Live) equals to 1.5 KN/Sq m; For 2 dimension research Member load, is the same as floor load for (ii) = (1.5 × 5 × 5)/4 equals to 1.87 KN/m.
c. Floor Load (Slab) equals to 4.44 KN/Sq m; For 2 dimension analysis, Floor load is considered as Member load, then Member load for (ii) equals to (4.44 × 5 × 5)/4 equals to 5.55 KN/m.
d. All Member Self Weight equals to −1.
   Seismic Definition:
e. Floor load equals to 100% of Floor load (Slab) + 50% of Floor load (Live) equals to 4.44 + (1.5/2) equals to 5.19 KN/Sq m. In terms of member load = (5.19 × 5 × 5)/4 KN equals to 6.48 KN/m.

5.3 *Loads on the Other Floors*

5.3.1 Dead Load

a. The slab's mass equals to (0.12 × 25 × 1 × 1) 3 KN/m^2.
b. Finishing Load equals to 1 × 1 × 0.06 × 24 = 1.44 KN/Sq m^2 Total = 3 + 1.44 equals to 4.44 kN/m^2.
c. Infill Brick Wall
   Weight of wall equals to 20 × 0.23 × (3.5 − 0.45) × 1 (@20KN/cum) = 14.03 KN/m.
   Imposed Load/Live Load-Live load at floor level = 3.5 KN/m^2.

5.4 *Loading for STAAD Pro*

- Member Load equals to 14.03 KN/m.
- Floor Load (Live) equals to 3.5 KN/Sq m; For 2 dimension analysis, Floor load is considered as Member load, then Member load for (ii) = (3.5 × 5x5)/4 = 4.37 KN/m.
- Floor Load (Slab) equals to 4.44 KN/sq m; For 2 Dimension analysis, Member load is the same as Floor load, then Member load for (ii) = (4.44 × 5x5)/4 = 5.55 KN/m.
- All Member Own Weight equals to minus 1.
- Loads for Seismic Definition: Floor load equals to 100% of Floor load (Slab) + 50% of Floor load (Live) = 4.44 + (3.5/2) = 6.19 KN/Sq m.

5.5 *Outcome Comparing Infill and Not Infill Structure*

See Tables 1, 2, 3, 4, 5, 6 and Figs. 4, 5, 6, 7, 8, 9.

Table 1 Maximal bending moments (kN-meter)

Serial number	Positioning	Not in infill	Infill
One	PHT 1	5.14193×10^2	7.5411×10
Two	PHT 2	4.54327×10^2	7.4545×10
Three	PHT 3	4.54009×10^2	7.4399×10
Four	PHT 4	4.54327×10^2	7.4545×10
Five	PHT 5	5.34267×10^2	7.4411×10
Six	PVT 1	4.84482×10^2	7.407×10
Seven	PVT 2	4.94303×10^2	7.3715×10
Eight	PVT 3	4.54281×10^2	7.3112×10

Table 2 Maximal shear force (kilo-N)

Maximal shear force			
Serial number	Position	Not in infill	Infill
One	PHT 1	3.20243×10^2	9.432×10
Two	PHT 2	2.8242×10^2	9.3645×10
Three	PHT 3	2.8025×10^2	9.3534×10
Four	PHT 4	2.8242×10^2	9.4081×10
Five	PHT 5	2.08493×10^2	9.2779×10
Six	PVT 1	3.02691×10^2	9.3931×10
Seven	PVT 2	3.09841×10^2	9.3563×10
Eight	PVT 3	2.75672×10^2	9.3505×10

Table 3 Maximum axial force (kN-m)

Maximal axial force			
Serial number	Position	Not in infill	Infill
One	PHT 1	32.89888×10^2	47.3555×10^2
Two	PHT 2	28.7581×10^2	38.93835×10^2
Three	PHT 3	28.81296×10^2	33.53677×10^2
Four	PHT 4	28.75813×10^2	38.93835×10^2
Five	PHT 5	34.8384×10^2	48.33904×10^2
Six	PVT 1	7.37118×10^2	34.03469×10^2
Seven	PVT 2	26.87605×10^2	29.43101×10^2
Eight	PVT 3	23.87488×10^2	18.93649×10^2

Table 4 Maximal dispersion in X-direction

Maximal displacement (X-trans)			
Serial number	Position	Not in infill	Infill
One	PHT 1	3.83748×10^2	0.9228×10
Two	PHT 2	2.0221×10^2	0.7800×10
Three	PHT 3	1.97922×10^2	0.5962×10
Four	PHT 4	2.0221×10^2	0.7800×10
Five	PHT 5	4.2347×10^2	1.2313×10
Six	PVT 1	3.66656×10^2	0.7759×10
Seven	PVT 2	3.0856×10^2	0.6980×10
Eight	PVT 3	2.53465×10^2	0.6134×10

Table 5 Maximal dispersion in Z direction

Maximal displacement (Z-trans)			
Serial number	Position	Not in infill	Infill
One	PHT 1	$1.87.802 \times 10^2$	4.409
Two	PHT 2	$2.11.277 \times 10^2$	6.990
Three	PHT 3	$2.07.555 \times 10^2$	6.542
Four	PHT 4	$2.11.277 \times 10^2$	6.990
Five	PHT 5	$2.30.188 \times 10^2$	7.896
Six	PVT 1	$2.31.521 \times 10^2$	5.936
Seven	PVT 2	$2.26.177 \times 10^2$	5.767
Eight	PVT 3	$2.16.129 \times 10^2$	5.659

Table 6 Maximal support reaction

Supported reaction			
Serial number	Position	Not in infill	Infill
One	PHT 1	33.00888×10^2	47.1655×10^2
Two	PHT 2	28.0681×10^2	38.95835×10^2
Three	PHT 3	28.02296×10^2	33.57677×10^2
Four	PHT 4	28.06813×10^2	38.95835×10^2
Five	PHT 5	34.4484×10^2	48.40904×10^2
Six	PVT 1	29.96814×10^2	34.09469×10^2
Seven	PVT 2	26.89605×10^2	29.49101×10^2
Eight	PVT 3	23.86488×10^2	27.79619×10^2

5.6 Comparison of the Infill and Without Infill Conditions for Vertical Position

See Tables 7, 8, 9, 10, 11, 12 and Figs. 10, 11, 12, 13, 14, 15.

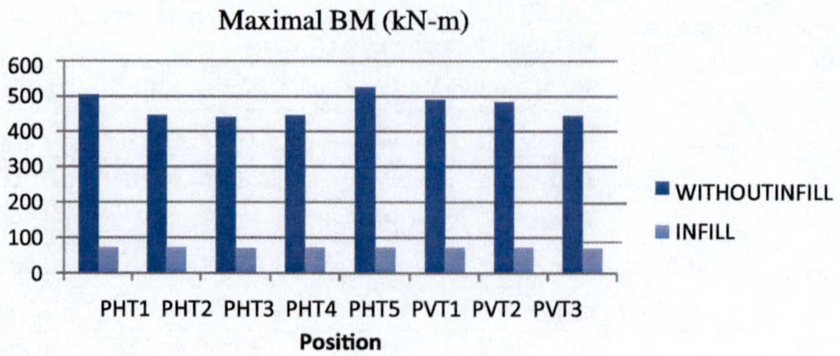

Fig. 4 Maximal bending moment

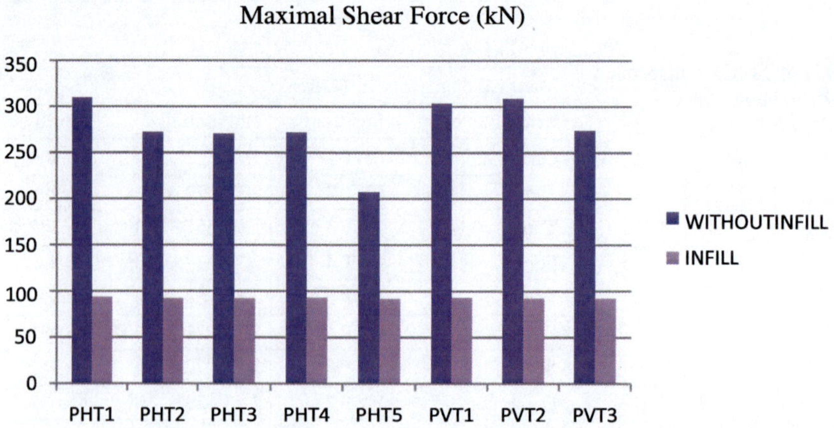

Fig. 5 Chart displaying the difference in CBR Value for soil mixtures with various percentages of waste broken glass powder and plastic granules. Differently with plastic granules and also in powder

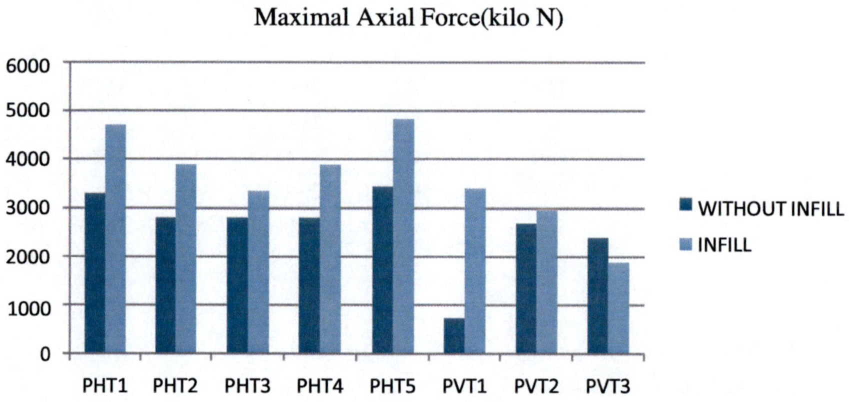

Fig. 6 Maximal axial force

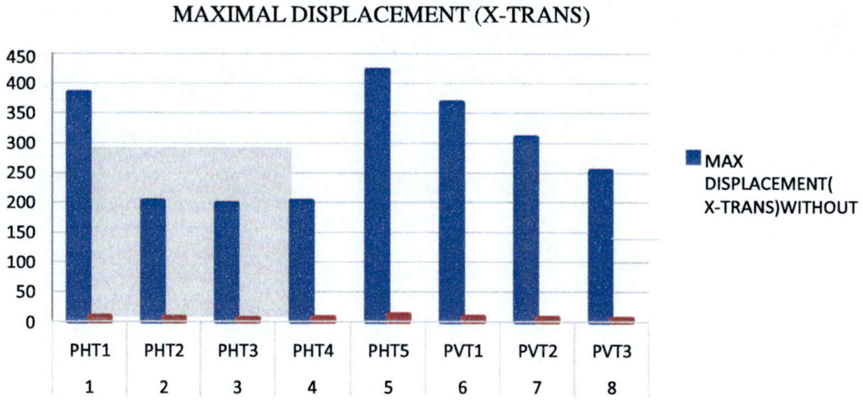

Fig. 7 Maximal dispersion in X direction

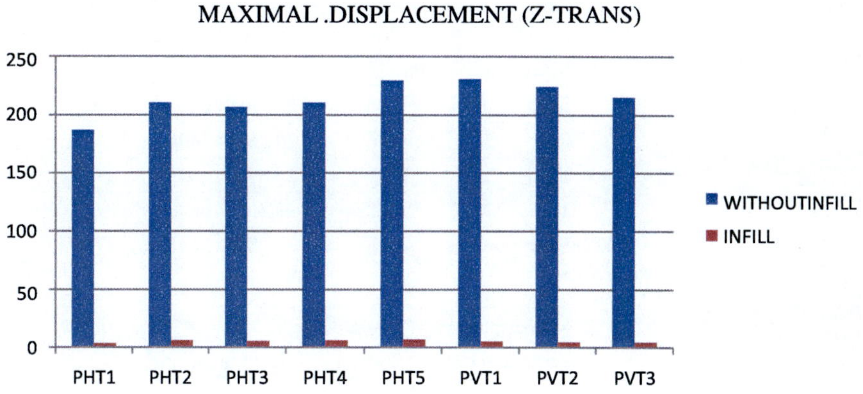

Fig. 8 Maximal displacement in Z direction

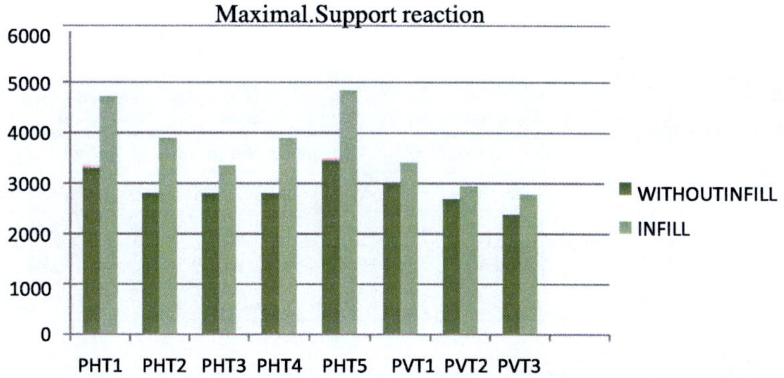

Fig. 9 Maximal support reaction

Table 7 Maximal bending moments (kilo N-meter)

Maximal BM			
Serial number	Position	Not in infill	Infill
One	PVT 1	4.93482×10^2	0.7407×10^2
Two	PVT 2	4.86303×10^2	0.72715×10^2
Three	PVT 3	4.47281×10^2	0.72112×10^2
Four	PVT 4	5.06193×10^2	0.73411×10^2

Table 8 Maximal shear force (kilo Newton)

Maximal shear force			
Serial number	Position	Not in infill	Infill
One	PVT 1	3.04691×10^2	93.931
Two	PVT 2	3.09841×10^2	93.563
Three	PVT 3	2.75672×10^2	93.505
Four	PVT 4	3.11243×10^2	94.32

Table 9 Maximum axial force (kilo Newton)

Maximum axial force			
Serial number	Position	Not in infill	Infill
One	PVT 1	29.96814×10^2	3409.469
Two	PVT 2	26.89605×10^2	2949.101
Three	PVT 3	23.86488×10^2	2779.619
Four	PVT 4	33.00888×10^2	4716.55

Table 10 Maximal displacement in X trans

Max displacement (X-trans)			
Serial number	Position	Not in infill	Infill
One	PVT 1	3.66656×10^2	8.759
Two	PVT 2	3.0856×10^2	6.98
Three	PVT 3	2.53465×10^2	6.134
Four	PVT 4	3.83748×10^2	9.228

Table 11 Maximal displacement in Z trans

Maximum displacement (Z-trans)			
Serial number	Position	Not in infill	Infill
One	PVT 1	2.31521×10^2	0.5936×10
Two	PVT 2	2.25177×10^2	0.5767×10
Three	PVT 3	2.16129×10^2	0.5659×10
Four	PVT 4	1.87802×10^2	0.4409×10

Table 12 Maximal support reaction

Supported reaction			
Serial number	Position	Not in infill	Infill
One	PVT 1	29.96814×10^2	34.09469×10^2
Two	PVT 2	26.89605×10^2	29.49101×10^2
Three	PVT 3	23.86488×10^2	27.79619×10^2
Four	PVT 4	33.00888×10^2	47.1655×10^2

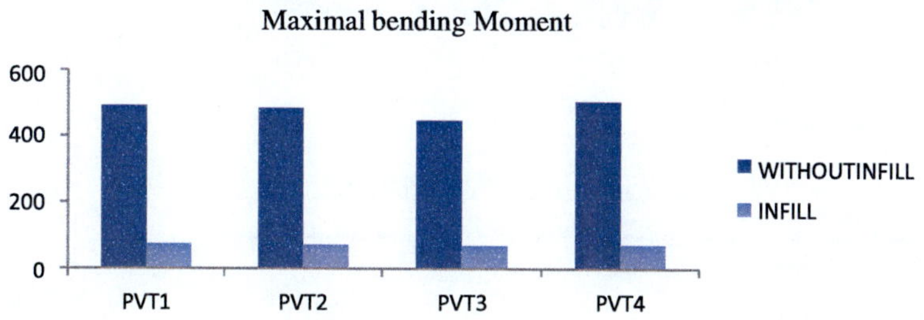

Fig. 10 Maximal bending Moment

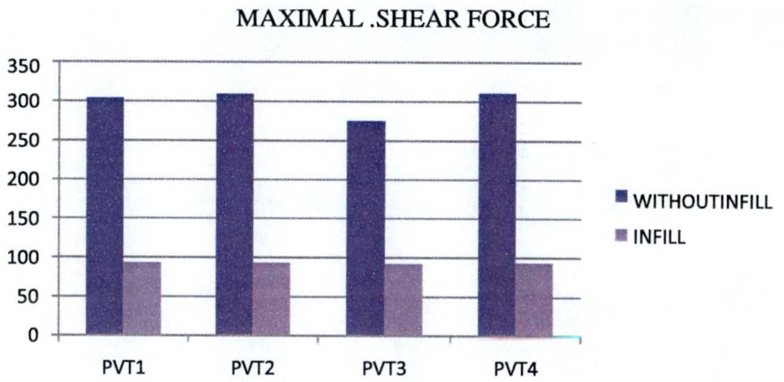

Fig. 11 Maximal shear force

5.7 Horizontal Position Infill and Without Infill Condition Comparison

See Table 13, 14, 15, 16, 17, 18 and Figs. 16, 17, 18, 19, 20, 21.

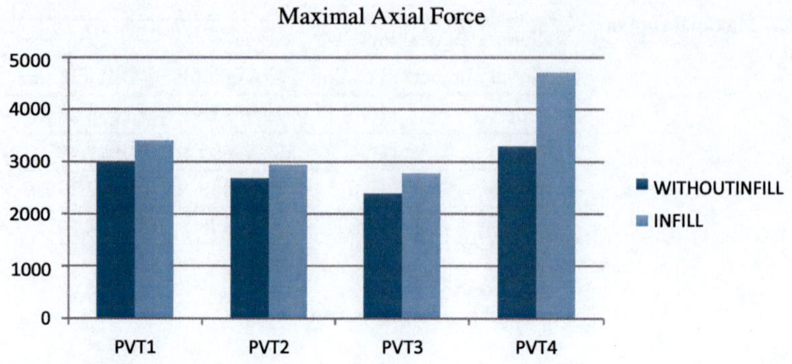

Fig. 12 Maximal axial force

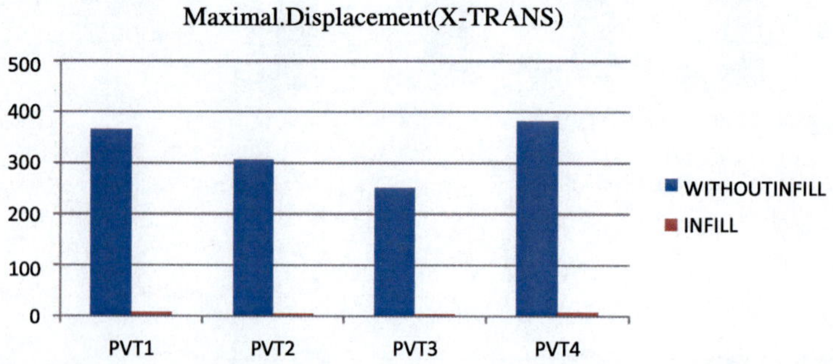

Fig. 13 Maximal displacement in X trans

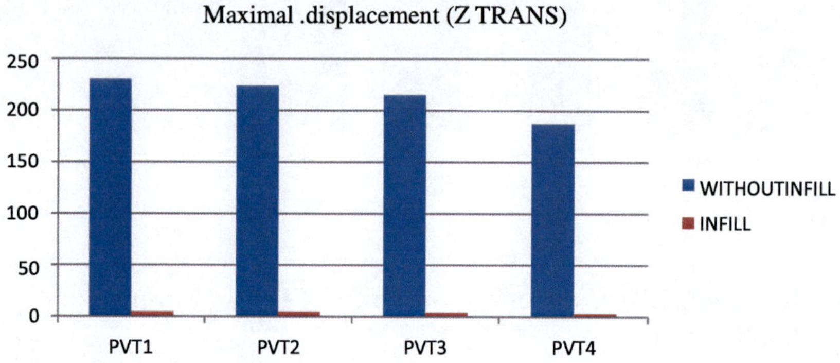

Fig. 14 Maximal displacement in Z direction

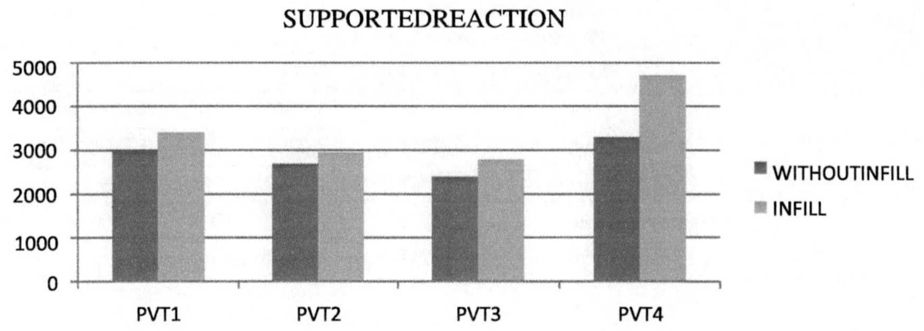

Fig. 15 Support reaction

Table 13 Maximal bending moments (kilo Newton meter)

Maximal bending moment			
Serial number	Position	Not in infill	Infill
One	PHT1	5.06193×10^2	7.3411×10
Two	PHT2	4.46327×10^2	7.3545×10
Three	PHT3	4.42009×10^2	7.2399×10
Four	PHT4	4.46327×10^2	7.3545×10
Five	PHT5	5.27267×10^2	7.3411×10

Table 14 Maximal shear force (kilo Newton)

Max. shear force			
Serial number	Position	Not in infill	Infill
One	PHT1	3.11243×10^2	9.432×10
Two	PHT2	2.7342×10^2	9.3645×10
Three	PHT3	2.7125×10^2	9.3534×10
Four	PHT4	2.7342×10^2	9.4081×10
Five	PHT5	2.08493×10^2	9.2779×10

Table 15 Maximal axial force (kilo-N)

Axial force			
Serial number	Position	Not in infill	Infill
One	PHT 1	33.00888×10^2	47.1655×10^2
Two	PHT 2	28.0681×10^2	38.95835×10^2
Three	PHT 3	28.02296×10^2	33.57677×10^2
Four	PHT 4	28.06813×10^2	38.95835×10^2
Five	PHT 5	34.4484×10^2	48.40904×10^2

Table 16 Maximal displacement in X-trans

Maximal displacement (X-trans)			
Serial number	Position	Not in infill	Infill
One	PHT 1	3.83748×10^2	0.9228×10
Two	PHT 2	2.0221×10^2	0.7800×10
Three	PHT 3	1.97922×10^2	0.5962×10
Four	PHT 4	2.0221×10^2	0.7800×10
Five	PHT 5	4.2247×10^2	1.2313×10

Table 17 Maximal displacement in Z-trans

Maximal displacement (Z-trans)			
Serial number	Position	Not in infill	Infill
One	PHT 1	1.87802×10^2	0.4409×10
Two	PHT 2	2.11277×10^2	0.699×10
Three	PHT 3	2.07555×10^2	0.6542×10
Four	PHT 4	2.11277×10^2	0.699×10
Five	PHT 5	2.30188×10^2	0.7896×10

Table 18 Maximal support reaction

Supported reaction			
Serial number	Position	Not in infill	Infill
One	PHT 1	33.00888×10^2	47.1655×10^2
Two	PHT 2	28.0681×10^2	38.95835×10^2
Three	PHT 3	28.02296×10^2	33.57677×10^2
Four	PHT 4	28.06813×10^2	38.95835×10^2
Five	PHT 5	34.4484×10^2	48.40904×10^2

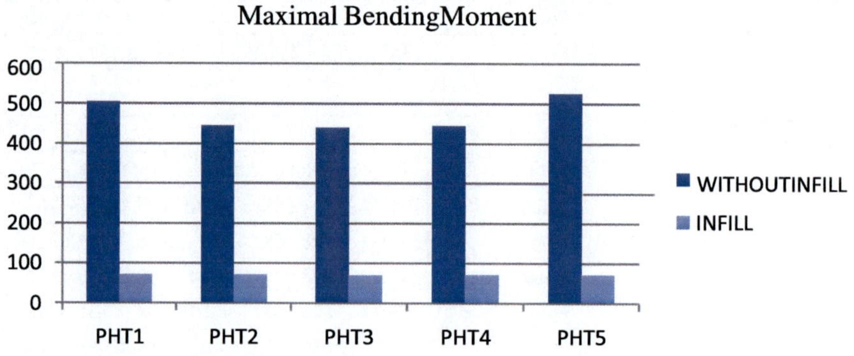

Fig. 16 Maximal bending moment

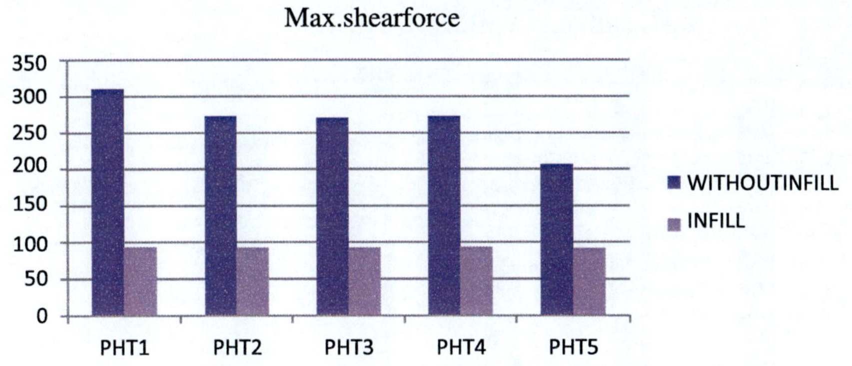

Fig. 17 Maximal shear force

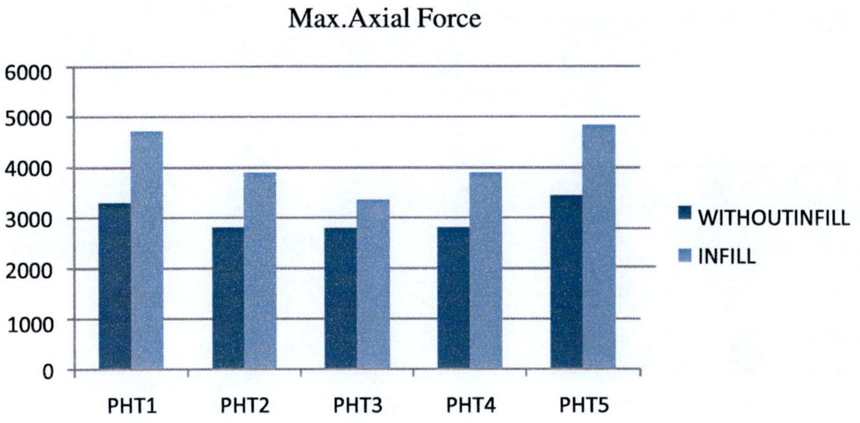

Fig. 18 Maximal axial force

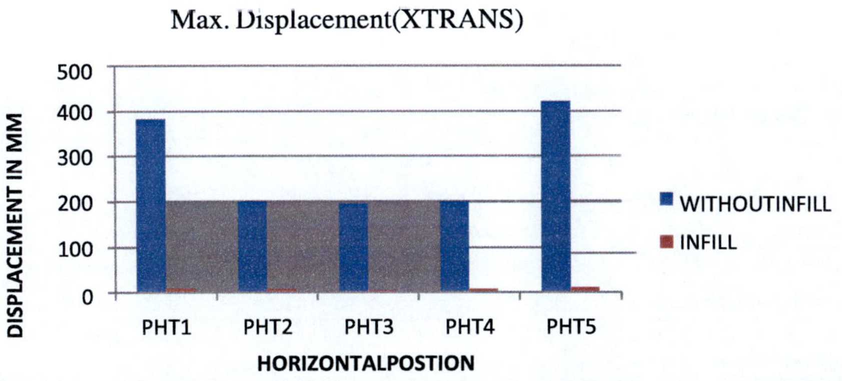

Fig. 19 Maximum displacement (X Trans)

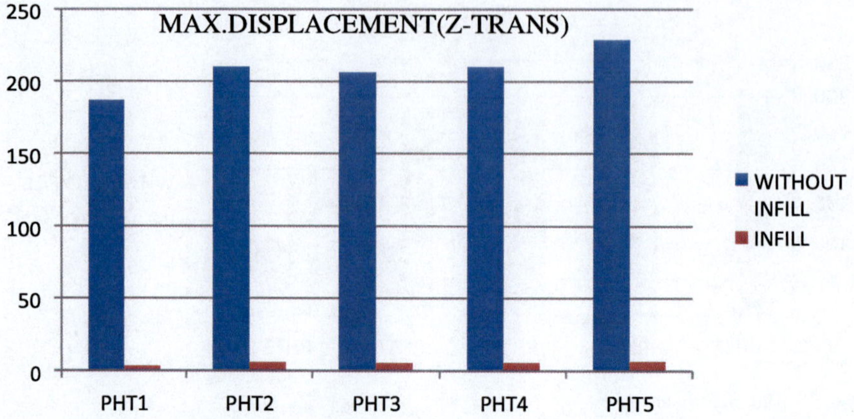

Fig. 20 Maximal displacement in Z trans

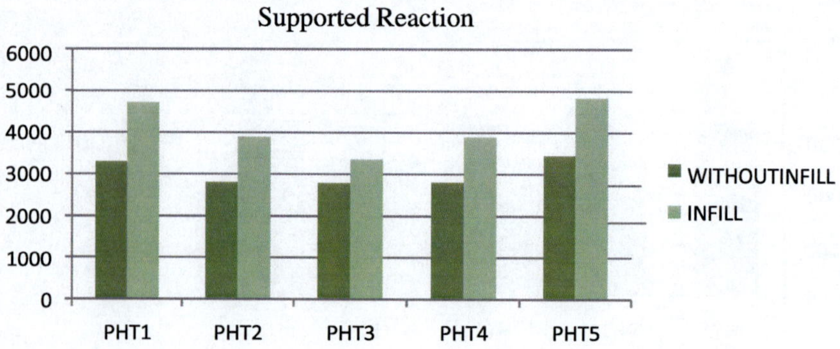

Fig. 21 Support reaction

5.8 *Cost Comparison*

See Tables 19 and 20.

Table 19 Material cost analysis according to S.O.R

Structure	Concrete (cubic meter)	S.O.R. rate	Amount
Bare frame	190	4200	Rs. 7,98,000/-
Floating column frame	165	4200	Rs. 6,93,000/-
Infill structure with floating column	177	4200	Rs. 7,43,400/-

Table 20 Maximal bending moments (kilo Newton-meter)

Structure	Rebar (tonne)	S.O.R. rate	Amount
Bare frame	20	39,000	Rs. 7,80,000/-
Floating column frame	15	39,000	Rs. 5,85,000/-
Infill structure with floating column	16.8	39,000	Rs. 6,55,200/-

6 Conclusions

6.1 General

From the present study, following conclusions have been drawn:

(a) The existence of floating columns affects the moments in beams and columns, according to the examination of several models. The position of the floating column inside the structure affects how beam and column moments vary. While going to design the construction which is experiencing the floating column, based on the location of the column, moments of beams and columns should be multiplied by a magnification factor.
(b) The top story columns have been shown to have the highest amplification factors in all situations of floating columns in any story caused by discontinuity of side columns.
(c) The comparison of the results with the floating column's location, both vertically and horizontally, demonstrating the variations in the results, will be highly useful for actual implementation to support the stability elements of the structure.
(d) After comparing structures with and without infills, it can be said that by using infill walls, one may lessen the forces and bending moments produced for a given amount of stress on beams and columns by dispersing them in infill walls as well.
(e) In terms of cost analysis, it is noted that the inclusion of the concept of a floating column can somewhat lower the cost of materials (concrete and rebar), although it has been noted that the cost of a wall can be increased in an infill construction.

6.2 The Works Scope

- More case studies are required to confirm the suggested outcomes. The building models used in this study are mid-rise building frame models that can be expanded to include more stories in the future and have a maximum height of eight levels.
- Currently, only linear analysis has been performed. For nonlinear analysis, the multiplication factor can also be examined.
- Future research may take into account a real-world scenario to demonstrate how it might work in practice.

- Since wind analysis was not taken into account in this study, it can be inferred for places with stronger winds in the future.
- In this structure, the cost of the materials is examined, but future costs for labor and machines may be taken into account.
- STAAD Pro software is utilized for analysis and designing in this study. While SAP2000, ETABS, or Tekla structure may be taken into consideration in the future. Another area that is the subject of extensive investigation is the construction of floating buildings.

References

1. Pankaj A, Manish S (2009) Earthquake resistant design of structures, PHI Learning Private Limited, New Delhi
2. Crisafulli FJ, Carr AJ, Park R (2000) Analytical modeling of infilled frame structures: a general review. Bull N Z Soc Earthq Eng 33:1
3. Hemanth G, Bhanupriya B, Ramakrishnaih A (2017) Earthquake analysis of multi-storeyed building with floating column. Int Res J Eng Technol (IRJET) 4(11)
4. Davis R, Menon D, Prasad AM 2008 Evaluation of magnification factors for open ground storey buildings. In: 14th world conference on earthquake engineering, Beijing, China
5. Choudhury T, Kaushik HB (2018) Seismic fragility of open ground storey RC frames with wall openings for vulnerability assessment. Eng Struct 345–357
6. Awkar JC, Lui EM (1997) Multistory seismic analysis and response. J Eng Struct 21(5):425–442
7. Criteria for earthquake resistance design of structures, Part: 1 General provisions and buildings, IS 1893:2002. Bureau of Indian Standards, New Delhi
8. Jain SK, Murty CVR, Dayal U, Arlekar JN, Chaubey SK (2001) The Republic day earthquake in the land of M. K. Gandhi, the Father of the Nation. Learning from Earthquakes (EERI)
9. Murty CVR, Goswami R, Vijayanarayanan AR, Mehta VV, Some concepts in earthquake behaviour of buildings. In: A handbook, Gujarat State Disaster Management Authority
10. Georgoussis GK (2009) An alternative approach for assessing eccentricities in asymmetric multistory structures 1: elastic systems. Struct Design Tall Spec Build 18(2):181–202
11. Sarkar P, Prasad AM, Menon D (2010) Vertical geometric irregularity in stepped building frames. Eng Struct 32(8):2176–2182
12. Arlekar JN, Jain SK, Murty CVR (1997) Seismic response of RC frame buildings with soft first storeys. In: Proceedings of the CBRI golden jubilee conference on natural hazards in Urban Habitat, New Delhi
13. Balsamo A, Colombo A (2005) Seismic behavior of a full-scale RC frame repaired using CFRP laminates. Eng Struct 27:769–780
14. Haridharan Kanthamani MK, Natarajan C (2015) Numerical simulation of damage in reinforced concrete slab subjected to elevated temperature. Int J Earth Sci Eng 8(1):242–248
15. Reyes G, Iman H, Kypros P (2010) Seismic behaviour of deficient RC frames strengthened with CFRP composites. Eng Struct 32(2010):3075–3085
16. Gilbert H, Ahmed A-A (1993) Analysis of building frames. J Struct Eng 119(2):468–483
17. Maison BF, Neuss CF (1985) Dynamic analysis of a forty four story building. J Struct Eng 111(7):1559–1572
18. Pujol S, Benavent-Climent A, Rodriguez ME, Smith-Pardo JP (2008) Masonry infill walls: an effective alternative for seismic strengthening of low-rise reinforced concrete building structures. In: The 14th world conference on earthquake engineering October 12–17, 2008, Beijing, China

19. Poonam AK, Gupta AK (2012) Study of response of structurally irregular building frames to seismic excitations. Int J Civil Struct Environ Infrastruct Eng Res Dev 2(2):25–31
20. Kirac N et al (2011) Failure of weak-storey during earthquakes. Eng Fail Anal 8:572–581 (Elsevier)
21. Choudhary HBK (2009) Effectiveness of some strengthening options for masonry-infilled RC frames with open first story. ASCE 8:733–745
22. Nelson et al (2010) Collapse modelling analysis of precast soft storey building in Australia. Eng Struct 1925–1936 (Elsevier)
23. Wilson JL et al (2015) Collapse behavior of precast soft storey building. Procedia Eng 10365–1042 (Elsevier)
24. Behera S (2012) Seismic analysis of multistorey building with floating column. Department of Civil Engineering, National Institute of Technology
25. Rohilla I, Gupta SM, Saini B, Seismic Response of Multi-Storey Irregular Building with Floating Column. IJRET Int J Res Eng

Analysis of Multi-Storeyed Structure by Using Design Softwares with Consideration of Lateral Loads

Danesh Khalotia, Saurabh Sairkar, and Pramod Gour

Abstract Every structure eventually fails during an earthquake because of its weak spots. This weakness results from irregular structures, which have discontinuities in their mass, stiffness, and geometry. Unusual structures make up a disproportionately huge proportion of urban infrastructure. One of the primary causes of building failures during earthquakes is vertical irregularities. For instance, buildings with flimsy storeys were the most notable buildings to collapse. As a result, the vertical irregularity significantly affects the structure's seismic performance. The dynamic behaviour of those building structures differs from a typical building due to height-wise modifications in stiffness and mass. STAAD, which stands for structural analysis, refers to the style of any object that is stable under a specific loading. The software called ETABS, also known as EXTENDED 3d Model OF BUILDING SYSTEMS, was created by Computer Systems or Structures, Inc. (CSI). The general-purpose civil engineering programme SAP2000 is perfect for the design and analysis of any kind of structural system. A practical and user-friendly object-based modelling situation that simplifies and simplifying the engineering process can be used to model, analyse, design, and optimise basic and sophisticated systems, ranging from 2 to 3D, of simple geometry to complex. Finding the best approximation software to generate the results of structural analysis is the aim of this research. The designer will then have a solid foundation from which to choose an analysis tool from STAAD, ETABS, or SAP 2000 [1, 2, 3, 6, 7, 8, 9]. A Ground+5, Ground+10, Ground+15, and Ground+20 building with unusual geometrical has been studied, designed, and the results were compared to determine the viability of these software programmes [8].

Keywords SAP 2000 · STAAD.PRO · ETAB · Seismic analysis · Structure analysis · Cost analysis

D. Khalotia (✉) · P. Gour
IES University, Bhopal 462044, India
e-mail: Dhaneshkhalotia@gmail.com

S. Sairkar
School of Advanced Sciences and Languages, VIT Bhopal University, Kotri Kalan, Bhopal-Indore Highway, Sehore 466114, India

K. K. Pathak et al. (eds.), *Latest Developments in Civil Engineering*, Lecture Notes in Civil Engineering 352, https://doi.org/10.1007/978-981-99-2676-3_50

1 Introduction

Three design software programmes, STAAD.PRO, ETABS, and SAP2000, can be used to design and analyse any type of structure using a static or dynamic approach. However, these software programmes will create different layouts and analytical outcomes for the same structural configurations. This is frequently because these programmes have different analytical mechanisms, which affects how they analyse the structure. This highlights the need for a comparison study of these two pieces of software to comprehend the key benefits and drawbacks. To encourage safe and cost-effective structures, it is important to compare the design output of different software programmes when studying and designing structures with geometric irregularities. This study compares the design outcomes produced by the software programmes ETABS and STAAD Pro while taking structural irregularities into consideration [1–3, 6–9, 13, 14, 17, 19].

The weak structural joints begin to fail first when there is an earthquake effect. The discontinuity in the structure's mass, stiffness, and geometry is what causes this weakness. An excessive amount of urban infrastructure is provided by irregular structures. One of the primary causes of building failures during earthquakes is vertical irregularities. The most notable structures to collapse, for instance, were those with soft storeys. These buildings' dynamic properties are different from those of regular buildings due to height-wise variations in stiffness and mass.

A Ground+10 [8] construction with unusual geometry has indeed been studied, designed, and the results compared to determine the viability of this software. Every type of structure, whether static or dynamic, can be designed and analysed using STAAD PRO [1–3], SAP 2000 [2, 3, 6, 9], and ETABS [1, 2, 6, 7], three design application programmes.

However, due to their various diagnostic components and the way they examine the structure, these products will provide a number of strategies and logical conclusion of the study for the same auxiliary arrangements. The requirement to compare the configuration outcomes of various programmings to safe in addition to conservative structures is significantly greater when trying to investigate and planning structures with geometric shape inconsistencies (Fig. 1).

1.1 Aim of the Study

The purpose of this study is to identify the best approximation software for generating results from Structural analysis. This can enable the designer to choose an analysis tool from among STAAD, ETABS, and SAP 2000 with confidence before conducting analysis [4, 5, 20].

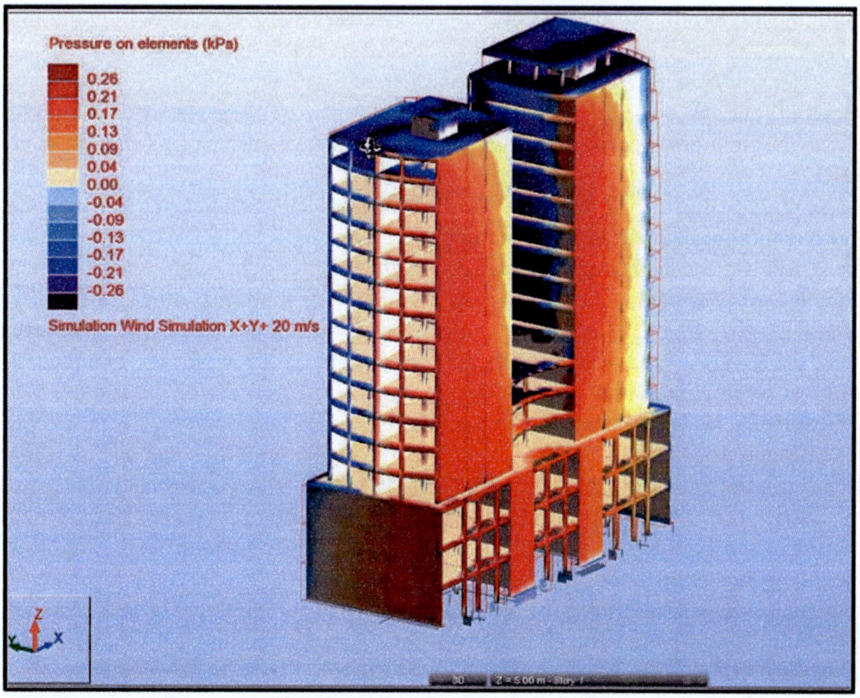

Fig. 1 Structure analysis

1.2 Objectives of the Study

(1) In order to model and analyse Ground+5, Ground+10, Ground+15, and Ground+20 storey Reinforced concrete (RC) framed structures with the aid of ETABS, SAP 2000, and STAAD-PRO [15, 16, 18].
(2) To create both a regular and an atypical multi-storey structure. As per IS. CODE—456 [21–23] and IS CODE—1893:2016 [21–23].
(3) To calculate the Shear forces (V), Bending moments (M), Reinforcement details, and results for the building's Structural components (Beams & Columns).
(4) To evaluate result of ETABS, SAP 2000, and STAAD-PRO [10–12].
(5) To determine which application produces results that are more accurate.

2 Literature Review

Richa Agarwal and Archna Tiwari (2017)—The research paper depicted a relative plan of three different structures as 5-storey, 10-storey, and 15-storey with various earthquake zones, namely, II, III, IV, V (as per IS. code: 1893 & 456—2000) of building, modelling and analysis of the structure were using structural programming STAAD.PRO and ETABS. In comparison to STAAD PRO, the design result showed a smaller area of steel to be used for the beam. In line with this, STAAD PRO software required less space for the column design result than ETABS. As a result, in both cases, the final product from ETABS provided less steel surface area than STAAD PRO.

S .Vijaya Bhaskar Reddy & V. Madhu (2018)—The research paper provided a thorough analysis of the simulation programmes ETABS and STAAD PRO, which were used for the evaluation and creation of rectangular plans with regular vertical heights and rectangular plans with vertical heights that were geometrically irregular. The goal of this study was to highlight the benefits of using ETABS over standard STAAD PRO practices. It was found that ETABS was more comprehensible, accurate, and compatible for design analysis. According to the results, load 1.5 (Self+Dead+Live) caused the maximum reaction produced to be 4572.12 kN in ETABS and 4624.92 kN in STAAD.-Pro. For irregular buildings, the displacement was along the x-direction and measured 106.25 mm in STAAD.-Pro and 53.47 mm in ETABS, respectively. As a result, ETABS produced more accurate results, which helped in the building's economical design. When an EQ length load is applied, Height decreases the storey overturning moment along the x-axis, and it is greater in regular buildings than in irregular ones. For beams and columns, the ETABS provided less reinforcing steel for RC building than for six regular buildings.

Mahmad Saber and D. GousePeera et al (2015)—The research paper provided a thorough explanation of how to analyse and design a structure using various applications, including STAAD PRO and ETABS. Using the static analysis method, a rectangular design with vertical regularity and a rectangle plan with vertical dimensional irregularity of a multi-storey building were analysed.

The result stated that because of the significant differences between how the results from STAAD PRO and ETABS were presented, it was challenging to draw conclusions about how to assign loading variables and design. When designing beams, ETABS provided less required steel surface area than STAAD PRO. Similar to both software, the steel required for the column section is required in these cases, but it is taken into account as a percentage between 0.30% and 0.50% From the column's design outcomes; the amount of steel determined by both software programmes was equal because the amount of steel needed to support the column forces in this specific problem would be less than the required steel limit for columns (i.e., 00.85%). As a result, it is impossible to compare the results in this case.

K Venu Manikanta and Dr. Dumpa Venkateswarlu (2016)—The primary goal of the research paper was to conduct a thorough analysis of the simulation programmes ETABS and STAAD PRO, which were used for the analysis and design of rectangular multi-storey buildings with vertical regular and irregular geometric patterns. The goal of this study was to highlight the benefits of using ETABS over standard STAAD PRO practices. Over STAAD.-PRO, it was found that ETABS was more user-friendly, accurate, and compatible for analysing design results.

Isha Bedi et al (2017)—The research paper suggested using STAAD.-PRO, ETABS, and SAP2000 to analyse and evaluate the results of RCC Frame Structures. The analysis of the proposed research led to the conclusion that STAAD.PRO was significantly more effective. When compared to ETABS and SAP, the values of the force derivative were low. The largest difference between both the values of STAAD-PRO, ETABS, and SAP2000 was caused by the Force derivative value at its maximum.

Sachin Patil et al (2017)—In the research paper, a defined structure's analysis and design was presented using ETABS and STAADpro, and the results were compared based on the amount of bending moment, base shear, axial force, and steel needed in each section. The application's affordable, secure, and user-friendly user interface was examined. According to the analysis of the columns and beams' results, ETABS provided higher values of shear stresses and bending stresses than SAP2000. Steel provided in ETABS has been larger than that provided in SAP 2000 based on design results for columns and beams. According to the results, SAP2000 is the more affordable and dependable design software when compared to ETABS.

3 Methodology

With the help of three different analysis tools—STAAD-PRO, SAP2000, and ETABS—we are comparing 12 cases in this study. The terms "Ground+5", "Ground+10," "Ground+15," and "Ground+20" refer to four different storey heights.

(1) **Step 1:** Plan selected for the study is of dimension 20.00 m × 25.00 m (Fig. 2).
(2) **Step 2:** To assign "sectional data and properties".
(3) **Step 3:** Assign "fixed end Condition".
(4) **Step 4:** Assign "seismic loading condition" & "Load combinations".
(5) **Step 5:** Perform Analysis to generate result Sheets.
(6) **Step 6:** Preparing Comparative Analysis results in "M.S.-Excel".
(7) **Step 7:** Providing conclusion as per results (Table 1).

3.1 Flowchart

See Fig. 3.

3.2 Load Calculation

DEAD LOAD (DL) (As per IS. Code: 875-1) [24]:

SELF-WEIGHT—Assigned in the downward direction on the Y -axis.

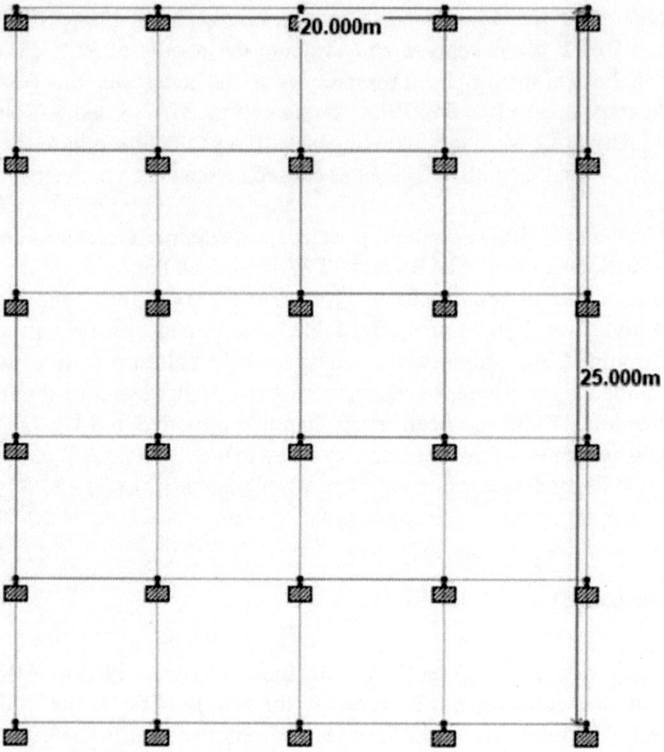

Fig. 2 Plan Of geometry

Table 1 Seismic description

Factor	Condition	Remark
Z (zone)	II (Bhopal Zone)	As per I.S.: code 1893—I: 2016 Table. 2
I-(importance factor)	"1. 50" (important structure)	Table.-6
R-(response reduction)	"5. 0" (S.M.R.F.)	Table.-7
Soil	Medium type soil	As per: CBR value < 3
Damping ratio	"0 0.50"	Damping effect

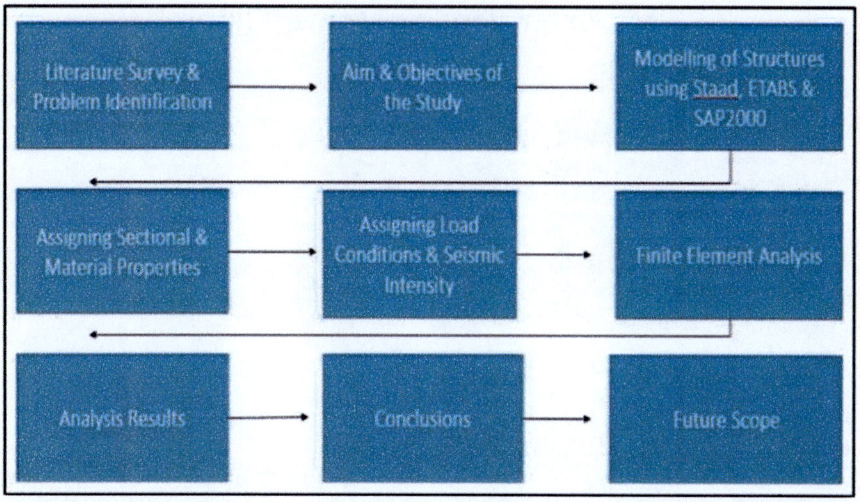

Fig. 3 Flowchart of the study

WALL LOAD—Thickness (t) × Height (h) × Density (d) (Unit weight of the material)

Wall Load = 0.230 m × 2.70 m × 20 KN/m^3 = 12.420 KN/m

PARAPET LOAD—Thickness × 1 m × Density (Unit weight of the material) 0.230 m × 1.000 m × 20.000 KN/m^3 = 4.60 KN/m

Slab Load—0.125 m × 25.0 KN/m^3 = 3.125 KN/m

LIVE LOAD (LL) (As per IS. CODE. 875-2) [25]—Floor Load (FL): Consider 5.0 KN/m^2 as given in IS CODE: 875-2 for commercial buildings.

SEISMIC LOAD (SL) (As per IS. CODE 1893—I:2016) [25]

Vb. = Ah × Weight of Building

Vb. = Base Shear,

Ah. = Horizontal Seismic Coefficient.

4 Results and Discussion

4.1 Case 1: Ground+5

See Figs. 4, 5, 6, 7, 8.

4.2 Case-2: Ground+10

See Figs. 9, 10, 11, 12, 13.

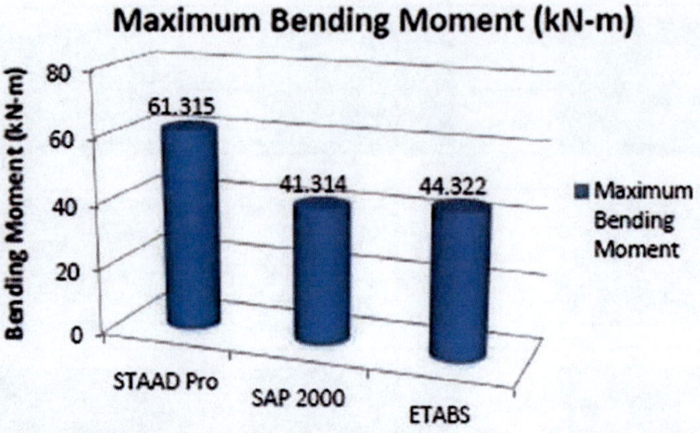

Fig. 4 Maximum Bending Moment (KN-m) for G+5 height

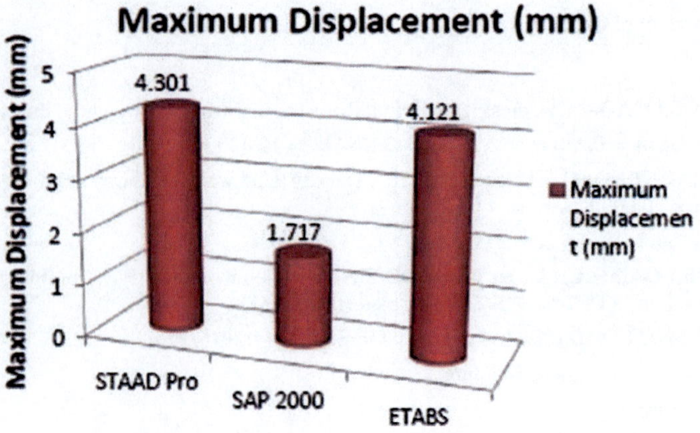

Fig. 5 Maximum Displacement (mm) for G+5 height

4.3 Case 3: Ground+15

See Figs. 14, 15, 16, 17, 18.

4.4 Case 4: Ground+20

See Figs. 19, 20, 21, 22, 23.

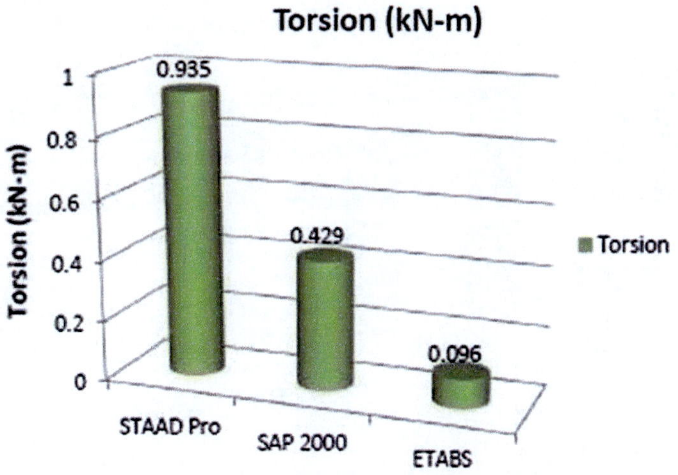

Fig. 6 Torsion (KN-m) for G+5 height

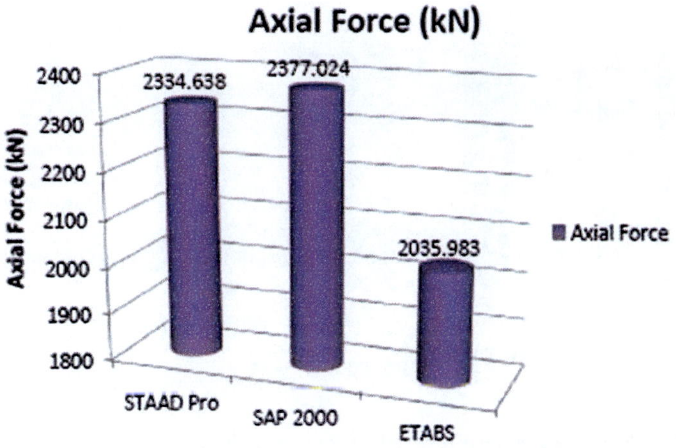

Fig. 7 Axial Force (KN) for G+5 height

5 Cost Analysis

5.1 Case I: G+5

See Table 2.

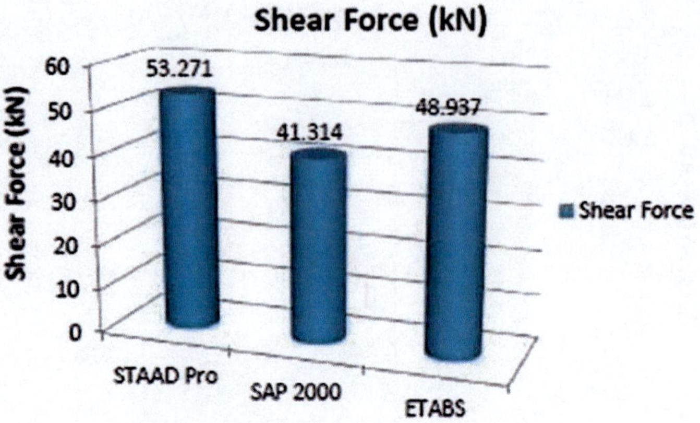

Fig. 8 Shear Force (KN) for G+5 height

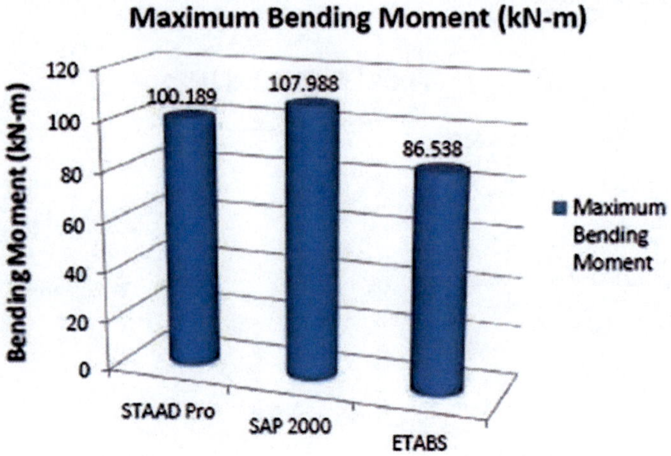

Fig. 9 Maximum Bending Moment (KN-m) for G+10 height

5.2 *Case II: G+10*

See Table 3.

5.3 *Case III: G+15*

See Table 4.

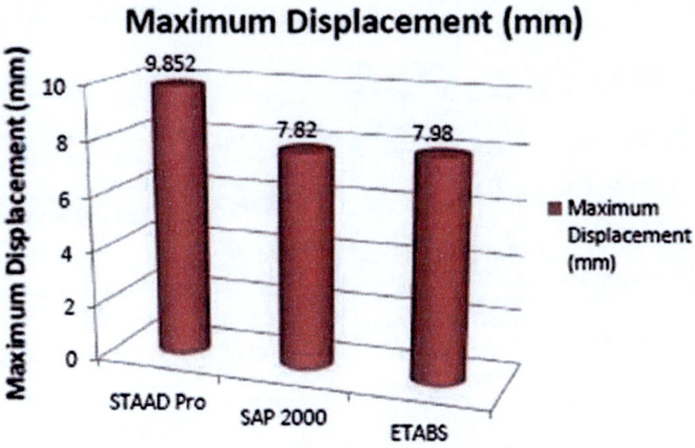

Fig. 10 Maximum Displacement (mm) for G+10 height

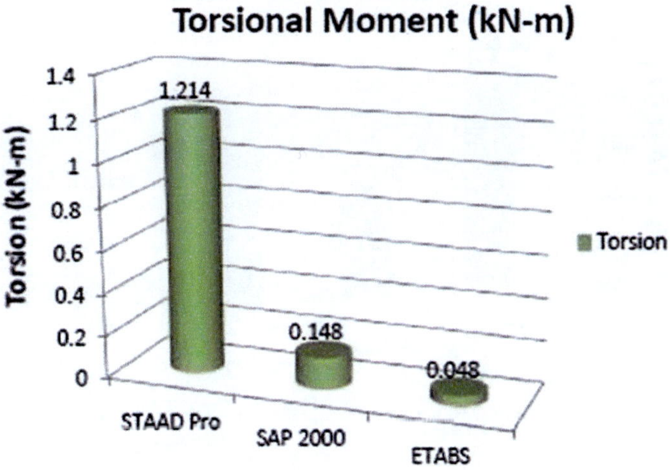

Fig. 11 Torsion (KN-m) for G+10 height

5.4 Case IV G+20

See Table 5.

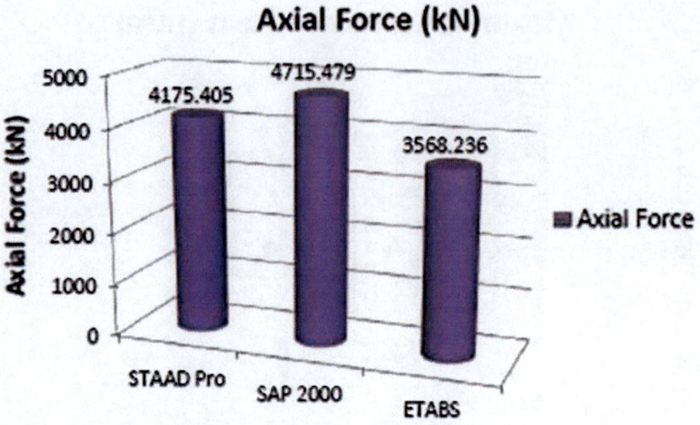

Fig. 12 Axial Force (KN) for G+10 height

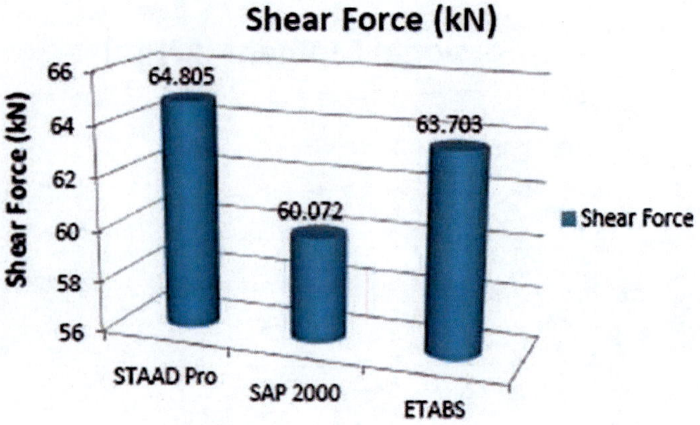

Fig. 13 Shear Force (KN) for G+10 height

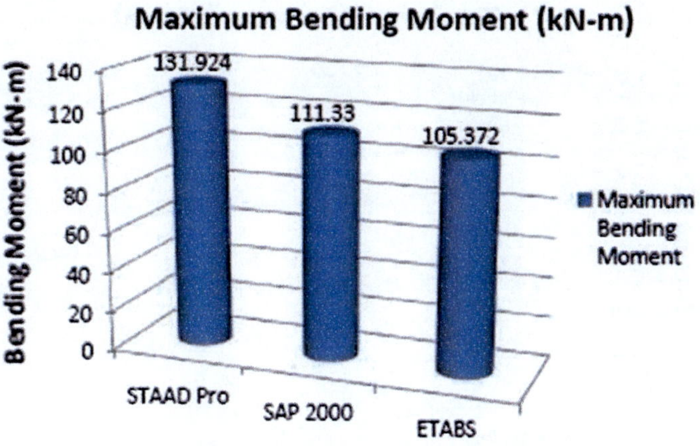

Fig. 14 Maximum Bending Moment (KN-m) for G+15 height

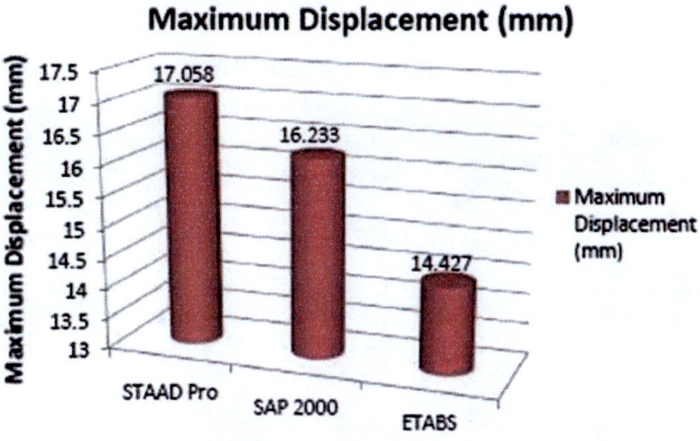

Fig. 15 Maximum Displacement (mm) for G+15 height

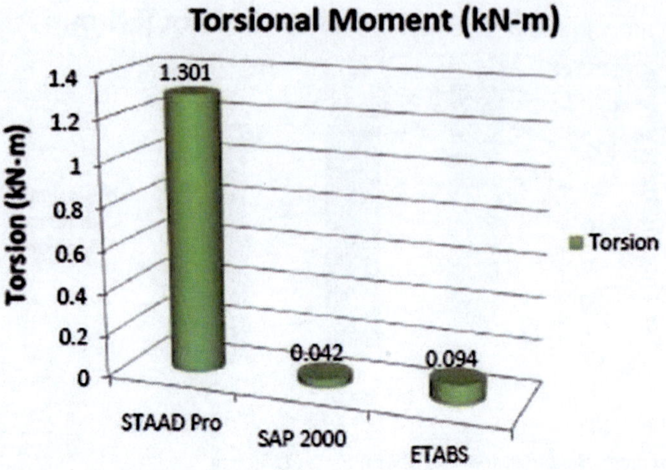

Fig. 16 Torsion (KN-m) for G+15 height

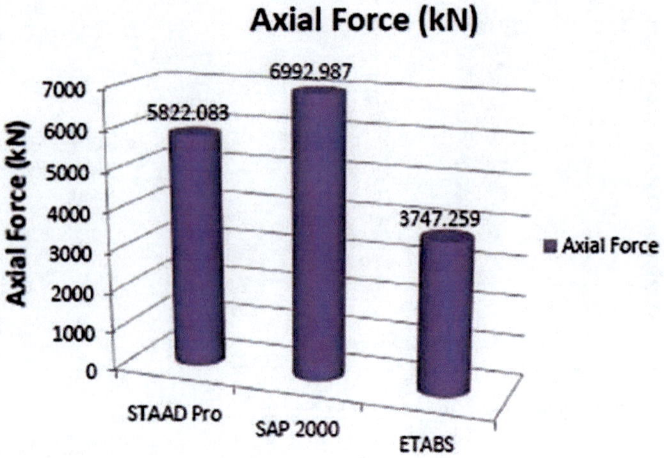

Fig. 17 Axial Force (KN) for G+15 height

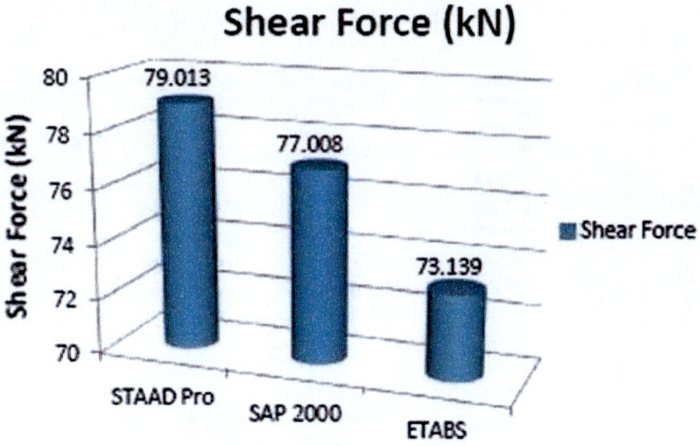

Fig. 18 Shear Force (KN) for G+15 height

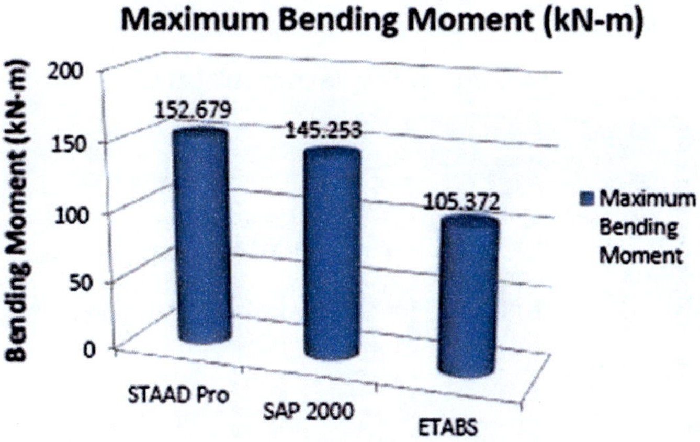

Fig. 19 Maximum Bending Moment (KN-m) for G+20 height

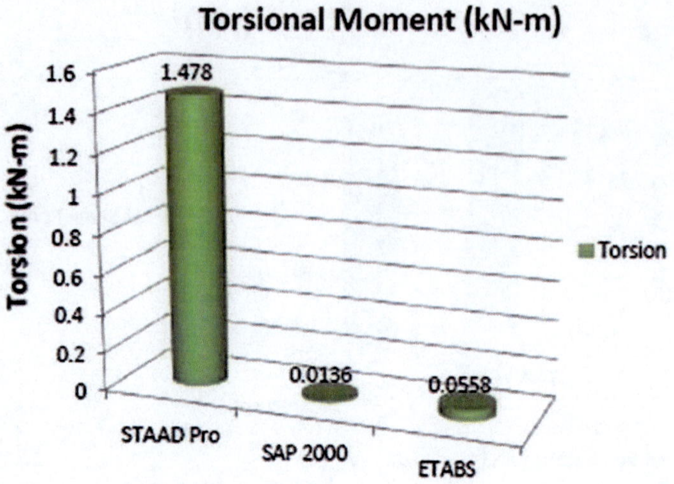

Fig. 20 Torsion (KN-m) for G+20 height

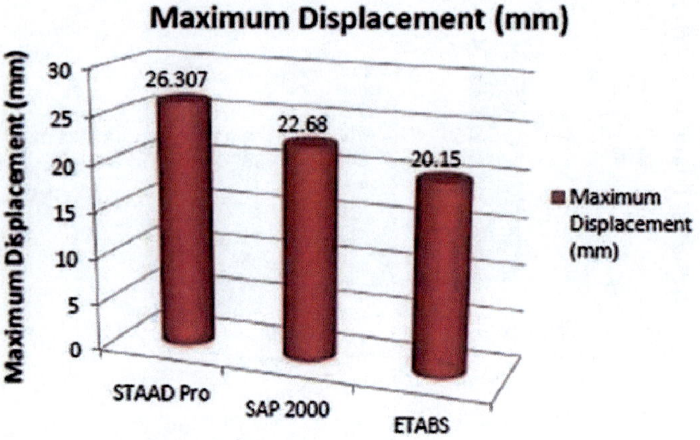

Fig. 21 Maximum Displacement (mm) for G+20 height

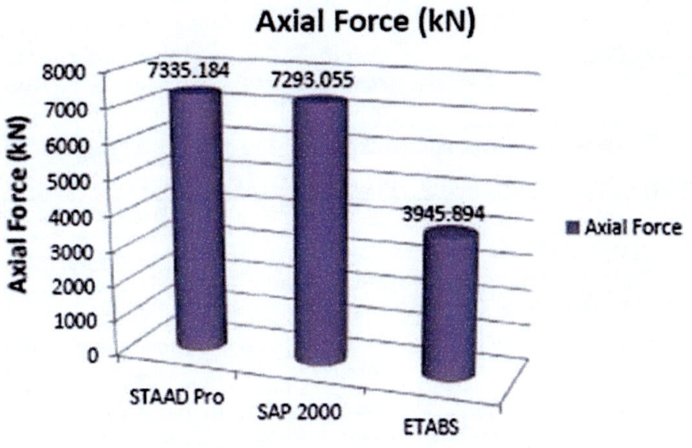

Fig. 22 Axial Force (KN) for G+20 height

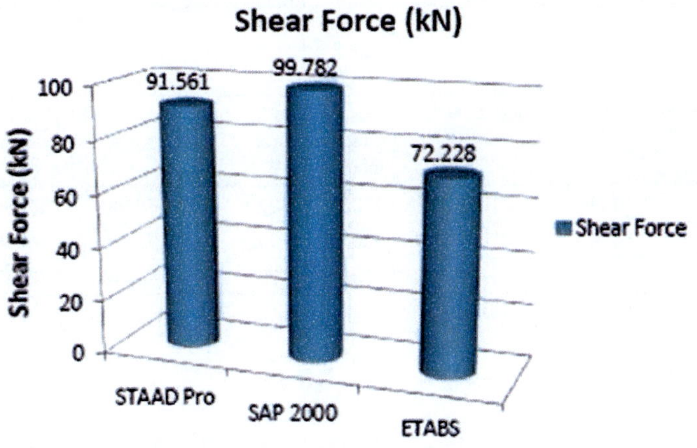

Fig. 23 Shear Force (KN) for G+20 height

Table 2 Cost analysis of G+5

Serial Number	Concrete	Rebar	Cost of concrete incu.m	Cost of rebar in Kg	Total cost of concrete	Total cost of rebar
ETABS	429	18086.763	4200	72.5	1801800	1311290.359
STAAD.PRO	429	25120.504	4200	72.5	1801800	1821236.61
SAP2000	429	16956.340	4200	72.5	1801800	1229334.712

Table 3 Cost analysis of G+10

Serial Number	Concrete	Rebar	Cost of concrete incu.m	Cost of rebar in Kg	Total cost of concrete	Total cost of rebar
ETABS	786.5	40761.062	4200	72.5	3303300	2955177.042
STAAD.PRO	786.5	47122.615	4200	72.5	3303300	3416389.644
SAP2000	786.5	50421.198	4200	72.5	3303300	3655536.919

Table 4 Cost analysis of G+15

Serial number	Concrete	Rebar	Cost of concrete incu.m	Cost of rebar in Kg	Total cost of concrete	Total cost of rebar
ETABS	1142.5	59472.122	4200	72.5	4798500	4311728.88
STAAD.PRO	1142.5	74526.469	4200	72.5	4798500	5403169.023
SAP2000	1142.5	62885.434	4200	72.5	4798500	4559194.022

Table 5 Cost analysis of G+20

Serial number	Concrete	Rebar	Cost of concrete incu.m	Cost of rebar in Kg	Total cost of concrete	Total cost of rebar
ETABS	1469.2	71043.436	4200	72.5	6170640	5150649.164
STAAD.PRO	1469.2	105109.39	4200	72.5	6170640	7620430.779
SAP2000	1469.2	99990.562	4200	72.5	6170640	7249315.8

6 Conclusion

In this study, the analysis output from three different analysis software tools—STAAD-PRO, ETABS, and SAP 2000—is being compared. Here, we have compared Ground+5, Ground+10, Ground+15, and Ground+20 storey structures while taking seismic zone II and medium soil condition into consideration.

The following findings from this study have been noted:

6.1 A. Ground+5 Storey

We found a variation of (32.62%) in the bending moment, where STAAD-PRO is (61.315 KN-m), ETABS is (44.3220 KN-m), and SAP 2000 is (41.3140 KN-m).

Small changes in analysis output, in terms of Forces, STAAD Pro [53.2710 KN], ETABS [48.9370 KN], and SAP 2000 [41.3140 KN] are the three programmes.

In terms of deflection, we observed almost similar values in STAAD Pro and ETABS, output whereas in SAP 2000. Comparatively speaking, the deflection is less.

6.2 B. Ground+10 Storey

STAAD PRO is (100.1890 KN-m), ETABS is (86.5370 KN-m), and SAP 2000 is (107.9880 KN-m), we discovered a variation of (19.86%) in the bending moment. STAAD Pro [64.805 KN], ETABS [63.7070 KN], and SAP 2000 [60.0720 KN] are the three programmes in terms of the forces' variations in analysis output. We observed nearly identical values for deflection in the output from STAAD.-Pro and ETABS, but less in SAP-2000.

6.3 C. Ground+15 Storey

In terms of bending moment, we noticed a variation of (20.12 %), where STAAD PRO is (131.9240 KN-m), ETABS is (105.3720 KN-m), and SAP 2000 is (111.330 KN-m). STAAD, ETABS, and SAP 2000 all produced analysis output with values of 79.0130 kN, 73.1380 kN, and 77.0080 kN regarding forces, minute variation. In terms of deflection, STAAD pro and ETABS output showed nearly identical values, whereas the deflection seen in ETABS (14.4270 mm) was significantly less.

6.4 D. Ground+20 Storey

We found a variation of 32.40% in the bending moment, where STAAD PRO is (152.679 KN-m), ETABS is (103.2080 KN-m), and SAP 2000 is (145.253 KN-m). The three software programmes STAAD PRIO [91.561 KN], ETABS [72.2280 KN], and SAP 2000 [99.7820 KN] have similar Forces variation in analysis output. In terms of deflection, we saw nearly identical values in the output from STAAD Pro and ETABS, whereas the deflection seen in ETABS [20.15 MM] is smaller compared to.

6.5 Future Scope

While wind pressure can be taken into account in the future, seismic analysis is taken into account in this study, While Indian-Standard provisions are taken into consideration in this study, similar research could be done in the future with European,

American, or British Codal provisions. While future analyses may take into account additional tall structures, Ground+20 Tall Structures are taken into consideration in this study.

In this study, STAAD. PRO, ETABS, and SAP 2000 are taken into account for analysis, whereas Tekla and mid as can be taken into account in the future.

References

1. Agarwal R, Tiwari A (2017) Comparison of design result of multi-story structure using ETABS and STAAD PRO software. Int J Eng Sci Comput 7(8)
2. Hu K, Yang Y, Mu S, Qua G (2012) Study on high-rise structure with oblique columns by ETABS, SAP2000, MIDAS/GEN and SATWE. Int Conf Adv Comput Model Simul, Procedia Eng 31: 474–480
3. Nura LN, Pandit JP (2019) Comparative study of structural software Sap2000 and STAAD Pro. Int J Eng Sci Inven (IJESI) 8(03) Series IV: 37–43
4. Manchalwar SA, Puri AS, Aswale V (2016) Comparative study of analysis and design of RC frame. Int J Sci, Eng Technol Res (IJSETR) 5(4)
5. Hiwase PD, Joshi A, Keshariya A (2018) Comparison between manual and software approach towards design of structural elements. Int J Eng Sci (IJES) 6: 54–56
6. Patil S, Javalge SC, Kurane RR (2017) Study of results of analysis and design of multistoried building by ETABS and SAP2000 software's. Int J Sci Res & Dev 5(02)
7. Venu Manikanta K, Venkateswarlu D (2016) Comparative study on design results of a multistoried building using Staad Pro And Etabs for regular and irregular plan configuration. Int J Res Sci Adv Eng 2(15): 204–215
8. Shukla R, Saha P (2017) Comparative study of a G+10 storied building using ETABS and STAAD. IJSRST 3(6)
9. Lallotra B, Singhal D (2017) State of the art report—a comparative study of structural analysis and design software—STAAD Pro, SAP-2000 & ETABS Software. Int J Eng Technol (IJET) 9(2)
10. Aman MN, Gajendra VT (2016) "Analysis and design of the multistory building by using STAAD Pro". Int Res J Eng Technol (IRJET) 03(06): 887–891
11. Patel MR, Singh RC (2017) "Analysis of a tall structure using Staad Pro providing different wind intensities as per 875 Part-III". Int J Eng Sci & Res Technol: 2018–2025
12. Anoop A, Hussian F, Neeraja R, Chandran R, Shabina S, Varsha S (2016) "Planning analysis and design of multi storied building by Staad.Pro.v8i". Int J Sci & Eng Res 7(4)
13. Deshmukh DR, Yadav AK, Supekar SN, Thakur AB, Sonawane HP, Jain IM (2016) "Analysis and design of G+19 storied building using Staad-Pro". 6(7) (Part-1): 17–19, ISSN: 2248–9622
14. Raju B, Rattaiah R (2015) 'Analysis and design of high-rise building (G+30) usingSTAAD.PRO'. Int J Res Sci Adv Eng 2(12): 50–54
15. Anoop A, Hussian F, Neeraja R, Chandran R, Shabina S, Varsha S, (2016) 'Planning analysis and design of multi storied building by STAAD.PRO.V8i'. Int J Sci & Eng Res 7(4): ISSN 2229-5518
16. Khan NM (2016) 'Analysis and design of apartment building'. Int J Innov Sci, Eng Technol 03(03): ISSN 2348-7698
17. Deshpande RD, Pai MN, Pawan N, Pednekar AP (2017) 'Analysis, design and estimation of basement+G+2 residential building'. Int Res J Eng Technol (IRJET) 04(06): e-ISSN: 2395-0056, p-ISSN: 2395-0072
18. Saleem SK, Ravi Kumar B (2017) 'Analysis and design of multi storeyed building by using STAADPRO'. Anveshana's Int J Res Eng Appl Sci 2(1): ISSN-2455-6300

19. Hugar A, Pujari SM, Pujari BG, Biradar AN, Gajendra (2016) 'Analysis and design of a commercial cum residential building by using STAAD Pro'. Int Res J Eng Technol (IRJET) 03(06): e-ISSN: 2395 -0056, p-ISSN: 2395-0072
20. Anup B, Venkateswarlu D (2016) 'Comparison between manual analysis and STAAD PRO.Analysis of multi storey building'. Int J Res Sci Adv Eng 2(15): 216–224
21. Madhurivassavai, Bhargavi V, Raghava Rao EV (2016) 'Analysis and design of multistoried building with G+8 floors by using Staadpro'. Int J Adv Technol Innov Res 08(02) ISSN 2348–2370
22. IS code 1893 and 456-2000
23. IS CODE-456 & IS CODE-875
24. Dead Load (As per IS CODE 875-1)
25. Live Load (As per IS CODE 875-2)
26. Seismic Load (As per IS CODE 1893-I: 2016)

Evaluation of Ductility of Hybrid Reinforced Concrete Beams Made with Glass Fiber-Reinforced Polymer Rebar

M. Naveen, K. Hemalatha, and V. Srinivasa Reddy

Abstract Concrete is a robust, adaptable, and moldable building material. Cement, fine aggregate, coarse aggregate, and water are the main ingredients. For concrete to perform better than traditional concrete, several researchers have added novel components. One of the best polymeric fibers for use in concrete is polyvinyl alcohol (PVA) fiber, although the design of the material is made more difficult by PVA fiber's particular microstructure. One of the most advanced composite material types that may be employed in a variety of civil engineering applications is GFRP rebar. Due to its advantageous characteristics, such as a high strength-to-weight ratio, GFRP reinforcement is becoming increasingly popular. Neutrality to magnetic fields and corrosion resistance. The capacity of reinforced concrete members to withstand significant deflection before failing is known as ductility. This property of structural members made of reinforced concrete is essential since it indicates when a component is failing and prevents catastrophic collapse. In the current study, M30-grade concrete was subjected to an experimental inquiry to establish its ductility. Two numbers of GFRP bars and conventional steel bars were used, and a comparison was done between concrete with and without 0.25% PVA fibers. Tensile strength measurements are made between GFRP and steel bars. The test results for the beams employing GFRP and PVA fibers are established during a 28-day curing period.

Keywords Polyvinyl alcohol · GFRP · PVA · Ductility · Fiber-reinforced concrete

M. Naveen (✉)
Structural Engineering, Gokaraju Rangaraju Institute of Engineering and Technology, Hyderabad, India
e-mail: m.naveen717@gmail.com

K. Hemalatha · V. Srinivasa Reddy
Civil Engineering, Gokaraju Rangaraju Institute of Engineering and Technology, Hyderabad, India

K. K. Pathak et al. (eds.), *Latest Developments in Civil Engineering*, Lecture Notes in Civil Engineering 352, https://doi.org/10.1007/978-981-99-2676-3_51

1 Introduction

1.1 General

The most common building material utilized worldwide in civil engineering projects is concrete. Every year, enormous amounts of various types of concrete are produced. One of the most fascinating topics for academics is fiber-reinforced concrete, which comes in many varieties. Cement, fine aggregate, coarse aggregate, and water make up traditional concrete [1]. Every new material that is added has some effect on the concrete, but the impact should be favorable and pleasant. This may be done by making certain adjustments to the concrete field.

Fiber-reinforced concrete (FRC) is created by organically incorporating and blending tiny, randomly distributed fibers into concrete. Polymeric fibers are known to be the most effective in preventing plastic shrinkage and cracking when added to concrete [2]. It is undeniable that ductility and post-ultimate behavior may be improved as a result of the inventive addition of polyvinyl alcohol (PVA) fibers to designed cementitious composites, according to the scant study that has been done thus far. When added to concrete, fibers are believed to boost its durability and capability to absorb energy. In terms of materials, concrete is thought to be brittle, weak in tension, and strong in compression [3]. By enhancing the tensile and flexural characteristics of concrete with fiber, the ductility of the material may be increased.

1.2 Hybrid Fibers

When compared to ECC with steel, PVA, and hybrid fibers, conventional concrete's flexural strength was shown to be lower [4]. Flexural strength increases by roughly 15% when the percentage of steel fibers is increased from 3 to 4, but it increases by 20% when the percentage of PVA fibers is increased from 1.5 to 2. Flexural strength in hybrid fibers was discovered to be within the average range of SECC 3 and PVA 1.5 while lowering the balling effect of the fibers [5].

When it comes to failure patterns, it is seen that the ECC only develops a single crack, reflecting its ductile behavior, as opposed to the CC, which breaks down into two pieces [6]. In this study, the impact of adding PVA fibers to concrete was examined. The optimal volume percent for geometric lengths, 6 and 12 mm, with an improvement of around 10% over plain concrete, is 0.25%. Concrete's compressive strength is only slightly boosted with increasing fiber content. With increasing fiber content, a similar trend for tensile and flexural strength is also shown.

The average strength improvement in flexure and splitting tensile of 20 and 30%, respectively, at a volume fraction of 0.25%, demonstrates the extent to which fibers may enhance the attributes of plain concrete [7]. Furthermore, compared to the control, FRCs demonstrated reduced flexural stiffness while also offering a larger load-bearing capability. This causes FRCs where the peak load values were maintained to have a greater ultimate deflection. According to his study's experimental findings, the use of PVA fibers improved the mechanical qualities of concrete.

By growing the extent percentage of PVA fibers, the workability of fiber-bolstered concrete diminishes. A 40% lower has been mounted for OPS concrete with 0.5% PVA fiber [8]. Additionally, the density of the concrete is reduced via way of means of the inclusion of PVA fibers with low unique gravities to the OPS mixes. PVA fibers, which assist barely lessening density, can't be disregarded. With a boom in PVA fibers, OPS concrete's compressive electricity has progressed for all age groups. Due to progressed adhesion on the fiber/matrix interface, the impact of PVA on the compressive electricity of lightweight concrete has been confirmed to be increasingly massive over time. Under no curing (AC) conditions, the inclusion of PVA fiber as much as 0.375% decreased the lack of compressive electricity [9].

Therefore, it is able to be concluded that PVA fibers may be utilized to reduce OPS concrete's sensitivity beneath neath subpar curing settings. OPS concrete turns into extra ductile due to the addition of PVA fiber, which increases the stress potential similar to height stress. In contrast, to manipulating concrete, the inclusion of PVA fibers substantially improves the splitting tensile and flexural strengths via way of means of as much as 30 and 32%, respectively. The modulus of elasticity is substantially tormented by the addition of PVA fibers to OPS concrete. This study's size of 16.1 GPa is more than that of in advance research. In mild of this, it is able to be stated that PVA fiber-bolstered OPS LWC confirmed the capacity and desirable overall performance for feasible use in growing inexperienced composite concrete structures. According to journal research, PVA fiber concrete outperforms conventional concrete in terms of flexural and split tensile strengths. Between 0.5 and 0.75% of the fiber added to the concrete has reached its maximum strength [10]. The tensile and flexural splitting strengths are dramatically increased by the inclusion of PVA fiber, up to 40%. PVA fiber is poisonous and degrades slowly in nature.

1.3 Ductility of the Concrete

The capacity of reinforced concrete members to withstand significant deflection before failing is known as ductility. This property of structural members made of reinforced concrete is essential since it indicates when a component is failing and prevents catastrophic collapse. In seismic regions, this is particularly important. We examine techniques for improving the ductility of reinforced concrete components.

Mix Design of Concrete

Concrete mix design is a process that involves choosing the right elements for concrete and their appropriate ratios with the goal of preparing concrete as affordably (value engineered) as feasible while maintaining a specific minimum strength, desirable workability, and durability.

Gather the following information before choosing a mix design, as some design requirements are predetermined based on this information.

Measures to Surge Ductility of RCC Structural Members

- It is advised to choose a straightforward and predictable structural layout.
- To avoid the failure of a column or the development of flexible hinges in columns, the weak beam and strong column concept must be applied in the design of a building.
- The avoidance of foundation failure may be utilized as another element to increase ductility.
- It is important to prevent undesirable brittle failures, such as those brought on by shear, anchoring, and concrete bond in compression failure.
- To allow for extensive rotation before the collapse, use beneath reinforced concrete sections.
- In order to make concrete suffer bigger deformations, it is advised to provide lateral confinement at crucial points of columns.

Utilizing the appropriate parts of the relevant codes, such as material specifications for ductility, ductile detailing of beams, columns, and frames, and employing suitable detailing for reinforcements at joints, it is possible to put the aforementioned precautions into practice.

1.4 GFRP Bar

The GFRP bar is defined as "Glass Fiber Reinforced Polymer which is a composite material of longitudinal glass fibers, polyester resins and matrix material" (Fig. 1).

Due to its properties and elastic behavior, which are extremely comparable to conventional steel rebars, it has currently replaced conventional steel rebars as the best alternative reinforcing material in RCC structures [11].

1.5 PVA Fibers

Vinyl alcohol (PVA) makes up at least 50% of the fiber-vinyl on the component in vinyl on fibers, which also contain at least 85% of the fiber's total weight in vinyl alcohol units as well as one or more acetal units. Another name for the fiber is vinal or vinyl.

The vinyl monomer is not stable; hence vinyl resin cannot be produced from it. In its place, polyvinyl acetate is hydrolyzed to yield polyvinyl alcohol (PVOH). The PVA polymer is then coagulated by spinning it either wet or dry through a nozzle into a coagulating solution. PVA fibers with good UV and chemical resistance, high tenacity, high Young's modulus, low elongation, and low creep (alkali, acid, oil, etc.).

Fig. 1 GFRP bars

This fiber can replace cotton in many applications since the fabrics manufactured from it feel like cotton, have an excellent affinity for water, don't release harmful fumes when burned, and are simple to wash and dry.

PVA fiber is an appropriate reinforcement material for cementitious composites due to its properties. An excellent environmentally friendly cement reinforcement material is polyvinyl alcohol fiber. It, by a reason of its distinctive molecular structure, has alkali and weather resistance, therefore it Enhances the concrete's resistance to frost. improving bending strength, impact strength, and fracture strength; improving permeability; and improving the impact and seismic resistance of concrete by forming a strong affinity with cement. Increased toughness is a result of concrete's improved brittleness, impact resistance, and bending strength. Strong bonding with cement matrix is one of PVA fiber's notable properties (Figs. 2 and 3).

1.6 Applications of PVA Fiber

PVA fibers' "stealth" quality is their main advantage. They can be cast all over the concrete and hardly or not at all be visible on the completed surfaces. This is true whether you are producing troweled finished concrete or concrete cast in a mold.

The ductility or flexibility of PVA-reinforced concrete is exceptional. Instead of having a quick catastrophic failure, the concrete section will flex substantially before it fractures. It only fails in a different way, which does not imply that it is more durable than AR glass fiber in use. In comparison tests, AR glass performs better, although it isn't always feasible to utilize it for a particular application.

Fig. 2 PVA fibers

Specification

Tensile Strength	>1500 MPa
Modulus	>37GPa
Elongation	6.5±1%
Fibre dimension	Length: 6mm / 12mm ; Diameter:20/ 40μm
Melt Point	230ºC
Density	1.3g/cm3
Acid&Alkali Resistance	Excellent
Reduction in 90ºC water	≤1%
Appearance	Faint yellow
Safety	Non-toxic, non-irritating, neutral

Fig. 3 PVA fiber properties

Concrete and PVA are said to "chemically bind" together. This assertion looks doubtful to ordinary observation. We won't give that assertion much credence and will just leave it alone. If it's accurate, we prevail. If not, we won't worry about it.

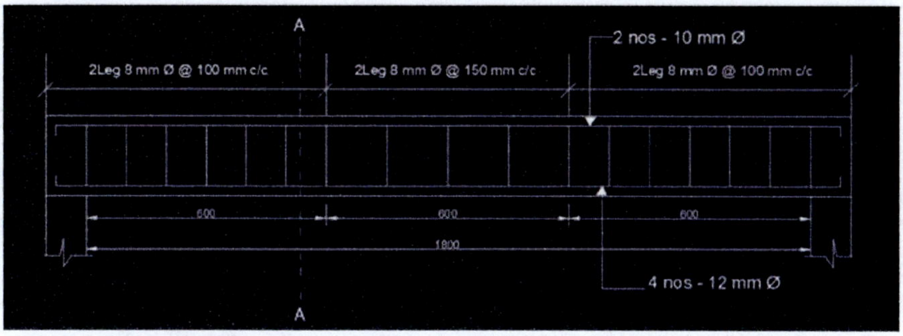

Fig. 4 Longitudinal section of the beam

1.7 Mix Proportions

- Cement = 394 kg/m^3
- Water = 197 kg/m^3
- Fine aggregates = 732 kg/m^3
- Coarse aggregate = 1139 kg/m^3
- Water–cement ratio = 0.50

Mix for M30-grade concrete is 1:1.86:2.89 at w/c of 0.50.

2 Beam Detailing

See Figs. 4 and 5.

3 Test Results and Discussions

From the above graph, we can observe that the stress–strain curve is initially linear up to the load of 61.5 kN (which is in the elastic limit), Then the curve has ascended up to the load of 63.1 kN then it has been instantly descended to 61.2 kN (plastic limit is formed) (Fig. 6). When the load keeps increasing the development of stresses has been constant, whereas strain keeps increasing up to 0.24–0.38, respectively. Then stress has been rapidly increased up to the load of 75.35 kN which is the ultimate load of the steel bar, When further increment of the load the steel bar has lost its resistance to the load so then the fracture point is observed at 52.7 kN.

From the above graph, the variation of Stress versus Strain is linear up to the load of 135 kN, Since the GFRP is brittle in nature. So, there won't be any warnings and it will lead to sudden failure (Fig. 7).

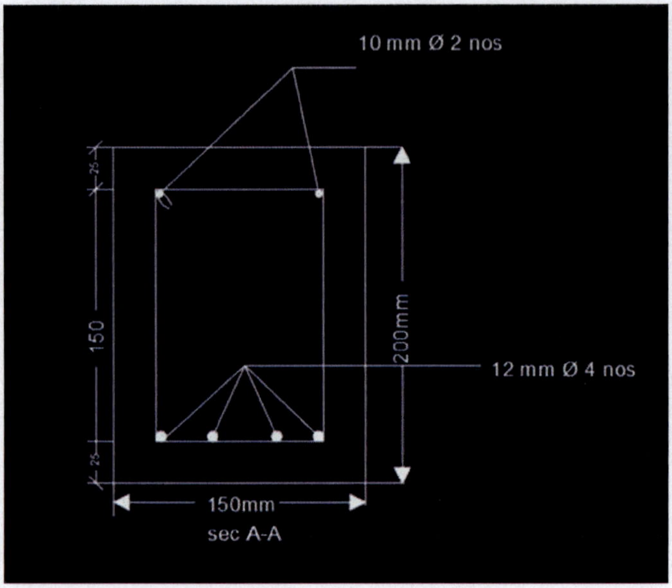

Fig. 5 Cross-section of the beam

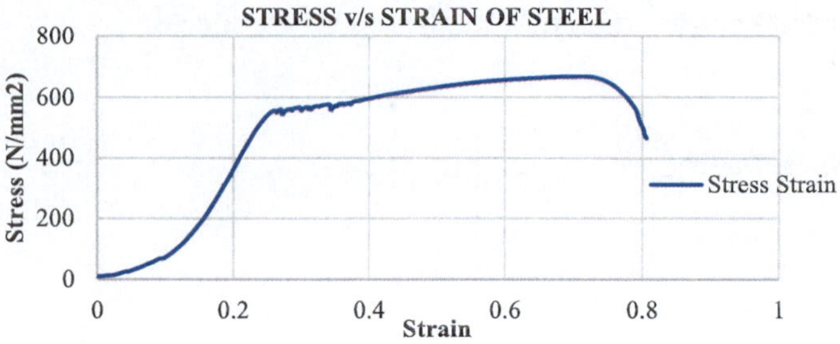

Fig. 6 Stress–strain curve of steel bar

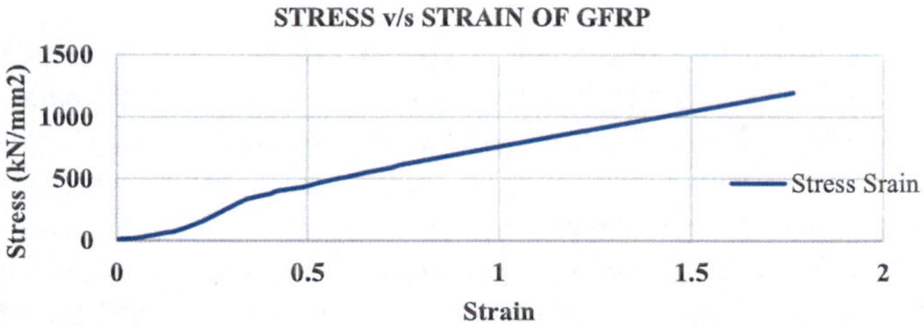

Fig. 7 Stress–strain curve of GFRP bar

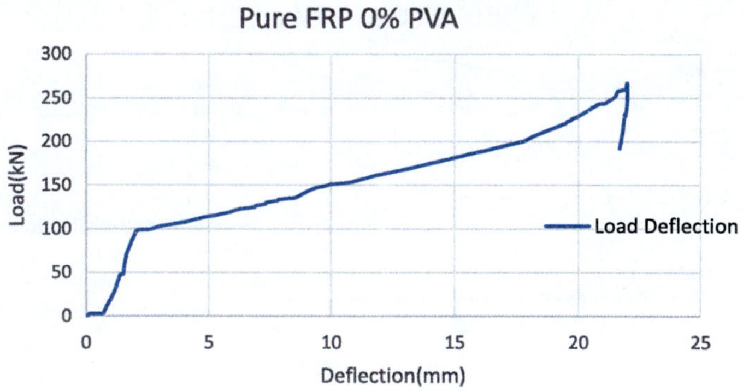

Fig. 8 Load–deflection curve of pure FRP 0% PVA

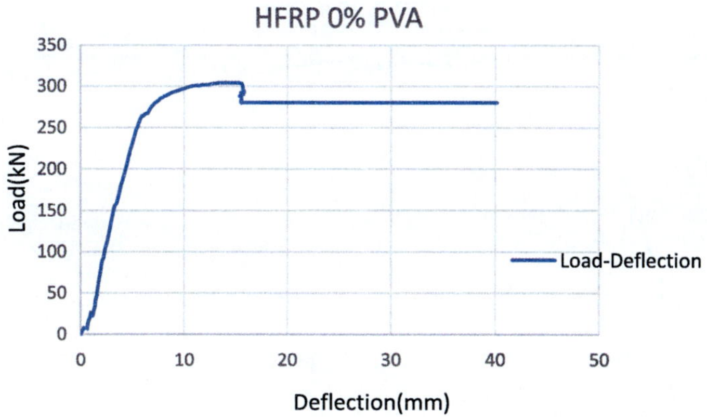

Fig. 9 Load–deflection curve of HFRP 0% PVA

3.1 Flexure Behavior of Reinforced Concrete Beams

The HFRP beam is attached to the testing equipment and the load is applied to it. The ultimate load and ultimate moment were used to investigate the flexural behavior (Figs. 8, 9, 10, 11 and Table 1).

3.2 Ductility Factor

Ductility refers to a material's capacity to stretch to its utmost extent before failing. The experiment is used to compute the ductility, which is given below (Fig. 12 and Table 2).

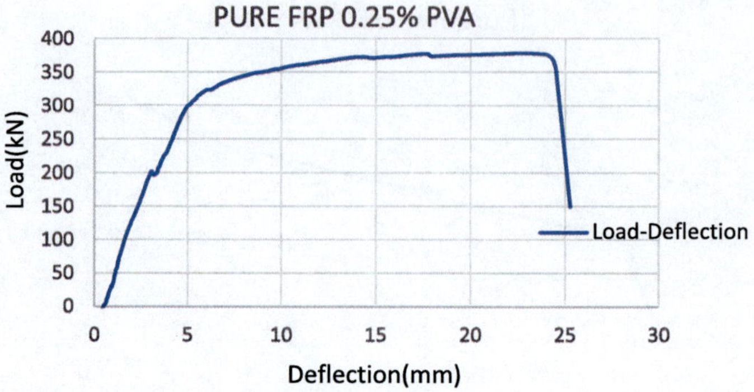

Fig. 10 Load–deflection curve of pure FRP 0.25% PVA

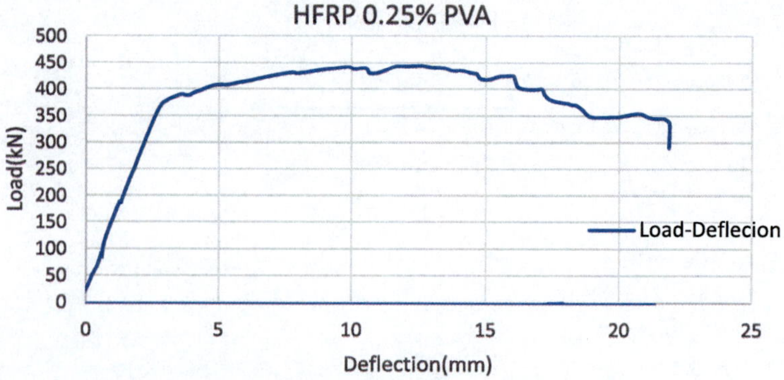

Fig. 11 Load–deflection curve of HFRP 0.25% PVA

Table 1 Test results from the experiment

Beam-type PVA (%)	Ultimate load (kN)	Ultimate moment (kN/mm)	% Change
FRP 0	257.8	64.25	–
HFRP 0	304	76	+18.28
FRP 0.25	379	94.75	+47.47
HFRP 0.25	445	111.25	+73.151

$$\text{Ductility Factor} = \frac{\text{Ultimate Deflection } (\Delta u)}{\text{Yield Deflection } (\Delta y)}$$

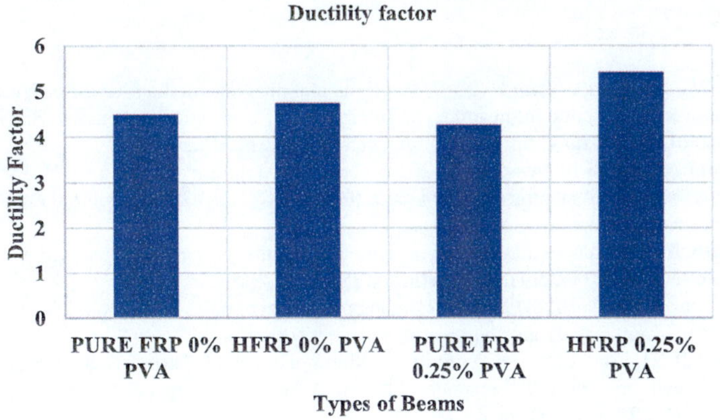

Fig. 12 Ductility factor

Table 2 Ductility factor of the beams

Sample	Δy (mm)	Δ*u* (mm)	*μ*	% Change
PURE FRP 0% PVA	3.18	14.27	4.48	–
HFRP 0% PVA	3.78	17.89	4.73	5.58
PURE FRP 0.25% PVA	3.78	15.89	4.26	4.91
HFRP 0.25% PVA	4.2	22.8	5.42	20.98

4 Conclusions

The findings that were drawn from research on beams were as follows:

- When comparing the tensile strength of steel and GFRP bars, GFRP is 1.8 times stronger than steel.
- The test findings reveal that when analyze to pure FRPRC beam and hybrid FRP beam without PVA fiber, hybrid fiber RC beam with PVA fiber behaves in a more ductile manner.
- The ductility factor increased by 20.98% at the optimum dosage of 0.25% of PVA fibers in hybrid made with RCC beams when compared with pure GFRP beams.

References

1. Du E, Wang Y, Sun J, Yang S (2019) Experimental analysis on ductility of polyvinyl alcohol fiber reinforced concrete frame joints. In: International information and engineering technology association, vol 43, no 1, pp 17–22. http://iieta.org/Journals/acsm
2. Noushini A, Samali B, Vessalas K (2013) Flexural toughness and ductility characteristics of polyvinyl-alcohol fiber reinforced concrete (PVA-FRC). In: VIII International conference on fracture mechanics of concrete and concrete structures, 2013
3. Velumani M, Menaka V (2018) Studies on the mechanical properties of polyvinyl alcohol fiber concrete review. JETIR 5(7). www.jetir.org. ISSN: 2349-5162
4. Du E, Dong S, Sun J (2018) Concrete bonding properties of polyvinyl-alcohol fiber in fabricated structures. Chem Eng Trans 66
5. El-Ghazaly H, Abd El-Azim A, Said M, Mahmoud Ali M (2019) Behavior of reinforced concrete columns with (PVA) under fire. Al-Azhar university civil engineering research magazine (CERM), vol 41, no 3
6. Devi M, Kannan L, Ganeshkumar M, Venkatachalam TS (2017) Flexural behavior of polyvinyl alcohol fiber reinforced concrete. SSRG Int J Civ Eng (SSRG–IJCE) 4(6)
7. Kamal MM, Safaan MA, Al-Gazzar MA (2006) Ductility of concrete beams reinforced with hybrid FRP rebars. HBRC J 2(3)
8. Thamrin R, Zaidir Z, Iwanda D (2022) Ductility estimation for flexural concrete beams longitudinally reinforced with hybrid FRP–steel bars. Polymers 14:1017. https://doi.org/10.3390/polym14051017
9. Li VC, Wang S (2002) Flexural behaviors of glass fiber-reinforced polymer (GFRP) reinforced engineered cementitious composite beams. ACI Mater J
10. Renić T, Kišiček T (2021) Ductility of concrete beams reinforced with FRP rebars. Buildings 11:424. https://doi.org/10.3390/buildings11090424
11. Renić T, Hafner I, Kišiček T (2021) Ductility of hybrid FRP—steel reinforced concrete sections. In: 2nd international conference on construction materials for sustainable future, 20–21 April 2021, Slovenia
12. Priyanka U, Siva Ramakrishnan S (2016) Hybrid reinforcement by using GFRP. Int Res J Eng Technol (IRJET) 03(06). e-ISSN: 2395-0056

Effect of Span Length on the Doubly Optimized Prestressed Concrete Beams

Ishan Jha and Krishna K. Pathak

Abstract The current work investigates the changes occurring in the optimized shape and cable layout of prestress concrete beams due to the change in span length. A single prestressed beam of fixed length is taken, and its span length is changed by placing supports along the length. The prestressed beam is then optimized for each case and the effect of the changed span over the optimized cable layout and concrete shape under different loading conditions is studied. The optimization of the prestressed beam is done using an indigenously developed software labeled "Simultaneous Shape and Cable Optimization" (SSCO). The software SSCO works on a fuzzy regulated integrated zero-order procedure where the concrete is simulated as a discretized continuum using nine noded Lagrangian elements, while the prestressing wire is represented as a curvilinear, three-noded bar element mapping a B-spline profile. The present objective of the optimization is to attain an optimized concrete shape and cable layout for a given predefined external load while removing the tensile stresses (< 0.5 N/mm^2) in the prestressed beams and to find out the optimum number of spans which will strike a perfect balance between the amount of material saved due to optimized shape and the number of spans needed to be given.

Keywords Prestressed concrete beams · B-spline · Shape optimization · Cable optimization · Fuzzy theory

1 Introduction

Prestressed concrete is a specific form of concrete which was brought into existence to counterpoise the vulnerability of concrete in tension [1]. Since then, prestressed concrete has been commonly employed in the field of construction for various

I. Jha (✉) · K. K. Pathak
Research Scholar, Department of Civil Engineering, Indian Institute of Technology (BHU), Uttar Pradesh, Varanasi, India
e-mail: ishan.jha.rs.civ18@iitbhu.ac.in

K. K. Pathak
e-mail: kkpathak.civ@iitbhu.ac.in

K. K. Pathak et al. (eds.), *Latest Developments in Civil Engineering*, Lecture Notes in Civil Engineering 352, https://doi.org/10.1007/978-981-99-2676-3_52

uses like the construction of railway slippers, bridge girders, slabs, beams, etc. As the adoption of prestressed structures increased, the demand for their more optimized state also increased and various methods were employed by researchers to obtain more optimized form [2–6]. The present study involves a new method for optimization of prestressed structures using an indigenously developed software, "Simultaneous Shape and Cable Optimization" (SSCO).

Greater span lengths are achievable with prestressed beams that are not possible with reinforced cement concrete beams. But, as the length of span increases, the deflection and bending moment also increases [7]. Also, the beams with higher span lengths require more depth, thus adding to the cost of the material. Few researchers [8–10] have researched the change in strength and ehaviour due to the number of spans over continuous prestressed concrete beams. However, there is by far no study on the ehavioural changes of optimized prestressed beams because of span length. Hence, to prevent any failure in real life, the optimal state of prestressed beams ought to be examined for the deflection and stress parameters.

The current work tries to fill up this gap by comparing the impact of the number of spans in prestressed beams on the overall optimized cable layout and concrete shape as obtained using the software SSCO. The software SSCO is an extension of the software GSO [11, 12], which was used to obtain the optimized shape of precast structures. The span length of a single prestressed beam is altered by adding supports along its length, and the effects on the optimal cable layout, concrete shape, and the percentage of material saved under various loading circumstances are investigated. The obtained optimized shape is also checked on the criteria for deflection and stress.

2 Finite Element Modeling in SSCO

In the present study, all the models have been modeled and optimized in SSCO (Simultaneous Shape and Cable Optimization), which is an indigenously developed software for optimization of prestressed concrete beam shape and cable layout. The software SSCO works on an integrated zero-order technique which is controlled using fuzzy membership functions. The SSCO optimizes the shape of concrete and the layout of the cable in a mutual way following an iterative process in order to satisfy the convergence criteria that are sourced by the user.

The concrete here is modeled as a discretized continuum using nine noded Lagrangian elements, and the cable is modeled as a three-noded curved bar element. The beam is then discretized into a certain number of design elements, with each design element bearing eight key nodes that can be used for the modification of concrete boundary, hence changing the underlying shape of concrete. The key nodes that are designated for changing the shape are called design nodes. The cable changes its layout following a B-spline profile so that proper continuity is maintained. The cable takes the majority of tensile force coming on the beam. Hence, the layout of the cable changes according to the changed shape of concrete at each iteration to take up the tensile forces coming on it. The change in the shape of concrete and the cable layout occurs simultaneously until the criteria for convergence are achieved. Three criteria are selected in the current study for achieving convergence, which are:

i. Removal of tensile stress i.e. $\sigma_{bottom/top} \leq 0.5$ N/mm$^{2.}$
ii. A concrete cover of a minimum of 40 mm should be maintained between the cable and concrete boundary.
iii. The maximum deflection (δ_{max})shouldn't be greater than$\delta_{limit} = Span/250$. . [13].

For the current work, the beam is segmented into five design elements, and the key nodes present at the bottom of the design elements are declared as the design nodes. The cable is initially laid straight, passing through the neutral axis of the rectangular beam. It is then made to follow a third-order B-spline curve, which moves up and down to counter tensile stress.

The flowchart in Fig. 1 shows a step-by-step approach followed by SSCO during the entire analysis.

3 Validation and Convergence Study

To prove the authenticity and reliability of the software SSCO, a validation study is performed. Through this study, the mesh size to be used is also determined. A single span simply supported prestressed beam having length = 6000 mm, depth = 600 mm, and width = 300 mm is taken with material properties of concrete being Young's modulus (E) = 3.5 × 10^4 N/mm^2, Poisson's ratio (ν) = 0.15, and density (ρ) = 2.5 × 10^{-5} N/mm^3 and the prestressing cable being Young's modulus (E) = 2.1 × 10^5 N/mm^2, Poisson's ratio (ν) = 0.3, and density (ρ) = 7.85 × 10^{-5} N/mm^3. An eccentricity of 75 mm is provided throughout the cable layout. The beam is imposed with an 18 N/mm load distributed uniformly across the length. The deflection (δ) and bending stress (σ_{bottom} and σ_{top}) are selected as the two benchmarks for this study. The δ, σ_{bottom}, and σ_{top} due to the levied loading and self-weight are calculated using SSCO for the prestressed beam and compared with results obtained for the same beam using the conventional formulae (F1) [14].

The beam is segmented into a single design element that is then divided into nine noded Lagrangian elements with element sizes ranging in Y-axis/X-axis from 600/1000 to 300/250. The results of the study are shown in Fig. 2. From the results, it is evident that a 600/500 element size is the point at which convergence is attained.

4 Results

A simply supported prestressed beam is taken having length = 36,000 mm, depth = 1300 mm, and width = 800 mm. The beam is made up of concrete having Young's modulus (E) = 3.5 × 10^4 N/mm^2, Poisson's ratio (ν) = 0.15, and density (ρ) = 2.5 ×

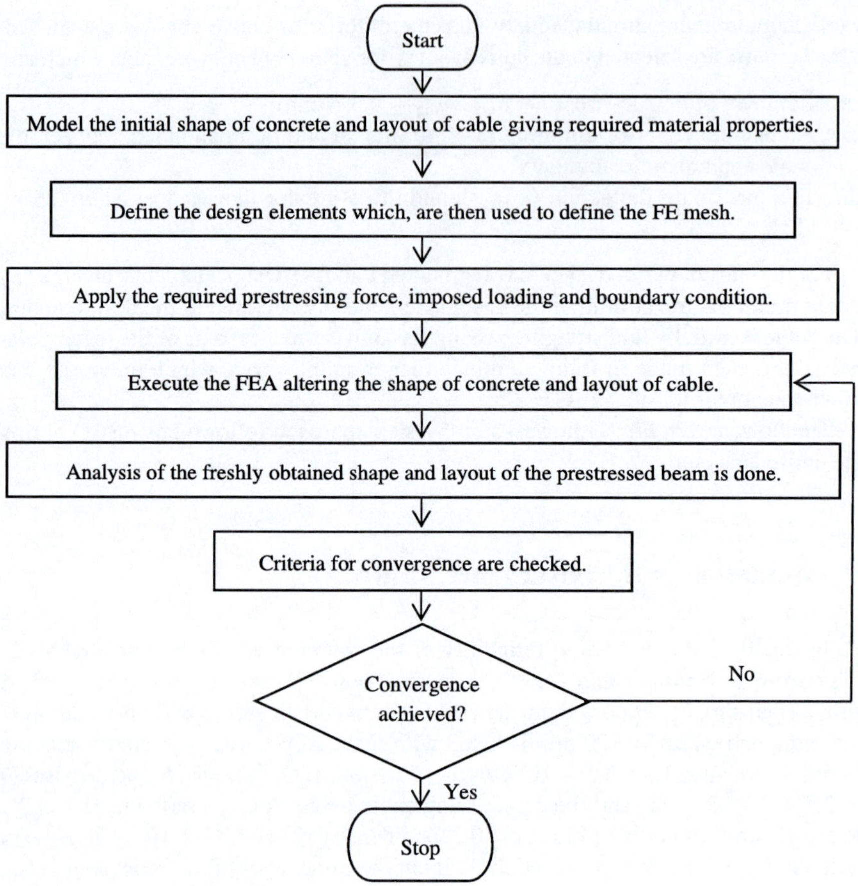

Fig. 1 Flowchart for SSCO procedure

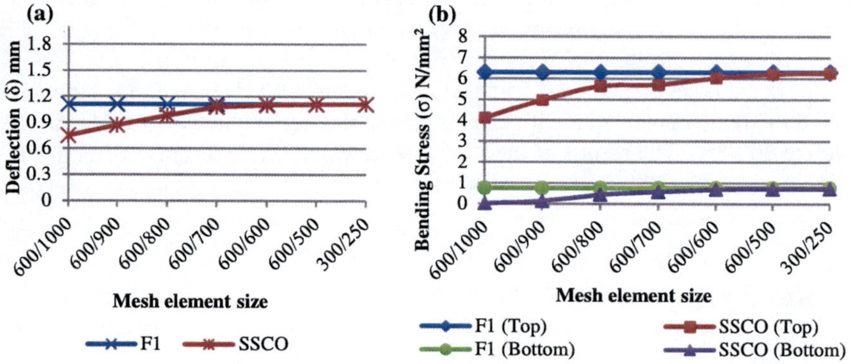

Fig. 2 Convergence study **a** based on deflection 'δ' **b** based on bending stress "σ_{bottom} and σ_{top}"

10^{-5} N/mm^3. The embedded prestressing cable has Young's modulus (E) = 2.1×10^5 N/mm^2, Poisson's ratio (ν) = 0.3, and density (ρ) = 7.85×10^{-5} N/mm^3. The beam is loaded under three different loadings, viz. (i) three-point load (3PL), (ii) uniformly distributed load (UDL), and (iii) uniformly varying load (UVL) with changing span length using intermediate supports by dividing the 36,000 mm length of the beam in (i) single span (B^1) having one full span of 36,000 mm length, (ii) two-span (B^2) having two spans of 18,000 mm length each, (iii) three-span (B^3) having three spans of 12,000 mm length each, (iv) four span (B^4) having four spans of 9000 mm length each, and (v) five span (B^5) having five spans of 7200 mm length each. The initial cross section in each case is kept constant, with the cable passing straight through the neutral axis. All the prestressed beams from B^{1-5} are then optimized for cable layout and concrete shape using SSCO for the three different load cases (3PL, UDL, and UVL) individually. The optimized state is then compared for the amount of material saved, change in deflection, and bending stress under each case from the initial unoptimized state. The results obtained under the 3PL, UDL, and UVL load case is compiled in Tables 1, 2, 3, respectively.

4.1 *Load Case 1: Three-Point Load (3PL)*

In load case 1, the prestressed concrete beam is imposed with three-point loads (i) 700 kN at 9000 mm from the beam's left side, (ii) 600 kN at 18,000 mm from the beam's left side, and (iii) 700 kN at 27,000 mm from the beam's left side acting in downward ('-'ve Y) direction. At the jacking ends, a 1000 kN prestressing force is applied while taking friction loss into account. The prestressed beam is then optimized using SSCO, and the results are compared with the beam's original unoptimized condition over the grounds of (i) bending stress (σ_{bottom} and σ_{top}), (ii) deflection (δ), and (iii) percentage material saved (M_{saved}) and are compiled in Table 1.

The optimized layout of cable and concrete shape obtained for all the prestressed beams B^{1-5} under the 3PL loading is illustrated in Fig. 3.

4.2 *Load Case 2: Uniformly Distributed Load (UDL)*

In load case 2, the prestressed concrete beam is imposed with a uniformly distributed load of 50 N/mm across the length of the prestressed beam acting in the downward ('-'ve Y) direction. At the jacking ends, a 1000 kN prestressing force is applied while taking friction loss into account. The prestressed beam is then optimized using SSCO, and the results are compared with the beam's original unoptimized condition over the grounds of (i) bending stress (σ_{bottom} and σ_{top}), (ii) deflection (δ), and (iii) percentage material saved (M_{saved}) and are compiled in Table 2.

The optimized layout of cable and concrete shape obtained for all the prestressed beams B^{1-5} under the UDL loading is shown in Fig. 4.

Table 1 Three-point load (3PL)

Prestressed beam	State	Section	Bending stress (N/mm^2)		δ (mm)	M_{saved} (%)
			σ_{bottom}	σ_{top}		
B^1	Initial	1–1	48.72	−51.22	−285.7	01.61
	Optimized		−02.54	−40.68	−067.9	
B^2	Initial	1–1	06.91	−08.74	−07.51	19.29
	Optimized		−03.64	−11.75	−01.02	
	Initial	2–2	06.94	−08.72	−07.52	
	Optimized		−09.95	−03.22	−01.92	
B^3	Initial	1–1	01.23	−03.13	−01.52	19.57
	Optimized		-04.79	−05.49	−00.87	
	Initial	2–2	01.83	−04.30	−04.59	
	Optimized		-03.02	−06.82	00.67	
	Initial	3–3	01.28	−03.08	−01.52	
	Optimized		−04.24	−05.36	−00.99	
B^4	Initial	1–1	−01.42	−00.46	00.15	20.26
	Optimized		−05.08	−00.23	00.85	
	Initial	2–2	−00.83	−00.98	−00.13	
	Optimized		−04.17	−01.29	00.22	
	Initial	3–3	−00.83	−00.90	−00.11	
	Optimized		−04.16	−01.03	00.28	
	Initial	4–4	−01.31	−00.33	00.16	
	Optimized		−04.63	−00.06	00.80	
B^5	Initial	1–1	−01.60	−00.29	00.17	20.33
	Optimized		−05.74	00.34	00.85	
	Initial	2–2	00.01	−01.89	−00.27	
	Optimized		−02.31	−03.36	−00.38	
	Initial	3–3	01.27	−03.68	−00.32	
	Optimized		00.15	−05.88	−00.41	
	Initial	4–4	00.02	−01.81	−00.26	
	Optimized		−02.22	−03.13	−00.35	
	Initial	5–5	−01.49	−00.18	00.17	
	Optimized		−05.29	00.47	00.81	

4.3 Load Case 3: Uniformly Varying Load (UVL)

In load case 3, the prestressed concrete beam is imposed with a uniformly varying load of 0 N/mm at starting point on the left-hand side and gradually increasing up to 150 N/mm towards the right-hand side, acting in the downward ('−'ve Y) direction. At the jacking ends, a 1000 kN prestressing force is applied while taking friction loss

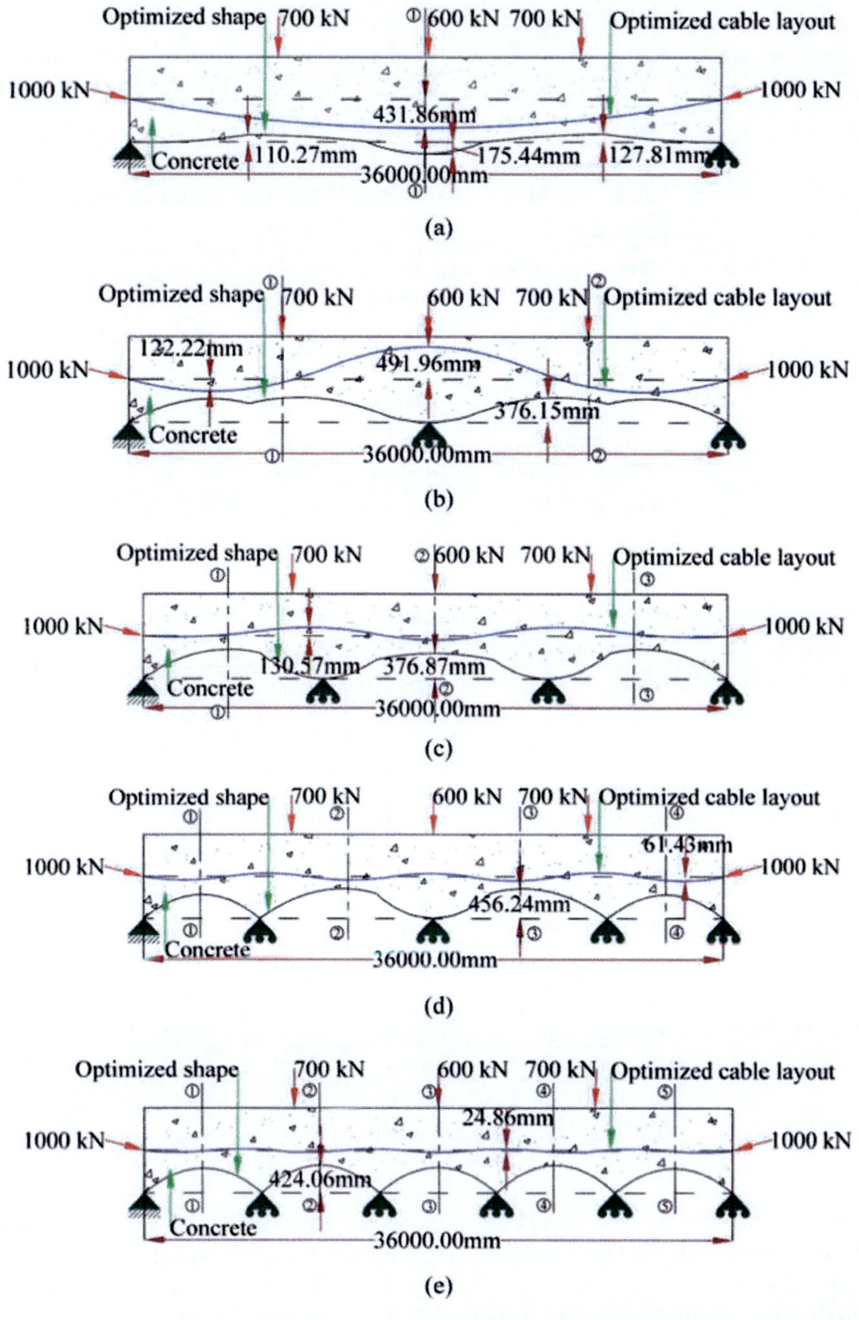

Fig. 3 Optimized state of prestressed concrete beam under 3PL for beam **a** B1, **b** B2, **c** B3, **d** B4, and **e** B5

Table 2 Load case 2: Uniformly distributed load (UDL)

Prestressed beam	State	Section	Bending stress		δ (mm)	M_{saved} (%)
			σ_{bottom}	σ_{top}		
B1	Initial	1–1	34.32	−36.21	−210.3	−3.46
	Optimized		−00.81	−30.76	−069.5	
B2	Initial	1–1	03.37	−05.27	−05.30	20.71
	Optimized		−05.54	−07.92	−05.28	
	Initial	2–2	03.40	−05.25	−05.31	
	Optimized		−05.01	−07.87	−01.15	
B3	Initial	1–1	01.25	−03.13	−01.34	21.31
	Optimized		−00.93	−04.48	−01.55	
	Initial	2–2	−00.40	−01.38	−02.35	
	Optimized		−02.82	−02.05	00.85	
	Initial	3–3	01.35	−03.06	−01.35	
	Optimized		−00.44	−04.51	−01.74	
B4	Initial	1–1	−00.04	−01.84	−00.29	22.76
	Optimized		−03.06	−02.54	00.01	
	Initial	2–2	−00.28	−01.53	−00.19	
	Optimized		−03.16	−02.34	00.09	
	Initial	3–3	−00.28	−01.45	−00.17	
	Optimized		−03.19	−02.05	00.17	
	Initial	4–4	00.06	−01.71	−00.28	
	Optimized		−02.40	−02.48	−00.11	
B5	Initial	1–1	−00.12	−01.77	−00.20	25.39
	Optimized		−04.39	−01.97	00.29	
	Initial	2–2	−00.61	−01.22	−00.05	
	Optimized		−03.45	−02.57	−00.01	
	Initial	3–3	−00.46	−01.33	−00.09	
	Optimized		−03.84	−02.07	00.15	
	Initial	4–4	−00.60	−01.14	−00.05	
	Optimized		−03.41	−02.27	00.03	
	Initial	5–5	−00.02	−01.66	−00.19	
	Optimized		−03.74	−01.95	00.21	

into account. The prestressed beam is then optimized using SSCO, and the results are compared with the beam's original unoptimized condition over the grounds of (i) bending stress (σ_{bottom} and σ_{top}), (ii) deflection (δ), and (iii) percentage material saved (M_{saved}) and are compiled in Table 3.

The optimized layout of cable and concrete shape obtained for all the prestressed beams B^{1-5} under the UVL loading is shown in Fig. 5.

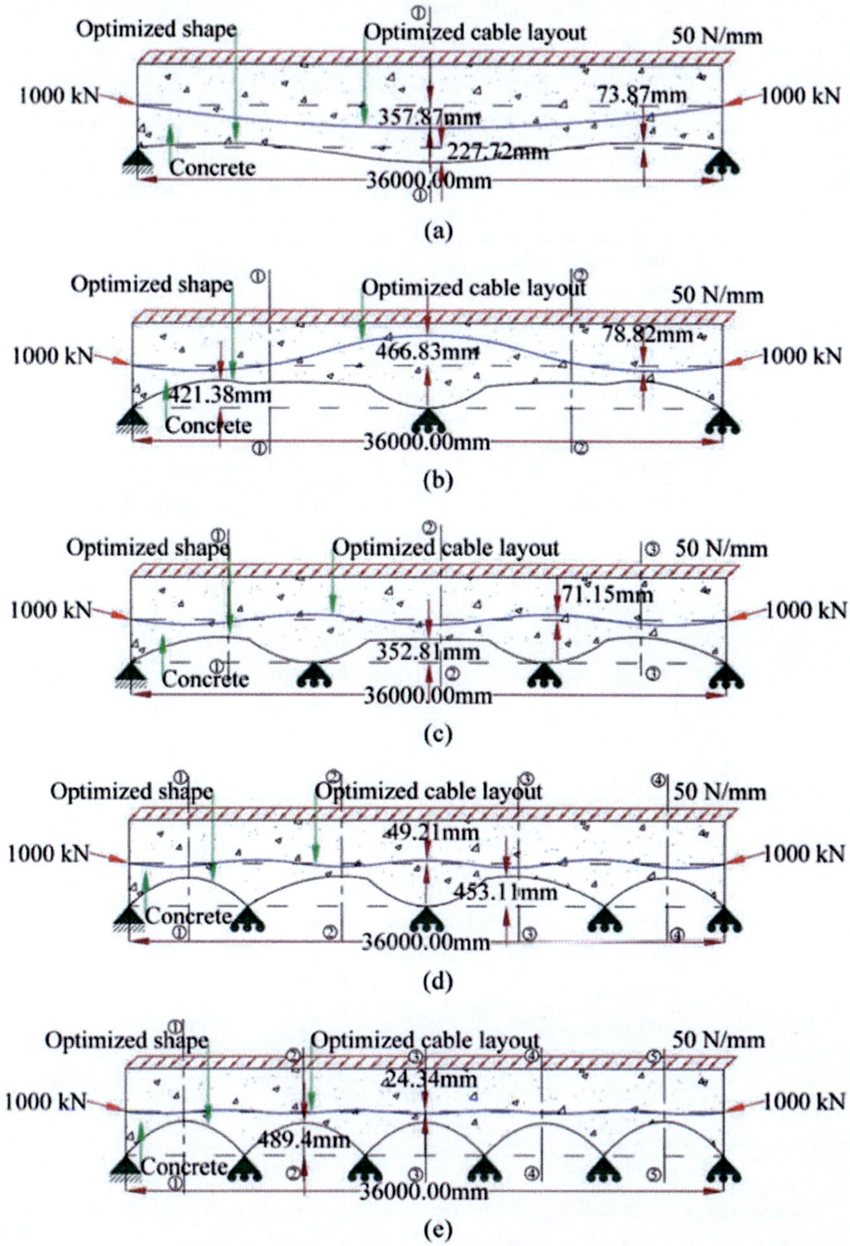

Fig. 4 Optimized state of prestressed concrete beam under UDL for beam **a** B1, **b** B2, **c** B3, **d** B4, and **e** B5

Table 3 Load case 3: Uniformly varying load (UVL)

Prestressed beam	State	Section	Bending stress		δ (mm)	M_{saved} (%)
			σ_{bottom}	σ_{top}		
B1	Initial	1–1	52.02	−53.90	−315.6	−4.42
	Optimized		−01.46	−39.45	−081.2	
B2	Initial	1–1	−00.81	−01.08	01.60	19.53
	Optimized		−07.31	−02.32	06.93	
	Initial	2–2	12.01	−13.85	−17.62	
	Optimized		−02.58	−16.55	−16.17	
B3	Initial	1–1	−00.13	−01.74	−00.46	19.74
	Optimized		−11.04	−02.51	02.36	
	Initial	2–2	−00.01	−01.79	−00.11	
	Optimized		−09.24	−03.45	02.26	
	Initial	3–3	05.11	−06.83	−03.64	
	Optimized		−03.13	−08.78	−02.82	
B4	Initial	1–1	−00.90	−00.98	−00.01	19.78
	Optimized		−06.15	−01.75	00.72	
	Initial	2–2	−0.09	−01.72	−00.26	
	Optimized		−05.12	−03.13	00.18	
	Initial	3–3	00.35	−02.09	−00.34	
	Optimized		−04.52	−03.22	00.29	
	Initial	4–4	01.93	−03.58	−00.88	
	Optimized		−00.74	−05.99	−01.14	
B5	Initial	1–1	−00.78	−01.11	−00.04	23.74
	Optimized		−05.07	−00.88	00.55	
	Initial	2–2	−00.58	−01.26	−00.06	
	Optimized		−03.33	−02.57	−00.03	
	Initial	3–3	−00.19	−01.60	−00.14	
	Optimized		−03.20	−02.68	−00.01	
	Initial	4–4	−00.23	−01.51	−00.11	
	Optimized		−03.21	−02.45	00.11	
	Initial	5–5	01.48	−03.17	−00.55	
	Optimized		00.04	−05.17	−00.81	

5 Discussion

The δ_{limit} for the beam of span 36,000 mm comes to 144 mm. Hence, δ_{max} at any section should not be greater than 144 mm. From the results obtained in Sect. 4, it can be seen that when the prestressed beam is loaded in a single span, the deflection

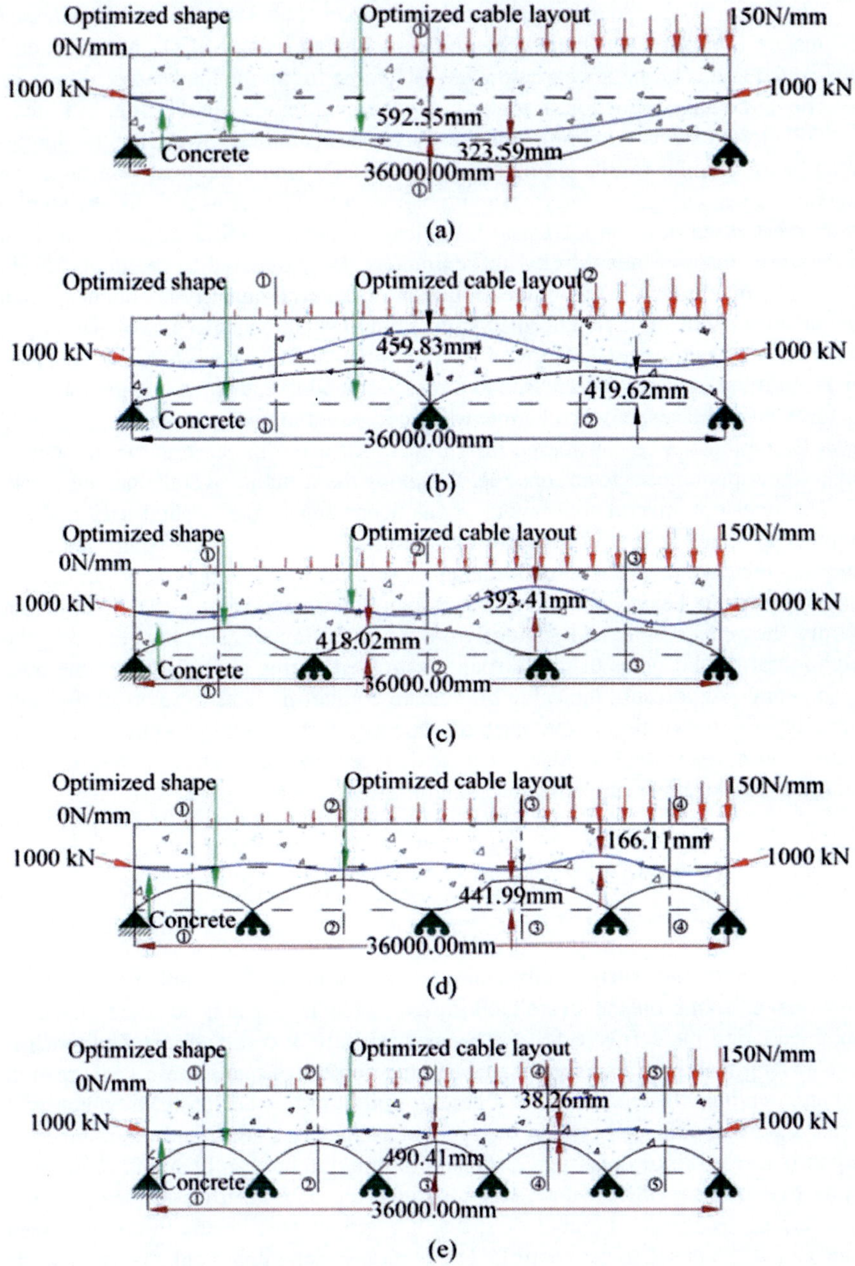

Fig. 5 Optimized state of prestressed concrete beam under UVL for beam **a** B1, **b** B2, **c** B3, **d** B4, and **e** B5

in Sect. 1 in all of the load cases is much more than δ_{limit}. But when the prestressed beams are optimized for their shape and cable layout using SSCO, the δ_{max} is also controlled and is brought well below the δ_{limit} value for all the load cases.

The percentage reduction in material for the prestressed beam optimum state in the 3PL load case is maximum for the five-span beam and lowest for the single-span beam. Similar results are obtained for the UDL and UVL load case, with the maximum percentage of material saving being up to 25%. However, for the single-span prestressed beam in UDL and UVL load cases, the optimized state shows an increase in material quantity by approximately 4% compared to the unoptimized state. The maximum change in the reduction of material quantity is obtained when the prestressed beam is transformed from a single-span beam to a two-span beam, where it can be observed that around 19% saving in the material is obtained. Further, as the number of spans is increased to three, four, and five, the percentage reduction in material increases, but the change, when compared to two-span beams, is around 1%–4% which can be considered insignificant for every increase in the number of spans the support needs to be constructed making the structure overall uneconomical.

The final optimized cable layout varies according to the applied load and the number of spans given to the prestressed beam. For a single-span beam, the cable shows a concave layout with coordinates of cable at mid-reaching near the bottom boundary of the beam. However, in five-span prestressed beams, the cable layout hardly shows any significant change from its initial layout. This is because of the high initial tensile stress in single-span prestressed beams as compared to the five-span beams. As a result, the cable in order to counter the tensile stress changes its initially straight position to the desired shape as per the loading and number of the span as observed in Sect. 4. Also, it can be deduced here that as the number of spans increases, the chances of having a tensile failure decrease.

6 Conclusion

The presented study successfully compares the effect of the number of spans in prestressed beams on the overall optimized cable layout and concrete shape as obtained using the software SSCO. From this study, it can be concluded that for a given length of the prestressed beam, a higher number of spans doesn't ensure high material savings. Instead, there is a certain optimum number of spans which will strike a perfect balance between the amount of material saved and the number of supports needed to be constructed. This optimum number can be found effectively using the software SSCO which will also give the optimized layout of cable and concrete shape for that number of spans. The prestressed beams in an optimized state can also be used to get controlled deflection which might not be possible in its unoptimized state. The present work can be used by industries in the designing of continuous shape-optimized prestressed beams and girders saving a lot of material and cost of construction. However, the one practical limitation of the present method is that it would be difficult to make formwork for the shapes obtained after optimization. But, as the research work progresses, it is hoped that the issue of having different formworks for different unorthodox shapes will be possible in the near future. In future, work on the optimization of cable force for a given imposed load can be introduced to the present method.

References

1. Masterman OJ (1952) Prestressed concrete. Nature 169:901–905
2. Erbatur F, Zaid RAl, Dahman NA (1991) Optimization and sensitivity of prestressed concrete beams. Comput Struct 45(5/6),881–886
3. Yoely YM, Amir O, Hanniel I (2018) Topology and shape optimization with explicit geometric constraints using a spline-based representation and a fixed grid. Procedia Manuf 21:189–196
4. Zelickman Y, Amir O (2021) Layout optimization of post-tensioned cables in concrete slabs. Struct Multi discip Optim 63:1951–1974
5. Mohammed AH, Taysi N, Nassani DE, Hussein AK (2017) Finite element analysis and optimization of bonded post-tensioned concrete slabs. Cogent Eng 4(1):1341288
6. Zelickman Y, Amir O (2022) Optimization of post-tensioned concrete slabs for minimum cost. Eng Struct 259:114132
7. Heyman J (1998) Structural analysis: a historical approach. Cambridge University Press, Cambridge
8. Aravinthan T, Witchukreangkrai E, Mutsuyoshi H (2005) Flexural Behavior of Two-Span Continuous Prestressed Concrete Girders with Highly Eccentric External Tendons. ACI Struct J 102(S40):402–411
9. Hawkins NM (1961) Strength and behavior of two-span continuous prestressed concrete beams. University of Illinois
10. Huber P, Huber T, Kollegger J (2020) Experimental and theoretical study on the shear behavior of single- and multi-span T- and I-shaped post-tensioned beams. Structural Con-crete 21:393–408
11. Jha I, Pathak KK (2021) Fuzzy-based integrated zero-order shape optimization of steel–concrete–steel sandwich beams. Curr Sci 121(7):941–949
12. Jha I, Pathak KK, Jha M, Ranjan A (2022) A comparative study of gradient descent method and a novel non-gradient method for structural shape optimiza-tion. Int J Math, Eng Manag Sci 7(2):258–271
13. IS 1343:2012: Prestressed concrete—code of practice. Bureau of Indian standards (2012)
14. Nawy EG (2009) Prestressed concrete a fundamental approach. 5th Edn, Pearson, New York (2009)

Evaluation of Flexure Performance of Beam Hybridized with PVA Fiber and GFRP Bars

Mohammed Yasir Hussain, K. Hemalatha, and V. Srinivasa Reddy

Abstract This study's goal is to present the use of PVA fibers, Poly Carboxyl Ether (PCE) superplasticizer, and glass fiber-reinforced polymer (GFRP) as reinforcement bars in concrete beams. There were four examples, each with a unique reinforcement and PVA fiber ratio. Two specimens, known as pure FRP beams since they only contained GFRP reinforcement and two hybrid fiber-reinforced polymer (HFRP) beams. Following dosing experiments with concrete, the volume fraction of PVA fiber in two specimens of the pure FRP beams and HFRP beams, respectively, was 0 and 0.25%. Pure FRP beams produced outcomes that weren't suitable for the beam considerations needed. In contrast, when compared to pure FRP beams, the HFRP beams performed better and produced satisfactory results. The results of this study, which examined the load–deflection capabilities of all the specimens described, demonstrated that HFRP beams had a higher load carrying ability and that a PVA fiber dosage of 0.25% produced favorable outcomes. Even though the HFRP beams had a strong capability for carrying loads, they deflected more than the others.

Keywords Hybrid fiber-reinforced concrete · Polyvinyl alcohol · Flexural load capacity · High-strength concrete

M. Y. Hussain (✉)
Structural Engineering, Gokaraju Rangaraju Institute of Engineering and Technology, Hyderabad, India
e-mail: yasirhussain1919@gmail.com

K. Hemalatha
Civil Engineering, Gokaraju Rangaraju Institute of Engineering and Technology, Hyderabad, India

V. S. Reddy
3Civil Engineering, Gokaraju Rangaraju Institute of Engineering and Technology, Hyderabad, India

K. K. Pathak et al. (eds.), *Latest Developments in Civil Engineering*, Lecture Notes in Civil Engineering 352, https://doi.org/10.1007/978-981-99-2676-3_53

1 Introduction

Steel being a tremendous construction element has its own limitations and drawbacks. It has undeniable properties for flexure strength in concrete structures. But, the irresistible nature towards corrosion and cost has pushed the construction industry for an alternative. Use of Fiber-Reinforced Polymer (FRP) has been one of the possible solutions but not the exact solution as per many literatures. The complete replacement of steel is not significant but hybridizing it with other polymers like FRP is getting its significance in this construction field as we know retrofitting is one of the many applications of hybridization. Moreover, the use of these useful polymers for High-Strength Concrete (HSC) is of great importance. This paper deals with the use of Polyvinyl Alcohol (PVA) and Glass Fiber-Reinforced Polymer (GFRP) rebars as reinforcement in HSC beam. Prerequisites are as follows.

1.1 Fiber-Reinforced Polymer

Goods made of fibers that are reinforced by polymer resin are known as fiber-reinforced polymer (FRP) products. Ideal fibers would be brittle, elastic, and stronger than the matrix. The composite's mechanical performance is influenced by the fiber type, direction, length, shape, volumetric ratio, matrix adhesion, and manufacturing method. As seen in Fig. 1, typical FRP reinforcing items include sheets, grids, bars, textiles, tendons, and ropes. There are numerous cross-sectional forms for the bars (strip, round, solid, and hollow). The composite's primary load-resisting element is made up of fibers [1]. They need to be strong, have a high elastic modulus, be sufficiently elongated after failure, and be resistant to the environment in which the structure will be used. The length, form of the cross section, and chemical makeup of the fibers all have an effect on how properly they work. Fibers are available in quite a few cross-sectional diameters and forms. The diameter of the fiber stages from 5 to 25 microns, relying on the form of fibers. Carbon, Glass, Aramid, and Basalt are the four maximum famous fibers utilized in FRPs, and the related composite merchandise is called CFRP, GFRP, AFRP, and BFRP, respectively [2]. Carbon Fibers: Carbon fibers provide incredible moisture and chemical resistance, an excessive elastic modulus (200–800 GPa), excessive energy and stiffness-to-weight ratios, and coffee susceptibility to fatigue loads [3, 4]. However, due to their low last strain, carbon fibers have a constrained effect resistance. Additionally, carbon fibers have excessive warmness and electric conductivity, which, relying on the designer, can be a gain or a drawback [14] (Fig. 1).

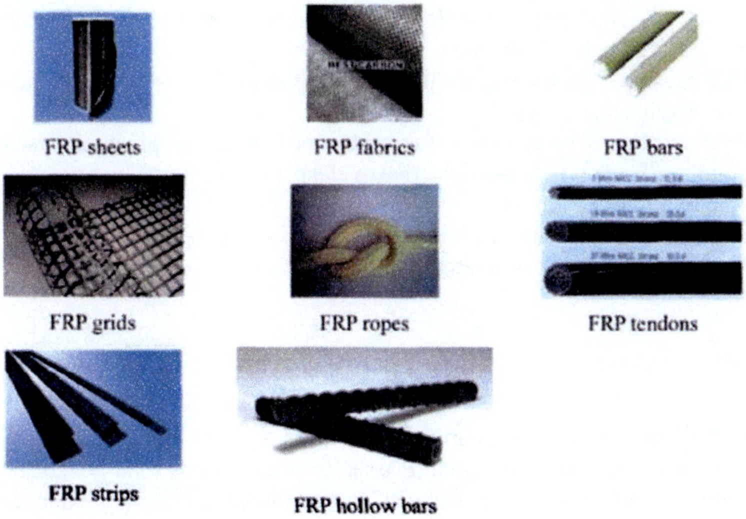

Fig. 1 Typical FRP reinforcement products

1.2 GFRP Rebar

Glass fiber reinforced polymer (GFRP) rebar is a type of marketed bar that includes additives and the thinnest glass fiber possible incorporated in a polymeric resin matrix. Corrosion of the reinforcing steel is the main factor contributing to the degeneration of reinforced concrete structures. One practical solution is to use non-corrosive glass fiber-reinforced polymer (GFRP) bars to strengthen concrete.

Glass fibers make up the majority of GFRP reinforcing bars. Due to the non-corrosive nature, GFRP bars may significantly increase the lifespan of reinforced concrete buildings while also lowering the expenses associated with maintenance, repair, and replacement. GFRP is a growing reinforcing option, but its design issues are distinct from those faced when designing traditional steel-reinforced concrete.

Advantages of GFRP Rebar

Rebar product of glass fiber-strengthened polymer is a fairly precious constructing fabric. Governments and different large-scale infrastructure companies an increasing number of realize that GFRP is a lower priced constructing fabric with the capacity to growth the lifespan of public centers wherein corrosion may have a massive poor effect on the economic system and the environment. The use of fiber glass reinforcing material has grown significantly as a result of the growth in corrosion caused by climate change. Future versions of these sophisticated composite materials would be able to more clearly exhibit their strengths and characteristics [5]. Here are some of the advantages of using GFRP rebar in various applications.

1. A high-quality vinyl ester resin that is corrosion-resistant and extends the life of a concrete structure is one of the components of GFRP.

2. Since GFRP rebar projects require no maintenance, construction companies may avoid paying for rehabilitation.
3. GFRP rebar has double the tensile strength of steel while weighing a quarter as much as typical steel rebar.
4. GFRP rebar is a great option for buildings like power plants and scientific institutions since it is non-conductive to electricity and heat.
5. Compared to epoxy-coated or stainless steel, GFRP rebar is a more affordable option when considering the long-term advantages of the material.
6. It has a greater weight-to-strength ratio than others.

1.3 PVA Fibers

Polyvinyl alcohol (PVA) makes up at least 50% of the fiber-forming component in vinylon fibers, which also contain at least 85% of the fiber's total weight in vinyl alcohol units as well as one or more acetal units. Another name for the fiber is vinal or vinylal.

Vinyl monomer is not stable; hence, vinyl resin cannot be produced from it. In its place, polyvinyl acetate is hydrolyzed to yield polyvinyl alcohol (PVOH). The PVA polymer is then coagulated by spinning it either wet or dry through a nozzle into a coagulating solution. PVA fibers with good UV and chemical resistance, high tenacity, high Young's modulus, low elongation, and low creep (alkali, acid, oil, etc.).

PVA fiber is an appropriate reinforcement material for cementitious composites due to its properties. An excellent environmentally friendly cement reinforcement material is polyvinyl alcohol fiber. It, because of its distinctive molecular structure, has alkali and weather resistance, therefore it enhances the concrete's resistance to frost. Improving bending strength, impact strength, and fracture strength; improving permeability; and improving the impact and seismic resistance of concrete by forming a strong affinity with cement [11]. increased toughness as a result of increased concrete's bending strength, impact resistance, and brittleness. Strong bonding with cement matrix is one of PVA fiber's notable properties [7, 8] (Fig. 2).

Fig. 2 Sample of PVA fiber

1.4 Applications of PVA Fiber

1. PVA Fibers' "stealth" quality is their main advantage. They can be cast all over the concrete and hardly or not at all be visible on the completed surfaces. This is true whether you are producing troweled finished concrete or concrete cast in a mold [12, 13].
2. The ductility or flexibility of PVA-reinforced concrete is exceptional. Instead of having a quick catastrophic failure, the concrete section will flex substantially before it fractures. It only fails in a different way, which does not imply that it is more durable than AR Glass fiber in use. In comparison tests, AR Glass performs better, although it isn't always feasible to utilize it for a particular application.
3. Concrete and PVA are said to "chemically bind" together. This assertion looks doubtful to ordinary observation. We won't give that assertion much credence and will just leave it alone. If it's accurate, we prevail. If not, we won't worry about it.

1.5 Hybrid Fiber-Reinforced Polymer

The use of polymers in addition to others to delimit the drawbacks and improve the qualities of the specimen is hybridization [9]. In this paper, we have used PVA fiber and GFRP rebar in the specimen beam. PVA is used to improve the strength of concrete and GFRP to improve the flexural performance of beam.

2 Theoretical and Experimental Work

2.1 Mix Proportions of M70-Grade Concrete

As per IS 10262:2019 and IS456:2000, the mix design has been done and the following proportions have been achieved.

Cement	Water	Fly ash	Silica fume	Coarse agg	Fine agg
1	0.35	0.186	0.063	2.36	1.14

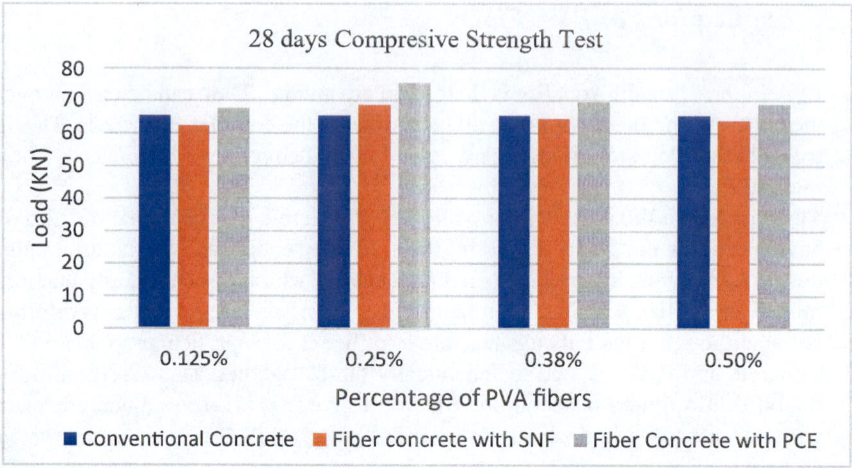

Fig. 3 Comparison of compressive strength

2.2 Comparison of PCE, SNF Plasticizer, and PVA Fiber Dosage for m70 Mix

The plasticizer and PVA fiber dosage were selected after the workability test with respect to slump and compression test and split tensile strength, respectively. The following is observed and the selection is done.

2.3 Compressive Strength Testing for 28 days

The compressive strength test was carried out for 28 days on cubes of 150 mm in length, 150 mm in width, and 150 mm in depth. The findings are listed below for the various combinations (Fig. 3).

2.4 Split Tensile Strength of 28 days

The split tensile strength test was carried out for 28 days on cylinders with a diameter of 150 mm and a length of 300 mm. The results are listed below for the various blends (Fig. 4).

PCE superplasticizer = 0.3% of cementitious content= 0.95 was selected. PVA fiber dosage was fixed at 0.25% volume fraction.

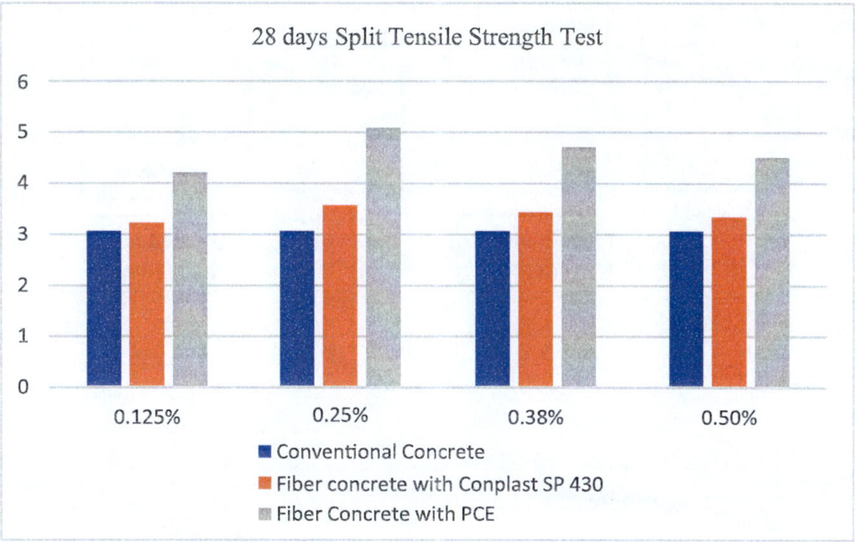

Fig. 4 Comparison of split tensile strength

2.4.1 Beam Design

The design of beam was done according to IS 456:200 and dimensions were fixed as (Figs. 5, 6 and 7).

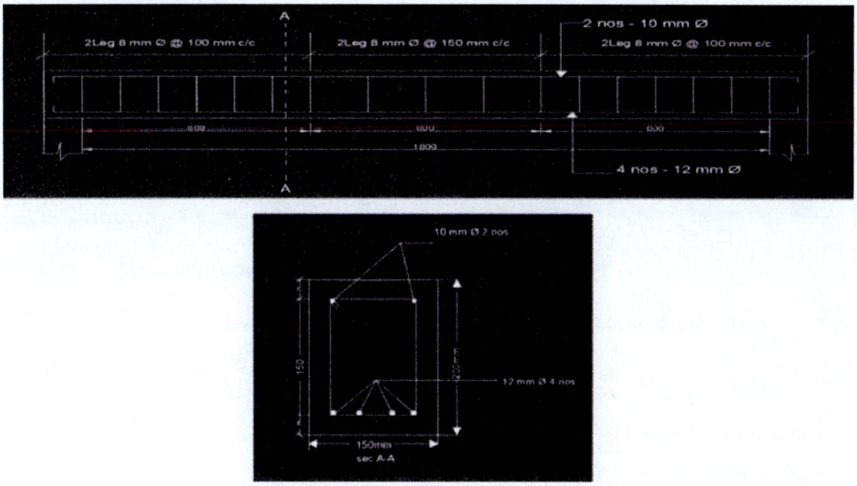

Fig. 5 Longitudinal section of beam

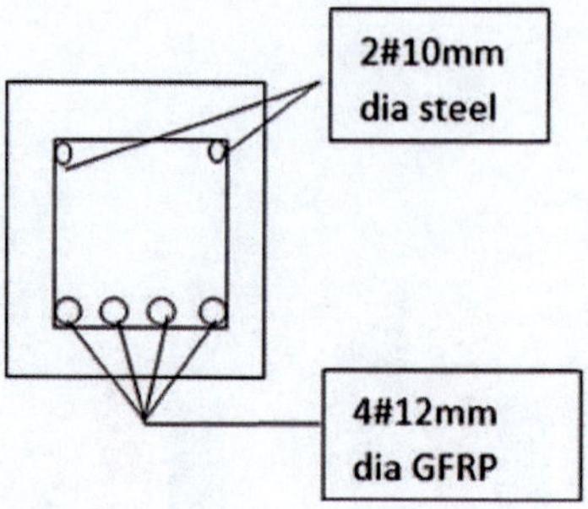

Fig. 6 Cross-sectional view of RCC beam

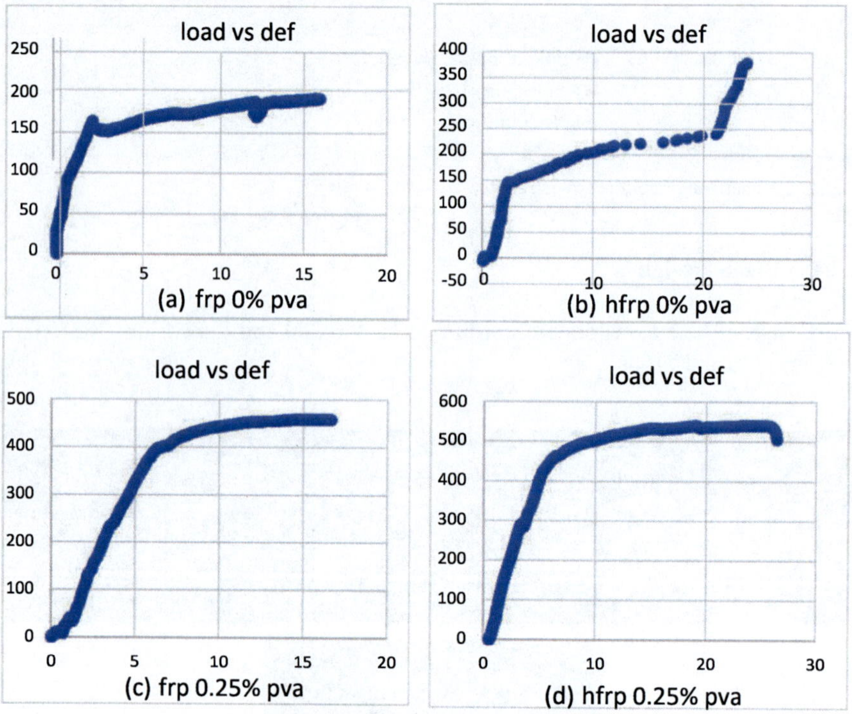

Fig. 7 A frp 0% pva, **b** hfrp 0% pva, **c** frp 0.25% pva, **d** hfrp 0.25% pva

Length l—2000 mm
Width b—150 mm Depth D—200 mm
Effective depth—175 mm

2.4.2 Flexural Performance of Beam Specimens

3 Discussion of Results

The test findings demonstrated that polycarboxylate ether-based superplasticizers were superior to SNF ones. PCE superplasticizer was used to obtain the desired slump in accordance with Table 1. The desired strength (80 MPa) was practically exclusively attained by PCE plasticizer; thus, we utilized 0.3% of PCE in the concrete mix since the compressive strength obtained after using PCE was superior to that of SNF, as indicated in Tables 2 and 3. Both cases attained high tensile strengths, but PCE plasticizer produced superior outcomes, as demonstrated in Tables 4 and 5. After 28 days, the testing was done (Table 6).

The PVA fiber percentages at 0, 0.125, 0.375, and 0.5% were studied. These similarities are shown in Fig. 3. These numbers demonstrate that the compressive and split tensile strengths of SNF-based specimens weren't very good. The PCE-based specimens, however, produced better outcomes, and the dosage of 0.25% PVA fiber was adequate in both situations of strength. The specimens with less than 0%

Table 1 Workability of different mixes

S. No	Sample	Plasticizer used	Plasticizer (%)	Slump value
1	PP-0	PCE-based	0.3	90 mm
2	PP-0.125	PCE-based	0.4	87 mm
3	PP-0.250	PCE-based	0.6	85 mm
4	PP-0.375	PCE-based	1.0	73 mm
5	PP-0.500	PCE-based	1.5	69 mm
6	CP-0	SNF-based	0.3	84 mm
7	CP-0.125	SNF-based	0.7	80 mm
8	CP-0.250	SNF-based	0.9	72 mm
9	CP-0.375	SNF-based	1.2	65 mm
10	CP-0.500	SNF-based	1.6	63 mm

0,0.125,0.25,0.375,0.500 are dosages of PVA fiber

Table 2 Compressive strength test results for SNF-based plasticizer (CP) of 28 days

Sample with PVA fiber dosage	Compressive strength of fiber concrete (N/mm^2)	% Change
CP-0	60.5	–
CP-0.125	62.3	+ 2.97
CP-0.250	68.6	+ 13.38
CP-0.375	64.5	+ 6.61
CP-0.500	63.9	+ 5.61

Table 3 Compressive strength test results for PCE-based plasticizer (PP) of 28 days

Sample with PVA fiber dosage	Compressive strength of fiber concrete (N/mm^2)	% Change
PP-0	65.8	–
PP-0.125	69.6	+ 2.73
PP-0.250	78.5	+ 14.58
PP-0.375	69.8	+ 6.07
PP-0.500	68.8	+ 4.55

Table 4 Split strength test results for SNF-based plasticizer of 28 days

Sample with PVA fiber dosage	Split tensile strength of fiber concrete (N/mm^2)	% Change
CP-0	3.05	–
CP-0.125	3.22	+5.57
CP-0.250	3.56	+16.72
CP-0.375	3.43	+12.13
CP-0.500	3.34	+9.50

Table 5 Split strength test results for PCE-based plasticizer of 28 days

Sample with PVA fiber dosage	Split tensile strength of fiber concrete (N/mm^2)	% Change
PP-0	3.55	–
PP-0.125	4.21	+18.37
PP-0.250	5.09	+43.94
PP-0.375	4.7	+28.14
PP-0.500	4.5	+20.02

Table 6 Comparison of FRP and HFRP beams

Specimen	Yield deflection (mm)	Ultimate Def (mm)	DF	Ult Load (kN)	Ultimate moment (kN-m)
FRP 0PVA	2.33	16.52	7.09	193.23	48.31
HFRP 0PVA	2.37	23.92	10.09	380.14	95.04
FRP 0.25PVA	7.6	18.38	2.41	458.13	114.53
HFRP 0.25PVA	6.24	26.16	4.19	525.81	131.45

PVA and more than 0.25% PVA weren't good enough. The dosage of 0.25% PVA complied with [15].

The beams were then cast with the M70 design mix, reinforced with pure FRP and hybrid-designed HFRP with dosages of 0 and 0.25% in both cases and tested for flexure response under 4-point loading. The results, which were plotted as shown in Figs. 7a, b, c, and d above, revealed that the deflection and load carrying capacity were dramatically altered. The findings of the beam with 0% PVA and pure FRP reinforcement were less than adequate, with yield deflection of 2.33 mm, ultimate deflection of 16.52 mm, deflection factor of 7.1, ultimate load of 193.23 MPa, and ultimate moment of 48.31 kN-m. The performance of the HFRP beam with 0% PVA was greater than that of the pure FRP beam (95.04 kN-m moment) [10]; however, the deflections were a little high (23.92 mm). The beams reinforced with pure FRP and 0.25% PVA performed better than FRP and HFRP with 0% PVA, with a moment of 114.53 kN-m, which is better than the preceding two cases. The HFRP 0.25% PVA demonstrated the best performance, with a moment capacity of 131.45 kN-m and a tolerable deflection. Compared to pure FRP beams, the HFRP specimens displayed higher yielding.

4 Conclusions

The study examined the flexural capabilities of pure FRP-reinforced and HFRP-reinforced beams having 0 and 0.25% PVA of volume fraction in both cases. The conclusions from the aforesaid experimental results were as follows.

- SNF specimens' compressive and split tensile strengths were lower than those of PCE specimens.
- Additionally, as compared to other percentages employed, the PVA percentage of 0.25% produced greater outcomes. The ability to achieve strength decreased when the fiber dosage was above 0.25%. Comparable to other dosages, the compressive strength and split tensile strength of the 0.25% dosage specimen were 78.5 MPa and 5.09 MPa, respectively.
- This demonstrates that PCE should be used in high-strength concrete mixes instead of SNF as the superplasticizer.
- Compared to previous specimens, the HFRP 0.25% PVA beam has a substantially higher load carrying capacity. Therefore, it is clear that employing only FRP reinforcement would not result in a considerable increase in strength; instead, for better outcomes, steel, and FRP bars should be used.
- The pure FRP beam with 0.25% PVA had 137.07% greater ultimate moment than those with 0% PVA. The HFRP beam with 0.25% PVA exhibited a 38.31% greater ultimate moment.
- Regardless of dosage variations with pure FRP beams, the yield strength was noticeably stronger in HFRP beams.
- Despite their greater load carrying capacity, HFRP beams exhibited greater deflections.

References

1. Saleem S, Hussain Q, Pimanmas A (2017) Compressive behavior of PET FRP–confined circular, square, and rectangular concrete columns. J Comp Const 21(3)
2. Suon S, Saleem S, Pimanmas A (2019) Compressive behavior of basalt FRP confined circular and non- circular concrete specimens. Constr Build Mater 195:85–103
3. Mourad AHI, Beckry Mohamed AM, El-Maaddawy T (2010) Effect of seawater and warm environment on glass/epoxy and glass/polyurethane composites. Appl Comp Mater 17:557–573
4. Raafat El-Hacha MA, Mark F Green, MA, Gordon R (2009) Wight, effect of severe environmental exposures on CFRP wrapped concrete columns. J Comp Const (1), 14
5. Hussain Q, Rattanapitikon W, Pimanmas A (2016) Axial load behavior of circular and square concrete columns confined with sprayed fiber-reinforced polymer composites. Polym Compos 37(8):2557–2567
6. Sen T, Reddy HNJ (2013) Pretreatment of woven jute FRP composite and its use in strengthening of reinforced concrete beams in flexure. Adv Mater Sci Eng 2013
7. Pimanmas A et al (2019) Axial strength and deformability of concrete confined with natural fibre-reinforced polymers. Mag Concr Res 71(2):55–70
8. Padanattil A, Karingamanna J, MKM (2017) Novel hybrid composites based on glass and sisal fiber for retrofitting of reinforced concrete structures. Const Buil Mater 133:146- 153
9. Ispir M, Dalgic KD, Ilki A (2018) Hybrid confinement of concrete through use of low and high rupture strain FRP. Compos B Eng 153:243–255
10. Wu G et al (2008) Structural performance of concrete confined with hybrid FRP composites. J Reinf Plast Comp 27(12):1323–1348
11. Rousakis TC (2013) Hybrid confinement of concrete by fiber-reinforced polymer sheets and fiber ropes under cyclic axial compressive loading. J Compos Constr 17:732–743
12. Rousakis TC (2014) Elastic fiber ropes of ultrahigh-extension capacity in strengthening of concrete through confinement. J Mater Civ Eng 26:34–44
13. Ankit Dasgupta et al. (2018) Retrofitting of concrete structure with fiber reinforced polymer. IJIRST –Intern J Innov Res Sci Techn 4(9):2349–6010, ISSN (online)
14. Sarker P, Begum M, Nasrin S et al (2011) (2011), Fiber reinforced polymers for structural retrofitting: a review. J Civil Eng (IEB) 39(1):49–57
15. Chetan Yalburgimath, Akash Rathod, S Bhavanishankar et al. (2018) Retrofitting of reinforced concrete beam using carbon fiber reinforced polymer (CFRP) fabric. Intern Res J Eng Technol (IRJET) 05(10), e-ISSN: 2395–0056
16. Noushini, Samali, Vessalas (2013) Flexural Toughness and ductility characteristics of polyvinyl- alcohol fibre reinforced concrete (Pva-Frc). Proceedings of the 8th International Conference on Fracture Mechanics of Concrete and Concrete Structure, pp. 1110–1121

Experimental Investigation on Flexural Strength of Bamboo Reinforced Concrete Beams

Rakesh Kumar Boodida and G. V. V. Satyanarayana

Abstract Day-to-day global warming issues lead to a challenge all over the World. To reduce global emissions, natural construction materials like bamboo as a reinforcement and other industrial by-products such as flyash, GGBS, etc., as cementitious materials in concrete may be used. Bamboo has been mostly used as a building material due to its low cost, flexibility, high strength, earthquake resistance, lightweight, availability and so on. The viability of using bamboo-reinforced concrete beams for rural and tribal construction is investigated in this study. Concrete beams reinforced with steel and bamboo both treated and untreated are casted and tested for flexural strength, deflection and fracture pattern. Finally, a comparison and analysis of the conventional steel-reinforced concrete beam over the treated and untreated bamboo-reinforced concrete beams is performed.

Keywords Bamboo · Bamboo reinforcement · Bamboo-reinforced concrete beams

1 Introduction

1.1 General

Many researchers are focusing on alternative materials such as bamboo as a reinforcement and other eco-friendly materials used to improve environmental circumstances. There is an ongoing search for non-polluting materials and energy-efficient manufacturing techniques. For engineering applications, academics and enterprises have resorted to materials such as vegetable fibres such as bamboo, soil and waste from industry, mining and agriculture.

Bamboo has been utilised as a building material since ancient times because of its low cost, great strength, practicality, lightweight, seismic resistance and so on

R. K. Boodida (✉) · G. V. V. Satyanarayana
GRIET, Bachupally, Hyderabad, India
e-mail: rakeshkumarrakhi10@gmail.com

K. K. Pathak et al. (eds.), *Latest Developments in Civil Engineering*, Lecture Notes in Civil Engineering 352, https://doi.org/10.1007/978-981-99-2676-3_54

[1]. Due to a lack of trustworthy technical knowledge regarding local materials, customers mostly utilise industrialised materials for which information is readily available. The biggest impediment to the use of bamboo as a structural composite is a lack of appropriate information on the ingredients of the composites and their endurance, as well as a lack of structural design data, exclusion from building codes and so on. The purpose of this study is to provide the knowledge accessible in the local area for creating concrete structural parts reinforced with bamboo.

1.2 Bamboo as an Engineering Material

Bamboos, contrary to popular belief, are gigantic grasses rather than trees. They are members of the Bambusoideae family. In general, the bamboo culm is a cylindrical shell separated at the nodes by transverse diaphragms. Bamboo shells are isotropic materials with a high parallel strength and a low perpendicular strength.

Bamboo is one material that will have a significant economic benefit since it reaches full growth in a matter of months and reaches maximum mechanical resistance in a matter of years [2]. Furthermore, it is abundant in tropical and subtropical parts of the world.

1.3 Bonding Strength

Bamboo has a very thin structure, it has a weak connection with concrete and must be treated to increase its binding strength [3].

Various treatments have been investigated to enhance the bond strength of bamboo with concrete, with varying degrees of effectiveness. A layer of epoxy coated to the bamboo surface followed by a coating of fine sand is one successful treatment [4]. Asphalt paints, tar-based paints and certain bituminous compounds can alternatively replace epoxy [5].

In our investigation, we attempted to improve the bond strength of bamboo by using a wire that was ribbed around the bamboo strips.

2 Selection of Best Bamboo

Bamboo is selected based on the test results conducted on different bamboo species, 1. *Bambusa balcooa* (Bhaluka), 2. *Dendrocalamus hamiltonii* (Koko), 3. *Dendrocalamus stocksii* (Manga) and 4. *Bambusa tulda* (Jati). Tests were conducted as per Indian Standard 6874:2008 [6].

Table 1 Moisture content of bamboo

Sample	Avg moisture content (%)
Bhaluka	6.36
Koko	10.58
Manga	11.70
Jati	0.53

Table 2 Shrinkage of bamboo

Sample	Avg shrinkage along length (%)	Avg shrinkage along diameter (%)	Avg shrinkage along thickness (%)
Bhaluka	5.285	5.640	12.970
Koko	3.995	1.710	2.930
Manga	0.930	1.510	2.750
Jati	2.055	2.210	10.620

2.1 Moisture Content of Bamboo

Moisture content test is conducted for the six bamboo samples of four different species as per IS 6874:2008 [6] and the results are as follows (Table 1).

2.2 Shrinkage of Bamboo

Shrinkage test is conducted for the six bamboo samples of four different species as per IS 6874:2008 [6] and the results are as follows (Table 2).

2.3 Compressive Strength of Bamboo

Compressive strength test is conducted for the six bamboo samples of four different species as per IS 6874:2008 [6] and the results are as follows (Table 3).

Table 3 Compressive strength of bamboo

Sample	Avg compressive strength (N/mm^2)
Bhaluka	41.47
Koko	61.39
Manga	60.85
Jati	79.91

Table 4 Tensile strength of Bamboo

Sample	Avg tensile strength (N/mm^2)
Bhaluka	239.72
Koko	401.83
Manga	317.48
Jati	312.05

Table 5 Shear strength of bamboo

Sample	Avg shear strength (N/mm^2)
Bhaluka	30.36
Koko	79.85
Manga	55.03
Jati	73.64

2.4 *Tensile Strength of Bamboo*

Tensile strength test is conducted for the six Bamboo samples of four different species as per IS 6874:2008 [6] and the results are as follows (Table 4).

2.5 *Shear Strength of Bamboo*

Shear strength test is conducted for the six Bamboo samples of four different species as per IS 6874:2008 [6] and the results are as follows (Table 5).

Based on the above test results, KOKO has high tensile and shear strength compared to other three types of bamboo, hence selected as reinforcement to cast beams. They were chopped into pieces to the requisite dimensions for the trial beam casting. The bamboo is so treated and ribbed with wire to increase the bond strength. Bamboo's compressive strength varies from 40 to 80 MPa. Bamboo has tensile and shear strengths ranging from 200 to 400 and 30 to 80 MPa, respectively.

3 Flexural Strength of Bamboo-Reinforced Concrete Beams

Specimen Details:

Beam 1 is made of size 1200 × 150 × 200 mm cast using conventional steel as reinforcement. The reinforcement in Beam is detailed and arranged as per Indian

Fig. 1 Reinforcement details of Beam 1

Standard 456 confined to 2000. The dia of steel bars for main and distribution reinforcement are 12 mm and 10 mm, respectively. Stirrups for shear reinforcement are made of 8 mm dia bars at a spacing of 230 mm c/c (Fig. 1).

Beam 2 is made of size 1200 × 150 × 200 mm cast using bamboo as main reinforcement and steel bars of 10 mm dia as distribution reinforcement. Stirrups for shear reinforcement are made of 8 mm dia bars at a spacing of 230 mm c/c (Fig. 2).

Beam 3 is made of size 1200 × 150 × 200 mm using bamboo as reinforcement for both main and distribution reinforcement. Stirrups for shear reinforcement are made of 8 mm dia bars at a spacing of 230 mm c/c (Fig. 3).

Beam 4 is made of size 1200 × 150 × 200 mm using wire-ribbed bamboo as reinforcement for both main and distribution reinforcement. Stirrups for shear reinforcement are made of 8 mm dia bars at a spacing of 230 mm c/c (Fig. 4).

All the four beams are of M30-Grade concrete as per Indian Standards and cured in water for 28 days before testing.

Fig. 2 Reinforcement details of Beam 2

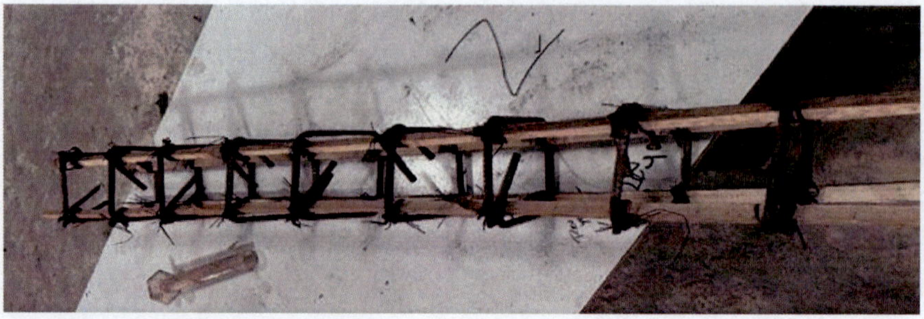

Fig. 3 Reinforcement details of Beam 3

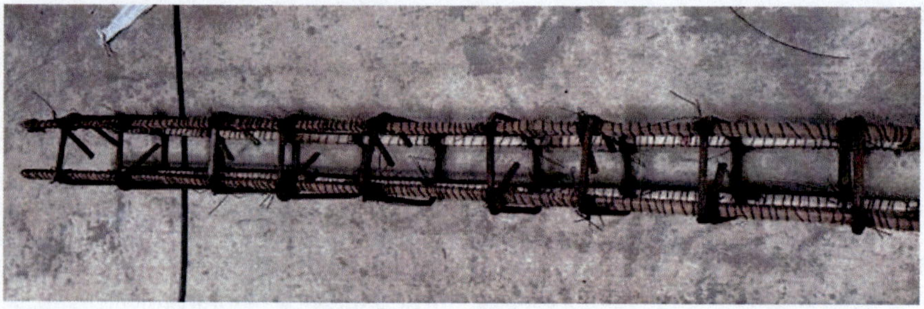

Fig. 4 Reinforcement details of Beam 4

Test Set-up

The test rig was set up in a loading frame lab with a capacity of 2000 tonnes. Casted specimens, i.e. beams, were loaded at a rate of 120 kg/cm^2/min. A specific set-up was created to record the deflection utilising a special instrument called LVDT (Linear Variable Differential Transducers), which can record sensitive and minute deflections. The LVDT has a range of 10 mm and its least count is 0.01 mm. Two LVDTs were attached to the specimen, which was set on a tripod and brought close to the specimen; an automated data collecting tool was utilised to capture the test results (Fig. 5).

4 Results and Discussion

The treated and untreated bamboo-reinforced beams, as well as the standard steel beam, were cast and cured for about 28 days. The specimen is next examined for flexural strength, deflection and crack pattern. The tests were conducted with two-point loading through computerised equipment. The load is applied ill the beam gets

Fig. 5 Loading frame with beam

failure. Deflections are noted through LVDT to get the load versus deflection graphs (Figs. 6, 7, 8, 9, 10 and 11; Tables 6, 7, 8 and 9).

It was observed that beams which have bamboo ribbed with steel wire have high flexural strength compared to beams with normal bamboo.

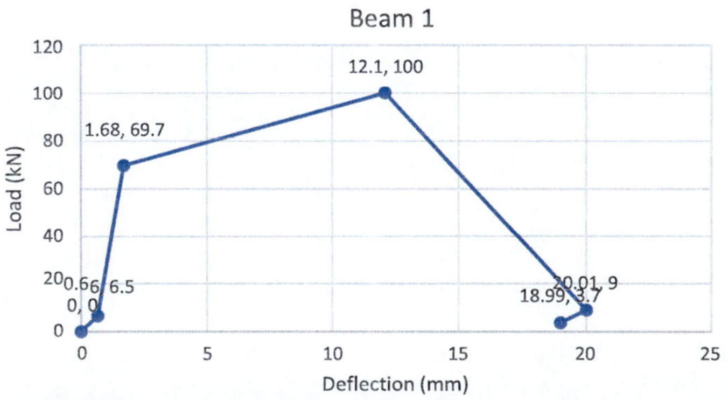

Fig. 6 Load versus deflection of Beam 1

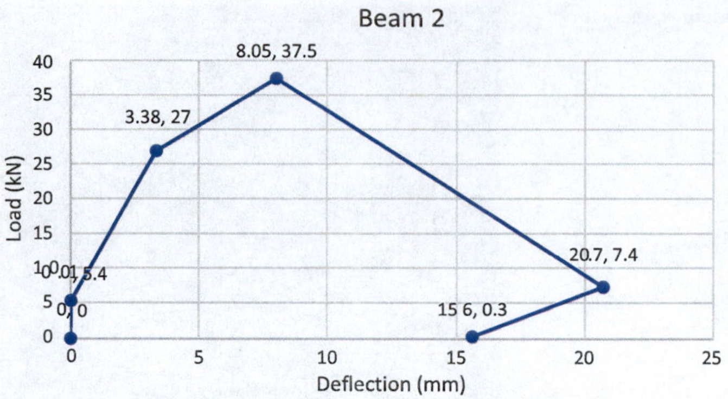

Fig. 7 Load versus deflection of Beam 2

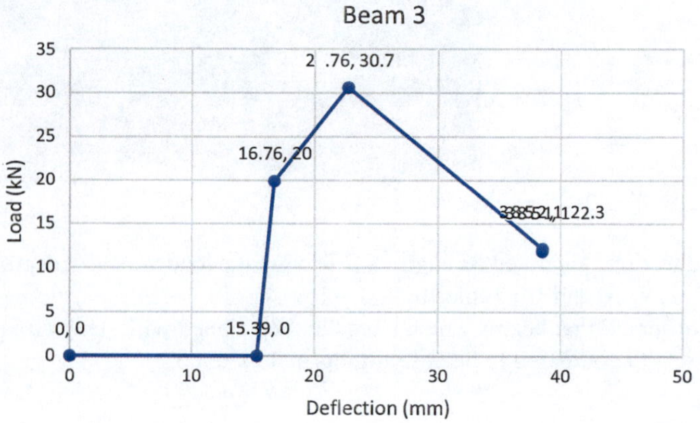

Fig. 8 Load versus deflection of Beam 3

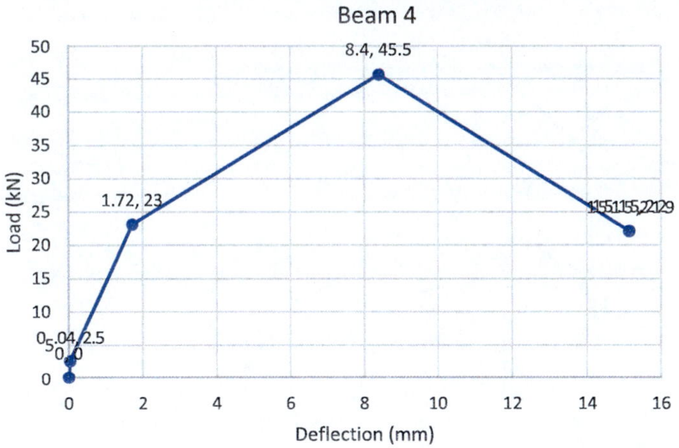

Fig. 9 Load versus deflection of Beam 4

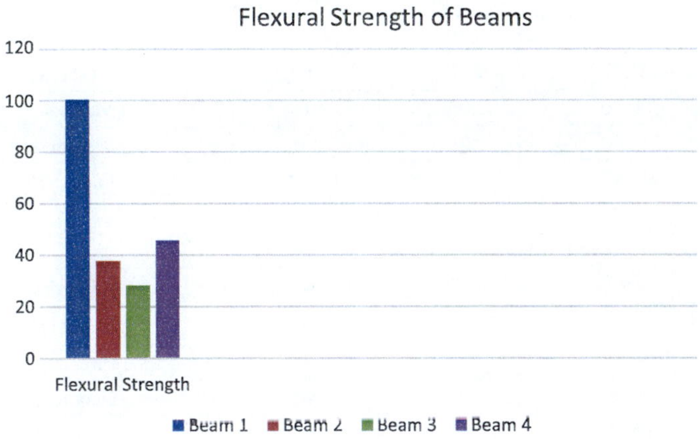

Fig. 10 Flexural strength of beams

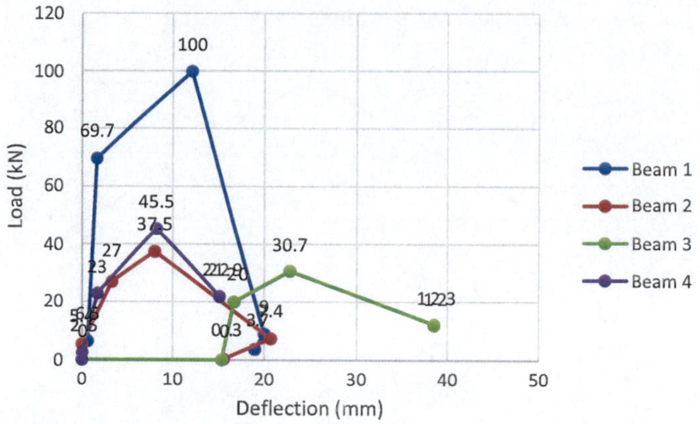

Fig. 11 Load versus deflection of beams

Table 6 Flexural strength results of Beam 1

S. No.	Time (s)	Load (kN)	Deflection (mm)	Remarks
1	0.10	6.50	0.66	Initial load
2	321.30	69.70	1.68	First crack load
3	409.90	100.00	12.10	Max load
4	516.00	9.00	20.01	Max deflection
5	521.90	3.70	18.99	Failure load

Table 7 Flexural strength results of Beam 2

S. No.	Time (s)	Load (kN)	Deflection (mm)	Remarks
1	0.10	5.40	0.01	Initial load
2	30.80	27.00	3.38	First crack load
3	48.00	37.50	8.05	Max load
4	95.20	7.40	20.70	Max deflection
5	201.20	0.30	15.60	Failure load

Table 8 Flexural strength results of Beam 3

S. No.	Time (s)	Load (kN)	Deflection (mm)	Remarks
1	0.10	0	15.39	Initial load
2	33.40	20.00	16.76	First crack load
3	61.20	30.70	22.76	Max load
4	114.10	12.30	38.52	Max deflection
5	116.60	12.00	38.51	Failure load

Table 9 Flexural strength results of Beam 4

S. No.	Time (s)	Load (kN)	Deflection (mm)	Remarks
1	0.10	2.50	0.04	Initial load
2	34.90	23.00	1.72	First crack load
3	75.80	45.50	8.24	Max load
4	99.60	22.00	15.15	Max deflection
5	100.20	21.90	15.15	Failure load

5 Conclusion

The following conclusions were drawn from the above investigation:

1. In the bamboo-reinforced concrete beams, the bamboo used in tensile zone failed at nodes due to non-continuity and uneven distribution of fibers.
2. In the bamboo-reinforced beams with wire ribbed, the strength of the beam increased due to the improvement in bond strength.
3. Bamboo reinforcement can be used for single-storied buildings in rural areas.

Acknowledgements We would like to gratefully acknowledge the support of Mr. Sravan Kumar of V.LIV.IN Bamboo, Hyderabad, for their continuous support in the research.

References

1. Van der Lugt P, Voglander JG (2015) The environmental impact of industrial bamboo products—life cycle assessment and carbon sequestration. INBAR technical report no. 35. https://doi.org/10.13140/RG.2.2.20797.46560
2. Kaminski S, Lawrence A, Trujillo D, King C (2016) Structural use of bamboo Part 3: Design values. Struct Eng 94(12):42–45. https://www.researchgate.net/publication/311282482_Structural_use_of_bamboo_Part_3_Design_values
3. Mali PR, Dattaa D (2019) Experimental study on improving bamboo concrete bond strength. Adv Concr Constr 7(3):191–201. http://www.koreascience.or.kr/article/JAKO201915540968236.page
4. Li T, Cheng DL, Wålinder ME, Zhou DG (2015) Wettability of oil heat-treated bamboo and bonding strength of laminated bamboo boar. Ind Crops Prod 69:15–20. https://doi.org/10.1016/j.indcrop.2015.02.008
5. Xing W, Hao J, Sikaro K (2019) Shear performance of adhesive bonding of cross-laminated bamboo. J Mater Civ Eng 31(9):201–216. https://doi.org/10.1061/(ASCE)MT.1943-5533.0002854
6. IS 6874 (2008) Method of tests for bamboo
7. FPL (2010) Wood handbook—wood as an engineering material. General technical report FPL-GTR-190: U.S. Department of Agriculture, Madison, WI, USA. https://www.precisebits.com/PDF/fpl_gtr190.pdf
8. Kute SY, Wakchaure MR (2013) Performance evaluation for enhancement of some of the engineering properties of bamboo as reinforcement in concrete. J Inst Eng 94:35–242. https://doi.org/10.1007/s40030-014-0063-1

9. Thwe MM (2003) Durability of bamboo-glass fiber reinforced polymer matrix hybrid composites. Compos Sci Technol 63(3):375–387. https://doi.org/10.1016/S0266-3538(02)00225-7
10. Yao W, Li Z (2003) Flexural behaviour of bamboo–fibre-reinforced mortar laminate. Cem Concr Res 33(1):15–19. https://doi.org/10.1016/S0008-8846(02)00909-2
11. Ma X, Jiang Z (2016) Flexural creep behaviour of bamboo culm in its radial direction. Jpn Wood Res Soc 62:487–491. https://jwoodscience.springeropen.com/articles/10.1007/s10086-016-1579-y
12. Wang Z, Wei Y (2020) Flexural behaviour of bamboo-concrete composite beams with perforated steel plate connection. J Wood Sci 66(4):1854–1859. https://jwoodscience.springeropen.com/articles/10.1186/s10086-020-1854-9

ANN-Based Model to Forecast the Compressive Strength Based on the Properties of Fresh Concrete

Yash Dangi and Sanjeev Kumar Verma

Abstract Researchers are working on determining the influence of fresh concrete properties on the characteristics strength and performance of concrete. According to different codes of practices, compressive strength reflects the quality of concrete significantly. In the present work, an artificial neural network model is developed to forecast the strength of concrete based on cement-to-aggregate ratio, water–cement ratio, and workability of the concrete.

Keywords Artificial neural network · Compressive strength · Fresh concrete properties · Hardened concrete

1 Introduction

Degradation and weakening of concrete structures are caused due to physical and chemical damages which results in the reduction in performance with time interval, physical damage occurs due to fire, expansion, and contraction stresses, while chemical damage occurs due to harsh environment.

In the present research, the influence of cement–aggregate ratio and water–cement ratio with slump value of fresh concrete over the compressive strength of concrete structures has been evaluated through laboratory testing of fresh and hardened concrete. Tests conducted are weighing the cement and aggregates, testing the properties of cement and aggregates, proportioning of ingredients, determination of water–cement ratio, determination of workability by slump cone method, and test of hardened concrete blocks after 28 days of curing for testing the condition of prepared concrete through compressive strength test.

Y. Dangi (✉) · S. K. Verma
SAGE School of Engineering and Technology, SAGE University, Bhopal, Madhya Pradesh, India
e-mail: yashdangii29@gmail.com

K. K. Pathak et al. (eds.), *Latest Developments in Civil Engineering*, Lecture Notes in Civil Engineering 352, https://doi.org/10.1007/978-981-99-2676-3_55

2 Literature Review

Researchers use several computational methods, including artificial neural networks, MATLAB, and fuzzy logic, to estimate the condition and strength of concrete.

Yeh [1] expressed the feasibility of accepting artificial neural networks to predict the compressive strength of (HPC) high-performance concrete. In order to forecast the compressive strength of concrete containing different amounts of fly ash at 7, 28, and 90 days of age, ANN, and fuzzy logic were utilized by Topcu and Sarıdemir [2]. A system has been developed by Lee [3] which gives in-place strength knowledge of the concrete to make easy concrete form removal and scheduling for construction. For this reason, the system is made with the help of artificial neural networks (ANN) that can learn cylinder test results as training patterns.

An attempt has been done by Kewalramani and Gupta [4] to use ultrasonic pulse velocity as a measure of compressive strength of concrete considering the advantages of NDT methods. This study was conducted for forecasting the crushing strength of concrete based on weight and UPV readings.

Yeh [5] describes artificial neural networks and nonlinear programming, and also how these computational methods can be used to optimize mixed concrete mixes (HPC) for specific workability and compressive strength.

Kasperkiewicz et al. [6] used artificial neural network for predicting strength properties of high-performance concrete (HPC) mixes. The 28-day compressive strength value was considered as the only aim of the prediction. Ghaboussi et al. [7] discussed the use of neural networks, developed by researchers in connectionism (a subfield of artificial intelligence) to model material behavior. According to Trtnik et al. [8], UPV test is one of the most accepted non-destructive techniques utilized in the evaluation of concrete characteristics.

However, it is not easy to accurately evaluate the compressive strength of concrete by this method. This is because the measured ultrasonic pulse velocity values are influenced by several factors that do not necessarily affect the compressive strength of concrete in a similar or identical manner.

3 Collection of Data

By varying cement–aggregate ratio and water–cement ratio, several samples of concretes were prepared. In order to determine the effect of fresh concrete properties over the compressive strength and to study the variation of tested parameters with respect to water–cement ratio and cement–aggregate ratio, experimental data is collected in the following format.

Total dataset is divided into training set, validation set, and test set. The training set is a part of input dataset used for neural network training, i.e., for adjustment of network weights. Validation set is a part of data used to tune network topology or

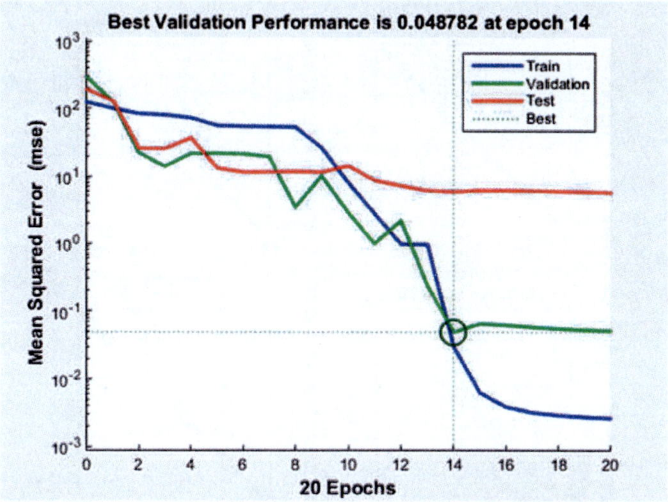

Fig. 1 MSE graph

network parameters other than weights. For example, it is used to define the number of hidden units to detect the moment when the neural network performance started to deteriorate. Here, validation set is used to calculate generalization loss and retain the best network. Test set is used only to test how well the neural network will perform on new data. Test set is used after the network is trained, to test what errors will occur during future network application.

The outputs produced by the model have been compared with the target outputs, which are the experimentally obtained compressive strength values. Generalization ability of the developed network is measured by the mean square error (MSE). Figure 1 presents the Mean square error (MSE) in all the three runs during training. It has been observed that MSE decreased after each run.

Figure 2 provides the results of all the three datasets using training, testing, and validation data with whole data considering the value of "r" correlation coefficient. Value of "r" indicates similarity between two data series, its value ranges from -1 to + 1. A larger value of "r" indicates a better correlation. It has been observed from Table 1, that value of "r" between the outputs from model and target values, for all the three sets of data, is reaching "1" which indicates a good correlation.

Experimental values of compressive strength have been compared with the values of compressive strength obtained through model and it has been revealed that both the values are comparable. Figure 3 presents the plot of error in each predicted value of strength.

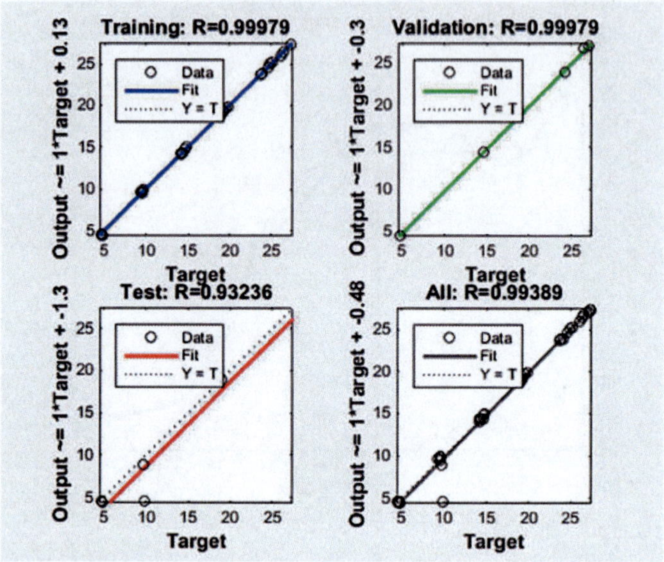

Fig. 2 Regression plot

Table 1 Experimental results

S. No	Weight (kg) of			Cement/ Aggregate (r1)	Water/ Cement (r2)	Slump height (h) in mm	Compressive strength (s) in MPa
	Cement	Aggregate (Fine + coarse)	Water				

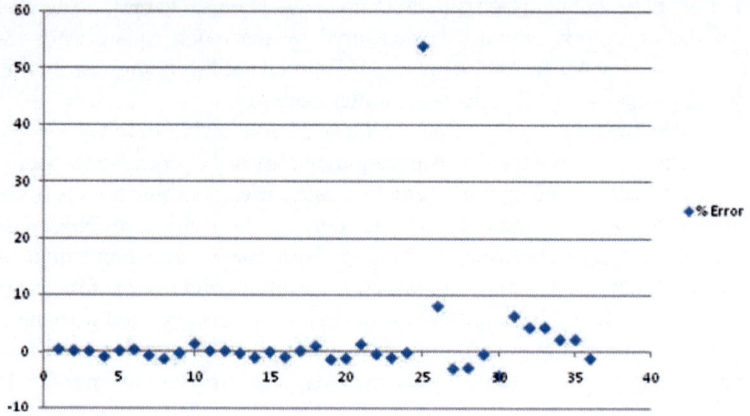

Fig. 3 Plot of error

4 Conclusions and Discussions

Following are the important points to be discussed:

1. Figure 1 shows the performance graph of model, which reveals that, in all the three states, mean square error is reducing. However, the least mean square error is observed at epoch 14.
2. Figures 2 and 3 show the statistical comparison between the forecasted results of ANN and experimental results.
3. Above table shows that samples are having low error less which indicates the accuracy of model.
4. Figure 3 shows the regression plot; in all three states regression coefficient is found almost equal to 1.
5. This study indicates the ability of the neural network tool as an excellent technique in order to model the predicted values of compressive strength of concrete.
6. Neural network tool has been used to develop a suitable method for predicting the compressive strength of concrete samples based on the properties of fresh concrete.

References

1. Yeh IC (1998) Modeling of strength of high-performance concrete using artificial neural networks. Cem Concr Res 28(12):1797–1808
2. Topcu IB, Sarıdemir M (2008) Prediction of compressive strength of concrete containing fly ash using artificial neural networks and fuzzy logic. Comput Mater Sci 41(3):305–311
3. Lee SC (2003) Prediction of concrete strength using artificial neural networks. Eng Struct 25(7):849–857
4. Kewalramani MA, Gupta R (2006) Concrete compressive strength prediction using ultrasonic pulse velocity through artificial neural networks. Autom Constr 15(3):374–379
5. Yeh IC (1999) Design of high-performance concrete mixture using neural networks and nonlinear programming. J Comput Civ Eng 13(1):36–42
6. Kasperkiewicz J, Racz J, Dubrawski A (1995) HPC strength prediction using artificial neural network. J Comput Civ Eng 9(4):279–284
7. Ghaboussi J, Garrett JH Jr, Wu X (1991) Knowledge-based modeling of material behavior with neural networks. J Eng Mech 117(1):132–153
8. Trtnik G, Kavčič F, Turk G (2009) Prediction of concrete strength using ultrasonic pulse velocity and artificial neural networks. Ultrasonics 49(1):53–60

A Comparative Analysis of Concrete-Filled Steel Tube and Aluminum Tube Columns Subjective to Axial Load

Vasu Malviya, Aslam Hussain, Amit Vishwakarma, and Aruna Rawat

Abstract Columns are the vertical members of the structure. A comparative study of two different types of confinement material for the concrete column has been carried out. This research adopted steel and aluminum tubes to confine the concrete column. Numerical analysis of confined columns has been done by a three-dimensional finite element model. A total of six confined concrete columns have been analyzed. The effect of increasing the area of hollow tubes increases the axial deformation capacity of concrete-filled tube column, load has been analyzed on the FEA model. A total number of six confined concrete columns has been analyzed. The analysis results have shown that the increment of the area of steel tube and aluminum tube decrees the axial deformation failure of the concrete column. FEA analysis results show that with the help of the concrete damage plasticity model different types of parameters like deformation, von Mises stress, logarithmic strain, and damage in tension can be plotted.

Keywords CFST · Axial load · FEA modal · CDP · Maximum axial deformation

1 Introduction

Steel and aluminum tubes are generally used to confine the concrete column. These confinement methods usually know as the composite structure and resulted in the behavior of the column as a single element. The performance of OPC concrete and steel or aluminum tube composite materials members of the building is excellent under vertical and lateral loading. It may also predict that the demand for OPC concrete, steel, or aluminum tube composite material will increase in the future due to the unmatched performance of composite material. Additionally, composite material will be the first choice for the construction industries for the important

V. Malviya (✉) · A. Hussain · A. Vishwakarma · A. Rawat
UIT RGPV, Bhopal, India
e-mail: fari_khan@rediffmail.com

K. K. Pathak et al. (eds.), *Latest Developments in Civil Engineering*, Lecture Notes in Civil Engineering 352, https://doi.org/10.1007/978-981-99-2676-3_56

Table 1 Performance of composite columns

Content	RC	Steel	SRC	CFST
Size of composite column section	Large	Small	Medium	Small
Seismic behavior of composite column	Satisfactory	Excellent	Good	Excellent
Fire resistance of the composite column	Excellent	Fair	Excellent	Good
Construction Capability	Poor	Excellent	Satisfactory	Excellent
Anti-corrosion ability	Good	Satisfactory	Excellent	Up to the mark
The behavior of composite members in long term	Poor	Good	Up to the mark	Good

structure. Therefore, the structural behavior and performance of OPC concrete and steel and aluminum tube confined material must be investigated.

Concrete is extraordinarily strong in compression and steel strong in tension. Hence, the composite material of steel tube and OPC concrete can withstand tensile and compression forces. Reinforced concrete structures are a vital example of the utilization of concrete compressive strength capacity and steel tensile strength ability as composite members. Furthermore, a composite structure of concrete and steel integrates the benefit of tensile and compressive strength at a low cost. Steel–concrete composite tubes manufacture of concrete-filled steel tubes (CFSTs), for instance, can be utilized to their fullest potential. Concrete is placed within a hollow steel tube to create a concrete-filled steel tube (CFST) column. In comparison to reinforced concrete and steel alone sections, this composite section has many structural advantages, including high strength, high ductility, and significant energy absorption capacity. Table 1 displays the performance of several column types that have been employed in the construction of high-rise buildings.

Because CFST buildings do not require the use of shuttering during concreting, the cost and duration of construction are decreased. The steel tube in CFST columns not only acts as formwork but also continuously confines the concrete core, improving the strength and ductility of the concrete. These benefits have been extensively utilized, which has resulted in the significant usage of concrete-filled tubular constructions in offshore structures, high-rise buildings, and bridges.

2 Literature Review

The experimental results of the CFST column were analyzed with the FEA method using ABAQUS [1]. The experiment validated that the performance of the CFST column for axial load is excellent to another system [2]. Comparative experimental studies on eccentrically loaded composite circular columns filled with OPC self-compacted concrete and conventional OPC concrete materials have been performed [3]. Nonlinear finite element code ABAQUS [7] is an extensive computational technique which helps them to investigate the behavior and effect of vertical load on circular cross-section concrete columns confined by steel tube filled [4]. Nonlinear finite element model of square cross-section steel tube filled by concrete column elements and checked the effect of confinement technique on local buckling [5]. The synergistic interaction between steel tube and filled concrete is driven by an equation for circular and square sections [7]. The behavior of centrally loaded concrete-filled steel tube short columns experimentally. The main objectives of the current study were to perform a parametric study on the circular CFST stub columns using the output data of the develop finite element numerical model.

Based on the above literature review objectives are listed below:

i. To develop an FEA model of square CFST columns.
ii. To study the behavior of steel and aluminum tubes filled with concrete.

- To study the load v/s deformation square CFST columns under axial compression.
- To present the failure pattern of square CFST columns under axial compression.

3 Methodology

Several researchers performed finite element analysis to examine the behavior of CFST with the use of commercial finite element software. Finite element analysis is a useful tool to predict the effect of several parameters on the performance of the composite structure under different loading conditions, which are very expensive or difficult tasks in the experimental study. The ABAQUS FE software is a quite famous tool for the numerical studies of the structure members. A numerical model has been developed in this study to predict and analysed the failure mode of a column.

3.1 FE Modeling Details

3.1.1 Materials

OPC concrete is confined by two confinement materials. Steel tubes and aluminum tubes have been used as confinement material for the concrete column in finite element modeling of the composite column. In FE modeling, a nonlinear compression and tension "concrete infill" and steel tube and aluminum tube are utilized to confine concrete columns. For concrete infill modeling, in FEA modeling, solid elements are utilized to modeled concrete infill, and shell elements are used to model steel tube and aluminum tube.

The concrete material behavior can then be defined in ABAQUS by defining the following three main sections:

ABAQUS can be used to define concrete material behavior.

Following are the three main sections:

General: In this section, the concrete density is required to be defined.

Elastic: The linear segment of the stress–strain curve is defined in this section. Young's modulus of confined concrete is defined in this part.

Plastic: ABAQUS provides material definitions for the behavior of concrete materials.

The plasticity model is used in this study.

A description of the function of each of these parameters can be found in ABAQUS help.

Dilation angle = 38 °

Eccentricity = 1.

The ratio of biaxial yield stress to uniaxial yield stress = 1.16.

$K_c = 0$.

Viscosity parameter = 0.

A tensile and compression behavior has also been assigned in the ABAQUS model.

Steel tube, aluminum tube, and the concrete material property also assign in the ABAQUS model shown in Tables 2, 3.

Table 2 Materials properties for IS code

Material	Density (Kg/m$^{3)}$	Modules of elasticity (MPa)	Poisson ratio	Thickness (mm)
Steel	7850	200,000	0.3	05, 15 and 25
Aluminum	2710	72,000	0.33	05, 15 and 25
Concrete	2400	32,786	0.2	100

Table 3 Load and deformation for confinement column

S. No	Dimension of composite (mm)	Confinement size (mm)	Confinement material	Load (N)	Deformation. (mm)
1	110 × 110x500	110 × 110x5	Steel tube	20,137	8
2	130 × 130x500	130 × 130x15	Steel tube	22,000	7
3	150 × 150x500	150 × 150x25	Steel tube	42,650	6.75
4	110 × 110x500	110 × 110x5	Aluminum Tube	20,158	5.98
5	130 × 130x500	130 × 130x15	Aluminum Tube	20,500	6.12
6	150 × 150x500	150 × 150x25	Aluminum Tube	30,000	3.66

3.1.2 Create Geometry of CFST

In this section, geometry of the steel tube and concrete in the form of the different parts has been created.

3.1.3 Interaction

Two different materials parts interacted with tangential and normal behavior.

3.1.4 Assigning Supports

Fixed support has been assigned at the bottom of the model base and a pressure load applied on the top surface of the column.

3.1.5 Run and Analysis

Post-processing of the program has been run and the outcome results get in the field output file.

4 Results and Discussion

A total of six beams have been analyzed and results are shown in the form of a table and figure.

1. The maximum load and displacement have been shown in Table 3, Figs. 1, 2, 5, 6.
2. Percentages of change in load and displacement have been shown in Table 4, Figs. 3, 4.
3. Various contour shapes for deformation, von Mises stress, damage in tension, and logarithmic strain (Fig. 7).

5 Conclusion

1. Based on load–displacement shown in Fig. 5 and Table 3, the maximum failure load resists by concrete-filled steel tube is 42650.00 N having confined by 150 mm × 150 mm x 25 mm.
2. Based on the load–displacement curve shown in Fig. 6 and Table 3, the maximum failure load resists by concrete-filled aluminum tube is 30,000.00 N having confined by 150 mm × 150 mm x 25 mm.

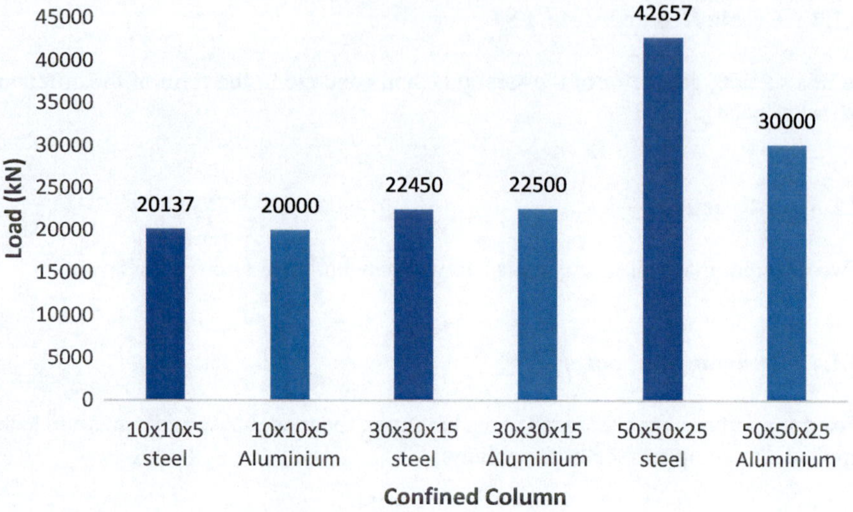

Fig. 1 Maximum load of concrete column confine by steel and aluminum tube

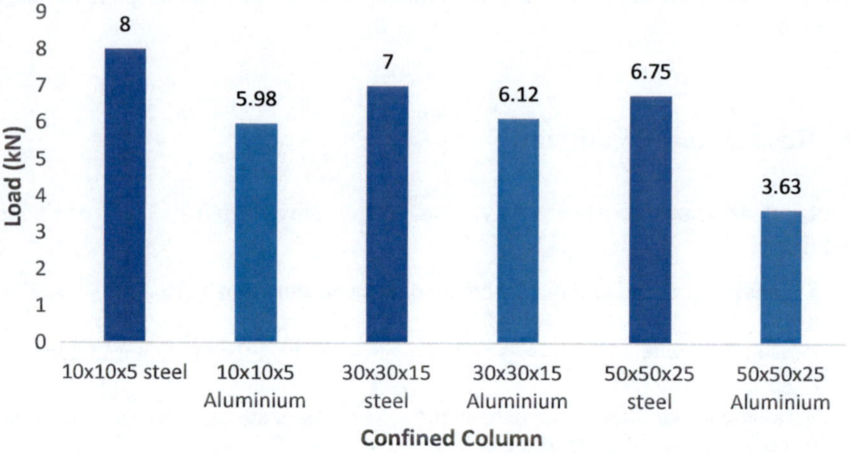

Fig. 2 Maximum deformation of concrete column confine by steel and aluminum tube

3. The maximum percentage of load increment in concrete-filled steel tube columns is 52.27 and for an aluminum tube is 32.27 has been shown in Table 4 and Figs. 3, 4.
4. The increment of the thickness of steel tube and aluminum increases the axial load-carrying capacity of the concrete column.
5. The increment of the thickness of steel tube and aluminum decrees the deformation failure of the concrete column.

Table 4 Load and deformation for confinement column

S. No	Dimension of composite (mm)	Confinement size (mm)	% of changes in load
1	110 × 110x500	110 × 110x5	0
2	130 × 130x500	130 × 130x15	10.03
3	150 × 150x500	150 × 150x25	52.27
4	110 × 110x500	110 × 110x5	0
5	130 × 130x500	130 × 130x15	1.42
6	150 × 150x500	150 × 150x25	32.74

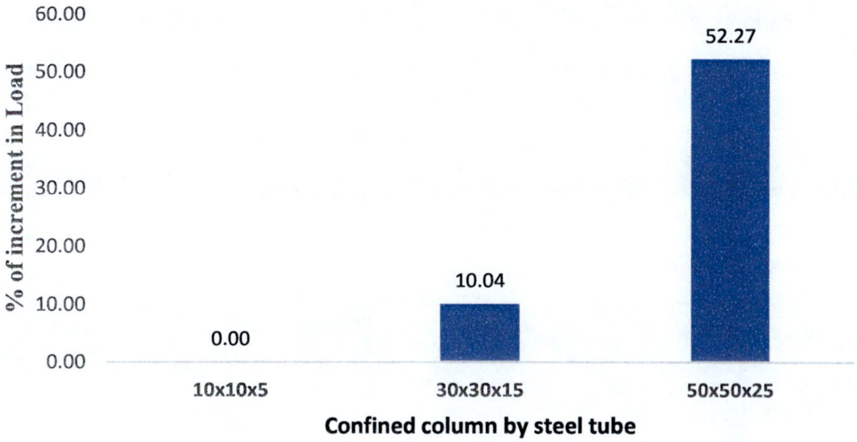

Fig. 3 % of incitement of load in concrete column confined by steel tube

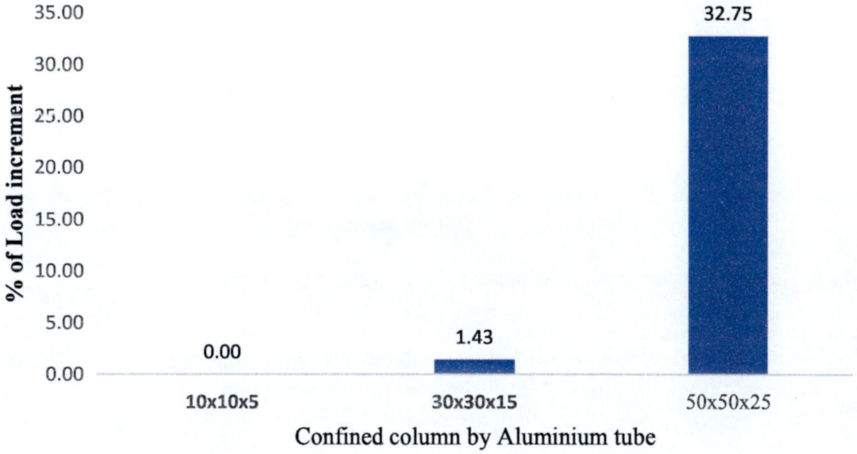

Fig. 4 % of incitement of load in concrete column confined by an aluminum tube

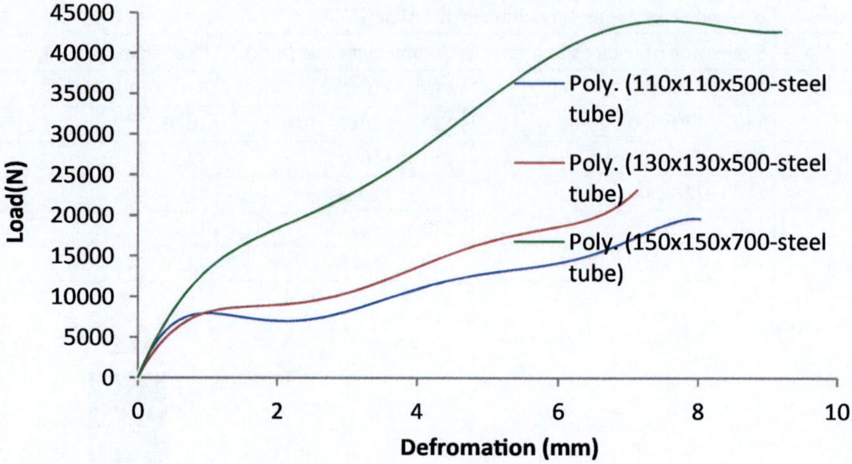

Fig. 5 Load deformation curve for concrete column confined by steel tube

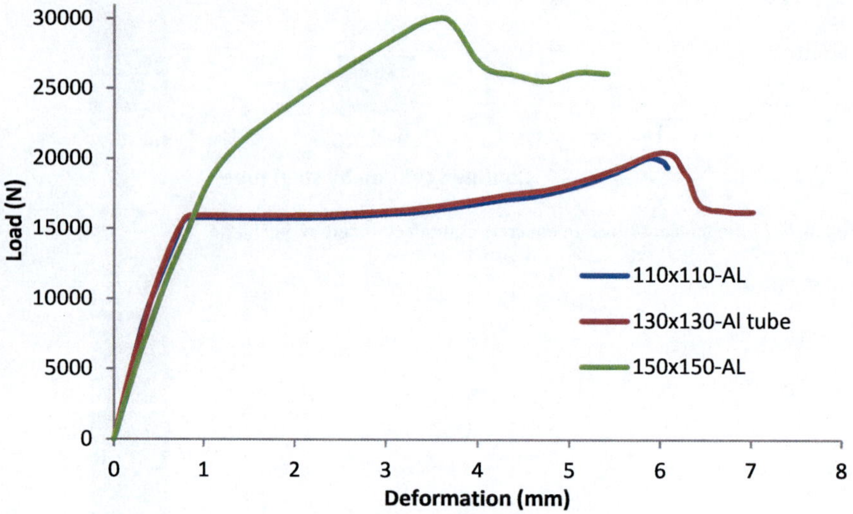

Fig. 6 Load deformation curve for concrete column confined by Aluminum tube

6. For future work, a concrete-filled steel tube will be analyzed for eccentric loading and flexural loading for different types of the cross section.

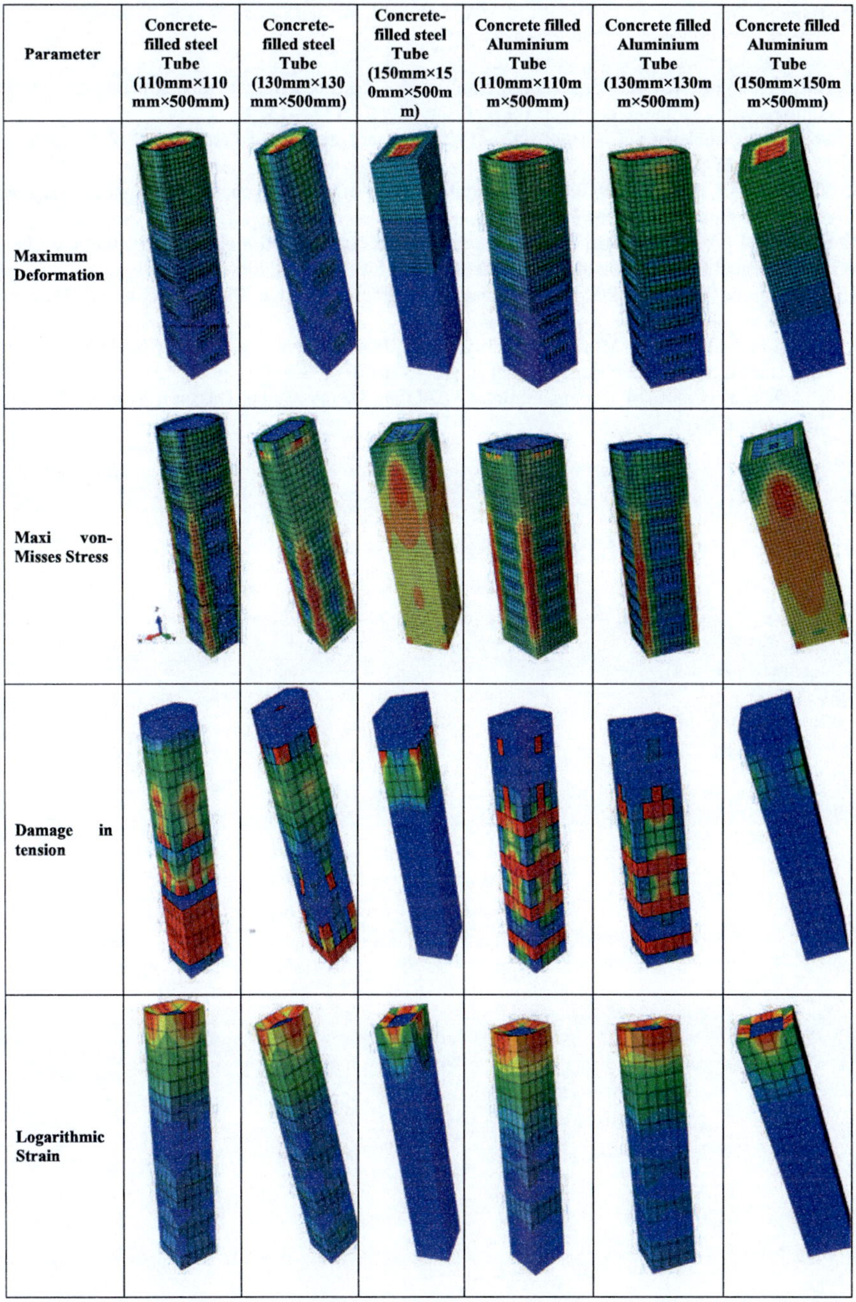

Fig. 7 Deformation, von Mises stress, damage in tension, and logarithmic strain for concrete column confined by steel and aluminum tube

References

1. Hu H, Huang C, Wu M, Wu Y (2003) Nonlinear analysis of axially loaded concrete-filled tube columns with confinement effect. J Struct Eng 129(10):1322–1329
2. Schneider SP, Huang C (1998) Axially loaded concrete-filled steel tubes. J Struct Eng 124(10):1125–1138
3. Yu Z, Ding F, Cai CS (2006) Experimental behavior of circular concrete-filled steel tube stub columns. J Constr Steel Res 63:165–174
4. Ellobody E, Young B, Lam D (2006) Behaviour of normal and high strength concrete-filled compact steel tube circular stub columns. J Constr Steel Res 62:706–715
5. Giakoumelis G, Lam D (2004) Axial capacity of circular concrete-filled tube columns. J Constr Steel Res 7(60):1049–1068
6. Sakino K, Nakahara H, Morino S, Nishiyama I (2004) Behavior of centrally loaded concrete-filled steel-tube short columns. J Struct Eng 2(130):180–188
7. ABAQUS Version 6.14 Documentation- ABAQUS Theory Guide, Dassault Systems Simulia Corporation, 2014.

Transportation Engineering

Evaluating the Ease of Access to Public Transport Systems in Urban and Suburban Centers

Thisaiveerasingam Thilakshan, Sabeen Sharic, Oshadhi Weerasinghe, and Saman Bandara

Abstract On par with the rapid motorization in the urban centers and the decline of public transport usage, fuel consumption, and travel delays have become major challenges in the context of transportation. Developers, investors, commuters, and residents tend to select locations with better accessibility for land use developments. The capability of public transport transfer points/stops in minimizing the level of passenger inconvenience to connect the trip attractors/land uses is important in retaining or attracting passengers to public transport in urban forms. Passenger inconvenience at public transport transfer points/stops can occur by increased waiting time to get the next public transport service or by the increased walking distance from the public transport transfer point/stop to the desired land use. Measuring the public transport connectivity to land uses is vital for public transport operators, passengers, regulators, and investors. Existing studies have focused on measuring public transport connectivity based on waiting time at public transport transfer points/stops. The research proposes a novel methodology in terms of "public transit connectivity index" and "coverage percentage" to compare urban centers in terms of public transport accessibility and to evaluate the accessibility to each land use category via public transportation utilizing Geographical Information Systems (GIS).

Keywords Public transit · Accessibility · Urban centers

T. Thilakshan · S. Bandara
Department of Civil Engineering, University of Moratuwa, Moratuwa, Sri Lanka

S. Sharic (✉)
Department of Management and Finance, General Sir John Kotelawala Defence University, Ratmalana, Sri Lanka
e-mail: sabeen@kdu.ac.lk

O. Weerasinghe
Urban Development Authority, Dambulla, Sri Lanka

K. K. Pathak et al. (eds.), *Latest Developments in Civil Engineering*, Lecture Notes in Civil Engineering 352, https://doi.org/10.1007/978-981-99-2676-3_57

1 Introduction

In the context of rapidly increasing population in the urban regions, it is expected that more than 60% of the entire population worldwide will be confined to the urban space from the 50% mark within the current decade. The economic growth of countries both developing and developed are dependent on these urban spaces to increase their economic revenues. Rapid development and urbanization, especially in developing countries, have led to unplanned developments and increase in the urban poor. The urban city concept has not only contributed to the GDP (Gross Domestic Product) in a national and domestic framework but also led to 70% of the global carbon emissions and 60% of the resource use [1]. These unsustainable outcomes have led to rethink the concept of urban city and to engage its' people with sustainable modes and practices. According to the United Nations, 68% of the global population of 9.8 billion will live in urban areas by 2050.

Road networks play an important role in facilitating increased access and improved connectivity in terms of travel, especially in urban areas. Transport mode choice depends on the availability of the specific mode and the availability of effective connectivity to reach a particular mode. The concept of sustainable transportation and usage of public transportation especially in the urban context is highly impacted by accessibility and connectivity. A study [2] discusses the different linkages in terms of targets of SDGs in relation to the five dimensions of sustainable transportation stated by the United Nations and provides an insight into the importance of accessibility as a front runner in terms of sustainable transportation via detailed analysis. There are many indicators available to evaluate connectivity individually and in relation to other components of transportation.

The capability of modes on enabling easy access to the desired trip attractions of passengers is one of the influential factors in passenger mode choice. Easy access represents the idea of direct connectivity or connectivity between the trip origin to the land uses through transfer points. Private vehicles can provide direct connectivity to the desired land uses at large. But, increasing private vehicle usage has led to adverse negative impact on the economy, society, and environment as a result of increased traffic congestion, increased stress level of passengers, and air/noise pollution, respectively. Therefore, public transport is encouraged as a viable alternative to private vehicles. The level of capability of public transport on providing direct connectivity or minimal inconvenience through transfer points in connecting the desired land uses of the passengers is one of the crucial factors for increased performance of public transport. In general, providing direct connectivity at large by public transport is not feasible due to operational issues. A feasible option in public transport is providing connectivity through transfer points. These transfer operations can consist of many modes such as private vehicle, public transport, and walking. A major criticism of transfer operations at public transport is its level of inconvenience in the form of waiting time to get the other desired mode or the level of walking distance to the desired land use from the public transport stop where passengers get down. This particular study investigates the latter aspect in detail. Minimizing the level of inconvenience caused to the passengers in reaching transfer point is important to attract passengers to public transport. Therefore, it is of utmost importance to measure the connectivity of the public transport modes to the desired land uses of passengers considering the tolerable level of the inconvenience of the public transport transfer points to the passengers.

Failure to measure connectivity to evaluate the ease of access to public transit systems in urban and suburban centers will not be favorable to the various stakeholders including passengers, public transport operators, public transport regulators, land use planners, and the corporate business sectors. If the walking distance from the public transport transfer stop to the desired land use is far more than the tolerable distance, passengers will avoid using public transport and will shift to private mode choices. This shift will reduce the overall public transit service, which is already showing steady decline in deviating from the sustainable transportation framework. Further, the purposes of individual land uses including attraction and generation of trips will not be achieved with poor public transit connectivity.

Therefore, the objective of this study is to develop a methodology to measure the level of connectivity to the public transport modes based on their capability to minimize the level of inconvenience of the walking distance from the transfer point to the passengers' desired destinations. The method/measure will be useful for service improvements that will help attracting more passengers to public transport resulting in increase in income of public transport operators, planning optimal public transport systems, planning appropriate land uses, and locating the suitable corporate business entities.

2 Literature and Concept Review

The concept of connectivity is highly complicated with other transportation concepts such as accessibility and mobility which needs to be differentiated and elaborated to understand the connectivity value of a certain region and compare with previous values in the time frame or with other towns with similar parameters. It is important to consider different levels of connectivity being applied to different modes (filtered permeability) [3]. Consistent connectivity is absolutely vital to the functioning of smart and sustainable cities [4]. The concept of sustainable cities can be narrated as being as good as their connectivity. The prevalence of forms of connectivity, the complex set of relationships between nodes, and their links imply an urban form that is unique in each case influencing sustainability in an urban setup. Many studies have been carried out measuring the public transport connectivity to the land uses based on the level of inconvenience at public transport transfer points focusing on waiting time. However, there isn't much literature that measures how well public transportation connects to passengers' preferred land uses based on how inconvenient it is for passengers to walk from the transfer point to their preferred land use.

There are different concepts in terms of defining connectivity in a particular road network. The study elaborates on the different definitions that have been used in research studies. The need to understand the idea of connectivity and differentiate the concept with accessibility and mobility is vital to establish individual and group relationships among the three parameters. In easy terms for understanding purposes, connectivity can be considered as the directness involved in a trip. Accessibility is basically a perspective of the number of places that can be reached in a certain timeframe whereas mobility is the distance that can be traveled in a given amount of time. A complete transportation network represents the ability to reach a particular location with a number of routes and modes available to select from depending on the travel requirements. Connectivity is defined as the density of connection in road or path networks and the directness of the links [5]. Further, a network with good connectivity consists of many short links, minimal dead ends (cul-de-sacs), and numerous intersections. The concept of connectivity is applicable both within a particular area (internally) or between two or more areas (externally) [5].

While assessing the concept of connectivity, the inclusion of mobility and accessibility is important to be considered. The need to consider CAM (Connectivity, Accessibility, Mobility) in transportation analysis and understanding the concept needed for a particular study considering the independent data available is important [6]. There are three models which help in adopting CAM and understanding their impact on each other (The Nested Model, The Snowman Model, and The Three-way Overlapping Model). Connectivity is a subjective concept and depends on other parameters such as speed, distance, modal choice, etc. It is important to note that increased connectivity means increased accessibility, but increased accessibility needs not necessarily represent a good connectivity network. The different street patterns play a major role in defining the connectivity of a particular area. Street connectivity reduces vehicle travel by reducing the travel distances between destinations and providing alternative modes for choice as well. Apart from accessibility, the resilience factor is increased in par with connectivity minimizing the hindrances when a particular road connection is closed and needs detour.

Definitions for connectivity include the degree of connection between vertices [7], whereas another study defines the concept as the degree of completeness of the links between nodes [8]. The importance of state street connectivity is a key element in neighborhood design [9]. The study encourages more grid-like networks where the high connectivity will have lesser distance and increase usage of non-motorized travel and access to public transit. It is important to decide on the method of connectivity measurement and the level of depth in the analysis depending on the nature of the analysis such as linking travel behavior to urban form and public policy/decision-making. A study [10] has been carried out on road systems and establishes their prime relationship with infrastructure affecting the economic development in a positive or negative way. Usually, a high connectivity network will cater to increased

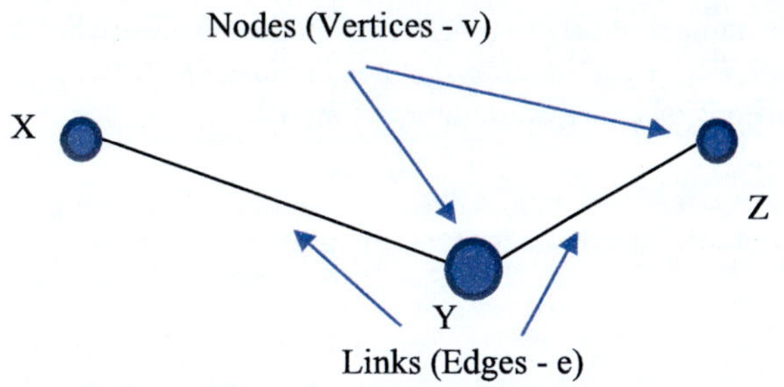

Fig. 1 Terms in connectivity analysis

economic and social development. There are three structural properties of a transport network.

1. Symmetry—All networks are connected in one direction
2. Connectivity—One vertex connected to another vertex in a network
3. Complement—In terms of A, B, and C in a triangular format network, if A is connected to B and B is connected to C; the hypothetical link of A to C becomes a compliment of the network ABC.

Usually, in the context of Transport Geography, a spatial approach is utilized in the analysis of transport networks, and a link is commonly analyzed in terms of the number of freights, the number of people, and the number of interactions along with the economic and social activities. Connectivity is not only analyzed in terms of road networks but also in terms of railway, air transport, and water (ships) as well. The Graph Theory mainly uses the nodes and links for analysis and connectivity also follows the same parameters for analysis in general. Considering a travel from X to Z via Y, the terms used in connectivity is depicted in Fig. 1. The nodes are the particular points where a travel starts, ends, or surpasses, whereas the links denote the path of travel.

These connectivity measures listed below are different in terms of the parameters and methodologies which can be adopted to measure connectivity as per the requirements for the required outcomes and analysis purposes.

- Block Length

(Usually, there is a maximum block length standard—300–600 feet) Shorter blocks represent more intersections which therefore reduces travel distance and increases the number of routes between locations. As the measure is easy in understanding, this is used as an overall performance measure.

- Block Size

It is measured by the area or perimeter with respect to the blocks.

- Block Density

It is the average number of census blocks per square mile. More block density represents higher connectivity. (Basically, higher number of blocks represents smaller blocks which means there are more intersections).

- Intersection Density

It is the number of intersections per unit area and is directly proportional to connectivity.

- Street Density

It is calculated by the number of linear miles per square kilometers and increases connectivity proportionately.

- Connected Node Ratio

The number of street intersections is divided by the number of intersections plus cul-de-sacs. Therefore, the increased ratio value represents less cul-de-sacs which means that the connectivity is high.

- Link Node Ratio (Connectivity Index CI)

The CI depicts how well the road network is connected and equals to the number of links divided by the number of nodes in a particular study area. The perfect grid will have a connectivity index and another study [11] proposes 1.4 as a good connectivity index value. The value does not reflect the length of the links and the calculation is simple and direct in terms of the network. The greater the CI value, more the connectivity. A maximum score is 2.5 for a completely connected network scenario, while a score above 0.75 is usually desired for an average road network scenario. The direct measure in a particular road network will ensure that the value depicts the real nature of the road connections including the number of links and intersections and the multiple number of route options available. A study [12] states reduction of speeds, reduction of crime rates, increase property values, and increase non-motorized travel, among the many pros of having a well-connected road network.

- Grid Pattern

Basically, measured in terms of both qualitative and quantitative approaches to see the grid-like pattern of the network.

- Pedestrian Route Directness

It is the ratio of the route distance to the direct straight-line distance (crow fly distance). A value closer to 1 depicted a more direct route in terms of the road network.

- Effective Walking Area

Capacity, speed, distance, and activity are the key factors for connectivity index [13]. The study further defines road connectivity as the level of transfer points and choice of routes to the users of connectivity that increases mobility as well. While assessing the concept of connectivity, the inclusion of mobility and accessibility is important to be considered. The need to consider CAM (Connectivity, Accessibility, Mobility) in transportation analysis and understanding the concept needed for a particular study considering the independent data available is important [6]. Public transport connectivity is the level of connectivity that is having the most efficient route that covers maximum locations increasing accessibility. In the context of the graph theory in public transportation analysis, a bus stop can be considered as a node and a bus network as links. There are a number of studies [14, 15] which analyze connectivity in the context of shortest distance, cost of travel, and delay optimization methods to provide an efficient public transport system. A study [16] has analyzed the levels of connectivity of selected nodes spatially and states high connectivity, high accessibility, and serviceability along with strong hierarchical structure enforce legibility and constant circulation in a road network.

Surveys and transport data are utilized to analyze and evaluate public transport connectivity [17]. Another study [18] adopted a framework to determine interconnectivity among public transport routes through Google online route planner information and to assess the overall connectivity of Auckland and compare cities for improvements. Travel time and number of transfers are impacted by the design of the network and public transport choice [19]. The quality of connectivity is defined as a function of spatial performance of connections between zones and further accessibility in terms of separate values for origins and destinations, while connectivity indicators are specific to the origin–destination pairs [20]. Thus, the importance of developing and adopting a connectivity measure to public transit points can be identified.

3 Methodology and Demonstration

The methodology was developed in two steps. As the first step, an initial analysis was conducted to compare the accessibility to public transportation network of urban and suburban centers. For the initial analysis, shape files of road networks, public bus route networks, bus stop locations, and boundaries of selected urban areas were required as the input data. In order to calculate the Public Transport Connectivity Index (PTCI), the ratio between no. of bus routes links and the no. of stops within the urban area boundary was estimated using available spatial data. The ArcMap tools and functions such as; "Select by location" and summarize option in attribute table, were utilized to count the required type of nodes and links. Further, the urban centers were ranked based on PTCI for comparison purposes.

As the second step, a detailed analysis was conducted to study the accessibility to the public transportation network considering different land uses of individual towns. This step was focused to evaluate further on urban centers which were initially screened through the PTCI. The road network, public bus route network, bus stop locations, urban area boundary, and categorized land use layer were used as the input data for the details analysis of the proposed methodology. The ArcMap software was utilized to develop the network analysis dataset by incorporating both road network and public bus route network. The developed road network database facilitates to

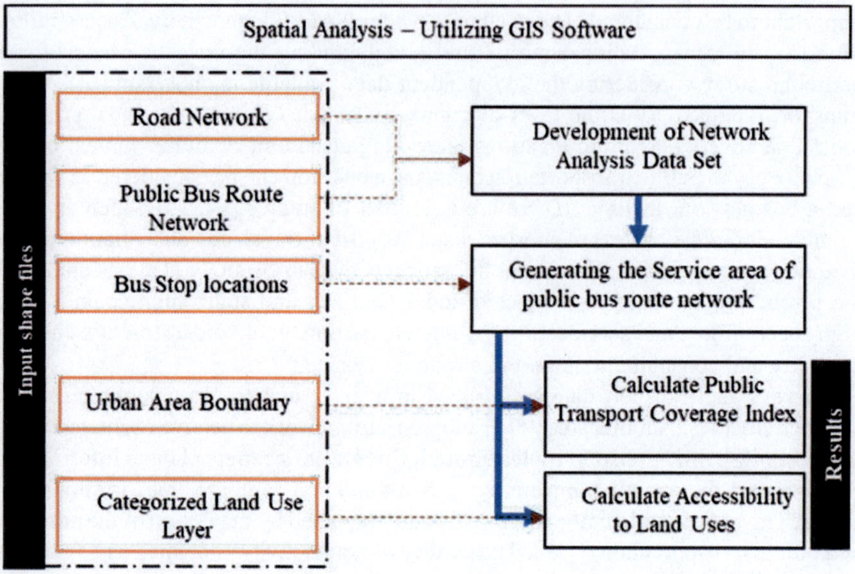

Fig. 2 Summarization of the methodology

analysis of the service areas, shortest routes, closest facility, etc. The service area analysis tool was used to demarcate the catchment of bus stops which can be also considered as the catchment of public transportation network. The catchment of bus stop was demarcated considering the walking distance which is approximately 400 or 500 m.

Based on the land uses within the selected urban area and land uses within the catchment area of public transportation system, the percentage of accessible land uses via public transportation system was calculated. The methodology was extended to evaluate accessibility to different land use types. The summarization of the proposed methodology is shown in Fig. 2.

3.1 Initial Analysis

The concepts of Road Connectivity Index (RCI) and Public Transport Connectivity Index (PTCI) are analyzed in a model study considering five urban and suburban towns in the Western province of Sri Lanka. The towns have been selected on the diversity in population, road connectivity, geographic positioning, and economic and social conditions. The three towns of Malabe, Homagama, and Padukka are in the Colombo district. Ingiriya belongs to the Kaluthara district and Bopitiya is located in the Gampaha district. All three districts are located in the Western Province (Fig. 3).

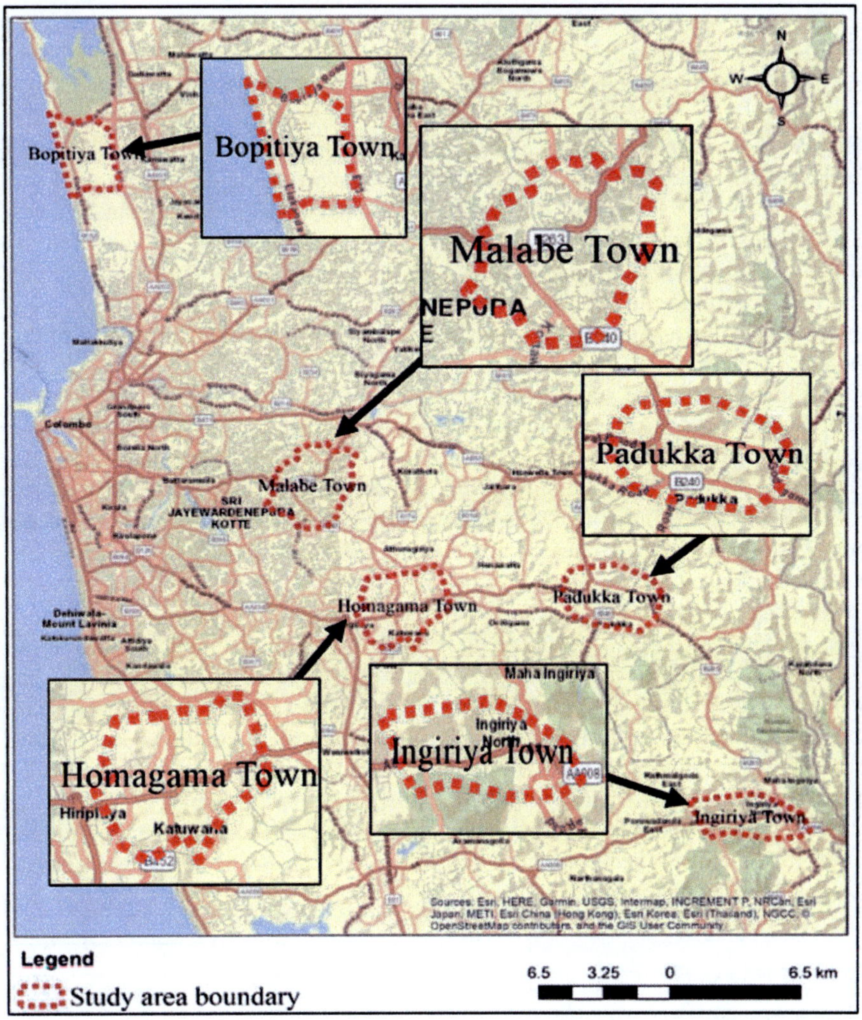

Fig. 3 Study area boundary with the five-town depiction

Bopitiya is a suburban town and is approximately 8 m above the sea level. It is in close proximity to the main airport of Sri Lanka (15–20 min travel time in average) and is facilitated for all main activities by the main towns of Jaela and Negombo. Malabe is located on the New Kandy Road (Kaduwela Road) approximately 10 km from the capital city of Colombo. The town and its vicinity had seen an immense boost in terms of trip attraction and generation due to the establishment of a number of educational institutions with high student volume and subsequent increase of land value. The Kaduwela interchange of the southern expressway has also contributed positively to the development of Malabe as an important transit destination. Homagama is an urban town on the A4 Highway and has access to both bus and train services in terms of public transit. It is also served by a number of bus services including inner and outer provincial routes. Padukka is a suburban town with a population of approximately 8000 and 33 km from the capital city. The town is served by a number of bus services and lies south to the main A4 Highway. There is a main station on the Kelani Valley line at Padukka. Ingiriya is an intersection junction of the Rathnapura Panadura, Ingiriya Padukka, and Ingiriya Bulathsinhala roads. Thus, the five towns will be analyzed in terms of their individual road connectivity and public transport connectivity to analyze their individual nature of connectivity and further compare and contrast in relation to other parameters.

The Connectivity Index "CI" method or the Link Node Ratio is utilized in this study to identify the road connectivity scenarios of the five identified townships. The values identified in the study will give an overall understanding of how behavior varies with geographical position and other parameters in general. Thus, the number of links and the number of nodes of the individual towns is analyzed using the GIS position to facilitate the identification of the number of nodes and links. The Road Connectivity Index (RCI) is calculated for the five towns as a ratio of the number of links and the number of nodes (for comparison and analysis purposes). In a similar approach, the Public Transport Connectivity Index (PTCI) is calculated by substituting the road nodes by the bus stops and the road links by the road network links as shown in Table 1. It must be noted that the bus stops on either side of the road had been amalgamated as a single bus stop on the network for analysis purposes using network analysis in ArcMap 10.5. Ingiriya is well connected in terms of its road connectivity (1.13), whereas Bopitiya is well connected in terms of its Public Transport Connectivity (1.38) highlighting the fact that good road connectivity does not always mean good public transport connectivity. Even though the values are individual representations of the road and public transport (bus) connectivity values, an overall idea of connectivity of the considered regions can be derived with ample scope for further analysis.

Table 1 PTCI index

Town	PTCI	Rank
Homagama town	1.133	4
Padukka town	1.070	5
Ingiriya town	1.143	3
Malabe town	1.278	2
Bopitiya town	1.375	1

3.2 Detail Analysis—Malabe (Application)

The methodology was developed to assess the accessibility to the main land uses located in an urban area via the public bus transportation system. The percentage of buildings which are located within walking distance from the bus stops was calculated as the indicator. The data collection was conducted using secondary sources. The shape file of road network was downloaded from Open Street Map. Public bus route network and the bus stop location data were collected from Google Maps. Further, the land use data in 2020 were collected from Urban Development Authority, Sri Lanka. The developed public transport coverage area for Malabe Town using ArcMap 10.5 is depicted in Fig. 4.

Basically, the detailed analysis was conducted utilizing spatial analysis techniques such as network analysis in ArcMap 10.5. The network database was developed by correcting the minor errors in downloaded road network. The developed road

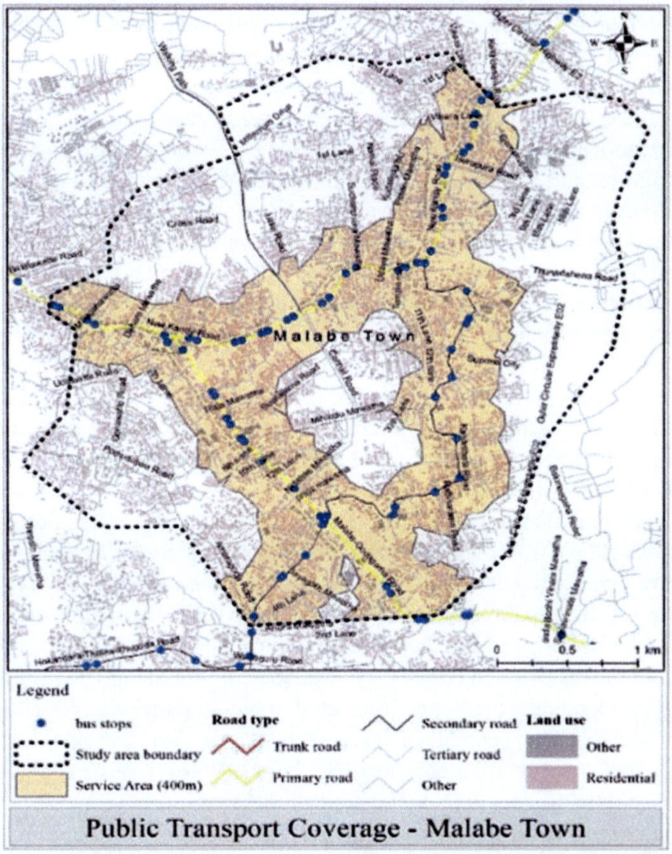

Fig. 4 Developed public transport coverage area for Malabe Town using ArcMap 10.5

network database facilitated the analysis of the service areas, shortest routes, closest facility, etc. In this research, service area demarcation tool was utilized to identify the catchment of public transportation routes. The bus stops were considered as the facilities for the analysis and catchment area was demarcated considering the 400 m distance along road network starting from the bus stop. The analysis demarcates, the 400 m service area for a bus stop on the public transportation network includes all the locations that can be reached within 400 m walking distance from that bus stop. Further, the demarcated service area facilitates to evaluate the accessibility. After determining the service area, that can be utilized to calculate land extent, demographical characteristics, and physical characteristics such as land uses within the covered area.

The number of buildings located in public transport catchment area by each land use category was calculated and the percentage of buildings located within 400 m distance from public transportation service out of the total building count was calculated. The proposed methodology was applied on Malabe urban area and results were evaluated. The study further considered to simplify the land use categories in Malabe town to understand the impact of every building type and its users in order to obtain services of public transit systems in urban and suburban centers.

Table 2 depicts an overall analysis of the Malabe area in terms of land use coverage in terms of the relevant percentages. The overall percentage via the land use analysis is 54%. The above analysis can be applied to the other four towns considered in the PTCI calculation and a ranking system can be derived using the outcomes as well. It is important to note that in the coverage percentage calculation considering the different land uses, a similar weightage has been vested upon all land use percentages. Considering the real case scenario, it is not the case and the study considers to provide weightages to every land uses considered in the regions to obtain a more reliable outcome utilizing an appropriate methodology. In order to reduce passenger inconvenience, less effort has been put into measuring public transportation connectedness based on how far people must walk from transfer points or stops to undesirable land uses. The goal of this study is to create an index that uses the walking distance between a public transportation stop and a land use to gauge the connectivity of the public transportation system.

Public Transport Connectivity Index (PTCI) is developed which indicates the ratio of the number of land uses located within 400-m walking distance from the available bus stops in a given study area and the total number of land uses available in the study area. Geological Information Systems (GIS) has been used to calculate the number of land uses in the buffer zone and the total number of land uses in the study area. This index was demonstrated to an Urban City "Malabe" in Sri Lanka. The results indicate that 54% of the land uses are better connected to bus stops within tolerable walking distance in Malabe. This percentage varies with separate land use categories such as residences (50%), commercial places (82%), banks (96%), industrial places (84%), public/semi-public officers (71%), and educational institutes (48%). These findings validate the existing form of land use categories in Malabe. This study concludes that, while majority of the important land uses are connected with bus stops, residential places and educational institutes have lesser public transport connectivity.

Table 2 Coverage percentage by land use type in Malabe

Main use	Building count within 400 m service area	Building count within study area	Coverage percentage by land use type (%)
Bank and allied	22	23	96
Com/residential	2	2	100
Commercial	931	1138	82
Educational	14	29	48
Filling stations	2	2	100
Health	19	22	86
Industrial	111	132	84
Institutional	2	3	67
Logistics	2	4	50
Other built-up	10	19	53
Private office	6	7	86
Public and semi-public	53	75	71
Religious	14	30	47
Residential	6069	12,027	50
Service	3	4	75
Socio-cultural	7	12	58
Sport—amusement	0	2	0
Stores—warehouses	2	3	67
Tourism	5	5	100
Under construction	322	668	48
Utility	2	4	50
Vacant building	68	114	60
Vacant urban	48	83	58
Total	**7714**	**14,408**	**54**

4 Discussion

The results indicate that there is a moderate level (54%) of connectivity to the land uses of the entire Malabe area from the public transport transfer points/stops. But the level of connectivity varies significantly between the different land use types.

As per the data collected, it shows that only 50% of the residents in this study area have easy access or tolerable access to public transport stops. This situation aligns with another study [21], where they have figured out less connectivity to bus stops in Western Province leading to lesser passenger satisfaction. These findings are eye-openers to bus transport regulators and operators to think for the introduction of feeder public transport services to take passengers to the main stops.

The analysis reveals that there is a higher connectivity to commercial places and banks (82 and 96%, respectively) from the public transport stops in Malabe. This context further justifies the findings of another study [22], where it is stated that the availability of transport infrastructure was one of the dominant factors for business location decision. These findings will be useful for the corporate sectors to think of expanding their services as their location are better accessible to the customers. This will also pave the way to relocate their offices if they are not in a tolerable access to customers. As both land uses commercial and banks attract more customers, lower percentage value for PTCI for these two land uses can interpret more private vehicle usage. Therefore, public transport regulators can think of locating public transport stops.

It is observed that there is better connectivity from the public transport stops to industrial places and Public/Semi-Public offices in Malabe with PTCI of 84 and 71%, respectively. This finding is strengthened as Malabe was the most rapidly industrialized area in Sri Lanka. It is further justified that availability of easy access is one of the main requirements for locating industrial places (factories) as per the guidelines of the Industrial Development Board of Sri Lanka. These land uses are trip attractors in the morning and trip generators in the evening. If PTCI yields lower values, it means these offices are far beyond the tolerable distance from the public transport stops. This would increase the level of inconvenience to the passengers or employees. At this junction, public transport regulators can go for new stop formations. Introducing staff transport and shared taxi services also could be suggested at this condition.

The analysis outlines very lower connectivity between public transport stops and educational institutes in Malabe with a PTCI value of 48%. This finding goes aligning with a study [23] which states that some of the main reasons for school trips to avoid public transport as the main mode in the Western province were higher travel time, longer waiting time, and poor accessibility. Introduction of feeder bus services can improve the public transport connectivity to the residence meanwhile the operation of Sisu-Sariya (Public Transport Service for school children), university bus service, and shared para transits can be recommended to increase public transport connectivity for educational institutes in Malabe. This study contributes to improving passenger attraction to public transport services through public transport connectivity to land uses. In terms of higher educational institutes, lower PTCI value means increase inconvenience or increased cost of transportation due to using para transits to get to destinations. Then, introduction of university bus services or improving shared para transit would be suggested options. Please note that the first paragraph of a section or subsection is not indented.

5 Conclusion

Inconvenience experienced by the passengers at public transport transfer points/stops to reach their trip attractions/land uses is a commonly known factor influencing public transport connectivity. Waiting time until getting the next public transport service and walking distance/time to the desired land uses are the main attributes that influence passenger inconvenience at public transport transfer points/stops. While there have been increased efforts to quantify passenger inconvenience based on waiting time, it is crucial to quantify it based on walking distance to comprehend the connectivity of urban public transportation.

To measure the passenger convenience of bus connection, easy access to land uses within a 400-m tolerance walking distance from bus stops in an urban setting has been taken into account. Public Transport Connectivity Index has been developed for this purpose. The city of Malabe in Colombo District in Sri Lanka had been considered for this demonstration. The evidence is clear: there are variations in the levels of easy access to land uses from the bus stops in the city of Malabe. Commercial places, banks, and industrial places have better connectivity from existing bus stops. Public and semi-public places have a moderate connectivity to bus stops. Residential places and educational institutes have lower connectivity to bus stops.

Corporate sectors or investors can further extend their business activities as their offices are easily accessible from the bus stops. But, feeder bus services should be introduced to make the residents access to bus stops as the residences are not located in close proximity to bus stops. Feeder bus services, shared paratransit operations, and educational passenger transport services should be initiated to make the school children and the higher education students to enable easy access to bus stops.

This study contributes to the domain of public transport connectivity through a novel index to measure passenger inconvenience at public transport transfer points/ stops. Future researchers can focus on walking time instead of walking distance as an implication of walking distance can be hampered by the road conditions. If the level of importance of each location can be weighed, this index can be further modified.

References

1. United Nations (2019) The sustainable goals report 2019, Statistics Division Development Data and Outreach Branch
2. Thilakshan T, Bandara JMSJ (2019) Identification of relevant sustainable transportation links to sustainable development goals (SDGs) in the national context. In: 113th annual IESL sessions 2019. Annual sessions of IESL. Colombo, Sri Lanka, pp 341–348
3. Cozens PM, Hillier D (2008) The shape of things to come: new urbanism, the grid and the cul-de-sac. Int Plan Stud 13(1):51–73
4. SmartCitiesWorld Homepage. https://www.smartcitiesworld.net/opinions/opinions/smart-cities-are-only-as-good-as-their-connectivity. Last accessed 21 Feb 2022
5. Victoria Transport Planning Institute, Todd Litman (2004) Evaluating accessibility for transport planning—measuring people's ability to reach desired services and activities
6. Labi S, Faiz A, Saeed TU, Alabi BNT, Woldemariam W (2019) Connectivity, accessibility, and mobility relationships in the context of low-volume road networks. Transp Res Rec 2673(12):717–727
7. Taaffe EJ, Gauthier HL (1973) Geography of Transportation. Prentice-Hall, Englewood Cliffs, NJ
8. Robinson H, Bamford C (1978) Geography of transport. Macdonald & Evans, Plymouth

9. Dill J (2004) Measuring network connectivity for bicycling and walking. Transport research board annual meeting
10. Weng J, Meng Q (2013) Estimating capacity and traffic delay in work zones: an overview. Trans Res Part C Emerg Technol 35:34–45
11. Ewing R (1996) Best Development Practices: Doing the Right Thing and Making Money at the Same Time. American Planning Association, Chicago
12. Gliderbloom JI, Riggs WW, Meares WL (2015) Does walkability matter? An examination of walkability's impact on housing values, foreclosures and crime. Cities 42(Part A):13–24
13. Kumar R, Parida P, Madhu E, Kumar AVAB (2016) Does connectivity index of transport network have impact on delay for driver? In: World conference on transport research—WCTR, 25, Shanghai, China
14. Derrible S, Kennedy C (2010) Evaluating, comparing, and improving metro networks application to plans for Toronto, Canada. Trans Res Board J Nat Acad, Washington, 43–51
15. Mishra S, Welch TF, Jha MK (2012) Performance indicators for public transit connectivity in multi-modal transportation networks. Transp Res Part A 46:1066–1085
16. Mathivathany V (2015) Accessibility of road network based on connectivity analysis technique in Moratuwa urban area of Colombo. In: 5th international symposium 2015
17. Ceder A, Net YL, Coriat C (2009) Measuring public transport connectivity performance applied in Auckland, New Zealand. Trans Res Record J Trans Res Board 2111(1)
18. Chodhury S, Cedar A, Velty V (2014) Measuring public-transport network connectivity using Google transit with comparison across cities. J Public Transp 17(4):76–92
19. Guo Z, Wilson NHM (2004) Assessment of the transfer penalty for transit trip. Transp Res Record 1872:10–18
20. Papaioannou D, Martinez LM (2015) The role of accessibility and connectivity in mode choice. A structural equation modeling approach. Trans Res Procedia 10:831–839
21. Ranawana HWMIS, Hewage D (2015) Factors affecting the service quality in public bus transportation in Sri Lanka. In: Proceedings of 8th international research conference, KDU
22. McQuaid RW (2004) Department for Transport—The importance of transport in business' location decisions
23. Damsara P, De Silva DI, Sirisoma N (2021) Analysis on transport mode choices of school children in Colombo District, Sri Lanka. Eng—J Inst Eng, Sri Lanka 54:17–26

Experimental Study on Aging of Different Grades of Bitumen for Bituminous Mixes Concrete

Raghvendra Pratap Singh Rajput and Rakesh Mehar

Abstract India consists of a vast network of highways and roads, and the whole span of those roads is more than 3.31 million kilometers. About 98% of the roads in India are flexible, so bitumen, used as a binder, is a key part of how well bituminous mixes work. The asphalt pavements are breaking down not only because of more traffic but also because of the extreme weather in the country. One of the primary causes of asphalt pavement damage is bitumen aging. The RTFOT rotates the bitumen thinner film, while the modified TFOT checks it for microfilm thickness. This research examines the usage of VG30, PMB40, CRMB55, and aggregate in the production of a bituminous concrete layer. The physical parameters of regular and aged bitumen were evaluated in the proposed research. After evaluating the quality, it was discovered that the aged bitumen outperformed the standard bitumen. This study evaluated the Marshall characteristics of various grades of conventional bitumen and RTFOT-modified bitumen. According to the study, all of the raw materials' physical parameters are in accordance with the MoRTH (Specifications for road and bridge works. Indian Road Congress New Delhi, India, 2013 [1]) criteria. We constructed the natural BC mix in the proposed research by employing aggregate and different grades of bitumen and performing Marshall testing. Again, the entire procedure was carried out with the old bitumen. Following laboratory experiments on various grades of bitumen and aged bitumen, it has been discovered that aged bitumen has greater strength than regular bitumen and has a lower Marshall flow value.

Keywords RTFOT · Binders · Marshall stability · VMA% · VFB%

1 Introduction

The most important job of the pavement is to support the weight of a vehicle, such as a truck or an airplane, without causing undue distortion. The pavement's layered architecture is intended to spread the weight under the wheel, resulting in a low

R. P. S. Rajput (✉) · R. Mehar
SATI, Vidisha, MP, India
e-mail: Bboyrajpot1612@gmail.com

K. K. Pathak et al. (eds.), *Latest Developments in Civil Engineering*, Lecture Notes in Civil Engineering 352, https://doi.org/10.1007/978-981-99-2676-3_58

enough stress at the lowest layer of the pavement, the sub-grade, to avoid damage. The most substantial load applied to a paved surface is applied by a car wheel or an aviation tire. The goal of an elastomeric pavement is to disperse the load such that the tension at the sub-grade soil level is modest enough that it can absorb the stress without serious deformation. The aging of bituminous binders is one of the most important variables affecting asphalt's lifespan. Age-related chemical and/or physical property changes often result in bituminous materials being harder and more brittle, which increases the risk of pavement failure. A pavement's cracking is one of the aging-related failure mechanisms (thermal or traffic-induced). Over the past few decades, researchers in the lab have examined the field aging of bituminous materials. The most important techniques for accelerating the aging of the materials are increasing temperature, decreasing layer thickness, laboratory and field aging effects on bitumen chemistry and rheology in porous asphalt mixture, and boosting oxygen concentration by raising airflow and pressure. Testing combines techniques for both short-term (mixing and constructing) and long-term (pavement service) aging.

The research is to study the physical properties of bituminous binders, both unmodified and modified regular and RTFOT bitumen. For the analysis of result, prepare a sample of bitumen by Marshall Apparatus. Now, compare the Marshall properties of different grades of bitumen in their normal and RTFOT modify stages.

The study of scope will define aging of bitumen is done using rolling thin film oven. The aging properties of bituminous binders are characterized using different bitumen tests. Compare the properties of normal bitumen and aged bitumen. The effect of aging on Marshall's characteristics of bituminous concrete mixes is investigated to understand the stability and flow value of modified and unmodified binders.

2 Methodology

The binder attributes of bitumen mix change over time while blending, batching, and presenting. In these phases, chemical and physical characteristics of bitumen change; in particular, oxidation occurs involving the increase in material's stiffness. This phenomenon is known as aging and it is one of the key factors affecting the lifetime of an asphalt pavement. Several methods of laboratory aging can be divided into two main categories, that is to simulate short-term aging and methodologies for simulating the long-term effects of oxidation process. The most common European method, which simulates the aging during storage, mixing with aggregates, transportation, and laying, is the RTFOT. Other tests, like Rotating Cylindrical Aging Tests (RCAT), are performed by following American standards. On the other hand, Pressure Aging Vessel (PAV) is the most recommended laboratory test for the simulation of long-term aging. All of them are based on the variation of three main factors: temperature, pressure, and time. Temperature is strongly related to the oxygen diffusion coefficient. The high pressure and the aging time can surely lead to greater aging, This speed up the process of oxidation and, consequently, results in major deterioration.

2.1 *Rolling Thin Film Oven Test (RTFOT)*

RTFOT was developed during the SHRP in an exertion to replicate bitumen's shortened lifespan aging bitumen which occurs especially during mixing in asphalt plan. Bitumen suffers high temperature of about 163 °C, combined with a constant and specific supply of air, following the standard procedure. It consists of an oven, which is equipped with an aluminum carriage with eight openings and spring clips for firmly holding the corresponding eight glass containers in a horizontal position. Moreover, an air jet is placed in a specific location. During the test, using a certain rate, the carriage revolves around the horizontal axis. This allows the air jet to enter each glass. In detail, the air shall blow along with central axis of each glass container. At the same time, high temperature leads the material to liquefy. The rotation together with the applied temperature leads bitumen to generate a thin coating that covers every square inch of each container, which also rotates following the carriage movement. Temperature is continuously monitored with a thermometer positioned inside the device. Fill the glass containers with 50 ± 0 5 g of bitumen, a specific amount that ensures that the material does not leak from the containers during rotation. The quantity of required glasses is determined by the operator's intended characterization tests [2].

3 Preparation of Crumb Rubber Modifies Bitumen

In order to produce the modified bitumen, 1500 g of bitumen is heated to a fluid state in a metal container with a 3-L capacity. As bitumen is heated to a temperature of about 160–180 °C and then crumb rubber is added, CRMB is created using a wet method with percentages of 5, 10, 15, and 20%, crumb rubber particles range in size from 0.300 to 0.150 mm and are added to ordinary bitumen. The mixture is manually stirred with the use of a gas burner for 15 min at 160 °C to create the 5% CRMB sample until there is a reaction between the bitumen and crumb rubber. Then, with the use of a heated oven that was also integrated with the stirrer, the 5% CRMB sample mixture was mixed for an hour at a temperature of 180–200 °C using a mechanical stirrer at a speed of 2000 rpm. The rubber particles expand as they take in oil, increasing the viscosity of the 5% CRMB sample over the first hour. The 5% CRMB sample is refrigerated to room temperature and properly maintained for testing [3].

This procedure for carrying out experimental work on bituminous concrete mixtures has been established by the suggested studies. Incorporating bitumen's aging qualities into the creation of bituminous concrete pavements is made possible by the proposed technique the suggested technique also makes an attempt to explain how bitumen of different grades may be utilized to make bituminous concrete. The goals and parameters of the current investigation were also attempted to be made clear using this technique. This technique validates the efforts in the current study to successfully investigate bituminous blends. This section includes a step-by-step overview of the methodology's framework mentioned above. It details the process used to build BC mixes and evaluate those utilizing marginal aggregates and a modified binder [3].

- Procurement of Material and Characterization.
- Assessment of Physical Properties of Raw Material.
- Gradation of Aggregates.
- Mix Design by using Aggregates and Binder.
- Analysis and Results.

The properties of aggregates have a big impact on the design of BC. A material that has enough strength, hardness, toughness, specific gravity, and shape is selected to the design mix; standard testing practices are implemented to follow IRC standards and MoRTH [1] as shown in the table below. Numerous tests on the aggregates were conducted as part of the proposed research work to ascertain their physical characteristics. The requirements for the physical properties of aggregates are shown in Table 1 [1].

In this experiment, bitumen samples PMB40 and CRMB55 are used. Bitumen is needed as a binding substance. The consistency, viscosity, temperature susceptibility, and safety of bitumen are evaluated using a variety of experiments. According to MoRTH [1], the following table details the standard test procedures that are used. The requirements of the physical properties of binders are tabulated in Table 2.

Table 1 Physical properties of aggregates [4]

S. No	Properties	Test methods	MoRTH [1]
1	AIV (Aggregate impact value)	IS:2386 (IV)	Max 24%
2	LA (Los Angeles abrasion value)	IS:2386 (IV)	Max 30%
3	Water absorption value	IS:2386 (III)	Max 2%
4	Specific gravity	IS:2386 (III)	2.5–3.0
5	Combined flakiness and elongation index	IS:2386 (I)	Max 35%

Table 2 Properties of modified bitumen [5, 6]

Properties	Test method	MoRTH specification [1]	Test method	MoRTH specification [1]
Penetration	IS 1203	30–50	IS 1203	30–50
Softening point °C	IS 1205	60	IS 1205	60
Ductility at 27 °C	IS 1208	+50	IS 1208	+50
Specific gravity	IS 1202	1.01	IS 1202	1.01
Flashpoint, °C	IS 1209	220	IS 1209	220
Fire point, °C	IS 1209	247	IS 1209	247

Table 3 Gradation for Grading-I of bituminous mix

S. No	Sieve size (mm)	Weight retained (gms)	Weight retained (%)	Cumulative weight retained (%)	Cumulative % by weight of total aggregate passing obtained values	Cumulative % by weight of total aggregate passing MoRTH [1]
1	19	00	0.0	0.0	100	100
2	13.2	470	9.4	9.4	90.6	79–100
3	9.5	380	7.6	16.0	84.0	70–88
4	4.75	1140	22.8	38.8	61.2	53–71
5	2.36	490	9.8	47.6	52.4	42–58
6	1.18	280	5.6	53.2	46.8	34–48
7	0.60	580	11.6	64.8	35.2	26–38
8	0.30	490	9.8	73.6	26.4	18–28
9	0.15	560	11.2	84.8	15.2	12–20
10	0.07	480	9.6	94.4	5.6	4–10

Table 4 Gradation of filler

S. No	IS sieve (mm)	Cumulative % passing by weight of total aggregate (obtained values)	Cumulative % passing by weight of total aggregate MoRTH [1]
1	0.6	100	100
2	0.3	100	95–100
3	0.075	98	85–100

4 Gradation of Aggregate

The aggregates are used in the construction of BC mixes, arranged through the sieves according to the MoRTH [1] for Grading-I of bituminous concrete. Dose gradation of aggregate manually without using a sieve shaker. The aggregate grading that satisfies the requirements of the MoRTH [1] specifications for midpoint gradation for Grading-I of BC Mix is selected and tabulated in Table 3 [5] (Table 4).

5 Results

The following physical characteristics of aggregates have been put to the test in the proposed research in accordance with the methodology provided. The findings of the physical characteristics of the specified aggregates are shown in Table 5 [5].

After conducting the tests on the bitumen as per the methodology in the proposed study, we performed the basic tests on bitumen. This table also puts the result of

Table 5 Result of physical properties of aggregates [4, 7, 8]

S. No	Properties	Test method	Natural aggregates	MoRTH specifications [1]
01	AIV (Aggregate impact value)	IS:2386 (IV)	15%	Max 24%
02	LA (Los Angeles abrasion test)	IS:2386 (IV)	21.78%	Max 30%
03	Water absorption value	IS:2386 (III)	1.012	Max 2%
04	Specific gravity	IS:2386 (III)	2.67 For 10 mm, 2.69For 6 mm, 2.83 For stone dust	2.5–3.0
05	Combined flakiness and elongation index	IS:2386 (I)	26.23%	Max 35%

the RTFOT abstract Bitumen and does a comparison of the properties. MoRTH [1], IRC SP55, and IRC SP53 standard test property limits are used. Tables 6 and 7 represents the physical properties of normal bitumen and RTFOT-modified bitumen [1] (Tables 6 and 7).

Table 6 Results of physical properties of modified bitumen (CRMB55) [9]

S. No	Properties	Test method	CRMB55 results	RTFOT CRMB55 results	MoRTH specifications [1]
1	Penetration	IS 1203	52	60	50–70
2	Softening point	IS 1203	77	106	45–50
3	Ductility at 27 °C	IS 1208	57	60	+40
4	Specific gravity	IS 1202	1.01	1.02	1.01
5	Flashpoint, °C	IS 1209	270	235	220
6	Fire point, °C	IS 1209	280	248	247

Table 7 Results of physical properties of UN-modified bitumen (VG30) [5]

S. No	Properties	Test method	VG-30 results	RTFOT VG-30 results	MoRTH specifications [1]
1	Penetration	IS 1203	66	69	50–70
2	Softening point °C	IS 1203	50	54	45–50
3	Ductility at 27 °C	IS 1208	53	60	+40
4	Specific gravity	IS 1202	1.01	1.02	1.01
5	Flashpoint, °C	IS 1209	240	260	220
6	Fire point, °C	IS 1209	260	275	247

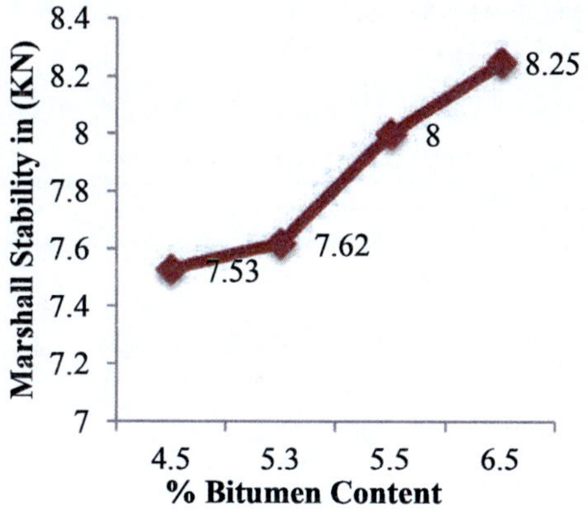

Fig. 1 Marshall stability vs. bitumen content (VG30)

5.1 Unmodified Bitumen VG30

The results of BC mixes with natural aggregates took some trial percentage of bitumen, i.e., 4.5, 5.3, 5.5, and 6.5. First of all, check the results of the V30-grade normal BC mix [5]. Figure 1 shows the relationship between the Marshall Stability Value and the VG30 bitumen content of BC mixes, observing that the increase in bitumen content increases Marshall Stability. Figure 2 shows the relationship between the Marshall Flow Value and VG30 Bitumen Content. It is observed that with an increase in bitumen content, the flow value of natural BC mixes is increasing. Figure 3 shows the relationship between the bitumen content and the unit weight of VG30 BC mixes. It is observed that with an increase in bitumen content, the unit weight increases up to a certain extent and then decreases with a further increase in bitumen content. The maximum unit weight was obtained at 5.3% of the bitumen content. Figure 4 shows the relationship between the percentage of void mineral aggregates and VG30 bitumen content. It is clear from the graph that there is an increase in bitumen content and the increase in %VMA. Figure 5 shows the relationship between VG30 bitumen content and the percentage of the void filled with bitumen. The graph shows the parabolic relation between these two parameters. As the increase in bitumen content in the mix, the increase in %VFB increases in bitumen content (Table 8).

5.2 RTFOT Unmodified Bitumen VG30

Table 9 shows the properties of Marshall of RTFOT bitumen (VG30). Figure 6 shows the relationship between the Marshall Stability Value and the RTFOT VG30 bitumen content of BC mixes. We observed that with an increase in bitumen content, the increase in Marshall Stability. Figure 7 shows the relationship between the Marshall

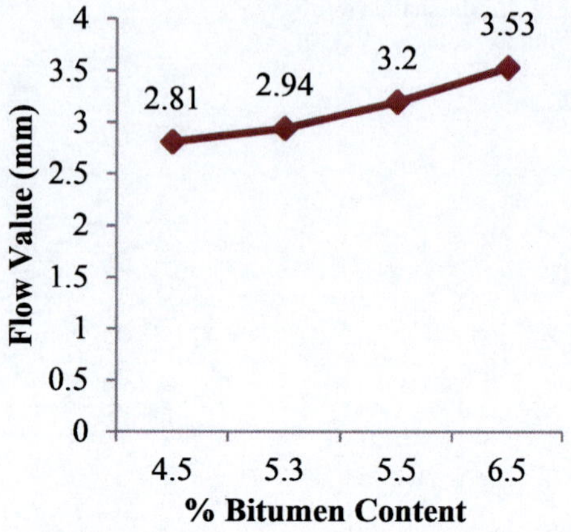

Fig. 2 Marshall flow value vs. bitumen content (VG30)

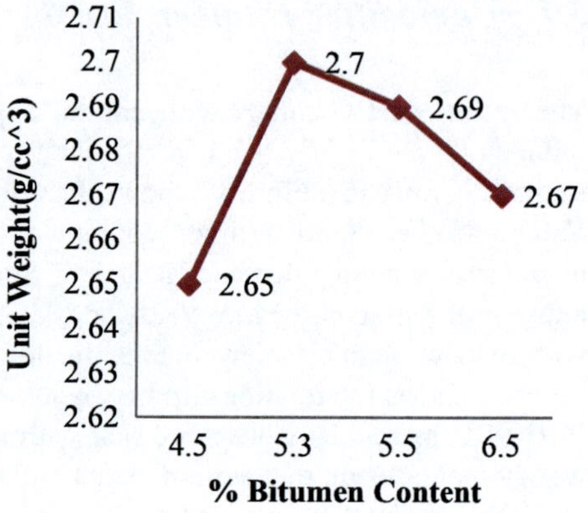

Fig. 3 Unit weight vs. bitumen content (VG30)

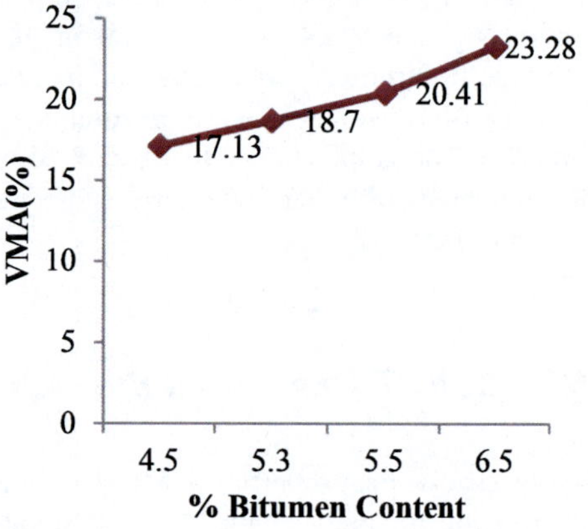

Fig. 4 VMA% vs. bitumen content (VG30)

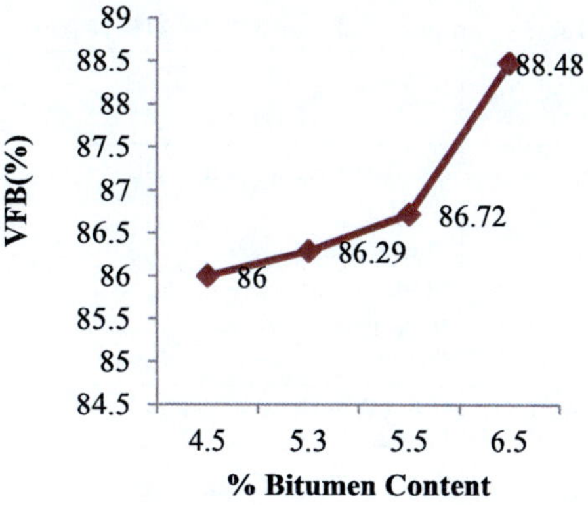

Fig. 5 VFB% vs. bitumen content (VG30)

Table 8 Shows the properties of Marshall of normal bitumen (VG30)

S. No	Property tested	Bitumen content by weight of aggregates (VG30)			
		4.5%	5.3%	5.5%	6.5%
1	MS (Marshall stability) (KN)	7.53	7.62	8.00	8.25
2	Flow value (mm)	2.81	2.94	3.20	3.53
3	Bulk density (g/cc)	2.65	2.70	2.69	2.67
4	VoV (Volume of voids) (V_v)%	2.45	2.51	2.69	2.76
5	VMA (Void in mineral aggregates) (VMA)%	17.13	18.7	20.41	23.28
6	VFB (Void filled with bitumen) (%)	86	86.29	86.72	88.48
7	Marshall quotient (KN/mm)	2.67	2.59	2.5	2.33

Flow Value and RTFOT VG30 Bitumen Content. It is observed that with an increase in bitumen content, the flow value of RTFOT BC mixes is increasing. Figure 8 shows the relationship between the bitumen content and the unit weight of RTFOT VG30 BC mixes. It is observed that with an increase in bitumen content, the unit weight increases up to a certain extent. Figure 9 shows the relationship between the percentage of void mineral aggregates and RTFOT VG30 bitumen content. One can see clearly through the graph that with an increase in bitumen content and an increase in %VMA. Figure 10 shows the relationship between the percentages of voids filled with bitumen and the content of RTFOT VG30 bitumen. The graph shows the relationship between these two parameters. As the increase in bitumen content in the mix, the increase in %VFB increases in bitumen content.

Table 9 Properties of Marshall for RTFOT bitumen (VG30)

S. No	Property tested	Bitumen content by weight of aggregates			
		4.5%	5.3%	5.5%	6.5%
1	MS (Marshall stability) (KN)	8.23	8.62	8.79	8.96
2	Flow value (mm)	3.01	3.24	3.40	3.53
3	Bulk density (g/cc)	3.67	4.26	4.11	4.04
4	VoV (Volume of voids) (V_v)%	4.46	5.35	5.51	6.11
5	VMA (Void in mineral aggregates) (VMA)%	24.1	27.35	29.40	39.42
6	VFB (Void filled with bitumen) (%)	81.32	97.78	92.95	94.65
7	Marshall quotient (KN/mm)	2.73	2.66	2.59	2.53

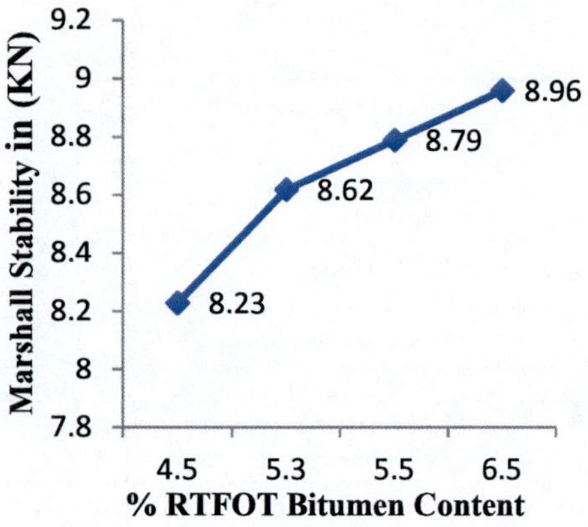

Fig. 6 Marshall stability vs. bitumen content (RTFOT VG30)

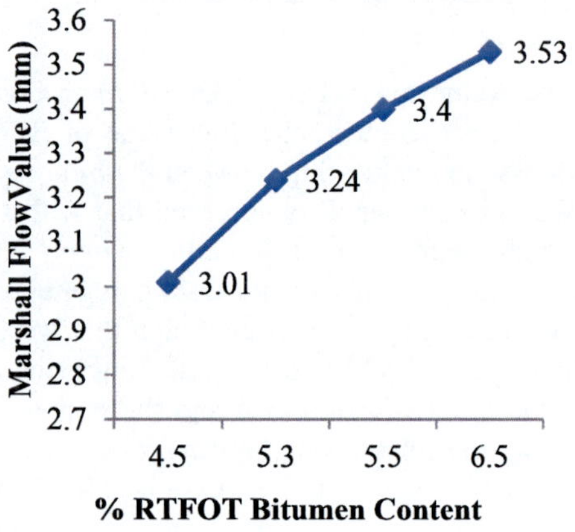

Fig. 7 Marshall flow value vs. bitumen content (RTFOT VG30)

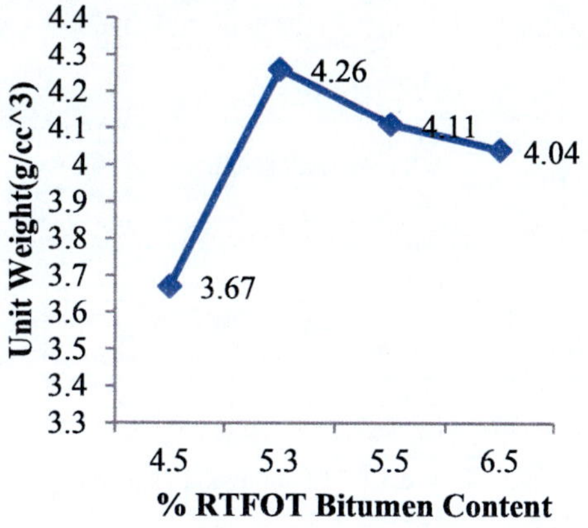

Fig. 8 Unit weight vs. bitumen content (RTFOT VG30)

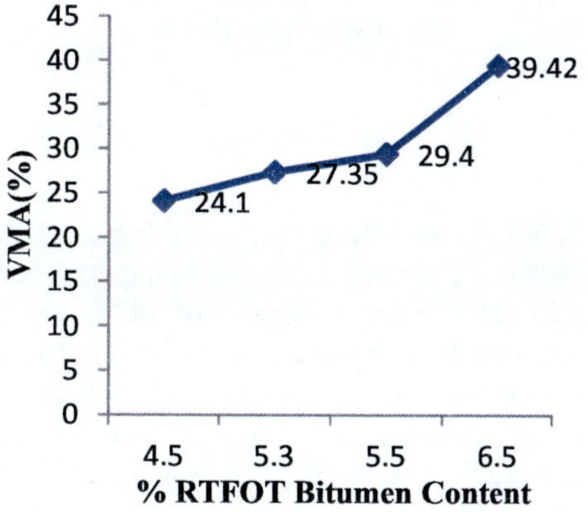

Fig. 9 VMA% vs. bitumen content (RTFOT VG30)

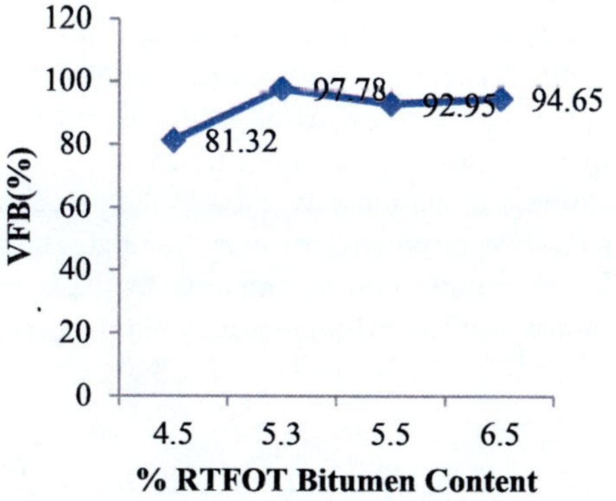

Fig. 10 VFB% vs. bitumen content (RTFOT VG30)

Table 10 Marshall results of modified bitumen CRMB55

S. No	Property tested	Bitumen content by weight of aggregates			
		4.5%	5.3%	5.5%	6.5%
1	MS (Marshall stability) (KN)	10.23	11.08	11.65	11.82
2	Flow value (mm)	3.05	3.13	3.19	3.27
3	Bulk density (g/cc)	2.55	2.43	2.5	2.65
4	VoV (Volume of voids) (V_V)%	3.13	3.85	4.22	4.6
5	VMA (Void in mineral aggregates) (VMA)%	15	17.08	19.01	20.05
6	VFB (Void filled with bitumen) (%)	64.06	65.56	68.32	72
7	Marshall quotient (KN/mm)	2.06	2.61	2.55	2.67

5.3 *Modified Bitumen CRMB55*

Table 10 shows the properties of Marshall for Results of Modified Bitumen CRMB55. Figure 11 shows the relationship between the Marshall Stability Value and the CRMB55 bitumen content of BC mixes. It is observed that with an increase in bitumen content comes an increase in Marshall Stability. Figure 12 shows the relationship between the Marshall Flow Value and CRMB55 Bitumen Content. We have observed that with an increase in bitumen content, the flow value of BC mixes is increasing and starts decreasing after the 5.5% of bitumen flow value. Figure 13 shows the relationship between the bitumen content and unit weight of CRMB55 BC mixes is broken. It is observed that with an increase in bitumen content, there is a decrease in optimum content. Furthermore, it is increasing after the optimum content. Figure 14 shows the relationship between the percentage of void mineral aggregates and CRMB55 bitumen content. We can see clearly through the graph that with an increase in bitumen content comes an increase in %VMA. Figure 15 shows the relationship between the percentages of void filled with bitumen and the CRMB55 bitumen content is depicted. The graph shows the straight-line relationship between these two parameters. As the increase in bitumen content in the mix, the increase in %VFB increases in bitumen content.

5.4 *RTFOT-Modified Bitumen CRMB55*

Table 11 shows the properties of Marshall Results for RTFOT-Modified Bitumen CRMB55. Figure 16 shows the relationship between the Marshall Stability Value and the RTFOT CRMB55 bitumen content of BC mixes. It is observed that with an

Fig. 11 Marshall stability vs. bitumen content (CRMB55)

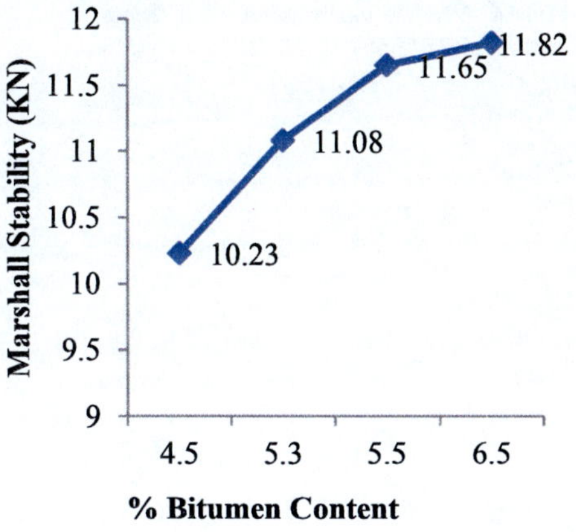

Fig. 12 Marshall flow value vs. bitumen content (CRMB55)

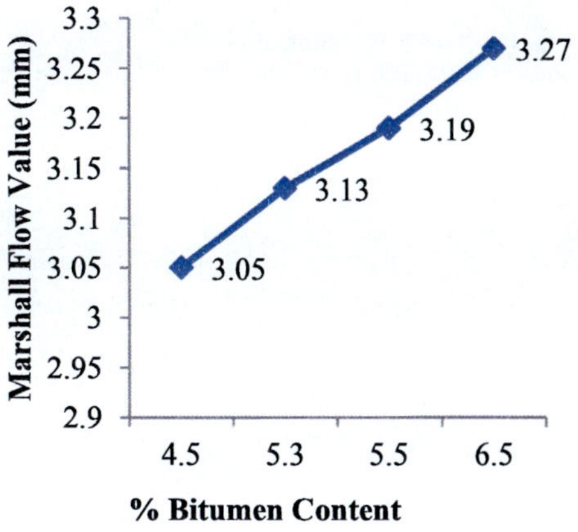

Fig. 13 Unit weight vs. bitumen content (CRMB55)

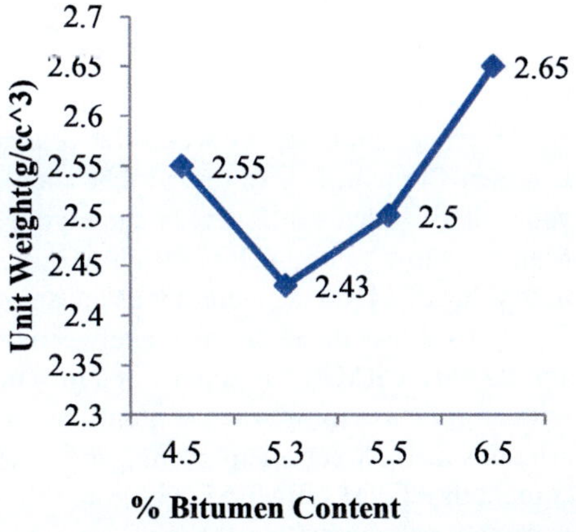

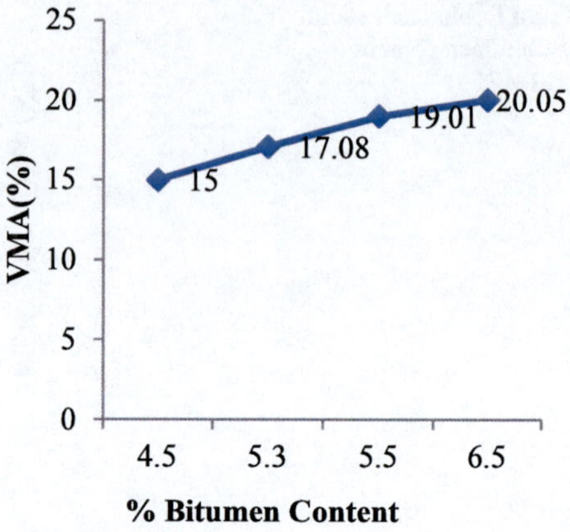

Fig. 14 VMA% vs. bitumen content (CRMB55)

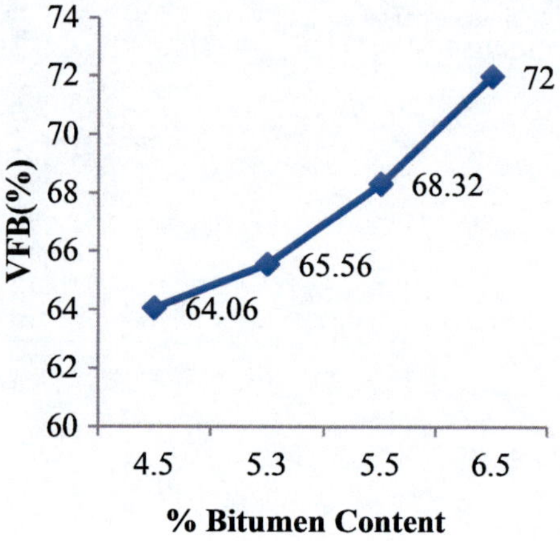

Fig. 15 VFB% vs. bitumen content (CRMB55)

increase in bitumen content, the graph is increasing in Marshall Stability. Figure 17 shows the relationship between the Marshall Flow Value and RTFOT CRMB55 Bitumen Content. It is observed that with an increase in bitumen content, the flow value of BC mixes are increasing. Figure 18 shows the relationship between the Marshall Flow Value and RTFOT CRMB55 Bitumen Content. It has been seen that the flow value of BC mixes increases as the amount of bitumen content in them increases. Figure 19 shows the relationship between the percentage of void mineral aggregates and RTFOT CRMB55 bitumen content. Through the graph, one can clearly see that with an increase in bitumen content, the increase in %VMA. Figure 20 depicts the relationship between the percentages of voids filled with bitumen and the bitumen content of RTFOT CRMB55. The graph shows the straight-line relationship between these two parameters. As the increase in bitumen content in the mix, the increase in

Table 11 Marshall of results of RTFOT-modified bitumen CRMB55

S. No	Property tested	Bitumen content by weight of aggregates			
		4.5%	5.3%	5.5%	6.5%
1	MS (Marshall stability) (KN)	10.98	11.38	11.9	12.32
2	Flow value (mm)	3.25	3.53	3.89	4.12
3	Bulk density (g/cc)	2.53	2.5	2.55	2.65
4	VoV (Volume of voids) (V_V)%	4.46	5.35	5.51	5.71
5	VMA (Void in mineral aggregates) (VMA)%	14.58	17.31	19	21.05
6	VFB (Void filled with bitumen) (%)	66.48	67.98	75.2	77.06
7	Marshall quotient (KN/mm)	2.35	2.32	2.22	2.22

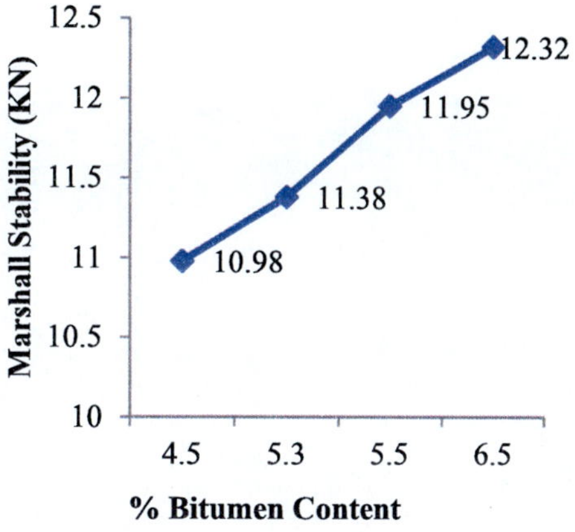

Fig. 16 Marshall stability vs. bitumen content (RTFOT CRMB55)

%VFB increases in bitumen content. The percentage of VFB decreases after 5.5% content bitumen.

6 Comparison Between the Different Grades of Bitumen

6.1 VG30

Table 12 comparisons of the properties of Marshall Stability between normal bitumen and RTFOT Bitumen. Figure 21 shows that after short-term aging, bitumen provides more strength compared to normal bitumen. It can bear more and provide great

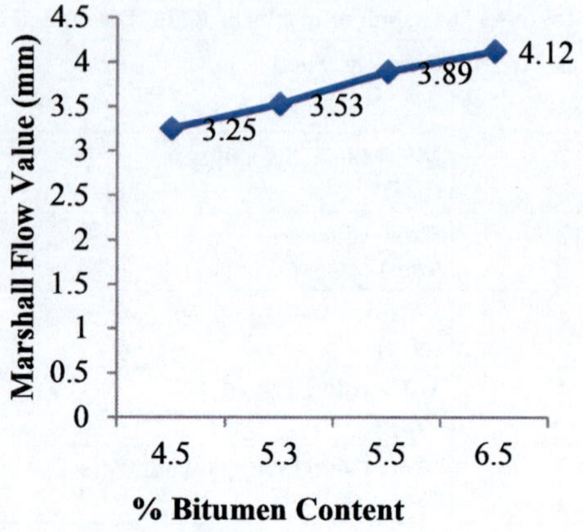

Fig. 17 Marshall flow value vs. bitumen content (RTFOT CRMB55)

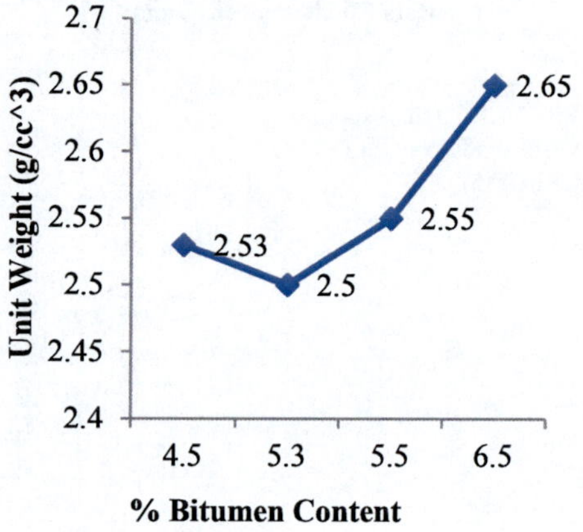

Fig. 18 Unit weight vs. bitumen content (RTFOT CRMB55)

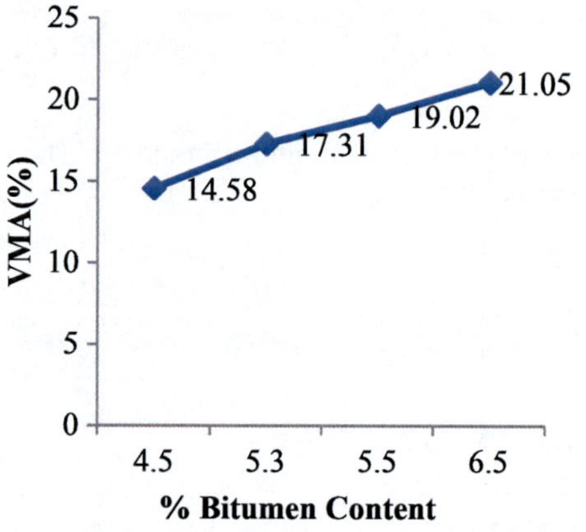

Fig. 19 VMA% vs. bitumen content (RTFOT CRMB55)

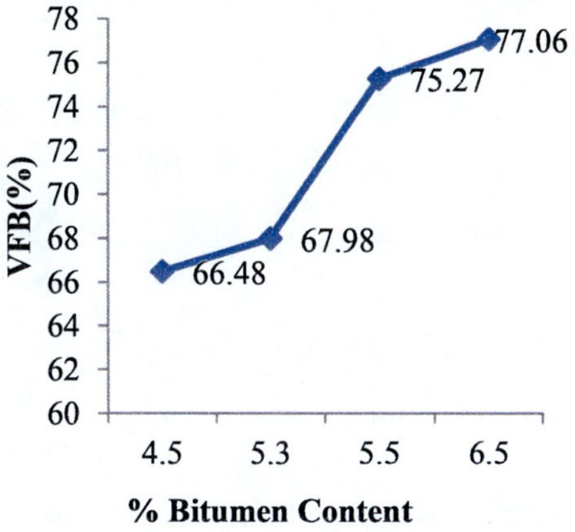

Fig. 20 VFB% vs. bitumen content (RTFOT CRMB55)

Table 12 Comparison between the properties of Marshall stability of normal bitumen VG30 and RTFOT VG30

S. No	Properties	Bitumen content (%)	Different types of bitumen	
			Normal VG30	RTFOT VG30
01	Marshall stability (KN)	4.5	7.53	8.23
02		5.3	7.62	8.65
03		5.5	8.00	8.79
04		6.5	8.25	8.96

strength against normal bitumen. Now compare the Marshall Flow value of normal bitumen VG30 with that of RTFOT-modified bitumen VG30. Figure 22 shows that after a short period of aging, bitumen provides more flow value to the load applied to its surface than normal bitumen. It can bear more and provide great strength against normal bitumen (Table 13).

6.2 *CRMB55*

Table 14 Comparison between the properties of Marshall Stability between normal bitumen and RTFOT Bitumen (CRMB55). Figure 23 shows that after short-term aging, bitumen CRMB55 provides more strength compared to normal bitumen CRMB55. It can bear more loads and provide great strength against normal bitumen. Figure 24 that after short-term aging bitumen CRMB55 provides more Flow value to wear the load apply on the surface of bitumen compared to the normal bitumen-CRMB55. It can wear load more and provide great strength against normal bitumen (Table 15).

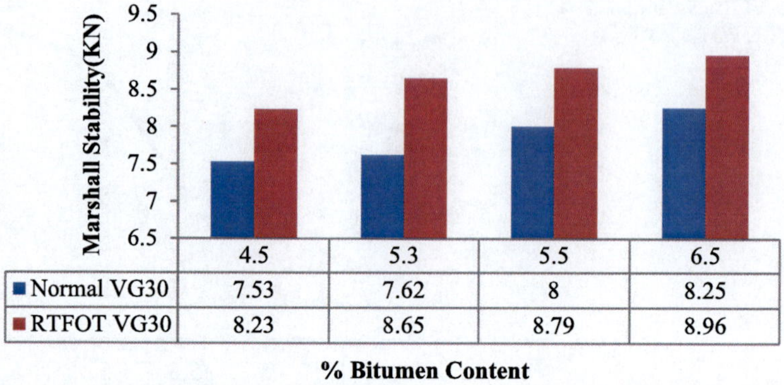

Fig. 21 Comparisons Marshall stability between normal VG30 & RTFOT VG30 bitumen

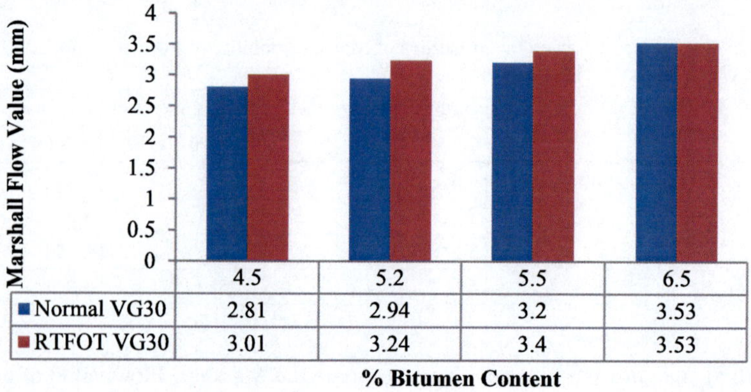

Fig. 22 Comparisons Marshall flow value between normal VG30 & RTFOT VG30 bitumen

Table 13 Comparison between the properties of Marshall flow value of normal bitumen VG30 and RTFOT VG30

S. No	Properties	Bitumen content (%)	Different types of bitumen	
			Normal VG30	RTFOT VG30
01	Marshall flow value (mm)	4.5	2.81	3.01
02		5.3	2.94	3.24
03		5.5	3.20	3.40
04		6.5	3.53	3.53

Table 14 Comparison between the properties of Marshall stability of normal bitumen CRMB55 and RTFOT CRMB55

S. No	Properties	Bitumen content (%)	Different types of bitumen	
			Normal CRMB55	RTFOT CRMB55
01	Marshall stability (KN)	4.5	10.23	10.98
02		5.3	11.08	11.38
03		5.5	11.65	11.95
04		6.5	11.82	12.32

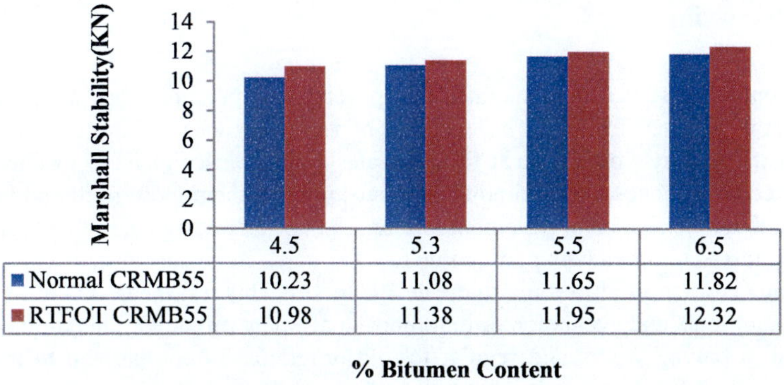

Fig. 23 Comparisons Marshall stability between normal CRMB55 & RTFOT CRMB55 bitumen

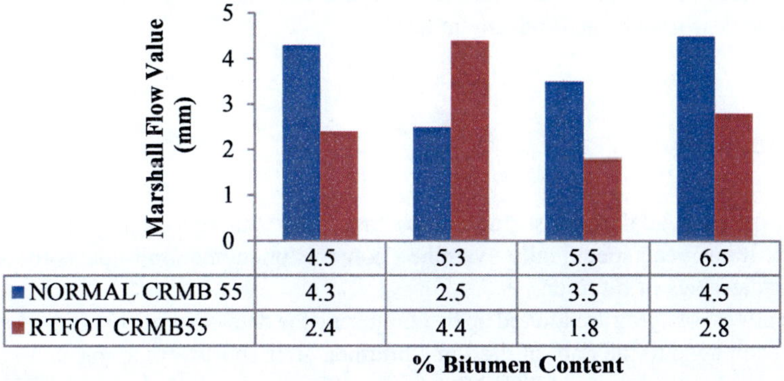

Fig. 24 Comparisons Marshall flow value between normal CRMB55 & RTFOT CRMB55 bitumen

Table 15 Comparison between the properties of Marshall stability of normal bitumen CRMB55 and RTFOT CRMB55

S. No	Properties	Bitumen content (%)	Different types of bitumen	
			Normal CRMB55	RTFOT CRMB55
01	Marshall stability (KN)	4.5	3.05	3.25
02		5.3	3.13	3.53
03		5.5	3.19	3.89
04		6.5	3.27	4.12

7 Discussion

The proposed research demonstrates that pavement temperature has a significant impact on the rate of aging. The viscosity of various grades of bitumen has been demonstrated to rise over time at variable rates Aging causes physical or chemical characteristic changes in bituminous materials, making them more rigid and fragile and reducing the chance of pavement failure. The hot-mix design uses this Marshall stability test of bitumen with particles up to 10, 6, and 2 stone dust in size. Marshall's technique is often used in India to create BC mixes. This test method is utilized in the development and evaluation of bituminous pavement mixtures. This test is often utilized in paving works and regular testing programs. In comparison to regular bitumen, RTFOT bitumen generates higher strength. The bitumen's short-term qualities have been altered, giving strength against the load applied to the road and a flow value of the load that allows quickly bearing the load and avoiding cracking on the bitumen's surface. The bituminous binder's age is one of the key factors that determine how long asphalt pavement lasts.

8 Conclusion

- The primary goal of this study is to investigate bitumen aging in the short term. It has been specifically examined how this phenomenon impacts the aging characteristics of bitumen.
- The effects of aging are looked up by comparing the data of single bitumen that had not been aged to the data of the same bitumen after short-term aging treatments.
- It shows how normal and RTFOT-modified BC behaved. It finds that the RTFOT-modified BC mix has better Marshall characteristics, as described in the results.
- At 5.2% bitumen content, the Marshall stability value begins to rise and continues to rise as bitumen content increases.
- The results show that bitumen aging is more severe for the top part of the pavement as it is in contact with the atmospheric air and the pavement temperature is higher. Therefore, a stiffness gradient develops along with the depth of the pavement, which plays a major role.

- CRMB55 of Marshall stability results in increased stability compared to pure bituminous concrete mix and is suitable for hot climate zones.
- The life cycle cost of a pavement structure for 20 years with polymer-modified bitumen showed a benefit of 1.36 times when compared to a pavement structure with unmodified bituminous mixes. Utilization of waste plastic and other modified bitumen is found to reduce the life cycle cost of pavement when compared to unmodified bitumen.
- The aging will reduce the cracks that develop on the pavement surface, and it will also increase the strength of the pavement surface.
- Though the aging can be reduced, the excessive hardening of bitumen can be reduced and the ability to support the traffic-induced stresses and strains increased.
- Through the aging process, the adhesion ability becomes stronger between the aggregate and asphalt.
- Due to heavy rainfall in India, our study conducted that the material may be used in the construction of rural roads.

References

1. MoRTH (Ministry of Road Transport and Highways) (2013) Specifications for road and bridge works. Indian Road Congress New Delhi, India. (Author)
2. ASTM D 2872-88 (2004) Standard test method for effect of heat and air on a moving film of asphalt (rolling thin-film oven test). American Society of Testing Materials (ASTM), 04.03, Pennsylvania, USA
3. Magar NR (2014) A study on the performance of crumb rubber modified bitumen by varying the sizes of crumb rubber. Int J Eng Trends Technol (IJETT) 14(2)
4. IS 2386 (1963) Method of test for aggregates for concrete, mechanical properties, part-IV
5. IS: 73 (2013) Paving bitumen–specification Indian Standard Code for unmodified bitumen
6. IRC: 37 (2018) Guidelines for the design of flexible pavements. Indian Road Congress, New Delhi
7. IS: 2386 (1963) Methods of test for aggregates for concrete–part 1: particle size and shape
8. IS 2386 (1963) Method of test for aggregates for concrete for determination of specific gravity, void, absorption and bulking, part-III
9. SP 53 (2010) Guidelines on use of modified bitumen in road construction
10. ASTM D 1075 (1979) Standard test method for effect of water on compressive strength of compacted bituminous mixtures. Philadelphia, American Society for Testing and Materials
11. AASHTO R-30 (2002) Standard practice for mixture conditioning of hot-mix asphalt (HMA). American Association of State Highway and Transportation Officials (AASHTO), Washington DC, USA
12. AASHTO T-283 (2003) Standard method of test for resistance of compacted asphalt Mixtures to moisture-induced damage. American Association of State Highway and Transportation Officials (AASHTO), Washington DC, USA

Threshold Enabled Measure to Evaluate the Travel Time Reliability Performance

Sabeen Sharic, Thisaiveerasingam Thilakshan, and Saman Bandara

Abstract The tolerance levels at both ends of a bus trip influence the perceived bus transport travel time reliability performance. Existing studies mainly consider the tolerance level at one end only. The objective of this study is to develop a method to evaluate the bus transport travel time reliability performance by defining tolerance levels at both ends of a trip. Threshold Enabled Departure Passenger Waiting Time (TEDPWT) is perceived as the half of Scheduled Departure Headway (SDH) at bus stops for frequent bus services that have uniform Scheduled Departure Headways. Threshold Waiting Time Enabled Departure Headway (TDH) is calculated as the difference between Actual Departure Headway and TEDPWT. By comparing TDH with TEDPWT, departure activity is categorized as Early Departure, Ontime Departure, and Late Departure. Similarly, Threshold Waiting Time Enabled Arrival Headway (TAH) is also defined, and arrival activity is categorized as Early Arrival, Ontime Arrival, and Late Arrival. Crosstabulation is conducted across departure activity levels versus arrival activity levels with respect to type of day (weekday and weekends) and time of the day. This crosstabulation analysis results in statistically significant percentage/probability value for the cross occurrence of the departure activity levels versus arrival activity levels. This percentage /probability value is used as the indicator to evaluate the travel time reliability performance. When applied to suburban feeder-bus routes, a major finding among many was, buses arrived at their destination late though they departed early on weekdays during peak hours. This study contributes to the domain of performance evaluation of bus transport travel time reliability.

Keywords Travel time reliability · Performance evaluation · Waiting time

S. Sharic (✉)
Department of Management and Finance, General Sir John Kotelawala Defence University, Ratmalana, Sri Lanka
e-mail: sabeen@kdu.ac.lk

T. Thilakshan · S. Bandara
Department of Civil Engineering, University of Moratuwa, Katubedda, Sri Lanka

K. K. Pathak et al. (eds.), *Latest Developments in Civil Engineering*, Lecture Notes in Civil Engineering 352, https://doi.org/10.1007/978-981-99-2676-3_59

1 Introduction

Passenger tolerance level is an important aspect in evaluating the travel time reliability performance. The existing studies have defined the tolerance level as the deviation of the passenger waiting time from the average travel time. Some studies have focused on the type of the arrival (early, on time, late) at the destination. Less studies have focused on the types of the departure activity at the origin and the type of the arrival activity (early, on time, late) at the destination. Failure to define tolerance level for both departure and arrival activities of trips in travel time reliability performance evaluations results in incorrect identification of passenger perceived satisfaction. Operation of bus routes without realizing the real perceived passenger satisfaction levels on travel time may lead toward lower passenger attraction. It also results in inefficient and unsuccessful transit planning and operations. The purpose of this study is to evaluate the travel time reliability of bus routes by defining the tolerance level for both departure and arrival activities of bus trips. This study paves the way for a better understanding of the effect on passengers due to travel time reliability performance. Meanwhile, both bus transport operators and the regulators can get better visibility of the travel time reliability performance.

2 Literature Review

Travel time is an important service quality element of bus transport systems that must be evaluated to measure the level of travel time being within the boundary expected by the passengers. Therefore, travel time should be reliable to satisfy the passengers. In general, there are three types of measures available in the literature to evaluate the travel time reliability performance. They are buffer measures, statistical range measures, and Tardy trip indicators.

Buffer time which is the difference between a given percentile of the travel time (e.g., 85%) and the average travel time for all the bus trips between origin and destination is a common parameter used for measuring the travel time reliability performance [1]. It denotes the extra time taken over the average travel time for bus trips between a particular origin and destination. The buffer time index is calculated as the ratio between the buffer time and the average travel time. Reliability buffer time which is calculated as the difference between the upper percentile, and an intermediate or lower percentile is the extra time that would be needed for the probability (equivalent to the value of the percentile) of arriving at the destination on time [2]. Excess reliability buffer time (ERBT) is calculated as the difference between the actual levels of reliability undergone by passengers and what they should have undergone if the transit systems worked according to the schedule which means 100% reliability [2]. However, buffer indexes generally assumed, for example, Weibull and normal distributions of travel time reliability, but no verification was made with real-life traffic data, which directly impacts the accuracy of predicting reliability. To lessen the underestimation of travel time reliability, it was recommended to define buffer indexes based on median rather than mean [3, 4]. Automatic Vehicle Location (AVL) systems have been widely used to collect data on travel time in the above. A main limitation of buffer indexes is their inability to meaningfully define both departure and arrival activities from origin and destination, respectively. Instead, these indexes consider merely the arrival activities at destination. To mitigate this problem, departure/arrival activities can be categorized such as early departure, on time departure, late departure, early arrival, on time, and late arrival as they are important for strategic decision-making.

In terms of statistical range measures of travel time reliability performance, mean and variability of travel times between two bus stops (time points) have been taken to monitor travel time variability during the same time intervals and among many bus routes [5]. Probability has been used as a measure of monitoring and evaluating the travel time variability of bus routes [6]. There, probability for a certain travel time between two bus stops to be greater than a particular value from the median travel time for that trip has been calculated using the lognormal probability distribution function. Another statistical measure used in travel time performance is the percent variation, which is the ratio between standard deviation and average travel time as a percentage. Variability index is another measure used in measuring travel time performance which is the ratio between the difference in peak period confidence intervals (upper 95% value—lower 95% value) and the difference in off-peak confidence intervals (upper 95% value—lower 95% value) as discussed by Lomax et al. [7]. These measures lack in accommodating threshold time either at origin or destination stops or at both. But these problems can be better solved if categories are developed for the levels of departure from the origin bus stops such as early, on time, and late as well as levels for arrival at the destination bus stop such as early, on time, and late. Cross tabulation using chi-square statistics test can help to identify statistically significant patterns available in its nature along with the provision of a percentage value for the occurrence of combinatorial categories of departure from the origin bus stop with arrival at the destination stop as suggested in White [8].

Tardy trip indicators are also the travel time reliability measures that represent the percentage of trips that would be too late relative to expectations out of some trips in unacceptable late intervals [7]. The Misery index which is one of the Tardy trip indicators is calculated as the ratio of the average travel time of the 20 percent worst trips minus the average travel time of all trips to the average travel time of all trips [9]. The limitations of this index are the subjective definition of the worst trips and the consideration of merely the arrival activities at the destination. This measure does not concern the departure activities at the origins.

All these measures (Buffer measures, statistical range measures, and Tardy trip indicators) are demonstrating the travel time reliability performance merely considering a deviation from the average travel time while considering the arrival activity at the destination. As they fail to consider what has happened at the origin, these measures cannot yield meaningful interpretations. For instance, an indicator to reveal merely a late arrival of a bus to the destination bus stop will yield less information on a situation if the bus has departed late from the origin bus stop and it has got late to the destination bus stop. Therefore, considering the levels of departure activities from the origin bus stop and the levels of arrival activities to the destination bus stops in developing an indicator to measure the travel time reliability performance of buses between bus stops is important. It will be useful in comparing and prioritizing a system of investments strategies to improve the travel time reliability performance of buses. In addition, there are contradictions in defining the expected travel time of the passengers. Passengers generally can tolerate a certain amount of time for departure from the original bus stop as well as they are tolerable of allocating a certain amount of time for the arrival at their destination. In addition, from the bus operators' point of view, their optimal operation should also be considered. Therefore, the allocation of threshold waiting time at departure stop and allowable perceived lateness at the destination of the trip is important in the development of performance measurement of travel time reliability.

3 Methodology

3.1 *Evaluation of Bus Transport Travel Time Reliability Performance Between Bus Stops*

Actual Departure Headway (ADH) is the difference between the departure times of two consecutive buses from the same stop. Scheduled Departure Headway (SDH) is the difference between the scheduled departure times of consecutive two buses at the same stop. Threshold Enabled Departure Passenger Waiting Time (TEDPWT) is perceived as the half of the Scheduled Departure Headway at bus stops for frequent bus services that have equal Scheduled Departure Headways [10–14]. Threshold Waiting Time Enabled Departure Headway (TDH) is calculated as the difference between ADH and TEDPWT. Similarly, Actual Arrival Headway (AAH), Scheduled Arrival Headway (SAH), Threshold Waiting Time Enabled Arrival Passenger Waiting Time (TEAPWT), and Threshold Waiting Time Enabled Arrival Headway (TAH) are also defined.

3.2 *Travel Time Reliability Performance Between Bus Stops*

Departure and arrival activities are categorized based on the concept of TEDPWT and TEAPWT as shown in Table 1. Nine levels of travel time reliability performances between origin and destination bus stops are developed based on the types of departure and arrival activities between two stops. Each trip between the identified two bus stops of a bus route is labeled with one of the levels of travel time reliability performance as shown in Table 2.

Table 1 Types of departure/arrival activities

Type of the departure/arrival activity	Definition
Early Departure from the origin stop (ED)	ADH < 2TDH
On time Departure from the origin stop (OD)	ADH = 2TDH
Late Departure from the origin stop (LD)	ADH > 2TDH
Early Arrival to the destination stop (EA)	AAH < 2TAH
On time Arrival to the destination stop (OA)	AAH = 2TAH
Late Arrival to the destination stop (LA)	AAH > 2TAH

Table 2 Levels of travel time reliability performance

Category	Departure from bus stop-origin	Arrival to bus stop-destination
01	ED	EA
02	ED	OA
03	ED	LA
04	OD	EA
05	OD	OA
06	OD	LA
07	LD	EA
08	LD	OA
09	LD	LA

3.3 Method of Analysis

Crosstabulation analysis using a chi-square test between departure activity types at the origin bus stop versus arrival activity types at a destination bus stop is carried out for the entire data set which includes both weekdays and weekends. Then the crosstabulation analysis is carried out between departure activity types at the origin bus stop, arrival activity types at the destination bus stop, and time intervals identified for both weekdays and weekends. Then the same analysis is carried out for both weekdays and weekends separately. This analysis yields the percentage or the probability of the occurrence of different types of departure/arrival activities. The percentage of occurrence or the probability is the indicator to evaluate the performance of travel time reliability between origin and destination bus stops. This analysis helps identify the patterns of occurrence among the nine categories shown in Table 2.

3.4 Validation of the Indicator

The Monte Carlo significance test is used when there are less than 5 occurrences for any possible combination of departure/arrival and time intervals as it compares the observed data with the random samples generated by the Monte Carlo simulation. This test yields a probability value (p) for the null hypothesis that there is no association between departure/arrival of buses with time intervals to be true. If the p-value is less than 0.05, then the null hypothesis is rejected [15]. The indicator on percentage of occurrence of the combinatorial types of departure/arrival activities is validated if and only if there is a statistically significant pattern identified using Monte Carlo significance test.

3.5 Data Collection

The proposed indicator is applied to a bus trip between Digana Junction bus stop and Katubedda bus stop for the frequent bus route 255 which runs between Kottawa and Mount Lavinia in the Colombo district, Sri Lanka. This route is 22.2 km in length and Digana Junction bus stop is located 1.5 km away from the Origin, Kottawa bus stop. The distance between Digana Junction bus stop and Katubedda bus stop is 11.1 km. Figure 1 illustrates the route map. The records on scheduled arrival time, scheduled departure time, actual arrival time, and actual departure time for both bus stops (Digana Junction and Katubedda) were collected from timekeepers at both the bus stops for 10 days including two weekend days. There were 594-weekday bus trips and 116-weekend bus trips for this analysis. The headways range between 05 and 10 min.

Depending on the data availability and uniformity of scheduled headways, different time periods can be considered. The time periods considered for this study for both weekdays and weekends are shown in Tables 3 and 4, respectively. Evening peak hours were not considered as the information on bus departure and arrivals were not recorded. Tables 5 and 6 illustrate data on the occurrence of origin-destination

Fig. 1 The route selection (Kottawa—Mt. Lavinia 255 route)

Table 3 Weekday time intervals and headways for the selected bus route and the stops

Category	Interval	Scheduled headway (min)
WD01	05:08–06:02 am	08
WD02	06:02–08:12 am	05
WD03	08:32–09:32 am	06
WD04	09:46–01:02 pm	07
WD05	01:07–03:12 pm	05

Table 4 Weekend time intervals and headways for the selected bus route and the stops

Category	Interval	Scheduled headway (min)
WE01	05.00–06.02 am	10
WE02	06.07–08.32 am	05
WE03	08.38–10.32 am	06
WE04	10.50–12.32 pm	06

travel time reliability performance (departure/arrival) along with the time intervals for weekdays and weekends, respectively.

Table 5 Occurrence (frequency) of combination of departure statuses and arrival types versus time intervals on Weekdays (WD)

Time intervals	Category (combination of departure status and arrival status)									
	1	2	3	4	5	6	7	8	9	Total no. of trips
WD01	24	15	1	4			1			45
WD02	46	37	29	12	5	12	9	1	16	167
WD03	6	5	11	5	3	11	1	3	30	75
WD04	1	5	8	4	7	19	3	4	34	85
WD05	68	78	15	18		8	4	2	9	202
Total no. of trips	145	140	64	43	15	50	18	10	89	574

Table 6 Occurrence (frequency) of combination of departure statuses and arrival types versus time intervals on weekends (WE)

Time intervals	Category (combination of departure status and arrival status)									
	1	2	3	4	5	6	7	8	9	Total no. of trips
WE01	7	7				1				15
WE02	23	23	6						1	53
WE03	2		8		2	1	1		10	24
WE04			8						16	24
Total no. of trips	32	30	22		2	2	1		27	116

4 Results

Table 7 outlines the results of the crosstabulation analysis carried out between departure activity types (from the origin bus stop) versus arrival activity types (at the destination bus stop) for types of the day (weekday and weekends). Tables 8 and 9 are the results of the crosstabulation analysis for the above but separately for weekdays and weekends, respectively.

5 Findings and Discussion

As per to the data collected and the results obtained (Table 7), it is observed that when a bus leaves the origin on time, there is a higher probability that the bus will come on time or early at the destination. This probability is relatively higher during off-peak hours in weekdays and lower in peak hours on both weekdays and weekends (Tables 8 and 9). As this is a feeder route which connects many trip attractors, the passenger demand for this route is high in the peak hours. This leads to higher boarding, alighting, and dead time per stop. These could be the possible reasons for lower travel time reliability during peak hours. There is a greater probability for a bus to arrive early to the destination when it leaves from the origin on time or early and this probability is higher in weekends compared to weekdays as there is lesser demand for travel in weekends (Table 7).

The probability for a bus to arrive late at the destination even when a bus leaves the origin on time or early is higher in weekdays compared to weekends (Table 7) and it is higher during the latter part of peak hours compared to early part of peak hours in weekdays (Table 8) as speed reduces with the traffic flow along the route. This probability is higher in weekends compared to weekdays during peak hours due to the low demand for this route in weekends and the average speed of the service is lower (Table 7).

When a bus leaves late from the origin, there is a higher probability that the bus arrives at the destination late (Table 7). This probability is higher in weekends than in the weekdays (Table 7). This probability is higher during peak hours during weekdays (Table 8). The probability for a bus arriving at the destination on time or early when it leaves the origin late is higher in weekdays compared to weekends (Table 8) and this probability is higher during off-peak hours as there is a possibility for higher speed due to less traffic during off-peak hours (Table 8).

The proposed indicator of this study can be used to identify the bus routes travel time reliability performance with respect to nine possible categories based on the departure and arrival time combinations. This information can be useful for both bus passengers and the operators to make decisions on making the trips and identifying the optimal operational plans. It also helps to identify the travel time reliability performance along the route with respect to time of the day, day of the week so that schedule updates can be done in an effective manner.

Table 7 Crosstabulation analysis on departure activities versus arrival activities by day type

Day	Departure status	Particulars	Arrival status			Total	Monte Carlo. Sig
			OA	EA	LA		
WD	OD	Count	145	140	64	349	0.00
		% within departure	41.5%	40.1%	18.3%	100.0%	
		% within arrival	70.4%	84.8%	31.5%	60.8%	
		% of total	25.3%	24.4%	11.1%	60.8%	
	ED	Count	43	15	50	108	0.00
		% within departure	39.8%	13.9%	46.3%	100.0%	
		% within arrival	20.9%	9.1%	24.6%	18.8%	
		% of total	7.5%	2.6%	8.7%	18.8%	
	LD	Count	18	10	89	117	0.00
		% within departure	15.4%	8.5%	76.1%	100.0%	
		% within arrival	8.7%	6.1%	43.8%	20.4%	
		% of total	3.1%	1.7%	15.5%	20.4%	
WE	OD	Count	32	30	22	84	0.00
		% within departure	38.1%	35.7%	26.2%	100.0%	
		% Within arrival	97.0%	93.8%	43.1%	72.4%	
		% of total	27.6%	25.9%	19.0%	72.4%	
	ED	Count	0	2	2	4	0.00
		% within departure	0.0%	50.0%	50.0%	100.0%	
		% within arrival	0.0%	6.3%	3.9%	3.4%	
		% of total	0.0%	1.7%	1.7%	3.4%	
	LD	Count	1	0	27	28	0.00
		% within departure	3.6%	0.0%	96.4%	100.0%	
		% within arrival	3.0%	0.0%	52.9%	24.1%	
		% of total	0.9%	0.0%	23.3%	24.1%	

Table 8 Crosstabulation analysis on departure activities versus arrival activities by time intervals on weekdays

Time interval	Departure status	Particulars	Arrival status			Total	Monte Carlo. Sig
			OA	EA	LA		
WD01	OD	Count	24	15	1	40	0.463
		% within departure	60.0%	37.5%	2.5%	100.0%	
		% within arrival	82.8%	100.0%	100.0%	88.9%	
		% of total	53.3%	33.3%	2.2%	88.9%	
	ED	Count	4	0	0	4	0.463
		% within departure	100.0%	0.0%	0.0%	100.0%	
		% within arrival	13.8%	0.0%	0.0%	8.9%	
		% of total	8.9%	0.0%	0.0%	8.9%	
	LD	Count	1	0	0	1	0.463
		% within departure	100.0%	0.0%	0.0%	100.0%	
		% within arrival	3.4%	0.0%	0.0%	2.2%	
		% of total	2.2%	0.0%	0.0%	2.2%	
WD02	OD	Count	46	37	29	112	0.001
		% within departure	41.1%	33.0%	25.9%	100.0%	
		% within arrival	68.7%	86.0%	50.9%	67.1%	
		% of total	27.5%	22.2%	17.4%	67.1%	
	ED	Count	12	5	12	29	0.001
		% within departure	41.4%	17.2%	41.4%	100.0%	
		% within arrival	17.9%	11.6%	21.1%	17.4%	
		% of total	7.2%	3.0%	7.2%	17.4%	
	LD	Count	9	1	16	26	0.001
		% within departure	34.6%	3.8%	61.5%	100.0%	
		% within arrival	13.4%	2.3%	28.1%	15.6%	
		% of total	5.4%	0.6%	9.6%	15.6%	

(continued)

Table 8 (continued)

Time interval	Departure status	Particulars	Arrival status			Total	Monte Carlo. Sig
			OA	EA	LA		
WD03	OD	Count	6	5	11	22	0.009
		% within departure	27.3%	22.7%	50.0%	100.0%	
		% within arrival	50.0%	45.5%	21.2%	29.3%	
		% of total	8.0%	6.7%	14.7%	29.3%	
	ED	Count	5	3	11	19	0.009
		% within departure	26.3%	15.8%	57.9%	100.0%	
		% within arrival	41.7%	27.3%	21.2%	25.3%	
		% of total	6.7%	4.0%	14.7%	25.3%	
	LD	Count	1	3	30	34	0.009
		% within departure	2.9%	8.8%	88.2%	100.0%	
		% within arrival	8.3%	27.3%	57.7%	45.3%	
		% of total	1.3%	4.0%	40.0%	45.3%	
WD04	OD	Count	1	5	8	14	0.146
		% within departure	7.1%	35.7%	57.1%	100.0%	
		% within arrival	12.5%	31.3%	13.1%	16.5%	
		% of total	1.2%	5.9%	9.4%	16.5%	
	ED	Count	4	7	19	30	0.146
		% within departure	13.3%	23.3%	63.3%	100.0%	
		% within arrival	50.0%	43.8%	31.1%	35.3%	
		% of total	4.7%	8.2%	22.4%	35.3%	
	LD	Count	3	4	34	41	0.146
		% within departure	7.3%	9.8%	82.9%	100.0%	
		% within arrival	37.5%	25.0%	55.7%	48.2%	
		% of total	3.5%	4.7%	40.0%	48.2%	
WD05	OD	Count	68	78	15	161	0.000
		% within departure	42.2%	48.4%	9.3%	100.0%	

(continued)

Table 8 (continued)

Time interval	Departure status	Particulars	Arrival status			Total	Monte Carlo. Sig
			OA	EA	LA		
		% within arrival	75.6%	97.5%	46.9%	79.7%	
		% of total	33.7%	38.6%	7.4%	79.7%	
	ED	Count	18	0	8	26	0.000
		% within departure	69.2%	0.0%	30.8%	100.0%	
		% within arrival	20.0%	0.0%	25.0%	12.9%	
		% of total	8.9%	0.0%	4.0%	12.9%	
	LD	Count	4	2	9	15	0.000
		% within departure	26.7%	13.3%	60.0%	100.0%	
		% within arrival	4.4%	2.5%	28.1%	7.4%	
		% of total	2.0%	1.0%	4.5%	7.4%	

The probability values discussed in this study as the indicator can be used to compare the travel time reliability performance of a bus trip before and after the introduction of service quality improvement strategies such as increasing the service frequency or change of bus size, etc. This will help assess the success of the implemented service quality strategy toward travel time reliability performance.

This indicator can also be used to monitor and evaluate the relevance of any service quality strategy implemented toward enhancing the travel time reliability performance such as bus priority lanes by benchmarking the probability value for a specific combination of statuses such as early departure and on time arrival.

High probability value for the occurrence of on time or early departure from the origin bus stop and early arrival to the destination bus stop and late departure from the origin but early arrival at the destination can refer to less demand for the trip or significantly less congestion along the route. This measure can help to identify amending schedule travel times or revise the frequency of the operation. Lower probability for early departure and on time arrival to the destination bus stop can mean medium or high demand. It can also refer to possibilities for higher disturbance such as traffic congestion and high stopping times at intermediate bus stops. Higher probability for early departure and late arrival to the destination bus stop can refer to higher traffic disturbance. Higher stopping times at intermediate bus stops or prevalent of higher traffic congestion can result higher probability value for the occurrence of on time departure and late arrival to the destination bus stop. Unprofitable bus route with lesser demand can be defined by the higher probability value for the occurrence of late departure and early arrival. Frequency of the route can be adjusted, or bus size can be reduced for making profitable route.

Table 9 Crosstabulation analysis on departure activities versus arrival activities by time intervals on weekends

Time interval	Departure status	Particulars	Arrival status			Total	Monte Carlo. Sig
			OA	EA	LA		
WE01	OD	Count	7	7	0	14	0.065
		% within departure	50.0%	50.0%	0.0%	100.0%	
		% within arrival	100.0%	100.0%	0.0%	93.3%	
		% of total	46.7%	46.7%	0.0%	93.3%	
	ED	Count	0	0	1	1	0.065
		% within departure	0.0%	0.0%	100.0%	100.0%	
		% within arrival	0.0%	0.0%	100.0%	6.7%	
		% of total	0.0%	0.0%	6.7%	6.7%	
WE02	OD	Count	23	23	6	52	0.126
		% within departure	44.2%	44.2%	11.5%	100.0%	
		% within arrival	100.0%	100.0%	85.7%	98.1%	
		% of total	43.4%	43.4%	11.3%	98.1%	
	LD	Count	0	0	1	1	0.126
		% within departure	0.0%	0.0%	100.0%	100.0%	
		% within arrival	0.0%	0.0%	14.3%	1.9%	
		% of total	0.0%	0.0%	1.9%	1.9%	
WE03	OD	Count	2	0	8	10	0.019
		% within departure	20.0%	0.0%	80.0%	100.0%	
		% within arrival	66.7%	0.0%	42.1%	41.7%	
		% of total	8.3%	0.0%	33.3%	41.7%	
	ED	Count	0	2	1	3	0.019
		% within departure	0.0%	66.7%	33.3%	100.0%	
		% within arrival	0.0%	100.0%	5.3%	12.5%	
		% of total	0.0%	8.3%	4.2%	12.5%	

(continued)

Table 9 (continued)

Time interval	Departure status	Particulars	Arrival status			Total	Monte Carlo. Sig
			OA	EA	LA		
	LD	Count	1	0	10	11	0.019
		% within departure	9.1%	0.0%	90.9%	100.0%	
		% within arrival	33.3%	0.0%	52.6%	45.8%	
		% of total	4.2%	0.0%	41.7%	45.8%	
WE04	OD	Count			8	8	-
		% within departure			100.0%	100.0%	
		% within arrival			33.3%	33.3%	
		% of total			33.3%	33.3%	
	LD	Count			16	16	-
		% within departure			100.0%	100.0%	
		% within arrival			66.7%	66.7%	
		% of total			66.7%	66.7%	

6 Conclusion

The contribution of this study is the method to evaluate the travel time reliability performance of bus routes with short headways. It differs from the existing literature on the evaluation of travel time reliability performance by accommodating tolerance levels for both departure and arrivals of buses at the origin and destination of a trip, respectively. This contribution helps better identifying the travel time reliability performance of the bus routes. This percentage value/probability value yielded through crosstabulation is helpful for the bus operators to understand their strengths and weaknesses. If the probability value indicates poor travel time reliability, it will help bus operators identify possible reasons. Meanwhile, regulators can make use of the findings of the study for bus scheduling and transport planning purposes. This index is more applicable for buses with a service headway of less than 10 min. This can be applied to most of the changing contexts of bus operations, such as the day of the week, stop levels and route levels, etc.

References

1. Chen C, Skabardonis A, Varaiya P (2003) Travel-time reliability as a measure of service. Transp Res Rec 1855:74–79
2. Uniman DL (2009) Service reliability measurement framework using smart card data: application to the London underground. Master thesis, Massachusetts Institute of Technology
3. Pu W (2011) Analytic relationships between travel time reliability measures. Transp Res Rec 2254:122–130
4. Russell MB (2014) Travel time reliability and level of service. Master thesis, Louisiana State University
5. Fosgerau M, Hjorth K, Brems C, Fukuda D (2008) Travel time variability definition and valuation. DTU transportation, Lyngby
6. Kieu LM, Bhaskar A, Chung E (2015) Public transport travel time variability definitions and monitoring. J Transp Eng 141(1):04014018
7. Lomax T, Schrank D, Turner S, Margiotta R (2003) Selecting travel reliability measures. Technical report, Texas Transportation Institute and Cambridge Systematics, United States of America
8. White DR (2004) A student's guide to statistics for analysis of cross tabulations. World Cult 14(2):179–193
9. Cambridge Systematics (2013) Analytical procedures for determining the impacts of reliability mitigation strategies. In: The second strategic highway research program. Transportation Research Board of the National Academies, Washington
10. Dial RB (1967) Transit pathfinder algorithm. Highw Res Rec 1855:67–85
11. Clerq FP (1972) A public transport assignment method. Verkeerstech Neth 23(6):268–274
12. Wirasinghe SC (1980) Nearly optimal parameter for a rail/feeder-bus system on a rectangular grid. Transp Res Part Gen 14(1):33–40
13. Tian Q, Yang H, Huang H (2012) Pareto efficient strategies for regulating public transit operations. Public Transp 3:199–222
14. Sun H, Wu J, Wu L, Yan X, Gao Z (2016) Estimating the influence of common distruptions on urban rail transport networks. Transp Res Part Policy Pract 94:62–75
15. Chatterjee S, Hadi AS (2006) Regression analysis by example, 4th edn. Wiley Interscience, New Jersey

Comparison Study on R.A.P with Old Asphalt Aggregate with Virgin Bitumen

Abhishek Kumar Soni, S. S. Goliya, and Rakesh Mehar

Abstract The cost of paving materials rises as the restoration and building of highways progresses, and there is a scarcity of good quality materials. One of the most essential answers to this problem is recycling. This procedure creates a sustainable pavement by milling old components from the pavement and mixing them with virgin materials to create recycled combinations. Economic savings, reduced environmental impact, and natural resource conservation are all significant advantages of recycling. The primary aims of this research were to examine the effectiveness of recycled materials in asphalt mixes, to examine the impact of recycled materials on mixture behavior, and to examine the impact of including progressively older components into virgin mixtures. Old materials processed from the field, filler and 12.5 mm nominal maximum size of virgin aggregate, and virgin bitumen of (40–50) penetration grade have been modified and utilized to make recycled mixes. There were a number of analyses performed on the mixes used in this study. The Marshall Stability (3 specimens). In the case of recycled asphalt pavement the recycled combinations including 100% virgin/old components and old with new aggregate and old aggregate with new bitumen, Recycled. The performance of mixtures with old aggregate content was good. When compared to blends with old bitumen, the percentages of variance for mixture properties were features of (Marshall Stability), and the reduction in mixture properties was in an acceptable range.

Keywords R.A.P · R.A.P Binder · R.A.P Aggregate

1 Introduction

Existing asphalt pavement materials are recycled to create new pavements that save time, money, and energy. Even though the aggregate and binder from old asphalt pavements have reached the end of their useful lives, they are still valuable. Pavement recycling is not a new concept; in the 1970s, an oil embargo drove up asphalt

A. K. Soni (✉) · S. S. Goliya · R. Mehar
SATI, Vidisha M.P, India
e-mail: soniabhi9826@gmail.com

K. K. Pathak et al. (eds.), *Latest Developments in Civil Engineering*, Lecture Notes in Civil Engineering 352, https://doi.org/10.1007/978-981-99-2676-3_60

prices, sparking interest in pavement recycling. Even though the aggregate and binder from old asphalt pavements have reached the end of their useful lives, they are still valuable. Significant breakthroughs in the creation of heavy-duty equipment and building processes have been made in the evolution of recycling. The use of recycled asphalt pavement (RAP) in asphalt mixtures has become all too widespread in many nations. Recycling asphalt pavements has proven to be very beneficial from a variety of perspectives, according to experience. Even though there is a great need for cost-effective pavement restoration and construction in all sections of the country, the recycling strategy has never been used. As a result, in the sector of road construction, the option of asphalt pavement recycling should be given. The purpose of this research is to encourage individuals to recycle by providing information on the best materials to utilize how recycled combinations function. There are only a few local studies on recycling [4].

2 Literature Review

As both the expense of new paving materials and the cost of resurfacing old ones continue to rise, the recycling method has become more attractive as a solution to both challenges.

Ramanujan [12]. To recycle is to find a new purpose for something already in existence; in this case, pavement materials that have outlived their usefulness to motorists. Milled materials still have significant value after the service life of the pavement mixture has been reached. It is possible to minimize the quantity of brand-new material required by adding milling materials, known as recovered asphalt pavement (RAP), to the virgin hot asphalt mixture. Old total and black-top fastener that has oxidized (matured) during administration in the field make up the majority of recuperated black-top asphalt (RAP), which is then eliminated and handled to create new asphalt. It is possible to create a recycled asphalt mixture by mixing RAP with fresh asphalt, new aggregate, and sometimes a recycling agent in a hot-mix recycling process. Recycling asphalt pavements is a technically sound, economically viable, and environmentally benign practice, as shown by its widespread adoption and the positive results of earlier recycling efforts by a wide range of organizations. Many state and local transportation departments throughout the United States are using hot-mix recycling of asphalt pavements as a primary technique of repair. Georgia, in the United States, has been recycling since 1991; most reused asphalts are worked with 10–25% recuperated black-top asphalt material.

Kandel [10]. Particularly popular as an alternative to the more commonplace milling and resurfacing or thin asphalt overlay restoration treatment, recycling is practiced in numerous nations. Recycling has been practiced over an area of around 2,000,000 m^2 in Queensland, USA, since the year 1990.

Because of weak economic conditions and a lack of updated innovative technologies, the reuse of removed asphalt pavement was not given more attention in India in the twentieth century.

Table 1 Table of reference

S. no	Author name	Topic	Year	Result
1	David A. Byrne	Recycling of Asphalt Pavements in New Bituminous Mixes	2005	Recovered binder 2.2
2	Ehsan Ali Hasan	Assessing Performance of Recycled Asphalt Concrete Sustainable Pavement	2010	Used recycling agent
3	Plescan Costel	Asphalt Pavement Recycling	2015	40% drop in quality

The cost of **Managua Sreenivasulu Reddy and Suvarna P** new asphalt mix was cheaper than that of recycled asphalt mix, and the lack of machinery was a constraint for not adopting this approach. At the conclusion of the twentieth century and the beginning of the twenty-first, Indians began researching asphalt pavement recycling. Authorized uses of asphalt After over a long time since its underlying preliminary in Nevada and Texas, RAP ends up being not just a beneficial option, but also a need to assure the future economic viability of flexible pavement construction (Table 1).

3 Proposed Methodology

In this study development of methodology for the assessment of characteristics of R.A.P and virgin material is based on a literature review of natural aggregates and old aggregates one unmodified bitumen binder and two old binders used as a pavement construction material. The proposed methodology of the present study is shown in the below flow chart. We have developed this methodology for conducting experimental work on the bituminous concrete mixes. The proposed methodology provides an idea of how we can incorporate the R.A.P Properties in the construction of bituminous concrete pavements. The proposed methodology also provides an attempt at how we can use RAP material in the construction of bituminous concrete. Through this methodology, we also tried to clear the objectives and scopes of the present study. This methodology clears our attempts made in the present study for successful investigations on bituminous mixes.

3.1 Asphalt Content Tester

Hot mix asphalt (HMA) and pavement samples may have their asphalt content determined with the use of an ignition furnace, also called an asphalt content furnace or binder ignition furnace, which removes the asphalt by sample heating rather than solvent. The applicable standards are ASTM D6307 [2] and EN 12,697-39. Compatible with a 415 V, 50 Hz, 20-A, three-phase AC power supply. Each specimen may

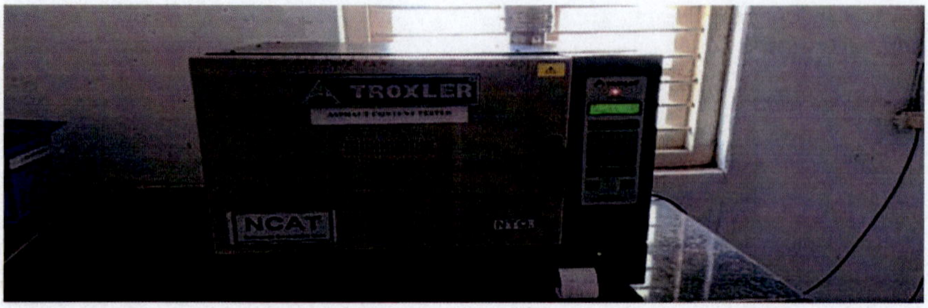

Fig. 1 Asphalt content tester

weigh no more than 4000 g. The recommended range for the specimen's weight is between one thousand and one thousand five hundred grams. The furnace can reach temperatures of up to 800 °C, whereas the industry norm is 538 °C and a 0.1-g balance is supplied. The duration of each test is twenty to thirty minutes. Furnace capacity: (350 × 430 × 300) measured in inches (L × B × H) (Fig. 1).

3.2 Aged Materials

The asphalt mixture was salvaged from a construction project along National Highway (46), which ran from Bhopal to Gwalior; the highway's exact location is shown in Fig. 2. This roadway was reconstructed by an Indian business. The roadway was in decent shape before the extraction began, with just a few shallow cracks and ruts; the milling depth for this job was 5 cm. It was assured that the recycled asphalt mixture wouldn't include any of the loam or hazardous substances that tend to collect on the top of roads. Binder and filler content, gradation, and aggregate characteristics were analyzed by extracting the recovered mixture using the ASTM D2172 [1] procedure. The extraction test results are shown in Table 5 for aged materials.

By randomly selecting ten samples from the milled material stack, submitting them to an extraction test to segregate cover from the total, and afterward sieving and isolating the total into different sizes to work out the degree for each example, we had the option to decide the old total degree as displayed, contrasts between tests were minor, and the typical degree of the ten examples acquired was the old total degree (Tables 2, 3, 4).

Typically less than 30%—a.
Typically less than 20%—b.
Typically less than 15%—c.
Typically less than 10%—d.

Fig. 2 Location of rehabilitation

Table 2 FHWA [3] typical particle size distribution for reclaimed asphalt pavement (percent by weight passing)

Screen size mesh	After processing or milling% finer
37.5 mm	100
27 mm	95–100
19 mm	84–100
12.5 mm	70–100
9.5 mm	58–95
7.5 mm	38–75
2.36 mm	25–60
1.18 mm	17–40
0.60 mm	10–35[a]
0.30 mm	5–25[b]
0.15 mm	3–20[c]
0.075 mm	2–15[d]

Table 3 Gradation for grading-1 of bituminous mix

S. no	Sieve size (mm)	Weight Retained (gms)	Weight Retained (%)	Cumulative weight retained (%)	Cumulative % by weight of total aggregate passing obtained values	Cumulative % by weight of total aggregate passing MoRTH, (2013) [11]
1	19	00	0.0	0.0	100	100
2	13.2	450	9.0	9.0	91.0	79–100
3	9.5	405	8.1	17.1	82.9	70–88
4	4.75	1150	23.0	40.1	59.9	53–71
5	2.36	500	10	50.1	49.9	42–58
6	1.18	310	6.2	56.3	43.7	34–48
7	0.60	490	9.8	66.1	33.9	26–38
8	0.30	450	9.0	75.1	24.9	18–28
9	0.15	540	10.8	85.9	14.1	12–20
10	0.075	430	8.6	94.5	5.5	4–10

Table 4 Gradation of filler

S. no	IS sieve (mm)	Cumulative % passing by weight of total aggregate (obtained values)	Cumulative % passing by weight of total aggregate MoRTH (2013)
1	0.6	100	100
2	0.3	100	95–100
3	0.075	98	85–100

4 Results

According to the given methodology, in the proposed study we have tested the physical properties of aggregates as mentioned below. Table 5 is representing the results of the physical properties of given aggregates (Tables 6, 7, 8).

5 Charts

CHART 1 This chart shows the relationship b/w marshal stability value and mix design the virgin material has high marshal stability value and the mix having new bitumen has high value and old bitumen and rap has low Marshall stability because the load carried by R.A.P is less (Fig. 3).

CHART 2 This chart shows the relation b/w mix design and the flow value of virgin is high as compared to others due to higher displacement and load (Fig. 4).

CHART 3 This chart shows the V.M. aggregate in different mix design (Fig. 5).

Table 5 Result of physical properties of Virgin aggregates

S. no	Properties	Test method	Natural aggregates	MoRTH specifications (2013)
01	Aggregate impact value	IS:2386 (IV) [6]	20%	Max 24%
02	Los Angeles abrasion test	IS:2386 (IV)	25.18%	Max 30%
03	Water absorption value	IS:2386 (III)	1.012	Max 2%
04	Specific gravity	IS:2386 (III)	2.67 for 10 mm, 2.69 for 6 mm, 2.83 for stone dust	2.5–3.0
05	Combined flakiness and elongation index	IS:2386 (I) [5]	26.23%	Max 35%

Table 6 Result of physical properties of old aggregates

S. no	Properties	Test method	Old aggregate	MoRTH specifications (2013)
01	Aggregate impact value	IS:2386 (IV)	18.14%	Max 24%
02	Los Angeles abrasion test	IS:2386 (IV)	22.%	Max 30%
03	Water absorption value	IS:2386 (III)	1.9	Max 2%
04	Specific gravity	IS:2386 (III) [7]	2.67 for 10 mm, 2.69 for 6 mm, 2.83 for stone dust	2.5–3.0
05	Combined flakiness and elongation index	IS:2386 (I)	18.14%	Max 35%

Table 7 Results of physical properties of old and virgin bitumen (VG30)

S. no	Properties	Test method	VG-30 results	Old VG-30 results	MoRTH specifications (2013)
1	Penetration	IS 1203 [8]	70	66	50–70
2	Softening point °C	IS 1205 [9]	50	66	45–50
3	Ductility at 27 °C	IS 1208	53	43	+40
4	Specific gravity	IS 1202	1.01	1.02	1.01
5	Flashpoint, °C	IS 1209	240	260	220
6	Fire point, °C	IS 1209	260	270	247

Table 8 Shows the properties of Marshall of normal bitumen (VG30)

S. no	Property tested	Bitumen content by weight of aggregates (VG30)			
		R.A.P	Virgin	New bitumen old aggregate	Old bitumen new aggregate
1	Marshall stability (KN)	7.3	8.72	8.20	7.00
2	Marshall Flow value (mm)	3.18	2.86	2.89	3.10
3	Bulk density (g/cc)	2.67	2.71	2.70	2.68
4	Volume of voids (V_v)%	2.50	2.62	2.66	2.70
5	Void in mineral aggregates (VMA)%	18.71	18.28	18.41	18.78
6	Void filled with bitumen, VFB (%)	85.9	86.4	86.3	86.2
7	Marshall quotient (KN/mm)	2.29	3.04	2.83	2.25

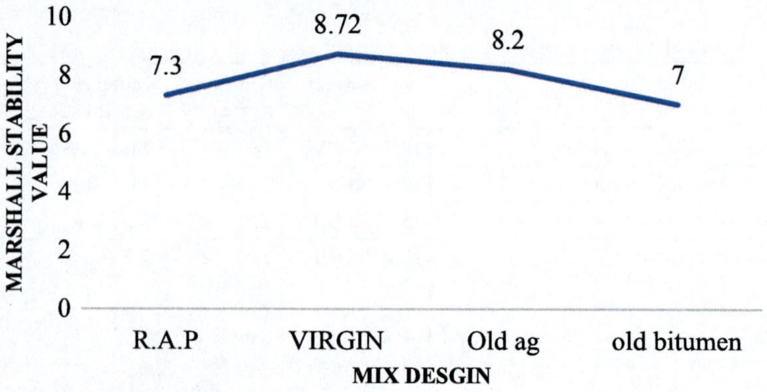

Fig. 3 Marshall stability and R.A.P design mix

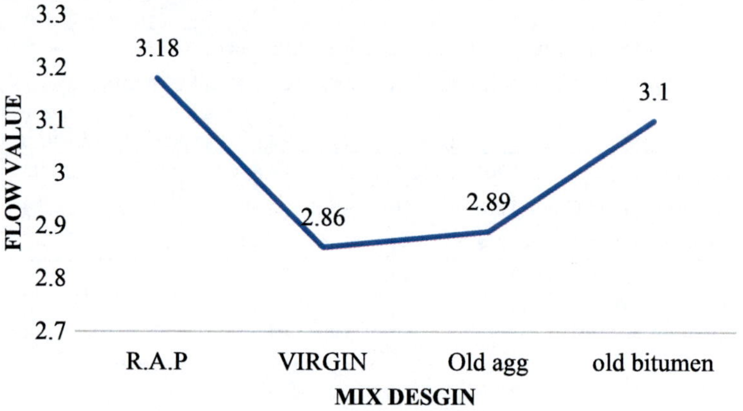

Fig. 4 Marshall flow value and R.A.P mix design

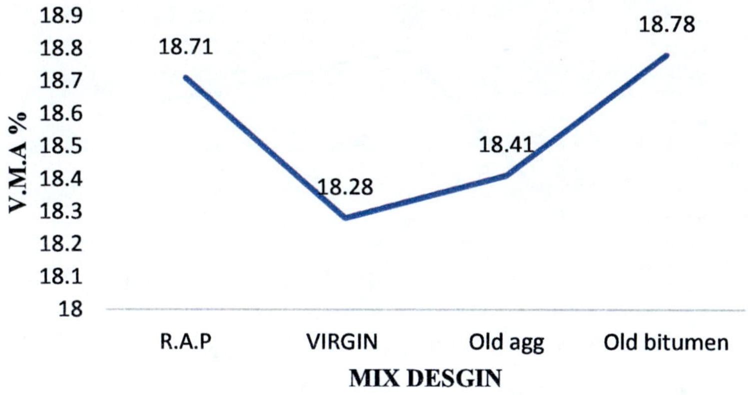

Fig. 5 V.M.A and R.A.P mix design

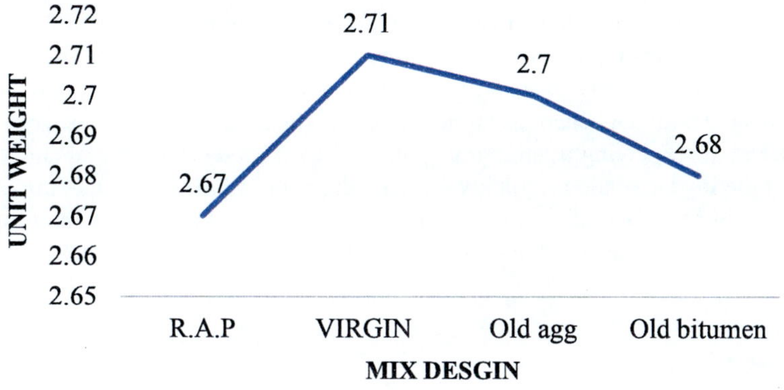

Fig. 6 Unit weight and R.A.P mix design

CHART 4 This chart shows bulk density of different mix and design (Fig. 6).
CHART 5 This chart shows V.F.B of different mix (Fig. 7).

6 Discussion

Proposed research demonstrates that pavement usage and thickness have a significant impact on the recycle asphalt pavement. It has been shown that the viscosity of various bitumen's increases at varying rates as time passes. Aging causes chemical or physical property changes in bituminous materials, making them harder and more brittle and reducing the chance of pavement failure. This Marshall Stability test is used in the hot-mix design of bitumen with particles up to 25 mm in size. The Marshall

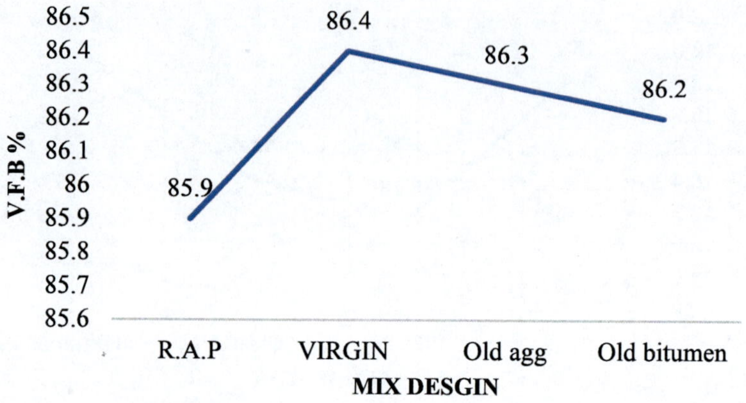

Fig. 7 V.F.B and R.A.P mix design

technique is often used in India to create BC mixes. This test method is utilized in the development and evaluation of recycle bituminous pavement mixtures. This exam is often utilized in paving work and regular testing programs. In comparison to recycle bitumen, new bitumen generates higher strength. The bitumen's short-term qualities have been altered, giving us strength against the load applied to the road and a flow value of the load that allows quickly wearing the load and avoiding cracking on the bitumen's surface. One of the most important things that affect is how long asphalt pavement lasts and how old the bituminous binders get as it affects R.A.P.

7 Conclusion

It is possible to draw the following inferences from the testing technique and its constraints:

- As per the behavior of R.A.P, we can use old material because its strength is not too less as coma paired to virgin so it can be used.
- If the thickness of the surface is less so we cannot use old material because of higher wear and tear b/w aggregate.
- If the surface is more aged likely to be more than 10 years, the surface used to higher oxidation of bitumen is more in age mix.
- R.A.P old aggregate can be used in new mix because its impact value is not that much affected. The difference is 3–5% in both impact and Los Angeles abrartion test.
- If the aggregate is extracted by burning process by asphalt content machine so due to higher temperature to burn bitumen the aggregate properties are affected.
- After extracting bitumen from benzene the bitumen properties get affected at some extinct Due to dust practical in bitumen.

- Bitumen can be used in the mix but at some percentage in the new mix.
- Due to wear and tear the fine particle percentage gets higher in the old mix so bitumen content or aggregate to be added in mix.
- After extracting aggregate from benzene there must be less bitumen % (min) in the new mix.

References

1. ASTM D2172 Standard Test Methods for Quantitative Extraction of Bitumen From Bituminous Paving Mixtures
2. ASTM D6307 Standard Test Method for Asphalt Content of Asphalt Mixture by Ignition Method
3. Federal Highway Administration (FHWA) (1997) User guidelines for waste and byproduct materials in pavement construction. Turner-Fairbank Highway Research Center, Publication Number: FHWA-RD-97-148
4. IRC: 120 (2015) Recommended practice for recycling of bituminous pavements
5. IS: 2386 (1963) Methods of test for aggregates for concrete–part 1: particle size and shape
6. IS 2386 (1963) Method of test for aggregates for concrete, mechanical properties, Part-IV
7. IS 2386 (1963) Method of test for aggregates for concrete for determination of specific gravity, void, absorption and bulking, Part-III
8. IS 1203 (1978) Method for testing tar and bituminous material Determination of Penetration
9. IS 1205 (1978) Method for testing tar and bituminous material Determination of softening point
10. Kandhal PS, Rao SS, Watson DE, Young B (1995) Performance of recycled hot mix asphalt mixtures. National Center for Asphalt Technology (NCAT), Report No. 95-1
11. MoRT&H (Ministry of Road Transport and Highways) (2013) Specifications for road and bridge works In Indian Road Congress New Delhi, India
12. Ramanujam JM (2000) Recycling of asphalt pavements. Department of Transport & Main Roads, Queensland, Internet link

Enhancement Penetration Resistance of Load by Using Coconut Coir Fiber and Copper Slag for Unpaved Roads

Garapati Venkata Sai Prasad

Abstract When an external load is applied on expansive soil it impacts on soil structure due to changes in swell pressure, volumetric swelling, swelling strain, shrinking strains, and shear strength. Use of geosynthetics as layers in expansive soil gives exceptional results in geotechnical applications since it is conversely useful in increasing the soil properties if correctly placed. The innate geo-textiles are used as a separator, reinforcement, and drainage layer for unpaved section. Several studies have exploited different materials to develop the bearing capacity of poor soil. The performance of Coconut Coir Fiber geo-textile on an unpaved road and paved road models are investigated using the California bearing ratio test setup in this study. The beneficial effects of coir geo-textiles (woven along with non-woven) on load penetration behavior and on sub-grade soil in unpaved road and paved road models are reported in this study. When these coir geo-textiles are employed, the CBR improvement is adequate for eliminating the need for a robust foundation course reached in both unpaved road and paved road. In comparison to non-woven geosynthetic, woven geosynthetic has been found to perform significantly better.

Keywords Expansive soil (ES) · Geo-textiles (GT) · Cyclic Plate Load Test (CPLT) · Geosynthetics (GS) · Coconut Coir Fiber geo-textile (CCFGT) · California bearing ratio test (CBRT) · Paved Road (PR) · Unpaved Road (UR) · Woven Geosynthetic (WG) · Non Woven Geosynthetics (NWG) · Copper Slag (CS)

1 Introduction

ES is also known as swell-shrink soils that expand and contract in response to variations in moisture content due to which considerable discomfort occurs in the soil, followed by damage to the overlying structures. These soils absorb water and expand

G. V. S. Prasad (✉)
Department of Civil Engineering, KG Reddy College of Engineering and Technology, Hyderabad, India
e-mail: gprasad@kgr.ac.in

K. K. Pathak et al. (eds.), *Latest Developments in Civil Engineering*, Lecture Notes in Civil Engineering 352, https://doi.org/10.1007/978-981-99-2676-3_61

during periods of high wetness in monsoons, losing their water-holding ability in summer. These soils lose moisture they have stored due to evaporation becoming harder, problems of ES allied to the civil engineering structures motivated several researchers to use efficient methods for expansive soil. Stabilization is one method by utilizing additives that help to reduce the volume change due to swelling. Synthetic and natural fibers are the two types of fibers available. Synthetic fibers such as Polypropylene, Nylon, plastic, glass asbestos, and others are more widely employed than natural fiber since they have better strength and resilience. Bhabar, coir, sisal, munja, bamboo, coconut fiber, and jute fiber are natural fibers accessible in India. Locally accessible fiber should be considered in the design to lower the cost of ply soil. At the same time, the structure's stability and longevity should be prioritized. The majority of these fibers have been shown to lose strength when wet and in a dry process. Natural fibers are not commonly utilized as reinforcement because of their poor strength and durability, but they are valued for their environmental affability and biodegradability. Even though assured natural fibers, such as coir, are tough and long lasting. They may be made more durable by properly treating them in cohesion-free soil as reinforcement and cohesive soils as filter fabric.

2 Review of the Literature

R. Tailor (Dec 2011) in the following experimental study geo-textile was utilized as reinforcement in the flexible pavement to control shrink and swell. It has been confirmed after two monsoons that it is capable of replacing a cushion of 600–1000 mm for expanding sub-grade.

Dr. P. Senthil Kumar (Dec 2012) in the following experimental study interfacing woven geo-textile and non-woven geo-textile in an unpaved road particularly with soft sub-grade, which increases the penetration resistance.

K. Meshram (Oct 2013) in the following experimental study, use of coir geo-textiles among sub-base and sub-grade layers is suitable which slows down the flow of water thins the layer. By adding a geo-textile layer or else membrane at the combined portion of granular sub-base layer and the sub-grade prevents sub-grade soil from entering the granular sub-base layer's interstices, allowing the granular sub-base layer to function properly as an evacuation layer which reduces the severity of sub-grade stress.

Radha Krishnan (2014) in the following experimental study, effects of magnesium chloride (Class-F) fly ash, in addition to fly ash-aluminum chloride, showed lower swelling properties on expansive soil.

3 The Materials that Were Used

3.1 Sample Soil Collected location

Expansive black cotton soil was obtained for this project from Allavaram, near Amalapuram, Andhra Pradesh. Top soil was removed up to 1000 mm and disturbed sampling was transported to the laboratory. A small amount of soil sample is sealed in a polythene bag to evaluate its natural moisture content. Later the soil is dried, crushed using wooden hammer, and sieved with 4.75 mm sieve.

3.2 Copper Slag (CS)

CS is a byproduct of the copper extraction process which involves smelting and refining copper ore. Refineries generate a significant amount of metallic dust and rock while they extract metal from copper ore, resulting in slag which may be utilized for a surprising variety of industrial uses (Table 1).

3.3 Coconut Coir Fiber

Coir fiber formed from natural fiber becomes more popular as an erosion control material, but not as a soil reinforcement material. Despite the fact that strong fiber with high lignin content, for example coir, can be used efficiently as a reinforcing material. It was recently discovered that adding fiber to the soil increases its ductility, preventing cracks from forming during shrinkage.

Table 1 Constituent elements of CS

S.No	Constituent elements	Content (%)
1	Silica as SiO_2	35
2	Iron as Fe_2O_3	44
3	Alumina as Al_2O_3	6
4	Calcium as CaO	1.5
5	Magnesium as MgO	2
6	Copper as Cu	0.9
7	Zinc as Zn	0.3
8	Cobalt as Co	0.13
9	Lead as Pb	0.04
10	Nickel as Ni	0.03

4 Methodology

Various laboratory experiments on soil samples treated with various quantities of admixtures as well as untreated soil samples were done in this inquiry and are described below according to the Indian standard code of practice.

4.1 Relative Density Test

Relative density is ratio of mass of unit per volume of soil to the mass of same volume of water.

The test was carried out in accordance with IS 2720 Part III.

4.2 Free Swell Index of Soil Test

Free swell, also known as differential free swell, is the increase in soil volume that occurs when exposed to water without any external limitation. The test was carried out in accordance with IS 2720 Part XL.

4.3 Grain Size Analysis Test

Grain size analysis is to determine the percentage size of each grain contained within a soil sample. The test was carried out in accordance with IS 2720 Part IV [1]

4.4 Liquid Limit and Plastic Limit

The Atterberg limit is used to calculate the moisture content at which fine-grained clay and silt soils has transition between phases. The test was carried out in accordance with IS 2720 Part VII [2]

4.5 Compaction Test

Compaction test is still one of the most important tools for predicting soil engineering behavior in the context of road construction to establish the OMC and MDD test has been done as per IS 2720 part IV [3].

4.6 California Bearing Ratio Test

The CBRT measures the load spreading ability of the pavement. This is only justifiable in the case of flexible pavements. CBRT was carried out in accordance with IS 2720 Part XVI [4].

4.7 Plate Bearing Test

The plate bearing test is used to measure the ultimate bearing capacity of soil in place. This test presupposes that the soil layers are generally uniform. For lab test we created a flexible pavement model in which we have used 20 cm of untreated or treated expansive soil as sub-grade 5 cm of gravel as sub-base and WBM–III as the base course.

As pressure increased, the deformation between cycles remained unchanged. To assess the ultimate load with and without the geo-textile as a reinforcement and separator, the test was sustained in anticipation of the model pavement collapsing. All of the model flexible pavements were subjected to these tests. All of the model flexible pavements were tested at OMC and under saturated conditions.

5 Discussion and Result

5.1 Mechanical Characteristics of Soil

For this investigation, soil was obtained from Amalapuram, village name as Allavaram in Andhra Pradesh. For laboratory testing, topsoil up to 1000 mm thick was removed, and disturbed soil was collected in bags and brought to the lab. A small amount of the sample was placed in a polythene bag to gauge its natural moisture content (Table 2).

5.2 Characteristics of Copper Slag

Copper extraction and refining produce copper slag as a byproduct. Refineries produce a significant amount of metallic dust, soot, and rock as they remove metal from copper ore, creating slag, which has a surprising variety of industrial uses (Table 3).

Table 2 Mechanical characteristics of soil

Sl. No	Properties	Value
1	Specific Gravity	2.65
2	**Atterberg's limits**	
	Liquid Limit (%)	72
	Plastic Limit (%)	32.55
	Plasticity Index (%)	39.55
3	**Soil compaction Values**	
	Percentage of OMC (Optimum Moisture Content)	29.04
	Percentage of MDD Maximum Dry Density (g/cc)	01.47
4	**Particle size distribution**	
	Percentage of sand	6
	Percentage of silt	30
	Percentage of clay	63
5	Indian Standard Soil Classification System	CH
6	UCS (Unconfined Compressive Strength) (kg/cm^2)	1.47
7	CBR (California bearing ratio) (%)	1.78
8	DFS (Free Swell Index) (%)	139

Table 3 Copper slag properties

Sl.No	Properties		Value
1	Specific gravity		3.7
2	Sieve analysis (particle distribution)	4.75 mm	100
		2.0 mm	40
		0.6 mm	42
		0.425 mm	17
		0.21 mm	4.5
		0.075 mm	2.0

5.3 *Coir Geo-Textile Properties*

As an erosion control material, coir fiber made of natural fiber is becoming increasingly popular. Although strong fiber with a high lignin concentration, such as coir, can be utilized efficiently as a reinforcing material. It was recently discovered that adding fiber to the soil enhances its ductility, preventing cracks from forming during shrinkage (Table 4).

Table 4 Coir geo-textile physical and chemical properties [5]

Chemical properties		Physical properties	
Description	Value	Description	Value
Percentage of lignin polymer	45.83	Length of Coir geo-textile in inches	6.7
Percentage of cellulose structure	43.43	Density (g/cc)	1.39
Percentage of hemi-cellulose structure	00.24	Tenacity (g/Tex)	10.01
Percentage of Pectin's and related compound	03.01	Percentage of breaking elongation	30
Percentage of water soluble	05.24	Diameter of Coir geo-textile in mm	0.1–1.4
Percentage of ash	02.21	Rigidity of modulus	1.8923 dyne/cm^2
Percentage of swelling in water (diameter)	5	–	–
Percentage of moisture at 65% RH	10.50	–	–

5.4 Compaction Test Values

See Table 5.

- Sample soil maximum dry density increased after exposure to 20% copper slag.

5.5 Soaked and Unsoaked CBRT Values

See Table 6.

Table 5 Expansive soil values treated with various percentages of copper Slag

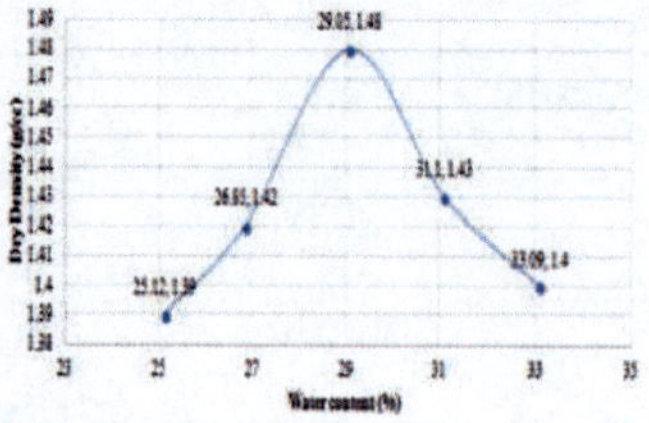

Expansive soil only

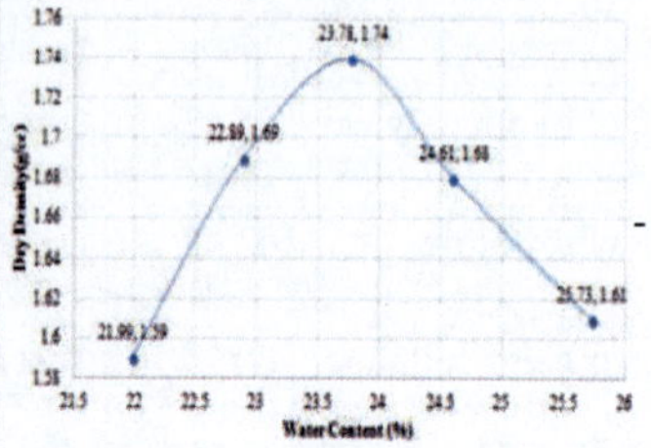

Expansive soil in addition of CS 19%

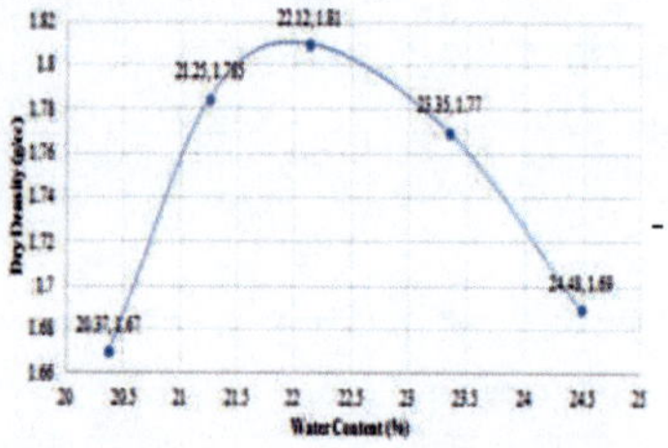

Expansive soil in addition of CS 20%

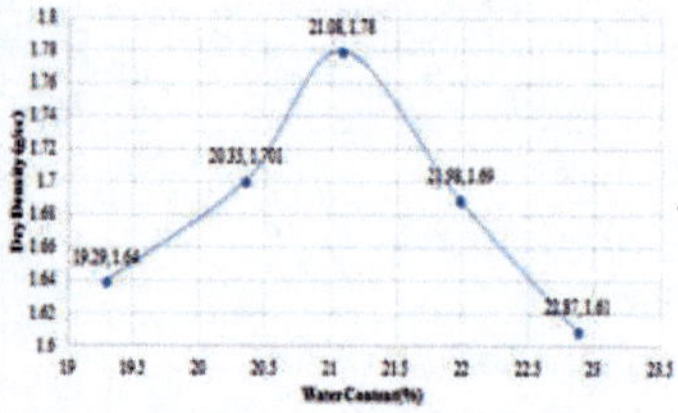

Expansive soil in addition of CS 21%

SI. No	Proportions of the mix	Maximum dry density (g/cc)	Optimum moisture content (%)
1	Origin soil	1.481	29.04
2	Expansive soil in addition of CS18%	1.68	23.73
3	Expansive soil in addition of CS19%	1.73	23.27
4	**Expansive soil in addition of CS20%**	**1.80**	**22.11**
5	Expansive soil in addition of CS21%	1.76	22.37

Table 6 CBRT copper slag treated soil and geo-textile reinforced soil values

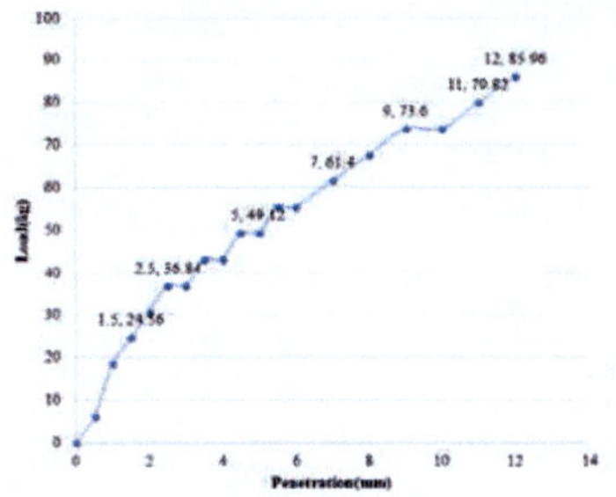 CBRT untreated soil value in unsoaked condition	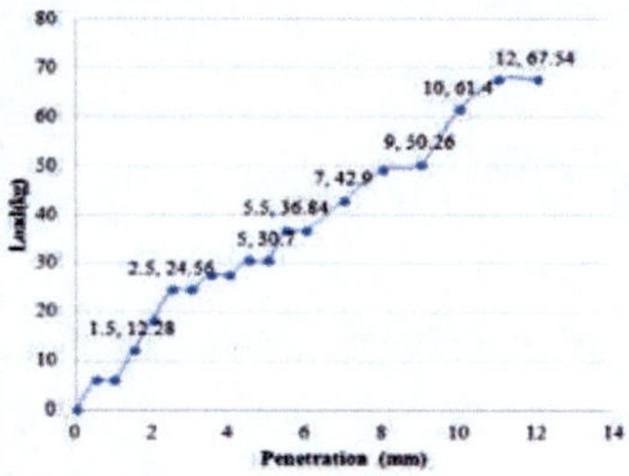 CBRT untreated soil value in soaked condition
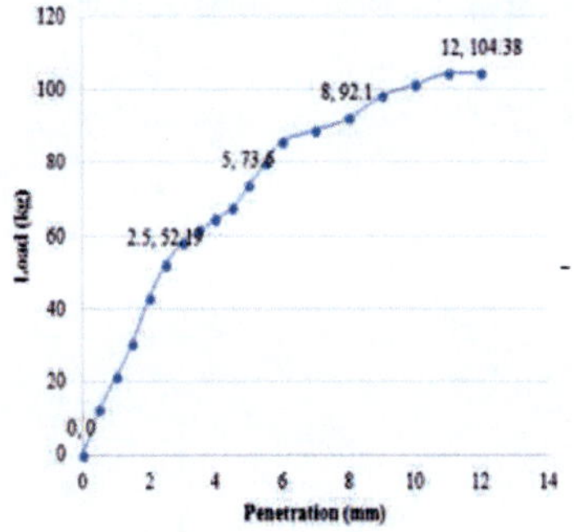 CBRT on sample soil in addition of CS 19% value in soaked condition	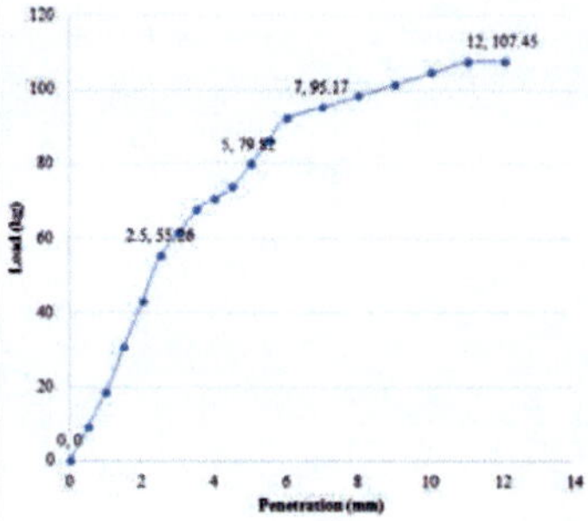 CBRT on sample soil in addition of CS 20% value in soaked condition
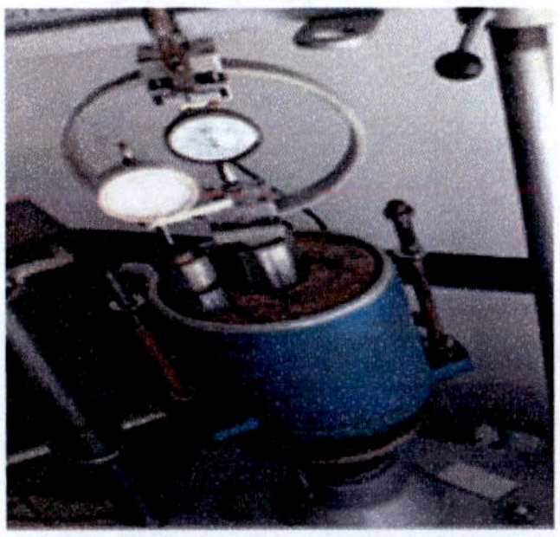 Unpaved road models experiment setup	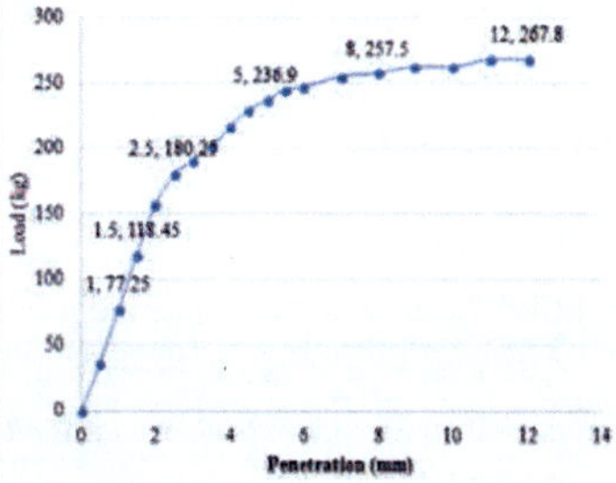 CBRT soil and sub-base value in soaked condition

(continued)

Table 6 (continued)

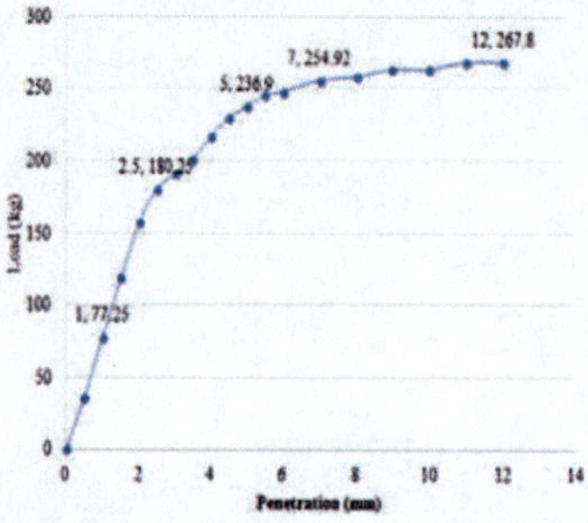

CBRT on Sample Soil in addition of CS 20% nonwoven geo-textile placed at middle of the sub-base

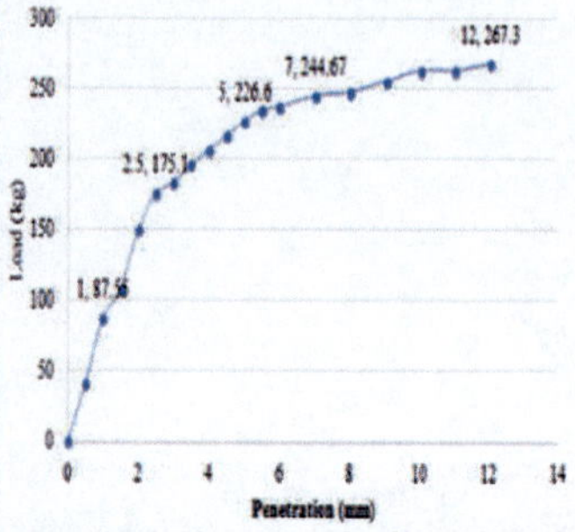

CBRT on sample soil in addition of CS 20% woven geo-textile placed at middle of the sub-base

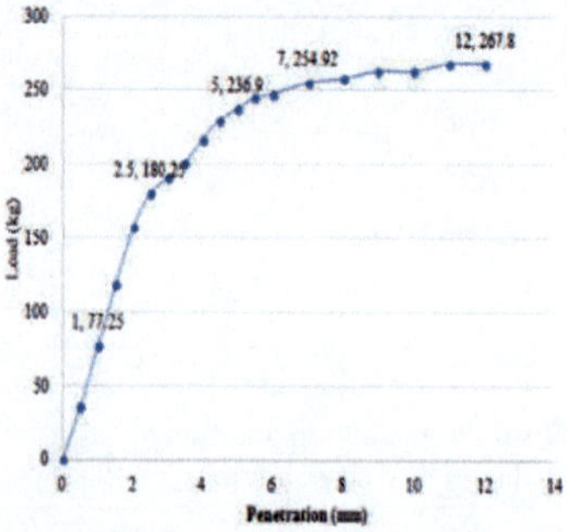

CBRT on sample soil in addition of CS 20% nonwoven geo-textile placed at edge of both layers in soaked condition

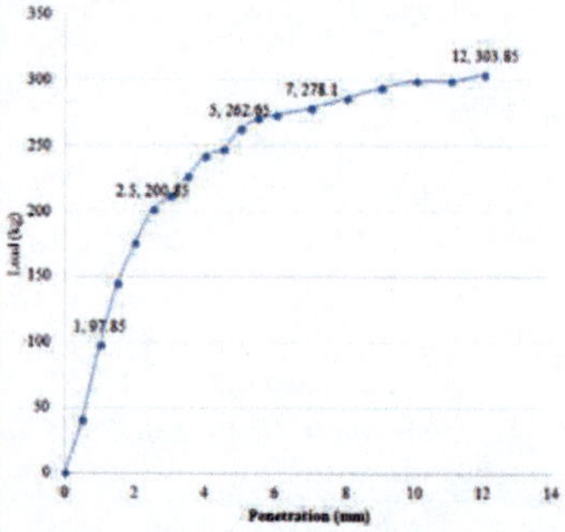

CBRT on sample soil in addition of CS 20% woven geo-textile placed at edge of both layers in soaked condition

S.No	Mix Proportion	CBR Value (%)
1	CBRT Untreated Soil Value in Unsoaked Condition	2.67
2	CBRT Untreated Soil Value in Soaked Condition	1.78
3	CBRT on expansive Soil in addition of CS18% Value in soaked Condition	3.81
4	CBRT on expansive Soil in addition of CS 19% Value in soaked Condition	4.02
5	CBRT on expansive Soil in addition of CS 20% Value in soaked Condition	4.47
6	CBRT on expansive Soil in addition of CS 21% Value in soaked Condition	4.24
7	CBRT on expansive Soil in addition of CS 22% Value in soaked Condition	3.81
8	Soil + Sub-base	4.71

(continued)

Table 6 (continued)

S.No	Mix Proportion	CBR Value (%)
10	CBRT on expansive Soil in addition of CS 20 Nonwoven Geo-textile placed at center of the Sub-base	11.26
11	CBRT on expansive Soil in addition of CS20 Woven Geo-textile placed at center of the Sub-base	12.77
12	CBRT on expansive Soil in addition of CS Nonwoven Geo-textile placed at edge of the both the layers	13.14
13	CBRT on expansive Soil in addition of CS Woven Geo- textile placed at edge of the both the layers	14.65

6 Conclusion

The succeeding conclusions have been generated based on the check outcomes gained from laboratory experiments:

- For the treated soil sample, 20% copper slag yields the most optimal moisture percentage.
- The soil treated with 20% copper slag has the highest CBR value.
- When compared to a soil sample that has not been treated, the treated sample soil's ultimate pressure increased from 650 to 3200 kPa.
- At the optimum moisture content, deformations of treated expansive soil for sub-grade flexible pavements have improved by 25% in cyclic plate load tests (Tables 6 and 7).

Table 7 Results of untreated and treated expansive clay for flexible pavements at optimum moisture content by cyclic plate load tests

Sl.No	Type of sub-grade	Sub-Base	Base course	Pressure (kPa)	Settlement(mm)
1	Origin Soil	– –	– –	181	2.07
2	Expansive soil untreated	Gravel	Water Bound Macadam-III	651	2.01
3	Expansive soil plus Copper Slag 20%	Gravel	Water Bound Macadam-III	1201	1.9
4	Expansive soil plus CopperSlag 20% as Single Woven Geo-textile used as separator and reinforcement provided in between Sub-Base and Sub-grade	Gravel	Water Bound Macadam-III	2201	1.8

(continued)

Table 7 (continued)

Sl.No	Type of sub-grade	Sub-Base	Base course	Pressure (kPa)	Settlement(mm)
5	Expansive Soil plus CopperSlag 20%plusdoubleWoven Geo-textile used as separator and reinforcement provided between Base Course and Sub-Base	Gravel	Water Bound Macadam-III	3200	1.6

References

1. IS:2720 part-4 (1975) Grain size analysis
2. IS:2720 part-5 (1970) Determination of Liquid limit and Plastic limit
3. IS:2720 part-6 (1974) Determination of Dry density and Optimum moisture content
4. IS:2720 part-40 (1977) Determination of Free Swell Index
5. https://www.sciencedirect.com/science/article/pii/B9780128183984000104 Coconut fiber: its structure, properties and application
6. Tailor R (2011) Geotextile as reinforcement in flexible pavement for swelling sub grade. In: Proceedings of Indian geotechnical conference, Kochi (Paper No.J-230.)
7. Kumar P, Ramasamy R (2012) Effect of geotextile on CBR strength of unpaved road with soft subgrade. Electron J Geotech Eng 17
8. Meshram K, Mittal S, Jain P, Agrawal P (2013) Application of coir geo-textile in rural roads construction on BC soil sub-grade. Int J Eng Innov Technol 3:264–268
9. Radhakrishnan G (2014) Swelling properties of expansive soils treated with chemicals and Flyash. Am J Eng Res (AJER) 03(04):245–250e-ISSN: 2320-0847 p-ISSN: 2320-0936
10. IS:2720 part-16 (1979) Determination of California bearing ratio
11. The use of natural geo-textiles in reinforcing the Unpaved roads. https://doi.org/10.13140/2.1.1079.3287
12. Geo-textile as reinforcement in flexible pavement for swelling sub-grade
13. Geo-textile strain in a full scale reinforced test embankment. https://doi.org/10.1016/0266-1144(94)00015-6
14. The effect of geo-textile layers and configuration on soil bearing capacity. https://doi.org/10.18280/mmep.080608

Land Use—Classification by Machine Learning Classifiers Using Landsat 8 Imagery

Reena Thakur and Prashant Panse

Abstract Through the twentieth and first decades of the twenty-first century, quick and abandoned population growth, in combination with industrial and economic development, enhanced the frequency of land-use-land-cover change (LULCC) for several times, particularly in evolving nations. Moreover, the most effective ways to manage and analyze land transformation are the quantifiable valuation of changes in LULC, it is important to evaluate the comparison of the accuracy of different Land Use-Land Cover mapping algorithms in order to choose the good classifier for future earth observation and its applications. The QGIS tool is applied to consider minimal distance, maximum likelihood, and spectral angle mapping in this paper. Various classes have been explored, and training is carried out based on these to determine the accuracy of each method.

Keywords Sand use · Land cover · Classification techniques · Accuracy assessment

1 Introduction

Due to converging causes, the remote sensing (RS) and data science fields have initiated to line up in recent years. The open-source, Kaggle [1] has shown that powerful machine learning (ML) algorithms may achieve high classification accuracy. Typically, Kaggle website provides open-source datasets for the researchers as many organizations used to post their data there to utilize as training and testing inputs for their models and achieve the outcome, with cash awards offered to the victors. Further, numerous new Landsat 8 and Sentinel-2, Earth-observing satellites, have newly been introduced, and their data has been made publicly available [2,

R. Thakur (✉)
Department of Computer Science and Engineering, Jhulelal Institute of Technology, Nagpur, India
e-mail: en19cs601002@medicaps.ac.in

P. Panse
Department of Information Technology, Medi-Caps University, Indore, India
e-mail: prashant.panse@medicaps.ac.in

K. K. Pathak et al. (eds.), *Latest Developments in Civil Engineering*, Lecture Notes in Civil Engineering 352, https://doi.org/10.1007/978-981-99-2676-3_62

3]. Finally, consumer computing capability has increased considerably while the cost of computing has decreased [4]. The machine learning importance in the Earth observation and RS communities has risen as a result of these reasons, especially in the sub-field of LCLU categorization. The ML algorithm like Random forests (RF) [5] is widely used. This is a widely used method that has shown effective in both regression and classification tasks, and hence with equally categorical as well as continuous variables [6]. RF has been employed in an extensive variety of Earth scientific applications, particularly modeling, due to its adaptability.

1.1 Inspiration and Objective

Few of the utmost significant environmental changes, such as deforestation, urbanization, residential development, agricultural expansion, and industrialization, take place at the massive amount and have a direct influence on the processes of the ecosystem. Furthermore, biological interactions can substantially transform landscapes and substantially spatial patterns turn to establish within them called as ecosystem engineering, a type of phenomenon (Hastings et al. 2007). As a result, mapping the LCLU at the landscape scale is critical for monitoring and managing these changes. The use of satellite data to classify objects is a critical first step in this project. For this research, Landsat 8 data was selected because of two factors: (1) its spatial resolution is relatively high, i.e., 10 m, and other (2) it uses three bands of red vegetation edge radiometry. Landsat 8 data is intriguing for LULC mapping because of these two qualities. Presently no studies use these data to evaluate the comparison based on the performance of traditional and developing ML methods in categorization and complicated land-scapes. Thus, using Landsat 8 data, the goal of this work is to assess the performance of three common ML algorithms on the basis of classification over an Indian terrain.

Steps for Classification as Shown in Fig. 1

1. Definition of the Classes is into Four Categories
   Dataset is divided into four classes for future classification
2. Training is performed on the classes to get
   a. Region of interest
      For the assessment of land region of interest is need to be identified
   b. Assessment
      Based on the region of interest, assessment is to be performed
   c. Classification of ROI
      Then classification is to be performed using tool
3. Refinement on classification is done
4. Assessment of the Accuracy of the Algorithms.

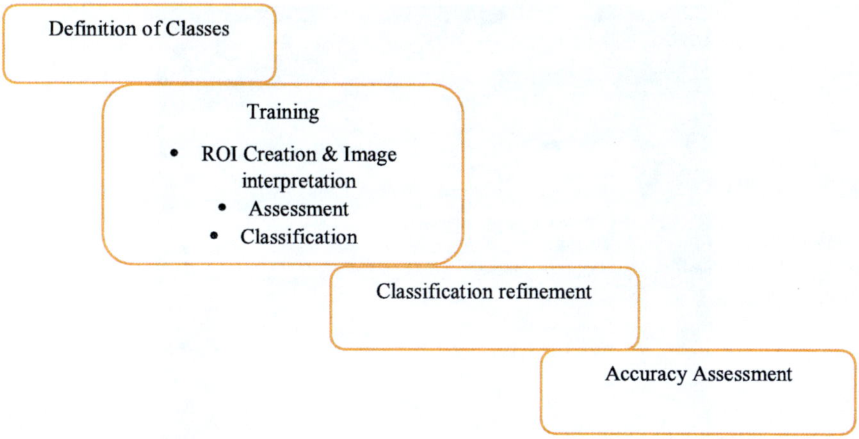

Fig. 1 Steps for classification

2 Study Area

The study area is the covering area of this study. The study area chosen for this study is a district in Nagpur, Maharashtra, India, as shown in Fig. 2. These regions were chosen to represent the various LULC groups. Land use/land cover (LULC) will be classified using standard QGIS Tool methods. The choice of a research area can aid in the right selection of a satellite image (Figs. 3, 4 and 5) respectively.

3 Data

The Landsat 8 Multispectral Instrument (MSI) consists of two satellites. These satellites data observe 10, 20, and 60 m, the Earth at spatial resolutions. Among freely available satellite products, the 10 m spatial resolution is the highly available. The existence of three red edge bands in the Landsat 8 data, as seen in Fig. 6, permits the data to capture the strong reflectivity of vegetation in the nearby infrared unit of the electromagnetic spectrum (EMS).

4 Detail Steps for Classification of Area

The technique structure in this work is separated into three stages, as shown in Fig. 7. The initial step is to process the satellite image, which includes image pre-processing and classification. The next step is the analysis, which involves comparing the data to ground fact data and determining the accuracy of each classifier. The final result

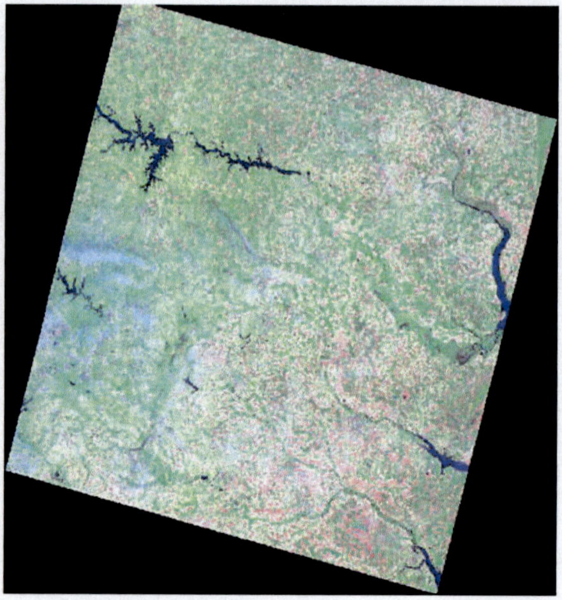

Fig. 2 Study area landsat image with the coordinates UL −79, 39 LR −76.9, 38.9 Product L8 OLI/TIRS

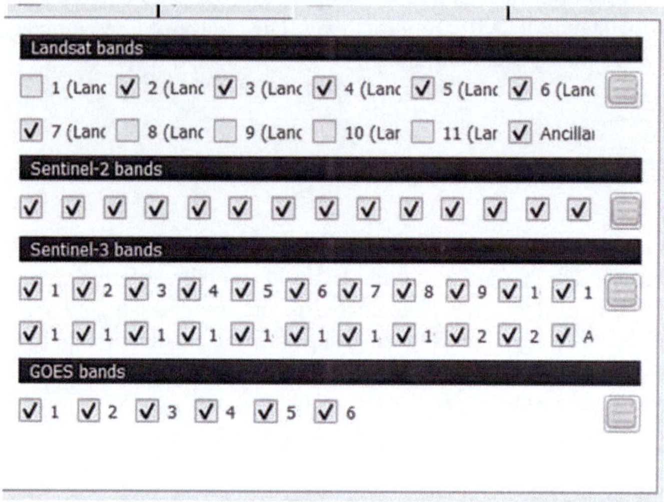

Fig. 3 Landsat band required

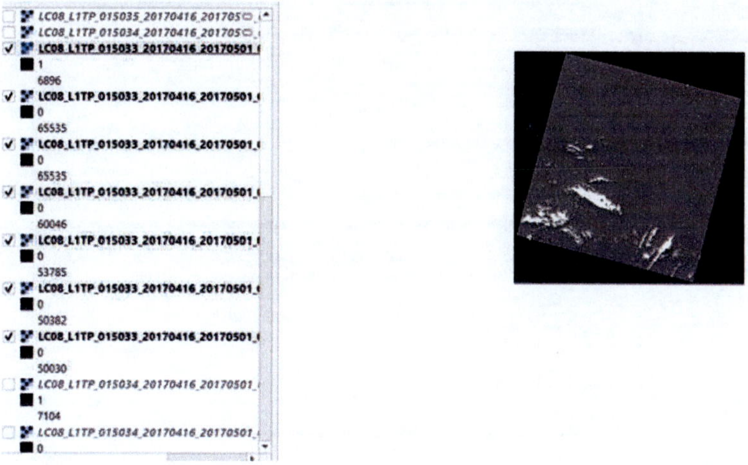

Fig. 4 Selection of bands

for image classification is a land as well as land cover map, which is produced in the third stage.

5 Empirical Result

The following Fig. 8 represents the graphical lustration of the four classes—water, built-ups, vegetation, and bare soil.

The following Fig. 9 represents the spectral distances of the four classes—water, built-ups, vegetation, and bare soil. Figure 10 shows the spectral details.

Figure 11 shows the classification preview consisting of a lake, buildings, trees, low vegetation, etc. (Table 1).

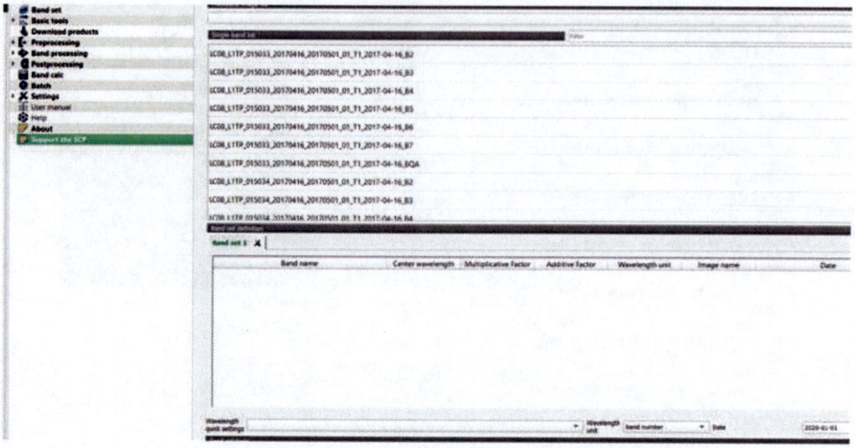

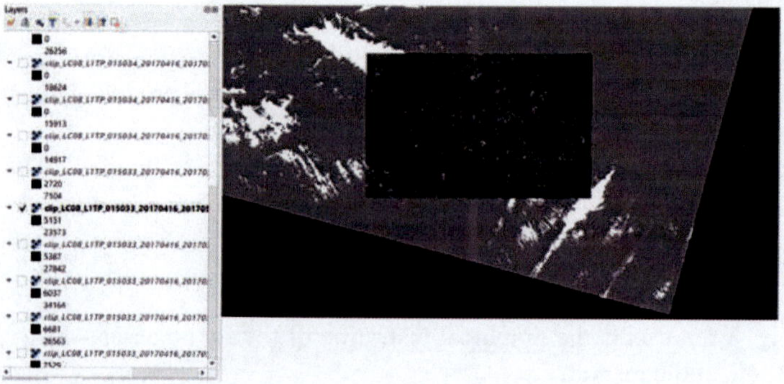

Fig. 5 Selection of clip data

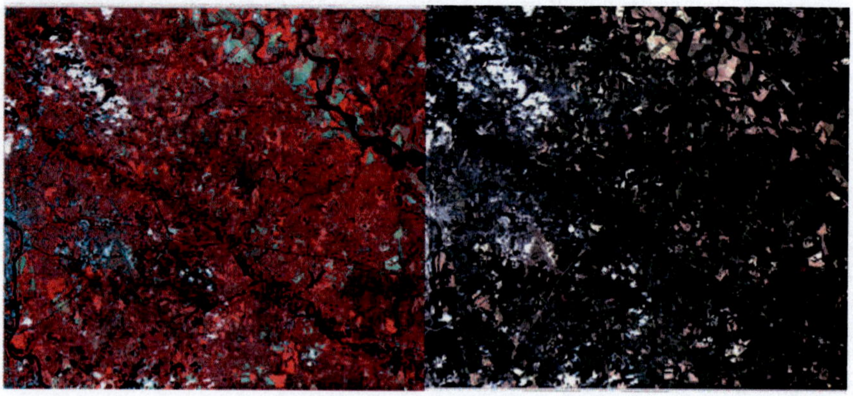

Fig. 6 Clip data

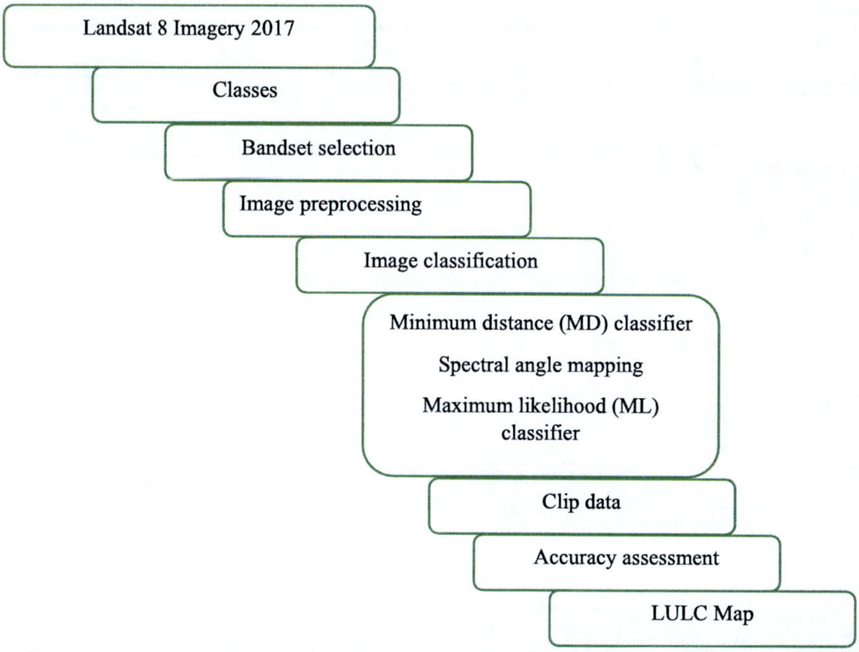

Fig. 7 Steps for classification of area

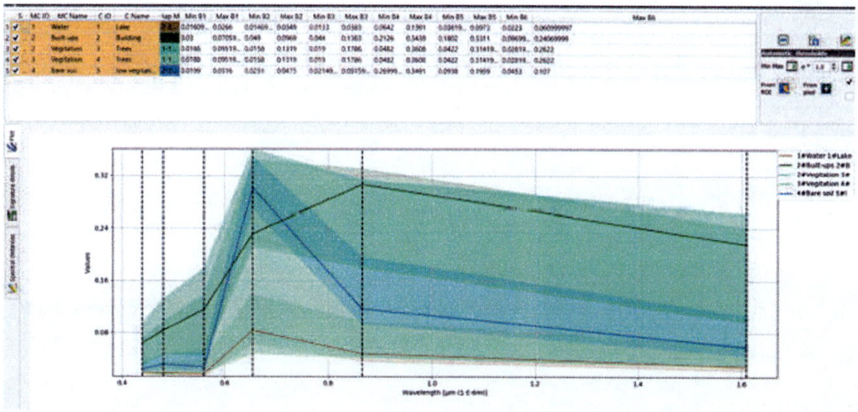

Fig. 8 Graphical representation

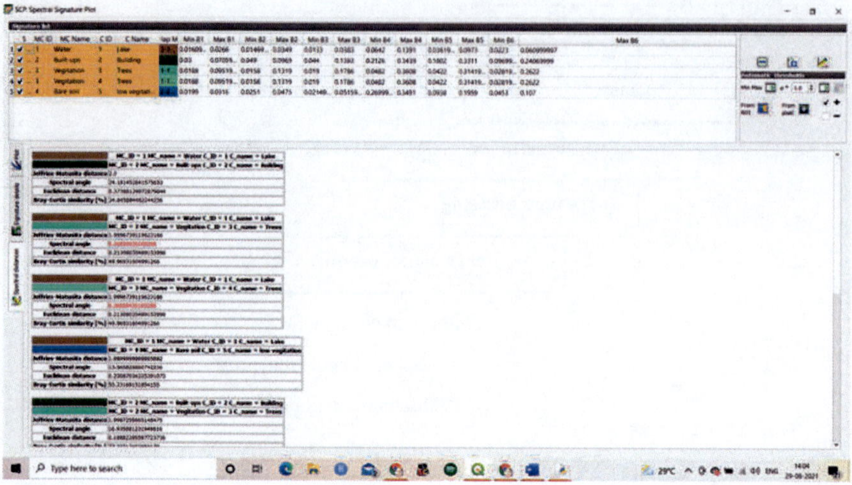

Fig. 9 Spectral distances

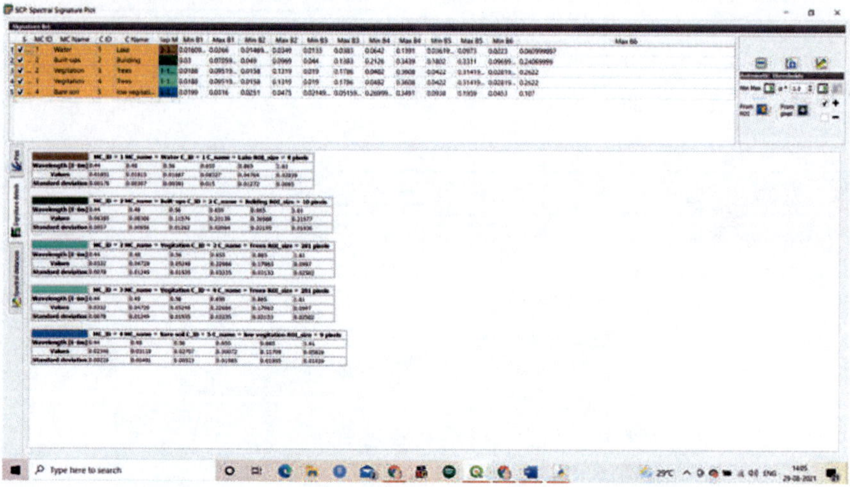

Fig. 10 Spectral details

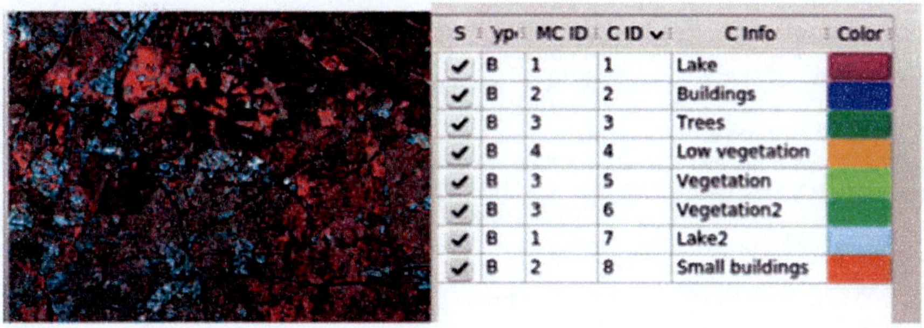

Fig. 11 Classification preview

Table 1 LULC classification efficiency of different methods

Techniques/ class name	Minimum distance classifier		Maximum likelihood classifier		Spectral angle mapping	
	Producer's accuracy (%)	User's accuracy (%)	Producer's accuracy (%)	User's accuracy (%)	Producer's accuracy (%)	User's accuracy (%)
Water (lake)	55.00	100.00	30.00	100.00	60.00	100.00
Built-up (building)	33.30	78.00	33.00	99.00	60.00	100.00
Vegetation	25.00	100.00	11.89	100.00	65.00	56.00
Bare soil	86.00	25.00	88.00	27.00	25.00	100.00
Overall classification accuracy	90%		64.83		74%	

6 Conclusion

The findings of image classification in this study illustrate the efficacy of the three approaches employed to categorize Land Use and Land Cover. Minimum distance generated the highest overall accuracy of the three classifiers, with 90% accuracy. The findings of image classification in this study illustrate the efficacy of the three approaches employed to categorize Land Use and Land Cover. Minimum distance generated the highest overall accuracy of the three classifiers, with 90% accuracy.

References

1. Google. Kaggle: your home for data science. https://www.kaggle.com/. Accessed 23 May 2019
2. Harris R, Baumann I (2015) Open data policies and satellite earth observation. Space Policy 32:44–53. https://doi.org/10.1016/j.spacepol.2015.01.001

3. Belward AS, Skoien JO (2015) Who launched what, when and why; trends in global land-cover observation capacity from civilian earth observation satellites. ISPRS J Photogramm Remote Sens 103:115–128. https://doi.org/10.1016/j.isprsjprs.2014.03.009
4. Waldrop MM (2016) The chips are down for Moore's Law. Nat News 530(7589):144
5. Fick SE, Hijmans RJ (2017) Worldclim 2: new 1-km spatial resolution climate surfaces for global land areas. Int J Climatol 37:4302–4315. https://doi.org/10.1002/joc.5086
6. Woznicki SA, Baynes J, Panlasigui S, Mehaffey M, Neale A (2019) Development of a spatially complete floodplain map of the conterminous United States using Random Forest. Sci Total Environ 647:942–953. https://doi.org/10.1016/j.scitotenv.2018.07.353

A Comprehensive Review on Travel Time Strategies for Performance Improvement of Public Transport System

Narendra Dudhe, Pradeep Kumar Agarwal, and Amit Vishwakarma

Abstract The study contains solutions for enhancing the public transportation system's travel time performance (PTS). The importance of a multimodal public transportation system resides in its travel time competence, service synchronization, and the heterogeneous nature of traffic, all of which needed improvements. The major decision influencer for commuter mode preference is PTS travel time competency versus personal trips. The identification of strategies based on previous research is an attempt to find a solution for improving travel time performance. Access time, in-vehicle and out-vehicle transit time, and egress time are all reduced by using travel time strategies. Access time strategies propose that getting to the public transportation network (PTN) is easier, that public transportation is more easily accessible, and that service coverage is better. The in-vehicle and out-vehicle travel time strategies show the key operational characteristics that influence travel time performance and how they can be improved. In-vehicle travel time plans emphasize the improvement and reorganization of areas responsible for enhancing travel time compatibility with personal trips. Interconnectivity and timely availability of cars at transfer sites were the focus of egress time initiatives. The long-term viability of public transportation is dependent on micro-level strategy enhancement.

Keywords Public transport system · Travel time performance · Travel time strategies · Performance evaluation · Commuters acceptance

N. Dudhe (✉) · A. Vishwakarma
Department of Civil Engineering, University Institute of Technology, Rajiv Gandhi Proudyogiki Vishwavidyalaya, Bhopal, India
e-mail: narendra.dudhe23@gmail.com

P. K. Agarwal
Department of Civil Engineering, Maulana Azad National Institute of Technology, Bhopal, India

K. K. Pathak et al. (eds.), *Latest Developments in Civil Engineering*, Lecture Notes in Civil Engineering 352, https://doi.org/10.1007/978-981-99-2676-3_63

1 Introduction

The strategies for improving public transportation travel time (PTTT) performance are an attempt to scrutinize the work done by researchers in order to identify the solution in terms of public transportation planning and operations strategies that are responsible for inefficient trips and the loss of commuter interest and reliability.

"The urban population is anticipated to expand to almost 600 million (40%) by 2031 and 850 million (50%) by 2051," according to the projections (handbook of urban statics 2016). To enable mobility to a large metropolitan population, an efficient PTS must be in place. PTS commuters have shifted to personal modes of transportation due to overcapacity and delays, resulting in traffic congestion and poor urban mobility. To provide excellent service and establish efficient PTS with travel time compatibility, strategic improvement is essential. The evaluation of journey time parameters serves as the foundation for strategy improvement.

Travel Time Ratio, Level of Service, Interconnectivity Ratio, Passenger Waiting Index, and Running Index are the parameters to evaluate for identifying inefficiencies in PTS [1]. Average passenger kilometers per vehicle-km, Annual ridership per bus, Passenger trips per effective vehicle-km, Operating cost per passenger-km, Revenue per passenger, Average fare per passenger-km [2]. Another essential criterion for evaluation is the time difference between personal automobile and public transportation trips (PTT).

To increase commuters' willingness to switch to PT, service reliability (SR) and level of service (LOS) must be improved. PTS competency can be achieved with personal car journeys by reducing vehicle time and increasing travel speed, resulting in SR. SR is another system effectiveness parameter that may be increased by maintaining punctuality and increasing the total serviceability index [3]. Travel time performance improvement techniques are separated into three parts: Access Time Strategies (ATS), Travel Time Strategies (TTS), and Egress Time Strategies (ETS) to recommend improvement approaches at different levels of public transportation trips (PTT). Based on the facts and conclusions in the literature, Table 1 represents ATS, Table 2 shows TTS, and Table 3 indicates ETS.

This study focuses on strategies for improving travel time performance that has previously been described in the literature to aid strategic decision-making in order to achieve competent travel time performance. Figure 1 shows a schematic diagram of a public transportation system with SAP, BRTS, or Metro as the main service.

2 Literature Review

The most important parameter for a successful PTS is travel time performance. Access time, PTTT, and egress time are the three major components of travel time. The average waiting time and travel time to reach a service access point (SAP), such

Table 1 Details of strategies for access time performance improvement

Strategy ID	Strategy	Empirical findings
ATS1	Reduction of walk time to reach SAP of BRTS stop or Metro station	Strong correlation exists between PTALs and access to services. The high PTAL areas generally have good access to services and low PTAL arcas have poor access to services
ATS2	Quick park and ride facility (having a own vehicle)	Transfer Time plays the most significant role in determining the respondent's preference. The next important factor is Parking Fair, and Public transport Travel Time (PTTT) has the least impact on the acceptance of P&R by travelers
ATS3	Availability of competent feeder network to reach service access point	i. Comfort level of public transport systems particularly depending on the crowd density in vehicles during peak hours ii. Commuters will adopt public transit and give priority for trip making, when the system is properly accessible from their trip origins and service is available at preferred travel times iii. A model for optimal trip distance of each hierarchy type of route is proposed based on features of passengers in the PTS
ATS4	Increase accessibility levels by providing higher frequency of service	The analysis helps to visualize and assess the level of transit services in the different regions. Every region, mapped LITA scores identify the sites with limited transit services relative to those sites with higher service. The calculated LITA could be used in a Geographic Information System (GIS) to identify the points where transit service is lacking in the studied regions
ATS5	Increase accessibility levels by higher FN service coverage	A well planned and systematic feeder network increases accessibility and facilitates commuters to access the public transport facility as it increases the service coverage

as a BRTS stop or a Metro station, are added together to calculate access time (Transport for London 2010). From the standpoint of commuters, trip planning involves everything from waiting periods to departure schedules, as well as the total route time and number of transfers. Access time, on the other hand, is frequently overlooked in PT accessibility research. To address this problem, a decision-support system was developed that allows any origin–destination pair to compute pre-trip waiting periods, travel durations, and the number of required transfers [6]. PTTT can be broadly divided into two categories In-vehicle travel time (IVTT) and Out-vehicle travel time (OVTT). The level of service is defined as the ratio of OVTT to IVTT. The OVTT/IVTT ratio is almost usually greater than 1. This reflects the fact that commuters spend more time outside than inside their vehicles. Longer excursions will always have a lower level of service due to the longer line-haul duration [1]. The performance evaluation procedure can help PTS improve its facilities in terms of

Table 2 Details of strategies for travel time performance improvement

Strategy ID	Strategy	Empirical findings
Strategies for travel time performance improvement: by minimizing out vehicle time		
TTS1	Minimize waiting time by strict adherence to bus schedules	Commuters experience highlighted issues like overcrowding discomfort and travel time unreliability that need to be evaluated as accessibility indicators. Implementation of bus schedule methodology to measure travel time variability in a congested PTN helps to improve accessibility levels and minimize waiting time
TTS2	Development of just in time approach by the availability of On-time running information of fleets	GPS-enabled vehicles can transmit current location information that can be available to commuters online, to plan their trips. The real-time information about trips can help to reduce waiting times and improve travel time performance
TTS3	Minimizing waiting time by quick availability of vehicles at transfer points	PTTT is the key factor of a multimodal trip for commuters to prefer PTS for longer trip distances ranging between 7.5 and 35 km. PTS is preferred if the access and egress distance is not too large. TT, waiting time and egress time are the significant and complex travel time factors. TT can be reduced by a higher frequency of service for improving efficiency and modal share
Strategies for travel time performance improvement: by increasing travel speed		
TTS4	Optimize number of bus stops	Stop locations can be placed at suitable intervals by considering walk access time so that commuters can have quick access to one of the adjacent stops. Adjacent Stops in one-kilometer vicinity can be relocated
TTS5	Minimizing number of traffic signals	In areas with lower pedestrian traffic, Actuated signals traffic control can be used along priority rapid transit corridors to increase the schedule reliability of transit service and escape unnecessary delays
TTS6	Optimizing stoppage time at bus stop	The Total Bus Stop Time (TBST) at stops is the time difference between the time bus pulls into the bus stop and when it returns to the main traffic stream. By considering variables for a dense urban area like methods of payment, time of the day, crowding level, the time lost serving a stop, and the location of the bus stops, improvements could be made in the optimization of TBST models
TTS7	Provision of the dedicated corridor	The key of successful BRTS is a dedicated corridor. The dedicated corridor provides optimum speed, safe, and scheduled trips which are essential for performance
TTS8	Strategies for vehicle waiting time minimization at intersections	For minimizing the number of vehicles from cycle to cycle inter green signal time model can be proposed. This model also incorporates restrictions for upper and lower bounds for green signal time and cycle time allocation which provides accurate and appropriate allocation for green signal time (Raviraj et al. 2018)

Table 3 Details of strategies for egress time performance improvement

Strategy ID	Strategy	Empirical findings
ETS1	Strategies for minimizing transfer time via interconnectivity improvement	A well-planned paratransit can prove means of interconnectivity and enhance and strengthen the PTS network
ETS2	Availability of Shared Vehicle Trips for Quick Service	An shared vehicle-based trips has the potential to provide door to door transport service with a high level of service and require less than 10% of today's private cars and parking places [4]
ETS3	Time table synchronization of feeder services with BRTS and Metros	The synchronization of timetables provides a more user-oriented, system-optimal, and well-connected public transport service which is more acceptable and attracts attention [5]

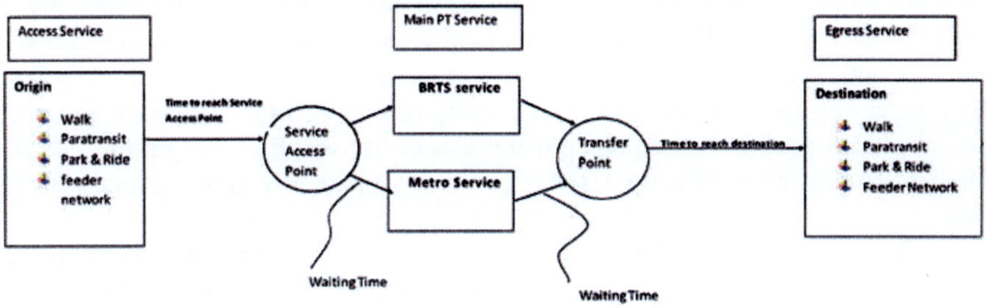

Fig. 1 Schematic diagram of public transport service

several operational elements [7]. The performance of the urban bus system was found to be about 70% when 29 evaluation parameters were categorized into eight indicator categories. The parameters for the 'passenger information systems' and 'social sustainability' sectors were deemed to be inefficient and in need of improvement [8].

According to the study, the city coverage index and city employment index of the BCLL and Mini bus systems are both less than 0.3, indicating that the systems are lacking in terms of city coverage and city employment [7]. The majority of research does not take into account all three segments of access time, PTTT, and egress time. As a result, we haven't been able to identify the flaws that are causing the inefficient transportation service. With personal vehicle trips, present public transportation options are inefficient and time-consuming.

The local Index for Transport Accessibility (LITA) index and a comprehensive spatial database on the condition of transit supply provides a practical approach to see and analyze the level of transit services in different regions. The mapping LITA scores show which sites have little transit service compared to others that have more.

The estimated LITA could be used in a Geographic Information System to identify areas with insufficient transit service.

Researchers have attempted to evaluate performance using various criteria such as the city coverage index, the LITA Index, the level of service, the interconnectivity ratio, and the passenger waiting index, but there is no sequential strategic framework for reducing travel time. This research proposes ways for improving PTS travel time performance.

3 Need of the Study

PTS must be available as a basic eminent facility in all cities with a population of at least 5 million people. It provides benefits to all members of society, "those who prefer to ride as well as those who have no other option." Over 90% of the population does not possess a four-wheel vehicle and relies on public transportation. PT saves money and resources for commuters because it uses less energy per capita per kilometer traveled than private travels [9]. Travel duration, stopping periods, entrance and egress timings, frequency of trips, comfort, and level of service safety are all areas in which PT can improve. The trip time element is the most significant and must be attended to first, since it reveals service reliability. According to the findings, traveling the PT takes 1.4–2.6 times longer than driving a personal vehicle [10].

The long-term sustainability of PTS is a difficult issue to address, as the percentage of commuters in PT does not rise at the same rate as the population. To make PT the preferred mode of transportation among commuters, significant improvements in quantitative parameters such as journey time between origin and destination are required. As a result, there is a genuine need to discover ways for improving travel time performance.

The cost of time spent traveling is referred to as the Value of Travel Time (VTT). Employees that commute spend important time in PT, which can increase "costs to businesses and services." The savings from reduced travel time and vehicle running costs are referred to as VTT savings [11]. Savings in travel time have a direct influence on vehicle running costs and commuter work hours. The reduction in travel time and the level of service provided by public transportation projects are the primary criteria for evaluation. As a result, methods must be developed whose implementation promotes travel time reduction by focusing on various travel time factors such as access and egress time, passenger waiting time at stops and intersections, frequency of service, and run-time information.

4 Strategies for Travel Time Performance Improvement

Travel time comprises three basic bifurcations access time, travel time, and egress time.

4.1 Strategy for Access Time Performance Improvement

The time it takes to go to a Service Access Point is known as access time (SAP). The whole travel and waiting time to get to SAP and choose the primary service, either BRTS or Metro, is the access time. Commuters have three primary options for getting to the SAP: walking to a BRTS stop or metro station, owning a vehicle with park-and-ride capability, or using a competent and efficient feeder network service.

ATS1: Reduction of walk time to reach SAP of BRTS stop or Metro station

The density of the PTN in any urban region is similar to the density of easily available PT. The Access time represents the distance walked from one's front door to the nearest public transportation SAP and has a direct impact on the PTTT between O and D. "*For buses the maximum walk time is defined as 8 min or a distance of 640 m. For rail, underground and light rail services the maximum walking time are defined as being 12 min or a walking distance of 960 m*" (PTALs, Transport for London 2010).

ATS2: Quick Park and Ride facility

The time it takes to drive to a parking lot, park, and walk to a bus stop or metro station must be less than the time it takes to walk to a bus stop or metro station. The amount of time spent in parking and getting to the next facility and the parking fair is the most important factor in a commuter's decision which influences whether or not to use P&R facilities [12].

ATS3: Availability of Competent Feeder Network to Reach Service Access Point

Great prospects of selecting PTS as a mode of transportation rely on frequent, uncrowded, comfortable, and efficient FN that can easily transport commuters to the SAP. Commuters must have access to PTS. The amount of comfort varies depending on a variety of conditions, such as crowd density in vehicles during peak hours. Commuters choose the private car for its convenience and comfort due to the higher level of discomfort in PTTs [13]. The PTS must be easily accessible geographically and to all sections of society. When the system is fully accessible and service is frequent, commuters will consider PT as a viable mode of transportation [14]. The density of FN should be larger and the length should be longer. The goal of FN is to broaden its service offering in order to improve accessibility and reduce walking distances [15].

ATS4: Increase Accessibility Levels by Providing Higher Frequency of Service

Increased frequency ensures that service is always available and improves accessibility. "Transit must provide an appropriate level of convenience, including more coverage and more frequent service to peripheral locations, in order to compete effectively with the personal vehicle trip" [16].

Yet Zhang [17] observed people will become less dependent on their vehicles if commuters have more access to transit options. To maintain accessibility levels, agencies must be able to analyze the regions that require more attention in order to enhance service frequency.

ATS5: Increase Accessibility Levels by Higher feeder Network Service Coverage

Increased service coverage lengthens routes per square kilometer, which has a direct impact on the number of commuters on each trip and the probability that commuters will use the service. LITA is regarded as a useful metric since it considers three key components of transportation system accessibility: frequency of service, capacity, and service coverage [16]. More riders would be attracted to a well-planned and scheduled regular FN from the origin to SAP. A competent FN ensures the shortest possible travel time, increases the number of commuters, and reduces the reliance on personal vehicle trips [18].

4.2 Strategies for Travel Time Performance Improvement

Travel time performance of BRTS or Metros can be improved by controlling two basic aspects first "reducing ideal time of commuters" and second "increasing travel speed." Ideal time is OVTT or the time spent in operational activities like boarding alighting, ticketing, waiting at intersections and transfer points, etc. The travel speed can be increased by improving travel speed. Systematic operations can reduce OVTT and increasing travel speed.

TTS1: Minimize Waiting time by Strict Adherence to Bus Schedules

Bus schedules that are strictly adhered to increase service quality and confidence among commuters by saving commuters valuable working hours Arbex and Cunha [19]. Due to crowding discomfort, researchers discovered a "population-weighted average reduction of 56.8% in accessibility to occupations in a standard workday morning peak due to crowding discomfort, as well as reductions of 6.2% due to travel time unreliability, and 59.2% when both are combined." Commuter reductions are due to user experience issues such as crowded discomfort and uncertainty about trip time. Maintaining punctuality can help to improve the predictability of travel times.

TTS2: Development of Just in Time Approach by Availability of On-time Running Information of Fleets

Commuters will be able to plan their trips more precisely if real-time fleet position information is available. It also cuts down on waiting time and, as a result, overall travel time. The public transportation system's reliability and performance will increase.

TTS3: Minimizing Waiting Time by Quick Availability of Vehicles at Transfer Points

To reduce transfer times, the schedules of PT modes such as BRTS, Metros, and feeder transit facilities have to be synchronized. It aids in the enhancement of travel time performance and the overall effectiveness of PTS. Improved transfer facilities, P&R facilities, and card access at public transit systems can minimize OVTT and transfer times (TT). By facilitating diverse modes at transfer locations, TT can be reduced, resulting in an Integrated Multi Modal Public Transportation System (MMTS) [1].

Strategies for Travel Time Performance Improvement: By Increasing Travel Speed

TTS4: Optimize Number of Bus Stop

Bus stops can be placed in strategic locations such as institutional areas, administrative buildings, CBDs, and intersections to connect people to other places. Short-distance bus stops can be combined into a single stop. A bus stop can be located within an 8-minute walking distance (PTALs, Transport for London 2010).

TTS5: Minimize Number of Traffic Signals or Operating Online to Reduce Crossing Time

Traffic signals (TS) control traffic while also increasing idle time and, as a result, total journey duration. Online traffic control solutions can reduce TS, and optimal time can be adjusted. Subways for pedestrian circulation, grade separators, and flyovers can be constructed in highly populated and narrow sections to reduce the need for TS. Controlling online traffic at signalized junctions has proven to be very effective at unsaturated intersections. Ceder and Reshetnik [20] concluded "An investigation was conducted in Israel on 30 most important critical intersections, which have heavy traffic resulting in 15-minute average delay. With a congested traffic flow saturation level of 8000 vehicle/h per intersection, the developed system can improve traffic flow by reducing queue length in oversaturated intersections, saving on the order of 1000 man-hours each rush hour per intersection."

TTS6: Optimize Stoppage Time at Bus Stops

Boarding and alighting (B&A) time, as well as passenger waiting time, influence bus stoppage time. Separate B&A doors, increased door width and number of doors, and low-floor buses can all help reduce B&A time. "The number of doors and B&A times have a significant effect on bus travel time to increase the level of service on the selected routes," according to a study conducted in Baghdad. Signal pre-emption for buses, on the other hand, reduces intersection delay and thus journey duration" [21].

TTS7: Provision of Dedicated Corridor

Because of the dedicated corridor, BRTS performs successfully. The average operating speed of Indian BRTS is between 18 and 24 km/hr. BRT stops of different systems are located before the intersection. For example, Ahmadabad BRTS has the maximum number of 127 stations in city BRT system of India with a spacing range between 525 and 710 meters [22]. Congestion pricing on Kuwait's network, which uses a genetic algorithm to improve traffic conditions, has resulted in significant reductions in travel time: "reductions in overall delay vary from 24.4 to 40.58%, and reductions in fuel used range from 36.76 to 60.89%." Fuel cost savings by users had a direct impact of 27.77% and 43.75%, respectively [23].

4.3 *Strategies for Egress Time Performance Improvement*

The egress time (ET) is a key factor in determining service mode. After alighting from main service, the availability of egress facilities must be frequent and convenient to reach the destination. It's a general observation in areas with adequate accessibility; ET is also lower when the FN is strong. Quick egress service enhances the chances of PT being selected as a service.

ETS1: Strategies for minimizing transfer time via interconnectivity improvement

Interconnectivity and, as a result, ET performance can be improved by a frequent and planned feeder service in the form of a small bus or paratransit. The length and placement of the service should be carefully planned, and feeders should be used in towns with an uniform population distribution [18].

ETS2: Availability of shared vehicle Trips for Quick Service

Localities with low population densities or suburbs in the early stages of development can benefit from shared vehicle trip service (SVT), which facilitates access and egress in the area. SVT service may prove cost-effective for users while also improving ET performance.

ETS3: Time Table Synchronization of feeder services with BRTS and Metros

A mathematical optimization model can be used to create an optimal timetable for a public transportation system that minimizes waiting times between service transfers. On the existing timetable, the weighted sum of transfer waiting times can be reduced by 11% using a heuristic technique [24].

5 Conclusion

Extensive research work has been done by researchers on various parameters like public transport accessibility levels, transfer time facility, significance of park and ride, frequency of service, network service coverage, travel time reliability, frequency of vehicle, real-time information of fleets, optimizing number of stops and stoppage time even though the efficiency of PTS is not noteworthy the reason is lack of synchronized efforts. To get remarkable results and increase commuter's acceptance of PTS we need to identify and improve the nonperforming and time-consuming parameters which makes PTS inefficient. After the review, the performance can be improved by implementing strategic decisions. Availability of on time fleets, service synchronization, on time running information of fleets, short accessibility, and egress time are some of the major indicators for PTS improvement. For improvement PTS is to be evaluated minutely at all suggested parameters and an action plan is to be proposed that can be implemented to achieve overall efficiency in public transport system. The major part of travel time is out-vehicle time due to unsynchronized public transport system. The OVT can be minimized by systematic planning and execution at certain levels, i.e., access to PT, boarding, and alighting, interconnectivity at transfer points, synchronized time table of feeder network with BRTS and metros, smart ticketing systems. To improve TTP, travel time monitoring is needed which is possible with GPS enabled and fleets. Hence it is recommended that GPS systems must be incorporated in PT fleets for having clear demarcation of travel time parameters. It is also recommended a committee to be formed with high rank officials which can do performance evaluation periodically.

References

1. Kumar PP, Parida M, Swami M (2013) Performance evaluation of multimodal transportation systems. In: Proceedings of the 2nd conference of transportation research group of India
2. TRIPP (2016) Indicators to measure performance efficiency of bus systems, a report. Transportation Research and Injury Prevention Programme, IITD
3. Jenelius E (2018) Public transport experienced service reliability: integrating travel time and travel conditions. J Transp Res Part A 117:275–291
4. Rigole PJ (2014) Study of a shared autonomous vehicles based mobility solution in stockholm. Royal Institute of Technology
5. Liu T, Ceder A (2017) Integrated public transport timetable synchronization and vehicle scheduling with demand assignment: a bi-objective bi-level model using deficit function approach. Proc 22nd Int Symp Transp Traffic Theory Transp Res Procedia 23:341–361
6. Kujalaa R, Weckströmb C, Mladenovic MN, Saramäki J (2018). Travel times and transfers in public transport: comprehensive accessibility analysis based on Pareto-optimal journeys. Comp Environ Urban Syst 67:41–54
7. Gurjar J, Agarwal PK, Jain PK (2019) A comprehensive methodology for comparative performance evaluation of public transport systems in urban areas. In: Proceedings of the world conference on transport research–WCTR 2019. Mumbai, India
8. Jasti PC, Ram VV (2019) Sustainable benchmarking of a public transport system using analytic hierarchy process and fuzzy logic: a case of Hyderabad, India. J Public Transp

9. Kenworthy JR (2020) Passenger transport energy use in Ten Swedish Cities: understanding the differences through a comparative review. J Energ 13:3719
10. Liao Y, Gil J, Pereira RHM, Yeh S, Vilhelm V (2020) Disparities in travel times between car and transit: spatiotemporal patterns in cities. J Sci Rep 10:1
11. Abrantes ALP, Wardman MR (2011) Meta-analysis of UK values of travel time: an update. J Transp Res Part A 45:1–17
12. Huang K, Liu Z (2019) Analysis of the acceptance of park-and-ride by users: a cumulative logistic regression approach. J Transp Land Use
13. İmrea S, Celebia D (2016) Measuring comfort in public transport: a case study for Istanbul. In: Proceedings of the world conference on transport research–WCTR. Shanghai, China
14. Shah J, Bhargav A (2016) Public transport accessibility levels for Ahmadabad, India. J Public Transp 19(3)
15. Jian G, Peng Z, Chengxiang Z, Hui Z (2012) Research on public transit network hierarchy based on residential transit trip distance. Hindawi Publishing Corporation, Discrete Dyn Nat Soc 2012. Article ID 390128
16. Wiley K, Maoh H, Kanaroglou P (2011) Exploring and modeling the level of service of urban public transit: the case of the Greater Toronto and Hamilton Area Canada. J Transp Lett 3(2):77–89
17. Zhang M (2006) Travel choice with no alternative: can land use reduce automobile dependence? J Plan Edu Res 25:311
18. Tabassum S, Tanaka S, Nakamura F, Ryo A (2016) Feeder network design for mass transit system in developing countries. In: Proceedings of the WCTR 2016 world conference on transport research. Shanghai, China
19. Arbex R, Cunha CB (2020) Estimating the influence of crowding and travel time variability on accessibility to jobs in a large public transport network using smart card big data. J Transp Geogr 85
20. Ceder A, Reshetnik I (2001) An algorithm to minimize queues at signalized intersections. J Oper Res Soc 52(6):615–622
21. Al-Jumailey SI (2011) Planning of operation policies for fixed bus routes in Baghdad city. In: Al-Mustansiriya University, Baghdad, Iraq. IEEE
22. Kathuria A, Parida M, Sekhar CR, Sharma A (2016) A review of bus rapid transit implementation in India. Cogent Eng 3:1241168
23. AlRukaibi F, AlRukaibi D, AlBurait A, Al-Mutairi A (2020) The impact of congestion charging technique on traffic flow and atmospheric pollution in Kuwait city. J Eng Res 9(1):51–62
24. Nielsen OA (2004) Behavioral responses to road pricing schemes: description of the Danish AKTA experiment. Intell Transp Syst 8:233–251
25. Ho CQ, Hensher DA, Wang S (2020) Joint estimation of mode and time of day choice accounting for arrival time flexibility, travel time reliability and crowding on public transport. J Transp Geogr 87

Engineering Properties of Paver Block Prepared by Incorporation of Stone Dust as Part Replacement of River Sand for Heavy Traffic Condition

Sarvesh P. S. Rajput, Amit Mandal, and Hemant Choudhary

Abstract Fierce growth in constructional exercises brings about the depletion of naturally accessible raw materials used to create concrete. Availability and accessibility of river sand, for example, the fine aggregate, right now is very costlier and getting scarcer due to non-accessibility of the river during the whole year, illegal digging, quick growth of constructional exercises, and so forth. On account of these causes the quality of sand has depleted. The analysis of stone dust emphasizes its properties and feasible usage in concrete. In this experimentation the concrete paver blocks containing increasing levels of stone dust as a part replacement of river sand is incorporated and enhanced properties were observed.

Keywords Stone dust · Alternatives to river sand · Sustainability · Waste management

1 Introduction

Sand is the third most utilized natural resource after air and water. It gets far more use than oil does. Food, glass, toothpaste, microprocessors, personal care items, paint, paper, and plastics are all made with it. In addition, it is used to create dams, houses, and other structures. The requirement for this resource has caused a boom in sand mining operations and companies. Rapid urbanization is the main cause of this dilemma.

People are migrating from rural to urban areas at an increasing rate. Urban areas are expanding faster and more dramatically than ever before across Asia, Africa, and Latin America. Despite being a profitable industry, sand mining significantly harms the environment.

S. P. S. Rajput · A. Mandal · H. Choudhary (✉)
Department of Civil Engineering, MANIT Bhopal, M.P., India
e-mail: spsrajput@manit.ac.in

S. P. S. Rajput
e-mail: spsrajput@manit.ac.in

K. K. Pathak et al. (eds.), *Latest Developments in Civil Engineering*, Lecture Notes in Civil Engineering 352, https://doi.org/10.1007/978-981-99-2676-3_64

The development of sand mining has given livelihood to the neighboring town. The disadvantages, however, far outweigh the advantages. Sand mining causes habitat in mined zones to become disturbed and eventually disappear. In other terms, the entire ecosystem is altered, including the physical habitats, food webs, and channel geometry. In addition, it intensifies the flow velocity, upsetting the flow regime and ultimately causing the river banks to erode.

The effects of sand mining on the ecology may ultimately result in:

1. Streambed shallowing due to channel expansion generates braided flow or subsurface inter-gravel flow in riffle locations, which prevents fish from moving between pools. As a result, fishing is no longer a viable way to make a living in the neighborhood.
2. The riverbed dries up due to exposure to solar radiation, which depletes both the surface and groundwater. Consequently, the river has progressively ceased to serve the nearby settlements as a supply of fresh water.
3. Sand erosion in streambeds causes the widening of river mouths and coastal inlets as well as the deepening of estuaries and rivers. Additionally, it causes saline water to infiltrate.
4. The environment above and below ground becomes worsened by the removal of vegetation and the degradation of the soil profile, which causes a fall in faunal populations.

The prime reason for excessive sand mining is the rapid constructional activities and due to this, there is an increase in the need of materials. The alternatives which have the potential to replace sand are stone dust, industrial slag, marble waste, ceramic tile waste, etc. There are numerous study, which have already used stone dust in concrete, mortar, paver blocks, and as filler material. The replacement of ingredients such as river sand in a mortar with stone dust has increased its performance [1]. The optimum replacement of sand with marble dust boosts the compressive strength of concrete. Also quick gain of split tensile strength and flexural strength is detected [2]. One of the applications of concrete is the use of it in making paver blocks, waste marble is also utilized to make the paving blocks. When marble waste is mixed as fine aggregate, on increasing the levels of replacement the water demand also increases which went on to increasing because the specific surface area increase. As the content of marble waste enhanced compressive strength of blocks decreases [3]. Nowadays researchers are greatly involved in finding different alternatives to fine aggregate due to the fact that have already been discussed above. Apart from these other materials such as walnut shell, ceramic waste, cenosphere, vermiculite, construction and demolition waste have also been utilized as a replacement for fine aggregate [4–8]. These discarded wastes have troublesome management as well as environmental issues, so using these wastes in making concrete greatly reduces the dumping part and also satisfies the gruesome river sand demand to make concrete. Cenosphere which is obtained from coal fly ash can replace up to thirty percent of sand without having any effect on the properties of concrete [6]. The incorporation of ceramic waste as fine aggregate in concrete enhances the compressive strength when it is substituted by the levels of 50% for controlled conditioned concrete but at the same

Fig. 1 Stone Dust obtained from nearby plants

time it decreases the workability [5]. Other studies relating the use of ceramic waste powder as part replacement of fine aggregate in self-compacting concrete enhance the compressive strength up to the level of 20% and beyond it reduces [9]. Overall by literature survey, it is observed that some of these aluminosilicate compounds greatly enhance the strength while the fineness results in lesser rheological properties.

A discarded product obtained from crushing units of stone is stone dust. Presently it is only being used as tile laying, martial for filling whether in pavement subgrade or foundations due to its low prices when balanced with other materials like murrum. When experimented it yields more fines as compared to natural sand while all other parameters remained the same. For this experimentation work an optimum amount of this material is obtained which can replace the rapidly exhausting river sand in concrete paver blocks (Fig. 1).

2 Material Used and Methodology

2.1 Cement

OPC 53 grade cement of mycem brand is used for this research. Table shows results of tests on cement's physical properties (Table 1).

Table 1 Properties of cement

S. no	Parameters	Obtained result	Codal provision
1	Specific gravity	2.88	3.15
2	Compressive strength	58 MPa	53 MPa
3	Fineness modulus	8.5	Less than 10%
4	Initial setting time	50 min	30 min (Min.)
5	Consistency	31%	–
6	Final setting time	350 min	600 min (Max.)

Table 2 Properties of fine aggregate

S. no	Parameters	Test value	Code referred
1	Fineness modulus	2.63	IS: 2386 (Part 1)-1963
2	Total Silt content	2.83%	IS: 2386 (part 2)-1963
3	Specific gravity	2.7	IS: 2386 (part 3)-1963
4	Water absorption	1.45%	IS: 2386 (part 3)-1963

2.2 *Fine Aggregate*

See Table 2.

2.3 *Coarse Aggregate*

See Table 3.

Table 3 Properties of coarse aggregate

S. no	Properties	Test value	Code referred
1	Fineness modulus	7.8	IS: 2386-part 1
2	Specific gravity	2.78	IS: 2386 (part 3)-1963
3	Density	1720 kg/m^3	IS: 2386 (part 3)-1963
4	Voids	37.30%	IS: 2386 (part 3)-1963
5	Water absorption	6.20%	IS: 2386 (part 3)-1963

Table 4 Properties of stone dust

Properties	Test value
Water absorption	1.12%
Specific gravity	2.66
Silt content	9.65%
Fineness modulus	3.32

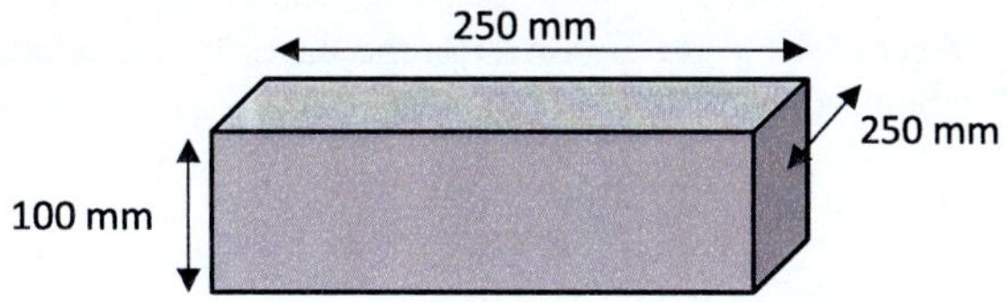

Fig. 2 Paver block

2.4 Stone Dust

See Table 4.

2.5 Paver Block Design

The paver block designed is based on heavy traffic conditions like, industrial and factory floors, bus stands, service stations, industrial pavements, etc. As per IS15658:2006 [10] the grade designation of a paver block for such traffic conditions is M50 and the recommended minimum paver block thickness of 100 mm (Fig. 2).

2.6 Mix Design

The mix design for the paver block is deduced from IS: 10262:2019 [11]. As per the heavy traffic conditions, M50 grade is selected for the mixes for which the target mean strength is given by

$$f'_{ck} = f_{ck} + 1.65\,S = 50 + 1.65 \times 5 = 58.25\ \text{N/mm}^2 \text{or}$$
$$f'_{ck} = f_{ck} + 6.5 = 50 + 6.5 = 56.5\ \text{N/mm}^2$$

Taking the maximum from the above two values, i.e., 58.25 N/mm^2 as the target mean strength for the M50 grade of concrete. For the mix 20% reduction of water content due to admixture is considered here.

Mix proportion for trial

Materials per cubic meter of concrete

Water = 149 kilogram

Cement = 480 kilogram

Coarse aggregate = 1314.47 kilogram

Sand = 607.72 kilogram

Water-cement ratio = 0.31

According to the mix design the paver block samples are prepared and checked for their workability, compressive strength, and flexural strength.

3 Result and Discussion

As per IS 15658:2006 ANNEXURE D [10], the apparent compressive strength of the specimen is calculated by dividing the load in N (maximum) by the plan area in mm^2. Now to calculate the corrected compressive strength this apparent compressive strength is multiplied by the correction factor from Table 5 of IS 15658:2006 [10]. For 100 mm paver block thickness, the correction factor for the plain block is 1.18.

3.1 Compressive Strength

From the Tables 6 and 7, it's clear that replacing sand with stone dust enhances the compressive strength of M50 grade concrete paver blocks by 40%, which is acceptable by Indian standards (Fig. 3).

Table 5 Mix proportion for different samples (per m^3)

Designation	Cement (kilogram)	Fine aggregate (kilogram)	Coarse aggregate (kilogram)	Replacement of cement (%)	Stone dust (kilogram)	W/C ratio
A	480	607.72	1314.47	0	–	0.31
B	480	546.95	1314.47	10	60.77	0.31
C	480	486.18	1314.47	20	121.54	0.31
D	480	425.40	1314.47	30	182.32	0.31
E	480	364.63	1314.47	40	243.09	0.31
F	480	303.86	1314.47	50	303.86	0.31

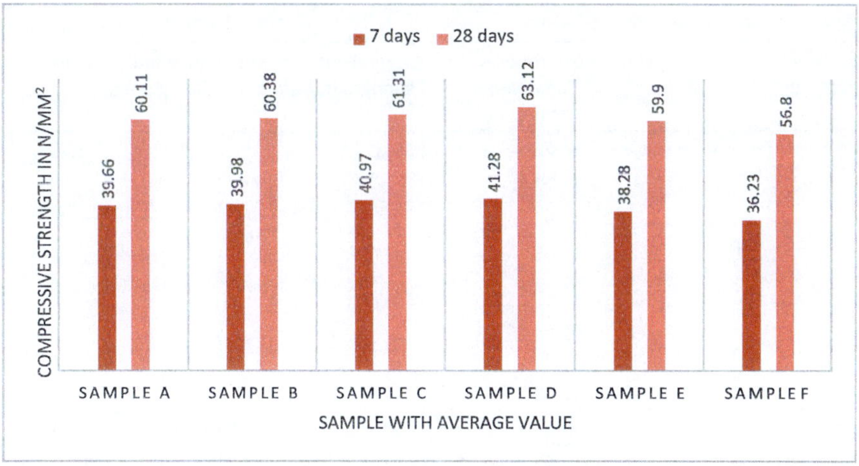

Fig. 3 Compressive strength of paver blocks after 7 and 28 days

3.2 *Flexural Strength*

The flexural test for paver block is conducted as per IS 15658:2006 [10], the mix containing 40% stone dust is selected because the optimum compressive strength was there when tested for it. The flexure strength after 7 and 28 days is given in the table below (Table 8).

We cannot calculate the direct tensile strength of concrete hence some indirect methods are used like flexural test, the empirical relation of tensile strength according to the Indian standard code [12] is given by *i*

$$f_t = 0.7\sqrt{f_{ck}}$$

where f_t, f_{ck} = tensile strength and characteristic compressive strength of cube, respectively, so

$$f_t = 0.7\sqrt{50} = 4.65\ \text{N/mm}^2$$

From the results obtained from the table flexural strength of the sample after 28 days of curing comes out to be 5.65 N/mm^2.

Table 6 7 days compressive strength

Substitution percentage	identification	Specimen's Compressive strength in MPa	Medial compressive strength in MPa	Standard Deviation	Variance
0%	A	36.25	39.659	3.391	11.498
		41.23			
		45.9			
		40.65			
		38.26			
		36.2			
		36.9			
		41.88			
10%	B	38.6	39.976	1.182	1.397
		39.56			
		40.98			
		41.87			
		38.65			
		39.15			
		40.25			
		40.75			
20%	C	44.23	40.956	2.301	5.297
		42.9			
		40.78			
		38.56			
		36.98			
		41.75			
		40.89			
		41.56			
30%	D	42.3	41.281	2.873	8.253
		45.6			
		40.8			
		36.55			
		39.48			
		42.72			
		43.65			
		39.15			
40%	E	38.65	38.276	2.023	4.092
		39.75			
		40.89			

(continued)

Table 6 (continued)

Substitution percentage	identification	Specimen's Compressive strength in MPa	Medial compressive strength in MPa	Standard Deviation	Variance
		36.34			
		39.56			
		37.85			
		34.56			
		38.61			
50%	F	36.75	36.226	2.851	8.128
		34.98			
		33.65			
		36.72			
		36.31			
		39.65			
		40.13			
		31.62			

3.3 *Water Absorption*

For conducting the water absorption test 3 specimen are taken as per IS 15658:2006 Table 4, these specimens are then completely immersed in water having room temperature of 24 ± 2 h. Then the specimens are removed and drained for 1 min after placing on coarser wire mesh (Table 9). After fully removal of water from surface the specimens are weighed and water absorption percentage is found as

$$\frac{W_w - W_d}{W\% = W_d}$$

3.4 *Abrasion Resistance*

For conducting the abrasion resistance test 3 specimen are taken as per IS 15658:2006 Table 4, these specimens are then tested via Cantabro abrasion test. The result of this test for paver block containing 40% stone dust as a part replacement of river sand in concrete is given by Table 10.

Table 7 28 days compressive strength

Substitution percentage	Identification	Specimen's Compressive strength in MPa	Medial compressive strength in MPa	Standard Deviation	Variance
0%	A	55.25	60.113	3.231	10.439
		59.63			
		58.13			
		60.56			
		63.9			
		65.42			
		59.23			
		58.78			
10%	B	56.96	60.378	2.691	7.241
		58.36			
		64.57			
		62.8			
		60.13			
		57.52			
		62.18			
		60.5			
20%	C	60.15	61.309	1.861	3.462
		59.26			
		59.7			
		63.56			
		64.2			
		62.5			
		60.32			
		60.78			
30%	D	63.25	63.120	2.406	5.787
		65.63			
		63.23			
		59.43			
		59.6			
		63.62			
		65.3			
		64.9			
40%	E	58.32	59.898	2.552	6.514
		59.62			
		56.5			

(continued)

Table 7 (continued)

Substitution percentage	Identification	Specimen's Compressive strength in MPa	Medial compressive strength in MPa	Standard Deviation	Variance
		60.31			
		65.25			
		58.63			
		59.65			
		60.9			
50%	F	60.12	56.796	1.995	3.978
		56.32			
		54.9			
		56.32			
		54.88			
		55.9			
		56.3			
		59.63			

Table 8 7 and 28 days flexural strength

Duration of testing after curing	Designation	Medial flexural strength in N/mm^2 (8 specimen)
7 days	FS7	3.75
28 days	FS28	5.65

Table 9 Water absorption data

Sample No	Ww	Wd	Water Absorption (%) for each specimen	Water Absorption (Average of 3 specimens)
1	8.88 kg	8.4 kg	5.7	
2	8.95 kg	8.56 kg	4.56	4.92%
3	8.6 kg	8.23 kg	4.49	

Table 10 Abrasion resistance data

Sample no	Cantabro loss (%)	Abrasion resistance value (average)	Variance
1	31.25	29.74	1.33
2	30.2		
3	28.6		

4 Conclusion

In this research work stone dust is used as a part replacement for river sand in making concrete paver blocks. Following were the results after experimentation;

1. The optimum percentage for M50 grade concrete paver blocks is 40%.
2. The compressive strength and flexure strength for these replacement levels are 63.12 KN/mm^2 and 5.65 KN/mm^2.
3. Water absorption for this mix is about 4.92%.
4. Abrasion resistance for this mix is 29.74%.

It can be concluded after the experimentation and results is stone dust can be effectively introduced to replace the river sand for making the paver blocks. It is a very costly as well as environment friendly alternative, or in other words we can say that it is a sustainable alternative to scarcer river sand.

References

1. "Crushed stone dust as a replacement for river sand in self compacting repair mortars—A sustainable solution. Elsevier Enhanced Reader"
2. "Development of green concrete using waste marble dust. Elsevier Enhanced Reader"
3. "Properties of concrete paving blocks made with waste marble. Elsevier Enhanced Reader"
4. Venkatesan B, Lijina VJ, Kannan V, Dhevasenaa PR (2021) "Partial replacement of fine aggregate by steel slag and coarse aggregate by walnut shell in concrete." Mater Today: Proc 37(2): 1761–1766, https://doi.org/10.1016/J.MATPR.2020.07.361
5. Torkittikul P, Chaipanich A (Jul.2010) Utilization of ceramic waste as fine aggregate within Portland cement and fly ash concretes. Cement Concr Compos 32(6):440–449. https://doi.org/10.1016/J.CEMCONCOMP.2010.02.004
6. Danish A, Mosaberpanah MA, Tuladhar R, Salim MU, Yaqub MA, Ahmad N (May2022) Effect of cenospheres on the engineering properties of lightweight cementitious composites: A comprehensive review. J Buildi Eng 49:104016. https://doi.org/10.1016/J.JOBE.2022.104016
7. Naveen Kumar K, Vijayan DS, Divahar R, Abirami R, Nivetha C (2020) "An experimental investigation on light-weight concrete blocks using vermiculite." Mater Today: Proc 22: 987–991 https://doi.org/10.1016/J.MATPR.2019.11.237
8. Shanko Y, Abera A (2022) "Performance of concrete materials containing recycled aggregate from construction and demolition waste." Results Mater 14: 100278, https://doi.org/10.1016/J.RINMA.2022.100278
9. Gautam L, Kumar Jain J, Alomayri T, Meena N, Kalla P (2022) "Performance evaluation of self-compacting concrete comprising ceramic waste powder as fine aggregate." Mater Today: Proc 61: 204–211, https://doi.org/10.1016/J.MATPR.2021.08.063.
10. IS: 15658 (2006) Precast concrete blocks for paving-specification
11. IS: 10262 (1982) Indian Standard Recommended guidelines for concrete mix design. Bureau of Indian standards, New Delhi
12. IS: 456 (2000) Plain and reinforced concrete—code of practice

Characterization of Paver Block Developed Using Rice Husk Ash (RHA) for Medium Traffic Condition

Hemant Choudhary, Sarvesh P. S. Rajput, and Amit Mandal

Abstract Typically, the wastes generated from the industrial and agricultural sector have raised problems related to waste management and pollution. However, using these wastes in addition to other conventional building materials has both functional and financial benefits. These wastes have no economic value and are easily available at lower cost including transportation. The use of these waste materials in construction has a significant contribution to protect the environment and conservation of natural resources. Silica fume, Fly ash, Corn Cob Ash, Rice Husk Ash, and volcanic ash are some of the waste items that have been investigated for usage in blended cement because they have pozzolanic qualities. Waste products from the rice industry, i.e., rice husk is usually dumped in open field or burned. It significantly affects the environment and also does not have any economical benefit. Rice husks generally have no constructional use. The husk ash was obtained from the mills which uses ash as fuel for thermal use because the burning temperature required is so high and we got the rice husk ash as residue or by-product by rice mills. This research paper uses partial replacement of cement in the construction of paver block by rice husk ash for medium traffic condition in levels of 0, 5, 10, 15, 20, and 25.

Keywords Rice husk ash · Blended cement · Sustainability · Waste management

1 Introduction

Concrete generation is a significant wellspring of cultural advancement as well as employment. With a characteristic development in population, the sum and sort of waste materials have expanded as needs be making in this way ecological problems. Traditionally, horticultural and industrial squanders have made waste administration and contamination problems. Different elective waste materials and modern results, for example, bottom ash, fly ash, recycled aggregates, sawdust, crumb rubber, and so forth were supplanted with regular aggregates [1–6]. Spite of the fact that all these

H. Choudhary (✉) · S. P. S. Rajput · A. Mandal
Department of Civil Engineering, MANIT-Bhopal, M.P., India
e-mail: 193111009@stu.manit.ac.in

K. K. Pathak et al. (eds.), *Latest Developments in Civil Engineering*, Lecture Notes in Civil Engineering 352, https://doi.org/10.1007/978-981-99-2676-3_65

materials are typically regarded as "crude" and therefore second rate compared to all the more exceptional forms as far as safety, durability, performance, occupant's wellbeing, and solace regarding ecological issue, consumption of natural items and vitality inside the development industry has made a huge interest for crude materials and for generation in this manner adding to the numerous ecological issues related with different biological system.

The squanders have commonly no economic value and are accessible locally at a negligible cost. Effective use of these has supplemented other customary materials in development and thus gives down-to-earth and monetary advantages. Also, legitimate usage of these squanders monitors the regular assets and helps in protecting the environment. Apart from the waste materials mentioned above, rice husk can likewise be utilized in concrete because of the accompanying focuses:

i. Rice is grown extensively in India and other Asian countries with coastal environments.
ii. It's a staple in most countries and produces megatons of husk every year.
iii. The natural environment is often tainted by the discarding of the husk and other debris that accumulate after rice has been harvested.
iv. While the local wildlife may find some usable nutrients in the husk, the vast majority of it goes to waste. Some part of it when blended in with other natural squanders and cow dung fills in as great fertilizer to the plants.
v. Helpful for low-rise structures and low-cost housing.
vi. Used as a building material that is good for the environment. With the goal of making housing more affordable for both urban and rural people in India and many other developing countries, huge ideas have been put forward that focus on lowering the cost of traditional building materials. Finding a substitution for the traditional material used today is a tedious task that should be thought about because it helps keep natural materials around for the future.

The production of rice husk in the world is approximately 31 million tons and, our country alone contributes approximately 20% of RHA every year [7]. As far as India is concerned, Punjab, West Bengal, Andhra Pradesh, Uttar Pradesh, and Odisha are extensive rice cultivators. In figure rice by-products are shown (Fig. 1).

It has been noted that the RHA was utilized in concrete by many researchers and it is observed that RHA is a sustainable material that has the property to add as an additive cementitious material to produce alkali-activated concrete [8]. Besides its use in concrete, we should be well aware that concentration and amount of the activating solution, water/cement ratio, and the curing conditions could also influence the durability as well as mechanical properties of the RHA utilized in concrete and mortars [9]. The utilization of RHA or any other likewise material can result in lower CO_2 emissions and also enhanced properties as far as conventional concrete is concerned [10]. The composition of RHA, i.e., chemical and mineralogical greatly relies on the origination of material and ways by which it has been produced. It usually has 85–90% of amorphous silica thus it is categorized as N-pozzolana in accordance

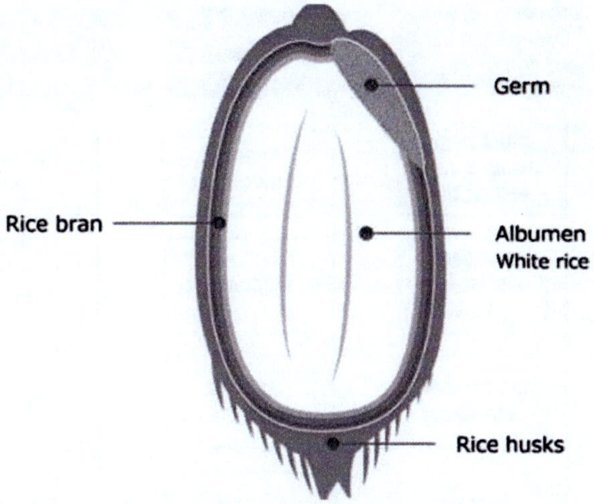

Fig. 1 By-products of rice

with ASTM norms. RHA can also be utilized as other construction material rather than concrete. It has been observed that this discarded material can also be utilized in the making of bricks. Lightweight cellular bricks having good insulating property as well as cost-effective new material can be produced by using RHA [11].

2 Methodology

The use of waste materials, specifically Rice Husk Ash, is the main emphasis of this research (RHA). Various experiments are utilized to assess the fundamental characteristics of the material adopted for the study. After that, the split tensile strength, compressive and flexural strength of cured concrete are assessed. This study uses a robust quantitative approach (Figs. 2 and 3). Figure below shows different tests that were done on materials and concrete.

3 Material Used

3.1 Cement

In this analysis OPC (Ultratech) 53 grade is used and the following are the properties (Table 1),

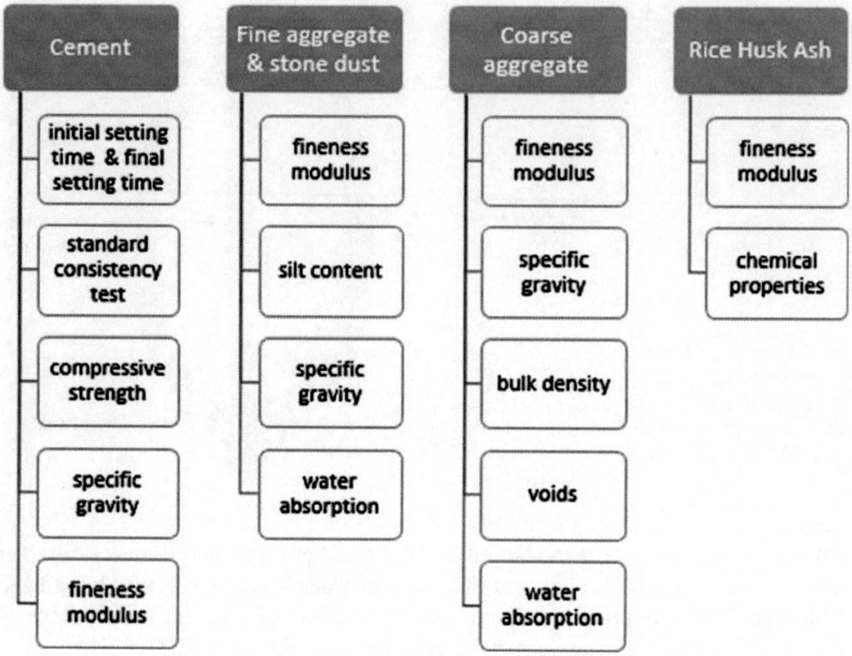

Fig. 2 Tests for evaluation of basic properties of material

Fig. 3 Tests on concrete

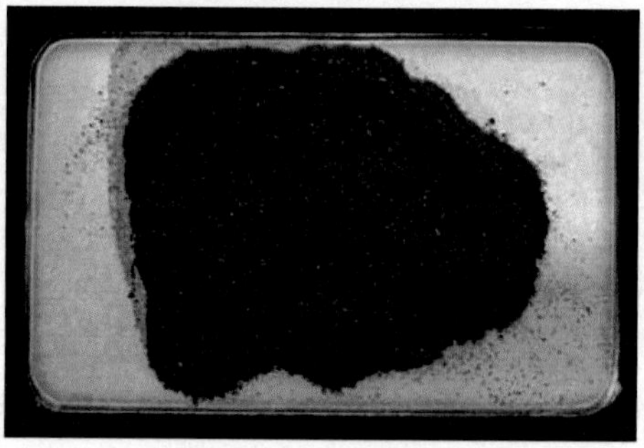

Table 1 Properties of cement

S.no	Cement properties	Result	Indian standard requirements
	Initial setting-time	52 min	Minimum of 30 min
	Final setting-time	370 min	Maximum of 600 min
	7 days Compressive strength	42 N/mm^2	22 N/mm^2
	28 days Compressive strength	60 N/mm^2	53 N/mm^2
	Standard consistency test	34%	–
	Specific gravity	2.92	Minimum 3.15
	Fineness modulus	7.2	Less than 10%

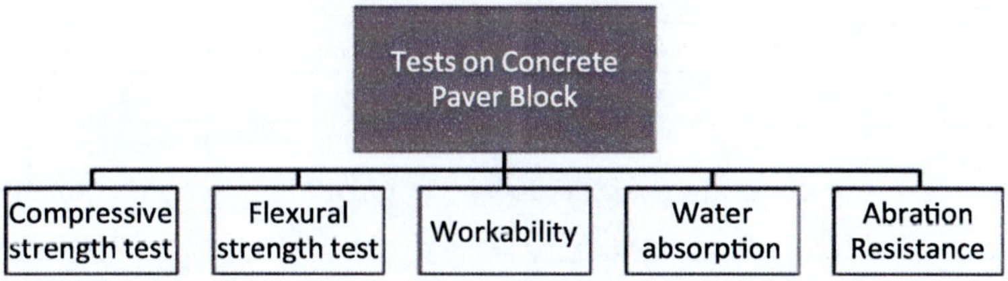

Fig. 4 Rice husk ash (RHA)

Table 2 Properties of stone dust

S.no	Parameters	Values (%)
1	Specific gravity	2.3
2	Fineness (<45 μm)	96
3	SiO_2	87.25
4	Al_2O_3	0.20
5	Fe_2O_3	0.17
6	CaO	0.57
7	MgO	0.35
8	C	5.95
9	SO_3	0.24
10	Ignition loss	5.44
11	Particles size (μm)	9
12	Pozzolanic activity	84

3.2 Rice Husk Ash (RHA)

For this experimental exercise, the RHA is been purchased from "Ambika Traders" and Gujarat (Fig. 4). The properties of RHA are as follows (Table 2):

Further the RHA is tested for moisture content and it is about 40% by mass. Blain's air permeability apparatus is used to find the average specific surface area and it comes out to be 643 m^2/Kg.

3.3 Fine Aggregate

See Table 3.

Table 3 Properties of fine aggregate

S.no	Properties	Test value	Code referred
	Fineness modulus	2.55	IS: 2386-part 1
	Specific gravity	2.58	IS: 2386 (part 3)-1963
	Silt content	3.4%	IS: 2386 (part 2) 1963
	Water absorption	1.15%	IS: 2386 (part 3) 1963

3.4 Coarse Aggregate

See Table 4.

4 Paver Block Design

The paver block is designed for medium traffic conditions like city streets, low volume roads, medium market roads, etc. As per IS15658:2006 paver block grade designation for such traffic conditions is M-40 and the recommended minimum thickness is 40 mm (Fig. 5).

Table 4 Properties of coarse aggregate

S.no	Properties	Test value	Code referred
	Fineness modulus	8.22	IS: 2386-part 1
	Specific gravity	2.7	IS: 2386 (part 3)-1963
	Bulk density	1580 kg/m^3	IS: 2386 (part 3) 1963
	Voids	38.50%	IS: 2386 (part 3) 1963
	Water absorption	7.60%	IS: 2386 (part 3) 1963

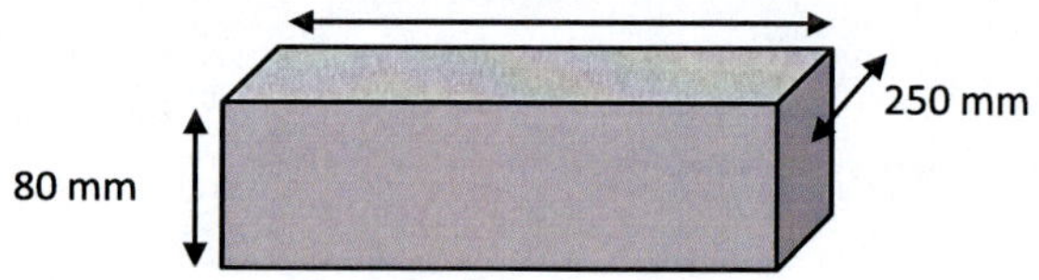

Fig. 5 Paver block

5 Mix Design

The paver block mix design is deduced from IS: 10262:2019 [12]. As per the medium traffic conditions M-40 grade is selected for the mixes for which the target mean strength is given by

$$F'_{ck} = F_{ck} + 1.65s = 40 + 1.65 \times 5 = 48.25 \text{ N/mm}_2$$

or

$$F'_{ck} = F_{ck} + 6.5 = 40 + 6.50 = 46.50 \text{ N/mm}_2$$

Taking the maximum from the above two values, i.e., 48.25 N/mm^2 as the target mean strength for M-40 grade concrete (Table 5).

- Mix proportion for trial

  Cement = 418.5 kg/m^3
  Water = 167 kg/m^3
  Coarse aggregate = 1155.06 kg /m^3
  Fine aggregate = 676.48 kg/m^3
  Water cement (w/c) ratio = 0.40

According to the mix design the paver block samples are prepared and checked for their strength (compressive and flexural) and workability.

6 Result and Analysis

6.1 *Compressive Strength*

Experimental procedure for determining compressive strength of paver block is adopted from IS 15658: 2006. As per this code for the average compressive strength value of 8 specimens of paver is taken as the compressive strength of that sample of paver. So, 8 specimens for each replacement are casted then it is tested for its compressive strength after 7 days and 28 days of curing. The apparent compressive strength of the sample is calculated by dividing the load in N (maximum) by the plan area in mm^2. Now to calculate the corrected compressive strength this compressive strength value which is also called as apparent compressive strength is multiplied by the correction factor provided in annexure D of IS 15658:2006. For an 80 mm thick plain paver block, the correction factor is 1.12. According to IS 456:2000 [13], the compressive strength of individual specimens should not be varied by ±15% of the average, the standard deviation shows those variations in the table (Fig. 6; Tables 6 and 7).

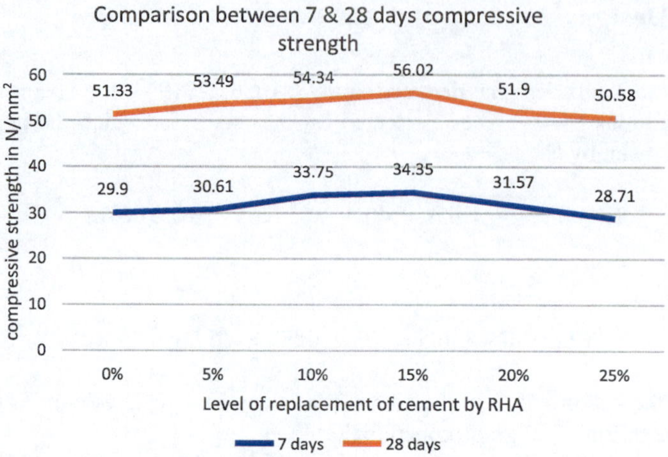

Fig. 6 Comparison between 7 and 28 days compressive strength

Table 5 Mix proportion for different samples

Identification	Fine aggregate (kg/m^3)	Cement (kg/m^3)	% Substitution of cement	Coarse aggregate (kg/m^3)	Rice Husk Ash (kg/m^3)	W/C ratio
PB1	676.48	418.5	0	1155	–	0.40
PB2	676.48	397.6	5	1155	20.9	0.40
PB3	676.48	376.65	10	1155	41.85	0.40
PB4	676.48	355.73	15	1155	62.78	0.40
PB5	676.48	334.8	20	1155	83.7	0.40
PB6	676.48	313.88	25	1155	104.63	0.40

6.2 *Flexural Strength*

The flexural strength is determined in accordance with the procedure given in IS 516. From 'Table 4 Sampling Requirements' of IS 15658: 2006 [13], for flexural strength/breaking load total 8 number of paver blocks are selected (Fig. 7). The flexural strength at 28 and 7 days for different mixes is shown in the Table 8 below.

6.3 *Workability*

At the time of casting, the workability of these blocks is an important parameter. It should not be too watery or not too dry or else the coherence can be greatly affected [14]. The decrease in the workability is observed when the amount of substitution is

Table 6 Days compressive strength

Designation	% Replacement of cement by RHA	Average Compressive strength	Standard deviation	Variance
PB1	0	29.9	1.29	1.65
PB2	5	30.61	0.83	0.69
PB3	10	33.75	0.94	0.89
PB4	15	56.02	1.8	3.23
PB5	20	31.57	0.91	0.83
PB6	25	28.71	0.72	0.52

Table 7 28 days compressive strength

Designation	% Replacement of cement by RHA	Average Compressive strength	Standard deviation	Variance
PB1	0	51.33	1.14	0.29
PB2	5	53.49	0.63	0.39
PB3	10	54.34	1.32	1.73
PB4	15	56.02	1.67	2.78
PB5	20	51.9	1.01	1.03
PB6	25	50.58	1.53	2.33

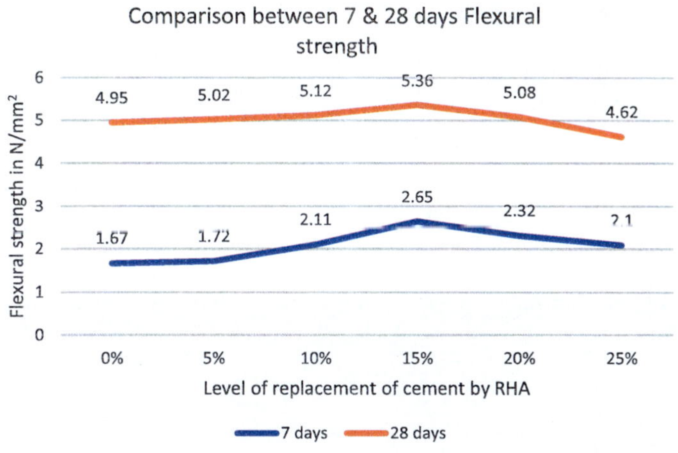

Fig. 7 Comparison between 7 and 28 days Flexural strength

increased, it is because of the fact that the greater the fineness, the greater the specific surface area which also results in a greater need for water requirements. In addition, due to the greater specific surface area of RHA, the reaction rate of it is also high which in turn reduces the rheological property of the mix (Fig. 8). Workability is

Table 8 Flexural strength after 7 and 28 days

Designation	% Replacement of cement by RHA	7 days Average flexural strength (KN/mm^2)	28 days Average flexural strength (KN/mm^2)
PB1	0	1.67	4.95
PB2	5	1.72	5.02
PB3	10	2.11	5.12
PB4	15	2.65	5.36
PB5	20	2.32	5.08
PB6	25	2.1	4.62

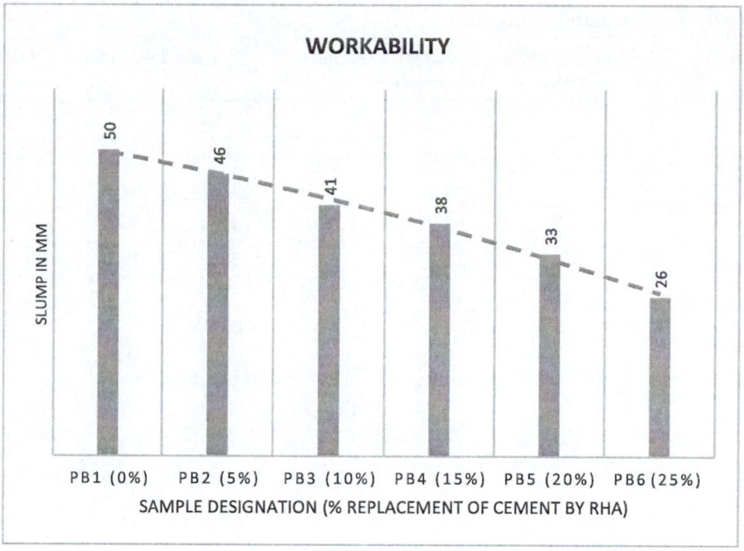

Fig. 8 Workability

determined by the slump cone test, the result for each designated sample is shown in the Table 9.

6.4 *Water Absorption*

The compressive strength of mix containing 15% RHA comes out to be maximum. So, for the water absorption test the samples having this particular mix are selected. For conducting the water absorption test 3 specimen are taken as per IS 15658:2006 Table 4, these specimens are then completely immersed in water having room temperature of 24 ± 2 h. Then the specimens are taken out and drained for 1 min after placing

Table 9 Workability

Designation of sample	Slump in mm	Percentage decrease
PB1 (0%)	50	–
PB2 (5%)	46	8%
PB3 (10%)	41	10.57%
PB4 (15%)	38	7.317%
PB5 (20%)	33	13.15%
PB6 (25%)	26	21.21%

Table 10 Water absorption data

Sample No	Ww	Wd	Water Absorption (%) for each specimen	Water Absorption (Average of 3 specimens)
1	8.657 kg	8.25 kg	5.15	
2	8.72 kg	8.264 kg	5.55	5.43%
3	8.69 kg	8.23 kg	5.6	

on coarser wire mesh. After fully removal of water from the surface the specimens are weighed and the water absorption percentage is found as Table 10:

$$W\% = \frac{W_w - W_d}{W_d}$$

6.5 *Abrasion Resistance*

For conducting the abrasion resistance test 3 specimen are taken as per IS 15658:2006 Table 4, these specimens are then tested via the Cantabro abrasion test. The result of this test for paver block containing 15% RHA as a substitute for cement in concrete is given by Table 11.

Table 11 Abrasion resistance data

Sample no	Cantabro loss (%)	Abrasion resistance value (average)	Variance
1	35.4	34.43	1.104
2	34.67		
3	33.23		

7 Conclusion

In this research work, RHA is used as a partial substitute for cement in making concrete paver blocks. Following were the results after experimentation;

1. The optimum percentage for M-40 grade concrete paver blocks is 15%.
2. The compressive strength and flexure strength for these replacement levels are 56.02 KN/mm^2& 5.06 KN/mm^2.
3. The slump value for a mix containing 15% RHA is 38 mm.
4. Water absorption for this mix is about 5.43%.
5. Abrasion resistance for this mix is 34.43%.

It can be concluded after the experimentation and results are RHA can be effectively introduced to replace the cement to a certain level for making the paver blocks. It is a very cost as well as environment-friendly alternative, or in other words, we can say that it is a sustainable alternative to cement.

References

1. Nath P, Sarker P (2011) "Effect of fly ash on the durability properties of high strength concrete." Proc Eng: 1149–1156
2. Kurama H, Kaya M (Sep.2008) Usage of coal combustion bottom ash in concrete mixture. Constr Build Mater 22(9):1922–1928. https://doi.org/10.1016/j.conbuildmat.2007.07.008
3. Tayeh BA, Al Saffar DM, Alyousef R (2020) "The utilization of recycled aggregate in high performance concrete: a review." JMR&T: 8469–8481
4. Xu J, Yao Z, Yang G, Han Q "Research on crumb rubber
5. Concrete: From a Multi-scale Review" (2019)
6. Siddique R, Singh M, Mehta S, Belarbi R (2020) "Utilization of treated saw dust in concrete as partial replacement of natural sand." J Clean Prod 261, https://doi.org/10.1016/j.jclepro.2020.121226
7. Kanchidurai S, Bharani G, Saravana Raja Mohan K (2017) "Strength and durability studies on concrete with partial replacement over burnt brick bat waste." IOP Conf Series: Earth Environ Sci 80(1). https://doi.org/10.1088/1755-1315/80/1/012018
8. Kaur JS, Ladhar A, Singh G, Singh R, Gill I, IT Assistance Dinesh Kumar, Pardeep Kumar E, Garg Er Mandeep Singh "Sponsored by Editorial Team Article Prepared by ConsultancyCell,PSCST."[Online]. Available: www.knowledgebank.irri.org
9. Kumar Das S, Adediran A, Rodrigue Kaze C, Mohammed Mustakim S, Leklou N (2022) "Production, characteristics, and utilization of rice husk ash in alkali activated materials: An overview of fresh and hardened state properties." Const Build Mater 345: 128341, https://doi.org/10.1016/j.conbuildmat.2022.128341
10. Jittin V, Bahurudeen A, Ajinkya SD (2020) "Utilisation of rice husk ash for cleaner production of different construction products." J Clean Prod 263. Elsevier Ltd, https://doi.org/10.1016/j.jclepro.2020.121578
11. Provis JL, Bernal SA (2014) Geopolymers and related alkali-activated materials. Annu Rev Mater Res 44:299–327. https://doi.org/10.1146/annurev-matsci-070813-113515
12. IS: 456 (2000) Plain and reinforced concrete—code of practice
13. IS: 15658 (2006) Precast concrete blocks for paving-specification

14. Ketov A, Rudakova L, Vaisman I, Ketov I, Haritonovs V, Sahmenko G (2021) "Recycling of rice husks ash for the preparation of resistant, lightweight and environment-friendly fired bricks." Const Build Mater 302, https://doi.org/10.1016/j.conbuildmat.2021.124385
15. IS: 10262 (1982). Indian Standard Recommended guidelines for concrete mix design. Bureau of Indian standards, New Delhi

A Comprehensive Analysis of Paver Block Using Ceramic Waste Powder (CWP) for Light Traffic Condition

Amit Mandal, Hemant Choudhary, and Sarvesh P. S. Rajput

Abstract Turbulent rise in construction activities resulted in excessive consumption of naturally available basic substance which are used for concrete manufacturing. The best possible route to get past this circumstance is to look for the alternatives. Construction activities and industrialization expanded and due to this the produced waste has also increased which are either landfilled or having no way to dispose or reuse. Some of them are fly ash, silica fumes, ground granulated blast furnace slag, limestone etc. This particular research paper emphasizes on the way of utilizing ceramic waste powder in concrete as paver block. The paver blocks are designed and strength criteria is satisfied according to the Indian standards for medium and heavy traffic conditions. The CWP is used in form of partial substitution of cement in concrete at the level of 0, 3, 6, 9. 12 and 15. The mechanical tests likewise compressive, flexural, split tensile strength, porosity and density etc. are performed to assess the feasibility of CWP in Concrete paver blocks.

Keywords Ceramic waste powder · Mechanical properties · Sustainability · Waste management

1 Introduction

The tile making industry and demolition activities are the two main sources which produce the ceramic waste powder (CWP). Basic raw substances like talk, potash, clay, dolomite, feldspar, and some chemicals likewise sodium silicate & tripolyphosphate are required to make ceramic tiles/any other ceramic products. Now, the next stage to grind or pulverise these raw substances, it is done by primary crushers and then particle size does the classification. Kiln with temperature 250–1250 °C is required to activate the pozzolana of the shredded raw material. Spray drying, forming, glazing and firing is then done. These processes provide many types of pollutants. It is estimated to generate about 15–25% waste during the manufacturing

A. Mandal (✉) · H. Choudhary · S. P. S. Rajput
Department of Civil Engineering, MANIT-Bhopal, M.P., India
e-mail: 213111009@stu.manit.ac.in

K. K. Pathak et al. (eds.), *Latest Developments in Civil Engineering*, Lecture Notes in Civil Engineering 352, https://doi.org/10.1007/978-981-99-2676-3_66

process. Production is approximately 200 million m^2 of tiles and at this moment very low amount of these waste is being utilized and rest of it is going to the dumping sites without any further possibility of reusing which creates pollution in considerable amounts. But if we talk about its properties, it is chemically and biologically resistant. Therefore, in all aspects of CWP utilization as a segment replacement (due to pozzolanic activity found) which in turns results in lowering the utilization of basic materials which further provides us a better environment with enhanced concrete properties and economical way of construction [1].

It has been noted that the wastes from ceramic industries is used in structural purpose earlier by replacing the coarse aggregate by the ceramic recycled aggregate which significantly increased the properties up to certain level of substitution [2]. The study shows that the durability performance increases while more substitution of cement replacement by CWP decreases the strength criteria [3]. The aggregate size greatly influences the strength properties of concrete pervious paver blocks. The increase in small pores reduces the compressive strength to a considerable amount [4]. The design of a pervious concrete is generally tougher due to the fact that the hydraulic properties are inversely proportional to the mechanical properties of the concrete [5]. Other elements such as very large parts of voids are mutually connected through each other which in turns reduces the strength parameter of the concrete, the quantity of those are approximately 20–30 percent [6]. So, while developing paver block all parameters are balanced in such a way that their proportion yields maximum requirement aspects.

One of the studies conducted by Messe Muenchen India in March 2018 by EAC (Euro Asia Consultancy) concluded that 26,000 crore present ceramic tile manufacturing will grow by 9% for the upcoming year. India is also the world's second largest manufacturer of ceramic tiles and produces about 1000 million m^2 of tiles per year. As per estimation approximately 20–22% of waste product is produced from the total production in India which is equal to 200–220 million square meters (Fig. 1).

2 Material & Methodology

The quantitative approach to find the required values regarding the proceeding experimentation is mentioned in the form of Fig. 2 as provided below.

2.1 Cement

In this research work Ordinary Portland cement 43 grade (Mycem branding) is used. The results of different tests on physical cement properties are given below in the Tables 1–3;

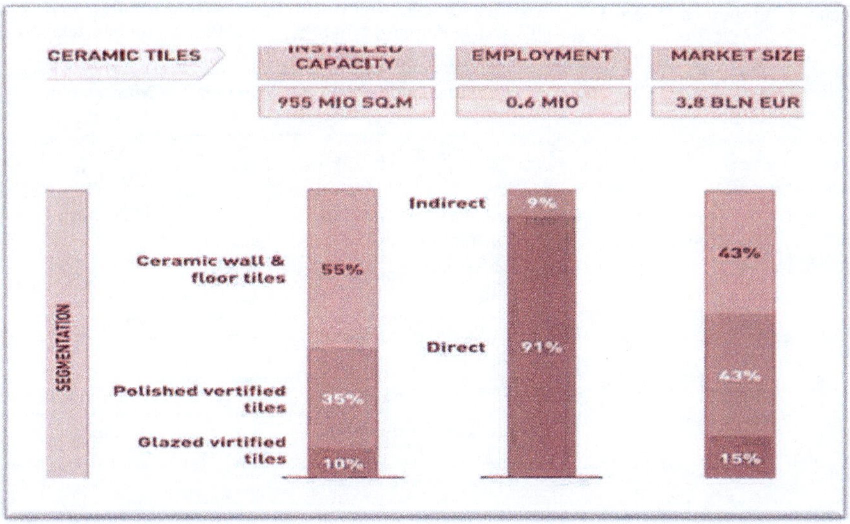

Fig. 1 Ceramic waste powder trend

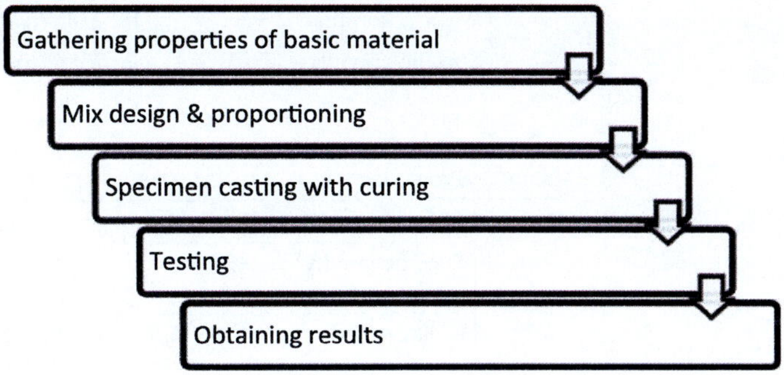

Fig. 2 Methodology

2.2 Fine Aggregate

See Table 2.

2.3 Coarse Aggregate

Ses Table 3.

Table 1 Properties of cement

S.no	Cement properties	Result	Indian standard requirement
1	Initial setting-time	50 min	Minimum of 30 min
2	Final setting-time	350 min	Maximum of 600 min
3	7 days Compressive strength	33 N/mm^2	43 N/mm^2
4	28 days Compressive strength	48 N/mm^2	22 N/mm^2
5	Standard consistency test	31%	–
6	Specific gravity	3.11	Minimum 3.15
7	Fineness modulus	8.3	Less than 10%

Table 2 Properties of fine aggregate

S.no	Properties	Test value	Code referred
1.	Fineness modulus	2.43	IS: 2386 (Part 1)-1963
2.	Silt content	3.8%	IS: 2386 (part 2)-1963
3.	Water absorption	3%	IS: 2386 (part 3)-1963
4.	Specific gravity	2.61	IS: 2386 (part 3)-1963

Table 3 Properties of coarse aggregate

S.no	Properties	Test value	Code referred
1	Fineness modulus	8.4	IS: 2386-part 1
2	Specific gravity	2.73	IS: 2386 (part 3)-1963
3	Bulk density	1650 kg/m^3	IS: 2386 (part 3) 1963
4	Voids	37.35%	IS: 2386 (part 3) 1963
5	Water absorption	9.20%	IS: 2386 (part 3) 1963

2.4 Ceramic Powder

Nearby tile making factories are the source for ceramic waste powder for this experimentation work (Table 4).

So, by this composition data it is quite clear that CWP is an aluminosilicate compound. We can deduce that greater silica content will result in greater cement strength but it will also delay the setting. The soundness will be also affected greatly due to lesser calcium oxides.

Table 4 Chemical properties of ceramic waste powder

Compounds in CWP	By weight percentage
Si-O_2	64.85
Al_2O_3	19.38
CaO	4.25
Fe_2O_3	4.11
K_2O	2.35
MgO	0.72
Na_2O	0.68
P_2O_5	0.15
Mn_2O_3	0.07
SO_3	0.11
Cl	0.003
Ignition losses	0.28

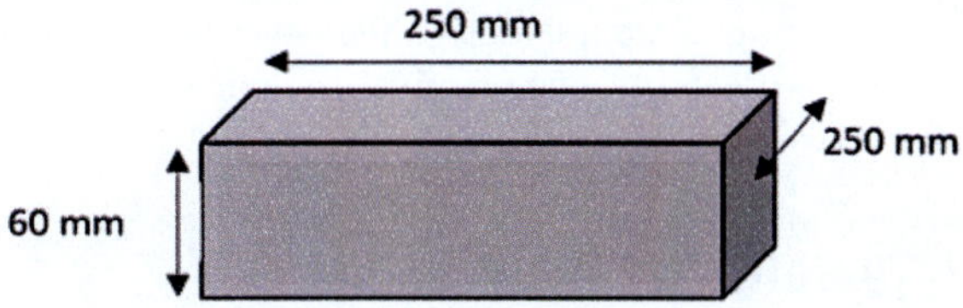

Fig. 3 Paver block

Further the Ceramic waste powder is tested for moisture content and it is about 38% by mass. Blain's air permeability apparatus is utilized to find the average specific surface area and it comes out to be 625 m^2/Kg.

3 Paver Block Design

The pervious paver block design is based on light traffic condition like, car parks, housing colonies, office complexes, farm houses, footways, beach sites etc. As per IS15658:2006 [7] the grade designation of the paver block for such traffic condition is M-35 and the recommended minimum paver block thickness is 60 mm (Fig. 3).

4 Mix Design

The mix design for the paver block is deduced from IS: 10262:2019 [8], According to the medium traffic conditions M-35 grade of concrete is selected for the mixes for which the target mean strength is given by;

$$f\prime_{ck} = f_{ck} + 1.65S = 35 + 1.65 \times 5 = 43.25\ N/mm^2$$

or

$$f\prime_{ck} = f_{ck} + 6.5 = 35 + 6.5 = 41.5\ N/mm^2$$

Taking maximum from above two values i.e. 43.25 N/mm^2 as the target mean strength for M-35 grade of concrete.

Mix proportion for trial;

Cement content = 416.05 Kg/m^3
Water content = 158.1 kg/m^3
Coarse aggregate = 1244.75 kg/m^3
Fine aggregate = 655.32 kg/m^3
Water-cement ratio = 0.38

According to the mix design the paver block samples are prepared and checked for their workability, compressive strength and flexural strength.

5 Result and Discussion

According to IS 15658:2006 ANNEXURE D, the specimen's apparent compressive strength is evaluated by dividing load in N (maximum) by the plan area in mm^2. Now to calculate the compressive strength (corrected), this compressive strength (apparent) is multiplied by the factor of correction from Table 5 of Indian standards 15658:2006. For 60 mm paver block thickness the correction factor for the plain block is 1.00 (Figs. 4 and 5).

Table 5 Mix proportion for different samples

Identification	Cement (kg/m^3)	River sand (kg/m^3)	Coarse aggregate (kg/m^3)	Percent substitution of cement (%)	Ceramic waste powder (kg/m^3)	W/C ratio
A	416.05	655.32	1244.75	0	–	0.38
B	403.57	655.32	1244.75	3	12.48	0.38
C	391.09	655.32	1244.75	6	24.96	0.38
D	378.61	655.32	1244.75	9	37.44	0.38
E	366.12	655.32	1244.75	12	49.93	0.38
F	353.64	655.32	1244.75	15	62.41	0.38

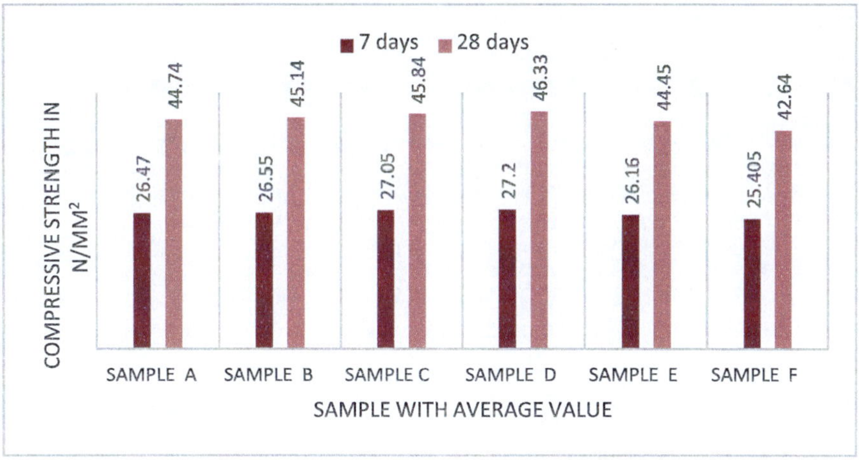

Fig. 4 Compressive strength of paver blocks

Fig. 5 Compressive and flexural test of paver block

5.1 Compressive Strength

See Tables 6 and 7.

Table 6 7 days compressive strength

Substitution level	Identification	Specimen's Compressive strength (N/mm^2)	Medial compressive strength (N/mm^2)	Standard Deviation	Variance
0%	A	27.1	26.471	1.105	1.222
		27.35			
		27.5			
		26.4			
		26.98			
		26.8			
		25.32			
		24.32			
3%	B	25.77	26.550	0.773	0.597
		27.32			
		27.98			
		25.98			
		26.32			
		26.77			
		26.41			
		25.85			
6%	C	27.42	27.045	1.092	1.192
		27.65			
		25.62			
		27.65			
		27.98			
		26.12			
		28.32			
		25.6			
9%	D	28.65	27.203	0.855	0.730
		27.95			
		27.41			
		26.45			
		27.32			
		26.41			
		26.11			
		27.32			

(continued)

Table 6 (continued)

Substitution level	Identification	Specimen's Compressive strength (N/mm^2)	Medial compressive strength (N/mm^2)	Standard Deviation	Variance
12%	E	26.32	26.156	0.406	0.165
		26.41			
		25.62			
		25.41			
		26.32			
		26.32			
		26.52			
		26.33			
15%	F	25.11	25.405	0.712	0.508
		26.25			
		26.41			
		25.32			
		24.65			
		24.65			
		24.88			
		25.97			

5.2 *Flexural Strength*

According to IS 456: 2000 [9], there are no direct ways to estimate the tensile-strength of concrete (Table 8). Because of that an indirect procedure is used i.e., flexural and slit tensile tests, and the approximate value of the tensile-strength is determined using compressive strength i.e.

$ft = 0.7\sqrt{35}$

Where,

f_t, f_{ck} = tensile strength & characteristic compressive strength respectively so,

$f_t = 0.7\sqrt{35} = 4.14\ N/mm^2$

From the results flexural strength is obtained prior to Twenty eight days of curing is 4.95 N/mm^2.

5.3 *Water Absorption*

For conducting the water absorption test 3 specimens are taken as per IS 15658:2006. In Table 4, these specimens are then completely immersed in water having room

Table 7 28 days compressive strength

Substitution level	Identification	Specimen's Compressive strength (N/mm^2)	Medial compressive strength (N/mm^2)	Standard Deviation	Variance
0%	A	45.62	44.741	0.714	0.509
		44.7			
		45.7			
		44.32			
		45.33			
		44.11			
		44.25			
		43.9			
3%	B	45.22	45.139	1.090	1.189
		45.65			
		46.98			
		44.82			
		43.26			
		44.32			
		45.74			
		45.12			
6%	C	46.5	45.841	0.741	0.549
		44.9			
		46.23			
		45.65			
		44.75			
		46.32			
		46.77			
		45.61			
9%	D	46.77	46.330	0.876	0.767
		46.85			
		47.12			
		45.62			
		46.32			
		46.16			
		47.2			
		44.6			
12%	E	45.6	44.449	1.258	1.582
		46.2			

(continued)

Table 7 (continued)

Substitution level	Identification	Specimen's Compressive strength (N/mm^2)	Medial compressive strength (N/mm^2)	Standard Deviation	Variance
		42.5			
		43.75			
		43.25			
		45.2			
		44.11			
		44.98			
15%	F	43.2	42.635	0.784	0.615
		41.65			
		42.35			
		43.98			
		42.77			
		42.5			
		41.65			
		42.98			

Table 8 7, 28 days flexural strength

Duration of testing after curing	Designation	Medial flexural strength in N/mm^2
7 days	FS7	3.2
28 days	FS28	4.95

Table 9 Water absorption data

Sample No	Ww	Wd	Water Absorption (%) for each specimen	Water Absorption (Average of 3 specimens)
1	8.88 kg	8.4 kg	5.7	
2	8.95 kg	8.56 kg	4.56	4.92%
3	8.6 kg	8.23 kg	4.49	

temperature of 24 ± 2 h. Then the specimens are removed and drained for 1 min after placing on a coarser wire-mesh (Table 9). After fully removing the water from the surface the specimens are weighed and water absorption percentage is found as;

$$W\% = \frac{W_w - W_d}{W_d}$$

Table 10 Abrasion resistance data

Sample no	Cantabro loss (%)	Abrasion resistance value (average)	Variance
1	35.2	34.1	0.95
2	33.6		
3	33.5		

5.4 Abrasion Resistance

For conducting the abrasion resistance test 3 specimens are taken as per IS 15658:2006 Table 4, these specimens are then tested via the Cantabro abrasion test (Table 10). The result of this test for the paver block is given by;

6 Conclusion

In this experimentation work ceramic waste powder as a part replacement for cement is used in producing concrete paver blocks (Fig. 6). Following were the results after experimentation;

1. The optimum percentage for M-35 grade concrete paver blocks is 9%.
2. The compressive strength and flexure strength for this replacement level are

$46.33\ \text{KN/mm}^2$ & $4.95\ \text{KN/mm}^2$.

3. Water absorption for this mix is about 4.92%.
4. Abrasion resistance for this mix is 34.1%.

From the results it can concluded evidently after the experimentation that ceramic waste powder can be effectively introduced to replace the cement to a certain extent for making the paver blocks. This is a very cost effective and environment friendly alternative or in other words we can say that it is a sustainable alternative.

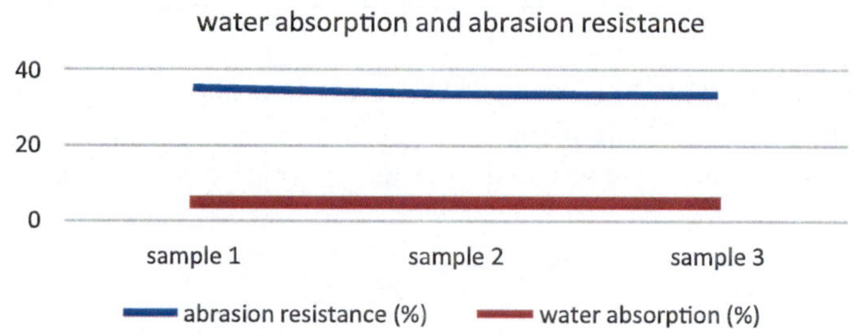

Fig. 6 Water absorption & abrasion resistance

References

1. Gautam L, Kumar Jain J, Alomayri T, Meena N, Kalla P (2022) "Performance evaluation of self-compacting concrete comprising ceramic waste powder as fine aggregate." Mater Today: Proc 61: 204–211, https://doi.org/10.1016/J.MATPR.2021.08.063
2. Ray S, Haque M, Soumic SA, Mita AF, Rahman MM, Tanmoy BB (Nov.2021) Use of ceramic wastes as aggregates in concrete production: A review. J Build Eng 43:102567. https://doi.org/10.1016/J.JOBE.2021.102567
3. Pacheco-Torgal F, Jalali S (Feb.2011) Cementitious building materials reinforced with vegetable fibres: A review. Constr Build Mater 25(2):575–581. https://doi.org/10.1016/J.CONBUILDMAT.2010.07.024
4. Yu F, Sun D, Wang J, Hu M (Jun.2019) Influence of aggregate size on compressive strength of pervious concrete. Constr Build Mater 209:463–475. https://doi.org/10.1016/J.CONBUILDMAT.2019.03.140
5. Kia A, Wong HS, Cheeseman CR (May2017) Clogging in permeable concrete: A review. J Environ Manage 193:221–233. https://doi.org/10.1016/J.JENVMAN.2017.02.018
6. ACI Committee 522 (2010) Pervious concrete (ACI 522R-10)
7. IS: 15658 (2006) Precast concrete blocks for paving-specification
8. IS: 10262 (1982) Indian Standard Recommended guidelines for concrete mix design. Bureau of Indian standards, New Delhi
9. IS: 456 (2000) Plain and reinforced concrete—code of practice

"Incorporation of Polythene for Improvement of Dense Graded Bituminous Macadam"

Dhanesh Khalotia, Shubham Tomar, and Rahul Saini

Abstract In the current observation explicit usage of non-bio degradable polythene is wastage from the household & pond liners. Polyethylene as Low Density Poly Ethylene & the cap of bottle is taken as the additional material in high density polyethylene in the construction of flexible pavement. A fixed % of zyco-therm will be added in this study. Reasons for utilisation of the mentioned material should be used earth non admissible unused material & to build up a high grade materials blend to anti-expanding traffic load & impact coming due to temperature alteration can damage the surface of asphalt layer. Examination of the proposed plastic waste will be cleaned & small size cutting with the end target through 2–3 mm strainer utilised the destruction machine. DBM Blends shall be presented to get ready for utilising the plain bituminous component as limited example & blended bitumen in Low Density PE = 2.0% & =4 .0% & High Density PE = 2%, = 3%, = 4%, = 5%) by weight, with permanent proportions of Zyco-Therm as 1.5% weight & 4.5% by weight for the above mentioned examples. Flow Test on Bitumen, mellowing pointing test & Penetration Test, were led to control & altered Dense Bitumen Macadam blends.

Keywords Marshall stability test · Low density PE · High density PE · Dense BM · Zyco-therm

1 Introduction

Bituminous altered polymer is emerging as a prominent developing material for stable asphalts. Since the majority of the polymer is plastic waste, the modified bitumen exhibits much better properties for street developments and waste made of plastics, which is generally considered a contaminated risk, and can be used in cycles. This helps to address the issue of contamination. When bitumen and other materials are combined and applied to a layer of granular, compacted material that is suitable for covering the subgrade, the result is known as asphalt. Polyethylene trash

D. Khalotia (✉) · S. Tomar · R. Saini
IES University, Bhopal 462044, India
e-mail: Dhaneshkhalotia@gmail.com

K. K. Pathak et al. (eds.), *Latest Developments in Civil Engineering*, Lecture Notes in Civil Engineering 352, https://doi.org/10.1007/978-981-99-2676-3_67

is frequently seen across the country and is responsible for a number of problems. Channels are blocked by polyethylene exempted items, which causes flash floods [1–4]. To solve this issue, non-biodegradable wastes must be used in the construction industry because they are long-lasting, don't decompose into manure, and cause severe environmental pollution when burned [5, 6].

(a) Low Density PE: Low Density Polyethylene (L.D.P.E) comprises of polythene from home & polyethylene ponder line which is collected from the nearby dump is sanitised & water & destroyed to the dimension of 2.00 mm to 3.00 mm [5–8].
   - 0.910–0.940 g/centimeter is the density.
   - 0.92 is the Specific Gravity.
   - 110 C is greater than the Melting temperature

(b) High Density PE: The High Density PE thickness is larger than the Low Density PE, High Density Polyethylene density is taken from 0.93 to 0.97 g/centimeter cubed melt temperature which could be higher than 130 Centigrade [6].

(c) Zyco-therm (ZT): The substances were added by the Zydex industry in Gujarat, India. When the humidity was stable at normal temperature and weight, the polymeric material was created and presented in the liquor. These substances are soluble in food.

65–75% of it is made up of (hydroxylalkyl) (alkoxy) (alkylsilyl) mixtures, 25–27% is benzyl alcohol, and 3–5% is ethylene glycol [5].

2 Gap Identification

The Technical authority and research scholars have worked hard to remove the un-environmental elements on the earth by using the non-biodegradable for the construction of a flexible pavement and by using the polyethylene in the bituminous mixture the following observations were drawn out [3]:

The experimental mixture made of the zyco-therm have workability higher than the conventional one. The problems like non recoverable deformation, crack, moisture damage are not prominently seen here. The marshal stability test and water resist property increases e because of it. In this Si-O-Si is the unique chemistry of the zyco-therm [5, 8].

3 Problem Identification

Through the following observations, the research team has attempted to determine the variation in bitumen samples caused by admixtures & the use of unusable materials:

When tested for performance, the mixture made with Zyco-Therm performed progressively enhanced than that of the basic specimen which is resistant to water abrasion, non-recover change, & cracks in the pavement. The specimen that was prepared using Zyco-Therm also demonstrated greater workability and compact ability than the orthodox mixture. A strong and intact chemical bond (Silicon O silicon native resemble bond) between the aggregate surface and the asphalt is now provided by Zyco-Therm chemistry of organo-silane. In a road that has been extended over time, the permanently present chemical bond activates superior moisture resistance, a complete coat of bitumen aggregates, compress equivalent, and blends a broad temperature zone [9]. Low Density Polyethylene reduces the voids that prevents bitumen from being oxidised by trapped air and absorbing moisture. It is aimed at demonstrating Marshall Stability Test numbers & larger water resistance. The polyethylene waste modified asphalt mix exhibits a higher ability to bind to the mix's properties, according to previous studies [6, 10–12].

4 Objectives

1. To research the effects of combining HMA with Low Density PE, High Density PE, and Zyco-therm.
2. In order to investigate the results of mixing HMA with low density, high density, and zyco-therm.
3. Make a decision to improve the properties of HMA using high- and low-density PE.
4. Calculate the mixture's usable percent of high- and low-density polyethylene [5, 13].

5 Literature Review

Mohd Rosliet Al 2017 It depicts the blacktop folio and mixing the silane-silane mixture Zyco-Therm. In this study, fasteners were blended with 0.1 wt% Zyco-Therm to improve blend creation at lower temperatures, purely to increase blend usefulness & compact ability. According to the entrance, rotational thickness, complex modulus, and stage edge, the characteristics of control black-top fasteners (60/

70 & 80/10 infiltration evaluation) and black-top folios consolidating 0.1% Zyco-Therm were taken into account. In this manner, barrel-shaped evidences made at specific tests require void of air and the temp to observe the presentation of fusing, Zyco-Therm with the control black-top blend. The evidence for Hamburg wheel following, roundabout elasticity (ITS), versatile modulus, dynamic downer, and moisture-induced harm was tested. When compared to the standard blend, examples organised using Zyco-Therm demonstrate higher usefulness & compact ability, according to compaction data obtained by compactor Servopak gyratory. According to the results of the blend performance tests, blends that were Zyco-Therm-arranged performed better in terms of protecting against dampness damage, lasting twisting, and breaking [12].

Dr. Malik Shoeb Ahmad 2014 Dense Graded Bitumen Macadam with LDFE Plastic jugs and recycled waste got (Low Density PE/P.W) convey packs that have been added to flexible asphalts as filler. The polyethylene waste has been cleared off and chopped into tiny size for the current examination, with the final goal being that it passes through a destroyer with a strainer of size 2–3 mm. As a limit, plain bitumen has been used to create the (DBM) blend in this experiment. Examples & BT blended varying degrees, such as 2, 4, 6, 8, 10 & 12% by weight (Low Density PE/PW). The Marshall Stability Tests were done using D.B.M blends with limits and adjustments. It's been taken into consideration that polyethylene unusable adjusted b/w blend exhibits better water resistance, soundness, thickness, and restricting properties. The current investigation will result in a reduction in street fixes and utilisation of plastic waste will assist in using less biodegradable unnecessary items. The test performance toy creator lead demonstrated the dependability of Dense BM blends to be improved altogether to expansion of poly-plastic squanders mixtures. The expansion of Low Density PE (P.W) reduces the air voids, preventing the ingestion of moisture. Marshall has improved as a result of this. Stability and respect. When compared to a blend mixed with plain bitumen, it has been observed that the strength estimates of the mixed altered show that plastic waste has been expanded fundamentally to the tune of 14% at 12% waste. It shows how the expansion of plastic waste has improved the blend's quality, which implies that taking plastic waste into account has increased the blend's thickness. Mass thickness for blending must also increase to account for plastic waste. The critical level of inactivity is estimated to be 12%, at which point the thickness must be at its maximum of 2.51 grammes per cubic centimetre, or about 25% larger than the thickness of the blend mixed with normal asphalt [1, 14].

Rohith N. J. Ranjitha 2013 Researchers have done work to make the observation to recognise the marshal property of Water Mix Asphalt that had been mixed with "Zyco-Therm" & Hot Mix Asphalt for (D.B.M.). The HMA received blending temperatures were 155 °C, 130 °C, and 115 °C, and the WMA received blending temperatures of 130 °C, 115 °C, and an additional material measure rate of one by 10% by weight of the fastener. The research facility study assumes that Marshall for the Water and Stability mix. The expansion added substance to a blend. The current investigation includes setting up and testing Marshall Research facility examples. HMA blend was tested at temperatures of one fifty five degree centigrade, one thirty degree centigrade, and one hundred fifteen degree centigrade, and Water Mix Asphalt blend was tested at one thirty degree centigrade and one hundred fifteen degree centigrade, with an added substance dose rate of 0.1 [15].

The Marshall Stability estimation of Hot Mix Asphalt examples produced at fifty five degree centigrade is highly valued. In contrast to H.M.A examples, the large Zyco-therm of the additional material measures the pace of point 1% y Weight of the folio that improved the solidity & Marshall Properties of Water Mix Asphalt examples set at one thirty degree Centigrade & one hundred fifteen degree Centigrade. The ideal folio proportion for the Hot Mix Asphalt blend at 155 °C was found to be 5.4% and for the WMA blend at 130 °C to be 0.1% Zyco-Therm should be found to be five point three seven percentage. This shows that the ideal folio element for a mixture of Hot Mix Asphalt and Wet Mix Asphalt varies depending on the temperature and added substance measurement rates.

Fernandez et al. (2008) the geometric evaluation of black closures modified with a polymer containing styrene- butadiene-styrene thermoplastic elastomeric (SBS) and modified folio properties were examined. By using fluorescence microscopy to determine the morphology, the rheological characteristics of PMB-SBS were investigated. The findings demonstrated how candy and shale oils affected the PMB's microstructure, accumulation quality, and adaptive viscosity. Shale oil could be utilised in place of sweet oil or even as a compatibiliser in this way without transferring ownership.

Sabina et al. (2009) studied how well modified plastic/polymer asphalt mixtures performed and found that delayed effects of balanced polythene's robustness and sustainability are 1.21 and 1.18 significantly greater in bituminous mixtures than standard mixtures. In order to test the introduction of 60/70 bitumen, 8 and 15% (by weight) of polyethylene bitumen is used. In contrast to the 8% modifier mixture, a balanced mixture of 15% polyethylene displayed only slightly reduced Marshall Stability properties.

Yousefi et al. (2009) It's been mentioned that polyethylene particles must combine to form larger particles because of surfaces, molecular forces of attraction, main tempering in the regulation process, and the presence of particles made using liquid bitumen segments. Polyethylene particles do not typically tend to break into asphalt. The author claims that the particles had sufficient energy at each location to move together, and hit the bitumen residual thin film that separated particles, beat the combination of polyethylene occurred particles, and resulted in the division of the polymer scale.

6 Methodology

The processes to be accompanied are in line the with Marshall Method A.S.T.M: D-1559. Marshall on 3 samples, tests are run for flow but also stability. For each sample of mix proportion with and without Zyco-therm and plastic, including both the entire specimen that has a constant bitumen content of 4.5%. Marshall Properties has lead with each mixed variable %, 2–5 of HDPE, and 2–4%. (Low Dense PE). A current investment will be made to observe how trying to add plastic (Low Density

Table 1 2% low density PE results

The value of Marshall stability test with 2% low density PE [6]		
Mix type	The value of Marshall stability (kilogram)	The value of Flow is (millimeter)
HMA mix with 2% (HDPE)	1560.00	= 3.5 mm
HMA-mix with 3% (HDPE)	1640.00	= 3.8 mm
HMA-mix with 4% (HDPE)	1700.00	= 4.1 mm
HMA-mix with 5% (HDPE)	1686.00	= 4 mm

Table 2 4% low density PE results

The value of Marshall stability test with 4r percent LDPE [5, 6, 10, 15–17]		
Type of mix	The value of Marshall stability (kilogram)	The value of flow is (millimeter)
Hot Mix Asphalt mix with 2% (HDPE)	1590.00	= 3.7
Hot Mix Asphalt mix with 3% (HDPE)	1685.00	= 3.9
Hot Mix Asphalt mix with 4% (HDPE)	1752.00	= 4
Hot Mix Asphalt mix with 5% (HDPE)	1690.00	= 4.2

PE & High Density PE) and merging it with Zyco-therm would then enhance the property of HMA [5, 8].

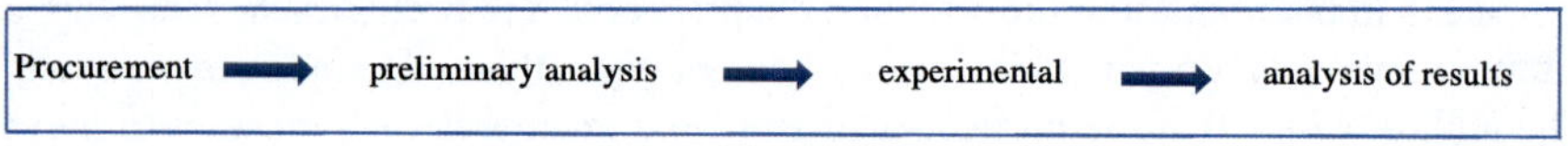

7 Results and Discussion

The finding of this research and the SF of a quantity of bituminous mixes are as follows (Tables 1–6). High Density PE% varies from two to five percent of weight, whilst Zyco-therm is also fixed at 1.5% and Low Density PE polythene at 2 and 4% of weight in large cases [5].

Table 3 Penetration test for 2% (low density PE)

The value of PENETRATION TEST 2% low density PE		
Sample-2	High density PE percent	Value (millimeter)
SAMPLE-1	2%	= 55.33
SAMPLE-2	3%	= 57.33
SAMPLE-3	4%	= 68.33
SAMPLE-4	5%	= 64.67

Table 4 Penetration test for 4% (low density PE)

The value of PENETRATION TEST 4% low density PE		
Samples	High density PE percent	Value (millimeter)
SAMPLE-1	2%	= 61.33
SAMPLE-2	3%	= 62.67
SAMPLE-3	4%	= 69.33
SAMPLE-4	5%	= 63

Table 5 Softening point (S.F) test

LDPE percent	HDPE percent	Marshal stability	The value of penetration (67.69)	The value of softening point (47.5 OC)
2% LDPE	2% HDPE	1560 kg	= 55.33 mm	= 54.1 OC
2% LDPE	3% HDPE	1640 kg	= 57.33 mm	= 54.7 OC
2% LDPE	4% HDPE	1700 kg	= 68.33 mm	= 55.1OC
2% LDPE	5% HDPE	1686 kg	= 64.67 mm	= 55.8 OC
4% LDPE	2% HDPE	1590 kg	= 61.33 mm	= 54.8 OC
4% LDPE	3% HDPE	1685 kg	= 62.67 mm	= 55.3 OC
4% LDPE	4% HDPE	1752 kg	= 69.33 mm	= 56.3 OC
4% LDPE	5% HDPE	1690 kg	= 63 mm	= 56.2 OC

Table 6 Cost analysis

Bitumen 100 percentage	Nos	Length (meter)	Breath (meter)	Depth meter	Quantity	Units	Rate (Rupees.)	Amount (Rupees.)
Providing & laying semi dense concrete with app batch type HMP using crushed aggregates of specified grading, premixed with bituminous binder @ four point five to five % of mix& filler, transport the hot mix to work site, laying with a hydrostatic paver finisher to the required grade, level & alignment, rolling with smooth wheeled, vibratory & tandem rollers to achieve the desired compaction as per MORTH specification clause No. 508 complete in all respects [5]	1	1000	5.50	0.02	110	Cumec		
Grading 2 layer Thick (25–30 mm) Bitumen (VG-30) [6]					= 17.38725	Cumec		
Quantity of Curves (As per AnnexureA1)					0			

(continued)

Table 6 (continued)

Bitumen 100 percentage	Nos	Length (meter)	Breath (meter)	Depth meter	Quantity	Units	Rate (Rupees.)	Amount (Rupees.)
Providing & laying semi dense concrete with app batch type HMP using crushed aggregates of specified grading, premixed with bituminous binder @ four point five to five % of mix& filler, transport the hot mix to work site, laying with a hydrostatic paver finisher to the required grade, level & alignment, rolling with smooth wheeled, vibratory & tandem rollers to achieve the desired compaction as per MORTH specification clause No. 508 complete in all respects [5]	1	1000	5.50	0.02	110	Cumec		
Extra Widen of CD portion					2.671875	Cumec		
Extra Quantity of Junction portion (As per Annexure-A3)					130.059125	Cumec	= 91880.00	= 1194983.241
As per 4% Low density PE & 4% High density PE					119.654395	Cumec	= 91880.00	= 1099384.581
Saving:-					10.40473	Cumec	= 91880.00	= 95598.65924

8 Conclusion

In this, we've demonstrated that using Low Density PE, High Density PE polythene, and a fixed amount of Zyco-therm enables us to boost the both specimen's property as well as its carrying capacity. The observation that can help in blocking the non-biodegradable materials is vital for road construction. Bitumen is flexible but macadam road life can waste shoot up [5, 13].

Test, done under given conclusions are:-

- The Marshall Test, which has been conducted on a bituminous mix with a mix of 4% Low Density PE, 4% High Density PE, as well as 1.5% Zyco-therm, does have a greater stability value of 1752 kg per equivalent to the number of flow 4.
- High Density PE is introduced into the mixture, growing it's own resistance to moisture and reducing its susceptibility to it.
- It is concluded that Zyco-therm has larger binding property of the mixed at a normal way [13].

This approach of economic construction is essential because it assists Zyco-therm decrease of BT in a mixture [13].

References

1. Dr. Malik Shoeb Ahmad (2014) Low density polyethylene modified dense graded bituminous macadam, Int J Eng Trends Technol (IJETT) 16(8)
2. Tariq Ali, Nouman Iqbal et al. (2014) Sustainability assessment of bitumen with polyethylene as polymer. IOSR J Mech Civ Eng (IOSR-JMCE) 10(5): 01–06, e-ISSN: 2278-1684, p-ISSN: 2320-334X
3. Gonzalez O, Munoz ME, Santamaria A (2006) Bitumen/polyethylene blends: using m-LLDPE to improve stability and viscoelastic properties. Rheol Act J 45(5):603–610
4. Navarro FJ, M., Partal, P., Garcia-Morales M., Martín-Alfonso M.J., Martinez-Boza F., Gallegos J.C., Gonzâlez N., Bordado M. and Diogo A.C. (2009) Bitumen modification with reactive and non-reactive (virgin and recycled) polymers: A comparative analysis. J Ind Eng Chem 15(4):458–464
5. Bindu Beena (2010) "Performance of low-carbon environmental warm mix asphalt". American Society of Civil Engineers"
6. Bindu C, Joseph Margret, Sibinesh P, George Shithin, Sivan Shyama (2020) Performance evaluation of warm mix asphalt using natural rubber modified bitumen and cashew nut shell liquid. Int J Pavement Res Technol
7. Albasheer et al (2011) "Evaluation of Aspha-Min zeolite for use in warm mix asphalt." National Center for Asphalt Technology Report 05-04. Auburn University, Auburn, Alabama
8. Alothman Dania, Gökçekuş Hüseyin, Ali Shaban (2022)
9. Gawande (2012) "Specifications for roads and bridge works"—2004. Fifth revision, Indian Roads Congress, New Delhi
10. Mutiu Akinpelu1, Bamidele IO Dahunsi, Oladipupo Olafusi, Olufemi S Awogboro (2013) Effect of polythene modified bitumen on properties of hot mix asphalt. ARPN J Eng Appl Sci 8(4), ISSN 1819-6608
11. Liliana MB Costa (2013) "Evaluation of the cecabase warm-mix additive". University Of Nevada Reno, in association with CECA Arkema Group, Nevada 89557

12. Mohd RosliMohd Hasan, TekSek Yee et al. (2017) "Performance characterizations of asphalt binders and mixtures incorporating silane additive ZycoTherm". School of Civil Engineering, UniversitiSains Malaysia (Engineering Campus), 14300 NibongTebal, Pulau Pinang
13. Singh H, Chopra T, Garg N, Singh M (2017) Effect of zycotherm additive on performance of neat bitumen and bituminous concrete mixes. Int J Civ Eng Technol 8:232–238
14. Malik Shoeb Ahmad (2014) Highway material testing (Laboratory Manual). Nemchand and Bros, Roorkee 1997
15. Rohith N, Ranjitha J (2013) "A study on Marshall stability properties of warm mix asphalt using zycotherm a chemical additive" Int J Eng Res & Technol (IJERT) 2(7), ISSN: 2278-0181
16. Khan Gundaliya (2012) "Laboratory evaluation of warm asphalt technology for use in Virginia." Virginia Transportation Research Council Report. In cooperation with the U.S Department of Transportation, VTRC 09-R11
17. Mohammed Sadequeet (2014) Mix design methods for asphalt concrete and other hot-mix types. Manual Series No. 2, 6th edn. Asphalt Institute, Lexington, Kentucky
18. Divya V, GyanenTakhelmayum (2017) "Evaluation of laboratory performance of sasobit and zycothermas an additives for warm mix asphalt on performance characteristic". Int J Adv Sci Res Eng (IJASRE) 3(1)
19. Johnson KwabenaAppiaha,*, Victor Nana Berko-Boatenga, Trinity AmaTagbor (2017) Use of waste plastic materials for road construction in Ghana. Council for Scientific and Industrial Research-Building and Road Research Institute, CSIR-BRRI, Kumasi, Ghana
20. Pe´rez-Lepe A., Martý´nez-Boza F.J., Gallegos C., Gonza´lez O., Mun˜o M.E. and Santamarý A. (2003) Influence of the processing conditions on the rheological behavior of polymer-modified bitumen. Fuel 82(11):1339–1348

Time and Cost Optimization of Unitized Façade Using Pull Planning Technique

Saimuddin Shaikh, Umesh Jadhav, and Fauwaz Parkar

Abstract Nowadays, the Unitized Façade system is becoming more important and popular due to its modern appearance and visual aesthetics in high-rise buildings. The entire building envelops quickly through unitized façade system which results in the external and internal environment getting separated. The Unitized Façade system has brilliant resistance to environmental factors such as general weathering, ultraviolet rays, and ozone. Utilizing the Unitized Façade system optimizes the number of labors required for the amount of manual handling at site in potentially hazardous conditions. Due to lack of technical knowledge and research, a limited number of construction professionals are familiar with their execution process. It has been observed that the project gets delayed due to improper execution work, lack of co-ordinations in labors, miscommunication between the stakeholders, consistent rework, mismatch of activity sequence, and overproduction of panels which directly affects the duration and cost of the project. The aim of this research is to optimize time and cost parameters by using the lean principle Pull Planning Technique in the Unitized Façade execution process. The pull planning technique is a practical approach of lean principles which focuses on the end goal rather than starting from the beginning and it works backward. This technique is a significant time and cost-saving process which reduces redundancy and eliminate unwanted delays. Lastly, Microsoft Project has been used for synchronizing all activities of the Unitized Façade execution process.

Keywords Unitized Façade System · Lean principles · Pull planning technique · Microsoft project · Baseline schedule · Critical path method

S. Shaikh (✉) · U. Jadhav · F. Parkar
Department of Civil Engineering, School of Engineering and Technology, AIKTC, Navi Mumbai, New Panvel 410206, India
e-mail: 20cem16@aiktc.ac.in

K. K. Pathak et al. (eds.), *Latest Developments in Civil Engineering*, Lecture Notes in Civil Engineering 352, https://doi.org/10.1007/978-981-99-2676-3_68

1 Introduction

In the past buildings, the entire structure load was supported to the exterior walls. The development of reinforced concrete and structural steel made it possible that small columns supported large loads, and the structural wall was no longer needed for structural support. Then the modern Façade system was born with the increased application of glass as an exterior façade. It is made up of glass and aluminum and silicon sealants act as a bond between them. As the Façade system is the exterior portion of the building which only carries their self-weight and wind pressure. Usually, the elements of the Façade system are fixed to the framed columns or building slabs with their primary role to separate internal and external environments. In high-rise buildings, transparent elements are becoming most important and popular due to their aesthetic appearance and superior visual. The Façade system is a huge investment in building construction which contributes about 20% of the total cost of the project [21].

Furthermore, the Façade system provides a protective cover between the building and the environment which is uncovered to environmental factors. These environmental factors comprise water, heat, fire, moisture, air, wind, sound, and seismic movements. There are various aluminum Façade systems concerning building envelopes such as Stick system and Semi-unitized system but due to the modern appearance and quick installation process especially in high-rise buildings, the Unitized Façade system is more prominent.

1.1 Unitized Façade Systems

The Unitized Façade system consists of a spandrel panel and glass vision panel attached to a prefabricated aluminum frame as shown in Fig. 1. As in this system, the entire panel is assembled in the factory under controlled working conditions with a high degree of precision and the process is inspected under deep supervision. Due to this, quality can be achieved in assembly and grant for fabrication lead-time. After that, these panels are transported to the construction site from a factory. Lastly, these panels are coupled together and installed with the least fieldwork. In this system, only 10% of the work is carried out at the site.

Furthermore, for installation, brackets are anchored to the slabs or columns after a detailed site survey; these glass panels are installed under deep supervision with minimum field labor and few joints. In this system, there are only three joints through each mullion and transom. There are two glasses for aluminum joints whereas a third joint is situated on the junction between half rails and half mullions. The manufacturer has to provide a serviceable joint system to evaluate the in-between panel joint of repair.

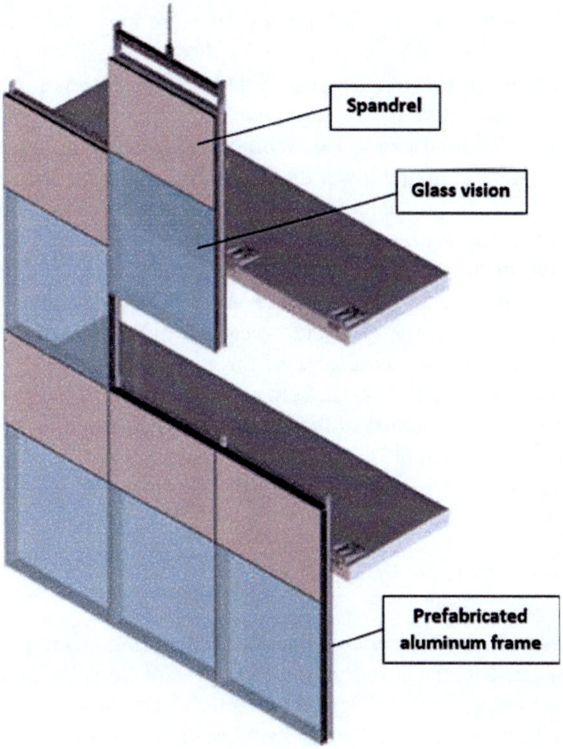

Fig. 1 Unitized Façade System [5]

2 Related Work

Pull scheduling minimizes delivery cycle times and improves cash flow [14]. The pull planning process assists the team members with collaboration, and commitment [18]. The collaborative pull planning technique supports giving feedback from the contractors to maintain the sequence of the schedule [20]. Lean principles refer to a technique to accept challenges in the installation process of a façade[10]. According toChung and Mutis, by applying the lean concept, the conventional quality control and quality assurance system in the preconstruction stage helps the façade detailing quality for the design team of façade[4]. The lean techniques improve schedule performance and the cost of projects by eliminating unwanted waste [1]. The Last Planner System reduces the construction complication of the project [17]. Manufacturing and design of adaptive facades, all stakeholders sign up for Life cycle assessment [11].

While for the Unitized façade mullion couple under wind actions, six full-scale tests are experimented [21]. Kinematic-based equations have derived to pinpoint key limits of Seismic Performance of Unitized Four-Sided Structural Glazing Systems [3]. Chen et al. state that the in-plane and out-of-plane distortion behaviour of

Unitized Curtain Wall has low-damage connection details [8]. MICMAC (Cross-Impact Matrix Multiplication Applied to Classification) method is used to determine various factors of the façade system for cost estimation [2]. Johnsa et al. experimented with the active rotary crane hook that gives high positional accuracy at the time of installation of panels [6]. While various installation methods such as Installation by the monorail, and tower crane are experimented [16].

A small excavator has robotic arm that holds the entire glass panel for installation [19]. Ilter et al. performed tests of wind resistance, water penetration, and air infiltration for design load, and seismic resistance requirements on the Unitized panel [9]. According to Kassem and Mitchel'sretrospective case studies, process mapping helped to identify issues in the early design phase of the façade system [12]. A model of the panel is demonstrated through Staad-pro to know the behavior of the wind load on panels [15]. Kassem et al. analysed a Decision Support System (DSS) that allows the selection of façade systems by examining the accuracy of the pre-selected façade system [13].

2.1 *Review of Previous Case Study on Critical Path Method*

In this study, the effectiveness of scheduling techniques of the Critical Path Method is used in the construction project. Specifically, in a project, the critical path is the longest path of dependent activities. Project managers can more easily arrange activities, create realistic schedules, keep projects within budget, and minimize delays by identifying the critical path. A critical path network has been made in which the start time and finished time of all the activities (A–V) are mentioned. The duration of each activity needs to be followed from start to finish. A duration of 61 days was planned for the project [7].

In Table 1, the duration of each activity was shown and the time estimates are calculated for the fabrication and construction of silo tanks. Totally, 23 activities have been required to complete the fabrication and construction of silo tanks i.e. starting from material inspection, allocation, and group project start-up meeting and it goes till load testing of signs, calibration of tanks, and handover. The earned value analysis based on an integrated time cost model had been applied to estimate the results for the Cost performance index (CPI), Schedule performance index (SPI), Cost variance (CV), and Schedule variance (SV).

As a result, it has been observed that 60% of project had been finished before the scheduled time, whereas 40% of the scheduled project was late from the scheduled time [7].

Table 1 Critical path table for fabrication and construction [7]

Critical path table for fabrication and construction of a 470 Barrel Silo Tank											
		Activity			Start and finish time (Days)				Floats		Critical
S/No	Job description	Code	Path	Duration	EST	EFT	LST	LFT	TF	FF	Path
1	Material Inspection, Allocation & Group Project Start-up meeting	A	1–2	3	0	3	0	3	0	0	x
2	Marking, Cutting/ Preparation of Plates for Rolling Shells, Domes & Cone	B	2–3	2	3	5	3	5	0	0	x
3	Movement of prepared plates to rolling mill	C	3–4	2	5	7	5	7	0	0	x
4	Marking, Cutting, Fitting & Tack welding of skid base	D	4–5	3	7	10	8	11	1	0	
5	Marking, Cutting, Drilling& Preparation of all Pad eyes	E	4–6	3	7	10	7	10	0	0	x
6	Full welding of Skid base	F	5–8	3	10	13	11	14	1	0	
7	Mark/Cut/Fit/Welding/ Prepare of skid unit of 4” air inlet Suction manifold	H	5–7	3	10	13	11	14	1	0	
8		–	7–8	0	13	13	14	14	1	0	
9	Mark, Cut, Drill, Fit, Weld & Preparation of 4 unit Air filter components	G	6–9	3	10	13	10	13	0	0	x

(continued)

Table 1 (continued)

Critical path table for fabrication and construction of a 470 Barrel Silo Tank

		Activity			Start and finish time (Days)				Floats		Critical
S/No	Job description	Code	Path	Duration	EST	EFT	LST	LFT	TF	FF	Path
10	Mark, Cut, Drill, Fit, Weld & Preparation of 1 unit of 24" air tight manhole	I	9–10	2	13	15	13	15	0	0	x
11	Mark, Cut & Prepare hinge pin & adjuster system with nuts & bolts	J	8–11	3	13	16	14	17	1	0	
12	Marking, Cutting, Drilling& Preparation of 6 unit of 4" clamps	K	10 –12	2	15	17	15	17	0	0	x
13		–	11 –12	0	16	16	17	17	1	1	
14	Mark, Cut, Fitting & Prefab Welding of 2 unit of 4" Inlet Charge line	L	12 –13	2	17	19	17	19	0	0	x
15	Delivery & tack weld of section of 2 dome, 1 cone & 4 cylindrical shells	M	13 –14	2	19	21	19	21	0	0	x
16	Fit/Tack/Full weld (Int. & Ext.) joints on big dome, 4 Cylindrical& cone	N	14 –15	15	21	36	21	36	0	0	x
17	Mark, Cut, Fit & Weld of 8" stanchion pipe & reinforcement to the body	O	15 –17	5	36	41	36	41	0	0	x
18	Marking, setting of skid accessories, welding & coupling onto stanchion	P	15 –16	3	36	39	38	41	2	0	

(continued)

Table 1 (continued)

Critical path table for fabrication and construction of a 470 Barrel Silo Tank

		Activity			Start and finish time (Days)				Floats		Critical
S/No	Job description	Code	Path	Duration	EST	EFT	LST	LFT	TF	FF	Path
19		–	16 –17	0	39	39	41	41	2	2	
20	Fit/Weld manhole, small dome, 4" discharge line, 4" Inlet, Filter, Nipples	Q	17 –18	5	41	46	41	46	0	0	x
21	Grind, Inspect/Correct weld flaws, install padeyes & clean debris	R	18 –19	3	46	49	46	49	0	0	x
22	Non-destructive testing of welded joints (MPI, DPI, Ultraflaw detection)	S	19 –20	3	49	52	49	52	0	0	x
23	Coupling accessories, pre pressure testing & final pressure testing	T	20 –21	3	52	55	52	55	0	0	x
24	Blasting, Coating (3 coat protection) & Inspection	U	21 –22	4	55	59	55	59	0	0	x
25	Load testing of Slings/ Shackle, Calibration of Tank & Handover	V	22 –23	2	59	61	59	61	0	0	x
	The Planned project duration is 61 days and the number of Nodes is 23										
	The table is based on activity—on—arc representation										

2.2 Complexities in the Traditional Critical Path Method (CPM)

The Critical Path Method (CPM) is used for scheduling construction projects. It is the traditional and frequent construction planning technique. Any construction project is required to be scheduled, and the CPM method does perceive in theory at the beginning of many construction projects. However, practically, CPM is not always a proper process for the project. Because the fact is that only one person decides the schedule for the entire team and project. Even for the most seasoned project managers, it is very difficult for the project manager to be the first stakeholder appointed to all schedules. It is tough to calculate the activity completion time and the critical path have to be estimated accurately but it is not distinct all the time in CPM. The critical path method cannot manage the schedule required for the resource allocation and usually, it does not give satisfactory results to the project. However, the pull planning technique is a substitute for the planning method which is achieving more grip in the construction industry. The main reason for the pull planning technique to have success is that it takes more input from all team members instead of one person.

3 Research Method

3.1 Lean Construction Principles

In construction, the term "Lean" introduces the combination of research and practical development to adopt Lean principles in the construction process. Lean also refers to improving and managing the process involved to complete any activity in construction by maximizing the value at minimal cost and taking into account the customer needs.

Lean construction is a method that helps construction industries to improve their process efficiency, productivity, and quality of work. This technique also reduces waste such as time, cost, labor hours, and rework.

Construction companies are more focused on the lean methodology which significantly maximizes customer value and minimize waste. The lean methodology can be used with ease in any industry where timespan, safety, and budget are all censorious, but as compared to the traditional construction methods, and project delivery is very different. Following are the 6 lean principles.

Identify Value from the Customer's Point of View. In the traditional perspective, construction focuses on identifying customer needs and whatever customer wants that is included in the specification and the plans. While in lean construction, the values of any customer are recognized which is deeper than that. In the planning phases of a project, a great different level of trust is established by a proper understanding of value from the customer's point of view.

In this principle, all stakeholders including the client, engineers, architect, contractor, and suppliers come to one platform. At that time all project teams also provide advice and helped to shape their expectations throughout the project.

Define the Value Stream. You can lay out all the essential processes when we get a clear understanding of the value from the customer's perspective. This process is known as the value stream. Necessary labour information, materials, and equipment are defined. Any unnecessary resources and steps which do not have value are removed.

Eliminate Waste. In lean principles, the first goal is to eliminate or minimize waste. There are eight types of waste in lean construction, such as unused talent, motion, defects, inventory, extra-processing, waiting, overproduction, and transportation.

Flow of Work Processes. In this principle, workflow is continuous and uninterrupted which is predictable and reliable. In the construction process, the sequence is the key; for example, the construction of any building cannot begin until the footings are laid. To achieve flow, clear communication is always essential between all parties. When some portion of the construction gets behind or ahead of schedule, then everyone must know for doing necessary adjustments to avoid excess inventory, waste of waiting, and motion.

Pull Planning and Scheduling. Based on downstream demand, reliable workflows depend on and are set according to that. Performance of work is recognized by lean construction and planning is made by pulling the schedule. Participants communicate and collaborate together by conducting meetings to determine the schedule network for tasks, materials, and labour.

Continuous Improvement. In this principle, continuously improving the processes and removing waste is the heart of the lean philosophy. Opportunities for improvement are picked out and pursued during the project and implemented in future projects.

The construction industry wants new methods for improvement, but it does not have the liability to stick to old ways and withstand change. Concerning focus end delivery of project which matches the need of the customer that is depending on the overall communication of team members. When projects come within budget and time and everyone are involved, that is better to fulfil the expectation of the customer.

3.2 Pull Planning Technique

The pull planning technique focuses on the end goal rather than starting from the beginning. Team members start with the end target of the project and work backward through the workflow schedule. In this manner, a well-defined workflow is developed and a clear plan is set for the pull planning. This technique substantially eliminates waste of rework [14].

The pull planning technique uses a handwritten project outline and coloured sticky notes to visually outline the workflow of the project. Each sticky note represents any necessary part of the project: "Installing the panels" for example. We require a huge space to perform the work on the timeline of the project. The use of notes is the simplest way for all team members to make their changes wherever needed. To know if there are any additional resources required to complete the activity. It will help to get the job done faster and more resourcefully. Pull scheduling can be controlled and found out Work In Progress (WIP) of the project by roughly multiplying the necessary output rate with respect to the overall project schedule [14].

In this technique, the timeline is drawn in a weekly manner and all activities must follow the sequence. A clear workflow is laid out, and team members should know what needs to be completed before the panel installation. Every person can easily determine the urgency of each activity and the current status of any activity involved in the project.

To identify any potential problems ahead of time, a collaborative environment allows the expertise of every member to make it possible. Because of this, the pull planning technique conserves both cost and time in the project by reducing waste and redundancy.

There are main three core rules in pull planning which use a straightforward format that assists in the process. Those three rules are "Pull," "Collaborate," and "Commit" continually used in this method. This system is known as pull planning and it will allow us to consistently be on schedule.

Pull. This is the primary step and it is most crucial because you will have to work backward from the end of the project and determine the sequence of each activity for avoiding unwanted delays. A Group of staff is gathered to ensure they will provide their input and collaborate seriously and work together on status updates along the way.

Pull scheduling is focused on the end goal instead of the beginning and identifies each predecessor of all activities in the proper sequence of events. This process helps to pinpoint any activity handoffs from the back end. To keep the project rolling, everyone on the project must know when and what needs to be done. It works only due to all team members understanding every project detail which results in a more fluid construction schedule.

Collaboration. The pull planning technique requires more collaboration than any existing construction technique. Every person works together as a team so that each stakeholder of the project acknowledges what is essential for the accomplishment of the work. The entire team members should think about what important steps to be taken next, as everyone are able to add their input at any time.

It is crucial to ensure any information added is comprehensive and detailed. Everyone can add their notes into the schedule, if you are not detailed and comprehensive, no one will understand what the note means. For example, a note with the "Quality control and Quality assurance" can have many meanings of different things. But, when the quality person guides the team on exactly which panels are going to be checked, they would efficiently learn whether or not the quality checking of panels is being checked on the 17 floors or another floor level.

This would grant all stakeholders to recognize what areas are not allowed to other workers for the time period of the quality checking of panels because of this everyone is doing their own jobs without clashing with one another. All stakeholders could permit the quality person if they want to delay the quality checking process for some days that are still in the process of completion in that particular area. This will ensure your project is moving smoothly and subcontractors are not trying to work in their way.

Commitment. Commitment is the third rule of the pull planning technique and it is the most essential of them all. In this rule, any of the stakeholders cannot move another note of the stakeholders. No one has the right to adjust or alter someone else's notes or ideas.

Usually, everybody thinks that their work should be completed first, but that cannot happen in the construction industry. Everybody has to follow the proper guidelines for that specific work that is in place and if it is done properly, then the project progress will not be affected and no work will have to be redone. As all stakeholders are involved and committed to flow the pull planning schedule and doing their particular work on the project within the stipulated time period. Successful application of pull planning requires commitment and discipline from every member [18].

The timeline of the project can be significantly delayed due to moving notes. Instead, everyone moves their notes, works together, and discusses which activity needs to be done before the rest and which is the best way that will improve the project's progress.

3.3 The Consistent Process Involved in Pull Planning Technique

In Fig. 2, the following process is involved which is consistently used during pull planning technique.

- To entail every important team member with the aim that all members shall contribute their relevant data regarding your project completion.
- Establish milestones which are primly important to the completion of various project phases and to guarantee proper and undisrupted delivery of your project. Set the milestones and follow them. When everything is in place, now you can manage workflows and plan your various phases including activities towards the attainment of those milestones.
- Schedule your activities in proper sequence. After recognizing your milestones, all confirmed activities must fix in your calendar and as per the planned activities, durations are set.
- Ascertain how you can itemise your activities into weekly work plans for getting a clear understanding to other team members.
- Weekly meetings are organized that gives the facilitator to review the weekly work plans and do proper adjustments when required. The facilitator is also able to scrutinize all finished activities to be right on course.

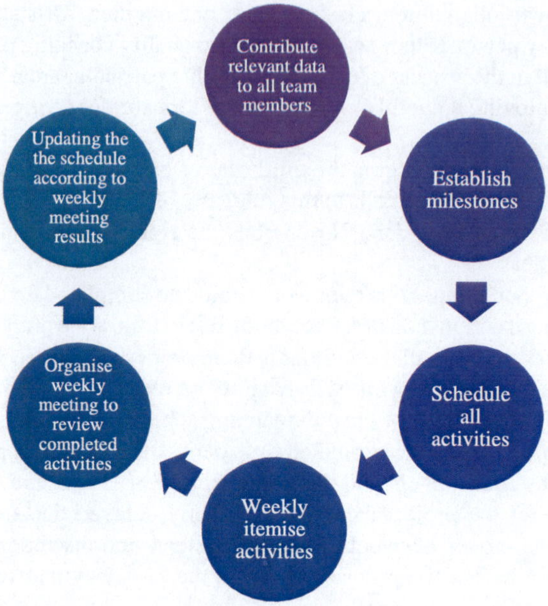

Fig. 2 Pull planning technique consistent process

- Updating the schedule according to weekly meeting results. It would provide you freedom for any adjustments to the overall schedule of the project.

4 Data Collection

4.1 Case Study

A Unitized Façade construction site has been selected for utilizing pull planning technique in the execution phase that is situated in Mumbai. The site is selected to know the issues related to the execution process of Unitized Façade work. Initially, the execution process of the Unitized Façade work has closely monitored. At the site, coordination has been done with the project manager, project planner, site engineer, and sub-contractors, and also attend weekly progress meetings by sharing the responsibility of the researcher.

A total of 5 floors of activity of Unitized Façade panels has been observed starting from site clearance up to the last activity quality check and all this work takes to 49 days to complete. Many meetings have been arranged with all stakeholders to give knowledge of how the pull planning technique works and as per the pull planning process, sticky notes were used to putting activity details of all stakeholders. Sticky

Table 2 Issues identified in the execution process

Execution process	Problems
Survey	Needs more persons to manage survey activities
Transportation	Access transportation flow and mismatch of panel sequence
Site logistic	Need to improve the unloading method and panels need to be stored nearly from where it is going to lift to their predetermined floor
Monorail and cradle installation	MS material not shifting on the floor as per the installation sequence, unavailability of labors, staff, safety person, and unnecessary movement for obtaining tools and machines
Panel preparation and Super Deck installation	Delay in panel preparation due to improper reading of drawing and miscommunication between site supervisor and site engineer
Panel shifting at level	A false sequence of panels shifting at the level and lack of co-ordinations between junior engineer/site supervisor and labors
Bracket and Panel installation	Consistently rework, need of assigning safety person and shortage of skilled labors
Quality control and quality assurance checking	Delay in checking the panels and documentation work

notes have been prepared for the workflow schedule on a weekly basis and was informed to all stakeholders to follow the same for the next five floors.

4.2 *Issues Identified in the Execution Process*

In Table 2, before implementing the pull planning technique, the following issues have been identified by observing every activity in detail. These issues are eliminated through this technique as the pull planning technique optimizes redundancy, waste of waiting, consistent rework, and reduced unwanted delays.

4.3 *Utilizing Coloured Sticky Notes in Unitized Façade Execution*

The main concern of the pull planning technique is to use coloured sticky notes for visualizing the workflow of the project and to see the big picture to all stakeholders of the schedule. This optimize redundancy, waste of waiting, and consistent rework as potential problems are identified upfront. In Fig. 3, the workflow schedule has been prepared on a weekly basis.

In this research, sticky notes are used to plot and identify specific activities of each stakeholder along a timeline on weekly basis. When the stakeholders identify and communicate their activity with other, each stakeholder can easily be able to visualize

Fig. 3 Workflow schedule showing weekly plan

the sequence and dependencies of these activities, as well as they can identify any possible overlap that could cause delays. This schedule is also considered the safety aspect to avoid any incident due to overlap of activity.

5 Data Analysis

5.1 Creating Baseline Schedule Before Use of Pull Planning Technique

Before preparing the baseline schedule, various meetings have been taken with each stakeholder, and also the site observation has been monitored of all activities. For the analysis of data, Microsoft Project software has been used. This baseline schedule has been prepared under the guidance of the project manager and project planner before using pull planning technique. In Fig. 4, all activities of the Unitized Façade are prepared for five levels i.e. Level 12 to 16.

ID		Task Mode	Task Name	Duration	Start	Finish	Predecessors	Resource Names	Cost
1	✓		**Execution**	**49 days**	**Thu 13-01-22**	**Thu 10-03-22**			**₹ 1,768,340.00**
2	✓		Project Manager	49 days	Thu 13-01-22	Thu 10-03-22		Project Manager	₹ 246,960.00
3	✓		**Site clearance form client**	**10 days**	**Thu 13-01-22**	**Mon 24-01-22**			**₹ 0.00**
4	✓		Level 12	2 days	Thu 13-01-22	Fri 14-01-22			₹ 0.00
5	✓		Level 13	2 days	Sat 15-01-22	Mon 17-01-22	4		₹ 0.00
6	✓		Level 14	2 days	Tue 18-01-22	Wed 19-01-22	5		₹ 0.00
7	✓		Level 15	2 days	Thu 20-01-22	Fri 21-01-22	6		₹ 0.00
8	✓		Level 16	2 days	Sat 22-01-22	Mon 24-01-22	7		₹ 0.00
9	✓		**Survey**	**15 days**	**Sat 15-01-22**	**Tue 01-02-22**	**4**		**₹ 12,960.00**
10	✓		Level 12	3 days	Sat 15-01-22	Tue 18-01-22	4	Surveyor,Surveyor Assist	₹ 4,320.00
11	✓		Level 13	3 days	Wed 19-01-22	Fri 21-01-22	10	Surveyor Assistant	₹ 1,440.00
12	✓		Level 14	3 days	Sat 22-01-22	Tue 25-01-22	11	Surveyor,Surveyor Assist	₹ 4,320.00
13	✓		Level 15	3 days	Wed 26-01-22	Fri 28-01-22	12	Surveyor Assistant	₹ 1,440.00
14	✓		Level 16	3 days	Sat 29-01-22	Tue 01-02-22	13	Surveyor Assistant	₹ 1,440.00
15	✓		**Transportation**	**17 days**	**Sat 22-01-22**	**Thu 10-02-22**	**14FS-9 days**	**Per Truck 7 Panels[50]**	**₹ 290,000.00**
16	✓		**Site Logistic (Panel unloading at stacking area)**	**17 days**	**Mon 24-01-22**	**Wed 16-02-22**	**15FS-16 days**	**Per Truck Mathardi Unloading fix charges + Site supervisor**	**₹ 154,700.00**
17	✓		**Monorail installation**	**20 days**	**Sat 15-01-22**	**Mon 07-02-22**	**4**	**Safety Person ,Electriciar**	**₹ 236,760.00**
18	✓		**Cradle installation**	**9 days**	**Fri 28-01-22**	**Mon 07-02-22**	**17FS-9 days**	**Electrician,Safety Person**	**₹ 85,320.00**
19	✓		**Panel preraration**	**10 days**	**Thu 27-01-22**	**Mon 07-02-22**	**18FS-10 days**		**₹ 67,400.00**
20	✓		Level 12	2 days	Thu 27-01-22	Fri 28-01-22	18FS-10 days	Jr. Engineer ,Site Supervis	₹ 16,160.00
21	✓		Level 13	2 days	Sat 29-01-22	Mon 31-01-22	20	Site Supervisor ,3 Skilled	₹ 12,160.00
22	✓		Level 14	2 days	Tue 01-02-22	Wed 02-02-22	21	3 Skilled Labour ,5 Unskill	₹ 14,360.00

Project: msproj11
Date: Fri 09-09-22

Task | Split | Milestone | Summary | Project Summary | External Tasks | External Milestone
Inactive Task | Inactive Milestone | Inactive Summary | Manual Task | Duration-only | Manual Summary Rollup | Manual Summary
Start-only | Finish-only | Deadline | Progress | Manual Progress

Page 1

Fig. 4 Baseline schedule

5.2 Creating an Actual Schedule After Utilizing Pull Planning Technique

The actual schedule has been prepared after utilizing pull planning technique for the next five levels i.e. Level 17 to 21. In Fig. 5, every activity is monitored precisely to know how much improvement could achieve at the site by following this technique.

6 Results

6.1 Comparison Between Baseline and Actual Schedule

In Table 3, duration and cost comparison has shown of all activities that are required to complete the execution process of Unitized Façade work. There is no change in site clearance activity in duration and cost, as site clearance has been done from the client side.

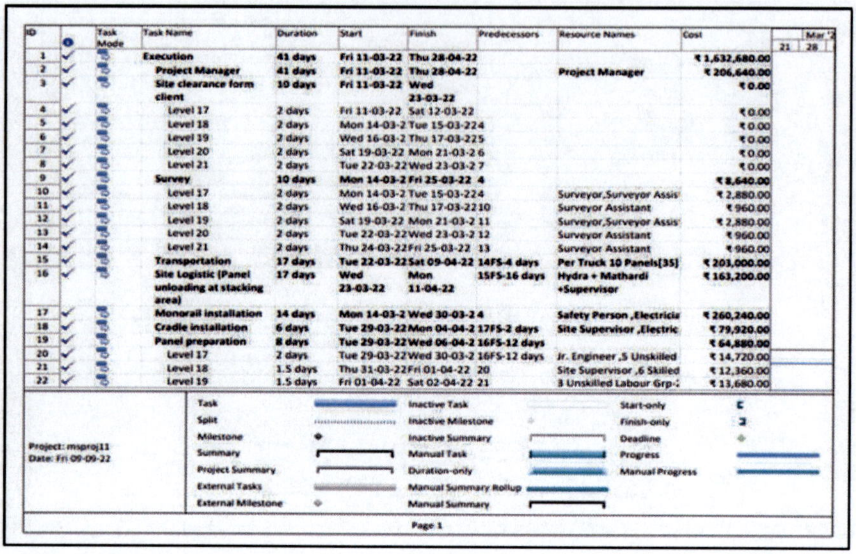

ID		Task Mode	Task Name	Duration	Start	Finish	Predecessors	Resource Names	Cost
1	✓		**Execution**	**41 days**	**Fri 11-03-22**	**Thu 28-04-22**			**₹ 1,632,680.00**
2	✓		**Project Manager**	**41 days**	**Fri 11-03-22**	**Thu 28-04-22**		**Project Manager**	**₹ 206,640.00**
3	✓		**Site clearance form client**	**10 days**	**Fri 11-03-22**	**Wed 23-03-22**			**₹ 0.00**
4	✓		Level 17	2 days	Fri 11-03-22	Sat 12-03-22			₹ 0.00
5	✓		Level 18	2 days	Mon 14-03-2	Tue 15-03-22	4		₹ 0.00
6	✓		Level 19	2 days	Wed 16-03-2	Thu 17-03-22	5		₹ 0.00
7	✓		Level 20	2 days	Sat 19-03-22	Mon 21-03-2	6		₹ 0.00
8	✓		Level 21	2 days	Tue 22-03-22	Wed 23-03-2	7		₹ 0.00
9	✓		**Survey**	**10 days**	**Mon 14-03-2**	**Fri 25-03-22**	**4**		**₹ 8,640.00**
10	✓		Level 17	2 days	Mon 14-03-2	Tue 15-03-22	4	Surveyor,Surveyor Assis	₹ 2,880.00
11	✓		Level 18	2 days	Wed 16-03-2	Thu 17-03-22	10	Surveyor Assistant	₹ 960.00
12	✓		Level 19	2 days	Sat 19-03-22	Mon 21-03-2	11	Surveyor,Surveyor Assis	₹ 2,880.00
13	✓		Level 20	2 days	Tue 22-03-22	Wed 23-03-2	12	Surveyor Assistant	₹ 960.00
14	✓		Level 21	2 days	Thu 24-03-22	Fri 25-03-22	13	Surveyor Assistant	₹ 960.00
15	✓		**Transportation**	**17 days**	**Tue 22-03-22**	**Sat 09-04-22**	**14FS-4 days**	**Per Truck 10 Panels[35]**	**₹ 203,000.00**
16	✓		**Site Logistic (Panel unloading at stacking area)**	**17 days**	**Wed 23-03-22**	**Mon 11-04-22**	**15FS-16 days**	**Hydra + Mathardi +Supervisor**	**₹ 163,200.00**
17	✓		**Monorail installation**	**14 days**	**Mon 14-03-2**	**Wed 30-03-2**	**4**	**Safety Person ,Electricia**	**₹ 260,240.00**
18	✓		**Cradle installation**	**6 days**	**Tue 29-03-22**	**Mon 04-04-2**	**17FS-2 days**	**Site Supervisor ,Electric**	**₹ 79,920.00**
19	✓		**Panel preparation**	**8 days**	**Tue 29-03-22**	**Wed 06-04-2**	**16FS-12 days**		**₹ 64,880.00**
20	✓		Level 17	2 days	Tue 29-03-22	Wed 30-03-2	16FS-12 days	Jr. Engineer ,5 Unskilled	₹ 14,720.00
21	✓		Level 18	1.5 days	Thu 31-03-22	Fri 01-04-22	20	Site Supervisor ,6 Skilled	₹ 12,360.00
22	✓		Level 19	1.5 days	Fri 01-04-22	Sat 02-04-22	21	3 Unskilled Labour Grp-:	₹ 13,680.00

Fig. 5 Actual schedule

Table 3 Comparison between baseline and actual schedule

		Baseline Schedule		Actual Schedule	
Sr. no	Activities	Duration (Days)	Cost (INR)	Duration (Days)	Cost (INR)
A	Project Manager	49	₹2,46,960	41	₹2,06,640
1	Site clearance from client	10	₹0	10	₹0
2	Survey	15	₹12,960	10	₹8,640
3	Transportation	17	₹2,90,000	17	₹2,03,000
4	Site Logistic (Panel unloading at stacking area)	17	₹1,54,700	17	₹1,63,200
5	Monorail installation	20	₹2,36,760	14	₹2,60,240
6	Cradle installation	9	₹85,320	6	₹79,920
7	Panel Preparation	10	₹67,400	8	₹64,880
8	Super Deck installation	15	₹25,200	10	₹16,800
9	Bracket installation	20	₹1,20,000	10	₹1,32,800
10	Panel shifting at level	15	₹2,46,960	10	₹2,19,680
11	Panel installation	20	₹1,98,080	15	₹2,09,520
12	QA/QC Checks	19	₹84,000	14	₹67, 360
13	Overall Execution	Total = 49	Total = ₹17,68,340	Total = 41	Total = ₹16,32,680

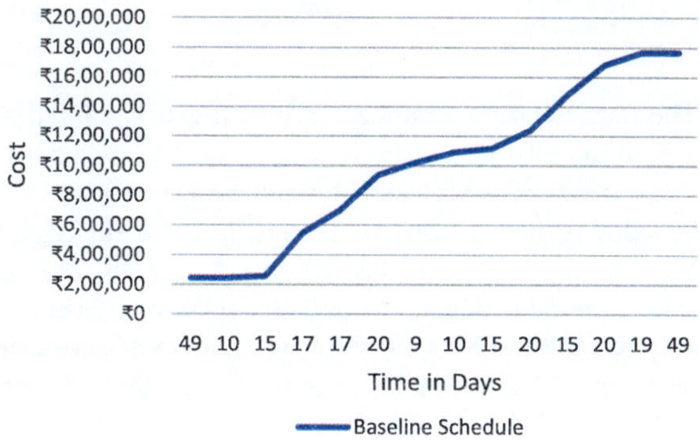

Fig. 6 Baseline schedule time and cost graph

6.2 *Baseline Schedule*

In Fig. 6, the graph shows the relation between time and cost of the baseline schedule activities. Cost of all activities ranges from ₹12,960 to ₹2,90,000.

6.3 *Actual Schedule*

In Fig. 7, the graph shows the relation between time and cost of the actual schedule activities. Cost of all activities ranges between ₹8,640 and ₹2,60,240.

From Table 3, Figs. 6, and 7,it has been observed that there is a drastically reduced time of 8 days whereas the overall cost of ₹1,35,660 of Unitized Façade execution work is significantly saved.

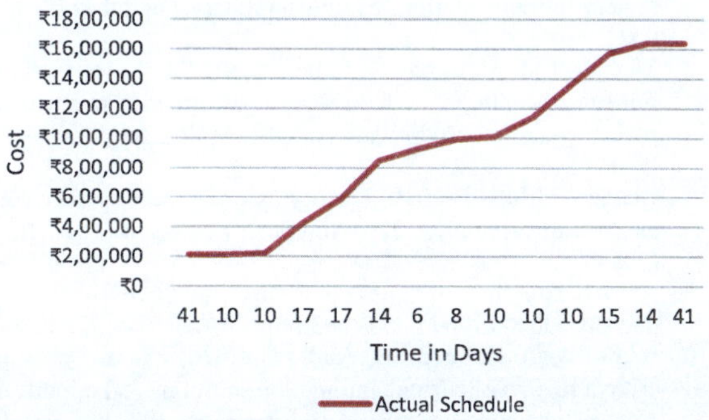

Fig. 7. Actual schedule time and cost graph

7 Discussion

The pull planning technique brings together all stakeholders on the same platform to pull the schedule, to collaborate, and have commitment with each other. All team members discuss between planned work and actual work [18]. Sticky notes help for visualizing the workflow of the schedule and identifying issues related to any activity. Work is managed by following a workflow schedule and resources are optimized in a well-defined manner to pull the schedule. Every team member is continuously engaged to do their work on time. Due to effective collaboration and commitment between all stakeholders, organizational culture becomes more powerful.

8 Conclusion

In this research, the pull planning technique has been utilized to optimize the time and cost of Unitized Façade execution work. This technique is a more accurate and systematic process with significant time and cost-saving. It effectively manages all activities, optimizes resources, and eliminates consistent rework, waste of waiting, and redundancy. This researcher has deeply monitored all activities and conducted weekly meetings to review completed activities. For analysis of data, Microsoft Project Software has been used for preparing the baseline and actual schedule. Results indicate that 16% duration has been reduced while up to 8% cost of Unitized Façade execution work has been drastically saved. It has been noticed that the pull planning technique is successful in minimizing the façade construction complexities during the execution of the project.

References

1. Deshpande AS, Filson LE, Salem OM, Miller R (2012) Lean techniques in the management of the design of an industrial project. Journal of Management In Engineering, ASCE 28:221–223
2. Le´sniak A, and Górka M (2020) Structural analysis of factors influencing the costs of facade system implementation. Multidisciplinary Digital Publishing Institute (MDPI), ISSN 1017–6021
3. Memari AM, Simmons N, Solnosky RL (2021) Derivation of kinematic equations based on full-scale racking tests for seismic performance evaluation of unitized four-sided structural sealant glazing curtain wall systems. Multidisciplinary Digital Publishing Institute (MDPI), ISSN 11–593 (2021)
4. Chung A, Mutis I (2019) Quality assurance and quality control of high-rise enclosure design using lean principles. The American Society of Civil Engineers, ASCE, ISSN 1084–0680 (2019)
5. Archi Expo by virtual expo group. https://www.archiexpo.com/prod/european-facade-products/product-149694-1711083.html
6. Johnsa B, Arashpourb M, Abdia E (2020) Curtain wall installation for high-rise buildings: critical review of current automation solutions and opportunities. International Symposium on Automation and Robotics in Construction (ISARC), pp. 393–400 (2020 Proceedings of the 37th ISARC, Kitakyushu, Japan, ISSN 2413–5844

7. Amade B, Achaka CH, Ubani EC (2013) An assessment of the effectiveness of scheduling techniques on the success of mechanical construction projects. In: International Journal of Emerging Trends in Engineering and Development, Issue 3, Vol.1, ISSN 2249–6149
8. Chen C, Xiang P, Jia L, Zhao X, Dhakal R, MacRae GA, Bhatta J, Clifton C (2020) Seismic behavior of unitized glazed curtain walls in a low-damage structural steel building. In: The 17th World Conference on Earthquake Engineering (2020)
9. Ilter E, Tavil A, Celik OC (2015) Full-scale performance testing and evaluation of unitized curtain walls. J Facade Des Eng, ISSN 2214–302/15 (2015)
10. Friblick F, Tommelein ID, Mueller E, Falk JH (2009) Development of an integrated façade system to improve the high-rise building process. In; 17th Annual Conference of the International Group for Lean Construction, pp. 359–370 (2009)
11. Crespi M (2021) Design and manufacturing of adaptive facades in a life cycle approach: a survey on challenges and solutions in the Italian Building Industry. Multidisciplinary Digital Publishing Institute (MDPI) 1(2):69–82
12. Kassem M, Mitchel D (2015) Bridging the gap between selection decisions of façade systems at the early design phase: Issues, challenges and solutions. J Façade Des Eng, ISSN 2214–302(15):165–183
13. Kassem M, Dawood N, Mitchell D (2012) A decision support system for the selection of curtain wall systems at the design development stage. J Constr Manag Econ, ISSN 0144–6193 print/ ISSN 1466–433:1039–1053 (2012)
14. Sacks R, Goldin M (2007) Lean Management Model for Construction of High-Rise Apartment Buildings. J Constr Eng Manag, ASCE, ISSN 0733–9364:374–384
15. Makwana RR (2013) Design consideration of façade wall system. Int J Eng Res Technol (IJERT), ISSN: 2278-0181 2(4):2449–2453
16. Trifonova R (2016) Unitized curtain walls installation and some health and safety aspects. International Balkans Conference on Challenges of Civil Engineering, 3-BCCCE, Epoka University, Tirana, Albania, pp. 190–199
17. Hussain S, Sekhar TS, Fatima A (2014) Analysis of lean construction by using last planner system. Int J Eng Res Technol (IJERT), 3(5):1059–1065
18. Tiwari S, Sarathy P (2012) Pull planning as a mechanism to deliver constructible design. In: 20th Annual Conference of the International Group for Lean Construction. San Diego, California, USA (2012)
19. Yu SN, Lee SY, Han CS, Lee KY, Lee SH (2007) Development of the curtain wall installation robot: Performance and efficiency tests at a construction site. Springer, Auton Robot, pp 281–291
20. Ghosh S, Reyes M, Perrenoud A, Coetzee M (2017) Increasing the Productivity of a Construction Project Using Collaborative Pull Planning. International conference of The American Society of Civil Engineers, ASCE, pp 825–836
21. Sivaprakasama T, Kesawana S, Mahendrana M, Stringfellowb J, Baleshan B (2020) Full-scale tests of aluminum mullion couples in unitized façades under wind actions", Elsevier, ISSN 0141–0296 (2020)

Traffic Estimation and Management Using GIS Mapping

Monika B. Khedkar, Prathamesh P. Gawde, and Rajendra Magar

Abstract Modern society is heavily dependent on effective transportation systems. The progress of a nation can be measured by the length of roads it has. The rapid growth in urban population and number of vehicles restrict the manoeurability on roads across the world. GIS applications have yielded excellent results in road traffic management not only in India but also throughout the world. Keeping in mind, this work addresses in providing solution to the traffic problem at Teen Hath Naka, a busy traffic intersection in the city of Thane, Maharashtra. Gram++—A GIS software developed by IIT Bombay, Mumbai has been used for graphical representation of the spatial and non-spatial data. The traffic data was collected through the local RTO office as well as manually carrying out traffic survey. Gram++ has facilitated the integration of this traffic data with the graphical replica of the intersection. The graphical model thus developed can depict the traffic data, scenario and solution to the same in a single framework. Continual modifications can be made by the traffic authority to this interface to make it more real-time.

Keywords Gram+ + · Geographic information system · Traffic congestion · Traffic volume · Urbanization

M. B. Khedkar (✉)
Construction Engineering and Management, Panvel, India
e-mail: khedkarmonika@gmail.com

P. P. Gawde
AIKTC, Panvel, India
e-mail: prathamesh.gawde@aiktc.ac.in

R. Magar
Civil Department, AIKTC, Panvel, India
e-mail: hod.ce@aiktc.ac

K. K. Pathak et al. (eds.), *Latest Developments in Civil Engineering*, Lecture Notes in Civil Engineering 352, https://doi.org/10.1007/978-981-99-2676-3_69

1 Introduction

It is aptly said that the progress of a country can be measured by the length of roads it has and further it is applicable to a city as well. Every city has its own topography and pattern of development. India is a country where speed of urbanization is enormous. There are very few examples of planned urban development in India. Most of the old cities are increasing in size like an octopus by encompassing the adjoining villages in its limits. Properly planned roads and equally efficient traffic management has become the need of the hour in most of the Indian cities.

Every city has its own traffic management system devised either in a planned manner or at times in a haphazard manner. As the size of the city increases horizontally as well as vertically, enormous amount of traffic as well as data related to it is generated. Management of this data to come up with feasible traffic solutions is a difficult task as monitoring information cannot be shared and utilized effectively. GIS is an effective tool to store and retrieve data. GIS provides us the freedom of compiling data in various forms and get a comprehensive output in the form we want. GIS empowers the decision making ability of its users. Traffic congestion management, which is otherwise a complex criterion, can be analyzed and provided with efficient solutions through GIS.

2 Brief Review of Literature

Many studies have been carried out for making traffic management easier by application of different software tools. Here is an overview of the literature reviewed in this regard.

Abraham et al. (1994) reported that the Congestion Management System (CSM) is a management system that focuses on congestion reduction and effective traffic flow, collecting information on accidents, highway geometry, and traffic volume. Road widening, turning lanes, and flexible scheduling are suggested to reduce peak travel [1].

Ogunbodede (2007) reported that a dynamic Traffic Information System (TIS) structure must be put in place to track municipal traffic jams and explore less congested routes. This will allow for a continuous free flow of traffic on urban highways [3].

Kasturia and Varma (2010) informed that PIS is a multi-objective generalised cost (GC) based design for a multi-mode transit system that incorporates GIS processing, network analysis, user interface, and database management [4].

Lu and Huang (2009) reported that the system combines GIS and traffic data to publish real-time visualizations [2].

Patnaik et al. (2016), this study categorized urban road networks according to the number of street classes and the LOS categories for average travel speeds. Speed data was used to calculate the ranges of different LOS categories, which were found to be 85 and above, 67–85, 50–67, 40–50, 30–40, and 30 and less [6].

Wang et al. (2018) informed that the quality of the bus and metro connection is key to attracting commuters for a metro system, and smart transportation systems are an integral part of any smart city [5].

Agyapong and Ojo (2018) The study found that poor road design, RTCs, and poor driver, pedestrian, and business behaviour are the biggest contributors to traffic congestion, which has a negative impact on productivity and sales [7].

Atta et al. (2018) informed that RFID technology can be used to dynamically change traffic signal behaviour by sensing the density of traffic and maintaining dynamic timings to reduce congestion [8].

Saleem et al. (2022) reported that FITCCS-VN is a fusion-based intelligent traffic congestion management system for virtual networks that uses ML approaches to gather traffic data and direct traffic on accessible routes. It has a 95% accuracy rate and a 5% miss rate, and offers drivers cutting-edge technology to reduce congestion while enhancing traffic flow [9].

3 Statement of the Problem

Thane is a city located on the Central railway's suburban train system located in North West of Mumbai. The city, known for its lakes, has become a major residential hub in the last two decades due to its proximity to all major areas in Mumbai Metropolitan region (MMR). As per 2011 census, the population of the city is approximately 32 lakhs. Inefficient public transport system has always been a major problem in Thane making traffic conditions in the city worse. Not more than 250 public transport buses serve this city with such a huge population. Thus it explains that the number of auto rickshaws, private buses, 2 wheelers and cars has increased enormously.

Traffic congestion is a day-to-day experience for the residents of Thane while travelling regularly from home to work/college and back. Construction of flyovers, construction of Metro, increased number and frequencies of buses are some of the major reforms happening in the transport sector of Thane city. However, the growth in the population of Thane and suburbs is enormous. This piece of work is an attempt to provide a solution to the traffic congestion at Teen hath Naka, one of the busiest intersection of Thane city.

Insights on how problems of traffic congestion in major cities throughout the world could be solved are already got through the literature review. Use of Gram++ software to prepare the traffic model for Thane city with emphasis on Teen hath naka in GIS environment would be the ultimate product of this work.

Aim

To devise Traffic Management plan for Thane city using GIS Mapping with emphasis on traffic reduction at Teen Hat-Naka.

Objectives

(1) To collect the available maps (hard & soft form) of Thane city from the relevant authority (TMC, MMRDA).
(2) Map Scanning: To import the maps in GIS environment.
(3) Digitizing: To digitize and create a replica in GIS environment.
(4) Geo Referencing and geometric rectification: To identify the various points of Origin and Destination (O & D) of traffic.
(5) Identification of new desire lines: To study existing desire lines. To identify & plan new ones to reduce traffic congestion.
(6) Developing solutions to tackle the problem: To suggest & check the feasibility of measures to improve road traffic conditions at teen hath naka, thane.

Scope of the Proposed Work

Development of a GIS Model for providing solution to road traffic issues in Thane city with special emphasis on teen hath naka. The model prepared can further be applied to various other busy intersections.

4 Methodology

The stages elaborate in the process were in this way:

(1) **Map Scanning**: Scanning of a physical map was done which was initially imported in raster data form which further it was converted into vector data form.
(2) **Geo Referencing and geometric rectification**: Geo referencing is the process of identifying an object's location in space. Identifying its position using coordinate systems or map projections.
(3) **Digitizing**: It is the process of coding geographic features as x, y coordinates in numerical form. It is done to create spatial data from recently printed maps and official documents.
(4) **Creation of databases**: This involves creating databases in Gram++ and adding various attributes to the digitized image of the map of the city.
(5) **Drawing of existing Desire lines**: It indicates construction of lines with some weightage to indicate the number of vehicles flowing in a particular direction. Thicker the line, more the number of users of that route.
(6) **Identification of new desire lines**: On the basis of the data collected and forecasts created from the same for the next 30 years, identifying new desire lines which may come up due to change in the land use or introduction of some new transport facility.
(7) **Developing solutions to tackle the problem**: Identifying solutions for the problem by making use of techniques learnt through literature review as well as by seeking guidance from the field & academic experts. Finally, checking the feasibility of the solutions devised (Fig. 1).

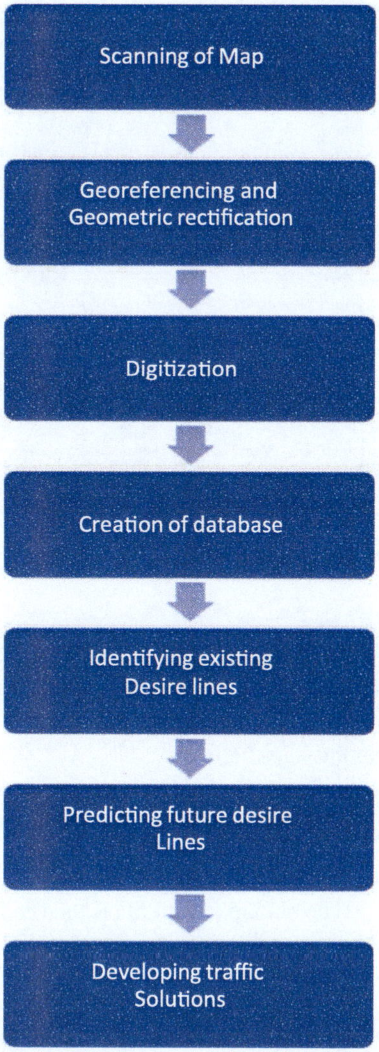

Fig. 1 Flowchart showing steps involved in the methodology adopted

5 Case Study: Replica and Database Creation

The software used is GRAM++ (Geo Referenced Area Management), developed by CSRE, IIT MUMBAI.MS-Access 2003 is used for handling the database. Use of GIS in traffic management can be helpful in deciding between alternatives and makes it possible to have a fool proof database of elements which do not change with time and to monitor those which change with time. One will then have to update the database at regular predefined intervals to monitor the completion of the project. This updation is done by the project managers/site engineers after he receives word from the site

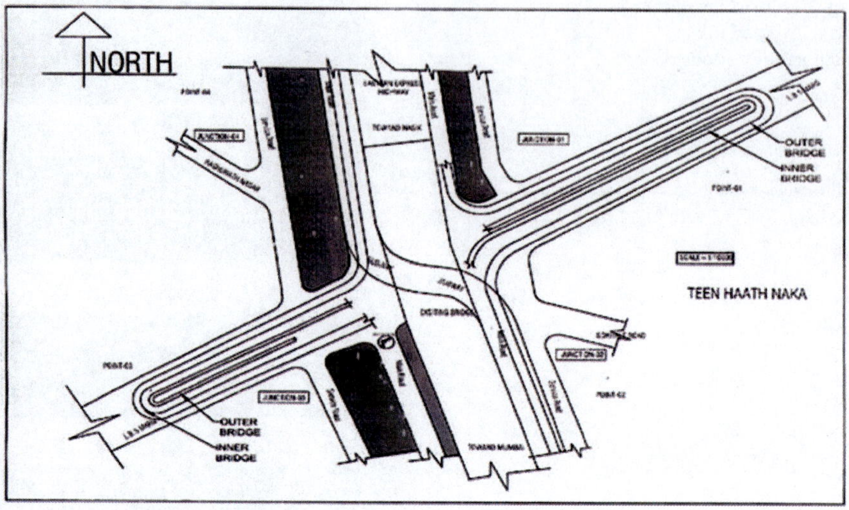

Fig. 2 Original CAD file converted to bitmap file

about the progress of various activities involved in the project. This application would automatically update the attributes of the various components. With the help of this application, in a single click, one can get all the relevant information related to the traffic improvement scheme at a particular junction. The work carried out is explained in the steps below.

Initially, the CAD file of the work is procured (final design plan on sept 12.dwg) and scanned and saved as a bitmap image. (scan.bmp) as shown in Fig. 2. With the help of Google earth, the point marks are fixed, co-ordinates for the same are known and accordingly are marked on the bitmap image as shown in Fig. 3.

The bitmap (.bmp) file is then converted to a vector (.vec) file in the Input–Output module of Gram++ as shown in Fig. 4. The entire map is then digitized and layers are created in the Map-Edit module of Gram++ (teen hath naka.vec) as needed, as demonstrated in Fig. 5. Once the map is digitized in point, segment and polygon layers, they need to be cleaned to remove off discrepancy if any.

Database creation is then started to attach non-spatial data to the image as shown in Fig. 6. Apart from it, if any additional image, graphs or videos are to be added to the layer-wise digitized image, and it can be done using child map feature as shown in Fig. 7.

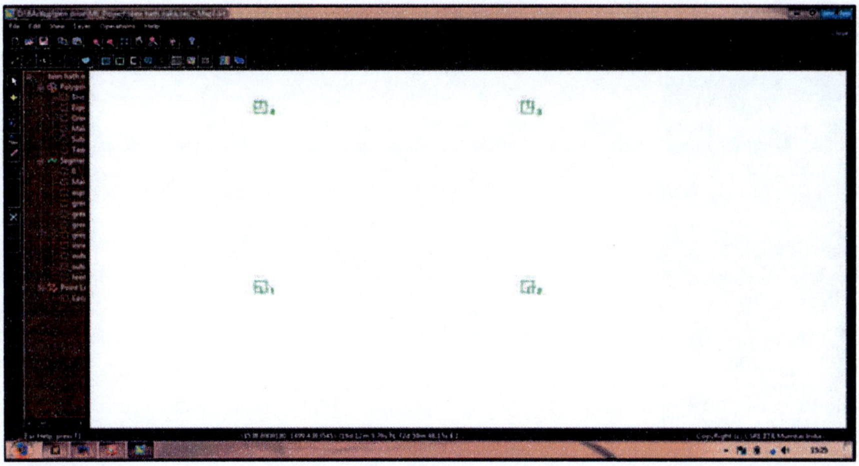

Fig. 3 Fixing of point marks on the scanned drawing

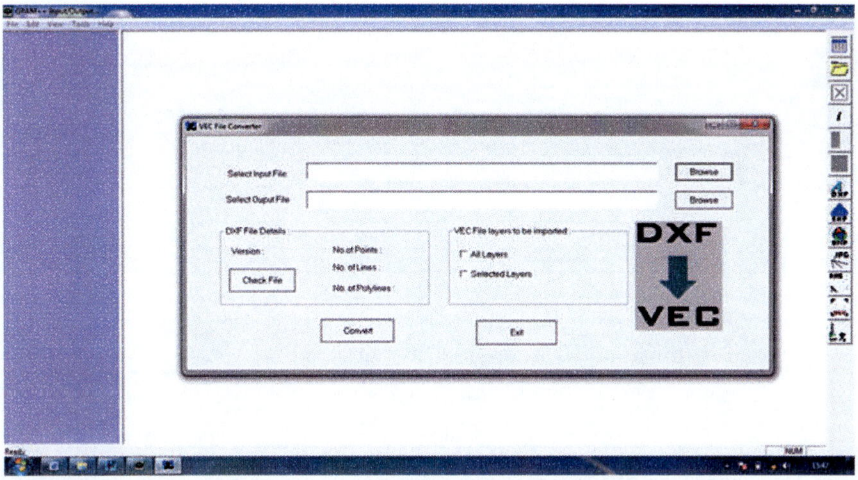

Fig. 4 Conversion of bitmap file into vector file in Input–output module of Gram++

6 Results and Discussions

Once the entire process as explained in the case study above is over, queries are run to extract the appropriate data from the database. The procedure, examples in addition to utility of few sample queries are mentioned below:

- Open Gram++. Select the module for Vector Analysis and open the file in.vec format (teen hath naka.vec).

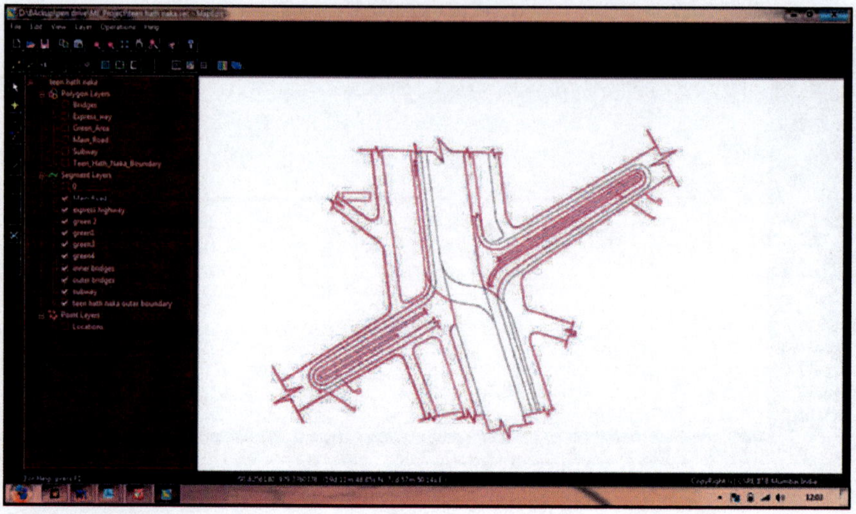

Fig. 5 Digitized form of vector file in Map-edit module of Gram++

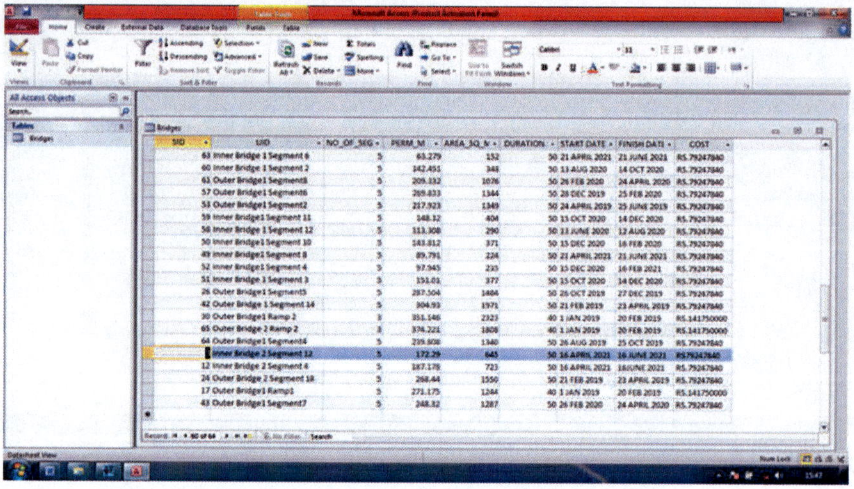

Fig. 6 Creation of database to attach non-spatial data

- Turn on the layer for which the database was produced. (Bridges, Express_ Way, Main_ Road, Subway, Green_ Area, Teen_ Hath_ Naka_ Boundary etc.).
- Go to the Tool menu in the top toolbar and choose "Query" for the chosen layer. Three alternatives are displayed in a pop-up menu. Click OK after selecting "Simple query." To perform the query for a specific field, select (double click) that field.

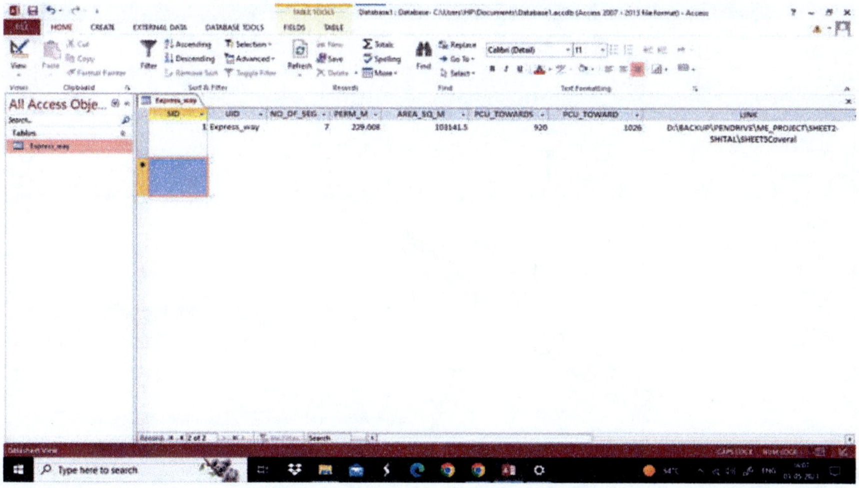

Fig. 7 Linking of child map

- The chosen field will be displayed in the "Enter Condition/Query" window, for example, "percentage completion." Type in the desired operators (, >, =, > =.=). In order to choose a value, double-click on the "Field Values" box. In the "Enter Condition/Query" window, this value will also be displayed. Choose "Ok."
- On the layer for which the query is executed, the query's results will be displayed in a highlighted format. A sample query run along with the result displayed is shown below for clear understanding.

Query: Show Points with PCU < 1000

See Fig. 8.

Procedure and Utility

Open gram++. Open the desired file (teen hath naka.vec) in vec analysis module. Go to Tools—Query—Simple query—Enter the relevant field out of the list and also enter the operators. In this case PCU is the field and <1000 is the operator and condition. Finally click on OK to get the output in Fig. 8. The utility of such a query related to PCU in a particular section can help us to know the amount of traffic occurring at that section during the busiest time of the day. Accordingly, decision can be made whether to install a traffic signal or a rotary island. At the same time, if bridge/flyover is to be constructed, the no of lanes to be provided will depend on the PCU$_=$s to be served.

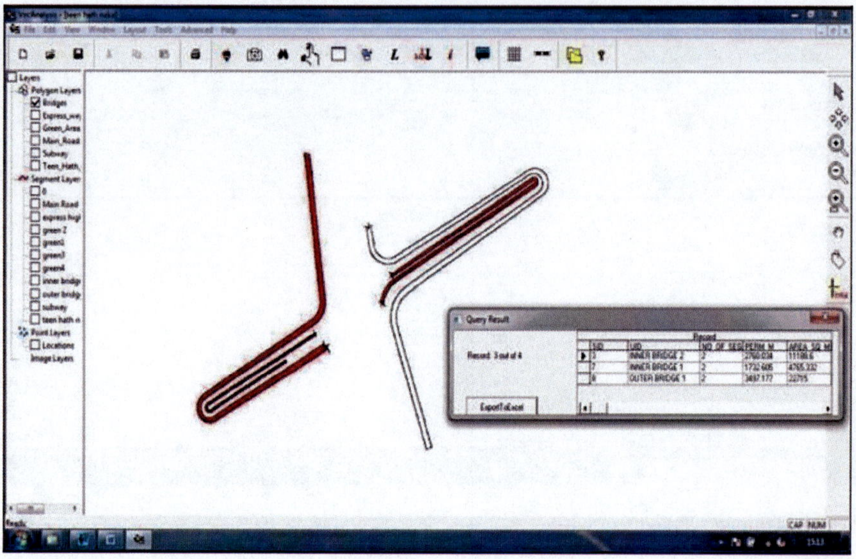

Fig. 8 Sections with PCU less than 1000 highlighted in red color

7 Conclusion

Quick and efficient urban infrastructure is an essential requirement of urban world. However; the pace of up gradation and growth in requirement for the same do no match many a times. Proper planning and timely execution of the infrastructure projects can help bridge this gap. This piece of work has explored Gram++ (An effective GIS tool) as a user friendly tool for planners, executors as well as monitors to make intelligent decisions; check the feasibility of different alternatives available and select one out of it which is most optimal. A query that was run for locations with less than or more than a particular value of PCU can help to know the spots of traffic congestion in the city on a GIS user interface. Similarly, the extent of work completed at a particular spot of a road section or bridge can be known and updated at any point of time. This can help the various stakeholders of the project to visualize the project status without physically visiting the site. Features like child map can provide depictive information on the actual view of the locations in the form of pictures and even videos. Efficient and timely decisions can help in arresting time and cost overrun.

This education will advantage project managers, site engineers and clients in the following manner.

Project Administrator

- Up-to date facts about the improvement of work.

- Graphical representation of each portion of the project is easily available making the decision making process faster and smoother communication among various stakeholders.
- Can effectively present the Project status to management and investors with a pictorial view of the project along with its schedule at the same time.

Site Engineer

- Control the progress of works at the site.
- Supports in updating the contractors beforehand about the start of their work or lacking of any.
- Can maintain a good record of all activities and works for final billing purpose.

Client

- Helps in knowing the particular status of the project.
- Has a view of the progress of work consequently knowing where enormous cost has been incurred.

Scope of Prospect Work

This work make the most of only three modules of Gram++ namely input–output, map edit and vector analysis. The remaining modules can be used for preparing various GIS based tools having applications in various civil engineering fields.

Scheme of the Chapterization

The proposed dissertation report shall comprise of the following chapters.

(1) Introduction.
(2) Brief Review of Literature.
(3) Statement of the Problem.
(4) Methodology.
(5) Case Study Replica and Database creation.
(6) Results and Discussion.
(7) Conclusion.

There shall be minor variation in the chapterization or contents or names there of when the final report will be compiled.

References

1. Hunt JD, McMillan JDP, Abraham JE (1994) Stated preference investigation of influences on attractiveness of residential locations. Transp Res Rec 1466:79
2. Lu B, Huang M (2009) A new measure for traffic data collection and processing. In: 2009 second international conference on intelligent computation technology and automation, vol 3. IEEE, pp 440–443

3. Ogunbodede EF (2007) Assessment of traffic congestions in Akure (Nigeria) using GIS approach: lessons and challenges for urban sustenance. In: Proceedings of conference on whole life urban sustainability, pp 1–25
4. Kasturia S, Verma A (2010) Multiobjective transit passenger information system design using GIS. J Urban Plan Dev 136(1):34–41
5. Wang ZJ, Liu Y, Chen F (2018) Evaluation and improvement of the interchange from bus to metro using smart card data and GIS. J Urban Plan Dev 144(2):05018004
6. Patnaik AK, Bhuyan PK, Rao KK (2016) Divisive Analysis (DIANA) of hierarchical clustering and GPS data for level of service criteria of urban streets. Alexandria Eng J 55(1):407–418
7. Agyapong F, Ojo TK (2018) Managing traffic congestion in the Accra central market, Ghana. J Urban Manag 7(2):85–96
8. Atta A, Abbas S, Khan MA, Ahmed G, Farooq U (2020) An adaptive approach: smart traffic congestion control system. J King Saud Univ—Comput Inform Sci 32(9):1012–1019
9. Saleem M, Abbas S, Ghazal TM, Khan MA, Sahawneh N, Ahmad M (2022) Smart cities: fusion-based intelligent traffic congestion control system for vehicular networks using machine learning techniques. Egypt Inform J 23(3):417–426

Time Cost Optimization Applied in Transit Camps

Chauhan Mohammed Shuaib Farooq and Girish Mahajan

Abstract This article is evaluation of transit camps (temporary structures) on the scale of time and cost. Construction activities are scheduled to be completed in specific time frame with some determined expenditure and required number of resources. Delay in construction projects leads to both excessive duration as well as expense which results in loss of stake holders and investors. The subsequent improvement in "transit camps", continues to develop the lifestyle of people residing in slum areas. This article examines the financial and time savings by planning, scheduling of activities, with qualitative outputs and ease for the people residing in these camps. According to the current status and available data by the literature reviews, it demonstrates the advantages and benefits of transit camps for a project. Additionally, this study focuses on the optimization of cost and time of activities which may reduce delays and indirectly help in economic savings with advanced construction management techniques.

Keywords Transit camps · Financial savings · Scheduling · Qualitative outputs · Management techniques

1 Introduction

A. Definition:

These are temporary structures constructed near the Construction sites, to provide shelter to the existing tenants residing in that location. These structures are constructed by assembling many different elements at the manufacturing unit and shifting those units to the construction site. It may consist of minor units in a facade

C. M. S. Farooq (✉) · G. Mahajan
Department of Civil Engineering, School of Engineering and Technology, AIKTC, New Panvel, Navi Mumbai 410206, India
e-mail: shuaib.chauhan.10@gmail.com

G. Mahajan
e-mail: girish.mahajan@aiktc.ac.in

K. K. Pathak et al. (eds.), *Latest Developments in Civil Engineering*, Lecture Notes in Civil Engineering 352, https://doi.org/10.1007/978-981-99-2676-3_70

or completely fitted bathrooms. It is basically used in typical projects having similar units (office rooms) and/or complex designs that is difficult to construct on site (special facade elements or steel frames). These projects are carried out in Public as well as Private sectors depending on various uses like military, residential, slum development projects etc.

B. Need of Work:

a. When the Construction site is at a very remote location or in a highly populated area.
b. Access to efficient offsite production on th econstruction site.
c. Site development is complex due to various stakeholders, limited space or time factor.
d. When the construction project is very huge, it is to be completed in short time due to climate, or is on very expensive land.
e. Saves economy of the developer on a large scale.

C. Objective of Research:

(a) To plan all the Operations precisely.
(b) To reduce Project cost by applying management techniques.
(c) To execute activities within scheduled time and with Quality Outputs.

2 Literature Review

Here are some reviews of the authors published below: Shamsul Haque Hamidullah et al., resulted as crash time reduces the cost after crashing increases i.e. its inversely related. It gives only a deterministic approach of activities, duration and cost [1]. Pornima M. Kashid et al., concludes that MS Project is very much suitable for planning and scheduling of activities due to its user-friendly nature software as compared to other softwares [2]. Cemil Akcay, illustrated in a paper its findings where the need of the genetic algorithm based on fuzzy time cost trade off was to minimize the project cost within the allotted project duration [3]. Uroš Klanšek et al., published a paper and its findings were that the project time and scheduling when performed with the Non-Linear Optimization process, the cost of the project is minimized [4]. Veludurthi Manoj Kumar et al., published a paper and its results were as follows: for a project Critical path method (CPM) technique is good for small networks (works) and can be majorly implemented when there is unavailability of softwares [5]. Anuja Rajguru et al., in their paper found that on reviewing different techniques for Optimization of the project, the minimization of the cost and duration is the most essential factor but while considering the cost parameter, cheap materials are suggested by maintaining the quality and strength of the project [6]. Mirko Pšunder et al., in an article concluded that the cost optimization process was done by the Non Linear Programming Approach. It avoids the need for linear approximation of non-linear expressions for a particular project, which is better than the Linear

Programming methods [7]. Mr. K. James Babu Raj et al., in a journal concluded that, two techniques which help in reducing the expense and to finish the project within the estimated time are: 1. Schedule Crunching. 2. Project Crashing. From both the techniques Schedule Crunching is the most effective for optimization of time and cost, but the quality of work is not guaranteed. Hence to achieve quality, crashing of the project seems to be a good technique [8]. Shanmugapriya Sundarrajan et al., in a paper found the components leading to time cost surplus were change in material prices, modifications in agreement, project location, transportation charges, standard of quality etc. [9]. Naveenkumar G. V. et al., published a journal which discusses the factors leading to escalation of cost that are delay in preliminary handling of project, increase in material cost, inappropriate selection of site, lack of planning, changes in designs, equipment issues etc. [10]. Bhushan V' Tatar et al., in a paper studied that Non Linear programming technique is better as compared to the deterministic approach as it decreases calculation time and improves speed and overall efficiency for scheduling and planning of the project [11]. Umesh Kamble et al., in a paper discussed the application of Microsoft Project for optimizing cost and time as it indirectly influences the project duration by decreasing the slack time [12]. It also concludes that the fluctuations also depend on seasonal variations, location of site which hampers the project cost and duration [13]. Dr. M. K. Trivedi et al., published a paper and its findings were, as many methodologies were applied to solve the optimization problem the most commonly adopted method is the genetic algorithm in the meta-heuristic method used for the optimization of cost and time problems [14]. It provides a number of solution to decision makers, and a better solution due to a random searching mechanism but it cannot deal with dynamic data [15].

3 Methodology

Methods

From the above critical literature review there are various methods for project planning. CPM method has been extensively used for the execution of construction projects while as the latter shall provide a breakthrough in project planning.

Critical Path Method (CPM)

Critical Path Method is a method where the tasks are scheduled according to their flexibilities to complete a project within the stipulated time frame. Critical path tends to have the largest path in the network which needs to be executed as per schedule for the completion of projects. Work breakdown structure (WBS) is made following which the interdependencies of the tasks and the resources required to execute the task are assigned. The project is based on the principle of early start time by which it is scheduled in the forward pass. Non-critical activities are protected by float where the activities can be postponed upto some extent to the latest start time without

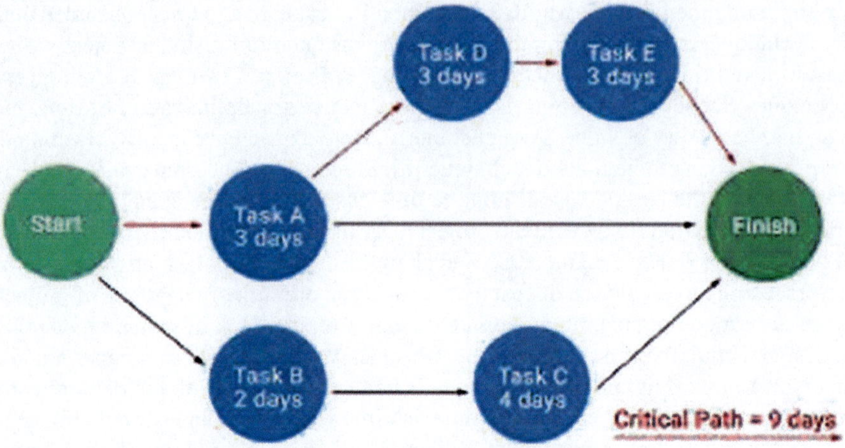

Fig. 1 CPM technique (*Source* www.construction-cpm.com)

hampering the project completion date. Critical activities associated with critical path have no float (Fig. 1).

Project Crashing

Project Crashing means to shorten the project completion time by adding extra resources to it. The project can be crashed by reducing the normal time of critical activities which is called crashing of activities. This can be achieved by increasing the utilities. Increase in utilities leads to increase in extra expenditure for completing the work. The increase in extra expense and utilities can be considered upto a certain extent that can be accepted by time cost constraint (Fig. 2).

Schedule Crunching

Schedule crunching refers to techniques, used when a project manager wants to minimize the duration of any activity in project without changing the scope of work. It is implemented when a project falls behind schedule and needs to catch up or to finish the project sooner than originally scheduled. In this technique increased duration and cost can be reduced upto some extent. But Crunching should be under an optimized manner, otherwise it leads to poor quality of optimization and also increase in project cost.

Methodology:

In order to fulfill objectives of the analysis the following tasks were performed:

1. Site investigation initial plans for the project. The aim of site inspection is to study factors affecting the project.
2. The factors affecting the project based on cost and duration is done through a survey from owners and engineers.

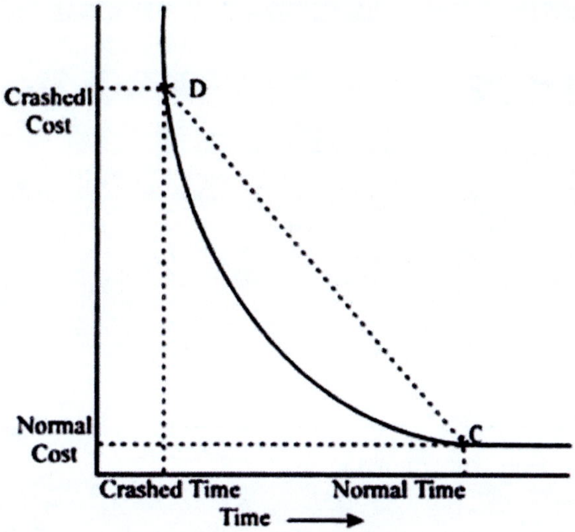

Fig. 2 Project crashing (www.project+crashing+graph.com)

3. The elements leading to delay and increase in expense is studied.
4. Once the factors are studied investigation is done for a structure by using MSP.
5. In this area, analysis of building starts with proper evaluation of expenditure and the time schedule of the structure is done.
6. Using the Microsoft project, scheduling, planning and costing is done.
7. Once the results are received, comparison is done based on actual and calculated values through MSP and accordingly the optimization is done for both constraints (Fig. 3).

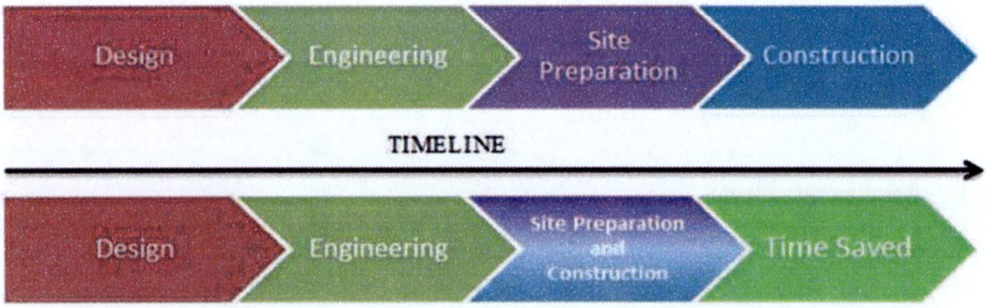

Fig. 3 Methodology of project using MSP

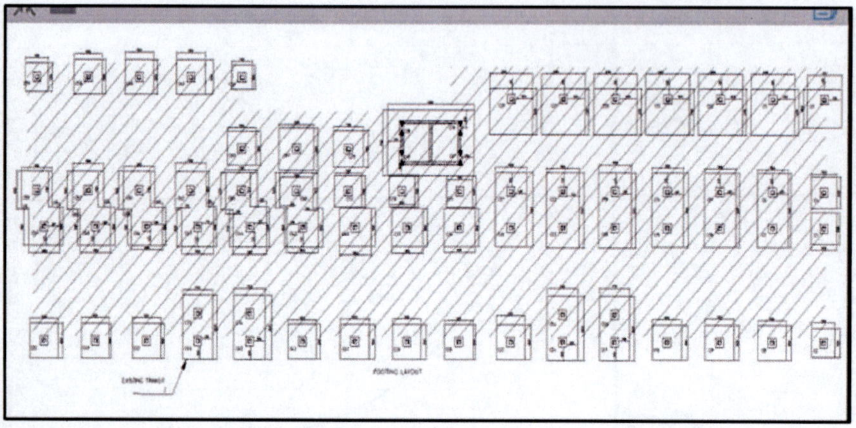

Fig. 4 RCC plan of transit camp (footing work)

SCHEDULE OF FOOTING

NUMBERS #COLUMN	FOOTING DIMENSION			FOOTING REINFORCEMENT			
				BOTTOM		TOP	
	L	B	D	ALONG B	ALONG L	ALONG B	ALONG L
RC1	REFER PLAN		500	T12 @100 c/c	T12 @100 c/c	-	-
RC2, RC59	REFER PLAN		450	T10 @100 c/c	T10 @100 c/c	-	-
RC9, RC63, RC60, RC8, RC58, RC16, RC25, RC35, RC36, RC41, RC43, RC32, RC49, RC52	REFER PLAN		500	T12 @100 c/c	T12 @100 c/c	-	-
RC3, RC4	REFER PLAN		500	T12 @100 c/c	T12 @100 c/c	-	-
RC6, RC7, RC10, RC11, RC14, RC15, RC19, RC22, RC26, RC62, RC18, RC23, RC47, RC61, RC57, RC55, RC51, RC44, RC48, RC66, RC82, RC65, RC40, RC17, RC50, RC24, RC54, RC45, RC74, RC46, RC75	REFER PLAN		700	T12 @100 c/c	T12 @100 c/c	T12 @100 c/c	T12 @100 c/c
RC34, RC37, RC40	REFER PLAN		700	T12 @100 c/c	T12 @100 c/c	-	-
RC64	REFER PLAN		450	T10 @100 c/c	T10 @100 c/c	-	-
RC67	REFER PLAN		700	T12 @100 c/c	T12 @100 c/c	-	-
RC69	REFER PLAN		600	T12 @100 c/c	T12 @100 c/c	-	-
RC70, RC71	REFER PLAN		500	T12 @100 c/c	T12 @100 c/c	-	-
RC72	REFER PLAN		500	T12 @125 c/c	T12 @125 c/c	-	-
RC30, RC31, RC38, RC39	REFER PLAN		750	T16 @125 c/c	T16 @125 c/c	T16 @125 c/c	T16 @125 c/c
RC5, RC12, RC13, RC20, RC21 RC28	REFER PLAN		500	T12 @100 c/c	T12 @100 c/c	-	-
RC29	REFER PLAN		600	T12 @125 c/c	T12 @125 c/c	-	-

Fig. 5 Footing reinforcement details

3.1 Site Layout Plans

Technical Site Plans (Auto-cad files) (Figs. 4, 5, 6, 7, 8, 9 and 10).

3.2 Site Images of Transit Camps (G+7)

A Slum Rehabilitation Project Behind Lotus Petrol Pump, Andheri-Mumbai (Figs. 11, 12, 13, 14, 15, 16 and 17).

SCHEDULE OF R.C.C. COLUMN

COLUMN NUMBERS		RC30, RC31, RC38, RC39	RC1, RC2, RC3, RC4, RC5, RC6, RC7, RC8, RC9, RC10, RC11, RC12, RC13, RC14, RC15, RC16, RC17, RC18, RC19, RC20, RC21, RC22, RC23, RC24, RC25, RC26, RC27, RC28, RC29, RC32, RC34, RC35, RC36, RC37, RC40, RC41, RC42, RC43, RC44, RC45, RC46, RC47, RC48, RC49, RC50 RC51, RC52, RC54, RC55, RC57, RC58, RC59, RC60, RC61, RC62, RC63, RC64, RC65, RC66, RC67, RC68, RC69, RC70, RC71, RC72, RC74, RC75
TERRACE TO PLINTH	SIZE STEEL LINKS	230 x 700 14 - 16 Φ 8 Φ @ 150 c/c	
PLINTH TO FOOTING	SIZE STEEL LINKS	230 x 700 14 - 16 Φ 8 Φ @ 150 c/c	450 x 450 16 - 12 Φ 8 Φ @ 150 c/c

Fig. 6 RCC column schedule upto footing

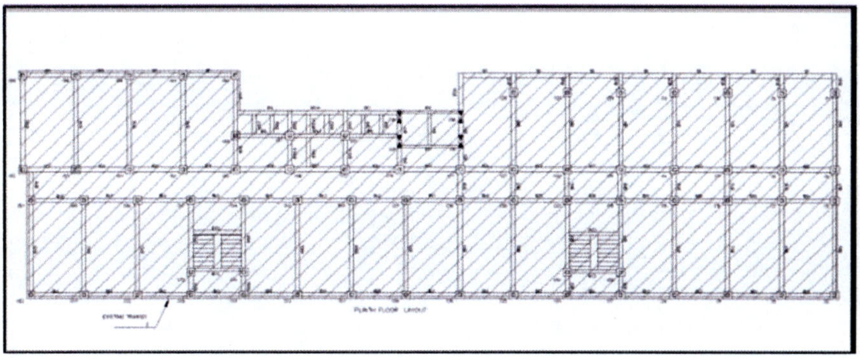

Fig. 7 Plinth floor layout

SCHEDULE OF TYPICAL FLOOR BEAM

BEAM NUMBERS	SIZE		BOTTOM REINFORCEMENT		TOP REINFORCEMENT			SHEAR STIRRUPS	REMARK
	B	D	STRAIGHT	CURTAILED	STRAIGHT	EXTRA LEFT SUPP	EXTRA RIGHT SUPP		
B12, B17	150	700	2-T16	-	2-T16	-	-	2L-T8@200	-
B92	230	700	3-T12	-	3-T16 +3-T16	-	-	2L-T8@150	-
B92A	230	700	3-T16	-	3-T16 +3-T16	-	-	2L-T10@100	CANT
B94	150	700	2-T12	-	2-T12	-	-	2L-T8@300	-
B97	230	700	3-T12	-	3-T12	-	-	2L-T8@175	-

Fig. 8 Lift reinforcement details

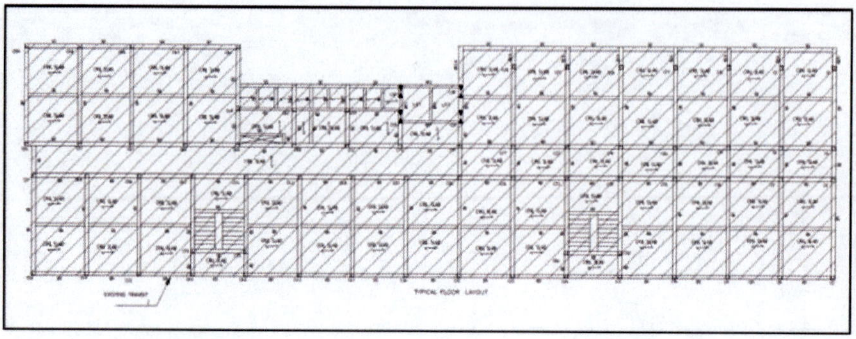

Fig. 9 Typical floor layout

SCHEDULE OF COLUMN

COLUMN NUMBERS		RC5,RC8,RC9,RC12, RC13,RC16,RC20,RC21,RC25, RC28,RC32,RC35,RC36, RC41,RC43,RC49,RC52, RC58,RC59,RC60,RC63,RC64, RC69	RC1,RC2,RC3,RC4,RC6,RC7,RC10, RC11,RC14,RC15,RC19,RC22,RC24, RC27,RC34,RC37,RC40,RC42,RC48, RC51,RC55,RC57,RC61,RC62,RC65, RC66,RC68,RC70,RC71,RC72	RC17,RC18,RC23,RC26,RC29,RC44, RC45,RC46,RC47,RC50,RC54,RC74, RC67,RC75
TERRACE TO 7TH FLOOR	STEEL	2ISMC 150	2ISMC 150	2ISMC 150
7TH FLOOR TO 6TH FLOOR	STEEL	2ISMC 150	2ISMC 150	2ISMC 150
6TH FLOOR TO 5TH FLOOR	STEEL	2ISMC 150	2ISMC 150	2ISMC 150
5TH FLOOR TO 4TH FLOOR	STEEL	2ISMC 150	2ISMC 150	2ISMC 200
4TH FLOOR TO 3RD FLOOR	STEEL	2ISMC 150	2ISMC 200	2ISMC 200
3RD FLOOR TO 2ND FLOOR	STEEL	2ISMC 200	2ISMC 200	2ISMC 250
2ND FLOOR TO 1ST FLOOR	STEEL	2ISMC 200	2ISMC 200	2ISMC 250
1ST FLOOR TO PLINTH	STEEL	2ISMC 200	2ISMC 250	2ISMC 250

SCHEDULE OF I SECTIONS TYPICAL

TYPE	DESIGNATION	WEIGHT KG / M
[illegible]	ISMB 250	37.3
[illegible]	ISMB 200	25.4
[illegible]	ISMB 150	14.9
[illegible]	ISMB 100	11.5

Fig. 10 Prefab MS column and beam details

Fig. 11 Actual photo of Prefab work, SRA-Andheri

Fig. 12 Actual photo of Prefab and laying of dack sheet, SRA-Andheri

Fig. 13 Actual photo of GF Prefab and channeling work, SRA-Andheri

Fig. 14 Actual photo of slab reinforcement work being done, SRA-Andheri

Fig. 15 Actual photo of internal finished rooms, SRA-Andheri

Fig. 16 Actual photo of kitchen and bath area in room, SRA-Andheri

Fig. 17 Actual photo of separate W/c units gents/ladies, SRA-Andheri

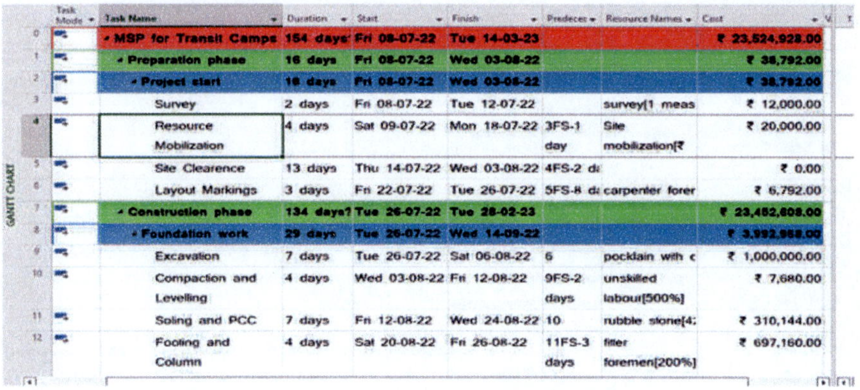

	Task Mode	Task Name	Duration	Start	Finish	Predeces	Resource Names	Cost
0		MSP for Transit Camps	154 days	Fri 08-07-22	Tue 14-03-23			₹ 23,524,928.00
1		Preparation phase	16 days	Fri 08-07-22	Wed 03-08-22			₹ 38,792.00
2		Project start	16 days	Fri 08-07-22	Wed 03-08-22			₹ 38,792.00
3		Survey	2 days	Fri 08-07-22	Tue 12-07-22		survey[1 meas	₹ 12,000.00
4		Resource Mobilization	4 days	Sat 09-07-22	Mon 18-07-22	3FS-1 day	Site mobilization[₹	₹ 20,000.00
5		Site Clearence	13 days	Thu 14-07-22	Wed 03-08-22	4FS-2 d		₹ 0.00
6		Layout Markings	3 days	Fri 22-07-22	Tue 26-07-22	5FS-8 d	carpenter forer	₹ 6,792.00
7		Construction phase	134 days?	Tue 26-07-22	Tue 28-02-23			₹ 23,452,608.00
8		Foundation work	29 days	Tue 26-07-22	Wed 14-09-22			₹ 3,992,968.00
9		Excavation	7 days	Tue 26-07-22	Sat 06-08-22	6	pocklain with c	₹ 1,000,000.00
10		Compaction and Levelling	4 days	Wed 03-08-22	Fri 12-08-22	9FS-2 days	unskilled labour[500%]	₹ 7,680.00
11		Soling and PCC	7 days	Fri 12-08-22	Wed 24-08-22	10	rubble stone[4:	₹ 310,144.00
12		Footing and Column	4 days	Sat 20-08-22	Fri 26-08-22	11FS-3 days	fitter foremen[200%]	₹ 697,160.00

Fig. 18 Gantt chart representing details of schedule and cost by CPM

Optimization Using Critical Path Method:

It is designed to accompany a project manager for developing a schedule, assigning resources to tasks, tracking progress, analyzing workloads and managing the budget. As the tasks are appointed to the resources the work schedule is evaluated, the cost data is calculated by the program, which completes the task part and any associated tasks and finally to the last level as per the schedule. Every activity can have its own calendar. The rates of the resources are used to calculate resource assignment costs. Every activity can be appointed to various different tasks and these tasks can be further appointed to multiple resources. Further based on the data obtained and the scope of work identified the construction activities are noted and logical linking of the same along with material and resource allocation was carried out. The estimated investment for the entire project considering delays, additional work will be obtained. The activities were scheduled to commence within the estimated time. The Critical Path has been identified by red colour whereas the non-critical path is denoted by bluc colour (Figs. 18, 19, 20, 21, 22, 23, 24 and 25).

4 Results

The work carried out under this study involves the outcomes of the advanced method of the CPM method for execution of transit camps (temporary structures). To accomplish the results, the objectives set were to schedule and assign utilities to the Transit structure by the CPM method. The results of cost and time were to be compared along the project lifespan with the data recovered from site and through MSP software. These objectives were successfully achieved and it was proved that CPM can be implemented to achieve optimization of time as well as cost of such kind of projects.

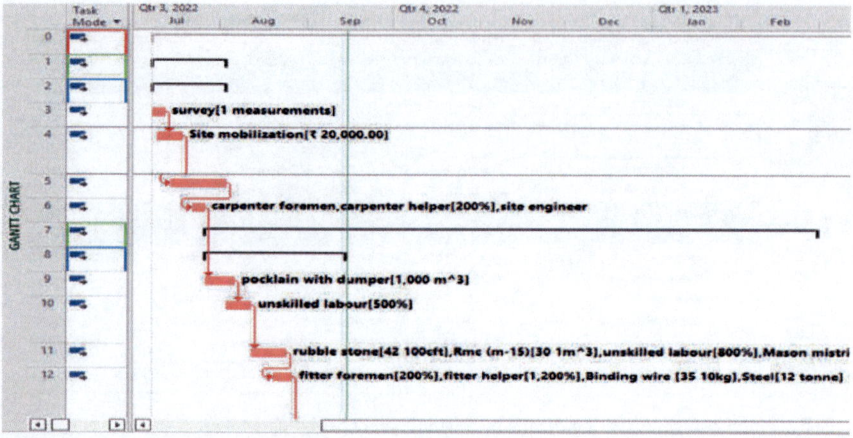

Fig. 19 Gantt chart representing graph (Fig. 18)

23	Plinth PCC	2 days	Sat 08-10-22	Tue 11-10-22	22FS-1	Rmc (m-15)[30	₹ 163,880.00
24	Curing work	7 days	Tue 11-10-22	Sat 22-10-22	23	unskilled labou	₹ 2,688.00
25	**Prefabrication Work**	**63 days?**	**Tue 18-10-22**	**Tue 24-01-23**			**₹ 9,308,576.00**
26	**GF C channels for columns (welded conections)**	2 days	Tue 18-10-22	Thu 20-10-22	24FS-3 days	Steel[4 tonne],Fabricat helper[700%],C	₹ 235,840.00
27	GF I channels for Beam work (welded connections)	3 days	Wed 19-10-22	Tue 25-10-22	26FS-1 day	Steel[6 tonne],Fabricat helper[700%],C	₹ 353,760.00
28	Laying of Dack sheet for slab work	1 day	Tue 25-10-22	Wed 26-10-22	27	Deck sheet[1 per slab]	₹ 200,000.00
29	Slab Reinforcement work.	1 day	Tue 25-10-22	Wed 26-10-22	28FS-1 day	Steel[2.5 tonne],fitter	₹ 151,860.00
30	Slab Casting.	1 day	Wed 26-10-22	Fri 28-10-22	29	Rmc (m-25)[38	₹ 219,424.00
31	Curing Work	7 days	Fri 28-10-22	Tue 08-11-22	30	unskilled labou	₹ 2,688.00
32	**1st floor C channels for**	2 days	Thu 03-11-22	Mon 07-11-22	31FS-3 days	Steel[4 tonne],Fabricat	₹ 235,840.00

GANTT CHART

Ready New Tasks : Manually Scheduled

Fig. 20 Gantt chart representing schedule and cost by CPM

The project time frame could be reduced by 31.55% and the overall cost expense on the project was brought down by 17.45% by using this technique.

- **Critical Path Method by Using MSP (Calculated Data)**

According to the methodology implemented after scheduling and application of resources the project commencement and completion date was obtained thereby obtaining the overall schedule for construction of 32,000 sq ft. area which tends to Rs. 735 per sq ft. Summing up all the activities' cost of the total project estimate that was obtained and has been described below (Table 1).

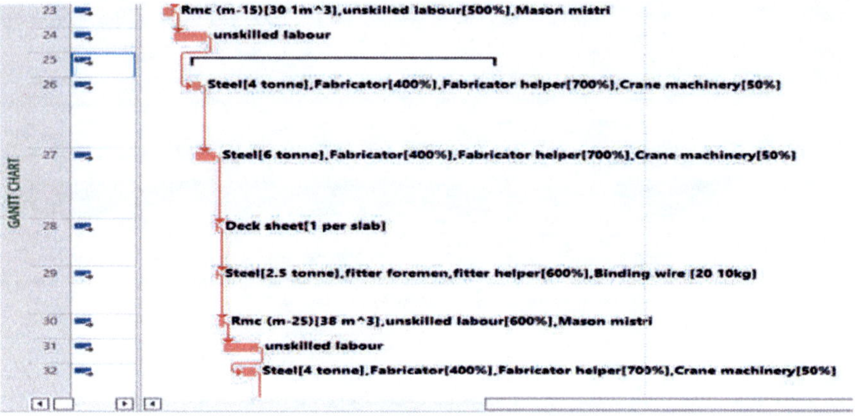

Fig. 21 Gantt chart representing graph (Fig. 20)

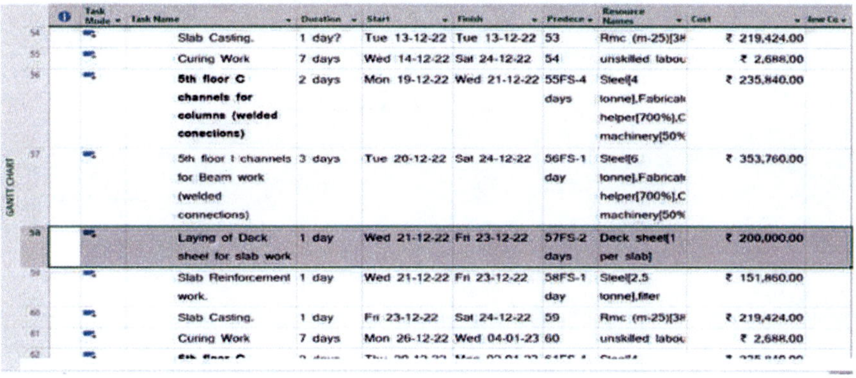

	Task Name	Duration	Start	Finish	Predecessors	Resource Names	Cost
54	Slab Casting.	1 day?	Tue 13-12-22	Tue 13-12-22	53	Rmc (m-25)[3	₹ 219,424.00
55	Curing Work	7 days	Wed 14-12-22	Sat 24-12-22	54	unskilled labou	₹ 2,688.00
56	**5th floor C channels for columns (welded conections)**	2 days	Mon 19-12-22	Wed 21-12-22	55FS-4 days	Steel[4 tonne],Fabricat helper[700%],C machinery[50%	₹ 235,840.00
57	5th floor I channels for Beam work (welded connections)	3 days	Tue 20-12-22	Sat 24-12-22	56FS-1 day	Steel[6 tonne],Fabricat helper[700%],C machinery[50%	₹ 353,760.00
58	Laying of Deck sheet for slab work	1 day	Wed 21-12-22	Fri 23-12-22	57FS-2 days	Deck sheet[1 per slab]	₹ 200,000.00
59	Slab Reinforcement work.	1 day	Wed 21-12-22	Fri 23-12-22	58FS-1 day	Steel[2.5 tonne],fitter	₹ 151,860.00
60	Slab Casting.	1 day	Fri 23-12-22	Sat 24-12-22	59	Rmc (m-25)[3	₹ 219,424.00
61	Curing Work	7 days	Mon 26-12-22	Wed 04-01-23	60	unskilled labou	₹ 2,688.00

Fig. 22 Gantt chart representing schedule and cost by CPM

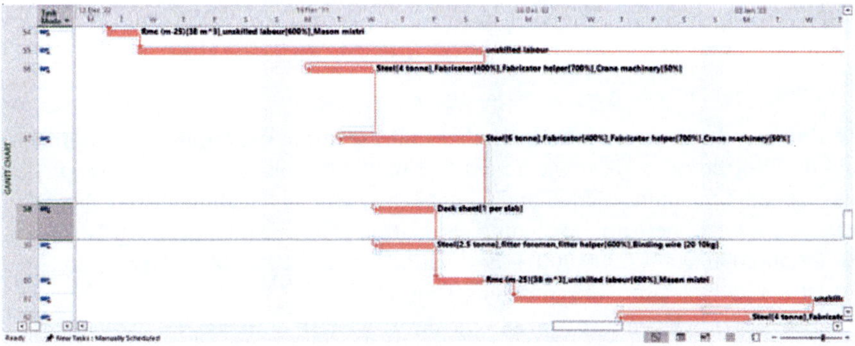

Fig. 23 Gantt chart representing graph (Fig. 22)

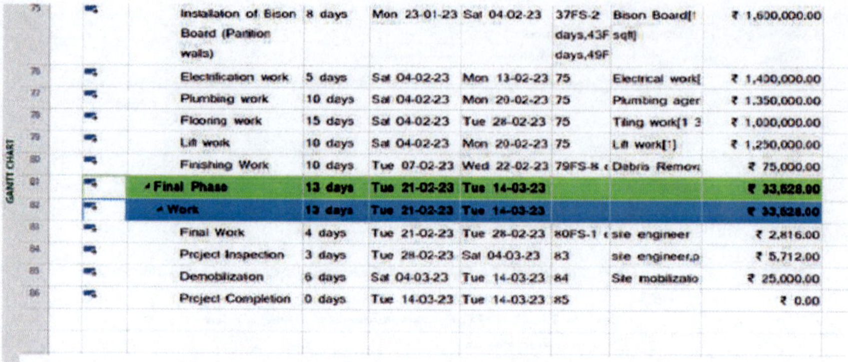

75		Installation of Bison Board (Partition walls)	8 days	Mon 23-01-23	Sat 04-02-23	37FS-2 days,43F days,49F	Bison Board[1 sqft]	₹ 1,600,000.00
76		Electrification work	5 days	Sat 04-02-23	Mon 13-02-23	75	Electrical work[	₹ 1,400,000.00
77		Plumbing work	10 days	Sat 04-02-23	Mon 20-02-23	75	Plumbing ager	₹ 1,350,000.00
78		Flooring work	15 days	Sat 04-02-23	Tue 28-02-23	75	Tiling work[1 3	₹ 1,000,000.00
79		Lift work	10 days	Sat 04-02-23	Mon 20-02-23	75	Lift work[1]	₹ 1,250,000.00
80		Finishing Work	10 days	Tue 07-02-23	Wed 22-02-23	79FS-8 c	Debris Remov	₹ 75,000.00
81		Final Phase	13 days	Tue 21-02-23	Tue 14-03-23			₹ 33,528.00
82		Work	13 days	Tue 21-02-23	Tue 14-03-23			₹ 33,528.00
83		Final Work	4 days	Tue 21-02-23	Tue 28-02-23	80FS-1 c	site engineer	₹ 2,816.00
84		Project Inspection	3 days	Tue 28-02-23	Sat 04-03-23	83	site engineer,p	₹ 5,712.00
85		Demobilization	6 days	Sat 04-03-23	Tue 14-03-23	84	Site mobilizatio	₹ 25,000.00
86		Project Completion	0 days	Tue 14-03-23	Tue 14-03-23	85		₹ 0.00

Fig. 24 Gantt chart representing schedule and cost by CPM

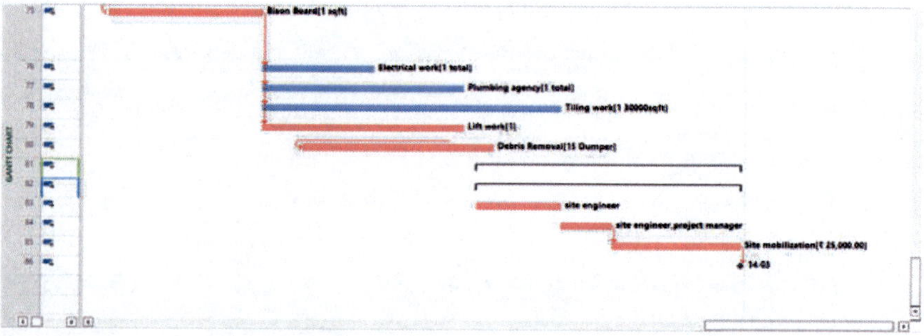

Fig. 25 Gantt chart representing graph (Fig. 24)

Table 1 Duration and cost in CPM by using MSP

Start	Cost	Finish	Scheduled duration
08-07-22	Rs. 23,524,928	14-03-23	154 days

- **Site Data**

Considering the observations and the data collected from site through the Contractor and the Project incharge, it was found that the total time period for its completion was around 7.5 months (220–225 days) and the cost of expenditure for 32,000 sq ft. area was Rs. 850 per sq ft. excluding excavation part which tends to a total sum of Rs. 28,500,000 (Rs. 27,200,000 + Excavation part Rs. 1,300,000). Hereby attached the letter requested from the contractor (Fig. 26).

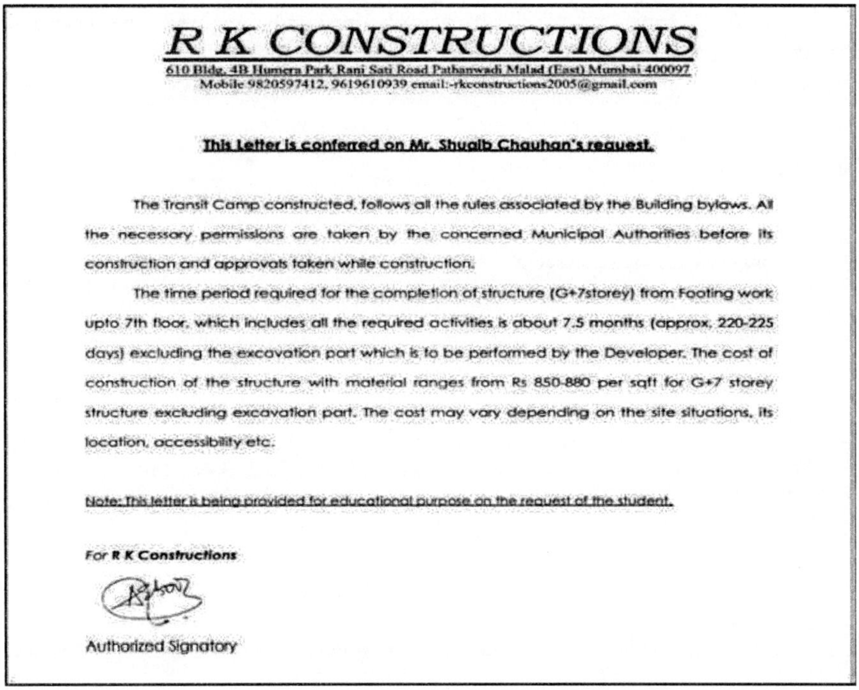

R K CONSTRUCTIONS

610 Bldg. 4B Humera Park Rani Sati Road Pathanwadi Malad (East) Mumbai 400097
Mobile 9820597412, 9619610939 email:-rkconstructions2005@gmail.com

This Letter is conferred on Mr. Shuaib Chauhan's request.

The Transit Camp constructed, follows all the rules associated by the Building bylaws. All the necessary permissions are taken by the concerned Municipal Authorities before its construction and approvals taken while construction.

The time period required for the completion of structure (G+7storey) from Footing work upto 7th floor, which includes all the required activities is about 7.5 months (approx. 220-225 days) excluding the excavation part which is to be performed by the Developer. The cost of construction of the structure with material ranges from Rs 850-880 per sqft for G+7 storey structure excluding excavation part. The cost may vary depending on the site situations, its location, accessibility etc.

Note: This letter is being provided for educational purpose on the request of the student.

For **R K Constructions**

Authorized Signatory

Fig. 26 Actual construction data from site

5 Discussion

Critical Path Method (CPM) technique provides a reliable and easy to monitor solution to the construction management problems thereby preventing excess cost and delays. The activities are properly scheduled thereby preventing any more delays during execution which makes the schedule robust and achievable. CPM does not change during the execution of projects that helps the higher management to prioritize the activities and the resources associated with the same. With the added benefits of duration and cost effectiveness it becomes practical for its application in the industry for such kind of projects.

6 Conclusion

A Transit camp project was identified and studied for its chronological activities with their inter-dependencies and the resource constraints.

Live project was scheduled by the (CPM) Critical path method by using MSP software.

The aspects of time, cost, flexibility was obtained and compared. These types of projects are mostly feasible for Slum Rehabilitation Projects.

References

1. www.construction-cpm.com
2. www.project+crashing+graph.com
3. Hamidullah SH et al (2007) Time cost optimization in project execution. IJARIIE. ISSN (O)-2395-4396
4. Kashid PM et al (2019) Time and cost optimization of construction projects: a review. IJESRT
5. Akcay C (2019) Time-cost optimization model proposal for construction projects with genetic algorithm and fuzzy logic approach. Institute of Graduate Studies in Science and Engineering, Istanbul University. Date of acceptance/reception: 02.12.2019/06.05.2019
6. Klanšek U et al. Cost optimization of time schedules for project management. J Econ Res Ekonomska Istraživanja 23(4):22–36. https://doi.org/10.1080/1331677X.2010.11517431
7. Kumar VM et al (2017) Comparative study of time-cost optimization. Int J Civ Eng Technol (IJCIET) 8(4)
8. Rajguru A et al. Effective techniques in cost optimization of construction project: an review. Int J Res Eng Technol (IJRET). eISSN: 2319-1163, pISSN: 2321-7308
9. Pšunder M et al. Cost optimization of construction project schedules. https://www.irbnet.de/daten/iconda/CIB16219.pdf
10. James Babu Raj K et al (2016) Time and cost optimisation in construction using M.S project. Int J Adv Res Trends Eng Technol (IJARTET) 3(Special Issue 2)
11. Sundarrajan S et al (2013) Investigation of significant factors influencing time and cost overruns in Indian construction projects. Int J Emerg Technol Adv Eng 3(10). ISSN 2250-2459. ISO 9001:2008 Certified Journal
12. Kamble U et al (2018) Implementing time cost trade optimization in commercial building using project management techniques in Microsoft project. Int Res J Eng Technol (IRJRET) 5
13. Naveenkumar GV et al (2016) Factors influencing time and cost overruns in construction projects. Int J Innov Res Sci Eng Technol 5(4). An ISO 3297: 2007 Certified Organization
14. Trivedi MK et al (2015) Use of optimization techniques in time cost tradeoffs (TCT) in civil construction: an overview. J Int J Civ Eng Mech 2
15. Tatar BV et al. Time cost optimization technique in construction project management. Int J Latest Trends Eng Technol 7(3)

WRE

Principal Causes of Soil Erosion in a Watershed from the Ganga Basin, India: Evidence from Land Use Land Cover Dynamics

Nikita Shivhare Mitra, Akansha Rupal Nath, Khushboo Pachori, Shyam Bihari Dwivedi, and Prabhat Kumar Singh Dikshit

Abstract Soil degradation is the primary issue faced by some of the countries including India. The main reason behind this problem is the soil erosion. So the primary objectives of this paper were to assess the potential impact of Land use land cover (LULC) dynamics and other principal causes of soil erosion on sediment yield. The evaluation of the impact of LULC dynamics was done with three scenarios (LULC of 2004 with climate data of 1995–2004, LULC of 2015 with climate data of 2005–2015, LULC of 2004 with climate data of 2005–2015). The soil and water assessment tool (SWAT) was used for all three scenarios to estimate the sediment yield. The calibration and validation of the SWAT model were done using the Sufi-2 algorithm. Multivariate linear regression technique was used to find out most dominating causes accountable for soil erosion. This study shows that due to the transition of Agriculture and Forest land to urban land and range land the average annual runoff is increased by 13% whereas the sediment yield is decreased by 26%. As the result of regression technique, we estimated that the coefficients of the variables for slope >40, Barren land type, agriculture land type and Soil type 3 (having Hydrological soil group C, with sandy Clay texture, and percentage of sand is dominant in this soil type) are the highest with values 2.95, 0.49, 0.25, and 0.13 respectively, which concludes that the slope is the dominant cause of soil erosion. Barren land and agricultural land are the most soil erosion prone LULC classes. Moreover, among the

Maps Disclaimer: The presentation of material and details in maps used in this chapter does not imply the expression of any opinion whatsoever on the part of the Publishers or Author concerning the legal status of any country, area or territory or of its authorities, or concerning the delimitation of its borders. The depiction and use of boundaries, geographic names and related data shown on maps and included in lists, tables, documents, and databases in this chapter are not warranted to be error free nor do they necessarily imply official endorsement or acceptance by the Publisher or Author.

N. S. Mitra (✉) · A. R. Nath · K. Pachori
Oriental College of Technology, Bhopal, India
e-mail: dr.nikitamitra.iitbhu@gmail.com

S. B. Dwivedi · P. K. S. Dikshit
Indian Institute of Technology (IIT), Banaras Hindu University (BHU), Varanasi, India

K. K. Pathak et al. (eds.), *Latest Developments in Civil Engineering*, Lecture Notes in Civil Engineering 352, https://doi.org/10.1007/978-981-99-2676-3_71

soil classes, soil type 3 was dominant causes of soil erosion. The R coefficient of the techniques was equal to 0.93 which shows the efficiency of the result. These results can be further utilized in land use planning and for applying effective measures for soil conservation.

Keywords SWAT · Multivariate linear regression · Soil erosion · Sufi-2 · LULC

1 Introduction

The soil degradation is one of the pressing issues in developing countries like India, where the agriculture and farming are the backbones of the country's economy. The primary cause of soil degradation is water erosion [1]. India is a rapidly developing country where the urbanization and industrialization are increasing exponentially. Also, the LULC pattern is changing very frequently. The rapid change in land use pattern is resulting in a change in the runoff, soil erosion, and other hydrological processes. So, it is imperative to estimate the impact of LULC dynamics for sustainability of the soil. Soil loss due to erosion is a global problem, on average yearly, about 0.9 mm of soil gets lost throughout the world. Soil erosion is a key ecological process that reflects specific characteristics, climate regimes, vegetation cover and land mismanagement depending upon natural and anthropogenic factors [2, 3] so it is essential to apply measures for reducing soil erosion. For applying effective means which can be utilized for the soil conservation, it is necessary to get the proper information in detail about the dominating causes of the soil erosion of that particular area. There are various methods developed to estimate soil erosion like universal soil loss equation (USLE), modified soil loss equation (MUSLE) and revised universal soil loss equation (RUSLE). Prioritization of soil erosion-prone areas [4, 5] and soil erosion estimation is done by various scientists using methods like USLE [6], MUSLE [7, 8] and RUSLE [9].

There are more than 50 physical models used for estimating soil erosion and sediment yield. Researchers have compared various soil erosion models [10, 11] and as a result it was revealed that soil and water assessment tool (SWAT), Water Erosion Prediction Project (WEPP), Agricultural Non-Point Source Pollution Model (AGNPS), Areal Nonpoint Source Watershed Environment Response Simulation (ANSWERS) & SHETRAN are most promising models [12]. There is much work done for application of SWAT for sediment yield response modeling under land use dynamics and other issues [13–16].

Soil erosion depends on many factors like rainfall, runoff, land type, soil type & slope. Assessing the impact of rainfall, slope gradient [17], land use [18], wind effect [19] it was found that, the more the amount of rainfall or runoff would be the more would be the soil erosion. The area which was mostly covered by wasteland and barren land would be more prone to soil erosion. When the soil gradient would be more the soil erosion would be more [18]. However, when these factors combined in different watersheds, it is difficult to find out which element is more dominant and which watershed would have more soil erosion, so it is imperative to determine the weight of these factors for effective soil management and to apply effective soil conservation measures. Wynn and Mastaghimi [20] used multiple linear regression analysis to determine the impact of vegetation and soil type on soil erosion.

In the present study, i.e., "Principal Causes of Soil Erosion in a Watershed from the Ganga Basin, India: Evidence from Land Use Land Cover Dynamics" can motivate the experts to formulate and implement effective and sustainable measures to minimize the undesirable effects of LULC dynamics and other causes of soil erosion. Here three scenarios (**Scenario 1**: LULC map of the year 2015 with climate data of the year 2005–2015. **Scenario 2**: LULC map of the year 2004 with climate data of the year 2005 to 2015. **Scenario 3**: LULC map of the year 2015 with climate data of the year 1996 to 2005) were created for modeling impacts of LULC dynamics on sediment yield. Twenty years of daily meteorological data were used to calculate the sediment yield of the study area using SWAT 2012. The area was divided into 46 subwatershed and 760 Hydrological response units (HRUs). The Sufi-2 algorithm was used for model calibration. The leading causes of the soil erosion were characterized and to find out principal causes of the soil erosion; multivariate linear regression statistical technique was used. The primary objectives of this study are:

1. To model the impact of LULC dynamics on sediment yield and calibrate the swat result using the Sufi-2 algorithm.
2. To find out the highest soil erosion impacted watersheds and find out the most soil erosion-prone soil type and land classes.
3. To categorize the causes of soil erosion and give the weightage of to each cause using multivariate linear regression.
4. To validate the result.

2 Study Area and Data Used

The Ganga basin in India covers five significant States Uttar Pradesh, Madhya Pradesh, Bihar Rajasthan and West Bengal. Full Uttar Pradesh comes inside Ganga basin. The study area is a part of Ganga basin situated in the south-east of Uttar Pradesh state as shown in Fig. 1, which covers four significant districts Varanasi, Balia, Babatpur, and Ghazipur. The latitude and longitude of the study area lie between latitude 82°22′7.511″E and 83°38′3.508″E, longitude 24°33′16.402″N and 25°51′47.728″N. The total area of the watershed is 15621.612 km^2.

There are two types of data required in the present study for modeling Spatial data (DEM, LULC map, Soil map) and Temporal data (Hydrological and climate data). For temporal data, 20 years of daily climatic data was procured from the Indian metrological department (IMD) from the year 1996 to 2015. The data consist of daily rainfall data solar, radiation data, minimum and maximum temperature data, relative humidity data and wind speed data. For spatial data, Shuttle radar thematic mapper (SRTM) Digital elevation model (DEM) of resolution 30 m was procured from USGS Earth Explorer website for watershed delineation. Soil data was collected

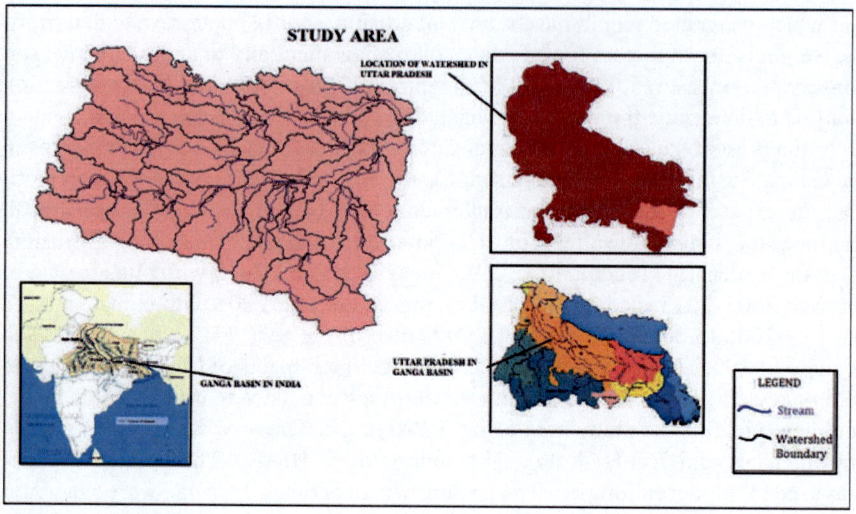

Fig. 1 Study area

from National Bureau of soil survey (NBSS). Also, some of the soil data was estimated by the lab experiments of soil samples. The observed data of the sediment yield for Varanasi station to be used for calibration was procured from Central Water Commission Varanasi.

3 Methodology

Assessment of the impact of LULC dynamics on sediment yield was done by creating three scenarios and comparing their results.

Scenario 1: LULC map of the year 2015 with climate data of the year 2005–2015. This scenario was prepared as a baseline from which the other scenarios would be compared for analysis and results.

Scenario 2: LULC map of the year 2004 with climate data of the year 2005–2015. This scenario was created to assess the effect of Land use land cover change on sediment yield.

Scenario 3: LULC map of the year 2015 with climate data of the year 1996–2005. This scenario was created to assess the effect of land use pattern along with the climate and other factors affecting soil erosion.

For preparing the LULC map the satellite imagery of Landsat 8 and Landsat 5 for the year 2015 and 2004 respectively were procured from USGS. Supervised image classification was done using the Erdas imagine tool. The LULC map for 2004 and 2015 are given in Fig. 2a and b respectively. After this, the change detection between two maps was done using ArcGIS, which analyzed the percentage transition of one

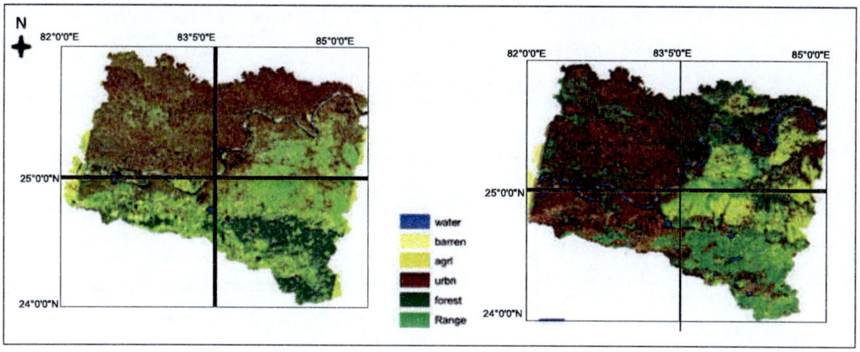

(a) LULC map of 2004 (b) LULC map of 2015

Fig. 2 **a** LULC map of 2004. **b** LULC map of 2015

land type to another. Table 1 Summarizes the change in LULC map for the different period after this, the sediment yield modeling was done using ArcSWAT for assessing the impact of LULC dynamics on sediment yield.

Methodology for estimating the principal causes of soil erosion and steps followed in this study is given in Fig. 3. At first, the spatial and temporal data were given as inputs in SWAT model using ArcSWAT. As an output, we got the calculated sediment yield values for the watershed. The calibration and validation step is done using Sufi-2 algorithm and Swat-Cup. Sediment yield per square kilometer for all the sub-watersheds are calculated, and the sub-watersheds were given rank according to these sediment yield values. After this step, all the parameters responsible for soil erosion were categorized.

For finding out the weight of these parameters, all the parameters were needed to be in the same unit. So for this sediment yield per square kilometer is estimated for all the soil type, land use land cover classes, and slope classes as shown in Table 2. Then using multivariate linear regression technique, the coefficient of each parameter was calculated using Matlab tool. After getting the coefficients of each parameter, these

Table 1 Summary of LULC pattern of a watershed for the different period

SWAT code	LULC type	Area referred to year (km^2)				% change
		2004	% of total area	2015 (%)	Total	
WATR	Water body	258	1.652152	314	2.010629	0.358478
BARR	Barren land	1256	8.043033	1268	8.119357	0.076324
AGRL	Agriculture	1678	10.74539	1071	6.857911	−3.88748
URBN	Urban	5098	32.646	6083	38.95114	6.305139
FRSD	Forest	3839	24.58376	2048	13.11391	−11.4698
RNGE	Rangeland	3487	22.32966	4833	30.94704	8.617383
	Total	**15,616**	**100**	**15,617**	**100**	

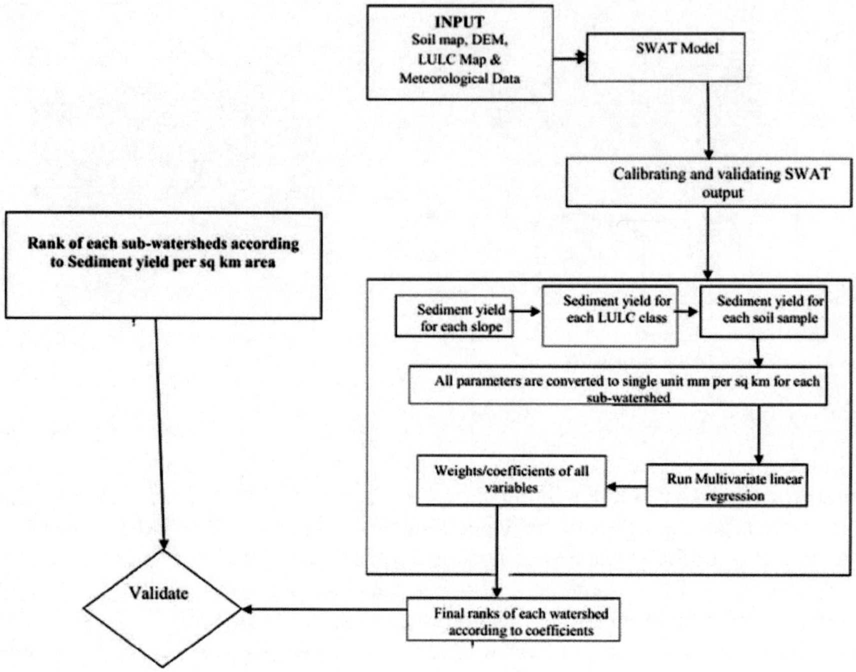

Fig. 3 Flowchart of methodology

Coefficients were multiplied to their respective values of each sub-watershed, and the ranking of each sub-watershed was calculated.

3.1 SWAT Modeling

The semi-distributed watershed model SWAT with ArcGIS interface was used for sediment estimation. In SWAT modeling we require DEM, LULC map, soil map, and meteorological data for processing. LULC was prepared using supervised image classification using Erdas imagine tool. The study area was classified into six land classes namely urban, range forest, agriculture, barren, and water body. For creating soil map NBSS data and field soil samples were taken, the soil map was divided into five classes, description is given in Table 3, at first watershed, was delineated using DEM in which 46 sub-watersheds were created, then HRU analysis was done using Slope, LULC map, and Soil map. The watershed was divided into 760 HRUs having unique properties. After this, by using HRUs and weather data, the SWAT model was run to calculate Sediment Yield.

Table 2 Sediment yield per square kilometer values for all parameters and Runoff per square kilometer for all Subwatersheds

SWS no.	Soil 1	Soil 2	Soil 3	Soil 4	Soil 5	Forest	Urban	Range
1	97	0	0	0	3	0	66.80853	27.82152
2	99.5	0	0	0	0.5	0	85.96467	9.652197
3	49	0	0	0	51	0.035806	55.00465	36.59376
4	40.86021	0	0	0	59.13979	0	68.42843	15.81991
5	21.55689	64.67066	0	0	13.77246	0	58.31792	21.47958
6	100	0	0	0	0	0	77.11304	19.6253
7	0	100	0	0	0	0	44.20087	24.04792
8	0	20.16575	0	0	79.83425	0.000291	73.90797	21.51272
9	53.94737	0	0	0	46.05263	0.000291	73.90797	21.51272
10	8.620689	0	0	0	91.37931	0.001986	69.64009	22.399
11	0	0	0	0	100	0.018086	66.93168	27.62285
12	69.3609	0	0	0	30.6391	0.012794	55.17699	27.25806
13	74.45652	0	0	0	25.54348	0.035964	43.99985	27.2349
14	0	0	0	0	99.9	0.053743	57.9113	33.79599
15	0	0	0	0	100	0.259348	51.09177	36.8988
16	0	0	0	0	0.1	0.103713	54.28334	39.00297
17	0	0	0	37.54266	62.45734	0.124595	33.31602	47.08508
18	0	0	0	0	100	0.679323	32.67078	54.24413
19	0	0	0	0	100	1.931718	54.68402	31.0285
20	0	0	0	0	100	0.45018	61.66894	30.67508

(continued)

Table 2 (continued)

SWS no.	Soil 1	Soil 2	Soil 3	Soil 4	Soil 5	Forest	Urban	Range
21	0	0	0	0	100	0.233247	63.68019	20.22738
22	0	0	0	0	100	1.638689	19.13898	60.77995
23	0	0	32	0	68	11.93744	16.98824	47.167
24	0	0	0	0	100	4.860466	13.09245	64.00867
25	0	0	3.298351	63.26836	33.43328	2.115549	23.92578	57.28991
26	0	0	6.115703	0	93.8843	3.400867	19.97968	64.63125
27	0	0	0	0	100	1.467889	58.3525	25.34156
28	0	0	0	0	100	9.713248	28.77396	21.9613
29	0	0	0	0	100	17.76799	22.06098	21.36708
30	0	0	0	0	100	7.004509	43.36769	24.91538
31	0	0	35.88957	0	64.11043	26.56368	19.87888	33.20519
32	0	0	46	0	54	10.51796	15.02628	56.5443
33	0	0	70	0	30	24.91066	14.7603	34.60749
34	0	0	0	0	100	3.443041	26.56206	28.79576
35	0	0	55.75	3.96	40.28	25.13263	16.20974	41.91578
36	0	0	76	18	6	7.562203	10.79214	55.308
37	0	0	55.71066	0	44.28934	21.80786	15.24055	29.16799
38	0	0	17.45562	0	82.54438	6.464401	28.77973	32.66317
39	0	0	65.79804	0	34.20195	22.29363	10.81614	37.77232

(continued)

Table 2 (continued)

SWS no.	Soil 1	Soil 2	Soil 3	Soil 4	Soil 5	Forest	Urban	Range
40	0	0	46.83698	0	53.16302	16.00889	19.49184	36.1876
41	0	0	33.75	0	66.25	22.69351	18.67113	35.36371
42	0	0	87	0	13	51	4	27
43	0	0	88	0	12	50.55837	4.954822	26.48421
44	0	0	35.43689	0	64.56311	33.95824	5.862145	34.49773
45	0	0	41.59292	0	58.40708	50.97362	1.007673	25.59616
46	0	0	45.08621	0	54.91379	47.08228	2.344216	26.26476
SWS no.	**Agriculture**	**Barren**	**Slope 1 0–10**	**Slope 2 10–20**	**Slope 3 20–30**	**Slope 4 30–40**	**Slope 5 >40**	**Runoff**
1	1.42838	2.614343	99.8	0.2	0	0	0	16.0495
2	0.396351	3.2878	100	0	0	0	0	8.113267
3	3.252676	3.898018	99.05	0.05	0	0	0	12.79222
4	3.054291	5.695412	97.56486	2.330938	0.104207	0	0	46.08896
5	3.840518	9.698103	98.38277	1.611267	0.005968	0	0	48.15576
6	1.026104	1.668914	99.5	0.5	0	0	0	9.556494
7	6.627581	7.17818	93.51598	6.30137	0.182648	0	0	161.8754
8	1.088247	1.528956	99.09	0.01	0	0	0	15.07964
9	1.088247	1.528956	91	9	0	0	0	14.90437
10	2.778771	1.346681	97.71184	2.27029	0.017876	0	0	97.8062
11	2.746352	2.271188	99.97578	0.024214	0	0	0	4.148779
12	3.157705	7.557644	97.08936	2.621843	0.286912	0.001888	0	19.56537
13	6.151137	12.74997	96.89715	3.09169	0.011161	0	0	35.13165

(continued)

Table 2 (continued)

SWS no.	Agriculture	Barren	Slope 1 0–10	Slope 2 10–20	Slope 3 20–30	Slope 4 30–40	Slope 5 >40	Runoff
14	4.311021	3.315601	100	0	0	0	0	7.098513
15	4.886012	6.367678	99.97519	0.024807	0	0	0	7.050139
16	2.841734	3.671438	100	0	0	0	0	191.7258
17	12.99861	5.478566	99.9	0.9	0	0	0	19.69125
18	1.289183	10.41119	99.91795	0.080403	0.001641	0	0	9.78696
19	6.269737	5.245166	99.99	0.01	0	0	0	4.530954
20	3.34624	2.191411	99.75927	0.232837	0.007893	0	0	5.624228
21	1.668779	6.748559	97	3	1	0	0	12.35488
22	4.44787	13.45781	100	0	0	0	0	70.76215
23	3.914441	19.05347	87.97462	5.861327	2.476566	2.523982	1.163512	75.10404
24	7.242787	9.560244	99.48255	0.517447	0	0	0	36.59072
25	11.34632	4.675402	100	0	0	0	0	14.24379
26	3.460178	7.847713	99.74391	0.249439	0.006652	0	0	12.84106
27	3.974104	3.911369	97.77029	1.706922	0.473128	0.049665	0	14.43394
28	20.20389	10.72474	96.0288	0.549913	0.001791	0	0	11.29506
29	20.90351	9.671475	92.40736	4.405286	1.58072	1.554807	0.051827	187.3673
30	7.455111	7.624897	96.5091	2.8615	0.616423	0.012977	0	19.97134
31	4.639314	14.0621	92.63258	5.557771	1.331166	0.426341	0.052142	48.96167
32	8.073002	8.693408	90.92982	2.094957	1.469726	2.018974	3.486529	39.4418

(continued)

Table 2 (continued)

SWS no.	Agriculture	Barren	Slope 1 0–10	Slope 2 10–20	Slope 3 20–30	Slope 4 30–40	Slope 5 >40	Runoff
33	6.202178	17.61733	90.99014	5.501881	2.074958	1.335754	0.097264	68.81961
34	27.19862	8.55849	97.30195	2.456946	0.241102	0	0	57.3152
35	10.64716	4.654725	70.71709	9.220831	6.57909	6.737883	6.745101	74.72728
36	12.06629	12.83493	86.1216	2.103377	2.103377	2.9	6.68796	90.93697
37	16.68063	13.01081	93.76973	4.659081	1.166318	0.402392	0.002476	22.66037
38	16.1967	9.190116	93.14838	4.980309	1.205105	0.609955	0.056258	47.84111
39	8.482526	19.71601	87.03614	8.371282	2.984207	1.579304	0.029067	46.79409
40	7.488992	16.61367	83.56728	10.19347	2.785289	2.390727	1.063228	26.20497
41	6.459953	14.63288	87.90092	6.734767	2.046646	1.976135	1.34153	38.21097
42	11	5	0	0	58	18	9	40.67212
43	8.353969	6.342959	51.76612	19.88242	11.19939	11.18463	5.967432	42.57624
44	12.00439	12.4086	86.09869	7.785459	2.79478	1.92368	1.39739	31.29254
45	11.23164	9.472391	62.56385	20.44895	9.627982	6.860388	0.498828	49.14521
46	10.03317	11.34751	73.8623	16.57925	5.477576	3.236434	0.844438	16.46584

Table 3 Soil properties used for SWAT modeling

SNAM (soil name)	Soil 1	Soil 2	Soil 3	Soil 4	Soil 5
HYDGRP (Hydrological soil group)	D	C	D	C	C
TEXTURE	L	L	CL	SCL	SL
SOL_Z1 (depth of soli layer 1)	300	300	300	300	300
SOL_BD1 (bulk density of layer 1)	1.5079	1.5583	1.5043	1.5847	1.5952
SOL_AWC1 (available water capacity of layer 1)	0.1441	0.1192	0.1327	0.0984	0.0893
SOL_K1 (hydraulic conductivity in mm/hr of layer 1)	4.1604	9.7772	4.0563	12.9805	30.1535
SOL_CBN1 (carbon content in layer 1)	0.6	0.8	0.8	0.7	0.6
CLAY1 (percentage of clay in layer 1)	28	22	30	21	3
SILT1 (percentage of silt in layer 1)	43	31	34	21	23
SAND1 (percentage of sand in layer 1)	30	47	36	58	64
SOL_ALB1 (soil albedo in layer 1)	0.3971	0.346	0.346	0.3707	0.3971
USLE_K1 (USLE erodibility factor in layer 1)	0.1719	0.163	0.1589	0.1569	0.1687
SOL_BD2 (Bulk density of layer 2)	1.535	1.5561	1.4814	1.5972	1.6026
SOL_AWC2 (available water capacity of layer 2)	0.1202	0.1158	0.1387	0.0957	0.0914
SOL_K2 (hydraulic conductivity in mm/hr of layer 2)	3.6147	10.3717	2.4191	12.8314	21.7927
SOL_CBN2 (carbon content in layer 2)	0.5	0.9	0.4	0.5	0.5
CLAY2 (percentage of clay in layer 2)	31	22	34	21	16
SILT2 (percentage of silt in layer 2)	26	28	36	20	22
SAND2 (percentage of sand in layer 2)	43	49	29	59	62
SOL_ALB2 (soil albedo in layer 2)	0.4254	0.323	0.4557	0.4254	0.4254
USLE_K2 (USLE erodibility factor in layer 2)	0.1561	0.158	0.1647	0.1583	0.1658

Note Soils 1–5 are explained in this table

3.2 Calibration and Validation of SWAT

As noted earlier, calibration is the modification of the elements influencing the SWAT yields, assisted by observed values and evaluated estimations of runoff, evapotranspiration, and other SWAT outputs. Validation compares the results of SWAT to the observed data without modification of the values of the influencing factors. Calibration is necessary for proper hydrological modeling. The twenty years of the dataset was divided into two halves from 1996–2005 and 2006–2015, where the first half is used for calibration and the second half is used for validation. Groundwater delay (GW_DELAY), Manning's "n" value for overland flow (OV_N), Saturated hydraulic conductivity (SOL_K), Base flow alpha factor (ALPHA_BF), Manning's "n" value for the main channel (CH_N2), Effective hydraulic conductivity

Table 4 Parameters used in the present study for calibration

Parameters	Best estimate values of parameters and their minimum and maximum range
GW_DELAY	191.07 (100.04, 300.00)
SOL_K	0.1 (0.58, 0.34)
ALPHA_BF	0.51 (0.23, 0.74)
OV_N	0.06 (0.00, 0.11)
CH_N2	0.2 (0,0.3)
CH_K2	83.95 (69.42, 150.00)
CN2	27.0 (29.00, 7.23)
USLE_P	0.3 (0,1)
SOL_AWC	0.07 (0.05, 0.15)
SURLAG	2.0 (0.05,23)
USLE_K	0.17 (0.01,0.63)

in main channel alluvium (CH_K2), SCS runoff curve number (CN2), USLE equation support parameter (USLE_P), Available water content (SOL_AWC), Surface runoff lag time (SURLAG), USLE equation soil erodibility (K) factor (USLE_K) are the parameters adjusted for calibration. Table 4 describes the parameters used in the present study for calibration with their best-estimated values and minimum–maximum ranges. The calibration was done using Sufi-2 and Swat-cup. The observed data were procured from CWC Varanasi for calibration.

3.3 *Multivariate Linear Regression Technique*

In the field of statistics, a linear regression is an algorithm of modeling a linear relationship between variables. One of the variables is the result or dependent variable which depends on one or more input variables or independent variables. The instance when only one variable is related linearly with the output variable is called as simple linear regression. When the case is to linearly relate one dependent variable with multiple explanatory variables it is called as multivariate linear regression technique. Here the computation would be complex due to added variables and multiple coefficients have to be determined. The equation defined for multivariate linear regression technique is:

$$Y = \alpha + \beta 1 X_1 + \beta 2 X_2 + \beta 3 X_3 \ldots\ldots\ldots \beta_n X_n \tag{1}$$

where

Y = output variable.

$X_1, X_2 \ldots\ldots\ldots X_n$ = independent or explanatory variables.

$\beta_1, \beta_2, \ldots\ldots \beta_n$ = Coefficients of explanatory variables and α is the constant.

3.4 Estimating the Weight/Coefficients of Parameters

To find out the impact of the causes of soil erosion, all the principal causes were categorized. Table 5 shows the land type wise classification that shows which land type class is more prominent to soil erosion. Table 6 shows the soil sample which is more prominent to soil erosion. For finding out the impact of runoff on soil erosion, the watershed was categorized from ascending order of runoff as shown in Fig. 4a and compared to the Fig. 4b which has given the watershed ascending order of the ranks according to the sediment yield per square kilometer of the watershed. By comparison, it could be seen that when the runoff is more the value of soil erosion is also high. However, in some cases where the values of runoff are comparatively less still the values of the soil erosion were high, which shows that run off is not the only cause of soil erosion. Logically it is understood that the steep slope will result in more soil erosion. However, it is difficult to find out which parameter among the steep slope, land type classes and soil samples are the most dominant parameter responsible for soil erosion. To find out the principal cause of soil erosion and the exact weight of these causes the multivariate linear regression analysis was done. For finding out the coefficient of all the parameters, we need to have all the parameters in the same unit.

For doing this, as the first step for each watershed the percentage of each soil samples, each LULC classes, and each slope classes were determined (calculation for one sub-watershed is shown in Fig. 5). In the next step, Table 1 was created including the values of all the parameters or 46 sub-watersheds. In the next step, multivariate regression analysis was applied using Matlab to find out the coefficient of all the

Table 5 Land type wise classification showing sediment yield for each LULC class

LULC type	Area in km^2	CN	AWC (mm)	USLE_LS	Precipitation (mm)	SED ton/ha
WATR	414.83	92	104.91	0.67	938.86	0
URBN	5683.33	83.21	102.13	0.18	917.93	3.45
RNGE	5233.71	80.43	103.27	0.26	936.31	4.71
AGRL	1071.01	84.28	105.53	0.48	959.55	11.92
BARR	1268.02	92.1	106.89	0.31	947.23	29.27
FRSD	1948.26	80.64	118.56	0.98	1008.96	1.38

Table 6 Soil type wise classification showing sediment yield for each soil sample

Soil	Area	Sed	Sed/km^2	Rank
Soil 3	8396.621	39,376.28	4.689539	1
Soil 5	8653.678	29,262.86	3.381552	2
Soil 4	15,434.71	42,474.61	2.751889	3
Soil 1	5540.322	1315.23	0.237392	4
Soil 2	912.8964	70.08	0.076767	5

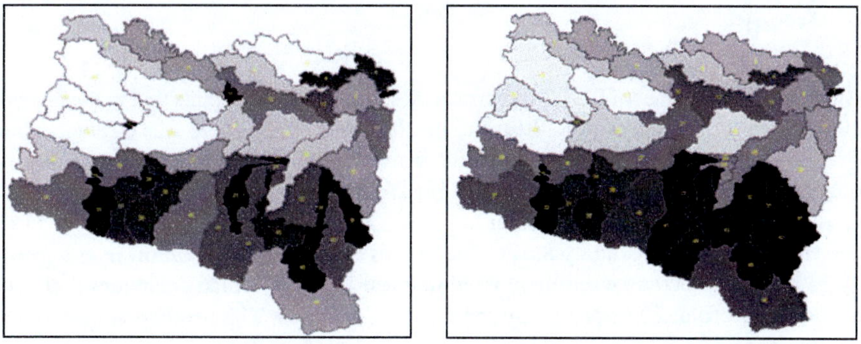

(a) watershed showing runoff, darker the watershed higher is the value of Runoff.

(b) watershed showing soil erosion, darker the watershed higher is the value of Soil erosion.

Fig. 4 **a** Watershed showing runoff, darker the watershed higher is the value of Runoff. **b** Watershed showing soil erosion, darker the watershed higher is the value of soil erosion

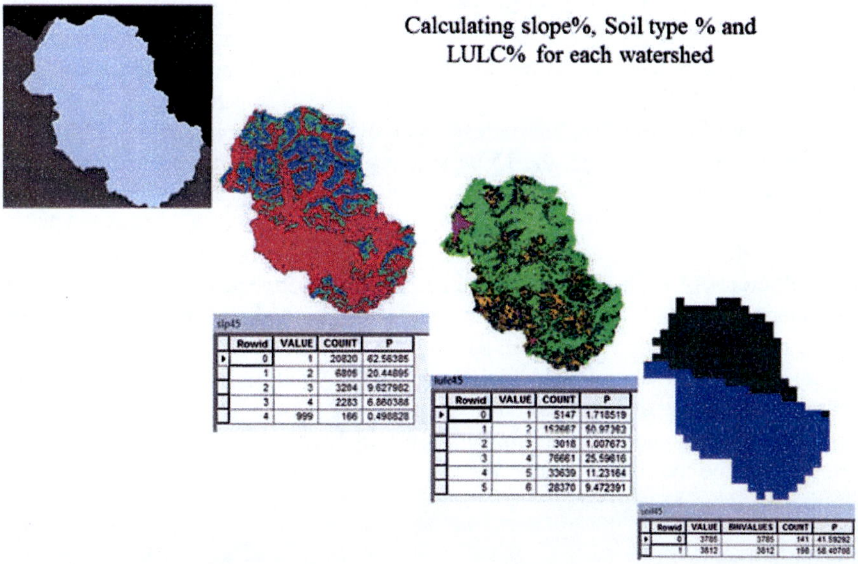

Fig. 5 Calculating the percentage of each slope, soil type and LULC class for one sub-watershed

parameters, as the last step the values of all the coefficient were multiplied with their corresponding parameter values and the compound value for each parameter was calculated. According to these valueseach sub-watershed was ranked. Moreover, this rank was then validated with the original rank of the sub-watershed according to their sediment yield values.

4 Results

Analysis of the impact of LULC dynamics on sediment yield depicts that in the period 2004–2015 (comparing the LULC map of 2004 and 2015), there was a remarkable change found in the land use pattern of the watershed due to urbanization and anthropogenic activities. The significant transitions observed were decreased in agricultural land by 3.8%, increase in the urban land by 6.3%, decrease in forest land by 11% and increase in range land by 8.6%. Due to these transitions, the runoff is increased by 13 and 26% decreases sediment yield in these 11 years, which concludes that soil erosion is profoundly impacted by the LULC dynamics. The detailed summary of the model response to land use land cover transition for all three scenarios is given in Table 7.

The results of the SWAT model for the year 2005–2015 revealed that the sub-watersheds 23, 7, 35, 29 and 36 are the most soil erosion prone sub-watersheds. Moreover, the maximum soil erosion occurs in July, August, and September. Also, it revealed that the soil erosion is more for soil sample 3 and 5 and land type barren and Agriculture and the slope greater than 40. The results of the Sufi 2 algorithm reveal that CH_K2, USLE_PCH_N2, SURLAG, and SOL_AWC are the most sensitive parameters. Figure 6 shows the results of calibration for sediment yield values from the year 2009–2015.

The result of the multivariate regression analysis is shown in Table 8. This table includes the coefficient of all the 15 parameters which are considered as the major causes responsible for soil erosion. They are soil samples (Soils 1–5), land type classes (urban, forest, Barren, agriculture and range), slope classes (0–10, 10–20, 20–30, 30–40 and >40), and runoff. This table also includes Standard error of all the coefficients of each parameter. The results conclude that the highest values of coefficient 2.5, 0.49, 0.25, 0.13 and 0.12 whose corresponding parameters are slope >40, Barren land, agricultural land, soil class 3 and soil 5. The R coefficient of the analysis i.e. the R^2 value is 0.89, which shows that the analysis has higher correlation. These results show that the slope and land type are the principal causes responsible for soil erosion. The standard errors are very low which show these values are efficient. After getting the coefficient values, these values were multiplied to the corresponding parameter giving equation as

Table 7 Model response to land use change

Temporal land use	Average annual precipitation (mm)	Average annual runoff (mm)	Annual sediment yield (ton/ha)
Scenario I	856.0	416.91	6.790
Scenario II	856.0	366.55	9.258
Scenario III	969.8	424.93	10.484
Change (I–II) (%)	0	13.7389	26.65803
Change (I–III) (%)	11.73438	−1.88737	35.23464

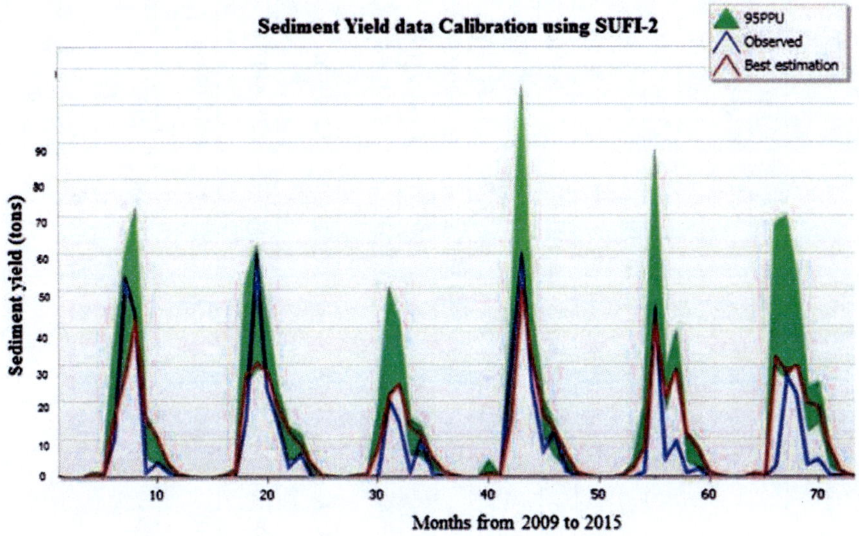

Fig. 6 The plot of estimated and observed values of sediment yield (mm)

$$
\begin{aligned}
Y = \alpha &+ 0.081\ \mathrm{X1} + 0.053\ \mathrm{X2} + 0.12\ \mathrm{X3} + 0.13\ \mathrm{X4} + 0.086\ \mathrm{X5} \\
&+ 0.14\ \mathrm{X6} + 0.21\mathrm{X7} + 0.13\mathrm{X8} + 0.25\mathrm{X9} + 0.49\ \mathrm{X10} - 0.036\ \mathrm{X11} \\
&+ 0.16\ \mathrm{X12} + 0.087\ \mathrm{X13} - 1.56\ \mathrm{X14} + 2.59\ \mathrm{X15} + 0.06\ \mathrm{X16}
\end{aligned}
$$

where,

Y = Sediment Yield, α = constant, X1 = Soil 1, X2 = Soil 2, X3 = Soil 3, X4 = Soil 4, X5 = Soil 5, X6 = Forest, X7 = Urban, X8 = Rangeland, X9 = Agriculture, X10 = Barren, X11 = Slope 1 (0–10), X11 = Slope 2 (10–20), X12 = Slope 3 (20–30), X13 = Slope 4 (30–40), X14 = Slope 5 (>40), X15 = Runoff.

After keeping these values in the equation, the compound value of each sub-watershed was calculated, and each sub watershed's rank was calculated. This rank was validated with the original ranks of the sub-watersheds. The result of the comparison is shown in the Fig. 7 as graph that was calculated as rank versus original ranks.

Table 8 Multivariate linear regression result

Variables	Coefficients	Standard error	T stat	*P*-value	Lower 95%	Upper 95%	Lower 95.0%	Upper 95.0%
Soil 1	0.08174197	0.04294113	1.903582	0.066931714	−0.0060825	0.169566439	−0.0060825	0.169566439
Soil 2	0.053207693	0.04682878	1.136218	0.265168036	−0.042567914	0.148983301	−0.042567914	0.148983301
Soil 3	0.123432684	0.06415191	1.924069	0.064209845	−0.0077727	0.254638069	−0.0077727	0.254638069
Soil 4	0.131382427	0.06242406	2.104676	0.044101419	0.003710886	0.259053969	0.003710886	0.259053969
Soil 5	0.086277888	0.0406273	2.123643	0.042353946	0.003185721	0.169370055	0.003185721	0.169370055
Forest	0.144347907	0.21259166	0.678991	0.502526718	−0.290450848	0.579146663	−0.290450848	0.579146663
Urban	0.213382437	0.2183652	0.977181	0.336561025	−0.233224553	0.659989427	−0.233224553	0.659989427
Rangeland	0.135574088	0.19883895	0.681829	0.500756964	−0.271097232	0.542245408	−0.271097232	0.542245408
Agriculture	0.251651926	0.27070548	0.929615	0.360244405	−0.302002951	0.805306804	−0.302002951	0.805306804
Barren	0.495353725	0.31777362	1.558826	0.129885206	−0.154566311	1.145273762	−0.154566311	1.145273762
Slope 1 0–10	−0.036294928	0.82277784	-0.04411	0.965116941	−1.719064546	1.64647469	−1.719064546	1.64647469
Slope 2 10–20	0.166832362	0.75803631	0.220085	0.82734699	−1.383525977	1.7171907	−1.383525977	1.7171907
Slope 3 20–30	0.087877065	1.07350471	0.08186	0.935320452	−2.107686588	2.283440718	−2.107686588	2.283440718
Slope 4 30–40	−1.569500815	1.09160328	-1.43779	0.16119658	−3.802080194	0.663078563	−3.802080194	0.663078563
Slope 5 >40	2.950917398	1.10581255	2.668551	0.012341833	0.689276783	5.212558014	0.689276783	5.212558014
Runoff	0.062831964	0.01541455	4.076146	0.000325212	0.031305669	0.09435826	0.031305669	0.09435826

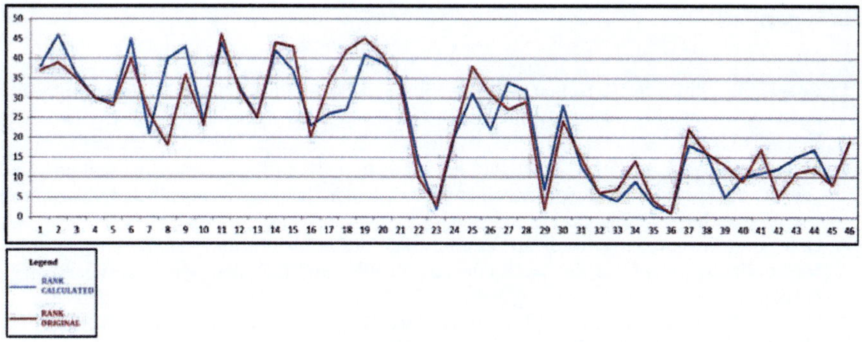

Fig. 7 Graph of calculated rank versus original rank

5 Conclusion

To model the remedies of the soil degradation problem, it is imperative to know the main parameters responsible for soil erosion; then only effective measures can be developed. So in this paper, the impact of LULC dynamics on sediment yield was analyzed and the statistics method multivariate linear regression and SWAT were both combined to find out the dominant cause of soil erosion. From the results, it can be concluded that there is a high impact of LULC dynamics on sediment yield due to land transitions, where 26% of sediment yield is decreased in 11 years. The land type and the slope are the principal reasons for soil erosion. So the management practices which can be directly applied to reduce the impact of the slope and land type would be most effective in the case of the present study area. The best management practices that can be applied as the remedies of soil erosion problem in this area could be (1) To convert the barren land into forest or rangeland. (2) To convert some percentage of barren and unused land to agricultural/forest land. Also, we can build some sediment filtration basin structures in the slope area for restricting sediment collection and soil erosion. (3) We can also model some runoff control hydraulic structures like bund, check dams, etc. (4) We can apply slope stability measures.

Acknowledgements We would like to thank the Indian Meteorological Department (IMD) in Pune, India for providing the daily meteorological data, and the National Bureau of Soil Survey and Land Utilization Planning in Nagpur India for providing the soil data.

References

1. Rahul AK, Shivhare N, Kumar S, Dwivedi SB, Dikshit PKS (2021) Modelling of daily suspended sediment concentration using FFBPNN and SVM algorithms. J Soft Comput Civil Eng 5(2):120–134
2. Mancino G, Nolè A, Salvati L, Ferrara A (2016) In-between forest expansion and cropland decline: a revised USLE model for soil erosion risk under land-use change in a Mediterranean region. Ecol Indic 71:544–550
3. Omar PJ, Shivhare N, Dwivedi SB, Dikshit PKS (2022) Identification of soil erosion-prone zone utilizing geo-informatics techniques and WSPM model. Sustain Water Resour Manage 8(3):66
4. Shivhare N, Rahul AK, Omar PJ, Chauhan MS, Gaur S, Dikshit PKS, Dwivedi SB (2018) Identification of critical soil erosion prone areas and prioritization of micro-watersheds using geoinformatics techniques. Ecol Eng 121:26–34
5. Rahul AK, Shivhare N, Dwivedi SB, Dikshit PKS (2020) Estimation of behavioral change of SSC of bed profile in the river using ADCP. Arab J Geosci 13(3):1–9
6. Bagarello V, Di Stefano C, Ferro V, Pampalone V (2017) Predicting maximum annual values of event soil loss by USLE-type models. Catena 155:10–19.S
7. Mondal A, Khare D, Kundu S, Mukherjee S, Mukhopadhyay A, Mondal S (2017) Uncertainty of soil erosion modelling using open source high resolution and aggregated DEMs. Geosci Front 8(3):425–436
8. Omar PJ, Shivhare N, Dwivedi SB, Gaur S, Dikshit PKS (2021) Study of methods available for groundwater and surface water interaction: a case study on Varanasi, India. In: The Ganga River basin: a hydrometeorological approach. Springer, Cham, pp 67–83
9. Ganasri BP, Ramesh H (2016) Assessment of soil erosion by RUSLE model using remote sensing and GIS—a case study of Nethravathi Basin. Geosci Front 7(6):953–961
10. Kinnell PIA (2016) Comparison between the USLE, the USLE-M and replicate plots to model rainfall erosion on bare fallow areas. CATENA 145:39–46
11. Li P, Mu X, Holden J, Wu Y, Irvine B, Wang F, Gao P, Zhao G, Sun W (2017) Comparison of soil erosion models used to study the Chinese Loess Plateau. Earth Sci Rev 170:17–30
12. Pandey A, Himanshu SK, Mishra SK, Singh VP (2016) Physically based soil erosion and sediment yield models revisited. CATENA 147:595–620
13. Guzha AC, Rufino MC, Okoth S, Jacobs S, Nóbrega RLB (2018) Impacts of land use and land cover change on surface runoff, discharge and low flows: evidence from East Africa. J Hydrol: Reg Stud 15:49–67
14. Ouyang W, Wu Y, Hao Z, Zhang Q, Bu Q, Gao X (2018) Combined impacts of land use and soil property changes on soil erosion in a mollisol area under long-term agricultural development. Sci Total Environ 613:798–809
15. Vanwalleghem T, Gómez JA, Amate JI, de Molina MG, Vanderlinden K, Guzmán G, Laguna A, Giráldez JV (2017) Impact of historical land use and soil management change on soil erosion and agricultural sustainability during the Anthropocene. Anthropocene 17:13–29
16. Welde K, Gebremariam B (2017) Effect of land use land cover dynamics on hydrological response of watershed: case study of Tekeze Dam watershed, northern Ethiopia. Int Soil Water Conserv Res 5(1):1–16
17. Fang H, Sun L, Tang Z (2015) Effects of rainfall and slope on runoff, soil erosion and rill development: an experimental study using two loess soils. Hydrol Process 29(11):2649–2658
18. Zhang Z, Sheng L, Yang J, Chen XA, Kong L, Wagan B (2015) Effects of land use and slope gradient on soil erosion in a red soil hilly watershed of southern China. Sustainability 7(10):14309–14325
19. Shivhare N, Rahul AK, Dwivedi SB, Dikshit PKS (2019) ARIMA based daily weather forecasting tool: a case study for Varanasi. Mausam 70(1):133–140

20. Wynn T, Mostaghimi S (2006) The effects of vegetation and soil type on streambank erosion, southwestern Virginia, USA. JAWRA J Am Water Resour Assoc 42(1):69–82
21. Rahul AK, Shivhare N, Kumar S, Dwivedi SB, Dikshit PKS (2022) Modelling suspended sediment concentration and discharge relationship using neural network and adaptive neuro-fuzzy inference system. Arab J Geosci 15(6):1–9

Optimisation of Irrigation Water Utilisation of Reservoir by Using Meta-heuristic Approach

Abhay Kumar Jha, R. S. Parihar, and S. M. Narulkar

Abstract The objective of irrigation reservoirs is to maximise crop productivity by efficient use of water supply. To tackle optimisation problems, conventional and meta-heuristic optimisation approaches are frequently used. In the current study, a meta-heuristic evolutionary algorithm called the genetic algorithm is used to solve the problem of reservoir operation for irrigation. In a situation with several crops, the main goal is to optimise crop output. The additive function for one formulation and the product function for another are solved. The study area chosen for the model's application is a reservoir in Madhya Pradesh which is a central state of India. The reservoir is filled in Monsoon season (June–September) and the storage water is used for irrigating crops in Rabi Season. The reservoir gets filled up to different levels of storage each year. For six separate storage levels, the optimization problem has been solved. On the MATLAB Platform's GA Solver, the issue is resolved. There are no deficits encountered until the fifth level of storage; however, a deficit is found at the sixth level. While the multiplicative model maximises the relative yield of other crops, the additive formulation maximises the relative yield of wheat. The result shows that the multiplicative model produces higher relative yield values.

Keywords Meta-heuristic evolutionary optimisation · Genetic Algorithm (GA) · Relative yield · Multiplicative model

A. K. Jha (✉) · R. S. Parihar
LNCT, Bhopal, India
e-mail: abhaykj@lnct.ac.in

R. S. Parihar
e-mail: hodce@lnct.ac.in

S. M. Narulkar
SGSITS, Indore, India

K. K. Pathak et al. (eds.), *Latest Developments in Civil Engineering*, Lecture Notes in Civil Engineering 352, https://doi.org/10.1007/978-981-99-2676-3_72

1 Introduction

Agriculture consumes the freshest water globally and is arguably the most mismanaged in its use. A nation's capability to sustain its social and economic systems depends on its ability to utilise water for irrigation effectively. Maximizing yield per unit of supplied water is the goal of efficient water use. Irrigation water management's primary goal is to increase agricultural production per unit of water. This is crucial to India's ability to sustain its social, economic, and environmental systems. In India, rainfall and river runoff have a fairly irregular relationship; most rivers have seasonal runoff. Additionally, the discrepancy between supply and demand shows marked variation. As a result, reservoirs on the rivers are required to close the supply–demand mismatch.

In India, many reservoirs have been commissioned and put into service. These reservoirs are multipurpose projects and are used for hydropower production, irrigation, municipal drinking water supply etc. However, most of the reservoirs were built exclusively for irrigation. The mainstays of water management systems are reservoirs, but they demand significant investments in terms of money, land, and the environment planning. Benefits from reservoir are realised only if the reservoir is planned and operated in an optimal way.

In the last three decades, the focus of researches on evolutionary algorithms that imitate the animal genetics or flocking behaviour of animals has increased significantly. A conventional optimisation technique has been obsolete since decades. Numerous studies have been published and Adeyemo [1], Ahmad et al. [2], and Neboh et al. have examined the methodologies [16]. Genetic algorithms (GA) have been used as a key tool in this category for reservoir operation and management. All of the GA's features are compiled in Goldberg's [12] textbook. The GA is a search heuristic tool that mimics natural evolutionary processes including inheritance, mutation, selection, and crossover. Numerous studies have helped to improve GA, particularly by merging it with other methods (hybrid) and evaluating performance [3, 11, 15] first applied GA to a four-reservoir problem in their paper. Nicklow et al. [17] have presented a state of the art paper on GA and applied its tool extensions to the water resources management problems. The problems of reservoir systems have been dealt extensively by many researchers [4]. Reddy and Kumar [18], Jotiprakash and Ganeshsan [13], Azamathulla et al. [7], Chen et al. [9], Mathur and Nikam [14], Scola et al. [20], Anand et al. [6]. The current study uses a GA model to develop a release programme that optimally distributes the reservoir's water storage across various crops over a range of fortnights. The Samrat Ashok Sagar Reservoir on the River Halali in the Madhya Pradesh districts of Vidisha and Raisen is the case study. The reservoir's command area is also known as the "bowl of wheat" which means that the wheat is produced in abundance in the command area. It produces Sharbati Wheat of the highest grade, which is well-known both in India and abroad.

2 Method

In the present work, the water management problem is optimised using the GA technique. The study's precise goals are to maximise yield and allocate water in a scenario with many crops and harvesting scenario. The issue is resolved in three steps. The initial step entails acquiring information on the cropping patterns and local meteorological data in the command area. The second stage involves calculating the fortnightly crop water requirements using the Penman Monteith method [5] and evaluating the relative yield ratios using [10] and other sources. The third stage consists of data preparation and solving the models to produce results, which are presented in the following section.

3 Computation of Demand

The CROPWAT-8.0 Software, which is an open access freeware, is used to estimate the evapotranspiration of the reference crop. The Reference Evapotranspiration (RET), also known as ET_o, is the rate of evapotranspiration from a reference surface with a hypothetical grass crop sown without any water scarcity. It makes use of a variety of inputs pertaining to the local area's climate, geography, soil types, and water quality. The software has the required tools to automatically relate the many existing formulae utilising the various inputs to generate the water requirements and maximum yield for each crop of interest. The software makes the assumption that all other agricultural production input elements (such as pesticides, knowledge, seeds, etc.) are maintained at their optimum levels. The equation used is the FAO Penman–Monteith Equation. The reference evapotranspiration value in *i*th season (ET_o) is calculated as per the Eq. 1) (1998)

$$ET_o = \frac{0.408\Delta(R_n - G) + \Upsilon\frac{900}{T+273}u_2(e_s - e_a)}{\Delta + \Upsilon(1 + 0.34u_2)} \tag{1}$$

where: ET_o = the reference evapotranspiration [mm/day], R_n = net radiation received at the crop surface [MJ m^{-2} day^{-1}], G = the soil heat flux density [MJ m^{-2} day^{-1}], T = the mean daily air temperature at 2 m height [°C], u^2 = wind speed at 2 m height [m s^{-1}], e_s = saturation vapour pressure [kPa] at T [°C], e_a = actual vapour pressure [kPa], Δ = slope of vapour pressure curve [kPa °C^{-1}], γ = the psychrometric constant [kPa °C^{-1}]. The crop evapotranspiration for the various crops sown in an area is necessary to determine each crop's seasonal water needs. The reference evapotranspiration (ET_o^i) values computed for different fortnights *i* were multiplied by the crop coefficient (k_c^i) to get the crop evapotranspiration (ET_c^i) (Eq. 2).

$$ET_c^i = k_c^i * ET_o^i. \tag{2}$$

4 Development of Model

The formulation and optimisation of the Samrat Ashok Sagar Reservoir's operation problem is described below along with the constraints of the objective function.

4.1 *Objective Function*

In the present study, the two different problems with different objective functions were solved.

4.1.1 Additive Model: Objective Function 1

The GA model's Objective function is to minimise the sum of the squared deviations of the mass balance equation and the irrigation demand deficit. (Eq. 1) The name "additive function" refers to this objective function. Reddy and Kumar [18, 19], Azamathulla et al. [7], Mathur and Nikam [14] are only a few researchers who have used this type of objective function in the past. The objective function was not weighed by the crop sensitivity factor to the water deficiency in different growth stages in the works of Reddy and Kumar [18], Mathur and Nikam [14]. The objective function in the current study is the squared deficit between the releases and demand of each crop in each season (Eq. 3).

$$\min f(x) = \sum_{nc=1}^{NC}\sum_{t=1}^{N} ky_t^{nc}\left(R_{t,nc} - D_{t,nc}\right)^2 + \sum_{t=1}^{N}\left(S_t(1 - B * e_t) - S_{t+1}(1 + B * e_t) + I_t - \sum_{nc=1}^{NC} R_{t,nc} - A_o * e_t\right)^2 \quad (3)$$

$R_{t,nc}$ = release in the period $t = 1, 2\ldots, N$ for crop $nc = 1, 2, \ldots, NC$ (MCM), $D_{t,nc}$ = demand to sustain the crop $nc = 1, 2,\ldots, NC$ in the period $t = 1, 2\ldots, T$ (MCM), ky_t^{nc} = yield response factor for time period $t = 1, 2\ldots, N$ of crop $nc = 1, 2\ldots, NC$, S_t = Storage at time period $t = 1, 2\ldots, N$ (MCM), S_{t+1} = Storage at time period $t + 1$ (MCM), I_t = Inflow at time period $t = 1, 2\ldots, N$ (MCM), A_o and B = Regression constant correlating surface area (Ha) and storage value, e_t = Rate of evaporation at each fortnight in (mm).

4.1.2 Multiplicative Model: Objective Function 2

The multiplicative model (Eq. 4) is more realistic and takes into account the effect of deficits to other seasons. In contrast, the additive model discussed in the previous paragraph has the limitation that it works in local time and the optimisation model does not have the effect of the deficits that occurred in other time periods.

$$\min f(x) = \sum_{nc=1}^{NC} \prod_{t=1}^{N} ky_t^{nc} \left(R_{t,nc} - D_{t,nc}\right)^2 + \sum_{t=1}^{N} \left(S_t(1 - B * e_t) - S_{t+1}(1 + B * e_t) + I_t - \sum_{nc=1}^{NC} R_{t,nc} - A_o * e_t \right)^2 \tag{4}$$

4.2 *Constraints*

Constraints of the Objective functions are two; Demand constraint and Capacity Constraint which are based on maximum irrigation demand and live storage capacity of the reservoir.

4.2.1 Demand/Release Constraints (Irrigation Demand)

To sustain the crops, irrigation releases must be more than or equal to zero. However, they must not exceed the maximum irrigation demand in order to achieve the desired yield.

$$0 \le R_{t,nc} \le D_{t,nc} \tag{5}$$

4.2.2 Capacity Constraints (Reservoir Storage)

Each fortnight, the reservoir's storage should not be more or less than its live storage or less than its dead storage.

$$\mathrm{S_{min}} \le \mathrm{S_t} \le \mathrm{S_{max}} \tag{6}$$

Reservoir data were subjected to a reservoir operation study based on the models discussed in the text. The results show that, even in situations with insufficient storage, the optimisation strategy can improve the annual net benefits. Under a multi-crop scenario there is a competition for the available water whenever the water available is less than the irrigation demands. The reservoir operation component releases water from the reservoir optimally, while the multi-crop water allocation component distributes water to various crops by evaluating how sensitive crop production is to moisture stress at various plant physiological growth phases. For operation throughout the Rabi (dry weather) Season, a single reservoir with multiple crop irrigation is taken into consideration in the current study. The evaluation and model formulation of various components of the study are presented in following paragraphs.

5 Data Collection and Discussion

The data used in the analysis was gathered from a number of sources including The Central Water Commission [8] and various state level offices of the M.P. The information regarding the cropping pattern and physical characteristics of the reservoir was provided by Water Resources Departments in Bhopal, the IMD website, the Water Portal website, and other sources. The irrigation during the Rabi Season is taken into account in the current issue, and in Table 1 the cropping pattern is displayed. The uncertainty of the monsoon season in India directly affects the runoff to the reservoir, which could cause variations in the reservoirs' filling levels. To investigate the issue in the current study, 6 alternative beginning storage levels in the reservoir were used. These are 204.2, 181.52, 158.53, 136.14, 113.45, 226.9 (full storage), 181.52, and the rest. The total command area is 37419 Ha, of which 32292 Ha, or 86.30%, will be covered by the reservoir system. Table 2 displays the crop duration used for the present study. Table 3 displays the evapotranspiration values for the chosen reference crop.

Table 1 Rabi season (cropping patterns)

S. no.	Crop	% Area	Area (ha)
1	Wheat	90	28,900
2	Barseem	0.3	100
3	Gram	7	2400
4	Vegetable	0.7	292
5	Moong	2	600
	Total	100	32,292

Table 2 The crop duration

S. no.	Crop	Duration	Days
1	Wheat	16 Nov–15 Mar	120
2	Barseem	16 Oct–15 Feb	120
3	Gram	16 Oct–28 Feb	135
4	Vegetable	1 Nov–31 Jan	90
5	Moong	15 Feb–15 April	60

Table 3 Evapotranspiration values (reference)

Fortnight	Duration	ET_o (mm)
1	16–31 Jan	49.00
2	1–15 Feb	51.82
3	16–28 Feb	56.13
4	1–15 Mar	85.72
5	16–31 Oct	53.98
6	1–15 Nov	48.95
7	16–31 Dec	40.00
8	1–15 Jan	40.04
9	16–31 Nov	40.04
10	1–15 Dec	40.00

6 Computation of Water Demand

The computation of a season reference Evapo-Transpiration (ET_o) values for the command area have been calculated for various growth seasons of Rabi Season (Dry Weather) crops using the local meteorological data and the CROPWAT 8.0 Software. The crop evapotranspiration coefficients were multiplied by the ETo values to create the crop evapo-transpiration (ETc) values for various crops (Eq. 2). Eventually, after the correct conversion and application of efficiency factors, the demands for the various seasons were calculated using the net sown area. In Table 4, sample findings for the Wheat crop are displayed. The demands for other crops have been calculated along similar lines.

7 Release Results at Various Initial Storages

The problem was approached using both the Additive and Multiplicative Models, according to the formulation described in the preceding discussions (Eqs. 3–6). For the Wheat crop, Fig. 1 calculates and displays the real demands, computed releases from the Additive Model and the Multiplicative Models in each week for various levels of beginning reservoir storage values.

Table 4 Computation of demand for wheat (fortnightly)

SowinG date	16-Nov	Crop	Wheat	Area (ha)	28,900	Computed demand
Days	Fort night	ET_o (mm)	Kc	Growth stage	$ET_C = Kc * ET_o$	Million Cubic Meters (MCM)
15	I	53.98	0.31	Ini	16.734	5.8033
30	II	48.95	0.44	Devp	21.538	7.4694
45	III	40.04	0.92	Devp	36.837	12.7750
60	IV	40.00	1.10	Devp	44.000	15.2592
75	V	40.01	1.10	Mid	44.011	15.2630
90	VI	40.04	1.10	Mid	44.044	15.2745
105	VII	48.95	0.92	Mid	45.034	15.6178
120	VIII	51.82	0.53	End	27.465	9.5247

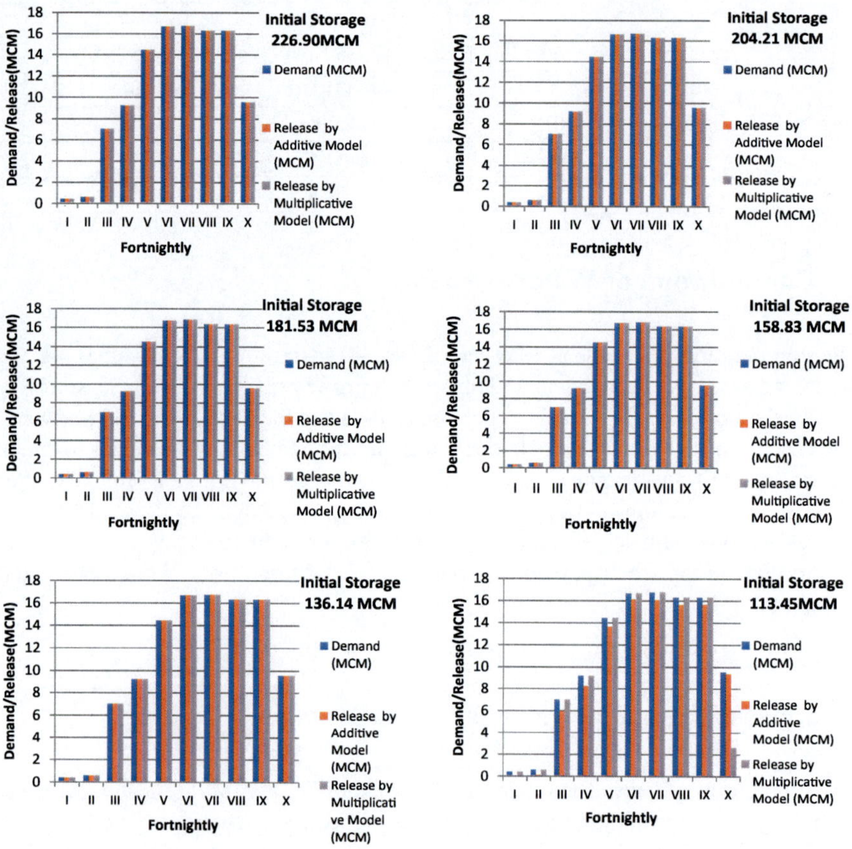

Fig. 1 Fortnightly demand/release computed by both models (for Wheat)

Both models estimated the releases for various crops in the command area in a similar way. The objective of the present study is to increase all crops' yield ratios in order to increase production. By dividing the total volume by the crop area and using the proper conversion factor, the releases calculated from the optimisation model for various crops in various fortnights at various levels of initial storage values in the reservoir are used to determine the amount of water to be supplied to the crop. The procedure comprises of solving three equations (Eqs. 7–9) at each time step to solve for the Actual Evapo-transpiration and the soil moisture storage in the root zone of each crop in various time periods. The relative yield is computed as per Eq. (10).

$$\mathrm{SM}_{t+1} * Z_{t+1} = \mathrm{SM}_t * Z_t + R_t + X_t + S_o(Z_{t+1} - Z_t) - \mathrm{AET}_t \tag{7}$$

$$Z_t = Z_{\max}\left[0.5 + 0.5\,\sin\left\{3.03\left(\frac{t}{t_{\max}}\right) - 1.47\right\}\right] \tag{8}$$

$$\mathrm{AET}_t = \begin{cases} 0; \ \mathrm{SM}_t \le \mathrm{WP}_t \\ \frac{\mathrm{PET}_t(\mathrm{SM}_t - \mathrm{WP})}{(1-p)(FC-\mathrm{WP})}; \ \mathrm{WP} < \mathrm{SM}_t \le (1-p)(\mathrm{FC} - \mathrm{WP}) \\ \mathrm{PET}_t; \ \mathrm{SM}_t \ge (1-p)(\mathrm{FC} - \mathrm{WP}) \end{cases} \tag{9}$$

$$R_t^*(X_t, \mathrm{AET}_t) = 1 - ky_t^{nc}\left(1 - \frac{\mathrm{AET}}{\mathrm{PET}}\right)_t \tag{10}$$

where, Z_t and Z_{t+1} are root zone depths in the periods t and $t + 1$ respectively, (cm), R_t is the rainfall in the period t (mm), X_t is the water depth allocated in period t (mm), S_o is the initial soil moisture content in the extended root zone (mm/cm), $t_{\max}$ is the time for the full development of root zone (days), $Z_{\max}$ is the maximum possible depth of effective root zone, ky_t^{nc} is the yield factor for period t of the crop nc, AET is the actual evapo-transpiration (mm) and ET_c is the crop evapo-transpiration for the crop *nc* in time *t*, SM_t and SM_{t+1} are the soil moisture content in depth units per unit root depth in period t (mm/cm), FC = Field capacity (mm/cm), WP = Wilting point (mm/cm), p = Crop water depletion fraction. The results of final yields for various crops at different storage levels using both the models have been presented in Table 5 and Fig. 2.

8 Conclusion

The present study is based on the operation of an irrigation reservoir namely the Samrat Ashok Sagar Reservoir (Halali Project) in Vidisha and Raisen Districts of Madhya Pradesh. The study is focussed on optimising irrigation water use for five

Table 5 Final yield by addition function and product function

Crop	Storage	Relative yield by addition function	Relative yield by product function
Wheat	226.9	1.000	1.000
	204.21	1.000	1.000
	181.52	1.000	1.000
	158.83	1.000	1.000
	136.14	1.000	1.000
	113.45	1.000	0.993
Barseem	226.9	1.000	1.000
	204.21	1.000	1.000
	181.52	1.000	1.000
	158.83	1.000	1.000
	136.14	1.000	1.000
	113.45	0.927	0.991
Gram	226.9	1.000	1.000
	204.21	1.000	1.000
	181.52	1.000	1.000
	158.83	1.000	1.000
	136.14	1.000	1.000
	113.45	0.992	0.998
Vegetable	226.9	1.000	1.000
	204.21	1.000	1.000
	181.52	1.000	1.000
	158.83	1.000	1.000
	136.14	1.000	1.000
	113.45	0.935	0.987
Moong	226.9	1.000	1.000
	204.21	1.000	1.000
	181.52	1.000	1.000
	158.83	1.000	1.000
	136.14	1.000	1.000
	113.45	0.911	0.997

crops in the command area during the Rabi season of arid weather. A Genetic Algorithm is used as the optimisation technique on the MATLAB platform. Two optimisation problems were formulated and solved using GA. The first is an additive model in which minimisation function is used and the weighted squared deviation of the demands/actual releases for all crops in all seasons is added. The weights represent the relative crop yield factors for the various time periods. The second is a multiplicative model that aims to minimise the weighted squared deviation addition for all

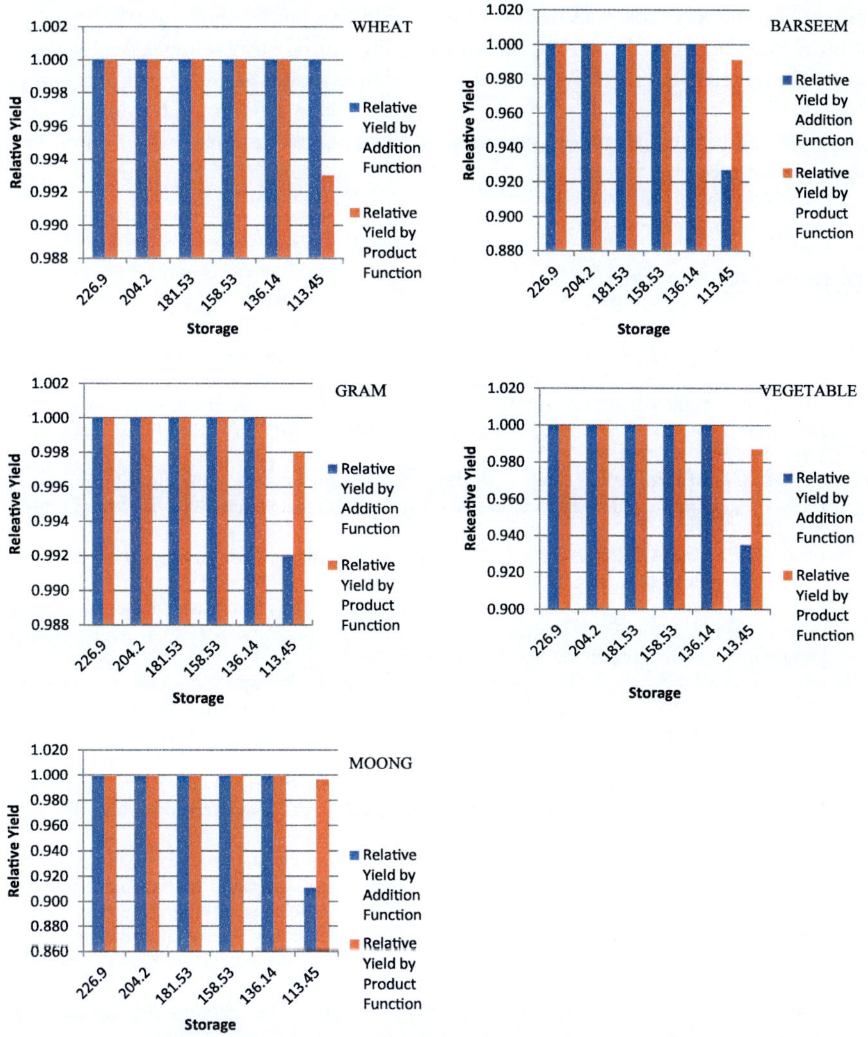

Fig. 2 Relative yields of different crops at different storage levels

crops during a season when they are multiplied together during the whole cropping season. Different levels of the reservoirs' initial storage values were applied to the models. The study's findings show that the model can be applied to optimally operate reservoirs for irrigation planning. The water is distributed among the various crops in various time periods in an optimal manner by the model created in the current study. There is competition among the crops at a reservoir storage level of 113.45 MCM inspite of all of the crops' demands for higher storage values having met fully. At this level, supply deficits and deficit allocations are made in order to maximise potential relative yields. The findings for the additive model show that the principal crop wheat receives all the required water at this initial storage level. Water shortages affect the other crops. When the multiplicative model is solved, the other crops benefit at the expense of a production shortfall of wheat. Even under deficit situations, the multiplicative model typically produces better relative yield figures than the additive model.

References

1. Adeyemo JA (2011) Reservoir operation using multi-objective evolutionary algorithms—a review. Asian J Sci Res. ISSN 1992-1454, https://doi.org/10.3923/ajsr
2. Ahmad A, Shafie AE, Razali SFM, Mohamad ZS (2014) Reservoir optimization in water resources: a review. Water Resour Manage 28(3391–3405):3391–3405. https://doi.org/10.1007/s11269-014-0700-5
3. Ahmad KN, Milhem MM (1989) Operation of a second steam stimulation pilot project in Kuwait. Paper presented at the middle east oil show, Bahrain, March 1989. https://doi.org/10.2118/17987-MS
4. Ahmed JA, Sarma AK (2005) Genetic algorithm for optimal operation policy of a multipurpose reservoir. Water Resour Manag 19:145–161
5. Allen RG, Pereira LS, Raes D, Smith M (1998) FAO irrigation and drainage paper no. 56 crop evapotranspiration (guidelines for computing crop water requirements). FAO, Water Resources, Development and Management Service, Rome, Italy
6. Anand J, Gosain AK, Khosa R (2018) Optimisation of multipurpose reservoir operation by coupling SWAT and genetic algorithm for optimal operating policy (Case Study: Ganga River basin). Preprints (www.preprints.org) Posted: 23 March. https://doi.org/10.20944/preprints201803.0199.v1 © 2018
7. Azamathulla HM, Wu FC, Ghani AA, Narulkar SM, Zakaria NA, Chang CK (2008) Comparison between genetic algorithm and linear programming approach for real time operation. J Hydro-environ Res 2:172–181
8. Central Water Commission (2006) Post project performance evaluation study of Samrat Ashok Sagar Project Madhya Pradesh. Final report, M.P. Water and Land Management Institute WALMI Hills Right Bank of Kaliasote Dam B No 538, Ravishankar Nagar PO Bhopal
9. Chen L, McPhee J, Yeh WWG (2007) A diversified multi-objective GA for optimizing reservoir rule curves. Adv Water Resour 30(5):1082–1093. http://repositorio.uchile.cl/bitstream/handle/2250/124937/Chen_Li.pdf?sequence=1
10. Doorenbos J, Kassam AH (1979) Yield response to water. FAO Irrigation and Drainage Paper No. 33, Rome, FAO
11. East V, Hall MJ (1994) Water resources system optimization using genetic algorithms. In: Hydroinformatics '94, Proceedings of 1st international conference on Hydroinformatics, Balkema, Rotterdam, The Netherlands, pp 225–231
12. Goldberg DE (1989) Genetic algorithms in search optimization and machine learning. Addison-Wesley, Reading, Mass
13. Jotiprakash V, Shanthi G (2006) Single reservoir operating policies using genetic algorithm. Water Resour Manag 20:917–929
14. Mathur YP, Nikam SJ (2009) Optimal reservoir operation policies using genetic algorithm. Int J Engi Technol 1(2):184–187. ISSN: 1793-8236. https://doi.org/10.7763/IJET.2009.V1.34
15. Mujumdar PP, Narulkar S (1993) Optimisation models for multi-reservoir planning and operation, vol 8, no 1. Indian National Committee on Hydrology, Roorkee, India, pp 29–52
16. Neboh N, Adeyemo J, Enitan A, Oludayo O (2015) A review on applications of evolutionary algorithms to reservoir operation for hydropower production. World Acad Sci Eng Technol, Int J Geol Environ Eng 9(9)

17. Nicklow J, Reed P, Savic D, Dessalegne T, Harrell L, Chan-Hilton A, Karamouz M, Minsker B, Ostfeld A, Singh A, Zechman E (2010) State of the art for genetic algorithms and beyond in water resources planning and management. J Water Resour Plan Manag 136
18. Reddy MJ, Kumar ND (2006) Optimal reservoir operation using multi-objective evolutionary algorithm. Water Resour Manage 20:861–878. https://doi.org/10.1007/s11269-005-9011-1C_Springer
19. Reddy MJ, Kumar ND (2007) Optimal reservoir operation for irrigation of multiple crops using elitist-mutated particle swarm optimization. Hydrol Sci–J–des Sci Hydrol 52(4):686–701
20. Scola LA, Neto OM, Takahashi RHC, Cerqueira SAAG (2010) Multi-objective optimal reservoir operation. In: WCCI IEEE world congress on computational intelligence, July 18–23, CCIB, Barcelona, Spain

Projection of Drought Indices Trend over the Lower Bundelkhand Region in Central India

A. Vishwakarma, M. K. Choudhary, and M. S. Chauhan

Abstract Frequent drought has emerged as the primary crisis in the Bundelkhand in recent years due to climate variation. It is essential to improve an early warning technique to predict the Bundelkhand drought. The latest generation global climate models (GCMs) and regional climate models (RCMs) are the better sources for accurately predicting long-term climate. The models are frequently updated and improved by an international body, the Intergovernmental Panel on Climate Change (IPCC). This study projects the future drought from 2021 to 2100 over Madhya Pradesh (MP) in Bundelkhand by using the two important drought indicators, the Standardized Precipitation Evapotranspiration Index (SPEI) and Standardized Precipitation Index (SPI). The study also found the non-parametric trend of projected drought on multi-scales (seasonal, non-seasonal, and annual). The increasing drought events for the RCP 4.5 scenario is observed as the carbon emission concentration is high until the year 2100. The rising trend of annual drought has been noticed over most of the MP in Bundelkhand. Besides, an insignificant trend has been observed on seasonal and monthly scales over the study region. The outcomes of the study will help policymakers for preparing drought alleviation and management strategies over the study region.

Keywords GCMs · Global climate models · MK · Mann-Kendall · RCMs · Regional climate models · RCP · Representative concentration pathway

Maps Disclaimer: The presentation of material and details in maps used in this chapter does not imply the expression of any opinion whatsoever on the part of the Publishers or Author concerning the legal status of any country, area or territory or of its authorities, or concerning the delimitation of its borders. The depiction and use of boundaries, geographic names and related data shown on maps and included in lists, tables, documents, and databases in this chapter are not warranted to be error free nor do they necessarily imply official endorsement or acceptance by the Publisher or Author.

A. Vishwakarma (✉) · M. K. Choudhary · M. S. Chauhan
Department of Civil Engineering, Maulana Azad National Institute of Technology (MANIT), Bhopal 462003, India
e-mail: ankur18vishwa@gmail.com

K. K. Pathak et al. (eds.), *Latest Developments in Civil Engineering*, Lecture Notes in Civil Engineering 352, https://doi.org/10.1007/978-981-99-2676-3_73

1 Introduction

A significant strategy for projecting potential future climatic scenarios is climate trend analysis [1]. When significant trends in hydrological variables, like a change in evapotranspiration, groundwater, precipitation, relative humidity, soil moisture, surface runoff, and temperature, are observed, non-parametric tests become more suitable than parametric tests.

The trend test can also be performed to evaluate the drought indices trend. The drought and its trends have been the subject of some recent global studies. Authors in [2] examined the trends and incidence of meteorological drought over western India, using the SPI at the six-month time-scale of 110 years from 1896 to 2005. The SPI trend was examined by the non-parametric MK trend test. The study also investigated the SPI trend using the split time steps as 1896–1931, 1932–1966 and 1967–2005. The study concluded that the trend of SPI for 110 years was found to increase over the subdivisions, western Rajasthan, Saurashtra, Kutch and Marathwada. Authors in [3] utilized the two drought indices, SPI and Standardized Soil Moisture Index (SSI), to assess the drought severity in Iran using the MK test. The study found that the northern and central parts of Iran were encountering increasing trends in droughts. Using the three, six and twelve months' time scales of SPI, authors in [4] utilized the MK test to observe the changes in the drought scenario throughout the Bundelkhand region. Majority of stations of Bundelkhand showed a diminishing trend of SPI at a 95% significance level. Authors in [5] examined the drought utilizing the MK and Sen's slope methods over India and concluded rising trends throughout the rainy season in the central and east regions. Authors in [6] evaluated the trend using the MK Z-statistic and Sen's method of SPI and SPEI meteorological drought indices, over semi-arid Botswana, Southern Africa, from 1960 to 2016. The study concluded that the wetting trend was shown for 1960–1979 and the drying trend for the remaining duration after 1979–2016. Authors in [7] assessed the effect of annual temperature variation on drought frequency in different parts of India. The study noticed a considerable correlation between temperature variation and drought occurrence. The places observed frequent droughts for temperature variation between 40 and 30 °C with once in 3–6 years. Authors in [8] introduced the new groundwater drought concept and evaluated the non-parametric trend of groundwater drought, i.e., Standardized Groundwater level Index (SGI) in the Ghataprabha river basin India. The Mann-Kendall trend of SGI was found to decrease, with over and above 61% of wells, average significantly declining by 0.21 m. Authors in [9] also evaluated the non-parametric trends of drought utilizing the SPI index with the help of the MK and Sen's methods over 566 stations throughout India. The study observed that droughts are becoming more frequent in the eastern, north-eastern and extreme southern regions, whereas droughts are becoming less common in the northern and southern parts of the country. Authors in [10] evaluated the trend of the meteorological drought index (simplified rainfall index, RIs) utilizing the modified MK test over the Narmada river basin. The outcomes found the rising trend of drought over the major districts of the Narmada river basin. Pandey and Narulkar [11] observed the meteorological drought over the various districts of Bundelkhand region using the SPI index. The present study utilizes the bias-corrected projected data to assess the future drought events happening in the future and assess their trend over the MP state in Bundelkhand and see the consequence of climate variation.

2 Study Area and data

The Bundelkhand region includes six districts in the state of MP in central India, which are shown in Table 2. The MP in Bundelkhand lies between 23°08′ N and 26°30′ N latitude and 78°11′ E and 80°41′ E longitude, which can be seen in Fig. 1.

Based on the daily historical observed maximum and minimum temperature of 0.5° × 0.5° resolution from NASA for 37 years (1982–2018), a bias-corrected CMIP5 GCM model has been found accurate for the study region. The description of the model is presented in Table 1.

Observed daily data of temperature have been collected for the rain-gauge (RG) stations of the study region that comes under the six districts. The description of the stations is displayed in Table 2.

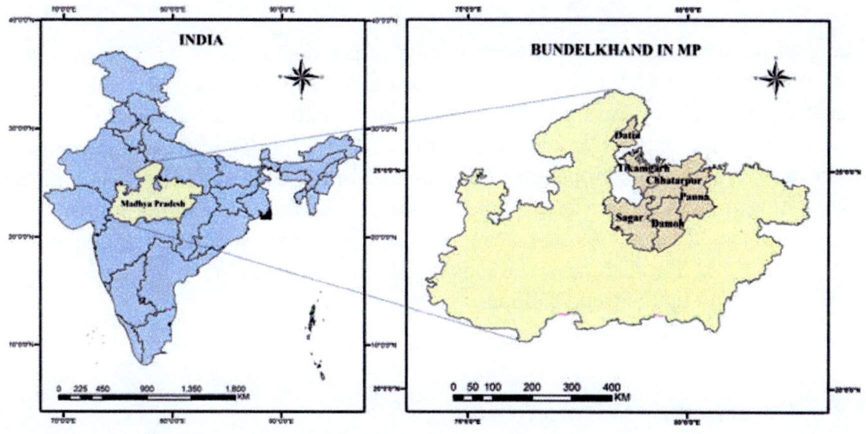

Fig. 1 Location of MP in Bundelkhand, India

Table 1 Detail of CMIP5-GCM model for the study region

S. no.	CMIP5 driving model	CORDEX model	Contributing CMIP5 modeling center
1	GFDL-ESM2M	GFDL-ESM2M-IITM-RegCM4	Geophysical Fluid Dynamics Laboratory (GFDL), USA

Table 2 Rain-gauge stations in MP Bundelkhand

S. no.	District	Location		RG stations
		Latitude	Longitude	
1	Sagar	23.839° N	78.738° E	Banda, Deori, Garhakota, Jaisinagar, Khurai, Malthone, Rahatgarh, Rehli, Sagar, Shahgarh
2	Chhatarpur	24.916° N	79.591° E	Bara Malhera, Bijawar, Buxwaha, Chhatarpur, Gaurihar, Khajuraho Aero, Laundi, Nowgong, Rajnagar
3	Tikamgarh	24.745° N	78.832° E	Baldeogarh, Niwari, Orchha, Prithvipur, Tikamgarh
4	Panna	24.718° N	80.187° E	Ajaigarh, Devendranagar, Gonour, Madla, Panna, Pawai, Shahnagar, Simaria
5	Damoh	23.832° N	79.439° E	Batiagarh, Damoh, Gaisabad, Hardua, Hatta, Jabera, Majhguwan Hansraj, Mala
6	Datia	25.684° N	78.566° E	Datia, Seondha

3 Methodology

3.1 Statistical Test for Trend and Variability Analysis

The goal of trend assessment is to anticipate future outcomes using data from the past. All independent meteorological indicators, for instance, monthly, yearly, and seasonal temperatures, rainfalls, etc., can be statistically evaluated in two steps to determine whether there is an increasing or declining trend. The MK method is used in the first, while Sen's slope estimator is used in the second [12]. The rising or diminishing trend has been investigated in the current study using normalized test statistics (Z) values at a 95% significant level. The pattern is stated to be rising when Z is positive and diminishing when Z is negative. The annual rate and direction of change are given by the trend's slope.

3.2 Assessment of Drought Indices

The drought risk management must include drought monitoring. It is frequently characterized broadly as a transient meteorological occurrence that results from a prolonged absence of precipitation relative to some long-term average stage. But droughts have many different aspects in each individual place, develop gradually, and are challenging to detect. The prompt dissemination of information on the drought's initiation, progression and geographic extent is crucial for the success of drought preparation and mitigation. Monitoring of the drought can provide this kind of information. Typically, monitoring is carried out using drought indicators. This study utilized two indicators, SPEI [13] and SPI [14]. SPI is a prominent drought index

Table 3 Levels of severity for SPEI and SPI

S. no.	Drought severity	SPI & SPEI ranges
1	Mild	−1.0 to 0
2	Moderate	−1.5 to −1.0
3	Severe	−2.0 to −1.5
4	Extreme	(less than) ≤ −2.0

that depends on precipitation. The benefits of the SPI incorporate its simplicity and flexibility. It can be computed for any time period and on different time scales. SPI evaluated using the probability of precipitation is computed as:

$$\mathrm{SPI} = (\mathrm{P} - \mathrm{Pm})/\sigma \quad (1)$$

where P is the precipitation at the station, Pm is the mean precipitation and σ is the standard deviation. When the SPI turns positive, the drought is over. The SPEI is derived from the SPI calculation and is relatively easy to calculate. The monthly (or any scale) precipitation is utilized as the input data to calculate the SPI. The change between P and PET is used by the SPEI. This is an example of a basic climate water balance that is used to calculate the SPEI at various time frames. Table 3 displays the SPEI and SPI values for the drought severity levels [15].

4 Results

4.1 Estimation of Future Drought

SPEI index has been noticed to be effective in indicating meteorological and agricultural drought compared to the SPI. Hence, future drought events have been assessed based on the SPEI index. The drought events have been assessed for the entire part of MP in Bundelkhand. The results for a future drought over the study region are shown in Fig. 2. The number of annual drought events has been observed to increase in mid-century (2041–2070) for the RCP 4.5 scenario, whereas found increased events in the third-century (2071–2100) for the RCP 8.5 scenario.

4.2 Trend Analysis of Future Drought

The trend of the projected drought has been analyzed using MK and Sen's slope methods. The trend has been observed on seasonal, non-seasonal, and annual scales for 2021–2100. Trend assessment of the drought has been estimated for both the scenarios RCP 4.5 and 8.5 and for both the drought indices (SPI and SPEI).

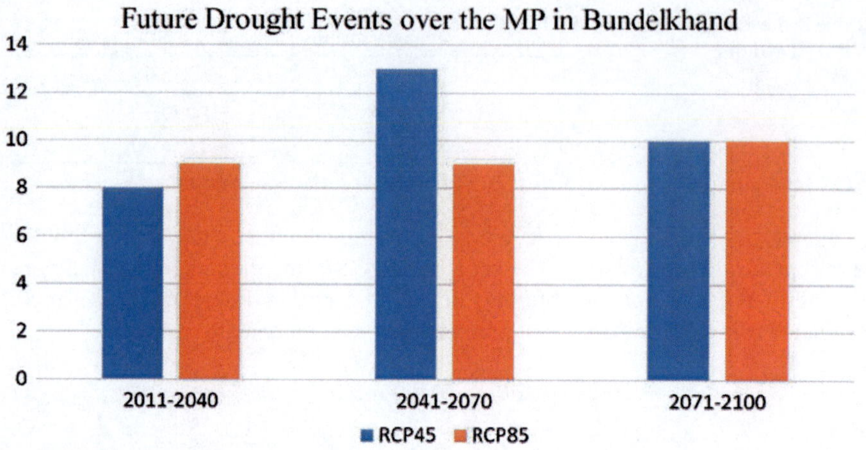

Fig. 2 Future drought events over MP Bundelkhand region

Trend Analysis of Drought for RCP 4.5

SPI has revealed a non-significant trend for all the months. The non-significant negative trend is seen for most of the part of MP Bundelkhand in April, June and November, whereas the insignificant increasing trend is seen for the remaining months presented in Fig. 3.

Similarly, a non-significant trend of SPEI has been seen for MP Bundelkhand in all of the months, as displayed in Fig. 4. A significant negative trend of SPEI has been seen for a few parts of Damoh and Panna in May.

The trend analysis has also been performed on seasonal (monsoon and non-monsoon) and annual time scales for both SPI and SPEI. A non-significant trend has been found for all the time scales (Seasonal, non-seasonal, and annual) taken for the study, as shown in Fig. 5.

Trend Analysis of Drought for RCP 8.5

SPI has revealed an insignificant trend for all the months except the western part of MP Bundelkhand in July. In the western part, a significant declining trend was found in some parts of Sagar and Tikamgarh in July. A significant negative trend has also been seen in August for a few parts of the Damoh and Sagar districts. Similarly, a non-significant trend of SPEI has been detected for the entire MP Bundelkhand in all the months except January. The significant negative trend of SPEI has also been seen for the entire part of MP Bundelkhand in January. Examination of the trend has been performed for RCP 8.5 on seasonal (monsoon and non-monsoon) and annual time scales for both SPI and SPEI. A non-significant trend has been noticed for all the time scales except for the annual trend of SPI in the lower part of the study region. A significant diminishing trend has been seen for most of the lower part of MP Bundelkhand, as displayed in Fig. 6.

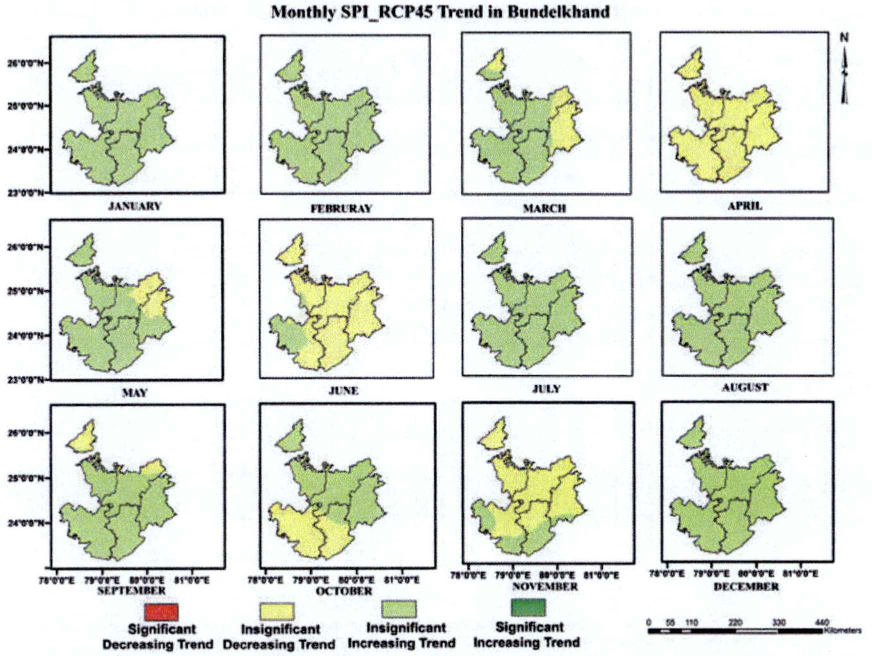

Fig. 3 Monthly SPI trend in MP Bundelkhand for RCP 4.5

5 Conclusions

This study utilized the future data to assess the future drought events and measure their trend over the MP Bundelkhand. The effect of climate variation on the future drought has been evaluated as compared to the historical observations. The future drought events have been assessed based on the SPEI index. The study found twenty-eight annual drought events in the future three-time periods (2021–2040, 2041–2070 and 2071–2100) for the RCP 4.5 scenario, while thirty-one drought events have been found for the RCP 8.5 scenario over the Bundelkhand region.

The conclusion drawn from both the non-parametric trend studies of RCP 4.5 as well as of RCP 8.5 is that there is no clear picture seen for the increasing and diminishing trend of drought in terms of SPEI and SPI on all time scales taken for the study at 95% significance level. Therefore individual drought events must be found for a better projection of the forthcoming climate over the MP in the Bundelkhand region.

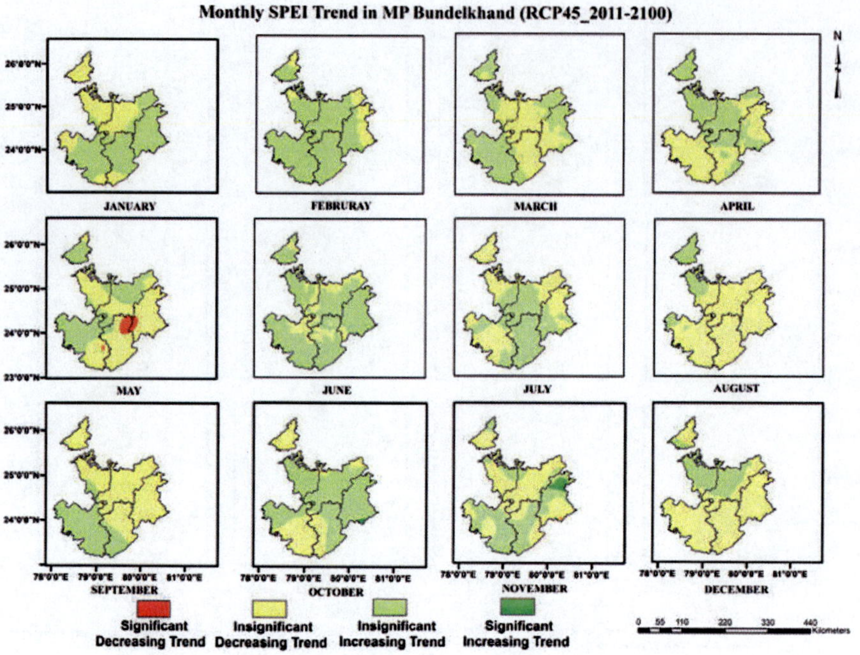

Fig. 4 Monthly SPEI trend in MP Bundelkhand for RCP 4.5

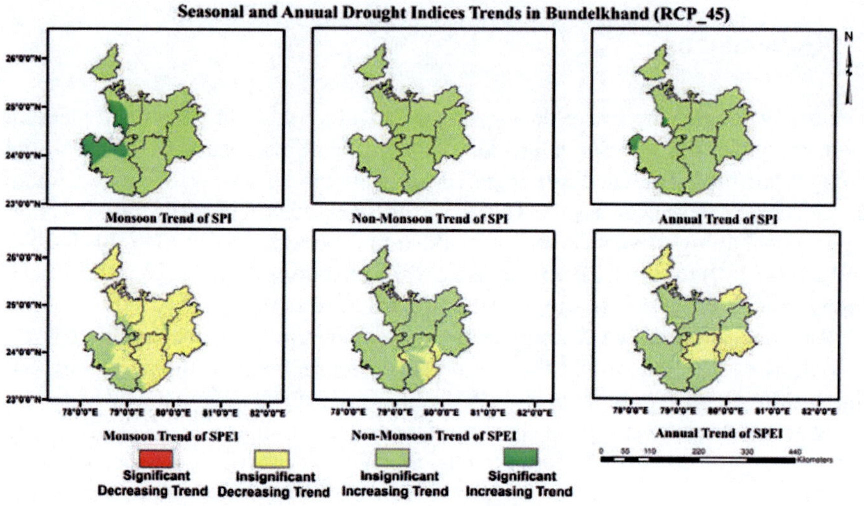

Fig. 5 Seasonal and annual trend in SPI and SPEI in MP Bundelkhand for RCP 4.5

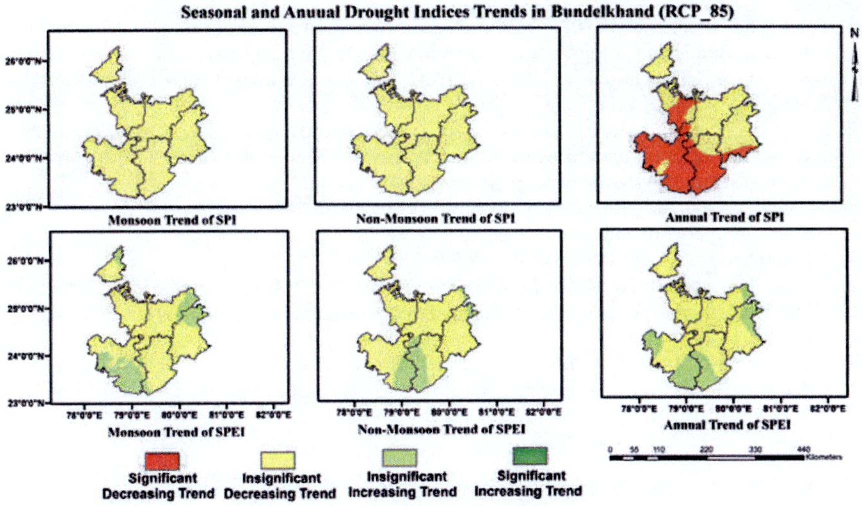

Fig. 6 Seasonal and annual trend in SPI and SPEI in Bundelkhand for RCP 8.5

References

1. Choudhary RR, Jhajharia D, Lal M, Jain JK, Lunayach A, Choudhary MK (2009) Climate and its variations over Bikaner since 1951–2008. J Indian Geol Congr 1(2):79–86
2. Ganguli P, Reddy MJ (2014) Evaluation of trends and multivariate frequency analysis of droughts in three meteorological subdivisions of western India. Int J Climatol 34(3):911–928. https://doi.org/10.1002/joc.3742
3. Golian S, Mazdiyasni O, AghaKouchak A (2015) Trends in meteorological and agricultural droughts in Iran. Theor Appl Climatol 119(3):679–688. https://doi.org/10.1007/s00704-014-1139-6
4. Thomas T, Nayak PC, Ghosh NC (2015) Spatiotemporal analysis of drought characteristics in the bundelkhand region of central india using the standardized precipitation index. J Hydrol Eng 20(11):05015004. https://doi.org/10.1061/(ASCE)HE.1943-5584.0001189
5. Das PK, Dutta D, Sharma JR, Dadhwal VK (2016) Trends and behaviour of meteorological drought (1901–2008) over Indian region using standardized precipitation–evapotranspiration index. Int J Climatol 36(2):909–916. https://doi.org/10.1002/joc.4392
6. Byakatonda J, Parida BP, Moalafhi DB, Kenabatho PK (2018) Analysis of long term drought severity characteristics and trends across semiarid Botswana using two drought indices. Atmos Res 213:492–508. https://doi.org/10.1016/j.atmosres.2018.07.002
7. Amrit K, Pandey RP, Mishra SK, Daradur M (2018) Relationship of drought frequency and severity with range of annual temperature variation. Nat Hazards 92(2):1199–1210. https://doi.org/10.1007/s11069-018-3247-6
8. Pathak AA, Dodamani BM (2019) Trend analysis of groundwater levels and assessment of regional groundwater drought: Ghataprabha River Basin, India. Nat Resour Res 28(3):631–643. https://doi.org/10.1007/s11053-018-9417-0
9. Sharma A, Goyal MK (2020) Assessment of drought trend and variability in India using wavelet transform. Hydrol Sci J 65(9):1539–1554. https://doi.org/10.1080/02626667.2020.1754422
10. Swain S, Mishra SK, Pandey A (2021) A detailed assessment of meteorological drought characteristics using simplified rainfall index over Narmada River Basin, India. Environ Earth Sci 80(6):1–5. https://doi.org/10.1007/s12665-021-09523-8

11. Pandey R, Narulkar SM (2021) Meteorological drought assessment using SPI for different district of Bundelkhand region. In: International conference on sustainable water resources development and management (SWARDHAM-2021), Government College of Engineering, Aurangabad, Maharashtra
12. Mondal A, Kundu S, Mukhopadhyay A (2012) Rainfall trend analysis by Mann-Kendall test: a case study of north-eastern part of Cuttack district, Orissa. Int J Geol Earth Environ Sci 2(1):70–78. http://www.cibtech.org/jgee.htm
13. Vicente-Serrano SM, Beguería S, López-Moreno JI (2010) A multiscalar drought index sensitive to global warming: the standardized precipitation evapotranspiration index. J Clim 23(7):1696–1718. https://doi.org/10.1175/2009JCLI2909.1
14. McKee TB, Doesken NJ, Kleist J (1993) The relationship of drought frequency and duration to time scales. In: Proceedings of the 8th conference on applied climatology, vol 17, no 22, pp 179–183
15. Tan C, Yang J, Li M (2015) Temporal-spatial variation of drought indicated by SPI and SPEI in Ningxia Hui Autonomous Region, China. Atmosphere 6(10):1399–1421. https://doi.org/10.3390/atmos6101399

A Case Study to Optimise the Operation Schedule for Irrigation Reservoir Using Harmony Search Algorithm

R. S. Parihar, Abhay Kumar Jha, and Sandeep M. Narulkar

Abstract Irrigated agriculture is widely regarded as the primary cause of water scarcity due to poor management. The key to management is the wise use of water to enhance agricultural yield. System optimisation and analysis are used to achieve this. The reservoirs are essential to the management of water in the majority of non-perennial rivers. Because building a reservoir is an expensive endeavor, it must be planned and run as effectively as possible to reap the greatest benefits. The irrigation reservoirs place a strong emphasis on maximising crop yield in its commanded area. Numerous mathematical programming strategies are used to achieve the best operation. Meta-heuristic optimisation strategies have gained popularity recently. Another meta-heuristic optimisation method that has been successful in reservoir problems is the Harmony Search Algorithm (HSA). The major goal of the paper is to show how HSA has the ability to improve regulation policies that allocate the reservoir's water storage across diverse crops in the best possible fortnights. An irrigation reservoir in Central India is used as a case study in order to increase crop yields during the dry season.

Keywords Water resources system · Reservoir · Optimal operation · Harmony search algorithm · Crop yield

1 Introduction

According to the World Resource Institute [9], extreme water stress affects 54% of the country's land area as a result of overuse of the resources. In poor nations, irrigation uses up to 90% of the available fresh water, and it is thought to be the main cause of water scarcity. The cause is improper management leading to wasteful water use.

R. S. Parihar (✉) · A. K. Jha
Civil Engineering, LNCT, Bhopal, India
e-mail: rajeevsinghparihar@gmail.com

S. M. Narulkar
Civil Engineering, SGSITS, Indore, India

K. K. Pathak et al. (eds.), *Latest Developments in Civil Engineering*, Lecture Notes in Civil Engineering 352, https://doi.org/10.1007/978-981-99-2676-3_74

India is not an exception. In India, where agriculture uses 78% of the country's water, a severe water shortage is imminent. The Central Water Commission [3] predicts that by 2050, the country's overall water demand will exceed availability, reducing the percentage of irrigation to 68%. According to the Planning Commission of the Government of India [8], efforts must be undertaken to raise irrigation efficiency from its current level of 35–40–60%. The key to management is the wise use of water to enhance agricultural yield. There are a lot of seasonal rivers in India that are not perennial. There are many reservoirs in India because the annual flow that needs to be stored is between 80 and 90% of the flow during the wet weather. The majority of them have irrigation as their primary objective. The irrigation reservoirs must be required to maximize crop productivity in the area under their control. Therefore, increasing water use efficiency ought to be crucial.

Because building a reservoir is an expensive endeavor, it must be planned and run wisely and ideally to reap the greatest benefits. Mathematical programming approaches have been used for decades to manage and operate reservoirs in the best possible way. Up to the 1990s, linear programming, dynamic programming, nonlinear programming, and their mixtures and variations were the most popular operation research techniques. These methods achieved the global optimum by resolving difficulties of a specific sort. However, in practice, the issues are extremely complicated and non-linear, making it impossible to find global optimum solutions. Numerous meta-heuristic methods have also been applied recently to handle reservoir operation optimisation issues. Heuristics are rules of the thumb for solving problems without applying a procedure exhaustively, according to the definition of the word in the context of computing [2]. Regarding their broad applicability, the techniques also have limitations. In order to solve this problem, meta-heuristic optimisation approaches based on repeated simulations have been developed. These techniques enable the discovery of suitable solutions with minimal memory and processing requirements [11]. The majority of meta-heuristic methods attempt to fundamentally emulate natural, physical, or biological principles by using a variety of operators [2]. Yoo and Kim's paper provides a thorough analysis of the meta-heuristic methods given by Bandaru and Deb [2] and their applications to hydrological sciences. The current work is based on the application of the Harmony Search Algorithm (HSA), a Meta-heuristic Technique, to reservoir operation studies. Geem et al. [5] created this interesting algorithm. The fundamental idea behind HSA was to draw a parallel between the pursuit of harmony in music and the optimal point in an optimisation process. Both procedures aim to deliver the best or optimum results. Despite being a relatively recent meta-heuristic algorithm, its benefits and effectiveness have been shown in a number of uses. Geem [6] used HSA to produce five distinct global optimal solutions with identical maximum benefits from irrigation and hydropower generation for the best operation schedule of a multiple dam system. Yang [10] examined new HAS's effectiveness in relation to meta-heuristic algorithms. Kougias and Theodossiou [7] used HSA to solve the well-known dam scheduling problem of maximising the operation of a four-reservoir system over a 24-h period utilising the Visual-Basic and MATLAB platforms. For the purpose of optimising reservoir operation during a flood, Atrabi et al. [1] solved the HSA.

The goal of the current study is to design a technique for the best agricultural water use at the Samrat Ashok Sagar Reservoir (SASR), also known as Halali Dam, in the Vidisha district of Madhya Pradesh. The reservoir is used to irrigate crops during lean periods when inflow is negligible in the reservoir. Multiple crops must be supported by the storage made accessible by wet season flows during the lean season. With the goal of optimising agricultural production, HSA is used in the current work to optimise irrigation over time among a variety of crops. The study's specific goal is to allocate water and maximum productivity in a multi-crop, multi-seasonal setting. Three steps are taken to resolve the issue. The fortnightly crop water requirements are first determined, and then the water is allocated using HSA optimisation.

2 Methodology

The present study has been framed based on an analysis of the literature reviewed in the previous discussion. The following lines provide the details.

2.1 *Computation of Demand*

One must investigate the crop-soil–water-atmosphere system in order to determine the necessary amount of water and the related crop yield. Typical estimates place evapo-transpiration (ET) as the primary consumer of water. Reference Applying the CROPWAT-8.0 Software described in FAO-56, evapo-transpiration (RET) is first estimated [4]. The programme makes use of inputs pertaining to the local area's meteorological conditions, topographical details, soil properties, and water quality. The software makes the assumption that all other agricultural production input elements (such as pesticides, knowledge, seeds, etc.) are maintained at their optimum levels. Using the Penman Monteith Method, the RET value for the project area for the ith season (RET^i) has been calculated (Eq. 1).

$$RET^i = \frac{0.408\Delta(R_n - G) + \gamma \frac{900}{T+273} u_2 (e_s - e_a)}{\Delta + \gamma(1 + 0.34u_2)} \quad (1)$$

where: RET^i = the reference evapo-transpiration [mm/day] in ith fortnight; R_n = net radiation received at the crop surface [MJ m^{-2} day^{-1}], G = the soil heat flux density [MJ m^{-2} day^{-1}], T = the mean daily air temperature at 2 m height [°C], u_2 = wind speed at 2 m height [m s^{-1}], e_s = saturation vapour pressure [kPa] at T [°C], e_a = actual vapour pressure [kPa], Δ = slope of vapour pressure curve [kPa °C^{-1}], γ = the psychrometric constant [kPa °C^{-1}]. For the various crops sown in an area the crop evapo-transpiration is required to assess the water requirement of individual crops in different seasons.

2.2 Potential Evapotranspiration

The Potential Evapo-Transpiration (PET) is calculated for the various crops sown in the command area of SASR canal system for various seasons *i*. The RET values computed from Eq. (1) were multiplied by the crop coefficient to get the PET values (Eq. 2).

$$\mathrm{PET}_c^i = k_c^i * \mathrm{RET}^i \tag{2}$$

k_c^i = the crop coefficient for *c*th crop in *i*th fortnight; PET_c^i values is the potential evapo-transpiration value for *c*th crop in *i*th fortnight This demand value is considered for each crop in each of the seasons.

2.3 Optimisation

Harmony Search algorithm is used in the present study for the optimisation. The flow chart of the optimisation process using HSA is shown in Fig. 1.

3 Model Development

Two objective functions were taken into consideration when formulating the reservoir operation optimisation problem for the SASR. One is the additive function while other is the multiplicative function. The first objective function is to minimise the total squared deviation of the deficits in irrigation demand and supply across all crops and seasons. This is added with the squared derivation of the storage balance equation (Eq. 3)

$$\min\ f(x) = \sum_{nc=1}^{NC} \sum_{t=1}^{N} ky_t^{nc} \left(R_{t,nc} - D_{t,nc}\right)^2 + \sum_{t=1}^{N} \left(S_t(1 - B * e_t) - S_{t+1}(1 + B * e_t) + I_t - \sum_{nc=1}^{NC} R_{t,nc} - A_o * e_t \right)^2 \tag{3}$$

The sum for all the crops of the product of the square deviation of releases and demands for a single crop for all the seasons added by the squared deviation of the storage balance equation is the second objective function.

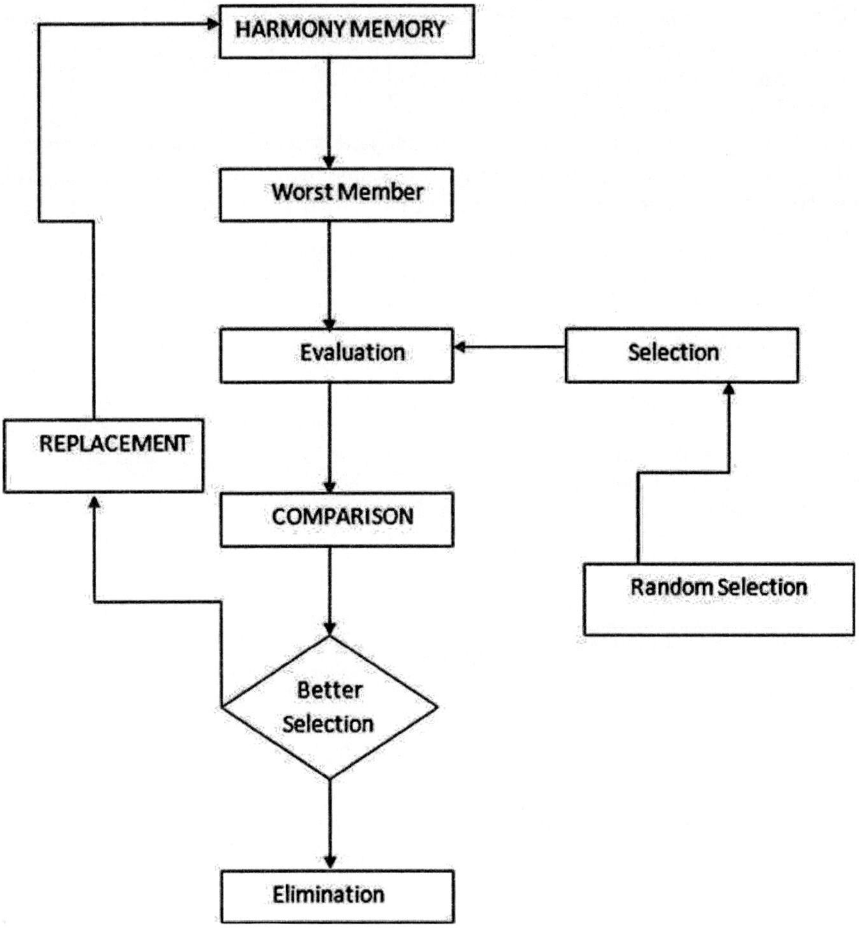

Fig. 1 Flow chart of HSA

$$\min\ f(x) = \sum_{nc=1}^{NC}\sum_{t=1}^{N} ky_t^{nc}\left(R_{t,nc} - D_{t,nc}\right)^2 + \sum_{t=1}^{N}\left(S_t(1 - B * e_t) - S_{t+1}(1 + B * e_t) + I_t - \sum_{nc=1}^{NC} R_{t,nc} - A_o * e_t\right)^2 \quad (4)$$

$R_{t,nc}$ = release in the period $t = 1, 2\ldots, N$ for crop $nc = 1, 2, \ldots, NC$ (MCM), $D_{t,nc}$ = demand to sustain the crop $nc = 1, 2, \ldots, NC$ in the period $t = 1, 2\ldots, T$ (MCM), ky_t^{nc} = yield response factor for time period $t = 1, 2\ldots, N$ of crop $nc = 1, 2\ldots, NC$, S_t = Storage at time period $t = 1, 2\ldots, N$ (MCM), S_{t+1} = Storage at time period $t +$

1 (MCM), I_t = Inflow at time period $t = 1, 2\ldots, N$ (MCM), A_o and B = Regression constant correlating surface area (Ha) and storage value, e_t = Rate of evaporation at each fortnight in (mm).

There are two distinct problems that are presented and solved. Although the first objective function is a very common adopted type, it is not a realistic choice. Although the second function is relatively infrequently used for optimisation, crop yield reduction makes it plausible. Both these formulations are subjected to following constraints.

3.1 Constraints Used for Irrigation Demand

The releases for irrigation should be more than or equal to zero to sustain the crops and also at the same time this should not exceed the maximum irrigation demand to produce the targeted yield (Eq. 5).

$$0 \leq R_{t,nc} \leq D_{t,nc} \quad (5)$$

$R_{t,nc}$ = release in the period $t = 1, 2\ldots, N$ for crop $nc = 1, 2, \ldots, NC$ (MCM), $D_{t,nc}$ = demand to sustain the crop $nc = 1, 2, \ldots, NC$ in the period $t = 1, 2\ldots, T$ (MCM).

3.2 Reservoir Storage—Capacity Constraints

The reservoir storage (S_t) in each fortnight should not be less than the dead storage, and should not be more than the live storage of the reservoir (Eq. 6).

$$S_{\min} \leq S_t \leq S_{\max} \quad (6)$$

$S_{\min}$ = Dead Storage of the reservoir in MCM, $S_{\max}$ = Maximum Capacity of the reservoir in MCM (Fig. 2).

3.3 Data Collection and Analysis

The information used in the analysis was gathered from a number of sources. The Central Water Commission [3], various state level offices of the M.P. Water Resources Departments in Bhopal, the IMD website, the Water Portal website, and other sources provided the information regarding the cropping pattern and physical characteristics of the reservoir. The irrigation during the Rabi Season is taken into account in the current issue, and in Table 1, the cropping pattern is displayed. The monsoon season in India is unpredictable and directly affects the runoff to the reservoir, which

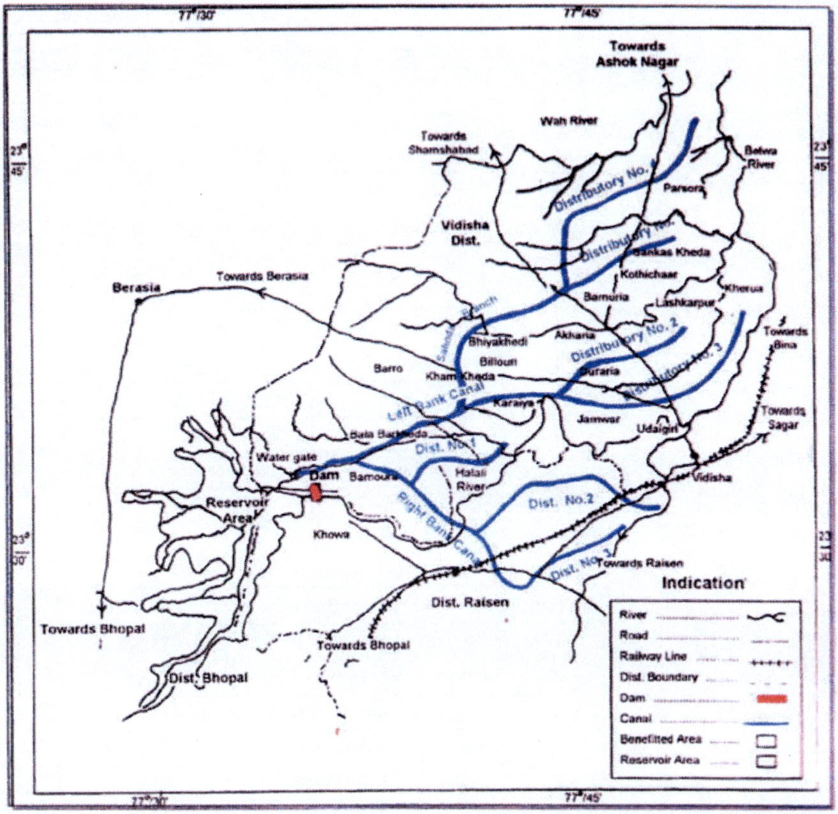

Fig. 2 Index map of Samrat Ashok Sagar Project. *Source* SMP_SAS_WRD Vidisha_20_03_12

could cause variations in the reservoirs' filling levels. To investigate the issue in the current study, 6 different initial storage levels in the reservoir were used. 226.9 (full storage), 204.21, 181.52, 158.53, 136.14, and 113.45 MCM are the numbers. The entire command area is 37419 Ha, of which 86.30%, or around 32,292 Ha, will be covered by the reservoir system. Table 2 displays the crop calendar used for the current study. Table 3 displays the evapotranspiration values for the chosen reference crop.

Table 1 Cropping patterns in the case study (Rabi season)

S. no.	Crop	% Area	Area (ha)
1	Gram	7	2400
2	Moong	2	600
3	Vegetable	0.7	292
4	Wheat	90	28,900
5	Barseem	0.3	100
	Total	100	32,292

Table 2 The crop calendar

S. no	Crop	Time duration	Days
1	Gram	16 Oct–28 Feb	135
2	Moong	15 Feb–15 April	60
3	Vegetable	1 Nov–31 Jan	90
4	Wheat	16 Nov–15 Mar	120
5	Barseem	16 Oct–15 Feb	120

Table 3 Reference evapotranspiration values

Fortnight	Time duration	ET_o (mm)
1	16–31 Oct	53.98
2	1–15 Nov	48.95
3	16–31 Nov	40.04
4	1–15 Dec	40.00
5	16–31 Dec	40.00
6	1–15 Jan	40.04
7	16–31 Jan	49.00
8	1–15 Feb	51.82
9	16–28 Feb	56.13
10	1–15 Mar	85.72

4 Results

The models described in the preceding sections have been implemented using SASR data for irrigation during the Rabi season. In the paragraphs that follow, along with discussions, the model's application results are shown.

Table 4 Computation of fortnightly demand for gram

Sowing date	16-Oct	Crop	Gram	Area sown	2400 (ha)	Demand	
Days	Fort night	ET_o (mm)	*Kc*	Growth stage	PET = $Kc * ET_o$ (mm)	Million Cubic Meters (MCM)	Mm
15	1	53.98	0.23	Ini	12.4154	0.358	14.898
30	2	48.95	0.28	Devp	13.706	0.395	16.447
45	3	40.04	0.69	Devp	27.6276	0.796	33.153
60	4	40.00	1.02	Devp	40.8	1.180	48.96
75	5	40.00	1.05	Devp	42	1.210	50.4
90	6	40.04	1.04	Mid	41.6416	1.199	49.970
105	7	49.00	0.89	Mid	43.61	1.256	25.332
120	8	51.82	0.63	Mid	32.6466	0.940	39.176
135	9	56.13	0.41	End	23.0133	0.663	27.616

4.1 Computation of Demand of Water

Using the CROPWAT 8.0 Software, the computation of fortnightly RET values is done using the local climate data. Additionally, Table 4 shows the results of PET computations for each crop across several fortnights and the subsequent demands.

4.2 Results of Releases at Various Initial Storages

According to the formulation, the Addition and Multiplication models were used to solve the issue. Figures 3, 4, 5, 6, 7 and 8 show the computed actual demands, computed releases from the Addition Model and the Multiplication Model for each of the fortnight for various reservoir initial storage levels.

4.3 Computations of Final Relative Yield Ratios for Different Initial Storages

The HSA model is used to optimise water allocation to different crops in order to maximise yield in the present situation. The original models of optimisation were producing results at the level of demand fulfilment. In order to calculate the relative yield ratios in various fortnights and the final relative yield ratios, these volumes based on the results were then used in a crop-water simulation model. The present

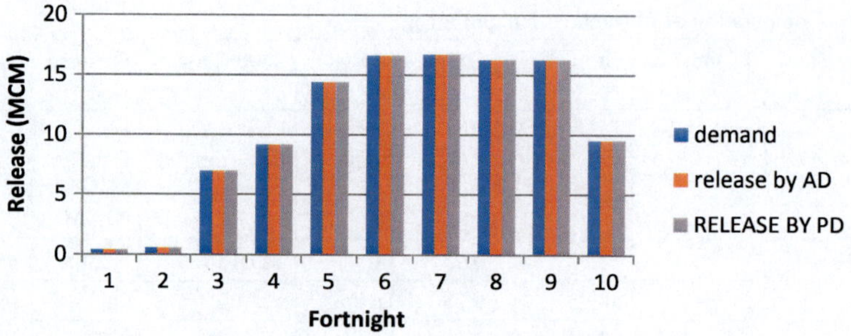

Fig. 3 Fortnightly demand versus Release from initial storage 226.90 MCM as per HSA

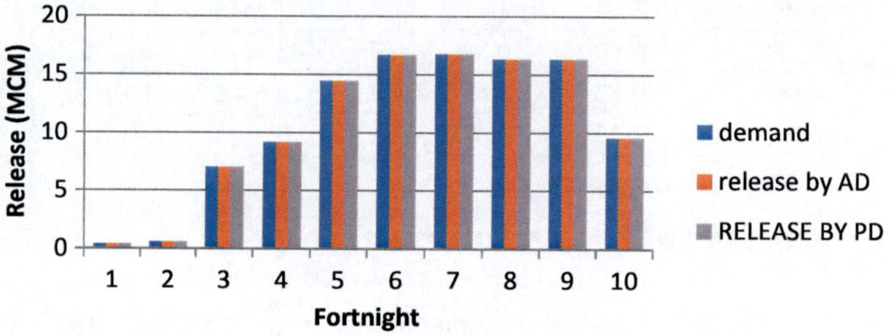

Fig. 4 Fortnightly demand versus Release from initial storage 204.21 MCM as per HSA

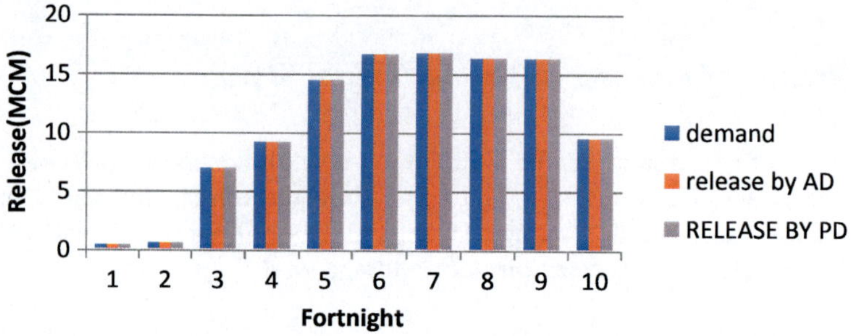

Fig. 5 Fortnightly demand versus Release from initial storage 181.52 MCM as per HSA

study's objective was to improve all crops' yield ratios in order to increase production. By dividing the total volume by the crop area and using the proper conversion factor, the releases calculated from the optimisation model for various crops in various fortnights at various levels of initial storage values in the reservoir are used to determine the depth of water supplied to the crop. Then, for each of the crops, Eq. (7) was used for soil moisture simulation model. In order to determine the Actual Evapotranspiration and the soil moisture storage in the root zone of each crop over time,

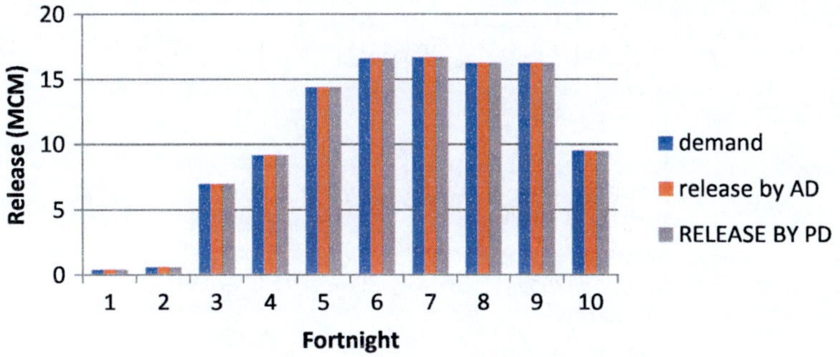

Fig. 6 Fortnightly demand versus Release from initial storage 158.53 MCM as per HSA

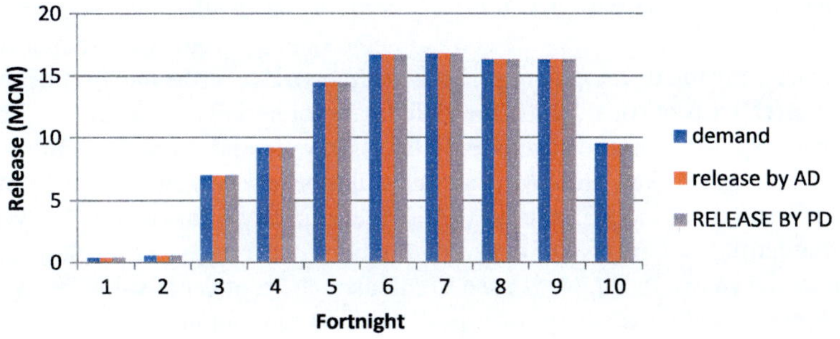

Fig. 7 Fortnightly demand versus Release from initial storage 136.14 MCM as per HSA

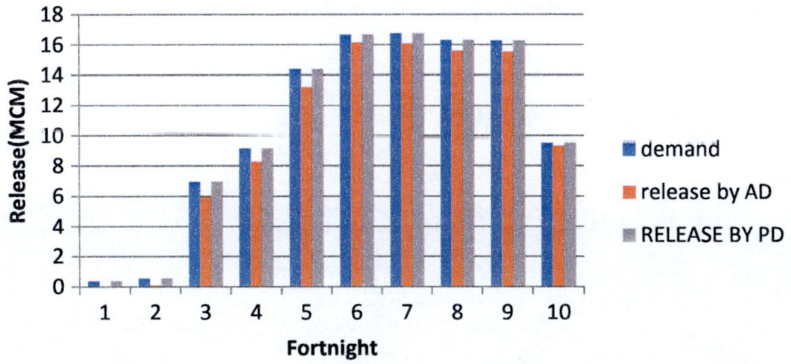

Fig. 8 Fortnightly demand versus Release from initial storage 113.45 MCM as per HSA

three equations (Eqs. 7–9) must be solved at each time step. Equation (10) is used to calculate the relative yield.

$$SM_{t+1} * Z_{t+1} = SM_t * Z_t + R_t + X_t + S_o(Z_{t+1} - Z_t) - AET_t \tag{7}$$

$$Z_t = Z_{\max}\left[0.5 + 0.5\sin\left\{3.03\left(\frac{t}{t_{\max}}\right) - 1.47\right\}\right] \tag{8}$$

$$\mathrm{AET}_t = \begin{cases} 0; \quad \mathrm{SM}_t \le \mathrm{WP}_t \\ \cdot \\ \frac{\mathrm{PET}_t(\mathrm{SM}_t - \mathrm{WP})}{(1-p)(\mathrm{FC}-\mathrm{WP})}; \quad \mathrm{WP} < \mathrm{SM}_t \le (1-p)(\mathrm{FC}-\mathrm{WP}) \\ \cdot \\ \mathrm{PET}_t; \quad \mathrm{SM}_t \ge (1-p)(\mathrm{FC}-\mathrm{WP}) \end{cases} \tag{9}$$

$$R_t^*(X_t, \mathrm{AET}_t) = 1 - ky_t^{nc}\left(1 - \frac{\mathrm{AET}}{\mathrm{PET}}\right)_t \tag{10}$$

where, Z_t and Z_{t+1} are root zone depths in the periods t and $t+1$ respectively, (cm), R_t is the rainfall in the period t (mm), X_t is the water depth allocated in period t (mm), S_o is the initial soil moisture content in the extended root zone (mm/cm), $t_{\max}$ is the time for the full development of root zone (days), $Z_{\max}$ is the maximum possible depth of effective root zone, ky_t^{nc} is the yield factor for period t of the crop nc, AET is the actual evapo-transpiration (mm) and PET is the potential evapo-transpiration for the crop *nc* in time *t*, $\mathrm{SM_t}$ and SM_{t+1} are the soil moisture content in depth units per unit root depth in period t (mm/cm), FC = Field capacity (mm/cm), WP = Wilting point (mm/cm), p = Crop water depletion fraction.

The model shown in Eq. (8) is used to calculate the root depth values for various crops. Since this was necessary for Eqs. (7) and (9) estimation of actual evapotranspiration and soil moisture. For Gram, the results are presented in Table 5. Results have been obtained for all other crops and are presented.

Equations (7), (9), and (10) are used to the data to compute the relative yields for each crop under various beginning storage conditions, as well as the soil moisture storage for various crops in various fortnights and actual evapotranspiration. Table 6 and Figs. 9, 10, 11, 12 and 13 show the findings of the final yields for various crops at varying storage levels using both models.

Table 5 Computation of fortnightly root depth values for gram

Fortnight	Time period (t days)	Root depth (Z_t cm)
Oct II	15.000	13.346
Nov I	30.000	41.704
Nov II	45.000	78.374
Dec I	60.000	114.202
Dec II	75.000	140.243
Jan I	90.000	149.996
Jan II	105.000	141.026
Feb I	120.000	115.573
Feb II	135.000	79.991

Table 6 Final yield for Gram by addition function and product function

Gram	Storage	Relative yield by addition Function	Relative yield by product function
	226.9	1.000	1.000
	204.21	1.000	1.000
	181.52	1.000	1.000
	158.83	1.000	1.000
	136.14	1.000	1.000
	113.45	0.993	1.000
Barseem	226.9	1.000	1.000
	204.21	1.000	1.000
	181.52	1.000	1.000
	158.83	1.000	1.000
	136.14	1.000	1.000
	113.45	0.927	1.000
Moong	226.9	1.000	1.000
	204.21	1.000	1.000
	181.52	1.000	1.000
	158.83	1.000	1.000
	136.14	1.000	1.000
	113.45	0.898	1.000
Vegetable	226.9	1.000	1.000
	204.21	1.000	1.000
	181.52	1.000	1.000
	158.83	1.000	1.000
	136.14	1.000	1.000
	113.45	0.937	1.000
Wheat	226.9	1.000	1.000
	204.21	1.000	1.000
	181.52	1.000	1.000
	158.83	1.000	1.000
	136.14	1.000	1.000
	113.45	1.000	1.000

5 Discussion on Results

According to the findings, there is no deficit for any of the crops up to an initial storage of 136.14 MCM. It was discovered that the deficit begins with 113.45 MCM of initial storage. In both cases, the objective function of the product type performs better than the objective function of the addition type.

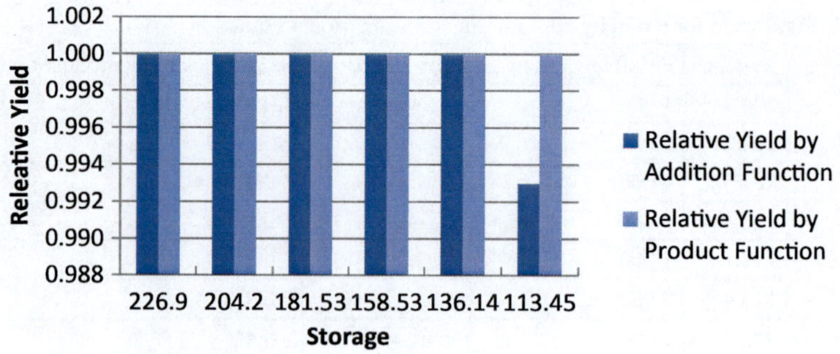

Fig. 9 Relative yield for Gram at different storage levels

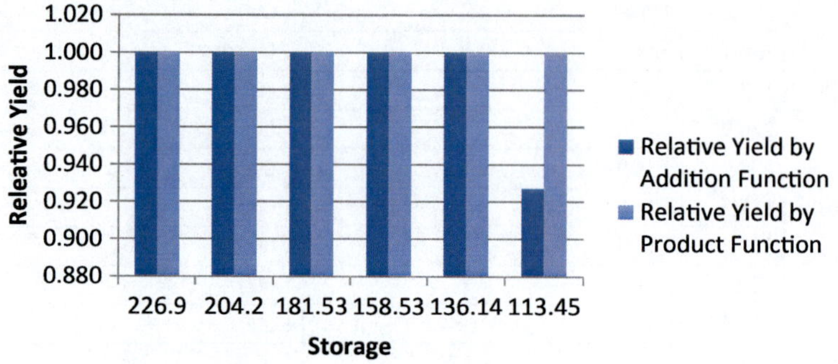

Fig. 10 Relative yield for Barseem at different storage levels

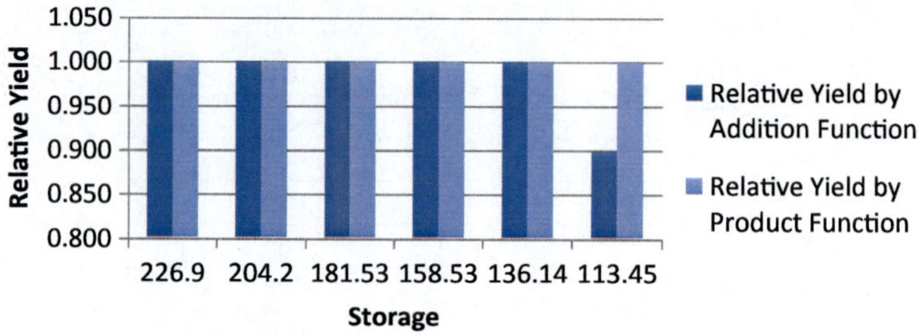

Fig. 11 Relative yield for Moong at different storage levels

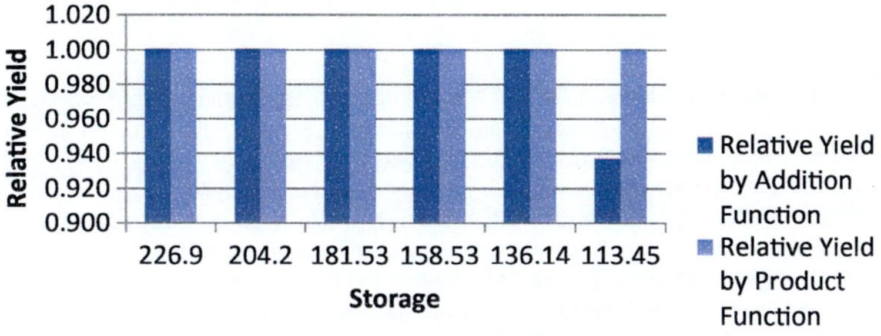

Fig. 12 Relative yield for Vegetables at different storage levels

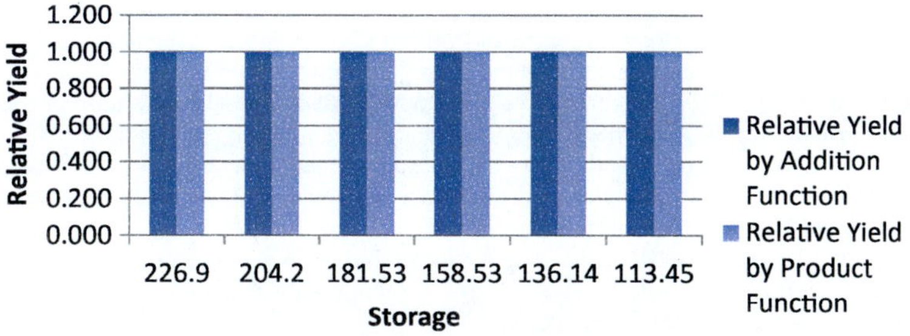

Fig. 13 Relative yield for Wheat at different storage levels

6 Conclusion

Utilising the HSA technique, the present study is being done on reservoir operation for irrigation using the Samrat Ashok Sagar Reservoir Project as a case study. To improve the yield in a multi-crop scenario, the water stored in a reservoir is optimised. The water is distributed among the various crops in various time periods in an optimal manner using the HSA model created in the present study. The distribution of deficits to achieve the highest feasible relative yields also takes into consideration supply deficits. It has been found that the model performs effectively and creates a series of releases to satisfy the demand for diverse crops across a range of fortnights. Two models are developed and applied. The squared deviations between the demands and the releases for all the crops in all the periods are added by the additive model. The product for all the crops is added by multiplying the squared deviation of each crop over all of its periods in the multiplicative model. Based on the final yield of the individual crops, the multiplicative model improves the additive model. According to the study, it is feasible to choose how best to operate a reservoir for irrigation using the model that has been presented. The deficient storage conditions in the beginning of the Rabi season have a negative impact on the water supplies and the crops suffer deficits. However the results of the present study indicate that the deficits are minimised in case of deficient storage.

References

1. Atrabi HB, Qaderi K, Rheinheimer DE, Sharifi E (2015) Application of harmony search algorithm to reservoir operation optimization. Water Resour Manag. Published Online September 2015. https://doi.org/10.1007/s11269-015-1143-3
2. Bandaru S, Deb K (2016) Metaheuristic techniques, 2016, a School of Engineering Science, University of Skövde, Skövde 541 28, Sweden b Department of Electrical and Computer Engineering, Michigan State University, East Lansing, 428 S. Shaw Lane, 2120 EB, MI 48824, USA COIN Report Number 2016029* Published in Decision Sciences: Theory and Practice, Edited by Sengupta Raghu Nandan, Gupta Aparna and Dutta Joydeep, CRC Press, Taylor & Francis Group Published October 26, 2016, pp 693–750
3. CWC (2014) Guidelines for improving water use efficiency in irrigation, domestic & industrial sectors. Performance Overview and Management Improvement Organization Irrigation Performance Overview Directorate, Central Water Commission, New Delhi, November, 2014, pp 1
4. Food and Agriculture Organization (FAO) (1998) Crop evapo-transpiration: guidelines for computing crop requirements. http://www.fao.org/docrep/X0490E/X0490E00.htm
5. Geem ZW, Joong HK, Loganathan GV (2001) A new heuristic optimization algorithm: harmony search. Simulation 76(2):60–68
6. Geem ZW (2007) Optimal scheduling of multiple dam system using harmony search algorithm. In: International work-conference on artificial neural networks, IWANN 2007: computational and ambient Intelligence, pp 316–323
7. Kougias IP, Theodossiou NP (2013) Application of the harmony search optimization algorithm for the solution of the multiple dam system scheduling. Published online: 4 January 2012 © Springer Science+Business Media, LLC
8. Planning Commission Government of India (2009) Report of the task force on irrigation, May 2009, pp 6. https://pdfs.semanticscholar.org/c28f/cda2bccc75b10e1d540031269e82a5d9c9b0.pdf
9. World Resource Institute (WRI) (2015). https://www.wri.org/blog/2015/02/3-maps-explain-india-s-growing-water-risks
10. Yang XS (2009) Harmony search as a metaheuristic algorithm. In: Geem ZW (ed) Music-inspired harmony search algorithm: theory and applications. Studies in computational intelligence, vol 191. Springer, Berlin, pp 1–14
11. Yoo DG, Hoon KJ (2014) Meta-heuristic algorithms as tools for hydrological science. Geosci Lett, Off J Asia Oceania Geosci Soc (AOGS), published 6 March 2014 in Springeropen